NATURAL DISASTERS
Hazards of the Dynamic Earth

NATURAL DISASTERS

Hazards of the Dynamic Earth

Stephen Marshak
University of Illinois
at Urbana-Champaign

Robert Rauber
University of Illinois
at Urbana-Champaign

Neil Johnson
Virginia Polytechnic Institute
and State University

W. W. Norton & Company
Independent Publishers Since 1923

W. W. Norton & Company has been independent since its founding in 1923, when William Warder Norton and Mary D. Herter Norton first published lectures delivered at the People's Institute, the adult education division of New York City's Cooper Union. The firm soon expanded its program beyond the Institute, publishing books by celebrated academics from America and abroad. By midcentury, the two major pillars of Norton's publishing program—trade books and college text—were firmly established. In the 1950s, the Norton family transferred control of the company to its employees, and today—with a staff of four hundred and a comparable number of trade, college, and professional titles published each year—W. W. Norton & Company stands as the largest and oldest publishing house owned wholly by its employees.

Copyright © 2022 by W. W. Norton & Company, Inc.

All rights reserved
Printed in Canada

Editor: Jake Schindel
Senior Project Editors: Thom Foley and Linda Feldman
Associate Production Director, College: Benjamin Reynolds
Production Manager: Richard Bretan
Assistant Editor: Mia Davis
Copy Editor: Norma Sims-Roche
Managing Editor, College: Marian Johnson
Managing Editor, College Digital Media: Kim Yi
Digital Media Editor: Michael Jaoui
Associate Media Editor: Arielle Holstein
Media Project Editor: Marcus van Harpen
Editorial Assistant, Media: Jasvir Singh
Marketing Managers, Geology: Marlee Lisker and Alex Ottley

Design Director: Rubina Yeh
Designer (interior): Jillian Burr
Designers (cover): Debra Morton-Hoyt and Eve Sanoussi
Director of College Permissions: Megan Schindel
Photo Editor: Stephanie Romeo
Photo Researcher: Fay Torresyap
Permissions Manager: Bethany Salminen
Text Permissions Specialist: Josh Garvin
Composition: Graphic World
Graphic World Project Manager: Gary Clark
Illustrations: Troutt Visual Services
Visualizing a Disaster art spreads by Stan Maddock, Craig Durant (Dragonfly Media Group), and Andrew Recher (Pens & Beetles Studios)
Manufacturing: Transcontinental—Beauceville, QC

Permission to use copyrighted material is included in the backmatter of this book.

Library of Congress Cataloging-in-Publication Data

Names: Marshak, Stephen, 1955- author. | Rauber, Robert M., author. | Johnson, Neil E., 1958- author.
Title: Natural disasters : hazards of a dynamic planet / Stephen Marshak, Robert Rauber, Neil Johnson.
Description: First edition. | New York, NY : W. W. Norton & Company, [2022] | Includes index.
Identifiers: LCCN 2021037137 | **ISBN 9780393921977** (paperback) | ISBN 9780393532623 (epub)
Subjects: LCSH: Natural disasters--Social aspects. | Natural disasters--Social aspects--Case studies. | Climatic changes. | Weather--Social aspects. | Weather--Social aspects--Case studies.
Classification: LCC GB5014 .M377 2022 | DDC 363.34--dc23
LC record available at https://lccn.loc.gov/2021037137

W. W. Norton & Company, Inc., 500 Fifth Avenue, New York, NY 10110
wwnorton.com

W. W. Norton & Company Ltd., 15 Carlisle Street, London W1D 3BS

To our families

BRIEF CONTENTS

Preface xxi
About the Authors xxvii

1. Our Planet Can Be Dangerous: Introducing Natural Disasters 2
2. Welcome to a Dynamic Planet: The Earth System and Plate Tectonics 30
3. A Violent Pulse: Earthquakes 72
4. The Wrath of Vulcan: Volcanic Eruptions 122
5. Unstable Slopes: The Danger of Mass Wasting 174
6. When the Sea Suddenly Rises: Tsunamis 216
7. Lost Ground: The Water Crisis, Sinkholes, Soils, and Subsidence 250
8. In Constant Motion: The Earth's Atmosphere 290
9. Local Tempests: Thunderstorms and Tornadoes 326
10. Cold and Icy: Hazardous Winter Weather 358
11. The Deadliest Storms: Hurricanes 382
12. Dangerous Deluges: Inland Flooding 420
13. Savage Seas: Hazards of the Shore 466
14. Dry, Hot, and Lethal: Drought, Heat Waves, and Wildfires 502
15. The Stealth Megadisaster: Climate Change 542
16. Dangers from Space: Space Weather and Meteorite Impacts 584

Appendix: Matter and Energy A-1
Glossary G-1
Credits C-1
Index I-1

CONTENTS

Preface xxi
About the Authors xxvii

CHAPTER 1

Our Planet Can Be Dangerous: Introducing Natural Disasters 2

1.1 Introduction 3

1.2 The Language of Danger 7
What's the Difference between a Hazard and a Disaster? 7
Classifying Hazards and Disasters by Their Source 7
Comparing the Time Frames of Natural Disasters 9

BOX 1.1 A CLOSER LOOK Contrasting Global Pandemics with Physical Natural Disasters 10

Distinguishing among Primary, Secondary, and Tertiary Disasters 12

1.3 Characterizing Disaster Size and Frequency 13
Comparing the Sizes of Disasters 13
Comparing the Frequency of Disasters 14
The Human Toll and Economic Impact of Disasters 14

1.4 Exposure, Vulnerability, and Risk 17
Characterizing Exposure and Vulnerability 17
Putting It All Together: Defining Disaster Risk 18

1.5 Factors Influencing Disaster Risk 19
Effect of Population Growth 19
Effect of Climate Change 20
Human Influence on Disaster Risk 20

1.6 Dealing with Disaster 21
Response, Recovery, Restoration, and Resilience 21
Who Responds to a Disaster? 22
Predictions, Forecasts, and Warnings 22
Hazard Preparedness, Mitigation, and Planning 25
The Cost of Disaster: Who Pays? 26

Chapter Review 28

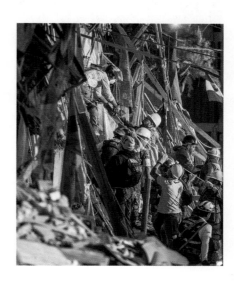

CHAPTER 2

Welcome to a Dynamic Planet: The Earth System and Plate Tectonics 30

- **2.1 Introduction** 31
- **2.2 A Quick Tour of the Earth System** 32
  - Welcome to the Neighborhood! 32
  - Introducing the Biosphere 33
  - Introducing the Atmosphere 33
  - Introducing the Hydrosphere and Cryosphere 34
  - Introducing the Geosphere 34
  - Energy in the Earth System 35
- **2.3 Earth Materials** 37
  - Forming Planet Earth 37
  - Introducing Minerals and Rocks 38
  - The Three Rock Groups 39
- **2.4 Journey to the Center of the Earth** 43
  - The Crust 44
  - The Mantle 45
  - The Core 46
  - Lithosphere and Asthenosphere 46
- **2.5 A Scientific Revolution: Discovering Plate Tectonics** 46
  - Continental Drift 47
  - Key Geologic Observations of the Mid-20th Century 48
  - The Proposal of Plate Tectonics 50
  - The Way the Earth Works: The Theory of Plate Tectonics 51
- **2.6 The Nature of Plate Boundaries** 52
  - Divergent Boundaries 52
  - Convergent Boundaries 54
  - Transform Boundaries 56
  - The Birth and Death of Plate Boundaries: Collisions and Rifts 56
  - Hot Spots 60
  - Plate Motion 61
- **2.7 The Origin of Topography and Geologic Structures** 61
  - Development of Topography 61
  - Geologic Structures 62
- **2.8 Geologic Time and Global Change** 66
  - Developing the Concept of Deep Time 66
  - Global Change 66
- **Chapter Review** 70

CHAPTER 3

A Violent Pulse: Earthquakes 72

- **3.1 Introduction** 73
- **3.2 What Causes Earthquakes?** 73

      Introducing Faults and Faulting   73
      Generating Seismic Waves: Elastic Rebound and Stick-Slip Behavior   76
      Defining the Location of an Earthquake   78
      Foreshocks, Mainshocks, and Aftershocks   78
      How Much Slip Happens during an Earthquake?   78

3.3 The Nature of Seismic Waves   79
      The Different Types of Seismic Waves   79
      Describing Seismic Waves: Wavelength, Period, and Amplitude   80
      How Seismic Waves Change with Distance   80
      Seismic Wave Reflection and Refraction   83

3.4 Measuring Earthquakes   83
      Seismographs and Seismograms   83
      Finding the Location of an Earthquake's Epicenter   85

3.5 Defining the Size of Earthquakes   86
      Modified Mercalli Intensity Scale   86
      Magnitude Scales   87
      Relating Intensity to Magnitude   89
      Energy Release by Earthquakes   90

3.6 Where Do Earthquakes Occur, and Why?   90
      Earthquakes at Plate Boundaries   90
      Earthquakes That Are Not at Plate Boundaries   93
      Induced Seismicity   95

3.7 How Do Earthquakes Cause Damage?   96
      Ground Rupture   96
      Ground Shaking   96
      Landslides   98
      Sediment Liquefaction   98
      Resonance Disasters   100
      Fire   100
      Tsunamis   100

BOX 3.1 DISASTERS AND SOCIETY The 2010 Haiti Catastrophe   102
      Disease   103

VISUALIZING A DISASTER Earthquakes   104

3.8 Can We Predict "the Big One"?   106
      Long-Term Predictions: Earthquake Probability   106
      Paleoseismicity Studies   107
      Seismic Gaps   108
      Short-Term Predictions   108
      Insight from Satellite Data   109
      Earthquake Early-Warning Systems   110
      Interpreting Earthquake Hazard Maps   110
      Regions of Seismic Risk in the United States   111

3.9 Prevention of Earthquake Damage and Casualties   112

BOX 3.2 DISASTERS AND SOCIETY The Megathrust Hazard of Cascadia   113
      The Challenge: Damage to Buildings   114

BOX 3.3 A CLOSER LOOK **Catastrophes Due to Building Collapse** 115
    Earthquake Engineering 115
    Zoning, Building Codes, and Common Sense 118
    Prevention of Casualties 119
Chapter Review 120

CHAPTER 4

The Wrath of Vulcan: Volcanic Eruptions 122

4.1 Introduction 123

4.2 Melting Rock 124
    Causes of Melting in the Earth 124
    What Are Magma and Lava Made of? 125

4.3 Forming Igneous Rocks 127
    The Rise of Magma 127
    How Do Melts Become Solid? 127
    The Variety of Igneous Rocks 128

4.4 Where Does Igneous Activity Take Place? 130
    Igneous Activity at Divergent Boundaries 130
    Igneous Activity at Convergent Boundaries 131
    Igneous Activity in Continental Rifts 132
    Igneous Activity at Hot Spots 133
    Large Igneous Provinces 134

4.5 Building a Volcano 135
    Lava Flows, Lakes, and Fountains 135
    Pyroclastic Debris 136
    Volcanic Edifices 138

4.6 Will It Flow or Will It Blow? 142
    Effusive Eruptions 142
    Explosive Eruptions 143
    Volcanic Explosivity Index 145
    Large Caldera Eruptions and Supervolcanoes 147

BOX 4.1 DISASTERS AND SOCIETY **The 1980 Explosion of Mt. St. Helens** 148

4.7 The Style of Volcanic Eruptions 150
    Hawaiian Eruptions 150
    Surtseyan Eruptions 150
    Strombolian and Vulcanian Eruptions 151

BOX 4.2 DISASTERS AND SOCIETY **The 2018 Eruption of Kilauea** 152
    Peléan and Plinian Eruptions (The Really Big Ones!) 153

4.8 Beware! The Many Hazards of Volcanoes 155
    Hazards Due to Lava 155
    Hazards Due to Ash Falls and Ash Clouds 156
    Hazards Due to Pyroclastic Flows 157
    Hazards Due to Lahars 157
    Hazards Due to Volcanic Gases 158

Hazards Due to Explosions 158
Volcanoes and Climates 160

4.9 Protecting Ourselves from Vulcan's Wrath 161
Active, Dormant, or Extinct? 161
Predicting Eruptions 163
Volcanic Observatories and Eruption Warnings 165

VISUALIZING A DISASTER Volcanic Eruptions 166
Mitigating Risk from Volcanic Hazards 168

BOX 4.3 DISASTERS AND SOCIETY Naples, Italy: A City Between Two Volcanoes 170

Chapter Review 172

CHAPTER 5
Unstable Slopes: The Danger of Mass Wasting 174

5.1 Introduction 175

5.2 Setting the Stage for Mass Wasting: Slopes 175
Uplift, Subsidence, and Slopes 175
Slope Steepness and Shape 176

5.3 Slope Stability 178
The Safety Factor 178
What Provides Resistance to Mass Wasting? 180
Failure Surfaces 182

5.4 Types of Mass Wasting 184
Creep and Solifluction 184
Slumps 186
Mudflows, Debris Flows, and Lahars 187

BOX 5.1 DISASTERS AND SOCIETY The Portuguese Bend Slump 188

BOX 5.2 DISASTERS AND SOCIETY Mudflows in Rio de Janeiro 190
Lateral Spreading 192
Rock and Debris Slides and Falls 193
Avalanches 196
Complex Landslides 197

5.5 Submarine Mass Wasting 198

5.6 What Factors Trigger Mass Wasting? 200
Shocks and Vibrations 200
Sediment Liquefaction 200
Changing Downslope Stress or Resistance Stress 201
Changing Slope Angle 203
Changing Slope Strength 204

5.7 How Can We Protect against Mass-Wasting Disasters? 205
Identifying Regions at Risk 205

VISUALIZING A DISASTER Mass Wasting 206
Detecting the Start of Mass Wasting 210
Preventing and Mitigating Mass Wasting 211

Chapter Review 214

CHAPTER 6

When the Sea Suddenly Rises: Tsunamis 216

- **6.1 Introduction** 217
- **6.2 The Origin and Movement of Tsunamis** 218
  - What Is a Tsunami? 218
  - Generation of Tsunamis 219
  - Why Does a Tsunami's Height Vary in the Open Ocean? 220
- **6.3 When a Tsunami Comes Ashore** 222
  - How Does a Tsunami Change as it Enters Shallow Water? 222
  - Describing a Tsunami That Reaches the Shore 223
  - How Does a Tsunami Differ from a Wind-Driven Wave? 223
- **6.4 How Do Tsunamis Destroy?** 226
- **6.5 Tsunamis Related to Earthquakes** 228

BOX 6.1 DISASTERS AND SOCIETY Ghost Forests of Cascadia 230

- **6.6 Landslide-Generated Tsunamis** 234
  - Subaerial Landslides 234
  - Submarine Slumping along Continental Margins 235
  - Flank Collapse of Volcanic Islands 236

BOX 6.2 DISASTERS AND SOCIETY The Enigmatic Papua New Guinea Tsunami of 1998 237

- **6.7 Other Causes of Tsunamis** 238
  - Tsunamis due to Explosive Volcanic Eruptions 238
  - Tsunamis due to Meteorite Impacts 239
- **6.8 Tsunamis and Society** 240
  - Secondary Disasters Related to Tsunamis 240
  - Tsunami Relief and Recovery 241

VISUALIZING A DISASTER Tsunamis 242
  - Tsunami Prediction and Mitigation 244

Chapter Review 248

CHAPTER 7

Lost Ground: The Water Crisis, Sinkholes, Soils, and Subsidence 250

- **7.1 Introduction** 251
- **7.2 A Most Precious Resource: Water** 252
  - The Hydrologic Cycle and Our Planet's Water Supply 252
  - The Global Water Crisis 253
- **7.3 Surface-Water Challenges** 255
  - Surface Freshwater Depletion 255
  - Surface Freshwater Contamination and Salinification 257
- **7.4 Groundwater Challenges** 258
  - Introducing Groundwater 258
  - Groundwater Depletion 260
  - Contaminated Groundwater 261

- 7.5 **Responding to the Water Crises** 263

  BOX 7.1 DISASTERS AND SOCIETY **The 2019 Water Crisis in India** 264

- 7.6 **Sinkholes** 266

  Origin of Karst Landscapes 266

  Formation of Sinkholes 267

  Dealing with the Risk of Sinkhole Formation 268

  BOX 7.2 A CLOSER LOOK **Liquefaction Sinkholes and Evaporite Sinkholes** 270

- 7.7 **Land Subsidence** 271

  Subsidence due to Pore Collapse 271

  Subsidence in Wetlands 273

  Subsidence in Delta Plains, Floodplains, and Coastal Plains 274

  Subsidence due to Mine Collapse 277

- 7.8 **Soil Hazards** 278

  How Does Soil Form? 279

  Soil Degradation 279

  BOX 7.3 DISASTERS AND SOCIETY **Dust Storms and the Dust Bowl** 282

  Expansive Soil 284

  VISUALIZING A DISASTER The Water Crisis, Land Subsidence, and Soil Erosion 286

  Chapter Review 288

CHAPTER 8

In Constant Motion: The Earth's Atmosphere 290

- 8.1 **Introduction** 291

- 8.2 **Atmospheric Composition** 292

  The Recipe for Air 292

  Air Pollution: The Danger of Dirty Air 294

- 8.3 **Describing Atmospheric Properties** 295

  Air Temperature 295

  Atmospheric Pressure 296

  Moisture and Relative Humidity 297

  Wind Speed and Direction 298

  Visibility, Cloud Cover, and Precipitation 299

  BOX 8.1 A CLOSER LOOK **Weather Radar** 300

  Measuring Atmospheric Conditions 300

- 8.4 **Vertical Structure and Heat Sources in the Atmosphere** 301

- 8.5 **Clouds and Precipitation** 302

  Cloud Formation 302

  Classification of Clouds 304

  Precipitation and Its Causes 304

  Clouds and Energy 305

- 8.6 **Winds of the World** 306

  Forces Acting on Air 306

The Effect of Heating and Cooling on Surface Air Pressure 309
What Controls Wind Direction? 310

8.7 The Atmosphere of the Tropics and Subtropics 311
Hadley Cells and the Intertropical Convergence Zone 312
The Southern Oscillation and El Niño 314
Monsoons 316

8.8 The Atmosphere of the Mid- and High Latitudes 317
Air Masses and Fronts 317
The Polar Front and the Jet Stream 318

8.9 Mid-latitude Cyclones 320

Chapter Review 324

CHAPTER 9

Local Tempests: Thunderstorms and Tornadoes 326

9.1 Introduction 327

9.2 Conditions That Produce Thunderstorms 328
What Is a Thunderstorm? 328
The Source of Moist Air 329
Lifting Mechanisms 330
Development of Atmospheric Instability 331

9.3 Types of Thunderstorms 333
Ordinary Thunderstorms 333
Squall-Line Thunderstorms 335
Supercell Thunderstorms 336

9.4 Hazards of Thunderstorms 340
Lightning 340
Downbursts 344
Hail 345

BOX 9.1 DISASTERS AND SOCIETY Hailstorms of Mendoza, Argentina 346

9.5 Tornadoes 347
Supercell Tornadoes 348
Non-supercell Tornadoes 350
Defining the Intensity of a Tornado 350

VISUALIZING A DISASTER Life Cycle of a Large Tornado 352

BOX 9.2 DISASTERS AND SOCIETY The Threat to Urban Centers 354

Chapter Review 356

CHAPTER 10

Cold and Icy: Hazardous Winter Weather 358

10.1 Introduction 359

10.2 The Dangers of Cold 360
Hypothermia, Frostbite, and Wind Chill 361
Cold Waves and the Polar Vortex 362

10.3 Hazardous Conditions during Winter Storms 363
Snowfall 363
Blizzards 363
Ice Storms 364

10.4 Causes of Winter Storms 366
Continental Mid-latitude Cyclones 366

BOX 10.1 DISASTERS AND SOCIETY **The Great Ice Storm of 1998** 367
Nor'easters and Oceanic Cyclones 368
Cold Air Damming 370
Lake-Effect Snowstorms 370
Mountain Snowstorms 371
Hazardous Winter Weather in Asia 372

BOX 10.2 DISASTERS AND SOCIETY **China's Worst Winter** 373

10.5 Dealing with Hazardous Winter Weather 374
Hazards to Personal Safety 374
Hazards to Aircraft 374
Hazards to Property, Infrastructure, and Agriculture 375

VISUALIZING A DISASTER Winter Storms 376

Chapter Review 380

CHAPTER 11

The Deadliest Storms: Hurricanes 382

11.1 Introduction 383

11.2 What Is a Hurricane? 384
Distinguishing among Tropical Cyclones by Wind Speed 384
What's Inside a Hurricane? 384

11.3 Hurricane Energy and Evolution 388
Where Does the Energy in a Hurricane Come From? 388
How Do Hurricanes Form and Evolve? 389
Hurricane Tracks and Dimensions 395

11.4 How Do Hurricanes Cause Damage? 397
Wind Damage 397

VISUALIZING A DISASTER Hurricanes 398
Storm Surge 400
Wave Damage 401
Rain, Inland Flooding, and Landslides 402

11.5 Hurricane Disasters: A World View 403
Caribbean and Central American Hurricanes 403

BOX 11.1 DISASTERS AND SOCIETY **Lasting Consequences of Hurricane Maria** 404
Atlantic and Gulf Coast Hurricanes 405

BOX 11.2 DISASTERS AND SOCIETY **Hurricane Katrina: Exposing the Vulnerability of Coastal Cities to Tropical Cyclones** 406
Pacific Hurricanes and Typhoons 408
Cyclones of the Indian Ocean and Southern Hemisphere 411

11.6 Hurricane Impacts and Mitigation 412
    Societal Disruptions 412
    Predicting Hurricane Tracks 413
    Preparing for the Storm's Arrival 414
    Hurricane Evacuations 415
    Hurricane Damage and Mitigation 417
Chapter Review 418

CHAPTER 12

Dangerous Deluges: Inland Flooding 420

12.1 Introduction 421
12.2 Draining the Land: Streams and Rivers 422
    Runoff and the Formation of Streams 422
    Describing Streams 422
    Erosion and Deposition by Streams 424
    Types of Streams 427
    Drainage Networks and Watersheds 429
    Stream Discharge and Stage 430
12.3 Natural Flooding in Streams and Rivers 432
    Flood Stages and Crests 432
    Slow-Onset Floods 432
    Flash Floods 434
BOX 12.1 DISASTERS AND SOCIETY Recent Mississippi Floods 436
12.4 Other Types of Floods 441
    Areal Flooding 441
    Urban Flooding 441
    Failures of Constructed Dams and Levees 442
12.5 Consequences of Flooding 445
    Danger due to Moving Water 445
    Damage due to Rising Water 445
12.6 Weather Systems That Cause Flooding 447
    The Rainy Season in a Monsoonal Climate 447
VISUALIZING A DISASTER Inland Flooding 448
BOX 12.2 DISASTERS AND SOCIETY The 2010 Monsoonal Floods in Pakistan 450
    Tropical Cyclones 451
    Mid-latitude Cyclones 451
    Mountain Thunderstorms 455
12.7 Flooding Risk 456
    Estimating Flooding Probability 456
    Flood-Frequency Graphs and Flood-Hazard Maps 457

BOX 12.3 A CLOSER LOOK Calculating Recurrence Intervals for Rare Floods   458
    Flood Watches, Warnings, and Advisories   460

12.8 Flood Mitigation   460
    Types of Flood Control   460
    Who Pays for Flood Damage?   462
    Flood Vulnerability   462

Chapter Review   464

CHAPTER 13

Savage Seas: Hazards of the Shore   466

13.1 Introduction   467

13.2 Circulation in the Ocean: Currents and Tides   467
    Currents: The Rivers of the Sea   467
    The Tide Comes In . . . The Tide Goes Out   469

13.3 Wind-Driven Waves   471
    Formation of Waves   471
    The Nature of Wave Motion   472
    When Waves Approach the Shore   473

13.4 Coastal Landforms   475
    Welcome to the Coast   475
    Sedimentary Coasts   477
    Rocky Coasts   478
    Organic Coasts   478

13.5 Hazards of Currents, Tides, and Waves   482
    Dangerous Currents   482
    Dangerous Tides   483
    Hazards of Waves   484

13.6 Coastal Challenges   487
    Nuisance Tides   487

BOX 13.1 DISASTERS AND SOCIETY Nuisance Tides on Pacific Islands   488
    Sea-Cliff Retreat   489
    Beach Erosion   490
    Protecting Coastal Property   491

BOX 13.2 DISASTERS AND SOCIETY Beach Restoration Following Hurricane Sandy   494
    Biological Hazards along Coasts   497

VISUALIZING A DISASTER Coastal Hazards   498

Chapter Review   500

CHAPTER 14

Dry, Hot, and Lethal: Drought, Heat Waves, and Wildfires 502

14.1 Introduction 503

14.2 Drought: When the Land Dries Out 504
  What Is Drought? 505
  The Relation of Drought to Climate Zones 506

BOX 14.1 DISASTERS AND SOCIETY Drought, Famine, and Infrastructure in Africa 507
  Monitoring and Predicting Drought 512

14.3 Heat Waves 514
  The Danger of High Temperatures 514
  When Is Hot Weather Considered a Heat Wave? 516
  Urban Heat Islands 517
  Weather Patterns Associated with Heat Waves 519

14.4 Wildfires: Flames of Destruction 521
  What Is Fire? 522
  Igniting a Wildfire 522
  Types of Wildfires 524
  The Spread of Wildfires 524
  The Special Challenge of Downslope Winds 528
  Climate Controls on Wildfires 529

BOX 14.2 DISASTERS AND SOCIETY The Human Cost of Downslope Winds in California 530

14.5 Consequences and Mitigation of Wildfires 532
  Impacts of Wildfires 532
  Fighting Wildfires 533

VISUALIZING A DISASTER Fire and Drought 534
  Mitigating Wildfire Hazards 537

Chapter Review 540

CHAPTER 15

The Stealth Megadisaster: Climate Change 542

15.1 Introduction 543

15.2 Controls on the Earth's Climate 543
  Heating the Atmosphere: Solar Radiation and the Greenhouse Effect 543
  Classifying Climates 545

15.3 Climate Change and Its Causes 546
  What Is Climate Change? 546
  Paleoclimate Indicators 547
  How Has Climate Changed Naturally over Earth History? 548
  What Causes Climate Change? 550
  Feedbacks Affecting Climate Change 553

15.4 Evidence of Recent Climate Change 554
  Discovering Recent Climate Change 554
  Observed Recent Temperature Change 554
  Observed Recent Sea-Level Change 556
  Melting Glacial and Sea Ice 557
  Biological Indicators of Climate Change 558
  Changes in Global Weather Patterns 559

15.5 Causes of Recent Climate Change 561
  Recent Changes in Atmospheric CO_2 Concentrations 561
  Are Rising CO_2 and CH_4 Concentrations Natural or Anthropogenic? 563

15.6 Impacts and Hazards of Climate Change 566
  Using GCMs to Predict the Future 566
  Future Sea-Level Rise 567
  Redistribution of Water Resources 567

BOX 15.1 DISASTERS AND SOCIETY New York City, Urban Flooding, and Sea-Level Rise 568
  Future Changes in the Arctic 569

VISUALIZING A DISASTER Consequences of Sea-Level Change 570
  Future Changes in the Biosphere 572
  Future Changes in the Oceans 573

BOX 15.2 DISASTERS AND SOCIETY The Global Poleward Spread of Disease and Pests 574
  Future Changes in Hazardous Weather 577
  Future Increases in Heat Waves and Droughts 577

15.7 Can Climate Change Impacts Be Mitigated? 578
Chapter Review 582

CHAPTER 16

Dangers from Space: Space Weather and Meteorite Impacts 584

16.1 Introduction 585

16.2 The Solar Inferno and Its Products 586
  The Sun's Energy Source and Structure 586
  The Sun's Magnetic Field and Sunspots 587
  Solar Storms 588

16.3 The Earth's Protective Shields 589
  The Earth's Atmosphere: A Shield against UV Radiation 590
  The Earth's Magnetosphere: A Shield against Space Weather 590

BOX 16.1 A CLOSER LOOK The Ozone Hole and the Montreal Protocol 591

16.4 Space Weather 593
  Consequences of Space Weather 593
  Mitigating the Consequences of Space Weather 594

16.5 Dangerous Objects Lurking in the Solar System 594

BOX 16.2 DISASTERS AND SOCIETY Intense Space Weather of the Past 595

  Asteroids 597
  Comets 597
  Meteors and Meteorites 598

16.6 Meteors, Fireballs, and Bolides 600

BOX 16.3 DISASTERS AND SOCIETY **The Tunguska and Chelyabinsk Events** 604

16.7 What Does an Impact Do to the Crust? 605
  Meteorite Craters 605
  Other Evidence of Large Impacts 606
  Examples of Impact Sites on the Earth 608
  Why Aren't There More Impact Sites on the Earth? 612

16.8 Extinction-Level Impacts 612
  Mass Extinctions of the Geologic Past 612
  How Might an Impact Cause Mass Extinction? The Chicxulub Example 613

VISUALIZING A DISASTER **Dangers from Space** 614

16.9 Mitigating Impact Risk 617
  Identifying Near-Earth Objects 617
  The Consequences and Frequency of Impacts 618
  Deflecting Near-Earth Objects 619

Chapter Review 622

APPENDIX: Matter and Energy A-1
A.1 Introduction A-1
A.2 The Nature of Matter A-1
A.3 States of Matter A-2
A.4 Force and Energy A-2
A.5 Heat, Temperature, and Heat Transfer A-4

Glossary G-1
Credits C-1
Index I-1

PREFACE

Narrative Themes

Natural Disasters: Hazards of the Dynamic Earth represents a 21st-century approach to describing physical dangers of the Earth System—-the various interacting realms of our planet, including the geosphere, atmosphere, hydrosphere, biosphere, and space. Toward this goal, the book encompasses six narrative themes.

1. Natural hazards are an integral part of the Earth System. Throughout geologic time, hazards have served a key role in the evolution of our planet and the life that inhabits it. Natural hazards become natural disasters when they cause significant damage and casualties to human communities.
2. Because the severity of a natural disaster is measured by how deeply it impacts people, in order to understand a disaster, we must first consider a community's vulnerability to such an event, and how vulnerability relates to broader societal issues such as poverty. For example, in countries with few resources, a given natural event can become a major disaster, while a similar event striking a more developed country with greater resources may only produce relatively minor consequences. Therefore, disasters must be studied from a global perspective.
3. Because natural disasters involve the solid Earth (the geosphere), surface water (the hydrosphere), and our planet's gaseous envelope (the atmosphere), a college course on natural disasters can benefit by covering all realms. For example, a complete discussion of floods should not only characterize the process and consequences of flooding, but also the meteorological causes of the event, and why certain parts of the world are at greater risk to such events than others.
4. While this book focuses on physical natural disasters, the COVID-19 pandemic, a biologic disaster, provides a context for understanding and visualizing the human and financial cost of disasters. Consequently, where relevant, the book compares the consequences of physical disasters with those of the COVID-19 biological disaster.
5. Different types of disasters happen at different rates. Some disasters—known as stealth disasters—develop so slowly that their consequences accumulate before society fully recognizes their impact. Examples include climate change, drought, and the water crisis. Because the consequences of stealth disasters are significant, this book discusses them in detail.
6. Many geological or geophysical disasters happen because the Earth System remains dynamic. Specifically, our planet's surface undergoes changes in response to the motion of plates (pieces of its outer, rigid shell), relative to one another. The theory of plate tectonics, a concept that explains these plate movements, helps determine why geological or geophysical hazards develop. Similarly, because climate change influences many surficial and meteorological hazards on the Earth, understanding climate change helps society make predictions about the risks associated with many types of hazards.

Pedagogical Approach and Special Features

Different students learn in different ways. Some students learn best by reading a clear narrative that lays out foundational concepts and illustrates them with examples. Other students learn best by analyzing visual images of features and processes. This book accommodates both types of learners.

The text in each chapter has been organized to follow a narrative arc. The arc begins with an example to set the stage, then continues with a description of key processes and phenomena, each illustrated with additional dramatic examples. We introduce appropriate vocabulary in context, so that phenomena can be discussed efficiently. We have also worked to ensure that the text provides sufficient background so that students, regardless of their previous knowledge, can completely understand a process. While this approach sometimes extends discussions, the result is greater clarity and a stronger impact.

The art in this book has been carefully designed to provide visual narratives. In many figures, a step-by-step illustration demonstrates how a phenomenon initiates and evolves. The lead authors have worked closely with experienced, scientific illustrators to ensure that each piece of art is technically correct, and that every detail, including textures and colors, provides students with the clearest and most realistic understanding of a process or phenomena. At the same time, illustrations are not cluttered or overly complex. The art has been tailored to fit with the book's written narrative, illustrating each new point when the topic first appears. The book includes hundreds of photos (many provided by author Stephen Marshak), carefully selected to illustrate specific points or examples. In addition to the individual figures and photos that occur throughout the text, every disaster-focused chapter contains a two-page narrative painting—Visualizing a Disaster—that synthesizes most of the chapter's key concepts into a unified illustration. These composite images help students to visually tie together the components of a chapter's narrative. They also provide a visual resource for study and review.

This book's combination of narrative text and narrative art provides a complete exposition of a chapter's subject. This thoroughness makes *Natural Disasters* an ideal textbook, not only for traditional lecture-based courses, because it fills in gaps and clarifies explanations that instructors might not have time to cover in class, but also for contemporary active-learning-style courses. When students read the book and engage in its pedagogical supplements, they will be well prepared to engage in classroom discussions and to work on projects during and beyond class sessions.

In addition to the primary narrative, many chapters contain two Disasters and Society boxes. This feature offers case studies of disasters, with a focus on society's responses to them. In the chapters that focus on a category of disaster, we present one example from North America and one from another part of the world, to demonstrate how disasters and society's responses to them can vary with locality. Some chapters also include a feature titled A Closer Look. These boxes provide a deeper understanding of certain topics. We have also included an Appendix that reviews basic scientific terminology for students who may be a bit rusty with their high-school level physics and chemistry. Notably, throughout the text, we use both metric and English units for measurements. This is to help American students without strong science backgrounds to visualize dimensions and learn conversions.

To encourage students to test their knowledge throughout, each chapter section ends with a Take-Home Message that highlights the major concept addressed in that section, and provides a Quick Question to challenge students to apply the knowledge that they just gained before moving forward. Each chapter also provides a concise two-page Chapter Review that includes a summary of key topics and numerous questions keyed to the Learning Objectives that open every chapter.

Organization and Coverage

Natural Disasters is a new book presenting an integrated and comprehensive view of physical natural disasters and their societal consequences. To develop this book, the lead authors researched topics in depth and tested ideas with disciplinary experts. They also visited locations of recent or ongoing disasters, to fully understand the nature of damage and society's response. In contrast with previous works on this topic, our book includes thorough coverage of meteorological disasters, and it integrates geological and meteorological concepts where appropriate. We also introduce categories of disasters that have not been covered by earlier books, because until recently, they have not been in the headlines. For example, we include a thorough discussion of the *water crisis*—the lack of clean fresh water where it's needed—that has already become a disaster for many communities. Similarly, we provide thorough coverage of drought, and not only do we provide a comprehensive chapter on climate change, but throughout the book, examples show how climate change has impacted the severity of other types of disasters.

The book begins with a general introduction to natural disasters. We then make the distinction between hazards and disasters and provide an overview of ways communities respond to such events. We also lay out the basic vocabulary used throughout the book to discuss natural disasters. Chapter 2 provides a concise synopsis of key geological concepts, as we introduce Earth materials, plate tectonics, and the Earth System concept—particularly for the benefit of students who have not had a previous introductory course in geology. This chapter provides the foundation for the first part of the book, which focuses on geological and geophysical disasters (earthquakes, volcanoes, landslides, tsunamis, the water crisis, and land subsidence).

In the second part of the book we focus on atmospheric phenomena that can threaten communities. To set the stage for this discussion, Chapter 8 provides a general review of atmospheric structure and behavior. This discussion includes an introduction to mid-latitude cyclones (weather systems that control most stormy weather in temperate regions of the Earth). Next, chapters discuss specific types of dangerous atmospheric phenomena (thunderstorms and tornadoes, winter storms, and hurricanes). This book provides a chapter dedicated to winter storms, because the danger of these extreme events is often overlooked. The position of Chapter 12, on inland flooding, after several chapters on meteorological topics, reflects the fact that understanding floods relies on a background of both geological and meteorological concepts. Chapter 13 follows, with a focus on coastal hazards, including coastal flooding. We then address the hazards and associated disasters that have grabbed so many headlines in recent years: drought, heat, and fire. The book then introduces an integrated treatment of climate change, utilizing topics covered previously. The book concludes by considering hazards that come from space, which includes expanded coverage of solar storms, a topic whose hazards are becoming progressively more worrisome as society develops increasing dependence on electronic technology.

Media and Interactive Learning Resources

GUIDED LEARNING EXPLORATIONS

Guided Learning Explorations are scaffolded activities that engage students in thinking critically and applying essential course concepts. Level 1 reinforces mastery of underlying core science concepts. Level 2 examines the science of how a hazard develops or is evaluated by Earth scientists. And Level 3 explores a real-world disaster case study, considering the impact and possible means of mitigation. Students receive feedback with every click and "unlock" each level question-by-question. The Guided Learning Explorations cover 14 major topics taught in introductory Natural Disasters courses. The Guided Learning Explorations are included at no extra cost with every new copy of the textbook, in print or ebook format, and can be set up to work within your campus LMS. They can also be purchased separately at digital.wwnorton.com/naturaldisasters.

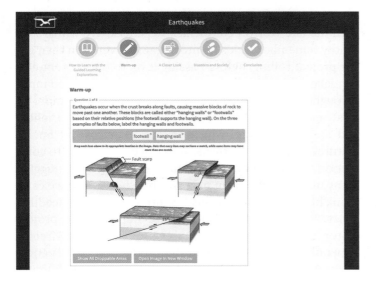

ANIMATIONS AND SIMULATIONS

Accompanying *Natural Disasters* at no cost to students or instructors is a rich collection of animations that illustrate the science behind hazards. These animations, developed by Alex Glass (Duke University), utilize a consistent style, applying a 3-D perspective to help students better grasp geologic and atmospheric phenomena. Some of these animations are simulations that allow students to control aspects and variables of an Earth process. Select animations have been integrated into the Guided Learning Explorations, and the full suite is available to students at digital.wwnorton.com/naturaldisasters. All animations and simulations can also be integrated into in an LMS via LTI integration. For help with integration, contact your Norton representative at https://wwnorton.com/find-your-rep.

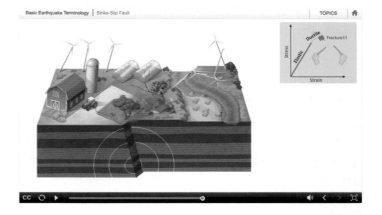

REAL WORLD VIDEOS. In addition to the suite of animations and simulations described above, instructors and students will have access to a robust suite of real-world videos showing the development of hazards and the occurrence and aftermath of disasters. These videos have been carefully selected by Mary Kosloski of the University of Iowa. Teaching notes to accompany the videos are available in the Instructor's Manual, and as with the animations, select videos have been integrated into the Guided Learning Explorations. The full suite of videos is available to students at digital.wwnorton.com/naturaldisasters, and all videos can be integrated into an LMS via LTI integration. For help with integration, contact your Norton representative at https://wwnorton.com/find-your-rep.

Additional Instructor Materials

LECTURE POWERPOINTS AND ART FILES. The art-focused lecture slides are optimized for presentation in live or online courses. Photos and art from the text, and linked animations and videos make these lecture slides engaging, while still covering the core concepts of each chapter. The Lecture PowerPoint images include alt text, and the slides are designed with the screen-reader experience in mind. Each set of lecture slides contains thought-provoking, multiple-choice classroom response questions (clicker questions), designed to help instructors measure student understanding in real time.

All the art from the book, sized for classroom display, is available in PowerPoint with text and in two JPEG formats—with and without labeling and balloon captions.

INSTRUCTOR'S MANUAL. The Instructor's Manual by Michelle Haskin and Ryan Sincavage offers helpful and creative resources for lecture planning and classroom activities in-class, online, and in hybrid classes. Every asset is tagged with the specific learning objective it covers. For each chapter, the IM offers:

- Teaching tips and common student struggles
- Class activities, including classroom procedures and online teaching and learning options
- Video and animation descriptions, with suggestions for using them effectively with their students
- Guided Learning Explorations descriptions, with suggestions for ways to incorporate them into lesson plans

TEST BANK. The test bank includes 800 questions written by instructors of natural disasters courses, utilizing art from the text, and tied directly to the book's chapter sections and learning objectives. The framework to develop our test banks, quizzes, and support materials is the result of a collaboration with leading academic researchers and advisers. Questions are classified by chapter, section, learning objective, Bloom's taxonomy, and difficulty, making it easy to construct tests and quizzes that are meaningful and diagnostic.

The Test Bank utilizes Norton Testmaker, which allows instructors to create assessments for their courses from anywhere with an Internet connection, without downloading files or installing specialized software. The format makes it easy to search and filter test bank questions by chapter, type, difficulty, learning objectives, and other criteria. Instructors can also customize test bank questions to fit each course, and easily export tests or Norton's ready-to-use quizzes to Microsoft Word or Common Cartridge files for the school's LMS.

LMS RESOURCES. It's easy to add high-quality Norton digital resources to online, hybrid, or lecture courses by accessing them through learning management systems (LMS). Graded activities can be configured to report to the LMS course grade book. The file below includes integration links to the following resources, organized by chapter:

- Ebook: An enhanced reading experience with embedded animations, interactive simulations, and narrative art videos
- Guided Learning Explorations
- Videos, animations, and simulations
- Flashcards

For these resources to function properly, instructors will need to ensure that LTI integration is enabled and supported on the school's learning management system (Blackboard, Canvas, Brightspace/D2L, or Moodle). If Norton is not yet an approved tool provider, the local Norton representative can provide assistance and can be contacted at https://wwnorton.com/find-your-rep.

Instructions for setting up these links with LMS integration can be found at https://wwnorton.knowledgeowl.com/help/lms-instructors.

Instructors can also add customizable multiple choice and short answer questions to their learning management system using Norton Testmaker. Explore a test bank of 800 questions to build quizzes and tests that can be easily exported as Common Cartridge files into an LMS.

Acknowledgments

Completion of this book would not have been possible without the continuing efforts of Kathy Marshak, who managed the flow of manuscript and proofs among the lead authors and the editors. At each stage, she read the text and checked the figures in detail to catch errors and inconsistencies, and to improve clarity. The lead authors also wish to profoundly thank all of our colleagues at W. W. Norton & Company, Inc. Initial discussion regarding this book began many years ago, with the late Norton editor Jack Repcheck. We will always fondly remember Jack's mentoring. Eric Svendsen then got the project rolling, before passing it on to our current editor, Jake Schindel, who has provided careful, detailed input throughout and helped to keep us on track and (more or less) on schedule. We are very grateful for Jake's friendly and steady guidance, which shaped many of the book's features, improved its focus, and drove the project to completion. Thom Foley, our primary project editor, juggled many moving parts from manuscript onward, and we can't thank him enough, especially his intense efforts at deadline times. Thanks to Linda Feldman for stepping in as project editor later in the process. Associate production director Benjamin Reynolds managed the production of this enormous project from its inception, and we are extremely grateful for his expert work in bringing it across the finish line.

Jan Troutt, at Troutt Visual Services, managed the Herculean task of managing the art program. Lead artist Joanne Brummett skillfully created visually dramatic images that convey concepts clearly. The willingness of the art team to work interactively with the lead author on developing and modifying figures is much appreciated. Stephanie Romeo and Fay Torresyap did a wonderful job of locating photos and obtaining permissions, and were more than patient with the lead author's repeated quests for just the right shot. Josh Garvin did similarly terrific work securing the line art permissions for this text. The two-page Visualizing a Disaster spreads were a delightful challenge to develop, in collaboration with talented artists—Stan Maddock, Andrew Recher, and Craig Durant. The lead authors greatly appreciate their efforts and patience.

This text is accompanied and enriched by a tremendous digital media suite, headlined by the terrific Guided Learning Explorations authored by Brian Zimmer of Appalachian State. His dedicated work on these resources make them excellent tools for students to hone and apply their concept understanding to real-world disaster scenarios. We are also very grateful to the group of digital media authors for their wonderful work on the suite of instructor and student tools that enhance the teaching and learning experience connected to this text: the Test Bank, Instructor's Manual, Lecture PowerPoints, and curated suite of real-world videos. Finally, we wish to thank the Digital Media team at Norton—Michael Jaoui, Arielle Holstein, Marcus van Harpen, and Jesse Singh—for their leadership and hard work bringing this robust media package to life, and ensuring that it works hand-in-hand with the textbook to give students the most impactful learning experience.

We are pleased to acknowledge discussions with our colleagues at the University of Illinois, including Susan Keiffer (author of *The Dynamics of Disaster*), Donald Wuebbles (an expert on climate change), Jeff Frame (an expert on severe weather), Patricia Gregg (an expert on volcanic hazards), Jesse Ribot (an expert on societal response to disasters), and Jim Best (an expert on river systems). We also greatly appreciate discussions of seismic hazards with Seth Stein at Northwestern University. This book has gained much from the participation of expert reviewers regarding specific topics, and from general reviewers and focus group participants regarding the book's content, structure, and interactive learning resources. We appreciate the input of all of these reviewers:

Alec Aitken, *University of Saskatchewan*
Joni Backstrom, *University of North Carolina—Wilmington*
Joseph Balta, *Texas A&M University*
LeeAnna Chapman, *University of North Carolina—Charlotte*
Ann Cook, *Ohio State University*
Jeremy Dillon, *University of Nebraska—Kearney*
Aley El-Shazly, *Marshall University*
Aurélie Germa, *University of South Florida*
Jacquelyn Hams, *Los Angeles Valley College*
Michelle Haskin, *University of North Carolina—Chapel Hill*
Amy Hochberg, *Utah State University*
Sarah Johnson, *Northern Kentucky University*
Mary Kosloski, *University of Iowa*
Brendan McNulty, *California State University—Dominguez Hills*
Claire O'Neal, *University of Delaware*
Mark Panning, *University of Florida*
Weisen Shen, *Stony Brook University*
Doug Shrake, *Capitol University*
Ryan Sincavage, *Radford University*
Seth Stein, *Northwestern University*
Lisa Tranel, *Illinois State University*

ABOUT THE AUTHORS

Stephen Marshak is a professor emeritus of geology at the University of Illinois, Urbana-Champaign, where he taught for 35 years. During this time, he also served as head of the department of geology and as director of the School of Earth, Society, & Environment. He holds an AB from Cornell University, an MS from the University of Arizona, and a PhD from Columbia University, all in geology. His research interests in structural geology and tectonics have taken him to the field on several continents. Steve, a Fellow of the Geological Society of America, has won the highest teaching awards at both the college and campus levels at the University of Illinois, and has also received a Neil Miner Award from the National Association of Geoscience Teachers for "exceptional contributions to the stimulation of interest in the Earth Sciences." In addition to writing research papers and *Natural Disasters*, he has authored or co-authored multiple textbooks: *Earth: Portrait of a Planet*; *Essentials of Geology*; *Earth Science: The Earth, the Atmosphere, and Space*; *Earth Structure: An Introduction to Structural Geology and Tectonics*; *Basic Methods of Structural Geology*; *Laboratory Manual for Introductory Geology*; *Laboratory Manual for Earth Science*; and *Geotours Workbook*.

Robert Rauber is a professor of atmospheric sciences at the University of Illinois, Urbana-Champaign, where he was department head for 12 years. He now serves as director of the School of Earth, Society, & Environment. Bob holds a BS in physics and a BA in English from the Pennsylvania State University, as well as an MS and PhD in atmospheric sciences from Colorado State University. He oversees a research program that focuses on the development and behavior of storms, and takes him on some rather exciting flights into the midst of severe weather. Bob has won campus teaching awards, is a Fellow of the American Meteorological Society, and served as the publication commissioner for the AMS. In addition to authoring research papers and *Natural Disasters*, Bob has co-authored *Earth Science: The Earth, the Atmosphere, and Space*; *Severe and Hazardous Weather—An Introduction to High Impact Meteorology*; and *Radar Meteorology, A First Course*.

Neil Johnson is a senior instructor and minerals curator at Virginia Polytechnic Institute and State University. He holds a BS from the Ohio State University, as well as an MS and PhD from Virginia Tech. He has taught the introductory course on Earth's natural hazards at Virginia Tech for many years. His research focuses on mineral characterization and economic geology.

NATURAL DISASTERS
Hazards of the Dynamic Earth

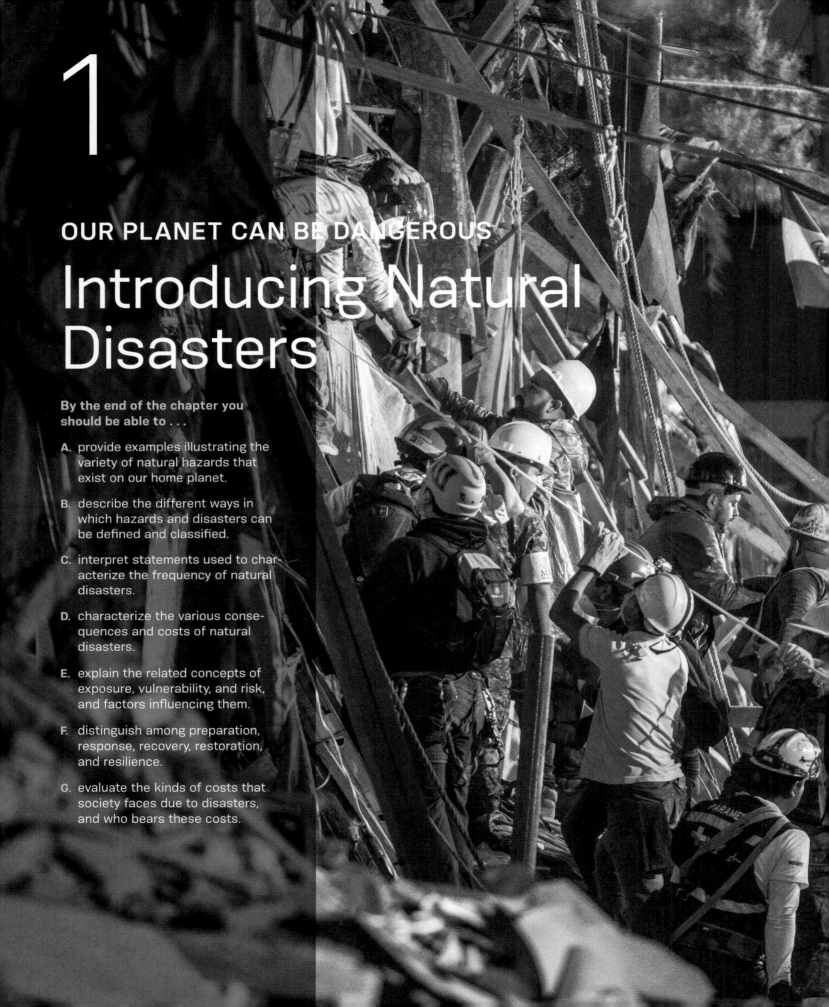

1

OUR PLANET CAN BE DANGEROUS
Introducing Natural Disasters

By the end of the chapter you should be able to . . .

A. provide examples illustrating the variety of natural hazards that exist on our home planet.

B. describe the different ways in which hazards and disasters can be defined and classified.

C. interpret statements used to characterize the frequency of natural disasters.

D. characterize the various consequences and costs of natural disasters.

E. explain the related concepts of exposure, vulnerability, and risk, and factors influencing them.

F. distinguish among preparation, response, recovery, restoration, and resilience.

G. evaluate the kinds of costs that society faces due to disasters, and who bears these costs.

1.1 Introduction

On March 11, 2011, in towns all along the east coast of Tōhoku, a region of northern Japan, people were following their normal daily routines . . . until 2:46 p.m. At that moment, the ground suddenly began to lurch up and down and back and forth, and for the next few minutes windows shattered, dishes flew off shelves, ceiling tiles fell, and building facades cracked. Disorientation and panic caused some people to freeze in place, some to dart outside, and others to dive beneath tables. Though terrifying, this **earthquake**—an episode of shaking caused by sudden movement of rock inside the Earth—was just the first stage of what would become a historical calamity.

The Earth movement that generated the Tōhoku earthquake also caused a broad region of seafloor east of Japan to lurch suddenly upward, sending a pulse of pressure into the overlying water. This pulse pushed the sea surface up slightly, producing a subtle mound of water that sailors on ships in the area didn't even notice. But because of its large area, the mound contained an immense volume of water. As soon as the water rose, gravity pulled it back down. This up-and-down motion of the sea surface generated very broad waves, known as **tsunamis**, which propagated outward at such high speed that the first one reached the coast of Japan only minutes later. As the waves approached the shore, they grew narrower and taller. In some locations, water rose as high as 20 m (65 ft) above normal sea level. The high water crossed beaches and spilled over seawalls, and in places submerged low-lying land up to a few kilometers inland (FIG. 1.1a). The moving water washed away communities and killed nearly 16,000 people (FIG. 1.1b).

When acting over millions of years, the movements that generate earthquakes can also lift mountains. But because of gravity, mountain slopes don't last forever. For example, though the south face of El Capitan, a towering cliff in Yosemite National Park (FIG. 1.2a), doesn't change for years at a time—allowing climbers to plan routes that use specific cracks in it for handholds—a rock block occasionally breaks free and falls. As it tumbles, it shatters into smaller fragments and dislodges other blocks, yielding a cascade of debris that roars toward the cliff's base (FIG. 1.2b). Generally, such **rockfalls** crush only trees, but sometimes they injure or kill people and damage buildings and roads. In populated regions, large rockfalls and other types of **landslides** (events during which rock and debris tumble, slip, or flow down a slope) have buried whole towns.

◀ Search and rescue team members look for victims beneath a collapsed building after an earthquake shook Mexico City in September 2017. When disasters strike, society must respond.

FIGURE 1.1 Tsunamis can devastate a coastal area.

(a) In 2011, tsunamis struck Japan. Here, the rising water of an incoming wave spills over seawalls.

(b) The tsunamis devastated many communities.

Earthquakes, tsunamis, and landslides remind us of our planet's dynamic nature. **Volcanic eruptions**, during which gas and molten rock rise from the Earth's interior and emerge from a *volcano*, a vent at the Earth's surface, provide another reminder. An eruption may produce a river of red-hot molten rock, called **lava**, that flows downslope, covering roads and homes (FIG. 1.3a), or it may blast debris skyward. In 1980, such an explosive eruption blew off the side of Mt. St. Helens, a volcano in Washington State, flattening forests over an area 10 times larger than that demolished by the Hiroshima atomic bomb (FIG. 1.3b, c).

FIGURE 1.2 The south face of El Capitan, Yosemite National Park, California.

(a) The 900 m (2,950 ft) rock wall of El Capitan presents a challenge to climbers.

(b) Sometimes, rock breaks free at a crack and tumbles as a rockfall.

FIGURE 1.3 The dangers of volcanic eruptions.

(a) Lava, at a temperature of about 1,100°C (2,000°F) flows across a street in Hawaii during a 2018 eruption.

(b) The explosive eruption of Mt. St. Helens, Washington State, 1980.

(c) The force of the blast from Mt. St. Helens flattened nearby forests.

FIGURE 1.4 Thunderstorms and tornadoes.

(a) Lightning flashes across the sky over Blackpool, England, causing everyone to head indoors.

(b) A tornado sweeps across the plains of Colorado.

(c) Damage left in the wake of a 2013 tornado in Moore, Oklahoma.

So far, we have focused on dangers associated with movements in or on the land. Society also faces peril from the Earth's atmosphere. At any given time, **storms**—episodes of strong winds and heavy *precipitation* (rain or snow)—are developing somewhere on the Earth. **Thunderstorms**, towering clouds that produce giant electrical sparks known as *lightning*, send people dashing for cover **(FIG. 1.4a)**. **Tornadoes**, narrow funnels of furiously spinning air, devastate the ground that they rake across **(FIG. 1.4b, c)**. **Hurricanes**, giant spiral-shaped storms, become headline news when they flatten buildings and swamp coastal towns or drench inland areas with torrential rains that trigger flooding and landslides **(FIG. 1.5)**.

Of course, without precipitation, life on land would not be possible, for all life requires water. But too much water produces **floods**, events during which water overtops the banks of rivers and streams and spreads over the surrounding countryside **(FIG. 1.6a)**. One of history's worst floods devastated large areas of China in 1931, when water from melting snow and record-breaking rains overtopped the banks of the Yellow and Yangtze Rivers and submerged villages and fields **(FIG. 1.6b)**. Tragically, many survivors of the flood later died due to lack of food, drinking water, shelter, and medicine. All told, the 1931 flooding and its aftermath took the lives of some 3,700,000 people and destroyed the property and livelihoods of millions more.

FIGURE 1.5 Hurricanes cause devastation both on the coast and inland.

(a) The immense spiral of clouds defining Hurricane Fran approaches the Bahamas and the southeast coast of the US. It caused 27 deaths and five billion dollars worth of damage.

(b) Damage along the coast of Florida caused by Hurricane Michael in 2018.

FIGURE 1.6 River flooding.

(a) A flooded community along the Russian River in California in 2019. Water submerged streets and moved buildings.

(b) A city in China flooded by the catastrophic rise of the Yangtze River in 1931.

Other atmospheric conditions, such as **droughts** (extended periods of little or no rain) and **heat waves** (prolonged times of unbearably high air temperature) are not only dangers themselves, but together with strong winds, they may lead to **wildfires** that consume forests, grasslands, and neighboring communities (**FIG. 1.7**). One of the most devastating wildfires of recent times, the 2018 Camp Fire in California, killed 85 people and consumed 19,000 buildings.

The destructive phenomena that we've described so far are unique to our planet, for in our Solar System, only the Earth has a dynamic interior, a circulating atmosphere and ocean, and life. But like all other planets and moons, the Earth faces dangers from space. These dangers include **solar storms**, during which the Sun blasts out energetic particles that can disrupt electronics on the Earth, and **meteorite impacts**, which happen when objects from elsewhere in the Solar System collide with the Earth. During our planet's long history, huge meteorites have struck the surface with devastating consequences. One such collision, which happened 66 million years ago at the site of what is now the Yucatán Peninsula, contributed to driving over 90% of living species—including all dinosaurs—to extinction.

FIGURE 1.7 California wildfires consume forests, grasslands, and homes.

Clearly, our beautiful home planet can occasionally be a dangerous place! The phenomena we've just described can damage portions of the Earth's surface, disrupt the environment, and cause significant harm to people and property. In order to decrease the peril that these phenomena represent to you and your community, it's important to understand why they exist, how they behave, how they can affect people, how they can be avoided, and how their consequences can be addressed. The purpose of this book, overall, is to introduce both the science behind dangerous natural phenomena and the ways they affect society. In this chapter, we introduce the basic vocabulary and ideas used to discuss such phenomena. We begin by distinguishing among natural hazards, natural hazardous events, and natural disasters. We then address the important related concepts of exposure, vulnerability, and risk. We also characterize the human dimensions of hazards and disasters and describe efforts that may help decrease their consequences.

1.2 The Language of Danger

What's the Difference between a Hazard and a Disaster?

When you see the word *hazard*, you might picture a substance or a phenomenon that poses a threat, meaning that it has the potential to cause injury or death or to destroy property. You wouldn't drink from a bottle labeled "Hazardous Chemical" or pour it on something you value. Hazardous chemicals represent an **anthropogenic hazard** (from the Greek *anthropos*, meaning human), one produced by the activities of people. Explosives, insecticides, and other objects or substances that people use, as well as smoking, drug use, and other activities that people do, are anthropogenic hazards because they are produced by people, caused by people, or undertaken by people. In contrast, a **natural hazard** is a phenomenon or process that exists independently of people but nevertheless has the potential to harm people or property, and therefore represents a threat to society.

Occasionally, natural hazards spawn a disturbance that damages and destroys features on a portion of the Earth's surface, disrupting the environment and perhaps injuring or killing plants and animals. Such disturbances, informally known as **natural hazardous events**, have happened since this planet formed, long before people appeared, and some, such as the meteorite impact 66 million years ago, changed the course of life's evolution. A natural hazardous event becomes a **natural disaster** only at the point when it causes significant numbers of human **casualties** (deaths or injuries), extensive destruction of property, and so much economic loss that its victims need outside help and resources to survive and recover. More precisely, the United Nations defines a natural disaster as "a serious disruption of the functioning of a community or society involving widespread human, material, economic, or environmental losses and impacts, which exceeds the ability of the affected community or society to cope using its own resources." Some disasters have both natural and anthropogenic components—researchers may refer to these events as *socio-natural disasters*.

Notably, the disaster caused by a natural hazardous event might not be immediately obvious. For example, destruction of a forest when tsunamis wash over an isolated island may cause the extinction of a plant species that contains a chemical that could have saved millions of lives. No one is directly injured, but future generations would certainly be affected.

Classifying Hazards and Disasters by Their Source

Our planet consists of many components, or *realms*—the atmosphere, the oceans, the land surface, the interior, and life—that constantly interact with one another. Researchers refer to all these realms, together with all their interactions, as the **Earth System** (see Chapter 2). We can distinguish among different types of natural hazards and disasters, in part, by the realm of the Earth System in which they develop **(FIG. 1.8)**. Those associated with features of the solid Earth are *geophysical* or *geological*; those involving water (generally on or under land) are *hydrological*; those involving large bodies of water and their shores are *maritime* and *coastal*; those involving weather (air pressure, rainfall, snowfall, temperature, and wind at a given time in a given location) are *meteorological*; those depending on atmospheric conditions over a longer period are *climatological*; and those resulting from interactions with objects or particles coming from outside the Earth are *extraterrestrial* **(TABLE 1.1)**. The categories in this table are broad, and some overlap.

All of the above phenomena are *physical*, in the sense that they are due to the action or behavior of nonliving objects or energy. Of course, many hazards and disasters on our planet are *biological*, meaning that they are caused by living organisms—examples include insect infestations and disease *epidemics*, called **pandemics** if they affect the entire globe **(BOX 1.1)**. The COVID-19 pandemic that began in 2019, for example, is a biological disaster that clearly caused global chaos, catastrophic numbers of deaths, and immense economic loss. This book focuses on physical hazards and disasters, and mentions biological ones only when they have been triggered by physical events. Note that wildfires involve multiple realms, so we classify them separately in Table 1.1. Generally, meteorological and climatological hazardous events cause most natural disasters in any given year.

FIGURE 1.8 Natural disasters can be classified by realm.

(a) A geological disaster, a volcanic eruption, can burn through neighborhoods.

(b) A coastal disaster, high storm waves, can batter buildings along the coast.

(c) A meteorological disaster, a blizzard, can shut down air travel.

(d) The landscape during a drought in Somalia, a possible indicator of a climatological disaster.

(e) An extraterrestrial disaster, such as a meteor striking the Earth, can flatten forests and cities or cause extinctions.

(f) A plague of locusts represents a biological disaster. The locusts eat crops and can cause starvation.

TABLE 1.1 Classes of Natural Hazards and Disasters Based on Their Realm

Geophysical/Geological	Earthquakes; landslides; volcanic eruptions; ground subsidence
Hydrological	Floods; water depletion; river erosion
Maritime and Coastal*	Storm waves; tsunamis, coastal erosion, currents; storm surge
Meteorological	Thunderstorms; tornadoes; hail; ice storms; hurricanes; mid-latitude cyclones; drought; blizzards
Climatological	Cold waves; heat waves; drought, climate change
Extraterrestrial	Meteorite impacts; solar storms
Biological	Epidemics; insect infestations
Wildfire	Brushfires; forest fires

*Maritime and coastal hazards are sometimes grouped with hydrological hazards.

While hazardous to people, the dangerous phenomena we discuss play important roles in the Earth System. For example, without the processes that cause volcanic eruptions, continents wouldn't exist, for volcanic activity produces new rock. Without wildfires, some plant species wouldn't reproduce, for heat causes their seeds to open. Without floods, rivers would become choked with debris, and floodplains and deltas would lose their fertility. And without storms, insufficient rain might fall on the land to foster life.

Comparing the Time Frames of Natural Disasters

Not all natural disasters develop at the same rate or last for the same amount of time. A **rapid-onset disaster**, such as a deadly earthquake, begins quickly, before people have had time to evacuate, whereas a **slow-onset disaster**, such as a destructive regional flood, may take days or weeks to develop, providing people with time to prepare (**FIG. 1.9**). Most rapid-onset disasters are of short duration, whereas most slow-onset disasters last for a while. For example, an earthquake lasts only seconds to minutes, but a regional flood may last for over a month.

Significantly, some disasters take so long to develop that we may not notice their onset, even after several months to decades. Such events have come to be known either as **stealth disasters** or as *creeping disasters*. Examples include sea-level rise due to climate change, leading to flooding of coastal cities and displacement of populations; drought, resulting in loss of crops; *groundwater depletion*, the pumping and removal of **groundwater** (water that resides in pores and cracks underground) at rates faster than it can be replenished, causing loss

FIGURE 1.9 Contrasting rates of natural disasters.

(a) An earthquake is a rapid-onset disaster. A 2018 quake in Anchorage, Alaska caused the sudden collapse of this road, trapping a vehicle that had been driving on it.

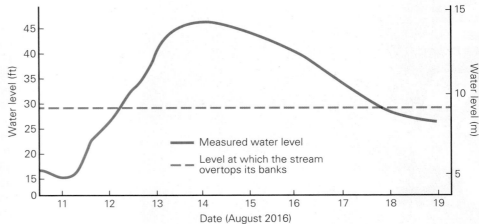

(b) This record of the water level (height above a reference elevation) in a Louisiana river shows the development of a slow-onset flood, during which floodwaters took days to rise and days to subside.

1.2 The Language of Danger • 9

BOX 1.1 A CLOSER LOOK...

Contrasting global pandemics with physical natural disasters

The year 2020 looked like a banner year for the world economy. Trade flourished, stock markets reached an all-time high, and global employment surpassed previous records. But in late 2019, a virus hosted by an animal somehow jumped across the species divide and infected a human in Wuhan, China. A virus is a microscopic infectious agent—10,000 times smaller than a grain of salt—that attaches to and infects a living cell. The virus then replicates in the cell. Eventually, new viruses burst out of the host cell, destroying it in the process, and infect other cells. The cell destruction caused by viruses, and the host's immune response to the virus, makes people sick.

The virus that appeared in 2019 is a type of coronavirus, meaning that its surface is covered by a "corona," an envelope of countless small protrusions. This virus causes a disease, now known as COVID-19, which starts as a fever and cough, but in about 1-2% of infected people it evolves into severe and potentially fatal pneumonia. The COVID-19 virus is highly contagious, for it transfers from person to person in exhaled droplets. Infected people can spread the disease even when they don't show symptoms. In our mobile society, seemingly healthy travelers soon carried it throughout many countries, in cars, trains, planes, and ships. By March 11, 2020, the World Health Organization declared the spread of COVID-19 to be a pandemic, for the disease had reached nearly the entire globe. By mid-April, at least 1.8 million people had contracted COVID-19 (inadequate testing means that the real numbers will never be known) and at least 110,000 people had died (hard numbers can't be known because of incomplete testing and reporting). In hot spots with high infection rates, hospitals became overwhelmed, and critical medical supplies ran out **(FIG. Bx1.1a)**. Although effective vaccines became available in 2021, the COVID pandemic continues, with a global death toll approaching 5 million by the year's end. Several causes contribute to spreading the disease: inadequate vaccine availability in some countries, the appearance of a particularly contagious version (the Delta variant), vaccination hesitancy, and avoidance of masking and social-distancing protocols.

A pandemic, as the world's experience with COVID-19 shows, can trigger an economic collapse. That's because, to stop the spread of the disease and limit casualties, most countries decreed that people should shelter in place by remaining indoors, away from others. So, over the course of only a couple of weeks, the world effectively shut down and went into a recession. Businesses large and small lost their revenue streams and went dormant or bankrupt, and stores, restaurants, and theaters closed. Consequently, millions of people suddenly became unemployed, buses and planes became empty, and even the busiest city streets were deserted **(FIG. Bx1.1b)**. To deal with the economic crisis, governments released massive loans and grants, thereby causing national debts to grow radically. By one estimate, COVID-19 cost the United States $15 trillion in 2020.

Catastrophic pandemics have swept across the world over the course of history. The Bubonic Plague, also known as the Black Death because hemorrhaging caused by the disease turned victims' skin black, remains the worst pandemic in human history. In the 14th century, this disease, caused by bacteria in fleas hosted by rats, led to the deaths of more than 130 million people out of a global population of about 475 million. It devastated Europe, where up to 60% of inhabitants died. The worst pandemic in the modern era, the 1918 influenza pandemic, infected one-third of the Earth's population (about 500 million people). Over 50 million people perished by its end, a death toll 30 times greater than all the fatalities caused by World War I (1914–1918).

FIGURE Bx1.1 Effects of the COVID-19 pandemic.

(a) The temporary field hospital set up in a convention and exhibition center in Madrid, Spain, in April 2020.

(b) An empty Times Square in New York City following the outbreak of COVID-19, during March 2020.

How do the casualties of physical natural disasters compare with pandemics? The 1931 flood of the Yangtze River in China may have directly killed over 400,000 people, and the famine that followed is estimated to have led to the deaths of 3.5 million more. The 1970 Bhola cyclone, a fearsome storm that struck the coast of what is now Bangladesh, was estimated to have killed up to 500,000 people. Estimated fatalities for each of the most deadly disasters over the past century (the 1920 Haiyuan earthquake, the 1976 Tangshan earthquake, the 2010 Haiti earthquake, and the 2004 Indian Ocean earthquake and associated tsunami) range from 230,000 to 280,000. Casualties from all of these natural disasters were less in number than those caused by the Bubonic Plague or the 1918 influenza pandemic. The COVID-19 pandemic has caused more deaths than the most dramatic rapid-onset natural disasters—such as earthquakes and volcanic eruptions—in written history. But by the end of 2020, it had not yet surpassed history's worst slow-onset disasters, such as immense regional floods or major droughts. Regardless, the local impact of casualties due to a natural disaster is very different from that of a pandemic. The former tend to concentrate in a smaller region, such as a city or a state. Within the region, the impact is catastrophic. Survivors not only find themselves surrounded by mountains of debris, but they may be left without life's essentials, such as clean water, food, and shelter. Pandemic casualties, however, are spread worldwide, and they do not necessarily interrupt the supply of life-essential materials.

In terms of economic cost, pandemics also differ significantly from physical natural disasters. Since written records began, most physical natural disasters have not produced global catastrophic economic impacts. For example, the 2010 Haiti earthquake devastated the country's national economy, and the country remains in crisis a decade later (see Chapter 3). However, the disaster did not result in a global recession. By contrast, even though the overall impacts of a pandemic don't take visible form—buildings still stand, trees still grow, and everything looks the same as the day before—its global economic costs are almost beyond comprehension. Due to COVID-19, individual and national debts accumulated at such rates that economists still can't say with any certainty how long a recovery will take.

Note that in our discussion, we compared pandemics to natural disasters that happened during human history. As you will see later in this book, the geologic record shows that giant meteor impacts and supervolcanic eruptions have altered the course of life evolution on our planet. If such events were to happen now, their consequences could dwarf those of even the worst pandemic. (The only event that could be comparable would be a major nuclear war.) While the probability of a supervolcanic eruption or a giant meteor impact in any given year, fortunately, is miniscule, society does face a near-term potential catastrophe in the form of climate change (see Chapter 15). Without slowing climate change, or mitigating its consequences, the human and economic toll of the droughts, floods, famines, storms, and loss of species that may result could exceed that of a pandemic. Clearly, both pandemics and natural disasters have the potential to alter the course of human civilization, but as you will see later, society can often blunt their consequences by taking appropriate action in advance.

of water supplies; and *soil erosion*, the loss of the top layer of the ground in which plants grow, reducing fertility.

We use Earth materials (such as rocks, water, soil, and metals) in all aspects of society. Some of these materials, unfortunately, can be dangerous to human health. Long-term contact with these materials can cause a **chronic health disaster**, a type of stealth disaster. Arsenic poisoning in Bangladesh serves as an example. In the 1980s, people in Bangladesh switched from drinking polluted river water to drinking groundwater pumped from deep wells, in order to avoid bacterial diseases. Unfortunately, some of the groundwater contained arsenic at a concentration that could cause illnesses. Chronic health disasters can also result from inhaling coal dust, rock dust (from rocks such as granite and sandstone), and certain types of asbestos dust (**FIG. 1.10**). When trapped in human lungs, these substances can cause black lung disease, silicosis, and lung cancer, respectively, and have been responsible for thousands of deaths.

FIGURE 1.10 Workers removing asbestos, a potentially hazardous mineral, must wear protective gear, and the work area must be sealed off.

Distinguishing among Primary, Secondary, and Tertiary Disasters

When TV news programs broadcast real-time images of a disaster, they're usually showing a **primary disaster**, casualties and destruction directly resulting from a natural hazardous event itself. Examples include buildings collapsed by an earthquake or homes destroyed by a flood (**FIG. 1.11a**).

Unfortunately, in some cases, a **secondary disaster**, a hazardous event triggered by the primary disaster, follows the primary disaster (**FIG. 1.11b**). For example, after an earthquake, piles of flammable debris may ignite, producing an urban *firestorm*, a huge blaze that generates strong winds. Such a secondary disaster happened after a 1923 earthquake devastated Tokyo—hot coals from cooking stoves ignited an inferno that incinerated 38,000 survivors of the earthquake.

Following a primary and secondary disaster, there may also be a **tertiary disaster**, a long-term disruption of society. The aftermath of a hurricane serves as an example. Wind and wave damage that happens during the storm represents the primary disaster. A secondary disaster could result from the collapse of dikes after the storm has passed, allowing large areas to be flooded. Loss of housing to decay and mold after the floodwaters recede represents a tertiary disaster. Tertiary disasters include long-term socioeconomic crises, such as job losses due to the destruction of stores and factories, or the disruption of government services and authority. And in cases where victims move into crowded refugee camps, disease can become a devastating tertiary disaster (**FIG. 1.11c**). Physical and mental health challenges arising from traumas during the disaster and rescue also represent a tertiary disaster, one that impacts some people long after other aspects of society have returned to normal.

FIGURE 1.11 Primary, secondary, and tertiary disasters.

(a) A major earthquake that collapses buildings and sets landslides in motion is a primary disaster.

(b) A fire triggered by an earthquake is a secondary disaster.

(c) The spread of disease among displaced earthquake survivors in refugee camps is a tertiary disaster.

> **Take-home message . . .**
>
> Features or processes of the Earth System that have the potential to cause harm to people or property are natural hazards. A natural hazardous event that harms people or property is a natural disaster, so natural disasters are defined by their effect on people. Rapid-onset disasters happen quickly, while slow-onset disasters develop slowly. Stealth disasters take so long to develop that we might not even realize that they are happening.
>
> **QUICK QUESTION** Distinguish among primary, secondary, and tertiary disasters.

1.3 Characterizing Disaster Size and Frequency

Comparing the Sizes of Disasters

When an earthquake struck China in 2008, landslides buried towns and countless buildings collapsed, resulting in almost 70,000 fatalities and $150 billion in damage. In contrast, an earthquake that struck Illinois the same year caused a few minor injuries and toppled a few chimneys. How can we compare these earthquakes in a meaningful way? Should both be called disasters?

As you will discover later in this book, researchers have devised scales to compare hazardous events of a given type. For example, geologists classify earthquakes using either the *Modified Mercalli Intensity scale*, based on the damage sustained and the perception of ground shaking by witnesses, or the *moment magnitude scale*, based on the amount of energy that an earthquake releases, regardless of its consequences. Atmospheric scientists classify hurricanes using the *Saffir-Simpson scale*, which assigns each storm to a category based on its maximum sustained wind velocity, and classify tornadoes using the *Enhanced Fujita scale*, based on the types of vegetation and human-built structures the tornado destroys.

Significantly, none of the above measures can give the full picture of the relative size or scope of a disaster. That's because the consequences of a hazardous event depend not only on the direct force applied during the hazardous event, but also on factors such as the dimensions of the area affected, the duration of the event, the number of people affected, and the economic cost of the event. A hurricane's Saffir-Simpson category, for example, provides no information about the storm's size, or about the extent of the damage it causes. To illustrate this problem, compare Typhoon Tip in 1979 (a *typhoon* is a hurricane in the western Pacific) with Hurricane Katrina in 2005. Both were Category 5 storms because of their wind velocities. Although Tip's area was much bigger than Katrina's **(FIG. 1.12)**, Katrina caused vastly more damage than Tip, because Tip mostly swirled over the ocean, while Katrina's *storm surge* (the bulge of water that the storm pushes landward) and waves wiped out coastal communities in Alabama and triggered the flooding of New Orleans. The consequences of a hazardous event also reflect human factors, such as quality of construction or the ability of people to prepare for and respond to the event, as we'll see shortly.

Since no all-purpose scale successfully compares natural disasters quantitatively, reports use informal adjectives to convey a sense of their consequences. Roughly speaking, a hazardous event is a *local disaster*, *minor disaster*, or *small-scale disaster* if it affects a relatively small area, causes relatively few casualties and relatively small economic losses, and results in

FIGURE 1.12 Contrasting sizes of disasters.

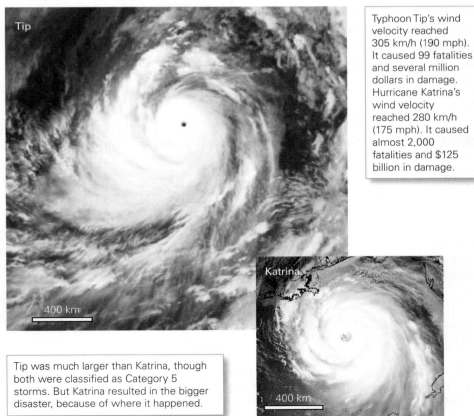

Typhoon Tip's wind velocity reached 305 km/h (190 mph). It caused 99 fatalities and several million dollars in damage. Hurricane Katrina's wind velocity reached 280 km/h (175 mph). It caused almost 2,000 fatalities and $125 billion in damage.

Tip was much larger than Katrina, though both were classified as Category 5 storms. But Katrina resulted in the bigger disaster, because of where it happened.

no long-term disruption of society. Events that affect a broad region, causing large losses of life and property, major economic losses, and long-term disruption of society are *major disasters*, *large-scale disasters*, *calamities*, or in extreme cases, **catastrophes**. Some researchers have suggested the term **megadisaster** for very rare, particularly immense catastrophes that have affected, or could affect, society worldwide. Use of such terms remains subjective, for no one has defined the specific range of casualty numbers or economic cost to distinguish among disasters in a consistent way.

Comparing the Frequency of Disasters

Natural disasters are not daily events, but several make the news in any given year. Researchers use two related terms to compare the frequency of disasters. The first, known as the **recurrence interval** (**RI**), indicates the average time between successive events of a given size (or, in some contexts, of a given size or larger). For example, if a long historical record is available, you can calculate the RI of an event of a given size or larger by taking the average of the time intervals between the events. To see how, imagine that a disaster struck in 1645, 1801, 1837, 1960, and 2017 **(FIG. 1.13a)**; the time gaps between the successive events are 156, 36, 123, and 57 years, respectively. The average of these numbers gives the RI. In this example, RI = 93 years, meaning that the average time between these disasters is 93 years. Significantly, an RI of 93 years does not mean that such an event happens every 93 years, like clockwork. Disasters are not periodic, for they can result from many phenomena, some of which happen randomly.

FIGURE 1.13 The concept of a recurrence interval for a disaster of a given type.

(a) A disaster of a given size (or larger) at location X has a recurrence interval of 93 years. Note that the disasters are not periodic.

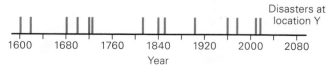

(b) A disaster of the same size (or larger) at location Y has a recurrence interval of about 30 years. Note that such a disaster is more likely to happen at location Y than at location X by 2080.

Unfortunately, news media and the public don't always recognize that the RI of a disaster is not the same as periodicity, and they may conclude incorrectly that if a disaster happened 50 years ago in a city, and if it has a recurrence interval of 93 years, then the city will be safe for another 43 years. Looking closely at the list of dates in the previous paragraph, you'll notice that two of the gaps are much less than 93 years. Indeed, the next disaster affecting the city could well happen in 100 years, in just 20 years . . . or even tomorrow! To avoid such confusion, researchers may represent the frequency of a disaster by specifying its **annual probability** (**AP**), a number that indicates the likelihood that an event will happen in a given year. (For some kinds of disasters, a different number, *annual exceedance probability*, or *AEP*, is used. It refers to the likelihood that an event of a given size or larger will take place.) RI and AP are related by the equation AP = 1 ÷ RI, so an event with a recurrence interval of 93 years has an AP of $\frac{1}{93}$, or about 0.011. You can multiply the decimal by 100 to convert it into a percentage. Therefore, the AP of our example is about 1.1%, or, put another way, slightly more than 1 in 100. The larger its AP, the more likely a disaster is to happen in a given year. For example, a disaster with an RI of 30 years has an AP of $\frac{1}{30}$, or 3.3%, and 3.3% is greater than 1.1% **(FIG. 1.13b)**. In Chapters 3 and 12, we will provide additional detail on RI, AP, and AEP.

Significantly, smaller hazardous events happen more frequently than do larger ones. For example, in a given region of earthquake activity, mild earthquakes of magnitude 4 happen roughly 40 times more frequently than do moderately strong ones of magnitude 6.

The Human Toll and Economic Impact of Disasters

Disasters make headlines, bluntly speaking, because of the shock and horror that they entail. To characterize the human toll of a disaster, officials ideally take into account not only casualties, but also the number of **displaced people** **(FIG. 1.14)**—including both *internally displaced people* (people who have left their homes but remain within the borders of their country; these people are also called *evacuees*) and *refugees* (people who have left their homes and traveled to another country)—as well as people who have not moved but have suffered significant economic loss or distress. The sum of the previous numbers defines the **total people affected** by a disaster. To establish the **economic impact** of a disaster, officials estimate the cost of replacing or repairing damaged buildings, property, and *infrastructure* (roads, power supplies, pipelines) and take into account the value of decreases in economic activity due to the loss of industrial production, agricultural production, and tourism. Unfortunately, statistics

FIGURE 1.14 Natural disasters may displace large numbers of people.

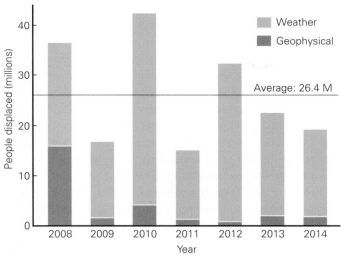

(a) Numbers of people displaced by natural disasters (both internally displaced people and refugees), 2008–2014. Most displacement over this 7-year period occurred due to weather-related events.

(b) Food being distributed to drought-displaced people in Somalia.

concerning disasters are difficult to tally, interpret, and compare for several reasons:

- News reports tend to highlight the number of fatalities due to the disaster **(TABLE 1.2)**, but the total number of people affected can be orders of magnitude more. In fact, the humanitarian crisis caused by a natural disaster may be huge, even if the number of fatalities was not. For example, 50,000–260,000 deaths were attributed to famine resulting from the East Africa drought of 2011, but the total number of people needing humanitarian aid approached 12 million.

- Reports of the economic impact of a disaster might not take into account broader losses due to associated secondary or tertiary disasters.

- The numbers reported aren't always accurate. In some cases, a country lacks sufficient emergency personnel or an adequate communication infrastructure to provide accurate data. In other cases, inaccuracy stems from political considerations that lead officials to intentionally report undercounts or overcounts.

- The impact on society that total casualty and economic loss numbers represent is different at different times in

TABLE 1.2 Notable Deadly Examples of Different Types of Disasters

Date	Location	Type	Death toll
1876–79	China	Drought	9,000,000–13,000,000
1931	China (Yangtze River)	Floods	1,000,000–4,000,000
1556	China (Shaanxi)	Earthquake	830,000
1970	Bangladesh/India (Bhola)	Tropical cyclone	500,000
2004	Indonesia (Sumatra) and other locations	Tsunami	228,000
1815	Indonesia (Mt. Tambora)	Volcanic eruption	71,000
1989	Bangladesh	Tornado	1,300
1871	United States (Wisconsin)	Wildfire	1,200-2,500

Note: The numbers reported are estimates. The amount of uncertainty in the numbers is based on when the events occurred (accounts are less accurate the further back in time the event took place) and the location of the event (less industrialized areas generally have less accurate statistics).

history because of population growth and inflation. For example, a 1601–1603 famine in Russia killed 2 million people. That number represented 30% of the region's population at the time, but would represent 1.4% of the region's population today. Similarly, unless economic loss estimates take inflation into account, they give a misleading sense of a disaster's financial burden on society. For example, $1 billion in 1919 represents about $26 billion in 2019.

- The value assigned to damaged or destroyed property depends on location. Specifically, property tends to have a greater assigned value in developed countries than in developing countries. For example, a hurricane that destroys the luxury condos of 1,000 people in an upscale resort may cause property losses with a higher dollar value than does a typhoon that destroys the shacks of 100,000 villagers living in a shantytown.
- In the case of socio-natural disasters, it may be impossible to separate the aspect of the disaster caused by society from the aspect caused by nature in a meaningful way. For example, the Indian famine of 1866–1867, which killed a third of the population of the state of Odisha, started with a drought, but became a disaster because colonial rulers continued to export agricultural products, leaving insufficient food for the local population.
- Reported disasters might not be classified consistently. For example, casualties due to inland floods and landslides might not be attributed to the hurricane that caused them.

Despite the limitations of statistics, numbers reported by government agencies, international aid organizations, and insurance companies give a general picture of which disasters, over time, cause the most casualties and the greatest economic losses. Between 150 and 250 million people are affected by disasters in a given year, on average. Of these victims, between 50,000 and 100,000 die, on average. Such numbers mean that in a given year, the number of people affected by natural disasters can exceed the number affected by military conflicts, but not necessarily a pandemic. Globally, meteorological, climatological, and hydrological disasters cause the most deaths. Geophysical and geological disasters generally cause fewer deaths, but in some years catastrophic earthquakes have topped the list.

In the United States between 1982 and 2019, deaths due to storms that caused more than a billion dollars in damage significantly exceeded deaths due to other types of natural disasters (**TABLE 1.3**). Tropical cyclones (hurricanes and typhoons), by far, produced the most damage because of their severity and frequency (**FIG. 1.15**). Roughly speaking, the inflation-adjusted cost of natural disasters over the last 37 years in the United States exceeded $1.7 trillion. Unfortunately, costs are rising. In the United States, disaster damage in 2020 totalled $95 billion, nearly double that of 2019. It's difficult to obtain accurate data on global economic losses due to disasters, but by some estimates, they represent about 0.15% to 0.50% of global gross domestic product (GDP), on average.

Fortunately, technological improvements, such as satellite monitoring of hurricanes and floods, have reduced casualties worldwide. For example, over 2 million people were evacuated before Cyclone Bulbul struck Bangladesh and eastern India in 2019, so fatalities were vastly fewer than they had been for a similar cyclone in 1970. But the increase in the number of people, the number of buildings, and the extent of

TABLE 1.3 Costs and Deaths from Billion-Dollar Disasters in the United States, 1982–2019

Disaster type	Number of events	Frequency (%)	Inflation-adjusted costs (in billions of dollars)	Percentage of total losses	Average event cost (in billions of dollars)	Deaths
Drought	25	10.0%	$216.0	12.5%	$8.6	1,733
Earthquake	2	0.8%	$65.9	3.8%	$32.9	120
Flooding*	31	12.4%	$123.9	7.1%	$4.4	548
Freeze	8	3.2%	$28.7	1.6%	$3.6	162
Severe storm	109	43.4%	$240.2	13.7%	$2.2	1,620
Tropical cyclone	43	17.1%	$936.3	53.9%	$22.8	6,489
Wildfire	16	6.4%	$80.4	4.6%	$5.0	344
Winter storm	17	6.7%	$49.2	2.8%	$2.9	1,048
All disasters	251	100%	$1,740.6	100%	$10.3	12,064

Source: Statistics as estimated by NOAA.
Note: Tropical Storm Imelda, Hurricane Dorian, and 2019 river flooding are not included in these statistics.
*Inland floods caused by tropical cyclones are included in the tropical cyclone category, not in flooding.

FIGURE 1.15 Deaths and economic impacts from natural disasters.

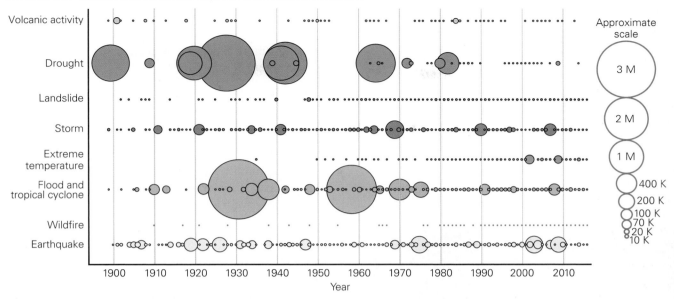

(a) Global deaths due to natural disasters since 1900. The diameter of the circle represents the number of deaths.

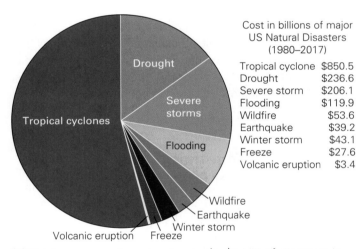

Cost in billions of major US Natural Disasters (1980–2017)	
Tropical cyclone	$850.5
Drought	$236.6
Severe storm	$206.1
Flooding	$119.9
Wildfire	$53.6
Earthquake	$39.2
Winter storm	$43.1
Freeze	$27.6
Volcanic eruption	$3.4

(b) Relative economic costs of major (> $1 billion) disasters in the United States for 1980-2017. The total cost of these disasters approaches $1.6 trillion.

infrastructure in harm's way has led to continuous increases in property losses.

How do people perceive the severity of disasters? To a large extent, this depends on coverage in the media. Researchers have found that coverage tends to emphasize the casualties from sudden-onset disasters or from disasters that are visually dramatic. For example, according to a study conducted in 2007, a famine must kill 40,000 people to warrant the kind of news coverage allotted to a single fatality caused by a volcanic eruption. Put another way, news coverage isn't in proportion to the total people affected by, or the economic impact of, a disaster, but rather reflects how spectacular or unusual the consequences of the disaster appear to be.

Take-home message . . .
Different scales have been devised for physically comparing natural hazardous events. But such scales don't characterize the full extent of disasters that happen due to these events because they don't take into account the dimensions or duration of the event or the human and economic costs. Statistics on disasters are difficult to interpret, but nevertheless emphasize that, globally, disasters impact millions of people and cost billions of dollars every year.

QUICK QUESTION What's the difference between the recurrence interval and the annual probability of a disaster?

1.4 Exposure, Vulnerability, and Risk

Characterizing Exposure and Vulnerability

In everyday English, we use the words *vulnerability* and *exposure* in a variety of different ways. In the context of natural disasters, each has a narrower meaning. Specifically, the term **exposure** refers to the potential casualties, together with the potential economic losses and social disruption, that a community faces because of its proximity to a natural hazard. **Vulnerability** refers to the characteristics of a community that affect its ability to cope with a hazardous event, such as the degree to which structures can resist damage, preparations can minimize the total number of affected people and the economic impacts, and resources are available to help victims and rebuild damaged infrastructure. Significantly, a community's wealth, traditions, and political considerations influence its exposure and vulnerability, for many reasons:

- Some communities, due to either architectural traditions or lack of financial resources, construct buildings out of less expensive or weaker materials that can't stand up to forces such as earthquake shaking or hurricane winds (FIG. 1.16).

FIGURE 1.16 Wealth and construction traditions may affect the quality of buildings. Poorly constructed buildings may collapse during earthquakes or storms.

(a) A typhoon destroyed an impoverished seaside town in the Philippines in 2013.

(b) Well-constructed buildings survived a hurricane in Myrtle Beach, South Carolina, in 1989.

- Due to the cost of land, poorer communities may have to build closer to known hazards, where land is cheaper.
- Wealthier communities lose more expensive buildings, containing more expensive objects, to a hazardous event than do poorer communities.
- Poorer people may live in more densely populated communities than do wealthier people.
- Funding for emergency services and personnel reflects the tax base of a community.
- Poorer communities are less likely to evacuate because transportation to safer locations may be scarce, and often they cannot pay for alternative housing. Also, lack of insurance may motivate residents to remain and try to protect their property.
- The political leanings of officials may influence decisions about investments in infrastructure, emergency-service funding and planning, zoning rules, aid distribution, and access to rebuilding funds.

In some cases, local customs also play a role in determining the response to a hazard or disaster. For example, in some communities, people trust officials and their warnings and may be willing to heed evacuation orders. In others, they may not move out of harm's way because they don't trust officials, or because they believe that their religious faith will protect them.

Putting It All Together: Defining Disaster Risk

In discussions of natural hazards, the term **risk** can be broadly defined as the probability that an individual or community will be subject to injury or loss due to hazardous events. We can relate risk for a given region to three factors: the annual probability of a hazardous event, the region's exposure, and the region's vulnerability. We've already discussed annual probability; now let's consider exposure and vulnerability.

Risk takes into account both the likelihood that a community will experience a natural hazardous event during a specified time interval, and the consequences of the event. The consequences, in turn, depend on exposure and on vulnerability, so we can represent risk by a simple formula:

$$\text{Risk} = \text{hazard} \times \text{exposure} \times \text{vulnerability}$$

In this equation, *hazard* encompasses both the nature of the hazardous events a community faces and the AP of those events.

To gain more insight into the concept of risk, let's contrast the risks faced over the course of a century in two areas.

The first is a small, wealthy farming community in central Canada. The second is a large, impoverished fishing community on a politically unstable, volcanically active tropical island in the western Pacific. The farming community will not experience a tsunami, typhoon, or volcanic eruption, and probably won't experience an earthquake, but the community may be subjected to blizzards, ice storms, floods, drought, or hail, which could reduce its income sources. Given the nature of the hazardous events that this community may face, and given the services available, relatively few casualties and relatively little financial loss would happen during a hazardous event. Therefore, the hazard and exposure components of the risk equation are low. Also, because affected residents could probably survive after the event by receiving insurance payments, living off savings, taking out loans, or receiving government subsidies, the community's vulnerability is low.

In contrast, the fishing community could experience a tsunami, volcanic eruption, typhoon, or earthquake, but the people won't endure blizzards or ice storms, floods, drought, or hail. Given the nature of the hazardous events that the community may face, and the lack of available services, many casualties and substantial financial losses could happen due to a hazardous event. Therefore, the hazard and exposure components of the risk equation are high. Due to lack of insurance, savings accounts, and government aid, many people would be destitute after an event. And because of its high population density, poor construction methods, and lack of financial resources and social services, the community's vulnerability is high. The information in this example indicates that the disaster risk for the tropical fishing community is significantly greater than that for the Canadian farming community.

Given a defined hazard, risk can be lowered by decreasing exposure, vulnerability, or both. For example, communities could decrease exposure to wildfire risk by enacting zoning laws that prohibit construction of buildings in places susceptible to fire, and could decrease vulnerability by improving the fire resistance of property, clearing brush, training emergency personnel, identifying water supplies, setting up a warning system, and educating the public on escape routes to follow in the event of a fire (see Chapter 14).

Nearly every place on the Earth faces some disaster risk, but some places face more than others. Governments and nongovernmental organizations analyze risk to determine where to invest in resources that could prevent disasters or improve responses to them. Insurance companies, as we will see, analyze risk to determine whether they can cover potential losses. Individuals, therefore, perhaps without realizing it, make a decision about how much risk they are comfortable living with when they choose a home. Such decisions about **acceptable risk** effectively balance an estimate of the likelihood of a disaster at a location against a subjective perception of a location's appeal. Questions about acceptable risk, therefore, are personal. Would you move to another location if your house had a 0.1% chance of being destroyed by a hurricane during the next 10 years? What if it had a 90% chance? What if it had a 30% chance, but your family lived nearby, you had a good job, and you liked the scenery and weather? What if you didn't have funds to move elsewhere? These aren't easy questions to answer.

> **Take-home message . . .**
> The vulnerability of a community to a natural disaster refers to the community's ability to prepare for, cope with, and recover from a hazardous event. Exposure takes into account a community's proximity to a natural hazard and the potential scope of a hazardous event arising from the hazard. Disaster risk depends on the annual probability of a hazardous event, and on the event's consequences, as determined by vulnerability and exposure.
>
> **QUICK QUESTION** How does the wealth of a community affect its vulnerability to natural disasters?

1.5 Factors Influencing Disaster Risk

Effect of Population Growth

Our planet's human population has doubled since about 1970 and surpassed 7.8 billion in 2020 (FIG. 1.17). This **population explosion** has major implications for the

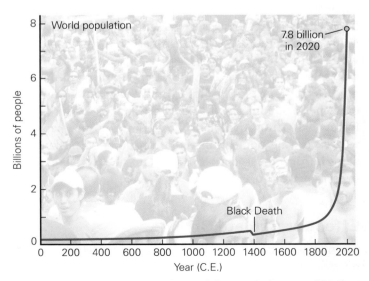

FIGURE 1.17 The growth of the world's population puts increasing numbers of people in harm's way.

FIGURE 1.18 Favelas (shantytowns) may be built on unstable slopes in Brazil, where heavy rains can trigger disastrous landslides.

disaster risk faced by society because it increases exposure to natural hazards. To see why, let's consider landslide risk in Rio de Janeiro, Brazil. In the mid-19th century, the city's small population only occupied the plains between smooth rock mountains known as sugarloafs. By the mid-20th century, however, the population had grown by so much that the plains were fully occupied, and land there had become expensive. Favelas, densely populated communities of poorly constructed houses, began appearing on the slopes rising toward the bases of the sugarloafs. Unfortunately, weak, slippery sediment underlies these slopes. During rains, the sediment mixes with water and turns into slimy mud, so the slopes sometimes fail, producing landslides that overwhelm buildings and result in casualties and property losses **(FIG. 1.18)**. A similar increase in exposure happens where growing populations build on flood-prone land or along hurricane-susceptible coasts, and where cities expand in earthquake-prone regions.

Effect of Climate Change

The number of natural disasters caused by geological and geophysical phenomena has stayed about the same over the past century, but the number of meteorological, hydrological, and climatological disasters appears to be increasing. Researchers are exploring the link between this trend and *global warming*, the gradual increase in atmospheric temperature that has been happening during the past two centuries as one manifestation of climate change (see Chapter 15).

Why might global warming influence disaster risk? Warming leads to an increase in the number of heat waves, droughts, and wildfires. Warming also causes an increase in seawater temperature (leading to the expansion of water) and causes glaciers to melt—both of these phenomena cause sea level to rise, worsening coastal flooding (see Chapter 13). And because warmer seawater nourishes hurricanes and kills corals (see Chapter 11), warming may result in more serious coastal damage by increasing the strength of hurricanes and by destroying reefs that protect coastlines from waves.

Human Influence on Disaster Risk

At the beginning of this chapter, we distinguished between anthropogenic and natural hazards, giving the impression that they are separate phenomena. In the modern world, however, it's often hard to completely separate the two because the actions of society have changed so much of our planet; as a result, human actions and interventions influence disaster risk. We've already mentioned how human decisions can turn a drought into a disastrous famine. Here, we examine some additional examples: urbanization, deforestation, and grassland destruction.

Urbanization, the migration of people from the countryside to cities and the consequent expansion of cities, can raise the risk associated with several types of disasters. For example, urbanization can increase flooding risk because the concrete and asphalt of cities prevent rainwater from infiltrating into the ground, so water from a downpour goes directly into streams, causing them to fill faster and rise higher (**FIG. 1.19a**). **Deforestation**, the removal of trees over a broad area, together with **grassland destruction** (the conversion of grasslands into farm fields or suburbs) also increases flooding risk, because without trees or grasses to absorb water, more rainfall flows into streams. Deforestation and grassland destruction also increase the risk of landslides, both because they allow more water to soak into the ground and because they decrease the presence of soil-strengthening roots. Rain that might not have triggered a landslide in a forested area or grassy prairie may cause one in a deforested area (**FIG. 1.19b**).

> **Take-home message...**
> Several anthropogenic changes that have taken place in the past two centuries influence disaster risk—namely, population growth, climate change, and activities such as urbanization, deforestation, and grassland destruction.
>
> **QUICK QUESTION** How might climate change affect disaster risk?

FIGURE 1.19 Human activities can worsen natural disasters.

(a) Rain on city streets can't soak into the ground, so it flows into drains that take it directly to a stream, causing flooding.

(b) Deforestation made this slope in Uganda more susceptible to a landslide. This 2018 event buried 30 homes.

1.6 Dealing with Disaster
Response, Recovery, Restoration, and Resilience

On August 13, 2017, winds began to strengthen and swirl in the tropical latitudes of the eastern Atlantic Ocean. Meteorologists from the **National Weather Service** (**NWS**), the government agency that monitors storms and other weather phenomena in the United States, took notice. On August 17, the storm had become strong enough to gain a name, Tropical Storm Harvey. When it moved over the western Gulf of Mexico, its winds accelerated, and it was renamed Hurricane Harvey (see Chapter 11). The NWS predicted that Harvey would make landfall on the coast of Texas on August 25. Given the low elevation of coastal Texas, it was clear that the storm would cause not only wind damage, but also flooding. So the **Federal Emergency Management Agency** (**FEMA**), in partnership with the Texas Division of Emergency Management and other officials, went to work. Supplies and vehicles were put in position, evacuation plans were reviewed, and emergency personnel were brought in. When Harvey arrived, it stalled and drenched the region with historic amounts of rain. Emergency personnel (together with other government agencies, private citizens, and nongovernmental organizations) *evacuated* 780,000 people (took them out of harm's way), *rescued* over 120,000 people (took them from the area after the disaster struck), provided 700 shelters that together could give 42,000 people a safe place to sleep, and worked frantically to restore essential services. The hurricane was a major disaster, but without emergency services, the human toll could have been much worse!

Clearly, when a natural disaster takes place, the affected communities, as well as outside agencies and organizations,

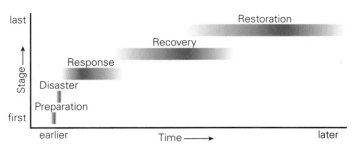

FIGURE 1.20 The relative times of the different stages of disaster relief.

must swing into action **(FIG. 1.20)**. For disasters that can be predicted, there is a **preparation stage** during which evacuations take place, extra emergency personnel are brought in, shelters are set up, and buildings are protected to the extent possible. During the **response stage**, which can begin even before disaster is over, officials and volunteers rescue survivors **(FIG. 1.21a)**, treat the injured, provide emergency shelter and security, and prevent the effects of the primary disaster from triggering a secondary disaster.

As the response stage concludes, affected communities enter the **recovery stage**. This stage involves cleaning up, finding or building longer-term shelter, re-establishing health care systems, and providing essential services such as access to drinking water, food, sanitation, electricity, and communication **(FIG. 1.21b, c)**. After these basic needs have been met and communities have been stabilized, officials turn their attention to rebuilding damaged buildings and infrastructure and re-establishing economic stability in the **restoration stage (FIG. 1.21d)**. Depending on the resources available, restoration may take months to years.

The ability of a community to respond, recover, and restore itself in a timely manner after a disaster characterizes its **resilience**. Resilience depends on the degree of devastation, the character and location of the damaged community, and its access to resources that can be used to repair damage and resettle people. Access to resources, in turn, depends not only on wealth, but also on the willingness and ability of government agencies or nongovernmental organizations to step in and help, and their efficiency in doing so. Consequently, different regions have different levels of resilience.

Who Responds to a Disaster?

Who is responsible for disaster response? Again, it depends on the size of the disaster. In the United States, local authorities bear the burden of a local disaster. If the response will require statewide resources, the state's governor issues a **state of emergency declaration**, which legally shifts the authority to coordinate aid to the state so that state police, the National Guard, and various state-employed emergency personnel can be mobilized to help. This declaration can be issued during the preparation or response stage. If the disaster looks like it will become too large for state authorities to manage, the governor asks the president to issue a **major disaster declaration**, which authorizes mobilization of federal funds and resources. FEMA can then deploy staff and supplies to assist victims **(FIG. 1.21e)**.

Other government agencies may play key roles during disasters. In the United States, for example, the *National Oceanic and Atmospheric Administration* (*NOAA*) may provide real-time descriptions and near-term predictions of weather disasters. The Coast Guard manages offshore rescue operations, and the US Navy may provide hospital ships. In cases where the National Guard does not have sufficient personnel, the regular army may be deployed to assist on land. When river flooding takes place, the *Army Corps of Engineers* monitors dams and levees. And in the case of wildfires, the *Forest Service* helps guide firefighting efforts. If public lands have been affected, the *Bureau of Land Management* (*BLM*) becomes involved in a variety of damage control and law enforcement activities. Private and public utilities also jump in to help, staging their equipment and personnel so as to have access to affected areas as quickly as possible. At an international scale, the United Nations and international aid agencies (such as the International Red Cross and Red Crescent), as well as the foreign-aid offices of wealthier countries, may provide assistance. And needless to say, losses during a disaster would be far worse were it not for volunteers—individuals, companies, faith-based organizations, contractors, and others—who help rescue neighbors, assist in efforts to confine the effects of the disaster, provide shelter and supplies to evacuees, remove downed trees, and repair damaged buildings.

Predictions, Forecasts, and Warnings

Imagine the days before satellites, radios, TV, radar, cell phones, and computers, not to mention modern scientific understanding of natural hazards. You might have had no idea that a hazard was about to generate a disaster. For example, if you were a sailor on a ship crossing the Atlantic in the early 16th century, you wouldn't have known what a hurricane was, let alone recognized warning signs that one was approaching. Today, of course, satellites perched in space send back real-time photographs of all kinds of storms as well as data on wind speed and rainfall distribution within. With these data and other information about global wind directions and atmospheric conditions, computer models can provide a **forecast**, an estimate of future weather. Such forecasts can help provide a **disaster prediction**, an estimate of the character, location, and timing of a disaster that allows government agencies and aid organizations to plan response and recovery.

FIGURE 1.21 Examples of response, recovery, and restoration after a hurricane.

(a) National Guard personnel, local authorities, and volunteers rescue victims trapped by floodwaters.

(d) Homes may be rebuilt, ideally to be less vulnerable.

(b) Once the waters have subsided, residents clear damaged materials from their homes to be hauled away.

(e) FEMA workers assist people impacted by the disaster.

(c) Inside damaged homes, wet drywall needs to be removed and mud cleaned out.

Long-term forecasting (more than 14 days out) of hazardous weather events, such as hurricanes, tornadoes, and thunderstorms, remains very uncertain. But in recent decades, *short-term forecasting* of such events (less than 3 days out) has become reasonably accurate, though surprises still occur. The shorter the time frame, the more accurate the forecast. When predictions indicate that conditions are right for a hazardous weather event to develop in or near the United States, the NWS issues a **watch**. If a hazardous weather event develops, the NWS issues a **warning** to areas that will be affected within a time frame of minutes to hours. Such warnings allow people to seek appropriate shelter in time **(FIG. 1.22)**.

Most people are used to hearing or seeing weather predictions, for they accompany every newscast and appear in every newspaper. Can we issue forecasts, or predictions, for other kinds of disasters? The broad-brush answer is . . .

FIGURE 1.22 An example of a tornado warning from the National Weather Service.

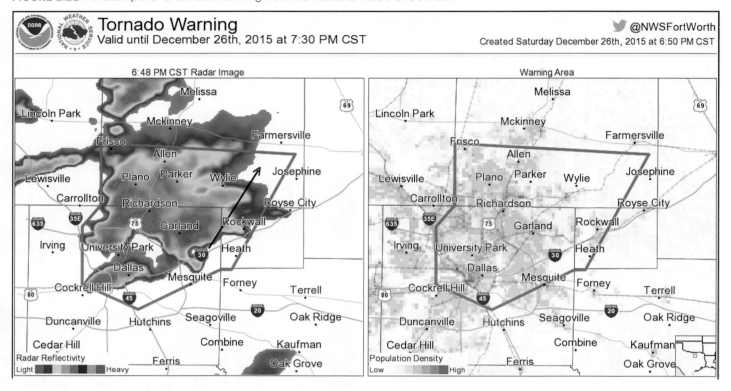

FIGURE 1.23 An example of a hazard potential map. This map shows where lava flows, ash falls, and mudflows (lahars) are likely to happen in the event of a volcanic eruption in Crater Lake National Park, Oregon.

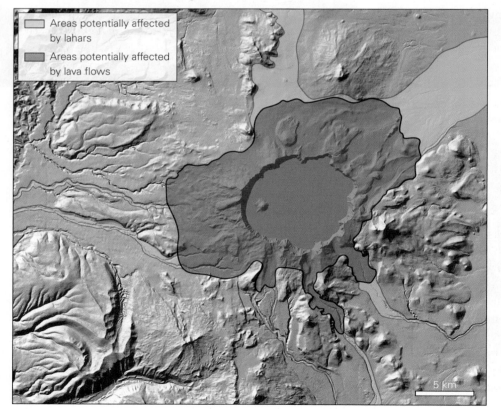

sometimes, but not always. That's because not all disasters have obvious **precursors**, events that indicate that a hazardous event is starting. In subsequent chapters, you'll see that observations can lead scientists to issue forecasts for some, but not all, volcanic eruptions, landslides, wildfires, and tsunamis. But a method for predicting the exact day that an earthquake will happen has not been devised, and may never be (though the AP of earthquakes can be estimated roughly from the historical record).

Planners can use data on the AP of hazardous events, along with assessments of preconditions that may make a locality susceptible to a given hazard, to draw a **hazard potential map**. Such maps show where disasters have a high probability of taking place (FIG. 1.23). A *flood potential map* highlights areas that are likely to flood if the water flow in a stream increases. An *earthquake potential map* delineates known *seismic belts*, places where earthquakes

24 • CHAPTER 1 • Our Planet Can Be Dangerous: Introducing Natural Disasters

FIGURE 1.24 Tornado warning sirens and refuge signs.

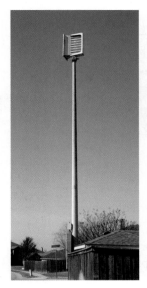

(a) This siren will warn that a tornado may be approaching.

(b) Signs can direct people to tornado shelters, such as a strong basement in a public building.

have happened in the past and are therefore likely to happen in the future. A *landslide potential map* takes into account variations in slope steepness, ground composition, and the occurrence of phenomena that can destabilize slopes. A *tornado potential map* delineates regions where weather systems that can generate tornadoes tend to develop. Such maps prove to be useful in designing building codes, setting insurance rates, motivating local governments to inform the public about what to do if a disaster should strike, and determining the amount of money that responders put into purchasing emergency equipment or hiring personnel. For example, communities in tornado-prone regions may invest in warning sirens **(FIG. 1.24a)** and require public buildings to post information on where to seek refuge if a tornado approaches **(FIG. 1.24b)**.

Hazard Preparedness, Mitigation, and Planning

Because natural disasters can be so devastating to communities, society needs to develop protocols and procedures to be ready for them. The push to develop **hazard preparedness** (or *disaster-risk reduction*) has been gaining global visibility over the past few decades. In fact, because of its importance, the United Nations has set up an entire office focused on disaster-risk reduction.

Hazard preparedness can include **hazard mitigation**, meaning the development of approaches that can prevent or decrease the consequences of a hazardous event. The consequences of the deadly 2004 tsunami in the Indian Ocean, associated with an earthquake off Sumatra, illustrate what happens when hazard mitigation has not been adequate. When the tsunami struck coastlines, people were caught entirely by surprise. There were no buoys offshore to detect the arrival of a tsunami before it reached the shore, and people on or near the beach didn't recognize the immediate precursor of a tsunami (a sudden drop in the sea surface along the coast). Even if a tsunami had been detected, there was no plan in place to contact authorities, and even if authorities had received a warning, there were no warning sirens or other facilities to send out broadcasts to alert the public. Even if the public had been warned, there were no marked escape routes to follow, and the public had not received training regarding what to do in the event of a tsunami. Finally, there were no accessible resources in place, either human or material, that could be called on to help with a response. As a result, the loss of life and property due to the event was staggering—an estimated 230,000 people died.

Modern hazard preparedness initiatives encompass several facets. First, communities and officials determine their exposure to a hazard and develop the ability to monitor the hazard and provide warnings of hazardous events. In addition, people tasked with responding to disasters develop plans to manage the response, including plans to maintain communication links in the case of power outages or destruction of cell-phone towers. They may also establish evacuation routes and develop plans for managing traffic, providing transportation, and sheltering displaced people.

Thoughtful *urban planning* measures, such as *zoning* that defines what can be built and where, and *building codes* that dictate how buildings should be designed to withstand a hazardous event, can help to mitigate the consequences of disasters. For example, communities might permit development only on land outside of flood-prone areas, or might require that buildings in earthquake-prone regions have adequate strength to withstand shaking **(FIG. 1.25a)**, or that homes constructed near coasts subject to storm surge be built on stilts **(FIG. 1.25b)**, or that roofs on homes in or near forests be made of fire-resistant materials.

Significantly, preparedness planning must be flexible in order to take into account societal conditions at the time of a disaster. For example, during the COVID-19 pandemic of 2020 (see Box 1.1), agencies such as FEMA rethought their responses and recommendations to natural disasters, for two key reasons. First, if emergency personnel were already deployed to address the needs of people afflicted by COVID-19, fewer personnel would be available to respond to the immediate needs of natural disasters. Similarly, if hospitals were already crammed with COVID patients, there might not be space to accommodate people injured by natural disasters. Second, because of *social distancing* requirements (the need for people to stay at least 2 m, or 6 ft, apart to avoid spreading

FIGURE 1.25 Efforts to mitigate the consequences of natural disasters.

(a) A design for an earthquake-resistant building being tested on a shaking table. Courtesy of Professor John W. van de Lindt, The NEESWOOD Project, Colorado State University.

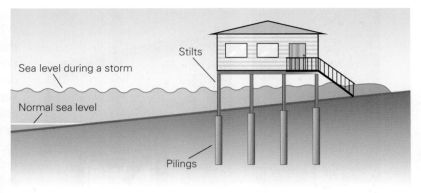

(b) By placing a house up on stilts, flooding might be avoided.

the contagion), normal strategies used by emergency personnel to rescue trapped or injured people, to shelter displaced people, and to distribute relief supplies, might put people too close to one another. The requirement to evacuate from the danger zones can conflict with the need to maintain social distancing, so facilities normally available for housing displaced people, such as school auditoriums and hotels, were closed.

Even after the response stage of the disaster is over (see Fig. 1.20), complications caused by a pandemic continue. For example, contractors sheltering in place cannot be safely deployed to repair damaged infrastructure and property. Because of the challenges resulting from realities of life in the time of a pandemic, officials continuously have to reevaluate strategies for coordinating relief efforts, and must encourage individuals to update their personal preparedness plans.

The Cost of Disaster: Who Pays?

On a global basis, the costs of natural disasters are staggering (FIG. 1.26). Who covers these costs? The answer to this question varies depending on location.

In many cases, especially in poorer communities and developing nations, it's the property owners themselves who pay—if they can. Destruction of farmers' crops may leave them in debt, and may even cause farmers to go bankrupt, unless they have a cash reserve in the bank or can borrow on the promise of next year's crop.

In developed nations, individuals and organizations turn to insurance companies to cover losses, if possible. An **insurance** policy is a contract between a company and a policyholder—the company receives payments from the

FIGURE 1.26 Costs of major disasters worldwide, 1995–2015, in billions of dollars.

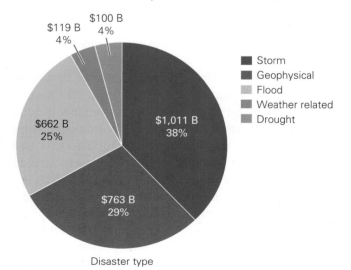

(a) Costs of disasters by type.

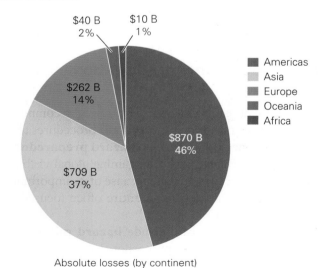

(b) Costs of disasters by region.

26 • CHAPTER 1 • Our Planet Can Be Dangerous: Introducing Natural Disasters

policyholder and, in return, agrees to cover the cost of the policyholder's lost or damaged property. Insurance companies cover such costs by selling policies to a broad segment of the population, for by doing so, they spread the risk. To set the rates for policies, and to determine how many policies they can write in a given community without becoming susceptible to bankruptcy if a disaster happens, insurance companies employ *actuaries* (people who examine statistics and calculate risk) to carry out *disaster risk assessments*. Since most policyholders, in a given year, won't suffer losses from disasters, the insurance company's cash reserves (money that policyholders pay in) exceed the amount the company pays out, so that the company remains solvent. However, losses can vary by a factor of five from year to year, and in bad years, an insurance company could conceivably run out of funds. To avoid this problem, some insurance companies buy policies from other insurance companies to help with catastrophic losses. Such **reinsurance** spreads risk to investors who intentionally take on high risk in the hope of a high return on their investment. On a global basis, however, insurance actually covers only a fraction of the losses due to disasters, and that fraction varies from year to year and location to location (FIG. 1.27).

Typical homeowners' insurance policies cover losses due to fire, wind, theft, and broken pipes—but in many flood-prone communities, they don't cover flood damage. This happens where actuaries have determined that their company cannot spread flood risk enough to make insurance affordable. Private flood insurance can't be obtained for some communities with significant exposure to flooding hazards. In the United States, the federal government may step in to provide affordable insurance for specific types of losses related to flooding through the *National Flood Insurance Program*. Similarly, the federal government may subsidize insurance to cover losses from meteorological or climatological disasters that impact farmers (FIG. 1.28). And, in places covered by a major disaster declaration, the federal government may provide low-interest loans to help people rebuild their homes,

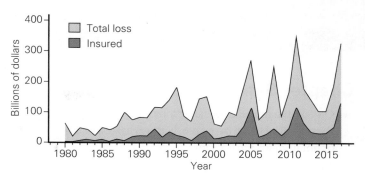

FIGURE 1.27 Total cost of disasters worldwide, 1980–2017, calculated in 2016 dollars. Note that insurance covered less than half of the cost, but nevertheless, disasters left insurance companies over $100 billion in bills during 2017.

and will foot the bill for some infrastructure repair. Who picks up the tab for losses covered by the federal government? The bill either goes to taxpayers or adds to the national debt. Therefore, rebuilding after some disasters becomes a political issue, for voters may ask, "Why should taxpayers support reconstruction in an area that has been destroyed by floods multiple times?" This question often arises when disaster strikes.

> **Take-home message . . .**
> Local disasters can be handled locally. If a disaster in the United States becomes severe enough, the governor declares a state of emergency, putting the state in charge. In appropriate cases, the US president issues a federal disaster declaration. Many entities, from the local to the national level, may become involved in response, recovery, and restoration. Efforts are underway to increase hazard preparedness by providing forecasts, developing response management plans, and enacting building and zoning codes.
>
> **QUICK QUESTION** Who pays for the costs of a regional disaster?

FIGURE 1.28 When floodwaters cover farmland and roads, as seen here near the Mississippi River in 2019, who pays?

1.6 Dealing with Disaster

Chapter 1 Review

Chapter Summary

- On our dynamic Earth, dangerous natural phenomena have the potential to harm people. Natural hazards exist whether or not humans are present.
- A natural hazardous event becomes a natural disaster when it causes significant human casualties and extensive property destruction, requiring outside assistance.
- Geophysical or geological disasters involve the solid Earth; hydrological hazards involve water; meteorological hazards involve weather; climatological hazards are due to long-term changes in atmospheric conditions; and extraterrestrial hazards originate in outer space.
- Physical disasters are distinct from biological ones such as the 2020 coronavirus pandemic.
- Some disasters take place very quickly, while others occur more slowly. We may not realize that a stealth disaster is happening.
- A primary disaster is the immediate result of a hazardous event. Secondary disasters are triggered directly by the primary disaster. Tertiary disasters may follow later.
- Different scales have been developed to compare the sizes of disasters of a given type. But no scales define the area affected or the extent of damage or casualties.
- Simplistically, the frequency of disasters can be represented by a recurrence interval or an annual probability.
- Natural disasters cause casualties and displace people. The economic impact of a disaster includes not only the costs of replacing buildings and property, but also the costs of lost industrial production, agricultural production, and tourism.
- Exposure refers to the potential human and economic cost that a community faces due to its proximity to a natural hazard. Vulnerability characterizes a community's ability to cope with a disaster. Poverty affects vulnerability.
- Risk can be defined by a simple equation: Risk = hazard × exposure × vulnerability. A number of factors, such as population growth, climate change, urbanization, and deforestation, influence risk.
- When a natural hazardous event approaches, people can try to prepare by evacuating or by securing property.
- Once a disaster strikes, there is an immediate response stage. After the response stage, the recovery stage begins. During the restoration stage, communities rebuild. Full recovery may take a long time.
- Response to a disaster can involve governmental and non-governmental agencies and groups. A governor can declare a state of emergency. A major disaster declaration authorizes use of federal resources.
- To help prepare the public for predictable disasters, agencies can issue watches and warnings. Many kinds of disasters can't be predicted.
- Hazard preparedness may involve creating hazard potential maps.
- Hazard mitigation efforts seek to decrease the consequences of a hazardous event.
- Affected individuals absorb the majority of costs in many disasters, especially in poorer communities. Insurance may not be available for places with significant exposure.

ANOTHER VIEW The COVID-19 pandemic, like other natural disasters, requires the public to change their behavior for the public good, until the crisis has passed. Making such changes often depends on a public relations effort. Authorities may take a serious approach, but sometimes they employ humor to catch the attention of the public and encourage collaboration.

Review Questions

Blue letters correspond to the chapter's learning objectives.

1. List several examples of natural hazards that could become natural disasters. (A, B)
2. Distinguish between anthropogenic and natural hazards, and provide examples of each. (A, B)
3. Why aren't all landslides considered to be natural disasters? (B)
4. Relate different kinds of natural disasters to the realm of the Earth System in which they occur. (B)
5. What is the difference between a rapid-onset disaster and a slow-onset one? Give an example of each. Which type does the photo show? (B)
6. Why do researchers refer to sea-level rise as a stealth disaster? (B)
7. Give examples of primary, secondary, and tertiary disasters. (B)
8. Why aren't the numbers provided by the Modified Mercalli Intensity scale or the Saffir-Simpson scale sufficient for defining the relative size of a disaster? (B)
9. Why might a statement of the recurrence interval of a natural disaster confuse people? What's another option for characterizing the frequency of natural disasters? (C)
10. What does "total people affected" during a disaster refer to? What costs does the "economic impact" of a disaster encompass? (D)
11. Does the global economic impact of natural disasters in an average year cost society tens of billions, hundreds of billions, or trillions of dollars? (D)
12. Define exposure and vulnerability as they pertain to natural disasters. What factors affect vulnerability? (E)
13. How do population growth, climate change, deforestation, and urbanization affect disaster risk? (E)
14. What factors may influence your decision about whether living at a certain location is an acceptable risk? (E)
15. What does a state of emergency declaration permit? (F)
16. How can communities prepare for a disaster? (F)
17. Distinguish among response, recovery, restoration, and resilience. What agencies and other organizations become involved in these activities? (F)
18. What is a major disaster declaration? (F)
19. Can warnings be provided for all kinds of disasters? (F)
20. Describe examples of hazard potential maps, and explain how such maps may be used. (E)
21. What steps can communities take to mitigate the consequences of disasters? (F)
22. Who pays for disasters? Why can the cost of paying for disasters become a political issue? (G)

On Further Thought

23. Imagine a town built on a low-lying plain bordering a river at the location where the river empties into the sea. How will urbanization affect the town's vulnerability to natural disasters? Could insuring homes in the town be problematic? (E, G)
24. Would a large meteorite impact on the Moon be considered to be a natural disaster by the United Nations? Explain your answer. (A, B)
25. List the natural hazards that your community may face. Can any of them be predicted? Describe any efforts at preparedness in your community. (A, F)
26. Examine the graph of global deaths due to natural disasters. Which disasters have produced the greatest numbers of deaths in the past century? (D)

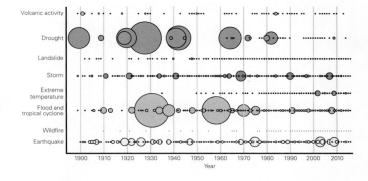

2

WELCOME TO A DYNAMIC PLANET
The Earth System and Plate Tectonics

By the end of the chapter you should be able to . . .

A. describe the realms of the Earth System, and give examples of how they interact.

B. explain the difference between a mineral and a rock, and characterize the three classes of rocks.

C. identify the internal layers of the Earth, and describe the nature of each.

D. explain the concept of continental drift, and outline observations that supported it.

E. explain the basic components of the theory of plate tectonics, and the evidence for the theory.

F. distinguish between collision and rifting, and how each relates to plate interactions.

G. identify basic types of geologic structures.

H. discuss the concepts of geologic time and global change, and provide examples of different kinds of change.

2.1 Introduction

It was December 24, 1968, and the *Apollo 8* spacecraft was silently orbiting above the stark, cratered surface of the Moon. For the first time in history, three humans—astronauts Frank Borman, Jim Lovell, and Bill Anders—were gazing directly at the Moon's far side. Just before their orbit brought them back to the near side, Borman started to rotate the spacecraft, as the mission plan required. Due to the maneuver, the view through Anders's window suddenly showed the Earth rising above the Moon's horizon. Anders gaped at the startling beauty of this image, and exclaimed, "Oh my God, look at the picture over there! There's the Earth comin' up. Wow, is that pretty!" Then, since it was the pre-digital age, he shouted to Lovell, "You got a color film, Jim? Hand me a roll of color, quick, would you? . . . Just grab me a color. A color exterior. Hurry up. Got one?" Alas, as the spacecraft continued its roll, the view started to disappear from Anders's window, and in disappointment he muttered, "Well, I think we missed it." Fortunately, that wasn't the case. Lovell exclaimed, "Hey, I got it right here! . . . Bill, I got it framed, it's very clear right here!" Anders leaned over and started snapping pictures. The best of those photos, now titled *Earthrise*, has become iconic (**FIG. 2.1**).

More than 50 years after *Apollo 8*'s mission, *Earthrise* still amazes, not only because it contrasts the beautiful blue of the Earth with the dull gray of the Moon, but also because it emphasizes that our planet truly is an island in space. All life, all land, all air, and all water that sustains humanity lies in, on, or near this isolated ball as it hurtles through a vacuum. The Earth is a unique planet, for unlike any other in the Solar System, it has remained dynamic, in many diverse ways, for all of the last 4.54 billion years. It constantly changes, sometimes quickly and blatantly and sometimes slowly and subtly, as its several components, or **realms**, exchange materials and energy. To emphasize the complexity of this interaction, Earth scientists now refer to these realms—the *atmosphere* (the gas surrounding the Earth's surface), the *geosphere* (the solid Earth from surface to center), the *hydrosphere* (water in its various forms distributed through other realms), the *cryosphere* (the portion of the hydrosphere that has frozen to become ice), and the *biosphere* (living organisms of all types)—and the interactions among them as the **Earth System** (**FIG. 2.2**).

◄ Pounding ocean surf ate into Australia's southern coast and left behind several sea stacks, known locally as the Twelve Apostles. These rock columns consist of limestone that formed from shells on the seafloor. The vertical movement of the rock relative to sea level, as well as the erosion of the rock today, emphasize that the Earth was and remains a dynamic planet.

FIGURE 2.1 *Earthrise*, the iconic photo by *Apollo 8* astronaut Bill Anders.

The dynamic character of the Earth System sets the stage for geophysical and geological hazards to develop and for natural disasters to take place. Therefore, to provide a foundation for our discussion of natural hazards and disasters in this book, we begin Chapter 2 with a whirlwind tour of the Earth System, highlighting the key features of each realm. We then focus in on the geosphere, first by describing the materials that it consists of, and then by describing its surface and interior. You'll see that the Earth's outer shell consists of many discrete pieces, called *plates*, which move with respect to one another at speeds of a few centimeters per year. These movements, explained by the *theory of plate tectonics*, produce earthquakes, volcanic eruptions, and mountains, and therefore generate the geophysical and geological disasters that we introduce in Chapters 3 through 7. In Chapter 8, we'll go into more detail about the atmosphere in order to provide a context for discussing hydrological, meteorological, and climatological disasters in succeeding chapters. Here in Chapter 2, you'll see that many features of the Earth System take a long time to develop and evolve. To help you understand this duration, we briefly discuss *geologic time*—the time since the Earth's formation. We conclude the chapter with a brief synopsis of the many ways that our planet has changed since its formation. Note that this chapter assumes a basic knowledge of matter and energy—if you're rusty on these topics, please see the book's Appendix for a concise review.

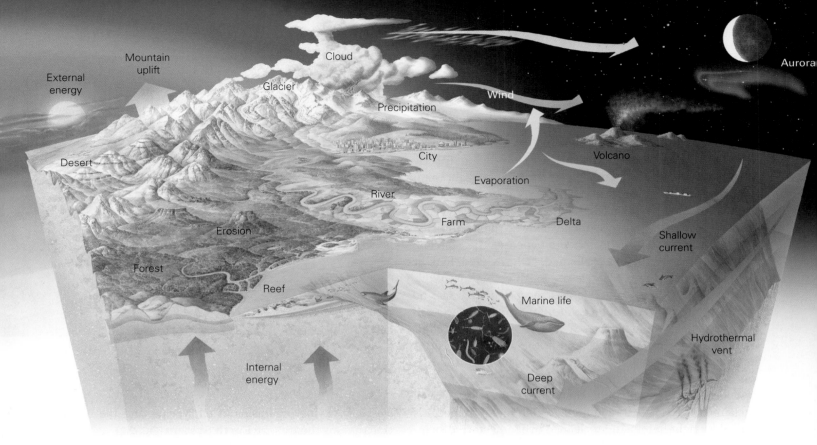

FIGURE 2.2 In the Earth System, many realms interact. Humans have modified many aspects of the system.

2.2 A Quick Tour of the Earth System

Imagine that you're an astronaut in a spacecraft that's returning from the Moon and goes into orbit around the Earth. What can you see and learn about the planet from such a vantage point? Certainly, you'll be able to detect the forces it applies to the space around it, and you'll be able to distinguish the realms of the Earth System. In fact, with measurements made from space, you'd be able to describe the basic character of these realms and even predict where natural hazards might develop within them. (You might even observe disasters taking place!) Let's look at your observations.

Welcome to the Neighborhood!

From a distance of 20,000 km (12,500 mi), the Earth appears as a bluish-glowing sphere with streaks of white and patches of tan and green (FIG. 2.3). Your instruments detect the planet's **magnetic field**, which can be symbolized by lines curving through space—these lines show the directions along which magnetic materials (such as compass needles) align (FIG. 2.4). The lines are parallel to the Earth's surface at the equator and point straight inward and outward at the

FIGURE 2.3 The Earth and the Moon as seen from space. The blue areas are oceans, the green and tan areas are land, and the white patches are clouds. The green areas are vegetated.

32 • CHAPTER 2 • Welcome to a Dynamic Planet: The Earth System and Plate Tectonics

FIGURE 2.4 The Earth's magnetic field. Magnetic field lines point in at the north magnetic pole, and out at the south magnetic pole.

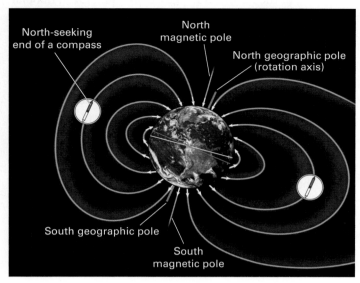

north and south *magnetic poles*, respectively. You can also detect the **solar wind**, a stream of dangerous charged particles blasting out of the Sun that warps the Earth's magnetic field into a teardrop shape. Fortunately, the field deflects most of these particles, but you realize that if the Sun were to send out a blast of particularly energetic particles, they could penetrate the field and disrupt electronics on the Earth's surface (see Chapter 16). The field doesn't stop your spacecraft, though, and you continue to speed toward the planet. As you get closer, the pull of gravity feels stronger. Carefully, you nudge your spacecraft into an orbit and start circling the planet.

Introducing the Biosphere

From orbit, you can see that the Earth hosts a **biosphere** composed of a vast variety of organisms, including plants, animals, plankton, and microbes. Most living organisms live on the surface of the land or in the oceans, but life exists from a few kilometers below the surface to a few kilometers above it. The distribution of land vegetation, indicated by green areas, serves as the most obvious indication of life (see Fig. 2.3). Your view also reveals the pervasiveness and consequences of human civilization, manifested by geometric patterns of vegetation (fields) as well as by cities and roads that light up at night **(FIG. 2.5)**.

Introducing the Atmosphere

A layer of gas, the **atmosphere**, surrounds the Earth. The atmosphere consists of a mixture of gases called **air**. Nitrogen (N_2) and oxygen (O_2) together make up 99% of dry air. The remaining 1% consists of argon (Ar), carbon dioxide (CO_2), methane (CH_4), ozone (O_3), and several other gases. Because these gases occur in such low concentrations, they are called *trace gases*. Oxygen in the atmosphere comes from the metabolism of photosynthetic organisms, such as plants and algae. In fact, if photosynthetic organisms were to vanish suddenly, other organisms would consume the O_2 of the air in less than 5,000 years. The density of air increases toward the planet's surface because the weight of overlying air pushes down on the air below it. In fact, 99.9% of the molecules in air lie within only 50 km (30 mi) of the surface! The weight of air at a location determines *air pressure*, the push that air applies to its surroundings.

Air contains water vapor—less in some places, more in others. Under certain conditions (discussed in Chapter 8), water vapor condenses or crystallizes to produce a layer of *clouds*, composed of tiny droplets or ice crystals, at elevations below about 12–16 km (7.5–10 mi), from which *precipitation* (rain or snow) may fall **(FIG. 2.6)**. Clouds move relative to the land or sea surface with the *wind*, the horizontal movement of air.

Atmospheric characteristics—cloud cover, wind velocity, precipitation rate, temperature, water vapor content, and air pressure—at a location at a given time define the location's **weather**. Weather can change rapidly. For example, the sky can be completely blue, then an hour later be filled with gray, rain-producing clouds. On occasion, a *storm*, in which the wind velocity and/or the rate of precipitation become particularly intense, takes place. Some storms are strong enough to cause natural disasters.

FIGURE 2.5 Lights at night reveal the reach of civilization.

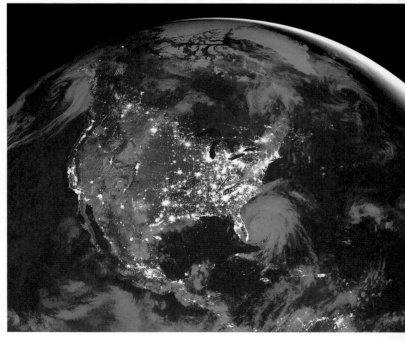

FIGURE 2.6 Clouds swirl in the lowermost part of the atmosphere, as seen from the International Space Station.

In addition to distinct cloud patterns, your view from the orbiting spacecraft reveals differences in the character and density of vegetation on land. Such variations depend on a region's **climate**—the weather conditions it experiences averaged over many years. Climate depends on latitude—it's hot near the equator and cold near the poles—and several other factors. Significantly, the concentration of trace gases, particularly CO_2 and CH_4, affects the amount of heat that the atmosphere retains.

Introducing the Hydrosphere and Cryosphere

You next turn your attention to the water of the Earth's surface. The Earth differs from all the other planets in the Solar System in that liquid water covers about 70% of its surface. Notably, this *surface water* of the Earth System constantly moves as it flows in rivers, bobs up and down in waves, and circulates in currents. About 97% of this water is salty and fills the **oceans** (see Fig. 2.3), where it circulates in vast currents. Close-up images indicate that waves on the ocean surface can become hazardous, particularly where they meet the land along coasts (see Chapter 13). *Freshwater*, the 3% of the Earth's water that doesn't contain salt, collects in ponds and lakes, and flows in streams and rivers, which sometimes overflow to produce hazardous floods (see Chapter 12). Were you to drill into the Earth's surface, you'd learn that most freshwater resides underground, as **groundwater**, filling pores (small open spaces) and cracks within rock and sediment. Even from space, you can see that life on land depends on freshwater, so an inadequate supply of freshwater could represent a hazard for this life (see Chapter 7).

Not all of the Earth's water exists in liquid form. In polar regions and at high elevations, you see white areas where water has frozen into solid ice. When ice builds up on the land surface in a thick enough layer to last all year, it becomes a **glacier**, and you can see many of these from space (FIG. 2.7). Glaciers flow very slowly. In very cold regions, near-surface land freezes into *permafrost*, and the sea surface freezes to form a thin layer of *sea ice*. The Earth's frozen regions together make up the **cryosphere**. The Earth's frozen and liquid water on land, in the oceans, and underground, as well as water vapor in the atmosphere, together make up the **hydrosphere**. Redistribution of water among the components of the hydrosphere, as happens when part of the cryosphere melts, may result in changes in **sea level** (the average elevation of the sea surface)—yet another hazard that the Earth's inhabitants have to deal with (see Chapter 15).

Introducing the Geosphere

The solid surface of the Earth and the planet's interior constitute the **geosphere**. The 70% of the geosphere's surface hidden beneath the oceans is the **seafloor** of ocean basins (FIG. 2.8). Instruments in your spacecraft allow you to map **bathymetry**—the shape of the seafloor—quickly, and you find that most of the seafloor forms flat plains at a depth of 4–5 km (2.5–3 mi) below sea level. The oceans become shallower over *continental shelves* along the fringes of continents and along submarine mountain ranges called *mid-ocean*

FIGURE 2.7 Glaciers are part of the Earth's cryosphere. This view from space shows a glacier in Alaska flowing to the ocean.

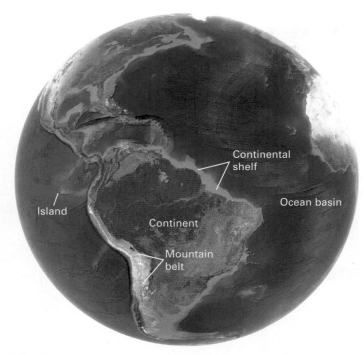

FIGURE 2.8 Land and sea on the Earth's surface. The darkness of green areas indicates the density of vegetation. Tan areas are largely unvegetated. The darkness of blue in the sea indicates the water depth—darker blue is deeper.

have been displaced along fractures. The presence of these *faults* indicates that dangerous earthquakes rattle the Earth's surface (see Chapter 3). The existence of faults, volcanoes, and mountains emphasizes that the geosphere hosts **tectonic activity**, meaning that it undergoes geologic movements and changes. In this regard, the Earth differs dramatically from its closest neighbor, the Moon, on which no tectonic activity now takes place. In fact, most features on the Moon's surface are craters from meteorite impacts billions of years ago. Tectonic activity, as well as **erosion** (the grinding away and removal of surface materials by moving water, air, and ice), constantly changes our planet's surface and has erased all but a few meteorite impact sites on the Earth. Rare, but potentially devastating, meteorite impacts can affect the Earth and therefore represent hazards (see Chapter 16).

Energy in the Earth System

The observations that you make from your spacecraft show that materials in all realms of the Earth System are moving and changing, though at vastly different rates. Roughly speaking, mountains rise at 1 mm (0.04 in) per year, glaciers creep at 25 cm (10 in) per day, ocean currents and streams flow at

ridges, as well as over isolated submarine peaks. The seafloor reaches its greatest depths—8–11 km (5–6.8 mi) below sea level—in elongate (long and relatively narrow) troughs called *trenches*. As we'll discuss later, bathymetric features that you can detect from space allow you to trace out the slowly moving plates.

The remaining 30% of the Earth's surface constitutes **land**, where the solid geosphere lies exposed to the atmosphere, except where covered by lakes, streams, or glaciers. Most land lies within **continents**, large blocks that are 2,000–11,000 km (1,240–6,800 mi) across. The remainder forms much smaller *islands*. **Topography**, the shape or form of the land surface, varies significantly. About two-thirds of the land surface resides at elevations of less than 1 km (0.6 mi) above sea level. Of the remainder, about 10% lies within **mountain belts** containing dramatic peaks rising from 3 to 8.85 km (2–5.5 mi) above sea level. The existence of **relief** (elevation differences between high and low points) means that slopes exist, and where there are slopes, the hazard of landslides exists (see Chapter 5).

You can see that some mountains and hills on the Earth are volcanoes, actively spewing out ash and lava (molten rock) **(FIG. 2.9)**. Because these eruptions can damage the land and temporarily diminish the amount of sunlight reaching the Earth's surface, they clearly represent a hazard (see Chapter 4). You can also see places where surface features

FIGURE 2.9 The ash cloud from a volcanic eruption as seen from space. The erupting volcano is Kliuchevskoi, on the east coast of Russia.

FIGURE 2.10 Energy in the Earth System comes from internal and external sources.

(a) The Earth's internal energy melts the rock that seeps from a volcano.

(b) External energy from the Sun heats the Earth's surface.

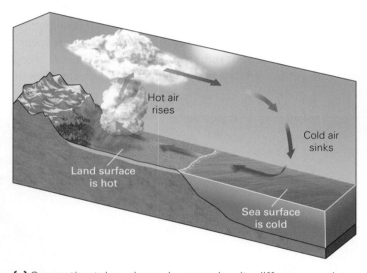

(c) Convection takes place whenever density differences exist in a fluid in the presence of gravity.

1–7 km (0.6–4 mi) per hour, lava flows out of a volcano at 0.1–30 km (0.06–19 mi) per hour, and the wind blows at velocities up to 200 km (125 mi) per hour. Where does the energy to drive this motion come from?

The Earth's **internal energy**, as we'll see, drives the movement of plates, and plate motion, in turn, generates most earthquakes and causes rock to melt and move upward beneath volcanoes (FIG. 2.10a). Internal energy exists because the Earth is hot inside—the temperature 20 km below the surface may reach 500°C (930°F), and temperatures at the center approach those found at the surface of the Sun. Some of this heat is left over from the formation of the Earth, and some comes from the decay of *radioactive atoms* (atoms that spontaneously release particles and energy). In contrast, motions in the atmosphere and hydrosphere are caused by **external energy**, which comes to the Earth from the Sun in the form of *radiation* (energy such as light that can travel across the vacuum of space) (FIG. 2.10b). External energy not only heats the surface of the planet, but drives photosynthesis.

All matter on the Earth feels the pull of **gravity**, the attractive force exerted by one body of matter on another. Gravity causes the downslope movement of land, the fall of rain, and the flow of rivers. Furthermore, in the presence of gravity, density variations in fluids (such as air and water) or soft materials (such as regions of the Earth's interior) can drive **convection** (FIG. 2.10c), during which less dense materials rise while denser materials sink. Convection plays a key role in atmospheric and oceanic circulation, influences plate motion, and plays a role in volcanic activity. The density variations that trigger convection in the Earth System are commonly a consequence of temperature variations. An increase in temperature causes a material (such as air or rock) to expand, and therefore, causes the material's density to decrease. Less dense materials feel a buoyancy force, causing them to rise above denser materials. Temperature variations in the oceans and atmosphere depend on input of external energy, and those within the geosphere depend on input of internal energy.

> **Take-home message . . .**
> The Earth System consists of several realms (geosphere, atmosphere, hydrosphere, cryosphere, and biosphere) that interact with one another. The surface of the geosphere, the solid Earth on which we live, has a wide range of elevations. 70% of the geosphere lies beneath the oceans. Internal and external energy, along with gravity, drive exchanges of materials among realms of the Earth System.
>
> **QUICK QUESTION** What is the overall relief between the peak of the highest mountain and the deepest spot in the ocean?

2.3 Earth Materials

Forming Planet Earth

The view from an orbiting spacecraft can provide a wonderful overview of the Earth's character, but can't provide details about its composition. To learn about the specific materials that make up our planet, researchers have spent more than three centuries exploring, investigating, and sampling, and in more recent years, performing laboratory analyses and creating computer models. As a result of this work, **geologists** (scientists who study the Earth) have developed a scientific model for how our planet formed, how it evolved, and what it's made of. According to this model, the Earth, like the other planets of the Solar System, formed from a *nebula*, a cloud of gas, ice, and dust. Gravitational attraction caused the cloud to flatten into a protoplanetary disk. The Sun grew at the center of this disk. Particles within this disk attach to one another, gradually producing tiny bodies called *planetesimals* (FIG. 2.11a). These then collided and combined into *protoplanets* until, eventually, eight survivors grew large enough to incorporate the rest of the matter in their orbits. The resulting spheres are the **planets** that orbit our Sun. The Earth and the three other *inner planets* (the ones closer to the Sun) are relatively small and consist mostly of rock and metal. The four much larger *outer planets* consist mostly of gas and ice. Hydrogen and helium

FIGURE 2.11 Formation of the Earth.

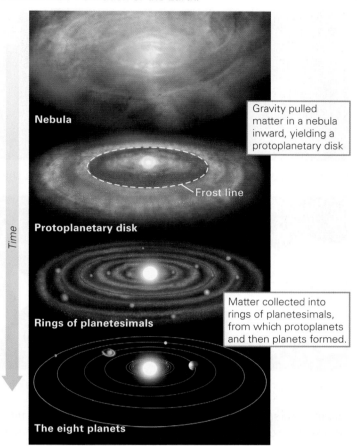

(a) Stages leading to the formation of planets.

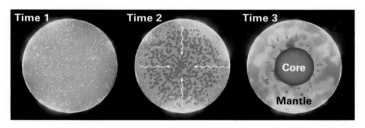

(b) Iron in the Earth's interior melted and sank to the center to form the core, surrounded by the mantle.

form almost all of the Sun. The process of planet formation concentrated certain other elements within planets. Four of those elements (iron, oxygen, silicon, and magnesium) account for about 91% of the Earth.

As it formed, the Earth, like the other inner planets, became hot enough inside to undergo *differentiation* (separation into layers)—dense iron sank to the center to form a metallic *core*, leaving behind a *mantle* of rock (FIG. 2.11b). Over time, portions of the mantle melted, producing molten rock that rose and solidified, gradually forming a thin *crust* of rock over the mantle—we'll describe these layers more thoroughly later in this chapter. Because a basic understanding of

FIGURE 2.12 The nature of minerals.

(a) These reddish quartz crystals display beautiful geometric shapes.

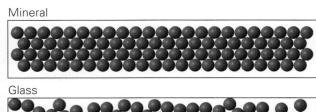

(b) In minerals, atoms are arranged in an orderly way, but in glass, atoms connect in a disordered way.

(c) Minerals grow by the attachment of atoms to the surface of a crystal.

the minerals, rocks, sediment, and soil that make up the natural surface of the Earth will help you to understand aspects of natural disasters, let's turn our attention to them briefly.

Introducing Minerals and Rocks

WHAT IS A MINERAL? Geologists define a **mineral** as a naturally occurring, generally inorganic, homogeneous, crystalline solid with a definable chemical composition. Let's examine the components of this definition. "Naturally occurring" means that real minerals form in nature, not in factories. "Generally inorganic" means that the vast majority of minerals are not *organic chemicals* (whose molecules contain chains or rings of carbon atoms)—so protein, sugar, tar, gasoline, and cellophane are not minerals. "Homogeneous" means that a piece of a mineral has the same composition and structure throughout. "Solid" means that a mineral can maintain its shape indefinitely. "Crystalline" means that the atoms within a mineral remain fixed in an orderly pattern known as a *crystal structure*, so a single, continuous piece of a mineral can be called a **crystal** (FIG. 2.12a). The crystal structure of a mineral controls the shape of a crystal, and also how a crystal breaks. Consequently, some crystals look like tiny bricks, some like pyramids, some like blades, some like needles, and some (known as *platy crystals*) like tiny sheets of paper. **Glass** isn't a mineral because the atoms within it are randomly arranged (FIG. 2.12b). The phrase "definable chemical composition" implies that we can write a chemical formula for a mineral to indicate the elements, and their relative proportions, that make up the mineral. **Quartz**, for example, always consists of SiO_2, and **calcite** always consists of $CaCO_3$.

Minerals form when atoms or ions come together and bond to one another in an orderly way (FIG. 2.12c). Such bonding can happen during the cooling and *solidification* of molten rock, during precipitation from a water solution (*precipitation* is the process by which dissolved ions attach to one another to form a solid), during precipitation from volcanic gas, or during *solid-state diffusion* (a process during which atoms move and rearrange themselves within a solid).

WHAT IS A ROCK? If you've ever struggled to move a large block of rock, you may think of rock only as a hard, heavy, solid mass. Geologists use a more precise definition: a **rock** is a coherent, naturally occurring solid consisting of an aggregate of mineral grains or, less commonly, a mass of glass. "Coherent" means that a rock holds together as a solid mass, so a pile of loose sand isn't a rock. "Naturally occurring" means that rocks form only by geologic processes, so manufactured materials such as concrete and brick aren't rocks. Finally, by "aggregate of grains," we mean a collection of many **grains** (rock fragments or crystals) that are attached to one another.

How does a rock hold together? Most rocks are coherent masses either because their grains are bonded to one another

FIGURE 2.13 Clastic and crystalline rocks.

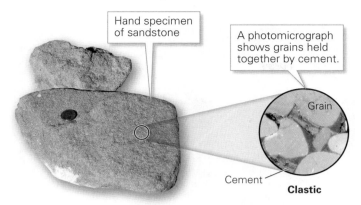

(a) Sandstone, a clastic rock.

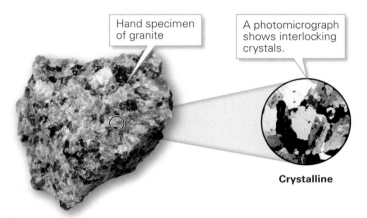

(b) Granite, a crystalline rock.

by **cement** (composed of mineral crystals that precipitated from water in the space between grains) **(FIG. 2.13a)**, or because their grains are crystals that grew together and interlocked with one another, like pieces in a jigsaw puzzle, as the rock formed **(FIG. 2.13b)**. Geologists refer to the former as **clastic rocks** and the latter as **crystalline rocks**. The way in which grains are held together in a rock, along with factors such as the size and relative orientation of grains, characterizes the **texture** of a rock.

Rock still attached to the Earth beneath it is called **bedrock (FIG. 2.14a)**. Geologists refer to an exposure of bedrock as an **outcrop**. A natural outcrop may be a knob of rock in a field or the woods, the bank of a stream, or a cliff **(FIG. 2.14b)**. People can produce outcrops by excavating road cuts, rail cuts, foundations, and mines **(FIG. 2.14c)**. When you see a loose block of rock, such as a boulder, keep in mind that it was once part of a bedrock outcrop, but at some time in the past it broke free.

The Three Rock Groups

Geologists distinguish among three rock groups—igneous, sedimentary, and metamorphic—on the basis of how rocks form.

IGNEOUS ROCKS. The word *igneous* comes from the Latin *ignis*, meaning fire—an appropriate name, because **igneous rocks** are formed by the solidification of very hot, molten rock (see Chapter 4). Molten rock underground is called **magma**. Rocks formed from cooling and solidifying magma are called *intrusive rocks* because they solidify after the magma has

FIGURE 2.14 Examples of outcrops.

(a) Bedrock forms the high cliffs at the top of this photo. Loose boulders litter the foreground.

(b) Bedrock cliffs along the coast of Maine. Boulders that have broken off lie at the base.

(c) Road cuts, like this one near Denver, Colorado, can expose large areas of bedrock.

FIGURE 2.15 Igneous rocks form by the solidification of molten rock.

(a) An intrusion of igneous rock (granite) into older, darker rock in California.

(b) A small lava flow on Hawaii. The red part is still molten, but the black part has solidified.

injected, or intruded, into pre-existing rock (**FIG. 2.15a**). Molten rock that emerges from a volcano and cools at the Earth's surface is called **lava** (**FIG. 2.15b**). Rocks formed from solidified *lava flows* (streams of molten rock) or from *pyroclastic debris* (fragments blasted explosively from a volcano) are called *extrusive rocks*. Most intrusive and extrusive igneous rocks that formed directly by solidification of molten rock, or *melt*, have a crystalline texture. Igneous rocks that formed from pyroclastic debris that has been welded or cemented together have a fragmental texture.

Geologists recognize many different types of igneous rocks, each of which has a name. These rocks differ from one another both by texture and by chemical composition. In crystalline igneous rocks, texture generally indicates how fast the rock cooled during its formation. Slow cooling, as happens when intrusive rocks solidify deep underground, tends to produce coarse-grained rocks, whereas fast cooling, as happens in lava that has flowed onto the Earth's surface, produces fine-grained rocks. Simplistically, the composition of an igneous rock depends on the proportion of the chemical compound silica (SiO_2) in the melt from which the rock formed. With this concept in mind, we can distinguish among three of the most common igneous rocks in the Earth's crust: granite (a coarse-grained rock containing relatively abundant SiO_2); basalt (a fine-grained rock containing relatively little SiO_2); and gabbro (a coarse-grained rock containing relatively little SiO_2) (**FIG. 2.16**). Significantly, the proportion of silica in an igneous rock affects the density of the rock: rocks with a high silica content are less dense than those with a low silica content. Therefore, a block of basalt weighs almost 10% more than an equal-sized block of granite.

SEDIMENTARY ROCKS. Exposure of rock to air and water at the Earth's surface causes it to undergo **weathering**, a process that breaks intact rock into pieces and can cause its minerals to undergo chemical changes. Geologists refer to the mechanical fragmentation of rock as *physical weathering*, and to the chemical alteration of minerals in a rock as *chemical weathering*. Physical weathering happens, for example, when frost or root growth breaks rock into pieces, and chemical weathering happens, for example, when minerals dissolve in water or when they react with water and air to form **clay** (a weak material composed mostly of extremely tiny platy crystals) or *iron oxides* (various minerals similar to those in rust). Physical and chemical weathering can work together to break rock down into loose fragments, or *clasts* (**FIG. 2.17**). The interaction of weathering products with rainwater percolating down into the ground, and with organisms, over a long time produces **soil**—the life-giving layer that hosts plants at the Earth's surface (**FIG. 2.18**) (see Chapter 7).

As we've just seen, weathering produces both solid clasts and dissolved ions. Clasts can be transported to a new

FIGURE 2.16 Examples of igneous rocks.

Granite

Gabbro

Basalt

FIGURE 2.17 Weathering of rocks produces sediment. In this example, granite, which is made up mainly of quartz, feldspar, and biotite, undergoes weathering to produce sand and clay.

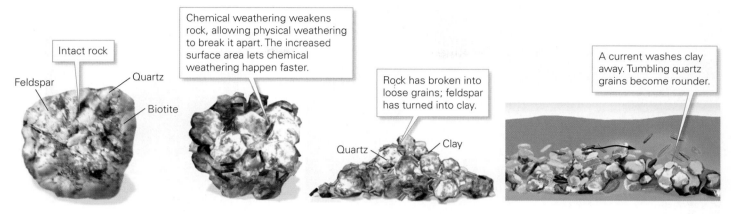

(a) Physical and chemical weathering acting together produce loose grains and ions in solution.

(b) The top of this granite outcrop has undergone weathering, so its grains are separating. The bottom is fresh (unweathered). An accumulation of loose clasts produced by weathering collects at the base of the outcrop.

location by moving water, ice, or wind. When the movement of these transporting agents slows or stops, the clasts undergo **deposition** (they accumulate). In some cases, ions in solution can precipitate directly to form a layer, as happens when salt precipitates on the floor of an evaporating salt lake. Ions can also be extracted from water by organisms such as plankton, corals, and clams to form shells. Geologists refer to an accumulation of clasts or shells, as well as to minerals precipitated from water onto the Earth's surface, as **sediment**. Deposition of sediment can build a succession of layers, known as **beds** (FIG. 2.19a). Each new sedimentary bed buries the one below it.

Eventually, layers at depth in a succession of sediment beds undergo *compaction* as the grains (clasts or crystals) within the beds are squeezed together by the weight of overlying layers. During this process, air or water that was once between grains gets squeezed out. As compaction takes place, minerals may precipitate from the groundwater that still remains between grains, and can cement the grains together. Compaction and cementation together result in **lithification**, the transformation of a bed of sediment into a bed of solid rock. In a general sense, therefore, we can define a **sedimentary rock** as a rock that formed either by lithification of buried sediments or by the precipitation of mineral crystals from water solutions onto the Earth's surface.

Different types of sedimentary rocks form from sediment deposited in different environments. For example, *sandstone* (a rock formed by lithification of quartz sand) and *shale* (a rock formed by lithification of clay) can form from the sediment deposited by rivers or along the coast (see Fig. 2.19a). *Limestone* (formed from calcite shells or precipitated calcite) can form from sediment deposited on reefs growing in warm, shallow seas (FIG. 2.19b). *Bedding*, the overall layering in sedimentary rocks, may be defined by contrasts in grain size and color, or because the rock splits more easily along bedding planes than within beds (see Fig. 2.19a). Geologists use the word **strata** for a succession of beds.

METAMORPHIC ROCKS. When a *protolith* (a pre-existing rock) endures changing environmental conditions without melting, it eventually becomes a **metamorphic rock** (from the Greek words for *meta*, meaning change, and *morphe*, meaning form). Most *metamorphism*, the process of forming metamorphic rock, takes place when the temperature and/or pressure in the environment increase substantially, as happens either when rock that formed near the Earth's surface ends up deep below the surface during the formation of mountains, or when a large igneous intrusion heats up its

FIGURE 2.18 Soil typically forms only a thin layer, like a blanket, over the substrate below. Here, we see a soil on top of chalk cliffs in southern England.

surroundings. Metamorphism can change the assemblage of minerals making up a rock as well as the texture of the rock. For example, during metamorphism, some crystals can grow bigger while others grow smaller, and new minerals can form as pre-existing ones chemically react with one another. Metamorphic rocks develop a crystalline texture, even if their protoliths were composed of cemented-together clasts. Changes during metamorphism can be so profound that a metamorphic rock looks as different from its protolith as a butterfly does from a caterpillar **(FIG. 2.20a)**.

Squeezing or shearing of rocks during metamorphism produces a type of layering called **foliation**. In some cases,

FIGURE 2.19 Examples of sedimentary rocks and sedimentary bedding.

(a) Sandstone, formed from cemented-together sand grains, and shale, formed from compacted mud, occur in distinct beds in this Utah road cut.

(b) Beds of limestone, formed from coral and other shells, exposed near Kingston, New York.

FIGURE 2.20 Examples of metamorphic rocks and metamorphic foliation.

Protolith (shale)

Metamorphic rock (schist)

(a) Metamorphism can change a rock's texture and the minerals within it. The protolith (red shale) could turn into a metamorphic rock (like schist).

(b) This rock was originally shale (gray) and sandstone (tan). The shale became slate, a metamorphic rock that splits on parallel foliation planes (dotted yellow line). Such foliation is called slaty cleavage.

(c) Compositional banding, manifested in this gneiss by the white and dark layers, is one form of foliation. The dotted yellow line shows where two layers meet.

foliation develops when crystals of minerals such as clay or *mica*, whose platy crystals resemble tiny sheets of paper, become aligned with one another. A rock with a foliation defined by aligned clay flakes is a *slate* (FIG. 2.20b), and one composed of aligned mica is a *schist*. In other cases, foliation develops when different layers consist of different minerals, so that the rock displays *compositional banding* (FIG. 2.20c). The presence of slippery clay mica along foliation planes can make the planes weak, so some metamorphic rocks split easily along foliation planes (see Fig. 2.20c).

Take-home message . . .

The Earth consists of a variety of different materials. Minerals serve as the building blocks of most rocks. Geologists recognize three groups of rocks—igneous, sedimentary, and metamorphic—distinguished by the way they form. Igneous rocks form by solidification of molten rock or accumulation of pyroclastic debris. Sedimentary rocks form at or near the Earth's surface, either by the compaction and cementation of clasts or by the precipitation of minerals from water, and occur in layers called beds. Metamorphic rocks form when other rock types are subjected to high temperatures and/or pressures, so that their minerals undergo changes without melting. Metamorphism commonly produces a type of layering called foliation.

QUICK QUESTION What is bedrock, and how does it differ from sediment?

2.4 Journey to the Center of the Earth

You can't see inside the Earth from space. But measurements of the planet's mass and shape provide a sense of the basic distribution of matter inside it. When geologists first made these measurements in the 19th century, they concluded that the Earth's interior resembles a hard-boiled egg, with three principal layers: the *crust* (the eggshell); the *mantle* (the white); and the *core* (the yolk) (FIG. 2.21a). The 20th-century development of methods to detect **seismic waves**, vibrations that travel outward from earthquakes and pass through the Earth, allowed researchers to refine this simple three-layer model. Seismic waves travel at different speeds through different materials, and they *refract* (bend) when they cross boundaries between layers. Significantly, some types of waves travel through both liquids and solids, whereas others travel only through solids. Therefore, by determining how long it takes for waves to pass from an earthquake to recording stations distributed around the globe, and by analyzing which types of waves arrive, researchers have produced a detailed image of the Earth's interior, depicting sublayers within the three main layers (FIG. 2.21b, c), and have determined which layers are

FIGURE 2.21 Looking inside the Earth.

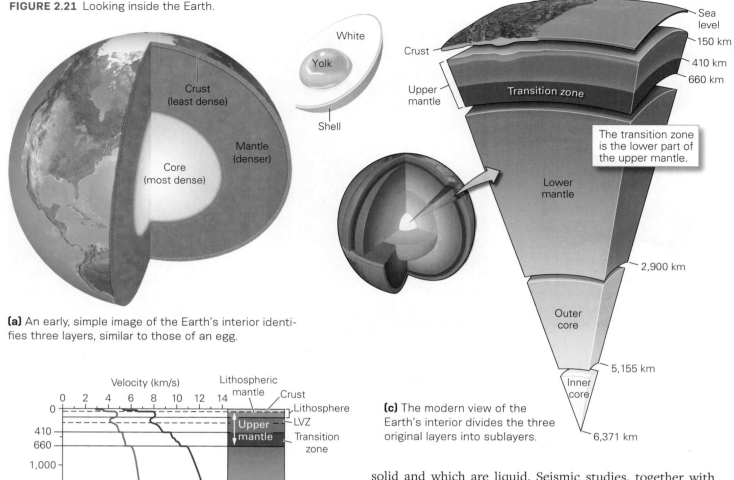

(a) An early, simple image of the Earth's interior identifies three layers, similar to those of an egg.

(b) A graph depicting how the velocities of two types of seismic waves (P and S) vary with depth. P-waves can travel through both liquids and solids, whereas S-waves cannot pass through liquids. Since S-waves don't pass through the outer core, it must be liquid. Note the abrupt changes in wave velocity that define layer boundaries.

(c) The modern view of the Earth's interior divides the three original layers into sublayers.

solid and which are liquid. Seismic studies, together with many other sources of information, ultimately allowed researchers to characterize the materials making up the layers. Let's examine each layer in turn.

The Crust

When you stand on the surface of the Earth, you are on top of its outermost layer, the **crust**. The crust consists of rocks whose composition differs significantly from that of the rocks in the underlying mantle. Because of this difference in composition, the density of the crust can be around 15% less than that of the underlying mantle. That difference causes seismic waves to refract abruptly at the crust-mantle boundary. That boundary is now called the **Moho**, an abbreviation of the name of the researcher who discovered this refraction.

The crust that underlies the ocean floor differs from the crust that makes up the continents in two ways (**FIG. 2.22**). First, *oceanic crust* has a thickness of only 7–10 km (4–6 mi), whereas *continental crust* has a thickness of 25–70 km (15–45 mi). But, compared with the radius of the Earth, even the thickest crust is so thin that if the Earth were the size of a balloon, the crust would be about the thickness of the balloon's skin. Second, oceanic crust and continental crust have

FIGURE 2.22 The continental and oceanic crusts differ markedly from each other in thickness and composition.

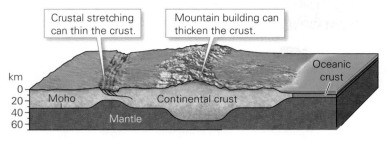

different compositions. Oceanic crust consists mostly of silica-poor basalt and gabbro. These rocks, in most places, have been buried by a thin blanket of sediment (less than 1 km thick) composed of clay and plankton shells that have settled like snow out of the sea. In contrast, continental crust consists of a great variety of igneous and metamorphic rocks, overlain in many places by up to several kilometers of sedimentary strata. Notably, the continental crust overall has an average chemical composition similar to that of granite.

The Mantle

The Earth's **mantle**, the thick layer of rock that lies between the crust and the core, accounts for most of our planet's volume. Popular media sometimes portray the Earth's mantle as a vast subsurface sea of molten rock. In fact, the mantle is almost entirely solid. Enough magma to generate volcanic eruptions forms only at specific locations, as we'll see in Chapter 4. The mantle, which consists entirely of a dark, very dense, very silica-poor igneous rock called **peridotite**

FIGURE 2.23 A sample of peridotite, the dense, dark rock that makes up the mantle.

(FIG. 2.23), has two sublayers: the *upper mantle* and the *lower mantle* (see Fig. 2.21). The lower part of the upper mantle is called the *transition zone*.

Temperature increases with depth in the Earth (**FIG. 2.24a**). Therefore, although it is solid, most of the mantle, below a depth of about 150 km (90 mi), is hot enough to undergo **plastic flow**, meaning that it can move and change shape very slowly without breaking. Consequently, convection can take place in the mantle, with hotter (and therefore less dense) mantle material rising in some localities and cooler (and therefore denser) material sinking at others. Keep in mind that this movement is extremely slow—it takes place at rates of only a few centimeters per year. But because of its importance, modern depictions of the Earth's interior emphasize the mantle's mobility (**FIG. 2.24b**).

FIGURE 2.24 The hot interior of the Earth.

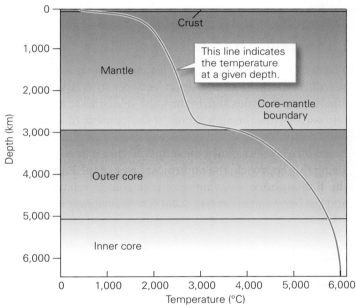

(a) A graph showing how temperature changes with increasing depth in the Earth.

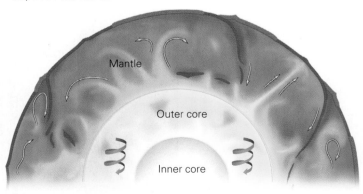

(b) Geologists envision the interior of the Earth as more dynamic than a simple layered model conveys. Mantle material is warm enough to flow plastically, and undergoes convection. The outer core, too, undergoes convection, but the inner core is solid.

2.4 Journey to the Center of the Earth • **45**

The Core

The Earth's **core** is a ball of metal at the center of the planet. Early calculations suggested that the core had the same density as gold, so people once held the fanciful hope that vast riches lay deep inside the Earth. Alas, geologists eventually concluded that the core consists of a far less glamorous material, *iron alloy*. This alloy consists of 90% iron and about 5% nickel. The remainder consists of oxygen, silicon, sulfur, and carbon. Studies of seismic waves led geologists to divide the core into two parts, the *outer core* and the *inner core* (see Fig. 2.24b). The iron alloy of the outer core is molten, so it can flow fairly rapidly; it is this flow that generates the Earth's magnetic field. The inner core is solid, even though its temperature approaches 6,000°C (10,830°F, comparable to that at the Sun's surface), because the immense pressure within the inner core—up to 3.5 million times the pressure at the Earth's surface—prevents atoms from breaking free of crystals.

Lithosphere and Asthenosphere

The division of the Earth into crust, mantle, and core reflects the differences in composition between the layers, which affect their densities and, in turn, the velocities at which seismic waves pass through them. In the context of discussing plate tectonics, however, geologists use a different approach based on a layer's rigidity, meaning whether it bends or flows. Using this approach, the outermost layer of the Earth, called the **lithosphere**, encompasses rock that is relatively rigid and can bend and break (FIG. 2.25). The lithosphere consists of the crust plus the top 100 km (60 mi) or so of the mantle (called the *lithospheric mantle*). The lithosphere overlies the **asthenosphere**, which consists of mantle that is hot enough to undergo plastic flow.

Take-home message . . .

The Earth has three main internal layers: the crust, mantle, and core. Each of these can be divided into sublayers. The mantle consists of solid peridotite, but most of it is warm enough to be able to flow slowly. The core consists of iron alloy.

QUICK QUESTION What is the difference between the inner and outer core? What is the difference between oceanic and continental crust?

2.5 A Scientific Revolution: Discovering Plate Tectonics

Using the terminology and concepts we've just introduced, we can turn our attention to the theory of plate tectonics. We start by describing how the theory arose from a melding of several ideas, including continental drift and seafloor spreading, and then we'll look at its key implications.

FIGURE 2.25 The lithosphere consists of the crust and the uppermost mantle, together. Lithosphere beneath continents is about 150 km thick.

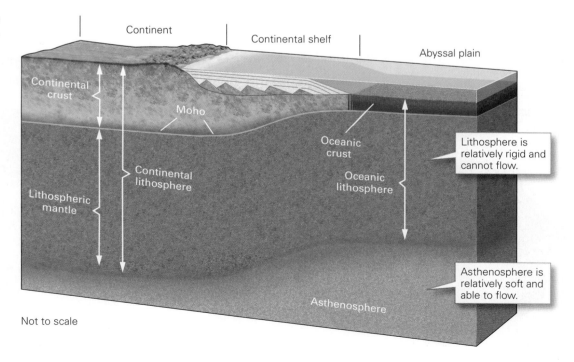

FIGURE 2.26 Some of Wegener's evidence for continental drift.

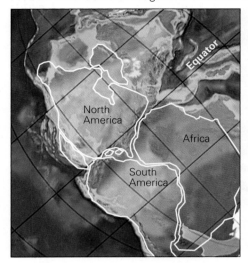

(a) The coastlines bordering the Atlantic match.

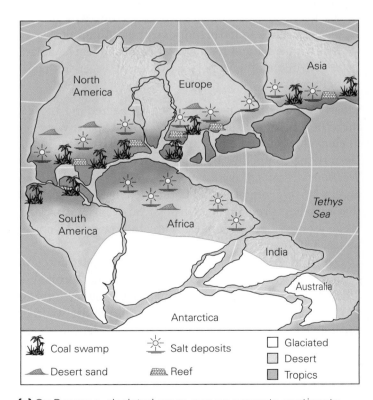

(b) Fossils of the same land animals and plants occur on all the continents.

Continental Drift

When the first good maps depicting the shapes of continents became available, it became apparent that the eastern coasts of North and South America look like they could nestle snugly against the western coasts of Africa and Eurasia (**FIG. 2.26a**). Similarly, the coasts of Antarctica, India, and Australia look like they could fit together and connect to the eastern coast of Africa. Were the continents ever actually together, like pieces of a giant jigsaw puzzle? Early in the 20th century, a German scientist named Alfred Wegener searched for geologic evidence to test the idea that that the continents were once connected to form a supercontinent that he named **Pangaea** (Greek for all land). The evidence that he found includes:

- On a map of Pangaea, mountain belts now separated by an ocean align with each other.

- Identical fossil species of land animals and plants can be found on continents now separated by oceans. Since the animals couldn't swim across oceans, their presence on now widely separated continents suggests that the continents were once together (**FIG. 2.26b**).

- The characteristics of sedimentary rocks provide clues to the climate in which the sediments in the rocks were deposited. For example, coal forms from plant debris that accumulates in a warm climate, and certain thick sandstone beds form from sand dunes that accumulate in deserts. Wegener found that the distribution of these rocks outlines climate belts, at appropriate latitudes, on his map of Pangaea (**FIG. 2.26c**).

(c) On Pangaea, glaciated areas now on separate continents were adjacent. Similarly, rocks indicative of tropical climates were aligned in a belt.

2.5 A Scientific Revolution: Discovering Plate Tectonics • **47**

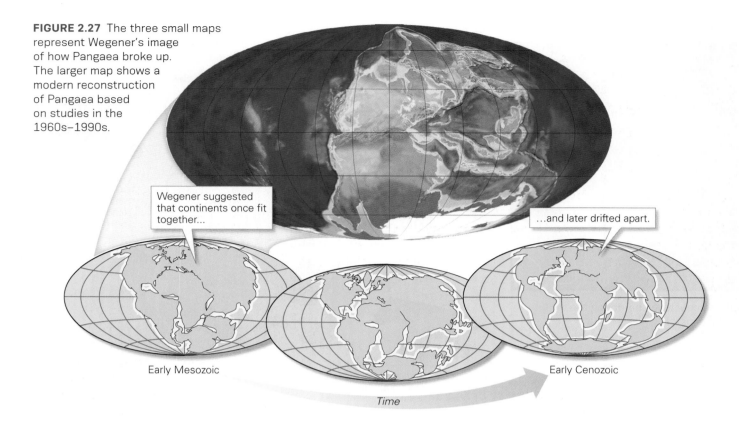

FIGURE 2.27 The three small maps represent Wegener's image of how Pangaea broke up. The larger map shows a modern reconstruction of Pangaea based on studies in the 1960s–1990s.

- Sedimentary rocks formed from sediments deposited by glaciers, as well as scratches in bedrock caused by moving glaciers, can be found in India, Australia, Antarctica, South America, and southern Africa, places that lay adjacent to one another at polar latitudes on Wegener's map of Pangaea.

Clearly, if the continents were once connected in Pangaea, then they must have subsequently moved apart. Wegener named this movement **continental drift** (FIG. 2.27). Despite the compelling evidence for continental drift that Wegener described, most geologists of his day refused to accept his interpretation, and preferred instead to assume that continents remain fixed in position over time, because Wegener could not adequately explain how or why continental drift operated.

Sadly, in 1930, Wegener—who not only contributed to geology, but was also a leading researcher in meteorology—died in a snowstorm after sledging across the icy interior of Greenland to resupply observers at a remote weather station. He would never know that three decades later, new observations would prove that he had been right—continents do move—and his ideas would be recognized as the first step in a *scientific revolution*, a complete rethinking of many ideas about the Earth System. But the next step would have to wait until new data about the Earth became available.

Key Geologic Observations of the Mid-20th Century

The discoveries that eventually changed geologists' minds about continental drift came from research, on both land and sea, in the three decades after Wegener's death. Here, we focus on three key discoveries.

SEAFLOOR BATHYMETRY. After World War II, geologists used sonar to record ocean depths all over the world. Two American geologists, Marie Tharp and Bruce Heezen, compiled the new depth data to depict the bathymetry of the Atlantic Ocean. Their work demonstrated that a submarine mountain range 2.5 km (1.5 mi) high, the *Mid-Atlantic Ridge*, runs down the center of the ocean, from north to south, for its entire length. Nineteenth-century studies had hinted at the range's presence, but Tharp and Heezen characterized its full extent. Tharp also noticed that a narrow valley follows the axis of the ridge (FIG. 2.28), and that this *median valley* resembles the East African rift, a linear belt of volcanoes, cracks, and frequent earthquakes along which Africa seems to be splitting into two blocks. She wondered if the Atlantic Ocean floor was actively splitting open right down its middle.

Eventually, Tharp and others collaborated to produce *bathymetric maps* showing the shape of the entire seafloor. Satellites have since made more accurate versions of these

FIGURE 2.28 One of Marie Tharp's bathymetric profiles across the North Atlantic Ocean. This profile shows the shape of the seafloor along a vertical slice. Note the median valley of the Mid-Atlantic Ridge. (Bermuda is a sea island formed by volcanic eruption.)

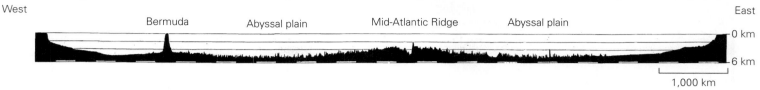

maps (**FIG. 2.29**). Tharp's maps revealed broad, flat surfaces, called **abyssal plains**, that make up most of the seafloor, as well as long belts of elevated seafloor, called **mid-ocean ridges**, whose crests lie about 2 km (1.2 mi) above abyssal-plain depths. (The adjective *mid-ocean* can be confusing because it gives the impression that all of these ridges run down the center of ocean basins. While the Mid-Atlantic Ridge does bisect the Atlantic Ocean, some ridges, such as the East Pacific Rise, lie closer to one side of an ocean basin than they do to the other.) In detail, a mid-ocean ridge consists of segments that do not align end to end. **Fracture zones**, narrow bands in which vertical cracks bounded by steep cliffs break up the oceanic crust, link the ends of ridge segments to each other. Along the margins of some, but not all, ocean basins are elongate troughs known as **deep-sea trenches**. And at hundreds of locations in the world's oceans, volcanic eruptions have produced peaks of igneous rock that have built up above the surrounding seafloor. If the top of a peak rises above sea level, it forms an *oceanic island*, whereas if its top lies below sea level, it's a **seamount**. Finally, along the edges of many continents are broad **continental shelves**, where the seafloor depth is less than 0.5 km (0.3 mi).

VARIATIONS IN HEAT FLOW AND SEAFLOOR SEDIMENT THICKNESS. From temperatures obtained by placing instruments in holes drilled into the seafloor, researchers calculated **heat flow**, the rate at which heat rises from the Earth's interior. By comparing temperatures from many locations, they learned that heat flow beneath mid-ocean ridges greatly exceeds that beneath abyssal plains. Researchers also found that the thickness of the sediment deposited on top of seafloor basalt increases progressively away from the axis of a mid-ocean ridge (**FIG. 2.30**). Specifically, virtually no sediment exists along the ridge axis, and the thickest sediment occurs on seafloor that is farthest from the axis. Eventually, researchers also

FIGURE 2.29 Topography and bathymetry of the Earth. Colors indicate elevation of land and depth of the seafloor.

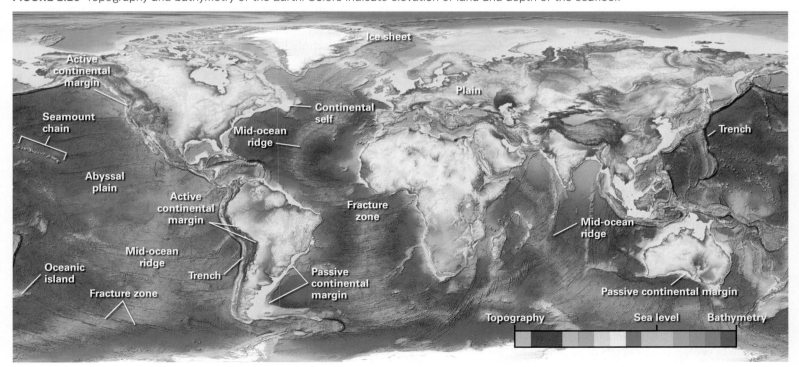

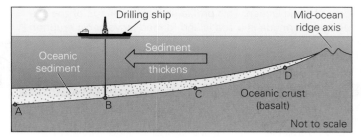

FIGURE 2.30 The thickness of seafloor sediment increases with distance from a mid-ocean ridge axis, as does the age of the sediment in contact with the basalt of the oceanic crust. The sediment at Point A is oldest, while the sediment at Point D is youngest.

learned that the age of the sediment directly in contact with seafloor basalt increases with distance from the mid-ocean ridge axis.

DISTRIBUTION OF EARTHQUAKES. In the early 20th century, researchers realized that most earthquakes occur along **faults**, fractures in rock on which sliding, or *slip*, takes place. By the 1950s, they had learned how to determine the locations at which earthquakes occur. This discovery led to the production of earthquake-distribution maps, which demonstrated that earthquakes are not scattered randomly around the Earth's surface, but rather concentrate in discrete bands, called **seismic belts (FIG 2.31)**. Some seismic belts occur along deep-sea trenches, some along segments of fracture zones, some along mid-ocean ridges, and some in mountain belts. These results imply that trenches, mid-ocean ridges, fracture zones, and mountain belts represent locations where slip on faults frequently takes place.

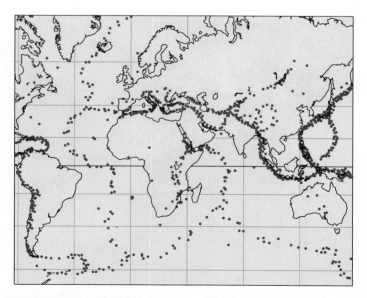

FIGURE 2.31 A 1950s-vintage map showing locations of earthquakes reveals the locations of seismic belts.

The Proposal of Plate Tectonics

When Marie Tharp suggested that the Atlantic Ocean was splitting down the middle, Bruce Heezen objected, because the concept seemed to support Wegener's model of continental drift, which Heezen didn't yet accept. But as new studies showed that earthquakes occur along the length of the Mid-Atlantic Ridge, he realized that Tharp had been right and began discussing the mid-ocean ridges with others, including Harry Hess, another American geologist. Hess thought hard about the implications of the new data on mid-ocean ridge bathymetry, the distribution of earthquakes, and variations in heat flow and oceanic sediment depths and ages, and came up with a way to explain them all. Hess suggested not only that the seafloor stretches apart along the axis of a mid-ocean ridge, but that new oceanic crust forms from magma that rises beneath the axis. Once formed, new seafloor moves away from the ridge. The stretching and breaking generates earthquakes, the presence of magma causes high heat flow, and the formation of new seafloor causes ocean basins to grow wider over time **(FIG. 2.32a)**. Another geologist, Robert Dietz, working independently of Hess, came up with a similar *hypothesis* (an idea that explains a phenomenon, but has not yet been proved) and coined the name **seafloor spreading** for the process.

If new seafloor indeed forms at mid-ocean ridges, then somewhere else old seafloor must be consumed, for otherwise the Earth would have to be expanding, and no evidence indicated that it was. Hess and Dietz both suggested that old seafloor sinks back into the mantle at deep-sea trenches, and that this movement generates the seismic belts observed along trenches. Subsequent work emphasized that it isn't just the crust that moves during seafloor spreading, but rather, the whole lithosphere, so modern images of seafloor spreading look different from Hess's original proposal **(FIG. 2.32b)**. It immediately became clear that the formation of new oceanic lithosphere by seafloor spreading, and the consumption of old oceanic lithosphere at trenches, provided the explanation for continental drift that eluded Wegener: continents move away from each other as new ocean basins form between them, and move toward each other as old ocean basins are consumed.

In science jargon, a hypothesis is just a viable idea. In contrast, a *theory* is an idea that has passed many tests, and has failed none. Consequently, scientists use the word "theory" for an idea in which they have a lot of confidence. In the early 1960s, geologists subjected the seafloor-spreading hypothesis and the continental-drift hypothesis to several tests and found that they passed with flying colors. For example, by studying *paleomagnetism*, the record (preserved in rocks) of the Earth's past magnetic field, they were able to confirm that not only do continents move relative to the Earth's magnetic

FIGURE 2.32 The concept of seafloor spreading. There are many differences between Hess's original image and a modern image of plate tectonics.

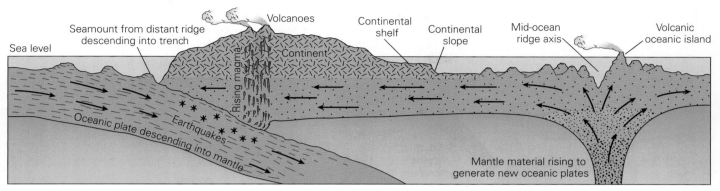

(a) Hess's original sketch illustrating seafloor spreading (ca. 1960).

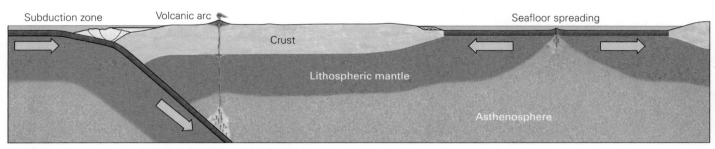

(b) A modern view of plate tectonics. We'll discuss the terms and concepts in this figure later in this chapter.

poles, but that they also move relative to each other. (Additional tests are described in geology textbooks.) Clearly, the map of the Earth's surface has changed slowly, but constantly over geologic time, and continues to change today.

The Way the Earth Works: The Theory of Plate Tectonics

By 1968, ideas revolving around seafloor spreading and continental drift had evolved into a comprehensive **theory of plate tectonics**, or simply *plate tectonics*. Geologists consider plate tectonics to be the "grand unifying theory" of geology because it provides the explanation for so many geologic features and phenomena. **Figure 2.33** provides an overview of plate tectonics that you can follow as we discuss its processes and related features.

According to this theory, the Earth's rigid outer shell, the lithosphere—which consists of the crust plus the uppermost part of the mantle (see Fig. 2.25)—isn't completely intact, but rather consists of separate pieces, known as **lithosphere plates**, or simply *plates*, that move relative to one another. Movement of plates can take place because the underlying part of the mantle, the asthenosphere, can flow like soft plastic—if the mantle were entirely rigid, plates could not move.

The rock making up the lithospheric mantle—the part of the mantle within the lithosphere—has the same composition as the rock making up the underlying asthenosphere. Why does lithospheric mantle behave differently from the asthenosphere? The answer comes from keeping in mind that temperature increases with depth in the Earth (see Fig. 2.24a). The behavior of rock depends on its temperature: hotter rock behaves more plastically than does cooler rock. (To see this difference, compare the behavior of a wax candle that you've just taken out of the freezer to one that you've just taken out of a warm oven.) The lithospheric mantle is relatively cool, so it behaves rigidly, while asthenosphere is hot enough to behave plastically.

Geologists refer to the divisions separating a plate from its neighbor as a **plate boundary** and to the area of a plate away from a plate boundary as a **plate interior**. Because almost all plate interactions take place at plate boundaries, major faults form at plate boundaries. And since slip on faults generates earthquakes, seismic belts delineate plate boundaries **(FIG. 2.34a)**. There are seven large plates and about a dozen smaller ones, as well as numerous tiny ones known as *microplates* **(FIG. 2.34b)**. Some plates have familiar-sounding names, such as the African Plate, while some have unfamiliar names, such as the Cocos Plate.

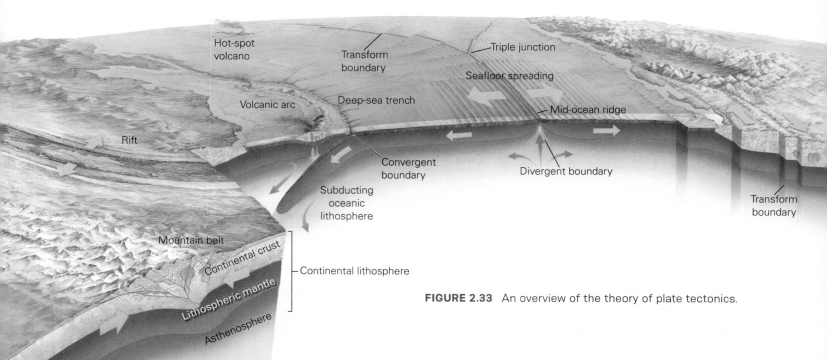

FIGURE 2.33 An overview of the theory of plate tectonics.

Note that some plates consist entirely of oceanic lithosphere, whereas others consist of both oceanic and continental lithosphere. Therefore, geologists distinguish between two types of continental margins: **active margins**, boundaries between continental and oceanic lithosphere that coincide with plate boundaries and host many earthquakes, and **passive margins**, which are not plate boundaries and rarely host earthquakes. The west coasts of North and South America are active margins, whereas the east coasts of North and South America are passive margins.

Take-home message . . .

Though many lines of evidence supported Wegener's continental drift concept, few geologists of his day accepted the idea. New evidence, particularly from studies of seafloor bathymetry, eventually led to the proposal of seafloor spreading, an idea that was subsequently proved by various observations. In the 1960s, ideas about the mobility of the Earth's lithosphere came together as the theory of plate tectonics, which states that the lithosphere consists of plates that move relative to one another. Interactions between plates take place along plate boundaries, while the interior of each plate remains intact.

QUICK QUESTION What is the difference between an active and a passive margin? Which hosts more earthquakes?

2.6 The Nature of Plate Boundaries

Geologists distinguish among three types of plate boundaries based on the relative motion of the plates: at *divergent boundaries*, plates move away from each other; at *convergent boundaries*, plates move toward each other; and at *transform boundaries*, plates slide sideways past each other (**FIG. 2.35**). As you will see in subsequent chapters, significant geologic hazards capable of producing natural disasters exist along plate boundaries. The descriptions of plate boundaries that we provide here will therefore provide context for discussing these natural hazards.

Divergent Boundaries

At a mid-ocean ridge, two plates move apart by the process of seafloor spreading. Because of this relative motion, geologists refer to mid-ocean ridges as **divergent boundaries**. During seafloor spreading, no open space actually develops between diverging plates. Rather, as the plates move apart, new oceanic lithosphere forms between them along the ridge axis (**FIG. 2.36a**). Overall, the seafloor slopes gently away from the ridge axis, reaching abyssal-plain depth at a distance of several hundred kilometers away.

How does new oceanic crust form? As plates move away from the ridge axis, hot rock of the asthenosphere rises from

FIGURE 2.34 The major lithosphere plates.

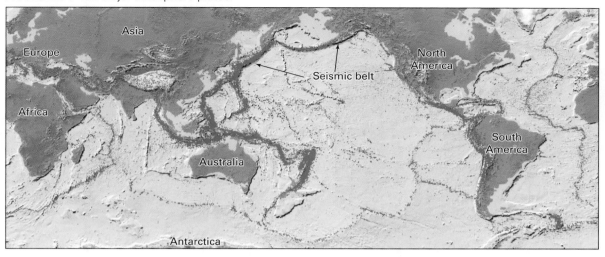

(a) This modern map of earthquake locations provides the basis for delineating plate boundaries.

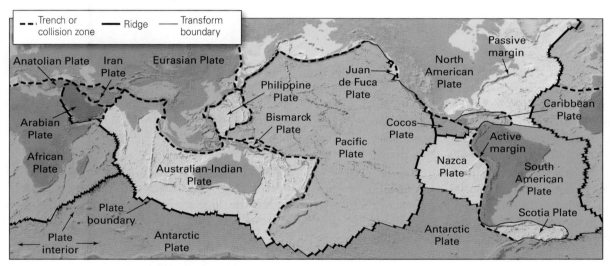

(b) Not all plates are the same size. Some are entirely oceanic, whereas others include both continental and oceanic lithosphere. This map does not show all the smaller plates, or the microplates.

FIGURE 2.35 The three types of plate boundaries. The yellow arrows indicate the relative motion of the plates.

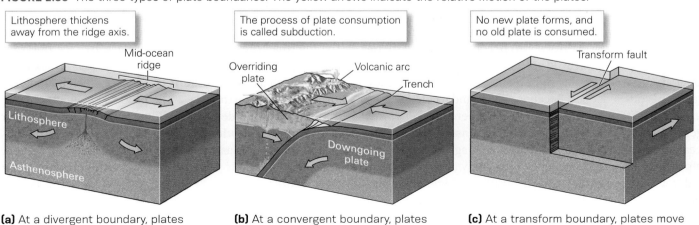

(a) At a divergent boundary, plates move apart.

(b) At a convergent boundary, plates move toward each other.

(c) At a transform boundary, plates move sideways past each other.

2.6 The Nature of Plate Boundaries • **53**

FIGURE 2.36 The nature of divergent boundaries.

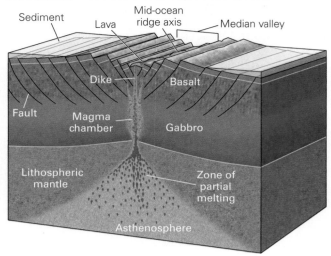

(a) The formation of crust along the ridge axis involves solidification at depth, intrusion of dikes, and extrusion of lava. Many faults occur along the ridge axis.

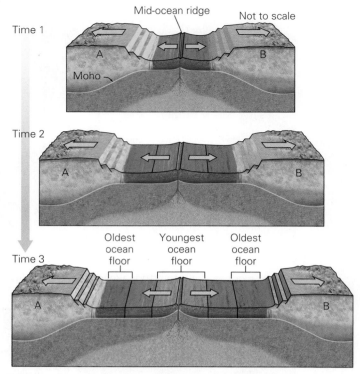

(b) New oceanic lithosphere forms at the ridge axis. Lithosphere thickens and ages as it moves away from the axis. The color bands indicate the relative ages of the lithosphere.

below. For reasons discussed in Chapter 4, this asthenosphere melts and produces magma. The resulting magma rises and accumulates in a *magma chamber* (a region containing a mush of crystals and molten rock) a few kilometers below the ridge axis. Simplistically, some of this magma solidifies into gabbro along the sides of the chamber, while some rises still higher, along vertical cracks, and solidifies into wall-like intrusive sheets, called *dikes*, made of basalt. The remaining magma makes it all the way to the surface of the seafloor, where it seeps out as lava from small submarine volcanoes and solidifies as a basalt lava flow. The entire layer of igneous rock formed at the ridge axis represents new oceanic crust. As it forms, the lithosphere stretches. Slip along a set of faults parallel to the axis accommodates some of this stretching.

At the ridge axis, a lithosphere plate consists only of new oceanic crust. As this crust moves away from the ridge axis, no more igneous rock is added to it, so the crust's thickness stays the same. But as the mantle directly underlying the crust moves away from the ridge axis, it progressively cools. Cooling of this mantle causes it to become rigid, and therefore to become part of the lithosphere. As the plate ages, the boundary between cooler rigid mantle and warmer plastic mantle becomes deeper. Because this boundary defines the base of the lithosphere, oceanic lithosphere thickens as it ages (FIG. 2.36b). Further, rock becomes denser as it cools, so oceanic lithosphere becomes denser as it ages. Like a cargo ship whose keel settles deeper into the water as workers load it with heavy cargo, aging lithosphere therefore settles deeper into the asthenosphere, and seafloor at the surface of older lithosphere lies lower than does the surface of young lithosphere of the ridge. This aging lithosphere, buried by sediment, underlies the abyssal plains. Because all seafloor forms at mid-ocean ridges, the youngest oceanic lithosphere borders the ridge axis, and the oldest oceanic lithosphere lies farthest from the ridge (FIG. 2.37).

Convergent Boundaries

Deep-sea trenches define most of the boundary of the Pacific Plate (FIG. 2.38a) and occur along margins of several other plates as well (see also Fig. 2.29). Trenches mark locations where the lithosphere of an oceanic plate bends, slides under the edge of another plate, and sinks into the asthenosphere (FIG. 2.38b). Due to the relative motion of the *downgoing plate* (the one that sinks) and the *overriding plate* (the one that doesn't), geologists refer to the plate boundary associated with a trench as a **convergent boundary**. Such boundaries host many large earthquakes. The process by which oceanic lithosphere slips under the overriding plate and sinks into the mantle is called **subduction**, so geologists also refer to convergent boundaries as **subduction zones**. Once subducted, pieces of old oceanic plates eventually settle down into the lower mantle.

As subduction takes place, seafloor sediment, as well as sand and mud that has washed into the trench from nearby land, gets scraped up and incorporated into a wedge-shaped

FIGURE 2.37 A map illustrating the age of oceanic lithosphere. Note that the youngest lithosphere is along the mid-ocean ridges.

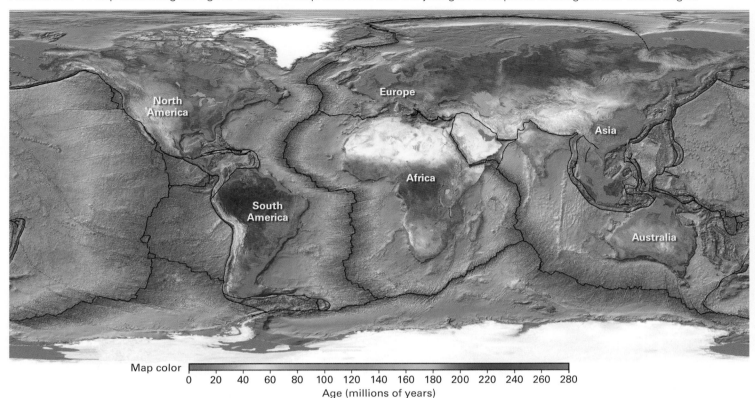

FIGURE 2.38 The nature of convergent boundaries.

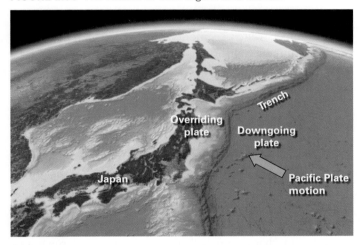

(a) Where the Pacific Plate subducts beneath Japan, a trench exists. The downgoing plate moves under the overriding plate. The arrow shows the motion of the Pacific Plate relative to Japan.

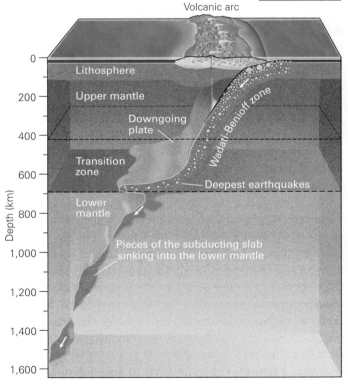

(b) The downgoing plate sinks into the mantle. Above a depth of 660 km (410 mi), earthquakes occur within the plate. This belt of earthquakes is called the Wadati-Benioff zone.

mass, known as an **accretionary prism**, along the margin of the overriding plate. In effect, an accretionary prism resembles a pile of sand that builds in front of a moving bulldozer (FIG. 2.39a). For reasons discussed in Chapter 4, a **volcanic arc** (or simply an *arc*), consisting of a chain of volcanoes, forms along the edge of the overriding plate. Subduction of

2.6 The Nature of Plate Boundaries • 55

FIGURE 2.39 Contrast between an island arc and a continental arc.

(a) An accretionary prism forms from sediment scraped up during subduction.

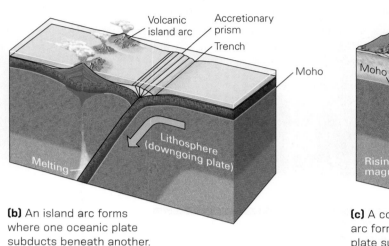

(b) An island arc forms where one oceanic plate subducts beneath another.

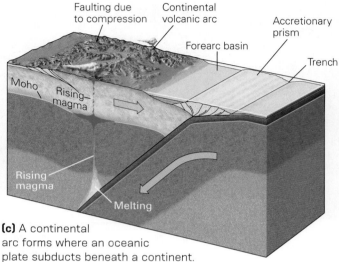

(c) A continental arc forms where an oceanic plate subducts beneath a continent.

one oceanic plate under another produces a volcanic **island arc** (**FIG. 2.39b**), whereas subduction of an oceanic plate beneath a continent produces a **continental arc** (**FIG. 2.39c**).

Transform Boundaries

In our discussion of seafloor bathymetry, we noted that mid-ocean ridges are segmented, and that fracture zones link the ends of the segments to each other (**FIG. 2.40a**). Originally, researchers incorrectly assumed that the entire length of each fracture zone was an actively slipping fault. But when information about the distribution of earthquakes along mid-ocean ridges became available, it became clear that slip occurs only on the segment of a fracture zone that lies between the ends of two ridge segments, and that the portions of the fracture zone that extend beyond the ends of ridge segments do not generate earthquakes (**FIG. 2.40b**). A Canadian geologist, J. Tuzo Wilson, introduced the term **transform fault** (or simply *transform*) for the actively slipping portion of a fracture zone. He emphasized that a transform fault represents a third type of plate boundary, now known as a **transform boundary**. At such boundaries, one plate moves horizontally relative to its neighbor along a vertical fault. In the case of many oceanic transforms, the fault remains the same length as the ocean gets wider (**FIG. 2.40c**).

Not all transforms link mid-ocean ridge segments, however—some connect trench segments, and some, such as the San Andreas fault of California, cut through continental lithosphere (**FIG. 2.41**). As we'll see in Chapter 3, earthquakes on continental transform faults cause many disasters.

The Birth and Death of Plate Boundaries: Collisions and Rifts

The configuration of plates and plate boundaries visible on our planet today has not existed for all of geologic history and will not stay the same in the future. When the ocean between two continents has been entirely subducted, continents once separated by that ocean can merge, and when a large continent breaks apart, a new ocean basin can form.

COLLISION. India was once a small, separate continent that lay far to the south of Asia. Over time, however, subduction consumed the ocean floor between India and Asia, and India moved northward. By 40–50 million years ago, all the ocean floor had been subducted, and India itself pushed into Asia. This type of event, leading to the merging of two landmasses after subduction of the intervening ocean floor, is known as a **collision**, and if the event involves two continents, it's a *continental collision* (**FIG. 2.42**).

56 • CHAPTER 2 • Welcome to a Dynamic Planet: The Earth System and Plate Tectonics

FIGURE 2.40 Transform boundaries at mid-ocean ridges.

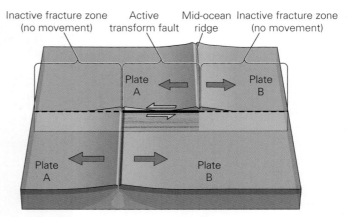

(b) Only the segment of a fracture zone between ridge segments is a transform boundary.

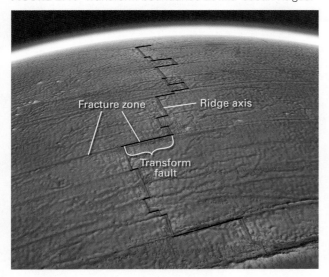

(a) The bathymetry of the Mid-Atlantic Ridge emphasizes that where fracture zones intersect a mid-ocean ridge, segments of the ridge are interrupted.

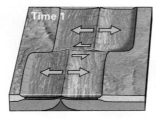 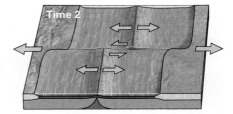

(c) The transform boundary can stay the same length as the seafloor spreads.

FIGURE 2.41 Transform faults that cut through continental crust.

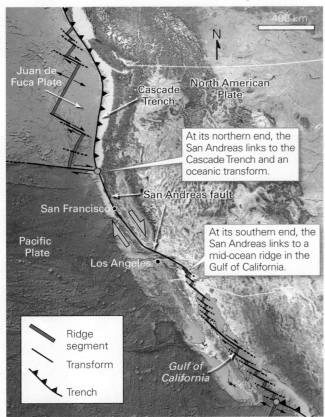

(a) The San Andreas fault is a transform that cuts through the continental crust of California.

(b) In places, the trace of the San Andreas fault can be clearly seen on the ground.

2.6 The Nature of Plate Boundaries • **57**

FIGURE 2.42 Consequences of collision.

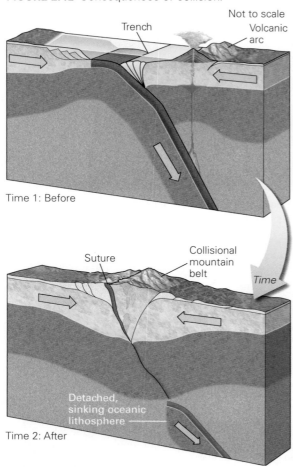

(a) When subduction consumes the oceanic lithosphere between two continents, the continents collide, and a collisional mountain belt develops.

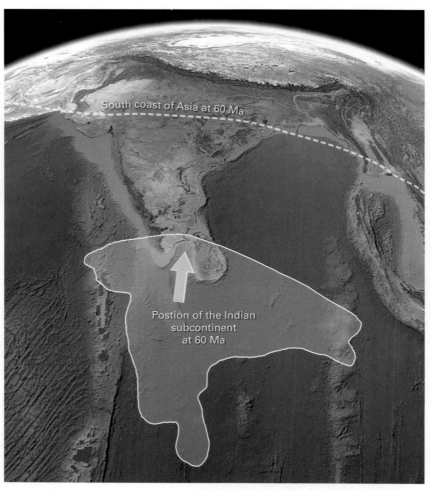

(b) The collision between India and Asia, which began about 40 million years ago, produced the Himalayas and led to the uplift of the Tibetan Plateau. At 60 million years ago, India was far to the south.

(c) The Himalayas' highest peak, Mt. Everest, as viewed looking west.

FIGURE 2.43 Rifting of a continent.

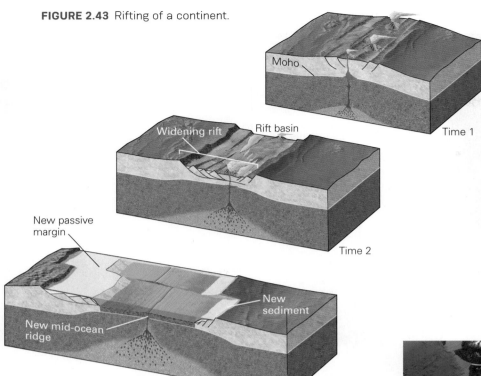

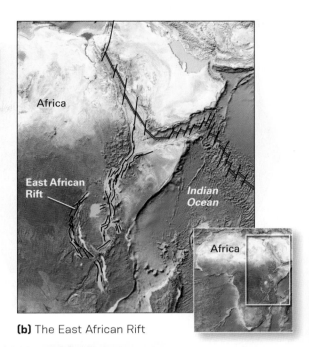

(a) As stretching starts to pull a continent apart, a rift forms. If rifting succeeds, a mid-ocean ridge forms, and the two continents are separated by a new ocean basin.

(b) The East African Rift

Collision happens when any two relatively buoyant pieces of crust (such a continent and an island arc) converge at a plate boundary, for such crust cannot be completely subducted. It may seem strange to think of rock composing continental crust as "buoyant," but the granite of continental crust is indeed about 20% less dense than the peridotite of the mantle, because it contains much more silica, so it can't sink into the asthenosphere. When collision happens, the convergent boundary that once lay between the buoyant pieces of crust ceases to exist, and rock and sediment that once lay along the margins of the two landmasses undergoes squeezing as if in a giant vise, producing a *collisional mountain belt*. The surface defining the boundary along which one crustal block attached to another is called a *suture*.

RIFTING. Geologists refer to the stretching and breaking apart of continental lithosphere as **rifting** (FIG. 2.43a). This process tends to be confined to a distinct belt, called a **rift**. Examples include the East African Rift in eastern Africa and the Basin and Range Province in the western United States (FIG. 2.43b, c). The manner in which continental lithosphere responds to rifting varies with depth—near the surface of the continent, stretching leads to the development of many faults, whose slip causes earthquakes, whereas deeper down, the warmer and softer rock of the continent stretches plastically.

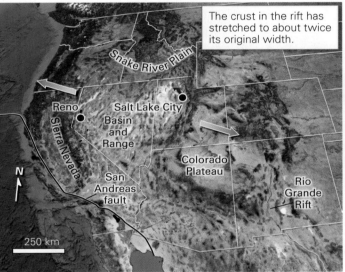

(c) The Basin and Range Province of the western United States is a rift that has been growing for the past 25 million years.

In rifts, blocks of crust tilt and slip downward as faulting progresses, producing low areas, called *rift basins*, that fill with sediment. Continental lithosphere thins during rifting, and hot asthenosphere rises beneath the rift and melts, producing magma that erupts as lava from volcanoes in the rift. As a result, rifts may host volcanic hazards (see Chapter 4).

If rifting succeeds in splitting a continent, a new mid-ocean ridge develops. Once seafloor spreading has started, the inactive remnants of the rift become the passive continental margins on either side of the new ocean basin. Over time, these rift remnants sink and become buried by thick accumulations of sediment washed in from the continent. The surface

FIGURE 2.44 The Big Island of Hawaii is a hot-spot volcano.

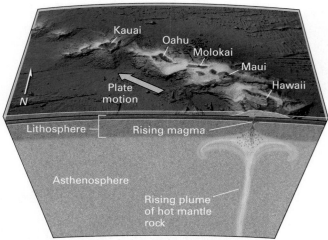

(a) The hot spot may exist because a plume of hot asthenosphere rises through the mantle beneath it.

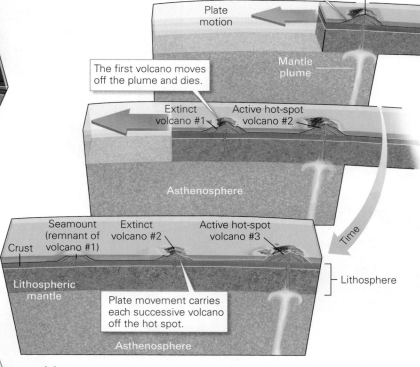

(b) As the plate moves with respect to a relatively fixed mantle plume, the volcano gets carried off the hot spot, goes extinct, and slowly sinks below the sea surface. Meanwhile, a new volcano forms over the plume.

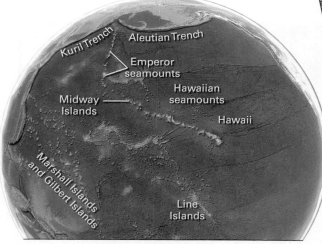

(c) The Hawaiian Islands and the Hawaiian-Emperor seamounts represent a hot-spot track.

of such a sediment accumulation forms a continental shelf. Not all rifts evolve into new mid-ocean ridges, however; in some cases, the process of stretching ceases before the continent splits, leaving the rift as a permanent scar in the continental crust.

Hot Spots

On a global basis, most *subaerial volcanoes* (volcanoes that rise into the atmosphere) occur along volcanic arcs bordering subduction zones or within rifts. Similarly, most *submarine volcanoes* (volcanoes hidden under the sea) occur along mid-ocean ridges. Geologists refer to such volcanoes resulting from the interaction of plates at a plate boundary as *plate-boundary volcanoes*. Not all volcanoes are plate-boundary volcanoes, however. Geologists have identified about a hundred places where volcanoes exist, but do not form as a direct consequence of plate interactions or rifting; they refer to these locations as **hot spots**, and the volcanoes at these locations are called *hot-spot volcanoes*. Some hot spots occur in the interior of a plate. Examples include Hawaii (FIG. 2.44a), which protrudes from the Pacific Plate thousands of kilometers from the nearest plate boundary, and Yellowstone National Park, located in the interior of the North American Plate. A few hot spots underlie mid-ocean ridges—volcanoes at these hot spots produce much more lava than normal seafloor spreading produces. Iceland serves as an example.

Why do hot spots exist? While researchers still debate this question, most favor a model in which a hot spot lies at the top of a **mantle plume**, a relatively narrow column of particularly hot asthenosphere that flows upward from deeper in the mantle. When peridotite in a plume reaches the base of the lithosphere, it melts, producing magma that seeps up through the lithosphere and erupts as a hot-spot volcano (see Chapter 4). Most hot-spot volcanoes lie at the end of a chain of *extinct volcanoes* (volcanoes that will never erupt again), because the position of a mantle plume stays fixed, more or less. Therefore, as a plate moves across the plume, active hot-spot volcanoes are carried off the hot spot and stop erupting (FIG. 2.44b). The resulting chain of extinct volcanic islands and seamounts defines a **hot-spot track** (FIG. 2.44c).

Plate Motion

Plates move in response to a combination of forces:

- Subducting lithosphere produces a *slab-pull force* that drags plates toward convergent boundaries because subducting plates are denser than the asthenosphere, so they sink like an anchor.

- The elevation of oceanic lithosphere at mid-ocean ridges produces a *ridge-push force* that causes plates to move away from the ridge axis because gravity makes elevated regions push outward toward lower elevations.

- Convection in the asthenosphere may cause the asthenosphere to move relative to the overlying lithosphere, and this movement applies a force to the base of the plate.

These forces combine to drive plates at rates of about 1–15 cm (0.4–6 in) per year, about as fast as your fingernails grow. Though small, such rates can yield large displacements over the Earth's long history. For example, in 100 million years, a plate moving at 2 cm (0.8 in) per year travels 2,000 km. Can we detect very slow plate motions by direct observation? Yes! By using the **global positioning system** (GPS), the same technology that drivers use to find their destinations, geologists can measure displacements as small as a few millimeters per year, and can easily "see" plates in motion (FIG. 2.45).

> ### Take-home message . . .
> Geologists distinguish among three types of plate boundaries, based on the relative movement of the plates involved: divergent, convergent, and transform boundaries. Seafloor spreading occurs at divergent boundaries, and subduction occurs at convergent boundaries. At transform boundaries, plates move sideways with respect to each other. Collisions occur when two buoyant pieces of crust converge, and rifts form where continents stretch and start to break apart. In some places, localized hot spots cause volcanoes to form independently of plate boundaries.
>
> **QUICK QUESTION** How fast do plates move, and how can their rate of movement be measured directly?

2.7 The Origin of Topography and Geologic Structures

Many geologic hazards exist because plate interactions not only cause horizontal movement of continents or islands, but can also result in vertical displacements. Such displacements, along with the consequences of erosion, yield topography that can include hazardous slopes, which host landslides, and valleys, which host floods. Plate movements can also produce geologic structures, such as cracks, faults, and foliation, that play a role in the development of hazards.

Development of Topography

The interaction of lithosphere plates with one another and with the underlying asthenosphere, as well as the occurrence of collision and rifting, can cause the land surface to move vertically. Geologists refer to upward movement as **uplift** and to downward movement as **subsidence**. Sometimes, these motions take place rapidly—for example, slip on a fault during a large earthquake can displace the Earth's surface by a few meters. But, averaged over time, uplift and subsidence take place at rates of 0.1–10 mm (0.004–0.4 in)—the thickness of your fingernail—per year. Although these rates seem slow, uplift can produce a mountain range in about 10 million years.

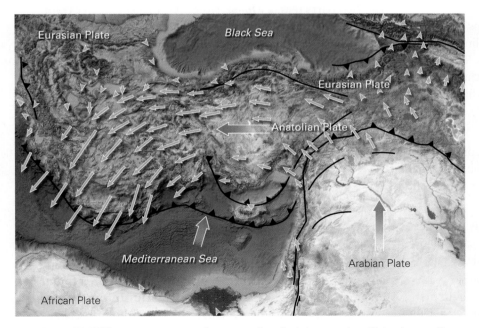

FIGURE 2.45 GPS measurements allow us to "see" plates moving. This observation is the ultimate proof of plate tectonics.

As uplift and subsidence take place, erosion and deposition begin to occur, resulting in the

FIGURE 2.46 Development of relief in mountain belts.

(a) Local relief of a mountain slope is the elevation difference between high and low points.

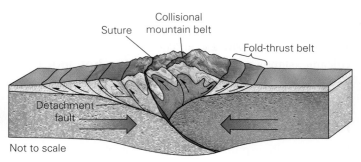

(b) In some places, relief develops as a consequence of continental collision.

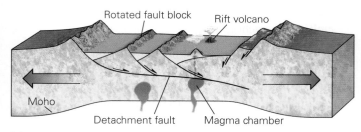

(c) Relief may also develop as a consequence of rifting.

development of distinctive shapes—such as valleys, cliffs, and ridges—on the Earth's surface, known as **landforms**. The elevation difference between the top and bottom of a landform determines the *local relief*—local difference in elevation (FIG. 2.46a). The detailed shapes of the landforms that develop depend on many factors, such as the nature of the eroding or transporting agent, the strength of the material being eroded, and the climate. As a result, we can find an incredible diversity of landforms at the Earth's surface.

The most extreme relief on the Earth develops as a consequence of **mountain building**, which produces linear belts of high, rugged topography called *mountain ranges*. The highest mountain range on the Earth, the Himalayas, reaches an elevation of 8.85 km (5.5 mi), and the steep slopes of its highest peak, Mt. Everest, have a local relief of over 3 km (1.9 mi). Mountains can be built by several processes:

- Collision squeezes the crust, causing it to become thicker, so that the land surface is uplifted and the Moho is pushed down (FIG 2.46b). The Himalayas are the result of the collision between India and Asia.

- Subduction at a convergent boundary not only produces a volcanic arc, but can lift the crust of the overriding plate. The Andes are forming at a convergent boundary.

- During rifting, a band of crust subsides relative to the rift margins, and the stretching of the crust causes faults along which crustal blocks slip downward and tilt (FIG 2.46c). The mountains of the Basin and Range Province of the western United States have been formed by rifting.

Geologic Structures

Because of the movements and interactions of plates, especially during mountain building, rocks are subjected to **stress**, defined as a force applied over an area (FIG. 2.47a). Geologists recognize three kinds of stress: *compression* squeezes materials together; *tension* stretches materials apart; and *shear* causes one part of a material to move past another part in a direction parallel to the boundary between the two parts (FIG. 2.47b–d). Note that in a stationary fluid, such as motionless air or water, stress is the same in all directions—such stress is called **pressure** (FIG. 2.47e). In solid rock, the magnitude of stress can be different in different directions, and this difference is what drives **deformation**: the process of breaking, bending, or modifying the shape and texture of rocks. Visible features that develop in rock due to deformation are called **geologic structures**.

The character of geologic structures depends on the conditions under which deformation takes place. At the relatively low pressures and temperatures found in the upper 15–20 km (9–12.5 mi) of the crust, rocks undergo **brittle deformation**, during which they crack or fracture to form pieces that no longer hold together (FIG. 2.48a)—you can see brittle deformation if you drop a plate on the floor and it shatters. At the higher pressures and temperatures found at greater depths, rocks undergo **plastic deformation**, meaning that they can change shape without breaking (FIG. 2.48b)—you can see plastic deformation by pressing on a ball of dough.

FIGURE 2.47 Stress can cause deformation.

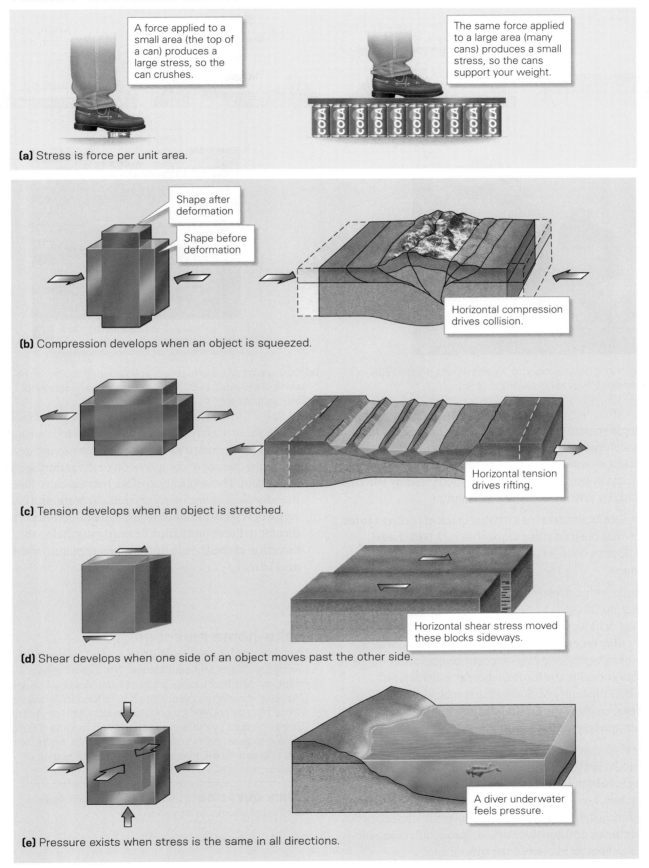

(a) Stress is force per unit area.

(b) Compression develops when an object is squeezed.

(c) Tension develops when an object is stretched.

(d) Shear develops when one side of an object moves past the other side.

(e) Pressure exists when stress is the same in all directions.

FIGURE 2.48 Contrasts between brittle and plastic deformation.

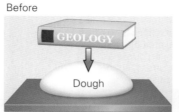
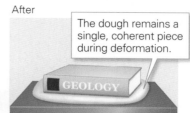

(a) During brittle deformation, rocks break into pieces. This sandstone ledge is breaking along cracks.

(b) During plastic deformation, rocks change shape without breaking. The "pancakes" in this outcrop were spherical clasts before deformation.

Geologic structures affect the strength of rock, which in turn may determine where geologic hazards develop. Geologists recognize several types of geologic structures, distinguished from one another by their geometry and by whether they formed by brittle or plastic deformation.

- *Joints:* Geologists refer to a natural crack in rock as a **joint** if there has been no shear along the crack (**FIG. 2.49a**). Rock doesn't connect across a joint, so joints are planes of weakness.

- *Faults:* As we've already mentioned, a *fault* is a fracture surface along which sliding occurs (**FIG. 2.49b**). The amount of movement that takes place across a fault is the fault's **displacement**. *Active faults* are those on which sliding has been occurring in recent geologic time and will probably occur in the future, whereas *inactive faults* ceased to slip long ago. Some faults, such as the San Andreas, displace the ground surface, but others are hidden completely underground.

- *Folds:* If you say that you have "folded a sheet of paper," you mean that you have bent the sheet. Geologists use the term **fold** in reference to a bend in a layer of rock (**FIG. 2.49c**). Folds form when stresses applied to layered rock during deformation cause the layers to undergo a shape change and become curved. Some folds resemble arches, and others resemble troughs.

- *Foliation:* As noted earlier, *foliation* refers to layering in metamorphic rock (**FIG. 2.49d, e**). This structure develops as a consequence of compression or shear during metamorphism. The application of such stress can cause brick-like, needle-like, and platy crystals to rotate, and can cause new platy minerals to grow in a direction perpendicular to the compression or nearly parallel to the direction of shear. Foliation can create planes of weakness in rock.

Take-home message...

Because of plate movements, the surface of the Earth undergoes uplift and subsidence. The resulting displacement, along with erosion, produces topography. The greatest relief develops in association with mountain building, which may be caused by collision, subduction, and rifting. Stress caused by plate interactions and mountain building produces geologic structures such as joints, faults, folds, and foliation. Some geologic structures are planes of weakness.

QUICK QUESTION Explain the difference between brittle and plastic deformation.

FIGURE 2.49 Examples of geologic structures.

(a) Joints cutting across beds of shale in New York.

(c) Folded beds of sandstone exposed in a cliff along the coast of Ireland.

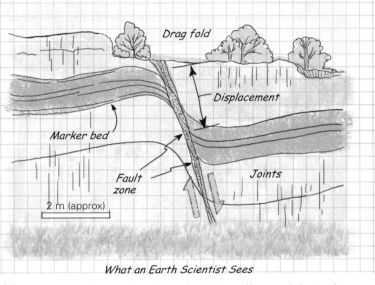

What an Earth Scientist Sees

(b) A photo and sketch showing a fault that offset rock layers in this road cut in Colorado.

(d) A metamorphic rock split on a foliation plane defined by aligned sheets of dark mica.

(e) Folded foliation in metamorphic rock. Note how the rock splits on foliation planes

2.7 The Origin of Topography and Geologic Structures • **65**

2.8 Geologic Time and Global Change

Developing the Concept of Deep Time

In the late 18th century, James Hutton, a Scottish farmer and scientist, began to wonder how rocks and landscapes formed, and how long it took for them to form. He realized that features within sedimentary rocks resembled features developing in modern depositional environments. These observations, and many others, eventually led Hutton to propose the principle of **uniformitarianism**, which states that physical processes that we can see operating in the modern world also operated in the past, at roughly the same rates, and produced the features we see in outcrops of ancient rocks **(FIG. 2.50)**. Put concisely, uniformitarianism means that "the present is the key to the past." Uniformitarianism will come into play, in a somewhat different way, when we discuss the frequency at which natural hazardous events take place. In such discussions, you'll see that "the past is the key to the present," in that by examining where and when hazardous events have happened in the past, we gain insight into where and when they might happen in the future.

In the context of uniformitarianism, Hutton deduced that the rocks, geologic structures, and landforms he observed couldn't all be the same age, and that the production of those features must have taken a very long time. Eventually, he concluded that **geologic time**, the time since the Earth formed, must be much longer than human history. When geologists first began to explore geologic time, they were able to determine only the *relative ages* of geologic features, meaning the age of one feature with respect to another. But beginning in the 1950s, researchers discovered how to measure the **numerical age** of rocks (their age in years) using a technique called **radioisotopic dating** (or *radiometric dating*). Using this technique, geologists determined the Earth to be 4.56–4.54 billion years old, so geologic time begins then. Over the past two centuries, studies of relative ages and numerical ages have led to the development of the **geologic time scale**, which subdivides geologic time **(FIG. 2.51)**. The largest subdivisions, *eons*, are named, from oldest to youngest, the Hadean, Archean, Proterozoic, and Phanerozoic. The first three eons together constitute the *Precambrian*. The eons were further subdivided into *eras*, *periods*, and *epochs*. Geologists abbreviate time designations as follows: Ga = billions of years before present, Ma = millions of years before present, and Ka = thousands of years before present.

Global Change

The Earth has changed in many ways during its long history. When it originated, our planet was molten, but it eventually cooled and formed a solid surface. The earliest oceans probably formed soon after, and a permanent ocean has existed since about 3.9 Ga. The volcanic activity that produced most of the rocks that became incorporated in continental crust took place between 4.0 and 2.5 Ga, and relatively large continental blocks had assembled by about 2.5 Ga. The fossil record indicates that life first appeared, in the form of single-celled microbes, by about 3.5 Ga, but complex multicellular life did not appear until about 570 Ma. Land plants didn't appear until about 470 Ma, so before that, the land surface consisted of barren rock and sediment.

FIGURE 2.50 The concept of uniformitarianism.

(a) Wave motion produces ripple marks (a pattern of small ridges and troughs) on the surface of this modern sandy beach.

(b) These ripple marks, on the surface of a tilted sandstone bed, are 145 million years old. Uniformitarianism states that they, too, were formed by water moving over sand.

FIGURE 2.51 The geologic time scale. (Note that each of the shorter columns is a blowup of the portion of the column to its left.)

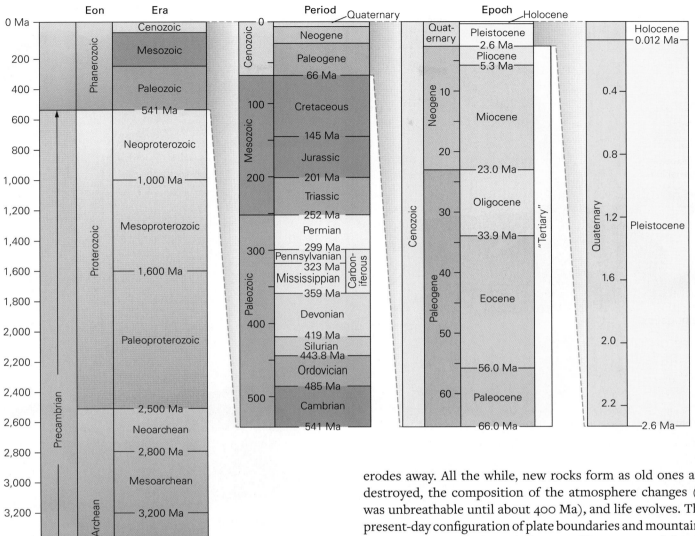

Plate tectonics has been operating in a form similar to what we see today, since at least 3.0 Ga, and since it began, continental blocks have waltzed around the globe, combining, breaking up, and recombining, and oceanic plates have formed by seafloor spreading and been subducted. Each collision and rifting event produces a mountain range, which eventually erodes away. All the while, new rocks form as old ones are destroyed, the composition of the atmosphere changes (it was unbreathable until about 400 Ma), and life evolves. The present-day configuration of plate boundaries and mountains has existed for only about the last 15 million years, and the climate belts of today for only about the last 10,000 years. What we see of the Earth today is just a snapshot in our planet's long biography.

Clearly, the geologic record shows that our planet constantly undergoes **global change**, defined as a significant modification of components in the Earth System over time. The Earth remains dynamic, because of continued inputs from internal and external energy sources, so it will continue to change in the future. We can distinguish among different types of global changes based on their rates. For example, some changes, such as mountain building and seafloor spreading, take place over the course of millions to hundreds of millions of years, whereas others, such as the impact of a meteorite or the retreat of a glacier, take place relatively rapidly (over seconds to millennia). We can also distinguish among types of change by the way in which the change progresses. For example, some changes, such as the evolution of the atmosphere and biosphere, proceed in one direction and

FIGURE 2.52 The rock cycle can take place because the Earth is dynamic.

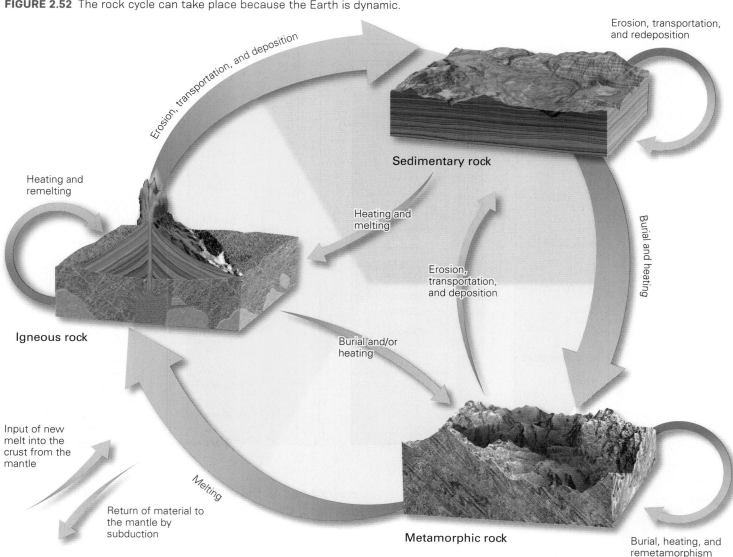

never repeat. But others are cyclic, in that they involve steps that occur over and over, though not necessarily with the same results or at the same rate. Examples include the **rock cycle**, during which atoms making up one type of rock may later become part of a different type of rock (**FIG. 2.52**); the **sea-level cycle**, during which the elevation of the sea surface relative to the land surface goes up and down by as much as 300 m (1,000 ft) (**FIG. 2.53a**); and the **supercontinent cycle**, during which blocks of continental crust collect into supercontinents and later break up to form smaller continents (**FIG. 2.53b**). As all these cycles take place, life on the Earth evolves.

Earth scientists refer to some types of cyclic change in the Earth System as **biogeochemical cycles**, meaning that they involve the movement of chemicals among various realms of the Earth System. A biogeochemical cycle can attain a *steady-state condition*, meaning that the proportions of a chemical in the different realms remain fairly constant even though its movement among realms continues. A global change in a biogeochemical cycle, therefore, is one that modifies the proportions of chemicals in different realms, causing a change from the steady-state condition. Examples of biogeochemical cycles include the **hydrologic cycle**—during which water moves from ocean to air to clouds to rain to land and plants, and eventually back to the ocean—and the **carbon cycle**—during which carbon moves among the atmosphere, hydrosphere, geosphere, and biosphere and can exist in many forms (such as CO_2, CH_4, oil, limestone, and coal). Global change in the hydrologic cycle happens, for example, during an ice age, when water transfers from the ocean into glaciers on land.

Particularly dramatic changes in the Earth System have accompanied the growth and modernization of human society. At the dawn of civilization, at 4,000 B.C.E, the human population was at most a few tens of millions, but in 2020, it surpassed 7.8 billion. As the human population grows and our

FIGURE 2.53 Sea-level rise and fall and the combination and break-up of continents are cyclic changes.

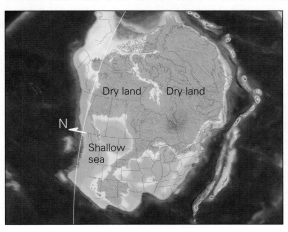

(a) During the Cambrian, a shallow sea covered much of North America.

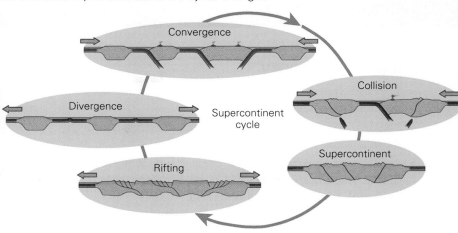

(b) The stages of the supercontinent cycle.

standard of living continues to improve, our use of the Earth's resources increases. Every time we move a pile of rock, plow a field, dig a mine, drain a wetland, pave a road, build a city, cut down a forest, plant a crop, or cause a species to go extinct, we change a portion of the Earth System. Clearly, humanity has become a significant agent of global change. Because of all the human-caused changes that have taken place, some researchers have suggested that the last few centuries, or millennia, should be considered a new geologic interval, named the **Anthropocene** (FIG. 2.54). In Chapter 15, we'll see how changes to the Earth's climate that are taking place now can influence natural disasters.

Take-home message . . .

The discovery of methods to determine the ages of rocks and other geologic features allowed researchers to develop the geologic time scale. The Earth has changed in many ways over geologic time. The geologic record indicates that atoms move among rocks of different types, continents come together and break apart, and sea level rises and falls. These changes are cyclic, but others, such as biological and atmospheric evolution, will never be repeated. In the past few centuries, humans have become a significant factor in driving changes in the Earth System.

QUICK QUESTION Describe examples of cyclic change.

FIGURE 2.54 During the Anthropocene, the land surface has been changed dramatically by human activity.

Chapter 2 Review

Chapter Summary

- The Earth System consists of several realms which exchange materials and energy.
- The Earth has a magnetic field that partially protects it from the solar wind.
- The biosphere is concentrated on the land surface and in the oceans. The atmosphere, of which 99.9% lies within 50 km of the Earth's surface, surrounds the Earth.
- The hydrosphere includes all surface and subsurface liquid water, most of which resides in the oceans. It also includes the cryosphere of frozen water.
- The geosphere extends from the surface of the solid Earth to the center of the planet. Most land lies within continents. Variations in land elevation define topography, and variations in seafloor depth define bathymetry.
- The Earth remains a dynamic planet due to inputs of internal energy from its interior and external energy from the Sun. Gravity also drives Earth movements.
- A mineral is a naturally occurring, generally inorganic, homogeneous, crystalline solid with a definable chemical composition. A rock, most commonly, is a coherent aggregate of minerals. Three rock groups can be distinguished by the ways in which they form.
- Igneous rocks form by solidification of a melt or by accumulation of erupted fragments. Extrusive rocks form at the surface; intrusive rocks form underground.
- Sedimentary rocks form by the accumulation and lithification of layers of sediment, or by precipitation of minerals from water onto the Earth's surface.
- Metamorphic rocks form by mineralogical or textural changes that take place when a pre-existing rock is subjected to elevated heat and pressure.
- The geosphere can be divided into three compositionally distinct layers.
- The crust, the outermost layer, is very thin. Continental crust is thicker and less dense than oceanic crust.
- The mantle is almost entirely solid, and consists of a very dense rock called peridotite.
- The uppermost part of the mantle plus the overlying crust together make up the rigid lithosphere. The lithosphere overlies the asthenosphere.
- The core consists of iron alloy. The outer core is liquid, and the inner core is solid.
- In the early 20th century, Alfred Wegener suggested that continents move with respect to one another, but few people believed him at the time.
- Data collection during the mid-20th century led to the discovery of seafloor spreading, by which new oceanic lithosphere forms at a mid-ocean ridge and then moves outward so that continents on either side move apart.
- Old oceanic lithosphere sinks back into the mantle at subduction zones.
- The ideas of continental drift, seafloor spreading, and subduction came together in the theory of plate tectonics, which states that the lithosphere is divided into plates that move relative to one another.
- The distribution of earthquakes delineates plate boundaries, the places where plates interact.
- The three types of plate boundaries are convergent, divergent, and transform.
- At convergent boundaries, subduction takes place, causing a trench and a volcanic arc to form. At divergent boundaries, seafloor spreading takes place, causing a mid-ocean ridge to form. Transform boundaries are actively slipping faults along which one plate slides sideways past another.
- At hot spots, volcanic activity happens independently of a plate boundary, probably in association with a plume of rising hot mantle.
- At rifts, continents stretch and pull apart. A rift may evolve into a divergent boundary. Collision happens when blocks of buoyant crust converge and form a collisional mountain belt.
- Plates move very slowly, at 1–15 cm per year. We can measure this motion with GPS.
- As uplift (upward vertical motion) and subsidence (sinking) take place, local relief develops and erosion and deposition occur.
- Stress acting on rock causes deformation, which results in development of geologic structures, such as joints, faults, folds, and foliation.
- The Earth formed from a planetary nebula surrounding the newborn Sun, about 4.54–4.56 billion years ago.
- The period since the Earth's formation is called geologic time. The geologic time scale is divided into eons, eras, and periods. The Earth constantly changes over geologic time.

Review Questions

Blue letters correspond to the chapter's learning objectives.

1. What does "the Earth System" refer to, and what realms does it include? **(A)**
2. What is the magnetic field of the Earth? What controls the shape of the field? **(A)**
3. What does the biosphere consist of? **(A)**
4. What does air consist of? Below what elevation do 99.9% of the molecules of the atmosphere reside? **(A)**
5. Describe the difference between internal energy and external energy in the Earth System. **(A)**
6. How does gravity contribute to movements in the Earth System such as convection? **(A)**
7. Provide the geologic definition of a mineral, and explain the components of this definition. **(B)**
8. Provide the geologic definition of a rock, and explain it. What is a rock's texture? **(B)**
9. How does weathering change rock, and what are the products of weathering? **(B)**
10. What are the three rock groups? Explain how rocks in each of these groups form. **(B)**
11. What are two key types of layering that can exist in rocks, and how does such layering form? **(B)**
12. Label the major internal layers of the Earth on the diagram. What does each consist of? How can researchers determine the depths of the boundaries between the layers? **(C)**
13. How does the crust beneath the ocean differ from the crust beneath the continents? **(C)**
14. How does temperature vary with depth in the Earth? As temperature increases, do rocks behave more rigidly, or more plastically? **(C)**
15. Describe the key sources of information that Wegener used to support his concept of continental drift. **(D)**
16. What other observations were made in the mid-20th century that led to the proposal of plate tectonics? **(D)**
17. How is seafloor spreading related to continental drift? **(D)**
18. Explain the difference between the lithosphere and the asthenosphere. **(E)**
19. How can the positions of lithosphere plate boundaries be identified on a map? **(E)**
20. At which type of plate boundary does seafloor spreading take place? At which type does subduction take place? Which type generally involves movement on a vertical fault? What type of plate boundary does the diagram show? **(E)**
21. Where do volcanic arcs form? How does a volcanic arc differ from a hot-spot track? **(E)**
22. What process leads to a continental collision, and what features forms as a result of such a collision? **(F)**
23. What is a continental rift? How does a rift differ from a mid-ocean ridge? **(F)**
24. Define the following terms: joint, fault, fold, foliation. **(G)**
25. How old is the Earth, according to the results of isotopic dating? What are the major divisions of geologic time? **(H)**
26. Give examples of ways in which the Earth has changed over geologic time. Why do some researchers refer to the past few centuries as the Anthropocene? **(H)**

On Further Thought

27. If people were to establish colonies on the Moon, would they face the same natural hazards as exist on the Earth? Explain your answer. **(A)**

3

A VIOLENT PULSE
Earthquakes

By the end of the chapter you should be able to...

A. explain what an earthquake is, and what processes cause earthquakes.

B. interpret a description of an earthquake's location, and explain how earthquake energy travels and changes away from that location.

C. describe how to measure earthquakes.

D. interpret the meaning of an earthquake's size as discussed in news media.

E. relate earthquakes to specific geologic settings in the context of plate tectonics theory.

F. distinguish among the different ways in which earthquakes cause damage.

G. understand the meaning of seismic hazard, and assess earthquake prediction and its limitations.

H. explain what features can help make a building earthquake resistant, and plan a strategy to help prevent earthquake damage and casualties.

3.1 Introduction

The children of Lisbon, Portugal, looked forward to November 1, the Feast of All Saints, when they could run from house to house and receive sweets. Sadly, November 1 of 1755 would not be a day of joy. At 9:40 A.M., residents of Lisbon felt a sudden jolt. Then, for the next few minutes, the ground rocked and swayed violently, terrifying and disorienting everyone. Dust billowed skyward as a cacophony of cracking, snapping, crunching, and clattering announced the collapse of buildings. Lisbon had endured a major **earthquake**, an episode of ground shaking generated when rocks break and scrape underground and send vibrations through the Earth.

Unfortunately, the calamity didn't end when the shaking stopped. About 40 minutes after the earthquake, panicked residents, who had clustered on quays along the harbor to avoid falling debris, saw a wall of water rising 6 m (20 ft) above their heads and approaching them at high speed. This wave—a *tsunami*—submerged the quays and rushed inland, washing away everything on low-lying land. Meanwhile, in the still-dry parts of the city, another secondary disaster was unfolding. The ground shaking had tipped over candles lit in honor of the day, igniting devastating fires. By day's end, over 60,000 people (20% of the local population) were dead, a great city of Europe lay in ruins, and almost half of Portugal's economy had vanished **(FIG. 3.1)**.

The *Great Lisbon Earthquake* had many repercussions. The human tragedy led European philosophers such as Voltaire to abandon the popular notion that "we live in the best of all possible worlds." It also alerted public officials to the need for emergency planning, and motivated the insurance industry to expand. Significantly, it also made scientists realize how little they understood about earthquakes, and consequently sparked modern research on earthquakes.

Prior to the late 18th century, **seismicity** (earthquake activity) had commonly been associated with the restless gyrations of mythical subterranean beasts, or to divine retribution. While there had been attempts to find natural, as opposed to supernatural, explanations for earthquakes, none were adequate. For example, ancient Greek philosophers attributed earthquakes to the movement of air between huge subterranean cavities, or to the swelling and shrinking of the Earth due to alternating floods and droughts. Not long after the Lisbon disaster, John Mitchell (1724–1793), an English scientist, proposed instead that earthquakes happen when rock breaks inside the Earth, the explanation still accepted today.

Over the next two centuries, the field of **seismology**, the scientific study of earthquakes (from the Greek word *seismos*, meaning shock or earthquake), became established. *Seismologists*, scientists who study the causes of earthquakes and the properties of **seismic waves** (vibrations generated by earthquakes), sought answers to key scientific questions about earthquakes, including: what geologic phenomena cause earthquakes; how seismic waves move through the Earth; how to compare earthquakes; and why earthquakes take place where they do. The first part of this chapter reviews answers to these questions by outlining the science of earthquakes. The later part of the chapter focuses on consequences of earthquakes and on steps that society can take to prepare for them and minimize their impact.

◄ People were crushed under this collapsed building in Mexico City, Mexico, during an earthquake in September 1985.

FIGURE 3.1 An illustration of the Great Lisbon earthquake of 1755, depicting ground shaking, fire, and the tsunami.

3.2 What Causes Earthquakes?

Several phenomena can produce earthquakes: slip on a fault; volcanic activity; landslides; meteor impacts; mine collapses; and explosions. Nearly all earthquakes, however, result from faulting, however, so this chapter will emphasize natural disasters associated with faulting.

Introducing Faults and Faulting

A **fault** is a fracture on which sliding, or *slip*, takes place. Faults develop when a portion of the Earth's crust undergoes *deformation* (movement or change in shape) due to the application of *stress*, the force applied over an area (see Chapter 2). Slip displaces rock or sediment on one side of the fault relative to that on the other side. If the surface of the fault—the *fault* *plane*—slopes at an angle, we can describe the slip direction by characterizing the movement of the *hanging wall*, the block of rock above the fault, relative to the *footwall*, the rock below the fault **(FIG. 3.2a)**. During slip on a **normal fault**, the hanging wall moves down the fault plane, relative to the footwall, whereas during slip on a **reverse fault**, the hanging wall slides up the fault plane relative to the footwall **(FIG. 3.2b)**. Geologists use another term, *thrust fault*, for a reverse fault with a gentle slope **(FIG. 3.2c)**. During slip on a **strike-slip fault**, the block on one side of the fault moves sideways, parallel to a horizontal line on the fault plane called a *strike line*, so that slip does not produce any upward or downward motion **(FIG. 3.2d)**. Normal faulting develops during stretching of the crust due to tension, a situation that occurs along a rift. In contrast, reverse or thrust faulting accommodates shortening of the crust due to compression, as occurs along a convergent boundary or a collision zone **(FIG. 3.2e)**. Motion along transform plate boundaries takes place along strike-slip faults.

What do faults look like? In an outcrop, a fault may stand out as a plane across which bedding, foliation, or the border of an igneous intrusion has been displaced **(FIG. 3.3a)**. On the land surface, a fault may look like a tear or rip along which landscape features have been displaced **(FIG. 3.3b)**. In cross section, a fault may appear as a band of broken-up or even pulverized rock **(FIG. 3.3c)**. The surface of a fault may be polished and grooved. If slip on a normal or reverse fault displaces the ground surface, a step, known as a *fault scarp*, develops **(FIG. 3.4a)**. Strike-slip faults that intersect the ground surface may offset features such as stream channels, fences, and roads **(FIG. 3.4b, c)**. At places where the trace of a strike-slip fault bends, the land surface may be pushed up or may sink down locally along the fault **(FIG. 3.4d)**. In some cases, the area that sinks will fill with water and become a *sag pond*. You can see faults at many locations, but don't panic! Not all faults serve as sources of earthquakes. Seismologists distinguish between **active faults**, which have slipped relatively recently and might slip in the near future, and *inactive faults*, which slipped in the distant geologic past and probably won't slip again. Do all active faults intersect the Earth's surface? No. Many are entirely underground, and these faults can generate earthquakes without ripping through a landscape.

FIGURE 3.2 Faults are classified according to the relative movement of blocks on either side and, in some cases, on the slope of the fault surface. Different types of faults accommodate different types of crustal deformation.

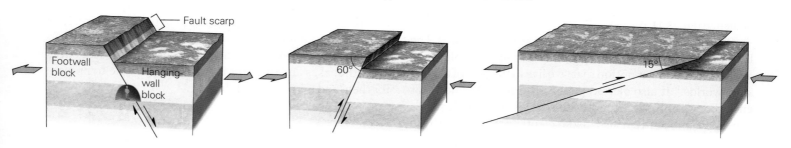

(a) On a normal fault, the hanging wall moves down. Note that if you stood in a tunnel straddling the fault, the hanging wall would hang over your head, and the footwall would be under your feet.

(b) On a reverse fault, the hanging wall moves up.

(c) A thrust fault is a reverse fault with a gentle slope.

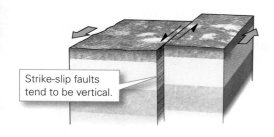

(d) On a strike-slip fault, blocks slip horizontally. This example depicts a left-lateral displacement, meaning that if you look across the fault, the block on the other side moves to your left.

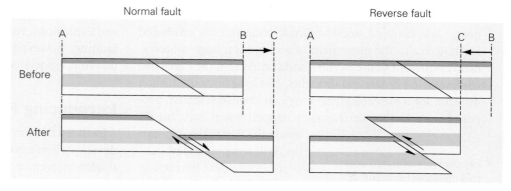

(e) This cross-section view emphasizes that a normal fault accommodates crustal stretching, whereas a reverse fault accommodates crustal shortening.

FIGURE 3.3 The consequences of faulting.

(a) A fault is visible in this road cut in Colorado.

The tan layers are volcanic ash. The dark layers at the top is a lava flow.

What an Earth Scientist Sees

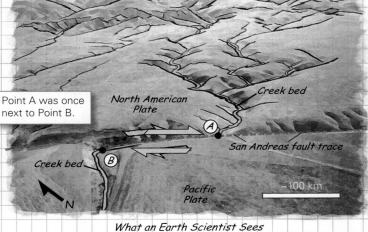
Point A was once next to Point B.

What an Earth Scientist Sees

(b) An aerial photo of the San Andreas fault in California, a strike-slip fault. The land in the foreground moved to the left relative to the land in the background, as indicated by the offset creek bed.

(c) Faulting can break up rock (left), and can polish or scratch surfaces (right). The orientation of the scratches indicates the direction of slip.

3.2 What Causes Earthquakes?

FIGURE 3.4 Some faults offset the ground surface.

(a) This distinct fault scarp formed during a large earthquake in Nevada in 1954. At this locality, about 3 m (10 ft) of slip occurred.

(b) Slip during the 1906 San Francisco earthquake offset this fence by about 3 m.

What an Earth Scientist Sees

(c) Slip on the San Andreas fault offset this dirt road, as seen in this aerial photo.

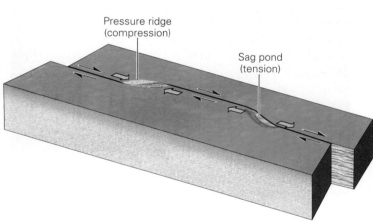

(d) If there are bends along the trace of a strike-slip fault, pressure ridges or sag ponds will develop, due to local compression or stretching, respectively.

Generating Seismic Waves: Elastic Rebound and Stick-Slip Behavior

Imagine an experiment during which you grip each side of a brick-shaped block of rock with a clamp, and then apply an upward push on one clamp and a downward push on the other, producing shear (see Chapter 2). Initially, the rock bends slightly but doesn't break. If you were to stop applying a push, the rock would relax, or *rebound*, and return to its original shape—the same phenomenon happens when you bend a stick and then let go, or stretch a rubber band and then release it. During such **elastic deformation**, an object changes shape for as long as stress is being applied, but recovers its original shape when the stress has been removed. Elastic deformation can happen because the chemical bonds that hold atoms together behave like tiny springs that can bend and stretch slightly without breaking when the stress is fairly small. Now repeat the experiment, but this time push harder, producing greater stress. At first

FIGURE 3.5 Producing a new fault by applying shear to a rock block. (This is a simplified illustration; real rock deformation machines are more complex.)

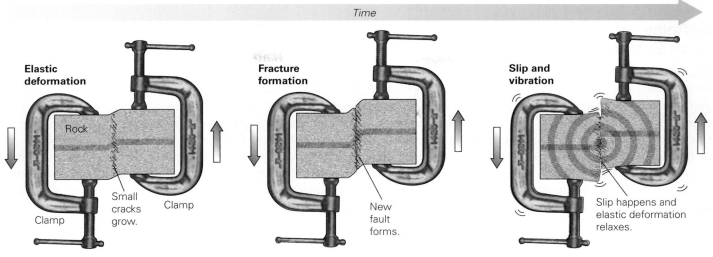

(a) Sufficient shear causes cracks to form after the rock has started to bend.

(b) Continued shear causes cracks to grow and link together.

(c) Slip takes place on the new fracture, and elastic rebound occurs.

the rock bends, but then cracks start to develop in the rock, because chemical bonds (see the Appendix) have been stretched beyond their limit and start to break (**FIG. 3.5a**). If you push hard enough, the cracks grow and intersect, until a through-going fracture suddenly develops and cuts across the entire block (**FIG. 3.5b**). At that instant, the block breaks in two, and rock on one side of the fracture slides past rock on the other side, so the fracture becomes a fault (**FIG. 3.5c**). When this new fault forms, the bent rock on either side of the fault elastically twangs back and forth momentarily like a snapped stick (**FIG. 3.6**), but it quickly relaxes and straightens out. Because of slip, however, rock layers on one side of the new fault no longer align with those on the other side. The fracturing and associated *elastic rebound*—the straightening out of bent rock—produce the energy of an earthquake, which propagates outward from the place where fracturing, slip, and rebound occurred as seismic waves. This overall explanation is called the **elastic-rebound theory** of earthquake generation.

Significantly, once a new fault forms, it doesn't continue to slip because **friction**—the resistance to sliding caused by bumps and irregularities on a surface that dig into an adjacent surface and act like tiny anchors—eventually slows and stops the movement. In addition, stress in the rock decreases when rock relaxes after slipping. Friction then prevents further sliding on the fault unless stress builds up again, overcomes friction, and causes another episode of slip. Commonly, the stress necessary to overcome friction on a pre-existing fault is less than the stress necessary to break intact rock and form a new fault. Therefore, most earthquakes happen on reactivated pre-existing faults. Geologists refer to

FIGURE 3.6 Elastic-rebound theory.

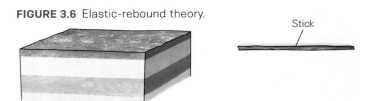

(a) Before deformation, layers of rock are unbent, like an unbent stick.

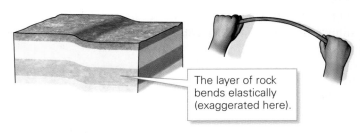

(b) Stress causes elastic deformation, like bending a stick.

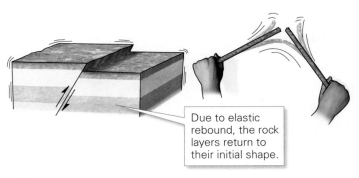

(c) When the rock fractures and the fault slips, the rock on either side rebounds, like a stick that vibrates after it snaps.

the cycle of stress buildup followed by stress release and sliding as **stick-slip behavior**.

Defining the Location of an Earthquake

Seismologists refer to the location where the generation of seismic waves begins as the **focus**, or **hypocenter**, of an earthquake (**FIG. 3.7a**). An earthquake whose focus lies at a depth of less than 70 km (43 mi) is a *shallow-focus earthquake*, one that occurs at a depth of 70–300 km (43–185 mi) is an *intermediate-focus earthquake*, and one whose focus lies at a depth of 300–660 km (185–410 mi) is a *deep-focus earthquake*. Because earthquake foci do not lie on the Earth's surface, we can't plot their positions directly on a map. The dot representing the position of an earthquake that you see on a map is the **epicenter** of the earthquake, the point on the surface of the Earth that lies vertically above the focus (**FIG. 3.7b**).

Foreshocks, Mainshocks, and Aftershocks

A significant earthquake, a **mainshock**, may be preceded and succeeded by smaller earthquakes. A cluster of smaller earthquakes, called **foreshocks**, sometimes precedes the mainshock, possibly due to development of cracks within the zone that will become the fault plane. A cluster of **aftershocks** always follows the mainshock because slip during the mainshock doesn't relax all the stress on the main fault plane and on nearby faults (**FIG. 3.8a**). The largest aftershock tends to be 10 times smaller than the mainshock—most are even smaller—and the frequency of aftershocks diminishes over time (**FIG. 3.8b**).

How Much Slip Happens during an Earthquake?

Generally, during an earthquake, only a portion of the fault surface slips. Seismologists can get a sense of the slip area involved in an earthquake by examining aftershock distribution (see Fig. 3.8). Not surprisingly, the bigger the earthquake, the larger the slip area. Small earthquakes involve a slip area of less than a hundred meters across. Destructive, but not catastrophic, earthquakes involve a slip area of several kilometers to tens of kilometers across. During the catastrophic 2011 Tōhoku earthquake (see Chapter 1), an area 300 km (186 mi) long by 100 km (60 mi) wide slipped, and during the catastrophic 2004 earthquake off Sumatra, an area 1,200 km (750 mi) long by 200 km (125 mi) wide slipped.

Similarly, the amount of displacement during an earthquake varies with earthquake size. The smallest earthquakes that people can feel result from displacements of only millimeters to centimeters. Damaging earthquakes

FIGURE 3.7 Distinguishing between the focus and the epicenter of an earthquake.

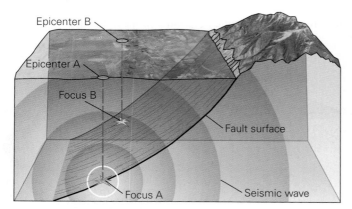

(a) The focus is the point on a fault where slip begins. The epicenter is the point on the Earth's surface vertically above the focus.

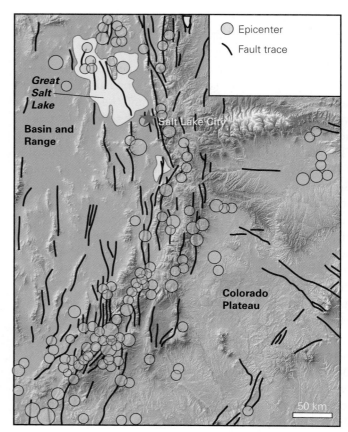

(b) The epicenters of earthquakes in Utah. The relative sizes of the circles represent the relative magnitudes (defined later) of the earthquakes. Note that earthquakes cluster along the faults that mark the eastern edge of the Basin and Range Province. The tan areas are higher elevation, and the green areas are lower elevation

FIGURE 3.8 Aftershocks of the 2011 Tōhoku earthquake.

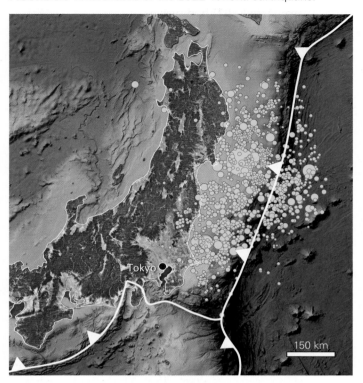

(a) Most aftershocks occurred along the thrust fault that defines the plate boundary. Some occurred on faults within the Pacific Plate, east of the plate boundary.

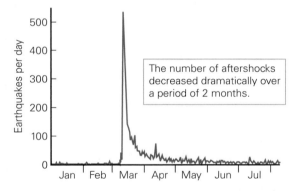

The number of aftershocks decreased dramatically over a period of 2 months.

(b) A graph depicting the frequency of aftershocks following the 2011 Tōhoku earthquake.

involve tens of centimeters, or more, of displacement. For example, the 1994 Northridge earthquake, which killed 60 people and caused $20 billion in damage in the Los Angeles area, resulted from about 50 cm (1.6 ft) of slip on a thrust fault. Displacement associated with catastrophic earthquakes can be meters to tens of meters. For example, almost 10 m (33 ft) of slip occurred during the earthquake that destroyed San Francisco in 1906 (see Fig. 3.4b), and up to 30 m (100 ft) of slip took place during the Tōhoku earthquake. The greatest displacement tends to occur

FIGURE 3.9 Variation in displacement on a fault plane. Generally, when an existing fault undergoes reactivation, only part of the fault slips. Most slip typically occurs near the focus (darkest color). It dies out away from that point.

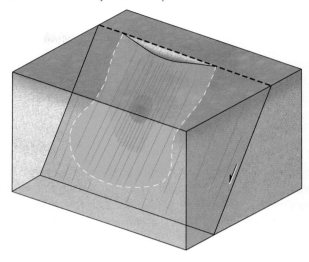

underground near the earthquake's focus. Consequently, the amount of slip observed at the Earth's surface is generally less than the maximum slip (**FIG. 3.9**).

> **Take-home message . . .**
>
> Most earthquakes happen when stress causes sudden formation of a new fault or slip on a pre-existing fault. When slip takes place, elastically deformed rock on either side of the fault rebounds and generates seismic waves. Faults commonly exhibit stick-slip behavior. The area of a fault that slips and the amount of displacement that takes place on a fault are greater for larger earthquakes.
>
> **QUICK QUESTION** What are aftershocks, and why do they occur?

3.3 The Nature of Seismic Waves

The Different Types of Seismic Waves

Seismic waves are vibrations that pass through rock much as sound waves pass through air. As seismic waves travel, chemical bonds in the rock's minerals elastically stretch and relax, and as each bond changes in length, it stimulates adjacent bonds to start stretching and relaxing. These movements are elastic, so after the waves have passed, atoms rebound to their original positions.

An earthquake generates different types of seismic waves (**TABLE 3.1**). Seismologists distinguish between two general

TABLE 3.1 Types of Seismic Waves

Name	Abbreviation	Motion
P-waves	**P**rimary	Compressional body waves (first to arrive)
S-waves	**S**econdary	Shear body waves (second to arrive)
L-waves	**L**ove	Surface waves that cause a back-and-forth, snake-like shimmying
R-waves	**R**ayleigh	Surface waves that cause up-and-down wave-like undulations

categories of waves based on where the waves travel: **body waves** pass through the interior of the Earth, and **surface waves** travel along the Earth's surface. Each category, in turn, includes two forms of waves. **P-waves** (*primary waves*) are compressional body waves, meaning that they cause back-and-forth motion parallel to the direction in which the waves themselves move (FIG. 3.10a). As a P-wave passes through a material, the material first contracts (squeezes together), then dilates (expands). You can see this kind of motion by pushing and pulling on the end of a spring—pulses of contraction (squeezing together) and dilation (spreading apart) move along the length of the spring. **S-waves** (*secondary waves*) are shear body waves, meaning body waves that cause up-and-down motion perpendicular to the direction of wave motion (FIG. 3.10b). The high point of the wave is its *crest*, and the low part is its *trough*. To see this kind of motion, jerk the end of a rope up and down and watch how the up-and-down motion travels along the rope. Surface waves include both *L-waves* (*Love waves*), which cause the ground surface to shimmy back and forth sideways like a snake, and *R-waves* (*Rayleigh waves*), which cause the ground surface to move up and down in rolling undulations (FIG. 3.10c).

When a fault slips, seismic waves propagate in all directions, so the *wave front* (the line tracing out the crest of the wave farthest from the source) can be thought of as having the shape of an oblong expanding bubble around the focus (FIG. 3.11). (Note that successive positions of a point on a wave as it moves define a curving line called a *seismic ray*.) When body waves reach the Earth's surface, the energy they transmit generates surface waves, which then propagate outward along the surface.

Different types of seismic waves travel at different velocities. P-waves move fastest, zooming along at 6 km/s (21,600 km/h, or 13,400 mph) in the crust, and even faster deeper in the Earth (see Fig. 2.21b). Since they travel the fastest, P-waves arrive first at a location distant from the focus, which is why they're called primary waves. S-waves travel at 60% of the speed of P-waves, so they arrive second, which is why they are called secondary waves. Both types of surface waves (L-waves and R-waves) are slower than body waves, so surface waves arrive after body waves.

The times at which the Loma Prieta earthquake reached different localities in the San Francisco Bay area give a sense of how fast seismic waves move. At 5:04 P.M. on October 17, 1989, a portion of the San Andreas fault 19 km (12 mi) beneath Loma Prieta peak (80 km, or 50 mi, to the southeast of San Francisco) slipped. It took 3 seconds for the P-waves to travel from the focus to the epicenter and cause Loma Prieta peak to start shaking. Two seconds after that, the nearby city of Santa Cruz felt the shock, and 14 seconds later, Candlestick Park in San Francisco, where the World Series baseball game was underway, began vibrating. Fortunately, the stadium remained standing, but other structures weren't so lucky. By the time the ground shaking stopped, it had cost the lives of 63 people, had injured over 3,700 more, and had left close to $12 billion in damage.

Describing Seismic Waves: Wavelength, Period, and Amplitude

When you throw a pebble into a pond, a series of waves, comprising a **wave train**, propagate outward. Seismic waves also move in a wave train (FIG. 3.12). The **wavelength** of a P-wave is the distance between two successive dilations or two successive contractions in a wave train. For other types of seismic waves, wavelength refers to the distance between two successive crests or two successive troughs. The *wave period* is the time it takes for a single wave to pass, and the *wave frequency* is the number of waves that pass in a second. To measure the amplitude of a wave, we first measure the *wave height*—the distance between the top of a crest and the bottom of a trough—and then divide by 2. In other words, the **amplitude** of a wave is half the wave height. Slip on a fault and associated elastic rebound send out a spectrum of seismic waves, each with a different speed, period, and amplitude. For a given wave type, longer-period waves travel more slowly, and carry more energy, than do shorter-period waves. The farther waves of different speeds travel, the more they separate.

How Seismic Waves Change with Distance

Imagine again the wave train produced when a pebble strikes a pond. The waves die out with increasing distance from the impact (FIG. 3.13a). That's because as waves travel through a material, they weaken, a phenomenon called *attenuation*. Such weakening of seismic waves happens for two reasons. First, waves spread out across a greater area as they travel away from the focus, so that the energy they carry also spreads out. Consequently, a given area at a location that lies farther

FIGURE 3.10 Seismologists recognize several different types of seismic waves.

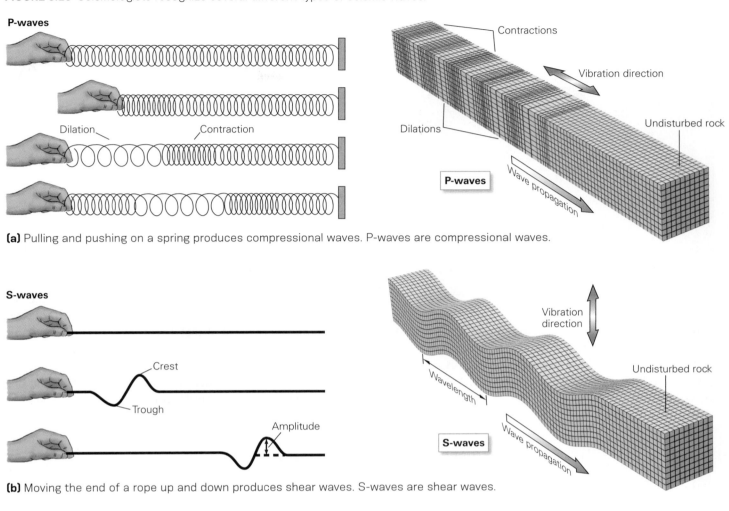

(a) Pulling and pushing on a spring produces compressional waves. P-waves are compressional waves.

(b) Moving the end of a rope up and down produces shear waves. S-waves are shear waves.

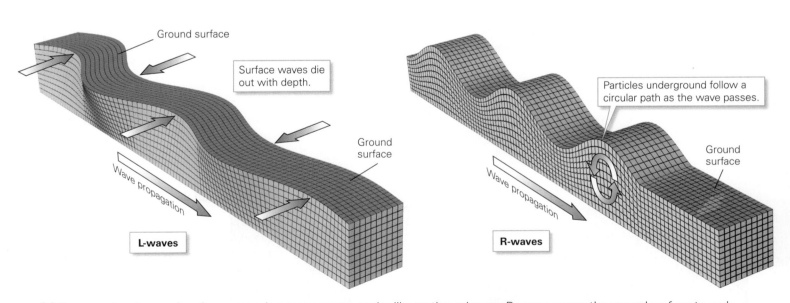

(c) There are two types of surface waves. L-waves cause a snake-like motion, whereas R-waves cause the ground surface to undulate up and down.

3.3 The Nature of Seismic Waves

FIGURE 3.11 Seismic waves start at an earthquake focus and propagate outward. Because seismic waves of any type travel faster with depth in the Earth, the wave front initially propagates faster downward.

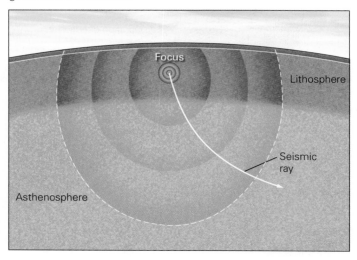

FIGURE 3.12 The terminology used to describe waves, here applied to an S-wave.

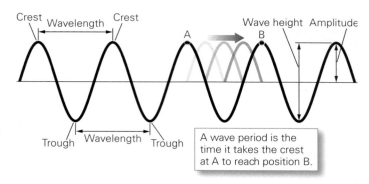

A wave period is the time it takes the crest at A to reach position B.

from the focus receives less energy than the same area at a location closer to the focus. Second, because rocks are not perfectly elastic, some of the waves' energy is absorbed by rocks, as if the rocks were shock absorbers. This energy converts into heat.

Waves also change as they pass from a stronger (harder) material (such as rigid bedrock) into a weaker (softer) material (such as soft sediment). Specifically, as the waves enter the weaker material, they slow down, so the energy they're carrying piles up, causing wavelength to decrease and wave amplitude to grow **(FIG. 3.13b)**. Because of this phenomenon, known as **amplification**, the 1989 Loma Prieta earthquake caused much more damage to buildings constructed on the soft sediment layers bordering San Francisco Bay than it did to buildings anchored in hard bedrock.

FIGURE 3.13 Seismic waves change as they travel, and these changes affect ground shaking.

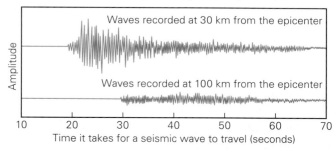

(a) Seismic waves undergo attenuation as they move away from the epicenter. These recordings represent seismic waves. The upper recording shows waves that traveled 30 km, and the lower one shows waves that traveled 100 km. Note that the waves arrived later at the location that is farther from the epicenter, and that the waves that traveled farther have smaller amplitudes.

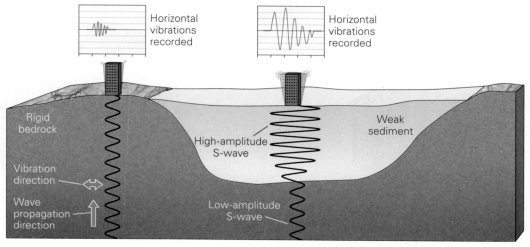

(b) Seismic waves are amplified when they pass from stronger material into weaker material. This example shows the amplification of S-waves passing from hard bedrock upward into soft sediment compared with S-waves moving entirely through bedrock. Buildings built on the sediment sway more violently.

Seismic Wave Reflection and Refraction

Imagine a train of seismic waves obliquely approaching the boundary between two different materials, such as the floor of a sedimentary basin (which separates soft sediment above from hard bedrock below), or a boundary between internal layers of the Earth. Some of the energy undergoes **reflection**, meaning that it bounces back from the boundary. Seismic energy that isn't reflected crosses the boundary, and when it does, its propagation direction changes, a phenomenon called **refraction**. To depict refraction, we can draw a seismic ray, which bends where it crosses a boundary. Where waves cross into a material through which they travel more slowly, the ray bends downward and away from the boundary, and where waves enter a material through which they travel faster, the ray bends upward and toward the boundary (FIG. 3.14a).

As we saw in Chapter 2, the composition of the Earth changes with depth, defining distinct layers: the crust, the mantle, and the core. At the crust-mantle boundary, seismic rays bend upward because their velocity in the mantle is greater than in the crust. Within the mantle, seismic waves travel progressively faster with depth so, overall, seismic rays follow broad curves that eventually bring them back to the Earth's surface at a steep angle (FIG. 3.14b, c).

> **Take-home message . . .**
> Earthquake energy travels as seismic waves. Body waves go through the Earth's interior, whereas surface waves move along the ground surface. Seismologists distinguish between two kinds of body waves (P-waves and S-waves) and two kinds of surface waves (L-waves and R-waves).
>
> **QUICK QUESTION** How do seismic waves change as they pass through the Earth?

3.4 Measuring Earthquakes

Seismographs and Seismograms

In order to detect earthquakes, seismologists use a **seismograph**, a device that can measure and record the ground motion, even if that motion is very tiny. A traditional *mechanical seismograph* uses a heavy weight suspended from a spring. The spring, in turn, hangs from a sturdy frame anchored to the ground. A pen extends from the weight and touches a revolving paper-covered cylinder attached to the frame. Seismographs can be configured in two ways: a *vertical-motion seismograph* detects and records up-and-down ground motion (FIG. 3.15a), whereas a *horizontal-motion seismograph* detects and records back-and-forth ground motion (FIG. 3.15b). Before an earthquake, when the ground is steady, the pen traces out a straight reference line on the paper as the cylinder turns, but when a seismic wave arrives and causes the ground surface to move, the seismograph frame, along with the paper-covered cylinder, moves with it. Though the frame moves, **inertia** (the tendency of an object at rest to remain at rest) makes the weight, with its attached pen, remain fixed. As the revolving cylinder moves with respect to the fixed pen, the pen traces a line on the paper that represents the ground motion (FIG. 3.15c). Note that if the cylinder were not revolving, the pen would simply go back and forth, or up and down, in place, but because the paper moves under the pen, the pen traces out a line that looks somewhat like a wave. Modern electronic seismographs work on the same principle, but the weight is a magnet that moves relative to a wire coil, thereby producing an electrical signal that can be recorded digitally. Such seismographs can record ground movements as small as a millionth of a millimeter (only 10 times the diameter of an atom).

FIGURE 3.14 Reflection and refraction of seismic energy.

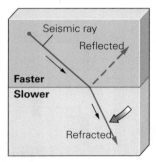
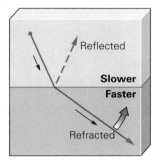

(a) The direction in which a seismic ray bends when it crosses a boundary between two materials depends on the relative speed of the waves in the two materials.

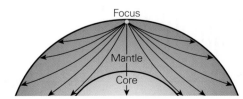

(b) Because density increases progressively with depth, a wave can refract enough to return to the surface.

(c) Overall, seismic rays follow curved paths through the Earth, due to refraction.

FIGURE 3.15 The operation of a traditional mechanical seismograph.

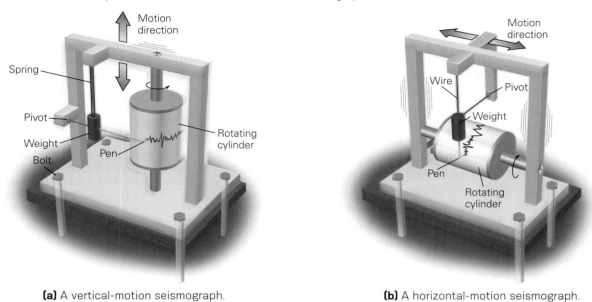

(a) A vertical-motion seismograph.

(b) A horizontal-motion seismograph.

(c) Before an earthquake, the pen traces a straight line. During an earthquake, the cylinder moves up and down while the pen stays in place.

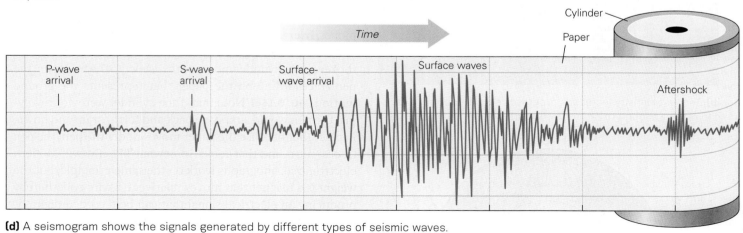

(d) A seismogram shows the signals generated by different types of seismic waves.

84 • CHAPTER 3 • A Violent Pulse: Earthquakes

The record of an earthquake produced by a seismograph is called a **seismogram** (FIG. 3.15d). At first glance, a seismogram looks like a messy squiggle of lines, but to a seismologist it contains a wealth of information. The horizontal axis on a seismogram represents time, and the vertical axis represents the amplitude of seismic waves. We refer to the instant at which a seismic wave appears at a seismograph station as the **arrival time** of the wave. The first squiggles on the record represent P-waves because P-waves travel the fastest and arrive first. Next come the S-waves, and finally the R-waves and L-waves. Typically, the surface waves have the largest amplitude and arrive over a relatively long interval of time.

Finding the Location of an Earthquake's Epicenter

Imagine two cars, one traveling at 80 km/h and one at 100 km/h. The distance between them increases the farther that they travel, so after 1 hour they will be 20 km apart, but after 2 hours, they will be 40 km apart. The same concept applies to seismic waves. Because P-waves and S-waves travel at different velocities, the difference between their respective arrival times at a seismograph increases with increased distance (FIG. 3.16a). We can use this difference, known as the $S - P$ time, to calculate the distance between a seismograph and an earthquake's epicenter by using a *travel-time*

FIGURE 3.16 Finding the epicenter of an earthquake.

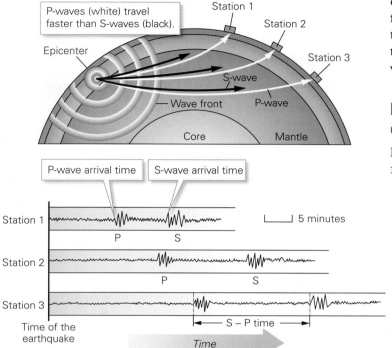

(a) P-waves travel faster than S-waves. The S − P time (pronounced "S minus P time"), therefore, represents the distance between the epicenter and a seismograph station.

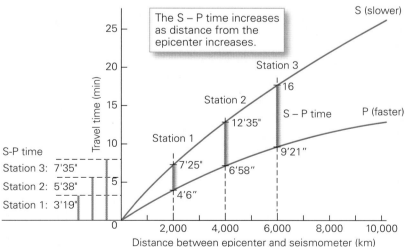

(b) To find the distance from a seismograph station to the epicenter, we draw a line segment representing the S − P time at each station at the scale on the left side of the travel-time graph. Then we slide the segment until we find where it fits so that each end of the segment lies on one of the travel-time curves. We then extrapolate the line down to the horizontal axis to find the distance from the station to the epicenter.

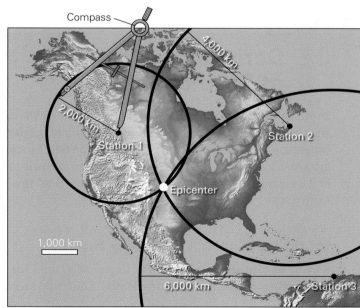

(c) Knowing the distance to the epicenter from three seismographs allows us to locate the epicenter.

3.4 Measuring Earthquakes • **85**

graph—**Figure 3.16b** describes this method. By repeating this calculation using seismograms recorded at three different seismograph stations, we can then triangulate to locate the epicenter. Simply draw three circles on a map—the center of each circle represents the position of a seismograph, and the radius of each circle represents the distance between that seismograph and the epicenter, at the scale of the map. The epicenter lies at the intersection of the three circles—this is the only point that has the appropriate measured distance from all three seismograph stations **(FIG. 3.16c)**. Seismologists have also developed methods that allow them to use seismograms to characterize the type of fault (normal, reverse, or strike-slip) responsible for an earthquake, the fault's orientation, and the depth to the focus.

> **Take-home message . . .**
> Seismographs detect and record ground motion. The record produced by a seismograph is called a seismogram. The wave-like lines on a seismogram represent seismic waves; the first one to appear is the P-wave. By measuring the difference in travel time for P-waves and S-waves recorded by seismograms from at least three different locations, we can determine the location of an earthquake's epicenter.
>
> **QUICK QUESTION** What role does a heavy weight serve in a traditional mechanical seismograph?

3.5 Defining the Size of Earthquakes

In our description so far, we've implied that some earthquakes are "large," in that they shake the ground violently, whereas others are "small," in that they can barely be felt. Seismologists have developed two different scales—an intensity scale and a magnitude scale—to define earthquake "size" in a uniform way.

Modified Mercalli Intensity Scale

Earthquake **intensity** refers to the degree of ground shaking at a locality. In 1902, an Italian scientist, Giuseppe Mercalli, devised a scale for defining earthquake intensity based on both an assessment of earthquake damage and people's perceptions of the shaking. A version of this scale, called the **Modified Mercalli Intensity (MMI) scale**, continues to be used today **(TABLE 3.2)**. Earthquake intensities on this scale are represented by roman numerals.

Earthquake intensity at a location depends on both the distance between the location and the earthquake's focus, and on the composition of the *substrate* (material beneath the ground surface) underlying the location. Distance matters because earthquake energy undergoes attenuation as it travels from the focus, and composition matters because weak substrates may cause amplification of seismic waves. Consequently, the intensity of a shallow-focus earthquake is

TABLE 3.2 Modified Mercalli Intensity Scale

MMI	Destructiveness (Perceptions of the Extent of Shaking and Damage)
I	*Not felt:* Detected only by seismic instruments; causes no damage.
II	*Weak shaking:* Felt by a few stationary people, especially in upper floors of buildings; suspended objects, such as lamps, may swing.
III	*Weak shaking:* Felt indoors; standing automobiles sway on their suspensions; it feels as though a heavy truck is passing.
IV	*Light shaking:* Shaking awakens some sleepers; dishes and windows rattle.
V	*Moderate shaking:* Most people awaken; some dishes and windows break, unstable objects tip over; trees and poles sway.
VI	*Strong shaking:* Shaking frightens some people; plaster walls crack, heavy furniture moves slightly, and a few chimneys crack, but overall little damage occurs.
VII	*Very strong shaking:* Most people are frightened; plaster cracks, windows break, some chimneys topple, and unstable furniture overturns; poorly built buildings sustain considerable damage.
VIII	*Severe shaking:* Many chimneys and factory smokestacks topple; heavy furniture overturns; substantial buildings sustain damage, and poorly built buildings suffer severe damage.
IX	*Violent shaking:* Frame buildings separate from their foundations; most buildings sustain damage, and some buildings collapse; the ground cracks, underground pipes break, and rails bend; some landslides occur.
X	*Extreme shaking:* Most masonry structures are destroyed; the ground cracks in places; landslides occur; bridges collapse; facades on buildings collapse; railways and roads suffer severe damage.
XI	*Extreme shaking:* Few masonry buildings remain; many bridges collapse; broad cracks form in the ground; most pipelines break; severe liquefaction of sediment may occur, causing the ground to fissure; many landslides develop; some dams collapse.
XII	*Extreme shaking:* Earthquake waves cause visible undulations of the ground surface; objects fly up off the ground; there is complete destruction of buildings and bridges of all types.

FIGURE 3.17 Representing perceptions of earthquake intensity.

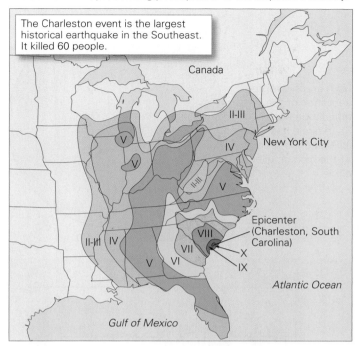

(a) A traditional intensity map, using the Modified Mercalli Intensity scale, for the 1886 Charleston, South Carolina, earthquake. Note that intensity decreases with distance from the epicenter.

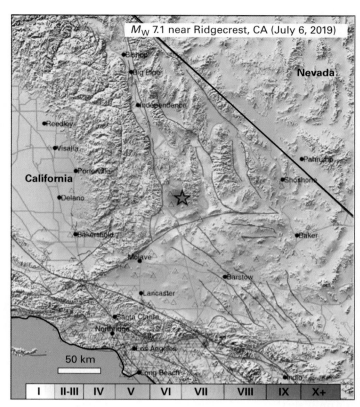

(b) A USGS ShakeMap for the 2019 Ridgecrest, California, earthquake. Though shaking was violent at the epicenter, it was light in Los Angeles. The red lines on the map are faults.

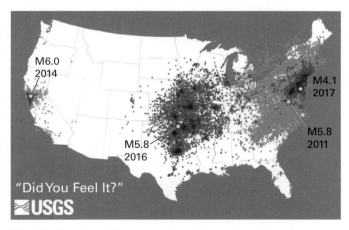

(c) A Did You Feel It? map for four different earthquakes of comparable energy release (indicated by magnitude, as we discuss later). Note that the midwestern and eastern earthquakes were felt over a broader area than the one in California.

greater than that of a deeper earthquake, and the intensity of an earthquake at a location near the epicenter is greater than one that is farther from the epicenter, unless there is local amplification.

Seismologists draw *intensity maps* to depict how earthquake intensity varies over a region (FIG. 3.17a). Contour lines on such maps separate regions in which the intensity differs. Traditionally, intensity maps were produced by documenting damage and by interviewing people in the days and weeks after an earthquake. In recent years, a computer program called *ShakeMap*, developed by the *US Geological Survey* (*USGS*), automatically takes magnitude data (described below) from seismographs and combines it with data about substrate composition to produce intensity maps (FIG 3.17b). These maps can be produced so quickly that emergency managers use them to plan deployment of personnel and equipment. The USGS has also devised a website that allows anyone to be a citizen scientist and contribute to the production of a *Did You Feel It? map* by logging onto the website and provide a *felt report*, a statement about the time and place that you observed shaking, along with details about the nature of the shaking. From the responses, a computer compiles and plots a map showing the area over which people detected the earthquake (FIG. 3.17c).

Magnitude Scales

When you hear a report of an earthquake disaster in the news, you'll probably encounter a phrase like, "An earthquake with a magnitude of 7.2 struck the city yesterday." What does this phrase mean? Earthquake **magnitude**

3.5 Defining the Size of Earthquakes • 87

represents the amount of energy released by an earthquake, as determined from a measurement of ground motion recorded by a seismograph.

In 1935, Charles Richter, an American seismologist, developed the first practical method for calculating earthquake magnitude. To make this scale, Richter assumed that the maximum ground motion observed at a location, as represented by the maximum deflection of a seismograph needle, represented the release of energy. He then created a scale that assigns an Arabic numeral to a particular magnitude of deflection. This scale is logarithmic, so each higher number represents a 10-fold increase in the amplitude of ground motion, relative to the number before (**FIG. 3.18a**). For example, a magnitude 8 earthquake results in ground motion with an amplitude 10 times greater than that of a magnitude 7 earthquake, and 1,000 times greater than that of a magnitude 5 earthquake (**FIG. 3.18b**).

Since seismic energy undergoes attenuation (weakening) as it travels away from the focus, the observed deflection of a seismograph needle depends on the distance between the seismograph and the epicenter. Richter decided to use the amplitude that would be observed at a seismograph located 100 km from the epicenter as the basis for his scale. But since it's unlikely that a real seismograph is located exactly 100 km from the epicenter, Richter designed a calibration chart that allows seismologists to use measurements from a real seismograph anywhere in the world to calculate earthquake magnitude consistently (**FIG. 3.18c**). The scale that Richter developed came to be known as the **Richter scale**, and a number on the scale was commonly called a *Richter magnitude*. A value of 0 on the Richter scale represents the smallest earthquake that could be measured with the seismographs available in 1935.

In practice, the calibration that Richter used when defining magnitude works best for data from seismographs located less than several hundred kilometers from the epicenter. To emphasize this limitation, Richter preferred to use the label *local magnitude* (M_L) for a number on his scale. Richter and others developed other magnitude scales that work better for estimating the magnitude of earthquakes that are farther from a seismograph station. For example, the *surface-wave magnitude scale* estimates magnitude (M_S) from the amplitude of specific surface waves, and the *body-wave magnitude scale* (m_B) estimates magnitude from the amplitude of specific body waves.

The above scales become unreliable for differentiating among larger earthquakes. A newer and more accurate way of determining magnitude comes from a direct calculation of earthquake energy release as represented by the *seismic moment*, a number that depends on three physical measures: the rigidity (resistance to shear) of the rock in which the fault formed; the area of the fault that slipped; and the

FIGURE 3.18 The Richter magnitude scale.

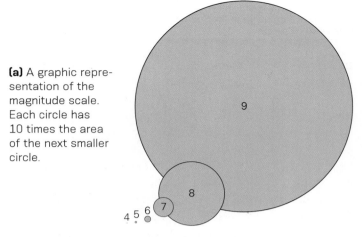

(a) A graphic representation of the magnitude scale. Each circle has 10 times the area of the next smaller circle.

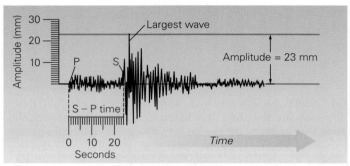

(b) To determine the magnitude of an earthquake, we measure the amplitude of the largest wave recorded by a seismograph.

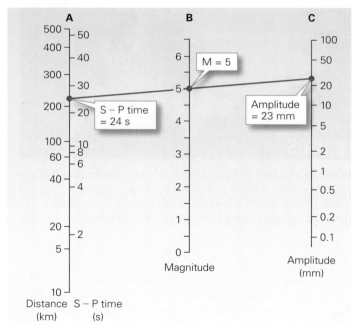

(c) Using this chart, we draw a line from the point on Column A representing the S − P time (calculated from the seismogram above) to the point on Column C representing the wave amplitude. We then read the Richter magnitude where the line intersects Column B.

average displacement on the fault during the earthquake. The resulting **moment magnitude scale** yields numbers (M_W) that match magnitudes determined from the previous scales up through magnitude 8, and is more accurate than the other scales for earthquakes with magnitudes greater than 8. For example, a 1960 earthquake along the coast of Chile registered as an M_L 8.3 on the Richter scale, but as an M_W 9.5 on the moment magnitude scale—this earthquake is now considered to be the strongest earthquake ever recorded. It may take a while before seismologists have all the information that they need to obtain a moment magnitude. As a result, the first report of an earthquake may provide a magnitude calculated quickly using one of the other scales. But when the moment magnitude has finally been determined, it becomes the magnitude of record.

To make discussion of earthquakes easier, seismologists commonly use familiar adjectives to describe an earthquake's magnitude (TABLE 3.3). Using these adjectives, we could call the 2011 Tōhoku earthquake (M_W 9.0) and the 2004 Sumatra earthquake (M_W 9.3) "great," and the Northridge (M_W 6.7) and the Loma Prieta (M_W 6.9) earthquakes "major."

News reporters sometimes incorrectly state that the moment magnitude scale "goes from 1 to 10." In fact, there are no defined limits to the moment magnitude scale. However, seismologists estimate that an M_W 9.5 earthquake is about as big as an earthquake can get, given the known dimensions of faults on the Earth. *Microearthquakes* (M_W 0 to M_W −2) can be detected by seismographs positioned close to the epicenter (the negative values are possible because modern seismographs are much more sensitive than the ones that Richter had in 1935).

TABLE 3.3 Adjectives for Describing Earthquakes

Adjective	Magnitude (M_W)	Intensity at Epicenter	Effects
Great	>8.0	X to XII	Total destruction
Major	7.0 to 7.9	IX to X	Extreme damage
Strong	6.0 to 6.9	VII to VIII	Moderate to serious damage
Moderate	5.0 to 5.9	VI to VII	Slight to moderate damage
Light	4.0 to 4.9	IV to V	Felt by most; slight damage
Minor	<3.9	III or smaller	Felt by some; hardly any damage

Note: These correlations between magnitude and intensity at the epicenter apply to shallow-focus earthquakes; deeper earthquakes of a given magnitude will have a lower intensity.

Relating Intensity to Magnitude

As we've seen, the intensity of an earthquake decreases with distance from the focus. Seismologists have provided a rough correlation between intensity and magnitude at the epicenter for shallow-focus earthquakes (see Table 3.3). For example, an M_W 7 shallow-focus earthquake yields intensity IX ground shaking near the epicenter. A deeper-focus earthquake of the same magnitude yields a lower-intensity earthquake at the epicenter because of attenuation between the focus and the ground (FIG. 3.19). Note that the same concept applies as regards distance from the epicenter—shaking far from the epicenter of a great earthquake can be the same as shaking close to the epicenter of a minor earthquake.

While we can get a sense of an earthquake's intensity at the epicenter by knowing its magnitude, magnitude alone does not define the felt area of the earthquake, as represented by the widths of intensity bands on a *ShakeMap* or by dots on a *Did You Feel It?* map. For example, the 2014 M_W 6.0 earthquake in Napa, California, was felt over a much smaller area than was the 2011 M_W 5.8 earthquake that struck central Virginia (see Fig. 3.17c). Such differences in felt area reflect differences in crustal rigidity: earthquake energy travels farther in rigid

FIGURE 3.19 The effect of focus depth on earthquake intensity.

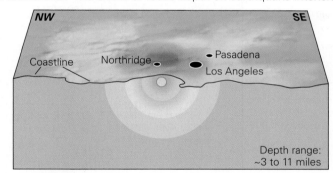

(a) The 1994 M_W 6.7 Northridge, California, earthquake occurred at a depth of 18 km (11 mi) and caused violent shaking at the epicenter.

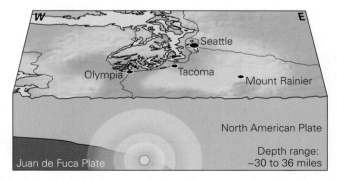

(b) The 2001 M_W 6.8 Nisqually, Washington, earthquake occurred at a depth of 57 km (35 mi) and caused only moderate to strong shaking at the epicenter.

crust because it undergoes less attenuation. Crust in the eastern United States is older, colder, and therefore more rigid than is California's crust because California lies along a plate boundary, so its crust is warmer and more broken up.

Energy Release by Earthquakes

To give a sense of the amount of energy released by an earthquake, seismologists compare earthquakes with other energy-releasing events. According to some estimates, an M_W 5.3 earthquake releases about as much energy as the Hiroshima atomic bomb, and an M_W 9.0 earthquake releases significantly more energy than the largest hydrogen bomb ever detonated. Notably, although an increase of one unit of magnitude represents a 10-fold increase in the maximum amplitude of ground motion, it represents about a 32-fold increase in energy release. Therefore, an M_W 8 earthquake releases about 1 million times more energy than an M_W 4 earthquake (FIG. 3.20). In fact, a single great earthquake can release as much energy as all other earthquakes on the Earth in a given year, combined. Fortunately, great earthquakes occur much less frequently than do small earthquakes.

> **Take-home message . . .**
> We can specify earthquake size by intensity (a representation based on perception of damage caused and shaking felt) or by magnitude (a representation of energy released by the earthquake, based on a measurement of ground motion).
>
> **QUICK QUESTION** Can you specify the size of an earthquake by giving just one intensity number? How about a single magnitude number?

3.6 Where Do Earthquakes Occur, and Why?

By plotting the distribution of earthquake epicenters on a map, seismologists have found that most, but not all, earthquakes occur in **seismic belts**, definable zones in which earthquakes happen fairly frequently (FIG. 3.21). Let's examine the geologic settings of seismic belts.

Earthquakes at Plate Boundaries

Plates move relative to their neighbors at rates of 1–15 cm (0.4–6 in) per year. So, over a period of decades to centuries, large stresses build along plate boundaries, and these stresses cause sudden slip on faults. Therefore, most earthquakes take place on plate boundaries. The fault type that occurs at a given location depends on the relative motion of plates at that boundary.

DIVERGENT-BOUNDARY SEISMICITY. Divergent boundaries (mid-ocean ridges) are broken into ridge segments linked by transform faults. Therefore, two kinds of faults develop at divergent boundaries: along ridge segments, stretching generates normal faults, whereas along transform faults, strike-slip faulting occurs (FIG. 3.22). Earthquakes on all these faults have shallow foci. These earthquakes have little impact on society because, with the exception of Iceland (a hot spot on the Mid-Atlantic Ridge), they occur far from any population centers.

TRANSFORM-BOUNDARY SEISMICITY ON CONTINENTS. At transform boundaries, one plate slides past another along a strike-slip fault. The majority of transform boundaries in the world link segments of mid-ocean ridges. But a few, such as the San Andreas fault in California, the Alpine fault in New Zealand, and the North Anatolian fault in Turkey, cut

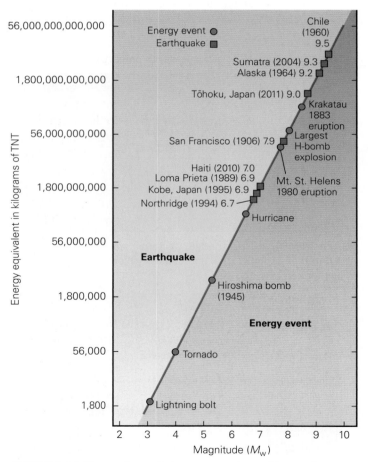

FIGURE 3.20 Comparison of the energy released by earthquakes with the energy released by other events. Note how dramatically energy increases with increasing magnitude.

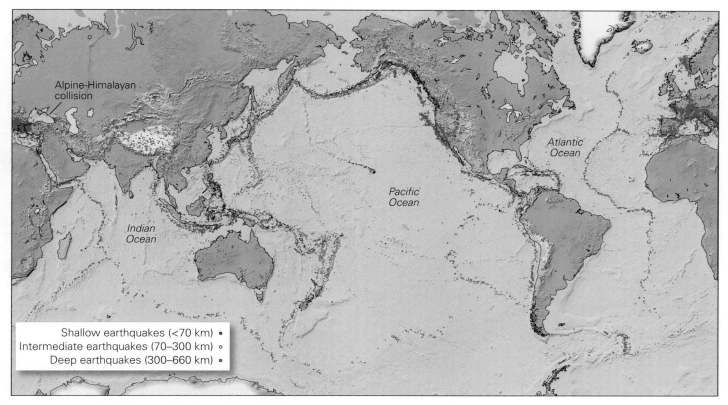

FIGURE 3.21 A map of the Earth's seismicity emphasizes that epicenters cluster in distinct seismic belts.

through continental crust. Large earthquakes on such *continental transform faults* can be very dangerous because many of these faults lie under or near population centers, and because these earthquakes have shallow foci, the seismic waves they produce have not undergone attenuation by the time they reach the ground surface.

The 1906 M_W 7.9 earthquake in San Francisco, an example of a transform-boundary earthquake, resulted from sudden slip on the San Andreas fault. This fault represents the plate boundary along which the Pacific Plate moves north at an average rate of 6 cm (2 in) per year relative to the North American Plate **(FIG. 3.23a)**. Because of stick-slip behavior, much of this movement happens in sudden jerks, each of which causes an earthquake. On April 18, 1906, a segment of the fault near San Francisco that was 15 km (9 mi) deep by about 470 km (292 mi) long suddenly slipped by as much as 8.5 m (28 ft) (see Fig. 3.4b). When the seismic waves generated by this slip struck the city, streets buckled and cracked, buildings swayed and collapsed, and nearby slopes slumped. Fires soon ballooned into a conflagration and destroyed much of what remained **(FIG. 3.23b)**. As a result, hundreds of people were killed, close to 300,000 were displaced, and the US Army was mobilized to keep the peace

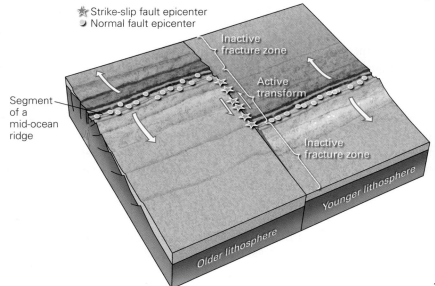

FIGURE 3.22 Seismicity at divergent boundaries. Normal faults occur along ridge segments, and strike-slip faults occur along transforms.

FIGURE 3.23 Seismicity on a continental transform fault.

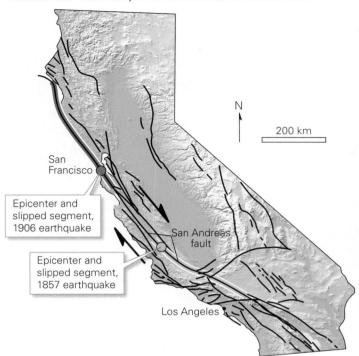

(a) The San Andreas fault system in California. Note that the system includes many faults in a band 100 km wide. The portions of the fault that slipped in the 1906 and 1857 earthquakes are indicated.

(b) A street in San Francisco after the 1906 earthquake.

(c) The 1989 Loma Prieta earthquake caused a two-level freeway ramp to collapse.

and to distribute aid. The largest earthquake on the San Andreas fault in the San Francisco area since that time took place at Loma Prieta in 1989. This M_W 6.9 earthquake caused extensive damage to bridges (FIG. 3.23c) and, as discussed earlier, to certain neighborhoods.

An even larger earthquake along this transform plate boundary took place in 1857, when 250 km (155 mi) of the southern San Andreas slipped. That event, the M_W 7.9 Fort Tejon earthquake, produced intensity IX ground shaking in areas north of Los Angeles (see Fig. 3.23a). Because of the sparse population of the area at the time, casualties and damage were not extensive, so the event is not as well known as the San Francisco earthquake. Should a similar-sized earthquake happen in the same area today, it would be catastrophic.

The San Andreas isn't the only active fault in California. Relative plate motion across the Pacific Plate–North American Plate boundary is distributed among many faults. Some of them are strike-slip faults, but some are not. For example, the Northridge earthquake, which we've mentioned already, occurred on a thrust fault along the north side of Los Angeles. Recall that thrust faults develop where the crust is being compressed. This happens along the San Andreas where the fault bends in such a way that it restrains plate motion (see Fig. 3.4d). During the past several million years, thrust faulting caused by the resulting compression has uplifted the Transverse Ranges that border Los Angeles (FIG. 3.24).

CONVERGENT-BOUNDARY SEISMICITY. Convergent boundaries are complicated regions where many earthquakes take place at a variety of depths (FIG. 3.25a). Seaward of the deep-sea trench, some earthquakes occur along normal faults, formed as the downgoing plate starts to bend. The most dangerous earthquakes, however, occur along relatively shallow thrust faults that delineate the boundary between the base of the overriding plate and the top of the downgoing plate (FIG. 3.25b). Because the foci of these earthquakes are so

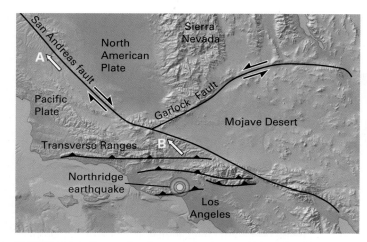

FIGURE 3.24 Generally, plate motion is parallel to the San Andreas fault (arrow at Point A). In the Transverse Ranges, the movement direction of the Pacific Plate (arrow at Point B) is oblique to the San Andreas fault due to a bend in the fault trace. Consequently, thrust faults have formed.

shallow, when the slip area is large, immense seismic energy reaches the ground surface before it has attenuated.

Notable examples of devastating convergent-boundary earthquakes include the largest recorded earthquakes: the 1960 M_W 9.5 earthquake in Chile; the 1964 M_W 9.2 Good Friday earthquake in Alaska; the 2004 M_W 9.3 earthquake off Sumatra; and the 2011 M_W 9.0 Tōhoku earthquake in Japan. Such earthquakes—those with magnitudes greater than M_W 8.6—are known as **megathrust earthquakes**. Later in this chapter, we'll see that the *Cascadia seismic zone*, along the coast of the northwestern United States (western Oregon and western Washington), was the site of an M_W 9.0 megathrust earthquake in 1700 and will host one again sometime in the future.

Unlike divergent or transform boundaries, convergent boundaries host intermediate- and deep-focus earthquakes as well as shallow-focus ones. A plot of the foci of such earthquakes on a cross section through the Earth defines a sloping band of seismicity called a **Wadati-Benioff zone** (named for the seismologists who first recognized it), which extends to a depth of 660 km (410 mi). Earthquakes in this zone occur within subducted lithosphere as it sinks through the asthenosphere (FIG. 3.25c) (see also Fig. 2.38b). Intermediate-depth earthquakes may be due to normal faults accommodating the bending, and consequent stretching, of the plate as it heads into the mantle. Very deep earthquakes may be due to sudden changes in the crystal structure of minerals as the plate sinks and becomes subjected to greater pressures.

Earthquakes That Are Not at Plate Boundaries

Not all earthquakes on continents are associated with plate boundaries. Here, we consider a few other geologic settings within continents at which earthquakes occur (FIG. 3.26).

FIGURE 3.25 Earthquakes at convergent boundaries.

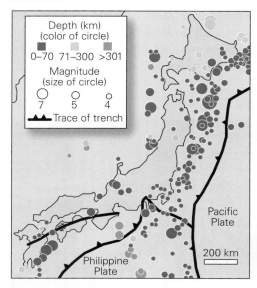

(a) This map of earthquake epicenters and depths in and near Japan shows how complex convergent-boundary seismicity can be.

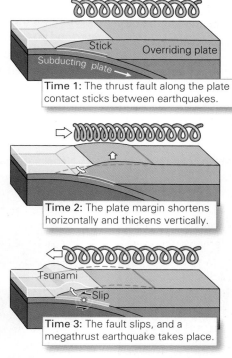

(b) A megathrust earthquake resembles the shortening, then rebound of a spring.

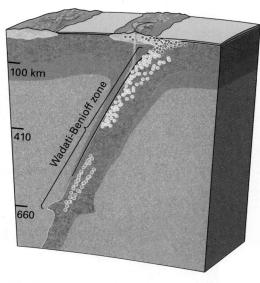

(c) At a convergent boundary, shallow-focus earthquakes (red dots) occur along the zone of contact between the downgoing plate and the overriding plate, as well as in the two plates themselves. Intermediate-focus (yellow dots) and deep-focus (blue dots) earthquakes define the Wadati-Benioff zone.

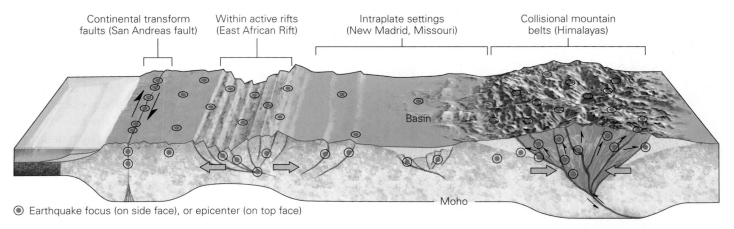

● Earthquake focus (on side face), or epicenter (on top face)

FIGURE 3.26 Seismicity within continents. Earthquakes can occur along continental transform faults, in rifts, in collisional mountain belts, and along ancient faults within the generally stable interior of a plate. An example of each type of earthquake is provided in parentheses.

RIFTS. The stretching of continental crust at a rift generates normal faults, and slip on these faults produces earthquakes. Active rifts today include the East African Rift, the Basin and Range Province in Nevada and Utah, and the Rio Grande Rift of New Mexico. In all these places, shallow-focus earthquakes occur, some of which cause major damage because they occur close to large metropolitan areas. The eastern and western boundaries of the Basin and Range rift, for example, remain seismically active. Salt Lake City, Utah, lies just west of the Wasatch fault, which defines the eastern edge of the rift. Many earthquakes take place at or near this boundary (see Fig. 3.7). Seismologists have found evidence that a few M_W 7 earthquakes have rocked the region during the past 10,000 years. An M_W 5.7 earthquake rattled the city in March, 2020.

CONTINENTAL COLLISION ZONES. In 2015, a large earthquake shook the ground in the mountainous landscape of Nepal **(FIG. 3.27)**. Monuments that had stood for centuries collapsed in Kathmandu, and throughout the remote countryside, landslides buried communities and roads. Rescue of survivors was very difficult because access to those communities was cut off by landslides. Earthquakes are fairly frequent in Nepal, as well as elsewhere along the Himalayas, because the range is actively growing due to the ongoing collision of India with Asia, and compression caused by this collision drives slip on thrust faults.

INTRAPLATE EARTHQUAKES. In a given year, 95% of the seismic energy produced on the Earth comes from plate-boundary, collisional, or rift-related earthquakes. The remaining earthquakes, which occur in the interiors of continents, are known as **intraplate earthquakes**. Though much less common, some of these earthquakes have magnitudes exceeding 7 and pose a hazard to cities. Intraplate earthquakes happen when ancient pre-existing faults slip in response to present-day stresses applied to continents. Most intraplate earthquakes have a shallow focus, typically no deeper than 30 km (19 mi).

Intraplate earthquakes are not uniformly distributed. In North America, most take place in southeastern Missouri,

FIGURE 3.27 Seismicity in a continental collision zone caused the 2015 Nepal earthquake.

(a) Towns throughout the Himalayas were destroyed.

(b) Around 2 million people were displaced.

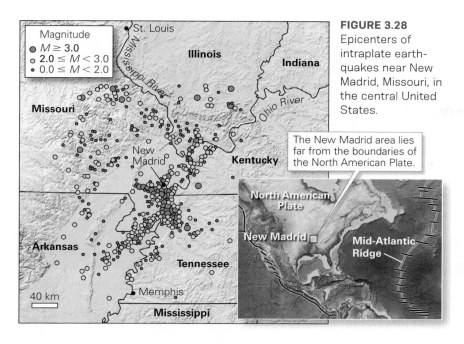

FIGURE 3.28 Epicenters of intraplate earthquakes near New Madrid, Missouri, in the central United States.

eastern Tennessee, eastern South Carolina (see Fig. 3.17a), and southern Quebec. The largest historical intraplate earthquakes to affect the continental United States struck near New Madrid, in southeastern Missouri. During the winter of 1811–1812, three M_W 7.0–7.4 earthquakes, resulting from slip on faults beneath the Mississippi valley, shook the region. The events probably represent slip on faults that first formed during Precambrian rifting. Displacement of the ground surface during the New Madrid earthquakes temporarily reversed the flow of the Mississippi River. Shaking toppled cabins, but because the area was sparsely populated, the earthquake caused no recorded deaths. The region remains seismically active today, a situation that concerns emergency planners because large cities, including St. Louis and Memphis, now lie within areas that endured intensity VI–VIII ground shaking during the 1811–1812 events (FIG. 3.28).

In the eastern United States, intraplate earthquakes are caused by reactivation of Precambrian or Mesozoic rift faults, or of thrust faults formed when Africa collided with North America. (This collision was the final stage of deformation that formed the Appalachian Mountains and united Pangaea.) Examples include the 2011 M_W 5.8 Mineral, Virginia, earthquake, which startled people in Washington, D.C., cracked the Washington Monument, and was felt in New York City and even as far west as Chicago (see Fig. 3.17c). The largest historical east-coast earthquake was an M_W 7.3 event that struck Charleston, South Carolina, in 1886, causing about 60 deaths and as much as $155 million in damage.

Induced Seismicity

Water pressure underground affects the stresses that can lead to sudden slip on a fault and generation of an earthquake. Simplistically, this relationship exists because water pressure decreases the push of one side of a fault against the other, and therefore effectively decreases the friction across a fault. Human activities can affect underground water pressure significantly. For example, filling a reservoir behind a dam built over a fault zone, or pumping water down into the ground at high pressure to dispose of it, can increase local water pressure enough to trigger small to moderate earthquakes. Such **induced seismicity**—seismicity caused by human actions—was first recognized when earthquakes happened every time waste fluid was pumped down wells at the Rocky Mountain Arsenal in Colorado by the US Army.

Induced seismicity has become an increasing problem in areas where many new oil or gas wells have been drilled. In addition to oil and gas, these wells also bring up salty groundwater that has been trapped in pores between the solid grains of rocks for millions of years. This salty water can't be disposed of in rivers or lakes, so it is pumped back underground into *injection wells* at high pressure. In Oklahoma, since about 2008, the development of hydrofracturing (fracking) methods made it economical to drill many new wells. But disposal of salty water in injection wells appears to have induced hundreds of earthquakes per year between 2014 and 2017 (FIG. 3.29). Reducing the pressure of the injected water can decrease the risk of induced seismicity.

FIGURE 3.29 The number of earthquakes per year increased dramatically during years when wastewater from oil wells was injected into the ground. The inset map shows the location of earthquakes and fault traces in Oklahoma.

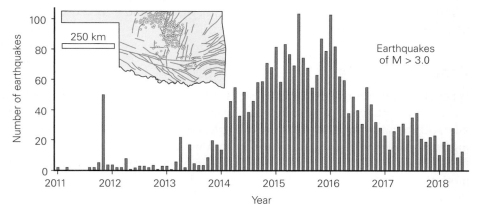

> **Take-home message . . .**
>
> Most, but not all, earthquakes happen along plate boundaries. Particularly catastrophic earthquakes occur at thrust faults associated with convergent boundaries and at continental transform faults because these earthquakes have relatively shallow foci. Large earthquakes also occur in rifts and continental collision zones, and a few take place along pre-existing faults within plate interiors.
>
> **QUICK QUESTION** Can human activity trigger earthquakes?

3.7 How Do Earthquakes Cause Damage?

As the description of the Lisbon disaster at the start of this chapter emphasizes, earthquakes can cause damage in many ways. The primary disaster associated with a major earthquake stems from the violent movement of the ground. But earthquakes, depending on where they occur, can also trigger secondary disasters, such as landslides, tsunamis, and firestorms, as well as tertiary disasters, such as disease. Here, we focus our attention on the various components of earthquake disasters.

Ground Rupture

Movies may give the impression that the ground opens up during an earthquake and swallows whole buildings. That impression isn't accurate. Where a fault intersects the ground surface, the surface tears and displacement develops, but generally a wide fissure doesn't develop. (As we'll see below, however, fissures can form due to landslides or to sediment liquefaction triggered by ground shaking.) Strike-slip faulting can crack buildings or roads by wrenching one side laterally with respect to the other **(FIG. 3.30a)**. Normal or reverse faulting can cause two sides of a building or road to move vertically relative to each other **(FIG. 3.30b)**. Clearly, it's not wise to build a structure that straddles an active fault.

Ground Shaking

When seismic waves reach the ground surface, they cause it to move. Different types of seismic waves generate different kinds of ground motion **(FIG. 3.31)**. For example, P-waves coming from below push the ground up and down, S-waves shear it from side to side, R-waves make it undulate in a wave-like motion, and L-waves make it shimmy like a snake (see also Fig. 3.10). But because different wave trains arrive at slightly

FIGURE 3.30 Damage caused by ground rupture.

(a) A strike-slip fault in New Zealand offset this road.

(b) A thrust fault in Taiwan displaced this running track during the 1999 Chi Chi earthquake.

different times, and because waves may interfere with one another, ground motion overall may feel chaotic.

The severity of the shaking at a given location depends on several factors, including: the magnitude of the earthquake, because higher-magnitude events release more energy; the distance from the focus, because seismic waves attenuate as they pass through the Earth; and the strength of the substrate, because seismic waves undergo amplification as they enter softer materials. Amplification may unexpectedly cause severe ground shaking, even at a distance far from the epicenter. In fact, ground motion over soft substrates can be 1.5–6.0 times greater than that over rigid bedrock nearby.

How much does the ground move during an earthquake? In a location subjected to intensity XII shaking, the ground surface may move by tens of centimeters or meters. The motion decreases away from the epicenter, but damage can still occur, even if it's only millimeters to centimeters. You'll even feel an earthquake that causes only 0.1 mm (0.004 in) of motion.

FIGURE 3.31 During an earthquake, the ground can shake in different ways, or in several ways at once, causing buildings and infrastructure to twist, sway, and bounce.

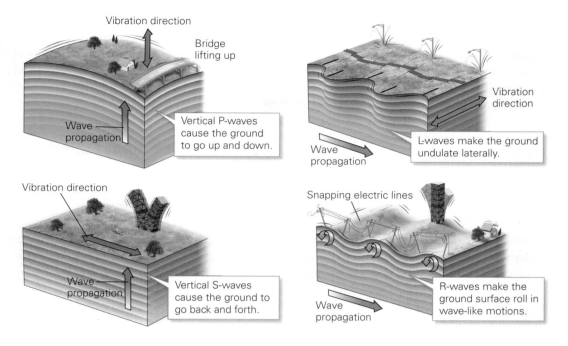

Regardless of the amount of ground motion, if it occurs very slowly, it's not an earthquake. To constitute an earthquake, the motion must happen very quickly, so that the ground accelerates significantly. Physicists define *acceleration* as a change in velocity in a specified time. To get a sense of how much acceleration happens during an earthquake, we can compare the maximum acceleration (known as the **peak ground acceleration**, or **PGA**) during the earthquake with the acceleration produced by the Earth's gravity, as defined by the gravitational constant, g, where $g = 9.8$ m/s². During a moderate earthquake (intensity VII), acceleration can approach $0.2\,g$, enough to knock you off your feet and cause significant damage, and during a great earthquake, acceleration can range between $1.7\,g$ and $3\,g$ (comparable to the acceleration you feel on a roller coaster) at the epicenter.

If you're out in an open field during an earthquake, you may be bounced around a bit, but your body is too flexible to break, so the ground motion won't kill you. Human-made structures, however, aren't so resilient **(FIG. 3.32)** (see also the chapter-opening photo). When ground motion takes place, roads, railroads, bridges, power lines, pipelines, canals, and sewers sway, twist back and forth, topple, lurch up and down, buckle, or rupture. As a result, windows shatter, roofs fail, building facades crash to the ground, and in some cases, walls and even whole buildings crumble.

How long does ground motion last during the mainshock? The answer for a given locality depends on two factors. First, it depends on the amount of time during which slip on the fault takes place and generates energy. Slip time depends largely on magnitude. For example, it took about

FIGURE 3.32 Ground shaking can damage buildings, railroads, and highways.

3.7 How Do Earthquakes Cause Damage? • **97**

15 seconds for the 20 km (12.5 mi) long fault surface that produced the M_W 6.7 Northridge earthquake to form. In contrast, it took 12 minutes for the 1,200 km long surface of the thrust fault that caused the 2004 M_W 9.3 Sumatra earthquake to slip. Second, the duration of shaking depends on the distance between the locality and the earthquake's focus—the greater the distance between a location and the focus, the greater the time delay between the fastest waves and the slowest waves, and thus the longer the duration of the earthquake.

Landslides

Seismic shaking can cause ground on steep slopes or ground underlain by weak sediment to give way. This movement results in a *landslide*, the tumbling or flow of soil and rock downslope (see Chapter 5). Earthquake-triggered landslides along the coast of California often make headlines, for when steep cliffs facing the Pacific Ocean collapse, expensive homes may tumble to the beach below **(FIG. 3.33a)**. Such events lead to the misperception that "California will fall into the sea" during an earthquake. Although small portions of the coastline do indeed tumble down to the beach, the state as a whole remains firmly attached to the continent, despite what Hollywood scriptwriters say.

Landslides are often a major cause of earthquake damage—they not only destroy access routes to isolated towns **(FIG. 3.33b)**, but in some cases, they bury towns. During the 2008 M_W 7.9 Sichuan earthquake in south-central China, an estimated 200,000 landslides of various sizes tumbled down hillslopes in the area. The landslides, together with building failures due to ground motion, led to the deaths of 70,000 people and caused over $150 billion in damage. Some of the landslides dammed rivers and causes *quake lakes* to form—if such a landslide dam later breaks, downstream regions can be flooded.

Sediment Liquefaction

In 1964, an M_W 7.5 earthquake struck Niigata, Japan, a city that had been partly built on land underlain by wet sand. During the shaking, the foundations of over 15,000 buildings sank into the ground, causing walls and roofs to crack. In fact, several buildings simply tipped over **(FIG. 3.34a)**! In 2011, an earthquake in Christchurch, New Zealand, caused a layer of wet sand underground to erupt out of cracks in the ground and produce small, cone-shaped mounds called *sand volcanoes* on the ground surface **(FIG. 3.34b)**. In some places where sand had erupted onto the surface, a pit-like depression—a type of *sinkhole*—developed nearby. Some of these sinkholes became large enough to swallow cars **(FIG. 3.34c)**.

These examples illustrate the process of **sediment liquefaction** (see Chapter 5 for more detail). Liquefaction of wet sand happens when shaking causes the tightly packed sand grains to move apart. As a result, what had been stable, load-bearing sand turns into a slurry of sand suspended in water that can't support weight. If cracks open up between the iquefied sand layer and the ground surface, the weight of the overlying sediment squeezes the wet sand upward. The

FIGURE 3.33 Landslides triggered by earthquakes.

(a) During the 1994 Northridge earthquake, a steep slope along the California coast collapsed, taking part of this house with it.

(b) Shaking during a 2011 earthquake in Gangtok, India, triggered landslides on steep mountain slopes, which blocked roads and inhibited rescue operations.

FIGURE 3.34 Consequences of sediment liquefaction.

(a) During the 1964 earthquake in Niigata, Japan, liquefaction of underlying sediment caused apartment buildings to tip over onto their sides.

(b) During a 2011 earthquake in Christchurch, New Zealand, fissures opened in the ground, and wet sand squirted up from below to build sand volcanoes.

(c) Drainage of the sand that fed sand volcanoes in Christchurch, New Zealand, formed sinkholes.

settling of overlying sediment layers into the liquefied layer can lead to formation of open fissures. A similar phenomenon happens in a special type of clay called *quick clay* (or *sensitive clay*): ground shaking transforms what had been a solid mass into slippery liquid mud by destroying cohesion among the grains.

In some cases, sediment liquefaction causes land to start flowing laterally, even on a very gentle slope, destroying hundreds of homes in the process, as happened during an earthquake that struck Indonesia in 2018 **(FIG. 3.35)**. This phenomenon also happened during the 1964 Good Friday earthquake in southern Alaska, sending the Turnagain Heights

FIGURE 3.35 Widespread sediment liquefaction occurred in the city of Palu (on Sulawesi Island, Indonesia), during an M_W 7.4 earthquake in 2018. A whole neighborhood was consumed as houses sank into the mud (see Chapter 5).

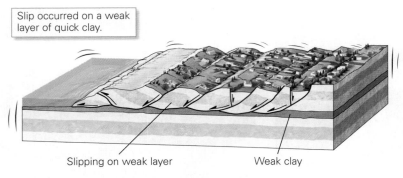

FIGURE 3.36 The 1964 Turnagain Heights disaster in Alaska. An earthquake caused sediment liquefaction, which led to slumping that destroyed the neighborhood.

neighborhood of Anchorage over a bluff and into the bay below (FIG. 3.36). We'll consider the mechanisms of sediment liquefaction, and see how it causes landslides, in Chapter 5.

Resonance Disasters

If you've ever played on a playground swing, you've experienced resonance. The swing goes back and forth with a specific period of motion that depends on the length of the chains that it hangs from—a swing suspended from long chains has a longer period than does one suspended from shorter chains. If you pump your legs at just the right time, you increase the amplitude of motion. The phenomenon that we've just described—increasing the amplitude of a periodic motion by adding energy at the same frequency—is called **resonance**.

While resonance adds to the fun in a playground, it's no laughing matter during an earthquake. That's because buildings begin to sway when the ground shakes. The period at which they sway, known as the *resonant period*, depends on the building's height. The resonant period for a 1-story building is 0.1 seconds; for a 4-story building, it's 0.5 s; for a 10-story building, it's 1–2 s; and for a skyscraper, it's 5–7 s. In an earthquake, the ground also vibrates with a specific period, which depends on the strength of the substrate—stronger materials (such as bedrock) vibrate with a shorter period than do weaker ones (such as soft sediment). If the resonant period of a building matches that of its substrate, the ground shaking adds energy to the building sway and amplifies it. For some buildings, the sway can become too much to sustain, and they collapse (FIG. 3.37a, b).

The pattern of earthquake-triggered building failure in Mexico City illustrates the danger of resonance. On September 15, 1985, slip on the convergent boundary along the west coast of Mexico produced an M_W 8.0 earthquake with an epicenter 400 km (250 mi) west of Mexico City. Because of the distance, its seismic waves had attenuated by the time they reached Mexico City, so it didn't seem as if they would cause great destruction. But the city is built on the flat ground of an old lake bed, underlain by soft sediment with a resonant period of 2 s. This period matched the resonant period of 6- to 15-story buildings, so the sway of those buildings was amplified, and many of them collapsed, leading to as many as 40,000 deaths (FIG. 3.37c). Older, shorter buildings survived the earthquake, even though they were constructed poorly, and taller buildings survived as well. Greater than expected damage due to amplification of building movements as a consequence of resonance is called a **resonance disaster**.

Fire

Ground shaking during an earthquake can make lamps, stoves, furnaces, and candles tip over, and it may break electric wires or topple power lines, generating sparks. As a consequence, areas already turned to rubble, and even areas that are not badly damaged, may be consumed by fire. Ruptured gas pipelines and oil tanks feed the flames, sending columns of fire erupting skyward (FIG. 3.38a). Debris-filled streets block firefighters from reaching the blazes, and broken pipes make fire hydrants inoperable. Once fires start to spread, they may merge into an unstoppable inferno—a *firestorm*—because as the hot air over a firestorm rises, the storm draws in air from the sides, generating strong winds—up to 160 km/h (100 mph)—that fan the flames (FIG. 3.38b). When a large earthquake hit Tokyo in 1923, coals from cooking stoves set the debris of wood-and-paper buildings alight, eventually producing a firestorm that incinerated 38,000 victims who had survived the earthquake itself. In some cases, property loss due to a firestorm can exceed that due to ground shaking. For example, most property loss due to the 1906 San Francisco earthquake was caused by fire after the ground had stopped shaking.

Tsunamis

When an earthquake occurs along a plate boundary in the ocean, the resulting movement can displace the seafloor, which in turn displaces the overlying ocean surface. The process generates a series of broad waves that travel rapidly outward—in the deep ocean, they move as fast as a jet plane! Though the sea surface rises only slightly at the site where the tsunamis form, the resulting waves may have a wavelength of tens to hundreds of kilometers, so they involve an immense volume of water. Large tsunamis build into tall plateaus of water when they reach the shore, submerging broad areas. Catastrophic tsunamis spread across the Indian Ocean following the 2004 M_W 9.3 earthquake in Sumatra and killed an estimated 230,000 people, and tsunamis triggered by the 2011

FIGURE 3.37 A resonance disaster happens when the resonant period of the ground and that of buildings of a certain size are the same.

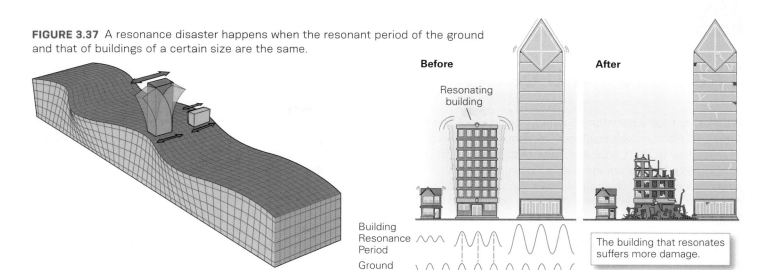

(a) If the fundamental resonant period of the ground is the same as that of a building, energy is added to the building's movement, and it starts swaying much more.

(b) In this example, the resonant period matched that of the middle building, so it collapsed, while taller and shorter buildings didn't.

The building that resonates suffers more damage.

(c) The smaller buildings in the foreground have collapsed, while skyscrapers remain standing in the background, in Mexico City after the 1985 earthquake.

FIGURE 3.38 Urban fires may be triggered by an earthquake.

(a) A fire ignited at a ruptured gas pipeline following the Tōhoku earthquake.

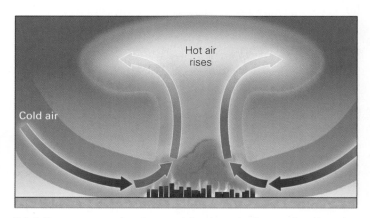

(b) A firestorm can develop as rising hot air above a burning city makes room for cool air to blow in from the sides. The resulting wind fans the fire.

3.7 How Do Earthquakes Cause Damage? • **101**

BOX 3.1 DISASTERS AND SOCIETY

The 2010 Haiti catastrophe

Haiti sits astride a transform boundary along which the North American Plate moves westward at about 2 cm (0.8 in) per year relative to the Caribbean Plate **(FIG. Bx3.1a)**. Therefore, earthquakes in Haiti are inevitable. But in 2010, people there weren't worrying a lot about earthquakes, because the last major one had happened over 200 years before, and because the impoverished, politically unstable nation had plenty of other issues to address. On the sunny afternoon of January 12, at 4:53 P.M., a 70 km (45 mi) long segment of a large strike-slip fault suddenly slipped by as much as 4 m (13 ft), releasing the stress that had been building for at least two centuries. The focus of this M_W 7 earthquake lay 25 km (15 mi) west-southwest of Port-au-Prince, the capital of Haiti, and 13 km (8 mi) beneath the ground surface **(FIG. Bx3.1b)**. Seismic waves reached the city in about 5 seconds, causing violent ground shaking that lasted for about 35 seconds.

The impact of an earthquake on a community depends not only on its magnitude, but also on the strength of the substrate, the steepness of slopes, construction practices, the quality of emergency services, and the capability of the population to respond to disaster. Much of Port-au-Prince sits on a basin of weak sediment, which amplified ground movements—in fact, beneath the harbor, sediment liquefied, causing wharves to sink into the sea. The buildings that were built on bedrock were fairly short, and their fundamental frequency matched that of their substrate, so they underwent a resonance disaster. Sadly, most of the city's buildings were constructed of weak masonry, and due to a lack of national building codes, most of that masonry was not reinforced, so many buildings collapsed **(FIG. Bx3.1c)**. To make matters worse, some of the city's neighborhoods were perched on steep slopes, which slid downhill during the quake, carrying whole neighborhoods with them **(FIG. Bx3.1d)**. When the shaking finally stopped, most of Port-au-Prince had become a rubble pile. The destruction killed 250,000 people, injured 300,000 more, and displaced 1.5 million people (out of a population of 10 million).

FIGURE Bx3.1 The 2010 Haiti earthquake.

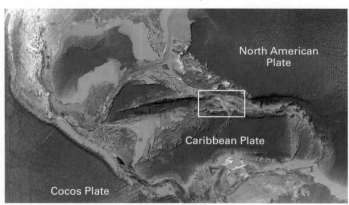

(a) Haiti lies along a plate boundary at the northern border of the Caribbean Plate.

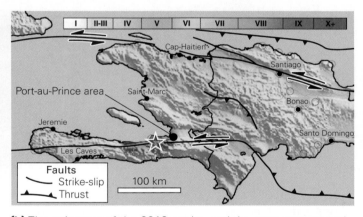

(b) The epicenter of the 2010 quake, and the strongest ground shaking, happened near Port-au-Prince, the capital city.

(c) Due to non-earthquake-resistant construction practices, many buildings in Port-au-Prince collapsed.

(d) Neighborhoods built on hillslopes slid downhill.

As a dense cloud of white dust slowly rose over the rubble, survivors began the frantic scramble to dig out victims, a task made more hazardous by aftershocks, over 50 of which had magnitudes of between 4.5 and 6.1. The aftershocks caused still-standing but weakened structures to collapse on rescuers. Due to lack of training and resources, emergency response immediately after the disaster was chaotic to nonexistent. In a short time, available services, such as hospitals and morgues, were overwhelmed. In the tropical climate, the bodies of victims began to rot, and the smell became overwhelming. Response was also slowed by lack of organizational communication, loss of key personnel, and confusion over which government or nongovernmental agency was in charge. Soon, though, the neighboring Dominican Republic stepped in to make airports and hospitals accessible. Eventually, the United States was given control of the Port-au-Prince airport so that air traffic could be managed, and the United Nations, several countries, the International Red Cross, Doctors without Borders, and other aid groups sent in planners, aid workers, and hospital ships. Unfortunately, no housing was available for displaced people, and it took a while for tent camps and other makeshift housing to appear. Camps for displaced people quickly became overcrowded, and they lacked adequate sanitation, so a tertiary disaster, an epidemic of cholera, began.

Two years after the 2010 earthquake, only about 40% of the international aid promised to Haiti had arrived, and 500,000 people who still had no permanent housing were living in uncomfortable and dangerous camps. Five years after the quake, tens of thousands of people remained in camps, and investigations revealed that substantial portions of relief funds had probably been misappropriated. Subsequently, Haiti has been struck by hurricanes, which have created a whole new set of problems. Sadly, an even larger earthquake (Mw 7.2) struck Haiti in 2021. Its epicenter was 50 km west of the 2010 epicenter, so it did not damage Port-au-Prince as badly as the 2010 earthquake did. Still, the 2021 event killed over 2,500 people and destroyed over 130,000 buildings. As Haiti continues to struggle with recovery and restoration due to the 2010 earthquake and subsequent hurricanes, its vulnerable population must cope with the response to still another disaster.

Tōhoku earthquake devastated the coast of Japan (FIG. 3.39). Because tsunamis can be so destructive, we investigate them more thoroughly in Chapter 6.

Disease

Once the ground shaking, building collapse, and fires have ceased (VISUALIZING A DISASTER, pp. 104–105), disease may still threaten lives in an earthquake-damaged region. Ground movement breaks water and sewer lines, cutting off supplies of clean water and exposing the public to pathogens, so diseases can reach epidemic proportions (BOX 3.1). Ground rupture, shaking, and landslides may not only destroy local medical facilities, but may also cut off transportation lines, preventing food and medicine from reaching stricken areas. Lack of refrigeration due to power outages and building collapses allows food supplies to spoil. The severity of such problems may exceed the ability of emergency services to cope, so diseases and medical problems may impact the population for months or years.

FIGURE 3.39 Examples of earthquake-generated tsunamis and their consequences.

(a) Damage caused by the 2004 Indian Ocean tsunami.

(b) The 2011 tsunami washing over the coast of Tōhoku, Japan.

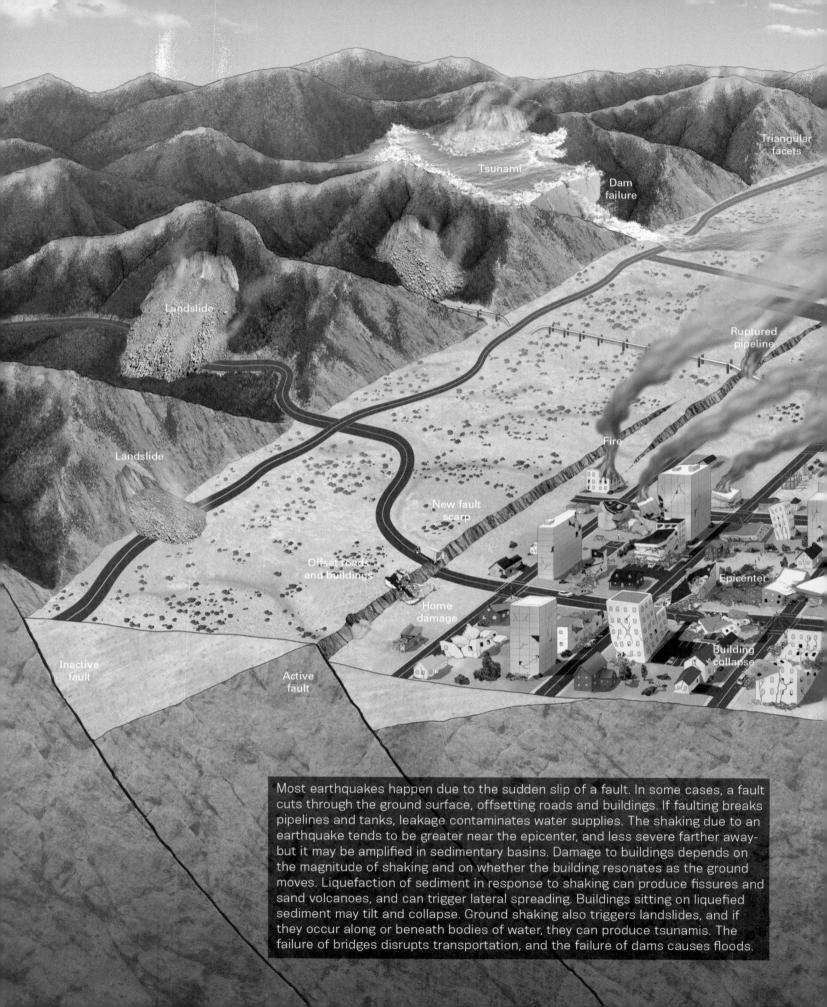

> **Take-home message . . .**
>
> Earthquakes cause damage in many ways. Ground shaking, landslides, and sediment liquefaction disrupt landscapes and cause buildings to crumble. Though ground shaking itself cannot kill people, falling debris can. In coastal areas, tsunamis may wash over broad areas of low land. Fire and disease may follow earthquakes, causing even more loss of life.
>
> **QUICK QUESTION** Why do epidemics sometimes follow in the wake of an earthquake?

3.8 Can We Predict "the Big One"?

Suppose someone predicts that an earthquake will happen near a certain city on a certain day, about a month in the future. What should local officials do? If the prediction proves to be correct, evacuating the city, or at least closing schools and bringing in medical supplies in advance, could save lives. But if orders come to evacuate the city, and the prediction proves to be wrong, the result will be lots of angry citizens and millions of dollars lost. Officials in southeastern Missouri faced this choice when a biologist and business consultant named Iben Browning predicted that the alignment of the Sun, Moon, and planets would trigger a major earthquake in New Madrid on December 2 or 3, 1990. Though seismologists responded that the prediction was *pseudoscience* (an idea that may superficially sound reasonable, but has no real scientific basis), New Madrid became a media circus, local authorities closed schools, and individuals invested in emergency supplies. December came and went and no earthquake happened.

The bogus 1990 prediction for New Madrid did help earthquake researchers better understand how to debunk pseudoscience and how to provide the public with better information. It also drew attention to the general question of whether earthquakes can ever be predicted. The real answer, as of now, depends on the time frame of the prediction. Seismologists can make *long-term predictions*, on the time scale of decades to centuries, of seismic activity. For example, it's fairly certain that during the next century, an earthquake will rattle Istanbul (which lies along a major active fault that has hosted many devastating earthquakes in the last century), but not north-central Canada (which lies far from a plate boundary and has no history of seismicity). Seismologists cannot, however, make accurate *short-term predictions*, on the time scale of hours to years. No one can say, for example, that an earthquake will happen in San Francisco 40 days from now, or 8 years from now. Here, we consider the basis of earthquake prediction and introduce early-warning systems.

Long-Term Predictions: Earthquake Probability

A long-term prediction estimates the *probability* (likelihood) that an earthquake will happen at a given location during a specified time period. For example, a seismologist may say, "The probability of a major earthquake occurring in the next 50 years in a given region is 20%." This sentence implies that there's a 1-in-5 chance that an earthquake of M_W 7.0 or above will happen before 50 years have passed. Seismologists refer to studies leading to long-term predictions as *probabilistic seismic-risk assessments*. Urban planners and earthquake engineers use such assessments to design building codes: codes requiring stronger buildings make sense for regions with greater seismic risk because the chance that a building will be shaken during its lifetime (generally 50–100 years) is greater.

Probabilistic seismic-risk assessment depends on two observations: First, a region in which earthquakes have occurred frequently in the past is likely to experience frequent earthquakes in the future. Therefore, seismic belts are regions of high seismic risk (see Fig. 3.21). This doesn't mean that disastrous earthquakes can't happen far from a seismic belt—they can and do—but the probability that an earthquake will happen outside of a seismic belt in any given time period is lower. Second, earthquakes of a given magnitude happen about 10 times more frequently than do earthquakes one unit of magnitude stronger. So, for example, there are about 10 times more M_W 4 earthquakes than there are M_W 5 earthquakes, and 100 times more M_W 2 earthquakes than there are M_W 4 earthquakes. This relationship applies generally to all seismic belts, regardless of the total number of earthquakes or the range of magnitudes recorded in a belt. So, if you know the frequency of M_W 4, 5, and 6 earthquakes, you can estimate the approximate frequency of larger events.

To provide a more specific sense of earthquake probability, seismologists try to determine the **recurrence interval (RI)** of earthquakes. (In discussing earthquakes, RI is the average time between successive events of a given magnitude.) We can calculate the RI during a specified time (t) in the past by counting the number of earthquakes (N) and applying the following equation: $RI = t \div N$. If we assume that earthquakes happen randomly, the annual probability (AP) of an earthquake of a given magnitude would be given by the equation $AP = 1/RI$. This assumption doesn't necessarily apply to earthquakes, however, for according to the stick-slip model, earthquakes happen when stress builds up enough to cause rocks to break or for friction on a pre-existing fault to be overcome, so the likelihood of an earthquake increases as more time passes since the previous earthquake. For example, if an earthquake of magnitude 7 on a fault has an RI of 200 years, then an earthquake might be more likely to occur on that fault after 180 years have passed than after 5 years have passed since the previous earthquake. But even that assumption doesn't always hold, because sometimes earthquakes happen in clusters.

As the vagueness of the above paragraph implies, estimating earthquake probability remains a challenge for seismologists. Even after decades of research, no one really knows why an earthquake begins at a given time and place, or why a particular quake ends up being minor or great. Put another way, in the context of the stick-slip model, no one can really predict the precise duration of the "stick" phase, nor can anyone predict how big the "slip" will be **(FIG. 3.40)**. Furthermore, because the historical record of earthquakes isn't very long, seismologists don't have a complete enough record of seismicity to determine recurrence intervals accurately, especially for seismic belts in which earthquakes have long recurrence intervals. Also, seismologists can't say for sure whether earthquakes are evenly spaced, or whether they cluster, with, say, three big events happening within a few years of each other, followed by a long period with no

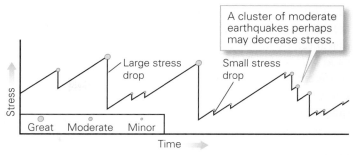

FIGURE 3.40 A simplistic stick-slip model, showing how the time since the last great earthquake may represent the amount of stress buildup. So, after a long time, there might be another large earthquake . . . or not. Sometimes earthquakes happen in clusters.

earthquakes. Because of these challenges, most recurrence intervals have a large uncertainty attached—for example, seismologists state the recurrence interval for major earthquakes in the New Madrid seismic belt as "500 to 1,200 years." Using more sophisticated statistical calculations, seismologists state probabilities for ranges of years, saying, for example, that a section of a fault has a 60% probability of hosting a major earthquake in the next 50 years **(FIG. 3.41)**.

Paleoseismicity Studies

As we've just discussed, to determine the recurrence interval for large earthquakes within a given seismic belt, seismologists must determine when previous earthquakes happened within that belt, but the historical record does not extend far enough back in time to date multiple large events. Studies of **paleoseismicity** (the geologic record of prehistoric seismicity) can add to the record of large earthquakes along a given fault. For example, in places where sedimentary strata have accumulated in a

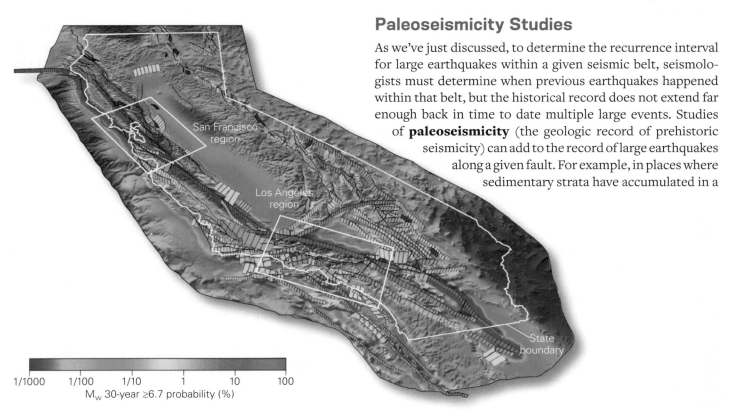

FIGURE 3.41 An earthquake hazard map for California. This map shows the probability of an earthquake of M_W 6.7 or greater happening in the next 30 years.

sag pond over a fault, researchers may dig a trench and look for buried beds that have been disrupted by liquefaction or beds that contain sand volcanoes. Each such bed, whose age can be determined by using radioisotopic dating of plant fragments, records the time of an earthquake (FIG. 3.42).

Seismic Gaps

As we've seen, the stick-slip model implies that at a location where an earthquake has happened recently, stresses are lower, so another earthquake is less likely to occur soon. It also implies that where an earthquake hasn't happened for a while, an earthquake is more likely to happen. Seismologists refer to a portion of an active fault that has not hosted a major earthquake for a long time, relative to adjacent portions that have, as a **seismic gap**. Let's consider two examples:

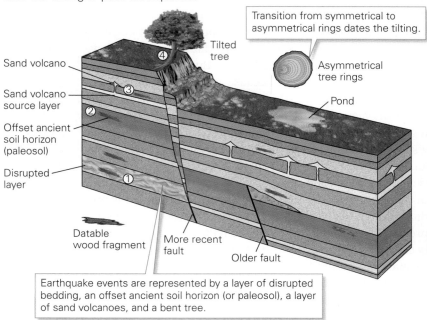

FIGURE 3.42 Studying paleoseismicity. Disrupted beds and other evidence indicate the timing of past earthquakes.

- *Loma Prieta gap (San Andreas fault):* A cross section of earthquake foci along the San Andreas fault for the 20-year period before the 1989 Loma Prieta earthquake shows that the portion of the fault beneath the Loma Prieta region had experienced much less seismicity than the region to the south (FIG. 3.43a). The Loma Prieta earthquake seemed to fill this gap.

- *Istanbul gap (North Anatolian fault, Turkey):* The North Anatolian fault is a large strike-slip fault along which Turkey slips westward (see Fig. 2.45). It has been the site of numerous devastating earthquakes in historic time. Since 1939, 11 significant earthquakes have taken place along the fault, each rupturing a different portion of the fault (FIG. 3.43b)—the largest of these, in 1939, was an M_W 7.8 event. Overall, there has been a general westward progression of seismic activity along the fault. Istanbul lies within a seismic gap near the west end of the fault, and if the seismic gap hypothesis is correct, the city might be the site of the next large earthquake.

Simplistically, seismic gaps may represent areas of greater seismic risk, for if the rate of displacement along the whole fault is assumed to be the same, then on average, all segments of the fault should host similar numbers of earthquakes over time. Whether this hypothesis is correct remains a subject of debate, and the recognition of seismic gaps has not yet provided clear forecasts of earthquakes. Seismologists don't know whether the lack of seismicity really means that a fault is sticking, or if fault creep is taking place along the fault instead. (**Fault creep** refers to a process by which slip takes place slowly, perhaps by plastic deformation, rather than in seismic pulses.) Recently, in studies of the Cascadia seismic zone, seismologists have learned that **slow earthquakes**, slip events that release energy over a time frame of hours to months, may be occurring on some faults. The slip rate of slow earthquakes is faster than that of fault creep, which accommodates slip over years, but is much slower than that of seismic slip, which takes place over seconds to minutes.

Short-Term Predictions

It would be great if somehow it were possible to predict that an earthquake will happen tomorrow, in the way that the evening weather report warns that a storm will arrive at 8:00 A.M. Unfortunately, such short-term predictions aren't yet possible, and they may never be. The absence of reliable methods for making short-term predictions based on scientific observations isn't for lack of trying. For example, scientists have tried to identify swarms of foreshocks, but swarms aren't always followed by mainshocks, and foreshocks can't really be recognized as such until after a mainshock has occurred. Other possible precursors have been explored—such as changes in electrical conductivity, the appearance of gases in wells, and even unusual animal activity—all attributed directly or indirectly to the opening of cracks in rock just prior to development of a large rupture. But these precursors either have not been fully tested or do not occur universally.

The history of seismicity at Parkfield, California, a town that straddles the San Andreas fault, illustrates the frustration

FIGURE 3.43 Examples of possible seismic gaps.

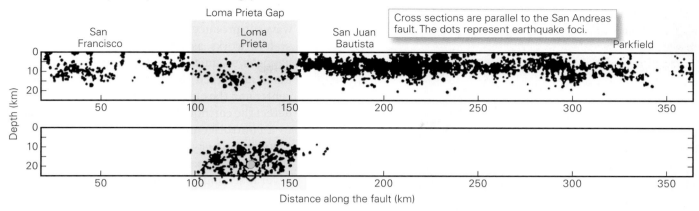

(a) The upper cross section shows earthquake foci along the San Andreas fault for the 20 years before the Loma Prieta earthquake. The lower cross section shows the Loma Prieta earthquake (yellow circle) and its aftershocks, which appear to have filled a seismic gap along the fault.

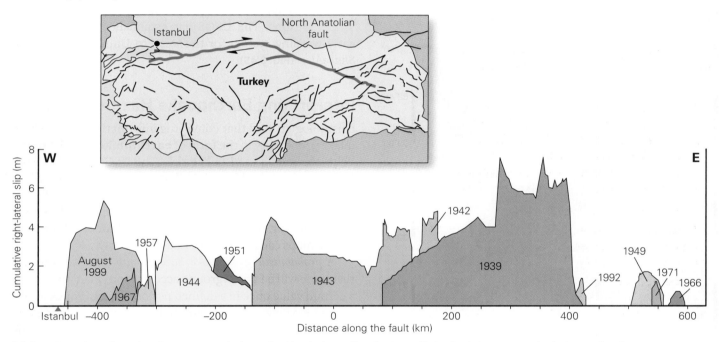

(b) A series of earthquakes has occurred along the North Anatolian fault, as if the fault is progressively fracturing from east to west. Seismic gaps exist at the west end of the fault and in the area east of the 1939 slip area. The colored patches indicate the amount of slip during the labeled earthquake. The horizontal scale represents distance, as measured from the left end of the fault trace (the red line on the map).

involved in making short-term predictions. In the 110 years between 1857 and 1966, six moderate earthquakes with an M_W of 5.5 or larger shook the town, and the events were assigned a recurrence interval of 12–32 years. Seismologists, therefore, predicted that the next earthquake after 1966 would happen between 1978 and 1998, and spent some $30 million setting up instruments to record the accumulation of elastic deformation and the fault slip. An earthquake finally did happen ... but not until 2004, years after the prediction window, and no precursors were identified.

Insight from Satellite Data

New satellite-mounted technologies are providing insight into seismic-risk assessment. For example, the global positioning system (GPS) allows geologists to measure relative motion of the crust—down to just a millimeter or two—on either side of a fault (see Chapter 2) **(FIG. 3.44a)**. If significant movements are happening, seismic slip is likely, unless the relative motion is being accommodated by fault creep. Another tool, **InSAR** (which stands for *I*nterferometric

FIGURE 3.44 Examples of how satellite data provide insight into where deformation of the crust is associated with seismicity.

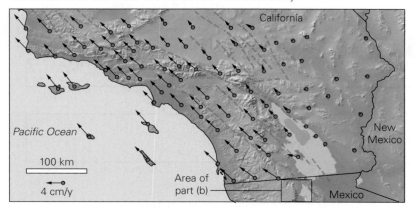

(a) A map showing just some of the GPS data for southern California. The arrows indicate the rate and direction of plate motion. Note that southwestern California is moving northwest relative to southeastern California. The difference in movement rates is accommodated by slip on faults.

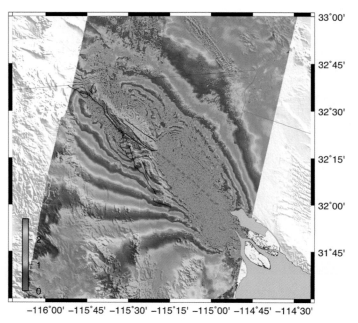

(b) InSAR data for the region just north of the northwestern end of the Gulf of California. Each band of colors indicates about 3 mm (0.12 in) of vertical displacement. This image represents displacements caused by the 2010 El Mayor–Cucapah earthquake.

Synthetic Aperture Radar), allows geologists to measure very subtle vertical movements of the ground (FIG. 3.44b). In some cases, the movements may be a clue that elastic deformation is building adjacent to a fault.

Earthquake Early-Warning Systems

The concept of short-term prediction should not be confused with the concept of an *earthquake early-warning system*, one that broadcasts an alert after an earthquake has happened, but before the damaging seismic waves it produces reach a city. An early-warning system works as follows: When an earthquake happens, seismic waves start traveling through the Earth and reach seismographs between the epicenter and the city. The instant these instruments detect the earthquake, a transmitter sends a signal to a control center, which automatically broadcasts emergency alerts to the city. The alerts travel at the speed of light, so they reach the city several seconds to a minute before the seismic waves do, and can trigger automatic shutdown of gas pipelines, trains, nuclear reactors, and power lines before shaking begins. The signals can also set off emergency sirens and alerts on TV, radio, or cellular networks, warning people to take precautions. The USGS has established an earthquake early-warning system for the western United States called *ShakeAlert* (FIG. 3.45).

Interpreting Earthquake Hazard Maps

Seismologists find that conveying information about earthquake probability in the form of **earthquake hazard maps** helps the public to visualize risk. The USGS provides such maps for the United States. At first glance, interpreting these maps seems simple: red areas have the highest risk, and other colors show areas with progressively less risk. In detail, understanding what these maps portray can be a bit tricky, for the maps can portray risk in different ways.

A *probabilistic earthquake hazard map* indicates the risk that there will be a peak ground acceleration (PGA) exceeding a certain value once in a defined period of time. For example, such a map may define regions that have a 2% chance of experiencing a given PGA within 50 years, which translates to one earthquake of this size, on average, in a 2,500-year period

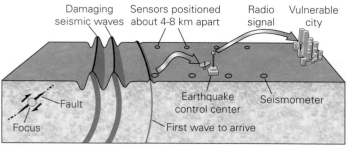

FIGURE 3.45 The *ShakeAlert* system, developed by the USGS, for the western United States. Seismographs detect P-waves and broadcast a radio signal before damaging surface waves arrive.

FIGURE 3.46 Different ways of portraying seismic risk in the continental United States.

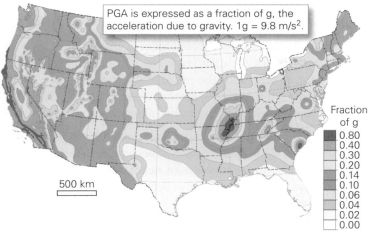

(a) This probabilistic earthquake hazard map outlines areas in which there is a 2% probability that PGA (peak ground acceleration) will exceed the indicated value during a period of 50 years. Red areas have higher risk of a strong earthquake.

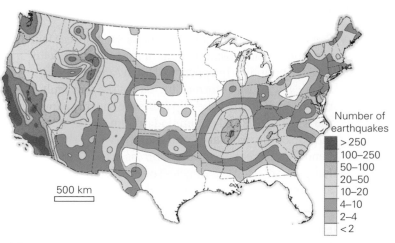

(b) This earthquake hazard map shows the number of damaging earthquakes that are likely to affect a region over a 10,000-year period.

(because 50 ÷ 0.02 = 2,500) **(FIG. 3.46a)**. Some researchers point out that this type of map overemphasizes risk for areas with longer RI. For example, if one area has an RI of 100 years, it has a higher probability in any given year of having an earthquake than does a region with an RI of 500 years, yet both appear with similar colors on this map. An alternative map type depicts the estimated number of damaging earthquakes that will happen within an area over a 10,000-year period

(FIG. 3.46b). Earthquake hazard maps in various forms have been developed for the entire world **(FIG. 3.47)**.

Regions of Seismic Risk in the United States

Despite the uncertainties involved in constructing earthquake hazard maps, such maps do outline areas of greater

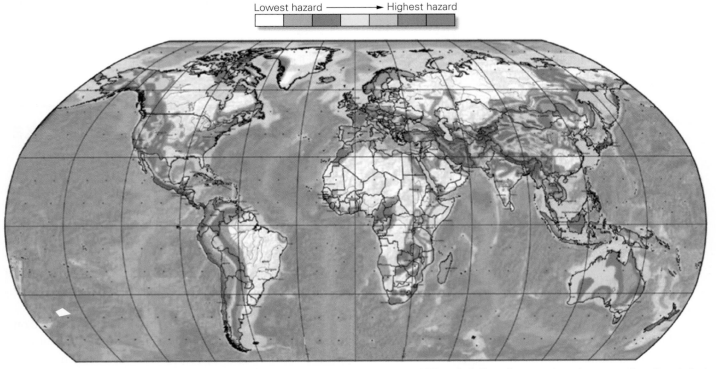

FIGURE 3.47 This global seismic hazard map portrays areas where there is a 10% probability of an earthquake exceeding the stated PGA in a 50-year period. Red and brown areas represent regions with the greatest seismic hazard.

3.8 Can We Predict "the Big One"? • **111**

seismic risk. Let's consider some examples of these areas in the United States:

- *San Andreas fault system:* The famous San Andreas fault is the longest and most notorious of the faults that delineate the boundary between the North American and Pacific Plates in California. But it's not the only fault there. Many other strike-slip faults are roughly parallel to the San Andreas and accommodate some of the relative motion of the two plates. And at places where the fault bends, local areas of compression (as in the Transverse Ranges bordering Los Angeles) or extension (as in the area around the Salton Sea) develop.

- *Cascadia:* North of Cape Mendocino, California, the western boundary of the United States is a convergent plate boundary. Subduction along this boundary has generated megathrust earthquakes, as well as smaller ones, affecting northern California and western Oregon and Washington (BOX 3.2). Some of the displacement takes place in the overriding plate. Such motion has produced earthquakes in the Seattle area.

- *Basin and Range Province:* Rifting has been taking place in the western United States for at least 25 million years. The location of active faulting associated with this rifting has shifted over time. Currently, the most active seismicity takes place along the western and eastern borders of the rift. Earthquakes along the boundary between the Sierra Nevada and the Basin and Range accommodate extension associated with rifting as well as strike-slip faulting due to the relative motion of the North American and Pacific Plates. The eastern zone of normal faulting poses a threat to Salt Lake City.

- *Intermountain seismic belt:* Earthquakes occur in western Wyoming, western Montana, and eastern Idaho. Some are associated with normal faulting accompanying the northward propagation of rifting and others with deformation around the Yellowstone hot spot.

- *Intracontinental regions:* Notable seismic activity currently takes place in Oklahoma; along the Mississippi valley in southeastern Missouri and adjacent areas; and in eastern Tennessee. This seismicity is associated primarily with the reactivation of ancient rift-related faults that have remained as weak spots in the continental crust since they were initiated. The most active of these intracontinental seismic zones is the New Madrid zone in the Mississippi valley (see Fig. 3.28b).

- *East coast:* An M_W 6.9–7.3 earthquake shook Charleston, South Carolina, in 1886; an M_W 5.8 earthquake rattled central Virginia in 2011; and seismicity also takes place along the Ramapo fault in New Jersey. These earthquakes result from the reactivation either of thrust faults formed during the collision of Africa with North America to form Pangaea, or of rift faults related to the breakup of Pangaea.

- *Aleutian Trench, Alaska:* A convergent plate boundary, delineated by the Aleutian Trench and the bordering volcanic arc, sweeps along the entire southern margin of Alaska. Megathrust earthquakes, including the 1964 Good Friday earthquake, have happened along this boundary.

- *Denali fault system:* A strike-slip fault system extends from southeastern Alaska and curves to become east-west-trending in south-central Alaska, near Denali, the tallest mountain in Alaska. An M_W 7.9 event happened on this fault in 2002.

- *Hawaii:* The Big Island of Hawaii is volcanically active. Normal faults cut downward through the volcano along the eastern side of the island, accommodating the seaward slip of that part of the island. One of the normal faults hosted an M_W 7.9 earthquake in 1868 and an M_W 6.9 earthquake in 2018.

> **Take-home message . . .**
> Seismologists can determine where earthquakes are most likely to take place, but cannot predict exactly when and where an earthquake will occur. Seismic risk is greater where seismicity has happened more frequently in the past and, therefore, the recurrence interval for earthquakes is shorter. Early-warning systems can provide seconds of warning by sending out signals that travel faster than seismic waves.
>
> **QUICK QUESTION** How can seismic risk be represented on a map?

3.9 Prevention of Earthquake Damage and Casualties

People who live in seismic zones can't be expected to just pack up and move to a region that doesn't have earthquakes to avoid an earthquake disaster. But contemplating the potential loss due to a major earthquake can be terrifying. If a major earthquake comparable to the 1906 event were to strike the San Francisco area, for example, there could be thousands of fatalities, over 150,000 displaced people, and by some estimates, $70–$400 billion in damage.

The consequences of an earthquake for a city depend on a number of factors, including the quake's magnitude; the city's distance from the focus; the style of building construction; the presence of nearby slopes; proximity to the coast; the composition of the underlying substrate; whether people are

BOX 3.2 DISASTERS AND SOCIETY

The megathrust hazard of Cascadia

On January 26, 1700, slip suddenly took place along the entire length of the thrust fault along which the Juan de Fuca Plate subducts beneath the North American Plate **(FIG. Bx3.2)**. This movement generated an earthquake with an estimated magnitude of M_W 8.7–9.2. Violent shaking rattled much of **Cascadia**, a region of the Pacific Northwest that includes Oregon and Washington. Also, as we'll discuss in Box 6.1, displacement of the seafloor caused by slip on the fault produced large tsunamis, which washed away many coastal Native American villages and reached the shores of Japan. Researchers calculated that, to cause such a large earthquake, the slipped area was about 1,000 km (600 mi) long by about 200 km (125 mi) wide, and that slip reached a maximum of about 30 m (100 ft). Clearly, the 1700 Cascadia event was a *megathrust earthquake*, very similar to the catastrophic 2011 Tōhoku earthquake and tsunami (see Chapter 1), and to the 2004 Sumatra earthquake and tsunami that killed 230,000 people.

The 1700 Cascadia megathrust earthquake happened over 100 years before the Lewis and Clark expedition first mapped the Pacific Northwest. Since then, the area's population has grown to over 9 million people, and its four major cities—Vancouver, Seattle, Tacoma, and Portland—host aerospace and other high-tech industries that account for about 5% of America's GDP. If another megathrust earthquake were to strike, more than 7 million people could be affected, and the death toll might exceed 10,000. Building codes requiring earthquake-resistant designs began to appear only in 1994, so engineers worry that as many as 75% of the buildings in the affected area could not withstand significant shaking. In fact, a million buildings—including thousands of schools, most hospitals, and probably all highway bridges—might fail. Clearing debris and repairing damage would cost hundreds of billions of dollars, and loss of strategic industries could impact the entire nation's economy.

What's the likelihood of another megathrust earthquake happening in Cascadia in the near term? GPS measurements show that the crust beneath Cascadia is actively deforming, and paleoseismicity studies indicate that the recurrence interval for a megathrust earthquake in the region ranges from 250 to 500 years. In light of these data, and taking into account the time since the last megathrust earthquake in the region (over 320 years), seismologists estimate that the probability of an M_W 8.2 or greater event happening in Cascadia during the next 50 years approaches 40%, and that the probability of an M_W 9.0 or greater event during that time approaches 10%–15%. Overall, these results imply that the annual probability of a great earthquake happening in Cascadia lies in the range of 0.2%–0.4%. Put bluntly, there is a significant chance of another Cascadia megathrust earthquake happening during the lifetime of this book's readers! Because the anticipated earthquake could be so large, journalists have taken to calling it "the really big one."

Although almost 10 earthquakes with magnitudes between M_W 6.0 and M_W 7.3 (that is, strong to major earthquakes) have affected Cascadia since 1870, officials and the general public did not fully appreciate that the risk of a great earthquake happening in Cascadia may exceed the risk of one happening in California, simply because the historical record of earthquakes in the region does not go back far enough to include megathrust earthquakes. The discovery of the 1700 megathrust earthquake (described further in Box 6.1), therefore, served as a call to action. New earthquake hazard maps are being developed, updated building codes are being enacted, and some buildings and bridges are undergoing seismic retrofitting. And, as we'll see in Chapter 6, tsunami evacuation plans have been implemented.

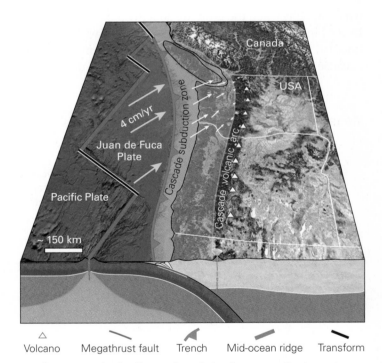

FIGURE Bx3.2 At the Cascadia subduction zone, the Juan de Fuca Plate slides underneath North America. The contact between the plates is a large thrust fault. Melting above the downgoing plate provides magma that rises beneath the Cascade volcanic arc.

at home or at work when it happens; and whether the government has the capacity to provide emergency services promptly. Given all these variables, let's see how people can strive to minimize the effects of an earthquake.

The Challenge: Damage to Buildings

What happens to a building when seismic waves shift the ground? To answer this question, stand a book on a sheet of paper and pull the paper to the right. The base of the book moves with the paper, but inertia keeps the top of the book where it was. Consequently, the book tips over **(FIG. 3.48a)**. Now picture a four-story building, attached to a solid foundation, on ground that suddenly shifts to the right. The foundation also shifts to the right, but because of inertia, the top of the building tries to stay where it was, and the building overall undergoes shear **(FIG. 3.48b)**. If the ground then lurches to the left, it suddenly undergoes shear in the opposite direction. What happens when the ground suddenly moves upward? The building accelerates upward with it, so the amount of stress pushing down on its support columns increases dramatically—in effect, the building momentarily weighs much more **(FIG. 3.48c)**. (You may have experienced such stress on a carnival ride that rapidly accelerates upward.) And what happens when the ground suddenly moves downward? Everything firmly attached to the ground drops with it. But if a floor isn't anchored securely, it gets left behind momentarily because of inertia. Gravity then pulls the floor down, so it lands on the support columns and applies a large stress to them **(FIG. 3.48d)**.

During an earthquake, many different seismic waves arrive, so a building experiences many cycles of up-and-down and back-and-forth movement, and it twists this way and that. What does this movement do? It depends on what the building is made of **(BOX 3.3)**. A building made of unreinforced masonry (adobe, bricks, or cinder blocks) or concrete can crack, and if enough cracks form and connect, the building crumbles and collapses **(FIG. 3.49)**. A reinforced concrete building or bridge (made of concrete containing steel reinforcing bars, or *rebar*) will be much stronger, but earthquake stresses can cause the concrete to flake, or *spall*, off the rebar. Spalling on a support column allows the rebar to bow outward and bend, so the column fails **(FIG. 3.50a)**. Such column failure can be a particular problem for buildings with "soft" first floors, meaning first floors used for parking or large, open commercial spaces, for such spaces contain few internal supports. In buildings or bridges constructed of multiple heavy concrete floors, the collapse of support columns holding up one floor causes the stress acting on other columns to increase, so those columns, too, may fail **(FIG. 3.50b)**. In quick succession, all the floors come crashing down, a process informally called *pancaking* **(FIG. 3.50c)**. Buildings with steel or wood frames have more flexibility, so they may accommodate earthquake motion better, though facades may break off, windows will break, and materials inside will be thrown around.

FIGURE 3.48 How ground motion causes building failure.

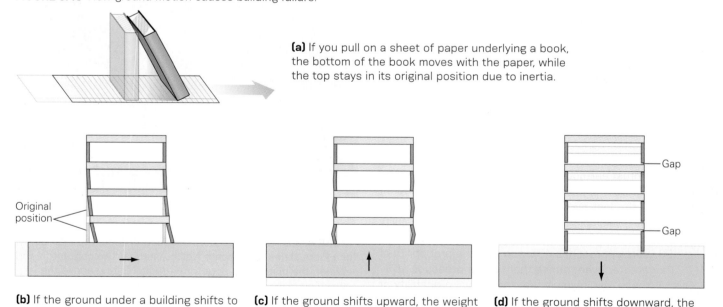

(a) If you pull on a sheet of paper underlying a book, the bottom of the book moves with the paper, while the top stays in its original position due to inertia.

(b) If the ground under a building shifts to the right, the building undergoes shear.

(c) If the ground shifts upward, the weight of the building effectively increases and pushes down on its support columns.

(d) If the ground shifts downward, the floors stay put momentarily and then fall downward onto the support columns.

BOX 3.3 A CLOSER LOOK...

Catastrophes due to building collapse

It wasn't yet 4:00 A.M. on a July morning in 1976, when an M_W 7.6 earthquake on a strike-slip fault struck Tangshan, an industrial city of 1 million people, that lies about 140 km (87 mi) east of Beijing, China. Because the focus of the earthquake was only 12 km (7 mi) beneath the Earth's surface, the intensity of the earthquake reached MMI XI within an area of about 55 km^2 (20 mi^2), but about 9,500 km^2 (3,668 mi^2) endured shaking of intensity VIII. Because the earthquake occurred in a region of relatively strong crust, its energy traveled great distances and shook Beijing with an intensity of VI. In fact, the earthquake was even felt in Mongolia, over 1,000 km (621 mi) away.

Unfortunately, most of the buildings in Tangshan were constructed of unreinforced brick. To make matters worse, the city was built on thick, sandy soil that underwent liquefaction. Because the construction was not earthquake resistant, this earthquake was one of the most deadly in history. Within the zone of highest-intensity shaking, 85% of the buildings collapsed, and the death toll was horrendous—the official count was close to 250,000, but some estimates placed it at over half a million. In addition, hundreds of thousands of people were injured. About a dozen aftershocks of M_W 6–7 rattled the area during the week following the mainshock, slowing rescue operations. At the time of the earthquake, China's Cultural Revolution was taking place, and the government insisted (incorrectly) that earthquake prediction was possible. Because the Tangshan earthquake was not predicted, many people believed that this natural disaster was a sign that the government was losing legitimacy. So the government put extra effort into relief operations. And in the aftermath of the Tangshan earthquake, China began the process of upgrading building codes.

In 1990, an M_W 7.4 earthquake generated by sudden slip on a strike-slip fault struck the mountainous south coast of the Caspian Sea, about 200 km (124 mi) northwest of Tehran. This area of Iran is part of the broad collisional zone that extends from the Alps through the Himalayas due to the northward movement of Africa, the Arabian Peninsula, and India into Asia. This region has suffered at least 14 major earthquakes during the past 1,200 years. Because the 1990 earthquake's focus was shallow (15 km, or 9 mi), the MMI reached X in the area around the epicenter.

Two cities and nearly 700 villages had been built near the epicenter of the quake. Most of the buildings in these population centers were constructed of unreinforced masonry. Because the earthquake happened just after midnight, when most people were sleeping, building collapse led to widespread casualties, including up to 50,000 deaths and 135,000 injuries. Virtually every *non-engineered building* (constructed without design input from professional engineers) in an area approaching 32,000 km^2 (12,355 mi^2) collapsed. Even *engineered structures* (designed professionally) fared poorly. Steel-frame structures, for example, collapsed because they had no lateral braces or *shear walls* (rigid walls that cannot undergo shear deformation), and connections and welds between beams were weak. Reinforced concrete buildings also failed because reinforcement was inadequate and because the low-quality concrete crumbled easily.

Because of the loss of housing, up to 400,000 people were displaced. Additional destruction came when an M_W 6.5 aftershock caused a dam to fail, producing a flood that washed away large areas of farm fields. Rescue was significantly hampered because landslides blocked roads leading into the devastated area, so many people who had been trapped by debris died of suffocation before they could be rescued. Because of the area's limited resources, the Iranian government declared a national disaster and requested international assistance. The United Nations Disaster Relief Organization (UNDRO) sent a coordination officer to help with relief efforts, and international aid agencies assisted in search and rescue operations.

Earthquake Engineering

Major efforts have gone into developing an understanding of how to make buildings less susceptible to earthquake damage. Such **earthquake engineering** can't prevent all damage, but **earthquake-resistant designs** can save lives and property. Earthquake engineers focus on the following goals: (1) preventing *structural components* (the columns, girders, beams, trusses, and load-bearing walls that support the building and fly about its roof) from failing; (2) ensuring that *nonstructural components* (non-load-bearing walls, partitions, ceilings, cabinets, and other features that make the building usable) do not detach from the structural components; and fly about (3) keeping the building attached to its *foundation* (the connection between the building and the ground), and keeping the foundation intact.

What features can make a building earthquake resistant? In a general sense, buildings stand up better during ground shaking if the stress caused by ground movement gets distributed evenly so that it doesn't exceed the strength of the building's structural components; if connections between parts of the building are reinforced so that they can't break;

FIGURE 3.49 Failure of masonry buildings.

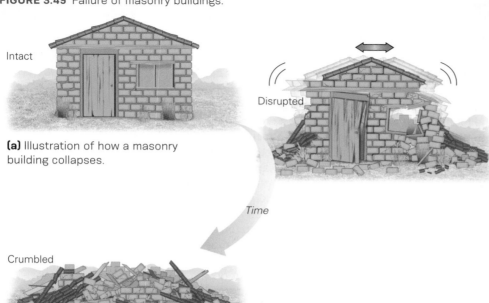

(a) Illustration of how a masonry building collapses.

(b) Collapsed buildings in Armenia following an M_W 6.8 earthquake in 1988.

if support columns can't bow outward; if the building can flex a bit, but not by enough to tip over or snap; and if the building's dimensions won't lead to a resonance disaster. In regions with significant seismic risk, certain kinds of construction materials, such as cinder blocks, bricks, and unreinforced concrete, should be avoided. For example, during an MMI IX earthquake, about 90% of adobe buildings, about 40% of brick buildings, and about 15% of wood-frame buildings will collapse, but less than 10% of properly reinforced concrete buildings will. Finally, avoid dangerous roofs—traditional heavy, brittle tile roofs can shatter and bury people inside, but lightweight sheet-metal roofs won't.

FIGURE 3.50 Failure of concrete columns can cause buildings and other structures to collapse.

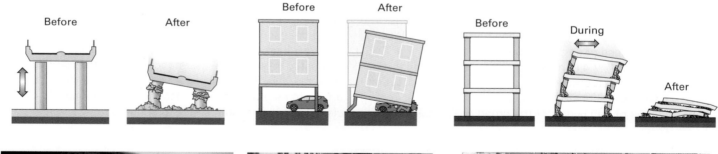

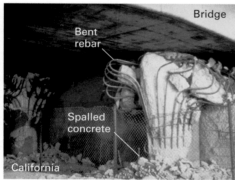

(a) Failure of columns holding up a bridge. Note the bent rebar and spalled-off concrete.

(b) Failure of a soft first floor. Here, an apartment building has crushed cars parked beneath it.

(c) Pancaking of a building.

Many features can be incorporated in the construction of large commercial buildings to improve their earthquake resistance. For example, cross braces or diagonal crossbeams can be added to the walls to keep them from twisting and shearing (**FIG. 3.51a**); strong bolts can be added to prevent beams from shifting off support columns and to prevent the building from shifting off its foundation; and support columns can be wrapped in steel cable or sheathed in steel plate so that they can't bow outward if the load pushing down on them increases (**FIG. 3.51b**). In recent years, some buildings have been designed with **base isolation**, meaning that the motion of the ground is decoupled from that of the superstructure (the portion of the building above the basement) so that the building sways less (**FIG. 3.51c**). This can be done by placing the building on rubber bearings, giant springs, or ball bearings, which absorb ground motion without transmitting it to the building. Other buildings have been designed with *motion dampers*, large masses (either blocks that can slide on horizontal rails, or a giant pendulum that can swing) that move with a period different from the resonant period of the building so that, like a person on a swing who intentionally moves their legs to slow it down, the motion damper decreases the amplitude and duration of swaying (**FIG. 3.51d**).

Earthquake-resistant design and construction can also improve the safety of private homes (**FIG. 3.52**). For example, to make the structural components of a wood-frame house withstand ground shaking better, metal connectors (framing anchors) can be added at all joints; horizontal or diagonal braces can be placed between the wall studs (the vertical framing members); *shear walls* made of strong plywood can be attached to several wall studs to prevent them from moving relative to one another; strong bolts can be used to attach the

FIGURE 3.51 Earthquake-resistance in the construction of commercial buildings.

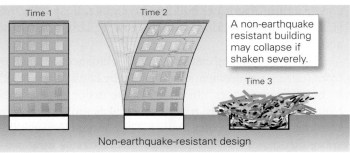

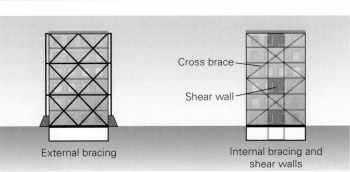

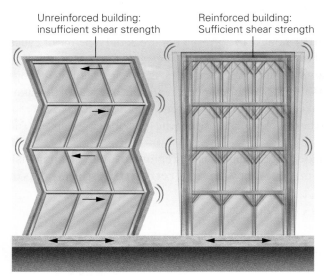

(a) External and internal bracing can make a non-earthquake-resistant building stronger against shearing.

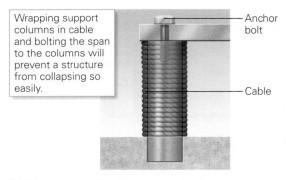

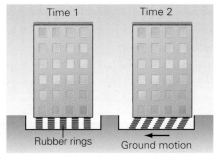

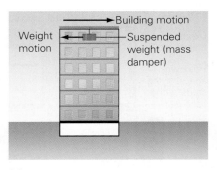

(b) Wrapping and bolting support columns can add to stability.

(c) Base isolation, by rubber rings or springs, can help prevent sway.

(d) Motion dampers can decrease the amplitude and duration of swaying.

3.9 Prevention of Earthquake Damage and Casualties • **117**

house frame to its foundation; and brick chimneys can be reinforced or replaced with lightweight metal chimneys. Injury to the residents due to the movement of nonstructural components can be prevented by bolting bookshelves to the wall, strapping the water heater and furnace to the wall, installing automatic gas shutoff valves, removing heavy wall hangings above beds or couches, and placing locking latches on kitchen cabinet walls.

The earthquake resistance of infrastructure (pipelines, power lines, roads, bridges, dams, and railroads) can be improved by incorporating the same features used in buildings, as well as some additional ones. For example, in the case of pipelines, special automatic valves can be installed; these valves contain a ball that, when shaken, falls into a hole and plugs the pipe to prevent leakage. Pipelines can also incorporate flexible components to prevent them from buckling or snapping. For example, the Trans Alaska Pipeline, which carries oil from wells on Alaska's north coast to shipping terminals on its south coast, crosses the Denali fault. At the crossing point, the pipeline has many intentional zigzags and sits on Teflon-coated "shoes" that allow it to slide on steel beams oriented parallel to the fault trace. The design works: the 7 m (23 ft) of strike-slip displacement that occurred during an M_W 7.9 earthquake in 2002 did not break the pipeline or cause any oil to leak. Adding joints that can expand and contract to bridge spans prevents them from snapping or falling off support columns. Such upgrades can prevent a repeat of the damage that occurred during the Loma Prieta earthquake of 1989, when the double-decker approaches to a bridge collapsed (see Fig. 3.23c).

Inadequate structures in regions of high seismic risk can be made safer by **seismic retrofitting**, the process of strengthening existing structures that were not originally designed to withstand an earthquake. Retrofitting methods can include adding support columns and sheathing existing ones with a steel casing, adding diagonal braces, adding an external or internal supplementary support frame, or adding base isolation (**FIG. 3.53**). Extra cables can be added to anchor the bridge span to its support columns.

Zoning, Building Codes, and Common Sense

Communities located in seismic belts can mitigate or diminish the consequences of earthquakes by enacting **earthquake zoning**. At a minimum, zoning should provide guidance that discourages construction on land underlain by weak mud or wet sand that could liquefy (**FIG. 3.54**), or adjacent to escarpments that could undergo landslides, or downstream of a dam that might fail. Zoning may also limit construction of critical buildings (schools, hospitals, fire stations, communications centers, power plants) on top of active faults.

Earthquake zoning can also provide guidance for *building codes* (design requirements established by law and regulated by building inspectors). In places where major earthquakes happen fairly frequently, it of course makes sense to require earthquake-resistant construction. The contrast between the consequences of two earthquakes demonstrates how construction techniques affect the stability of buildings during earthquakes. The M_W 6.8 earthquake that struck Armenia in 1988 caused 400 times as many deaths as the M_W 6.7 earthquake that struck San Fernando, California, in 1971 because many buildings in Armenia were constructed of unreinforced

FIGURE 3.52 Earthquake-proofing a home.

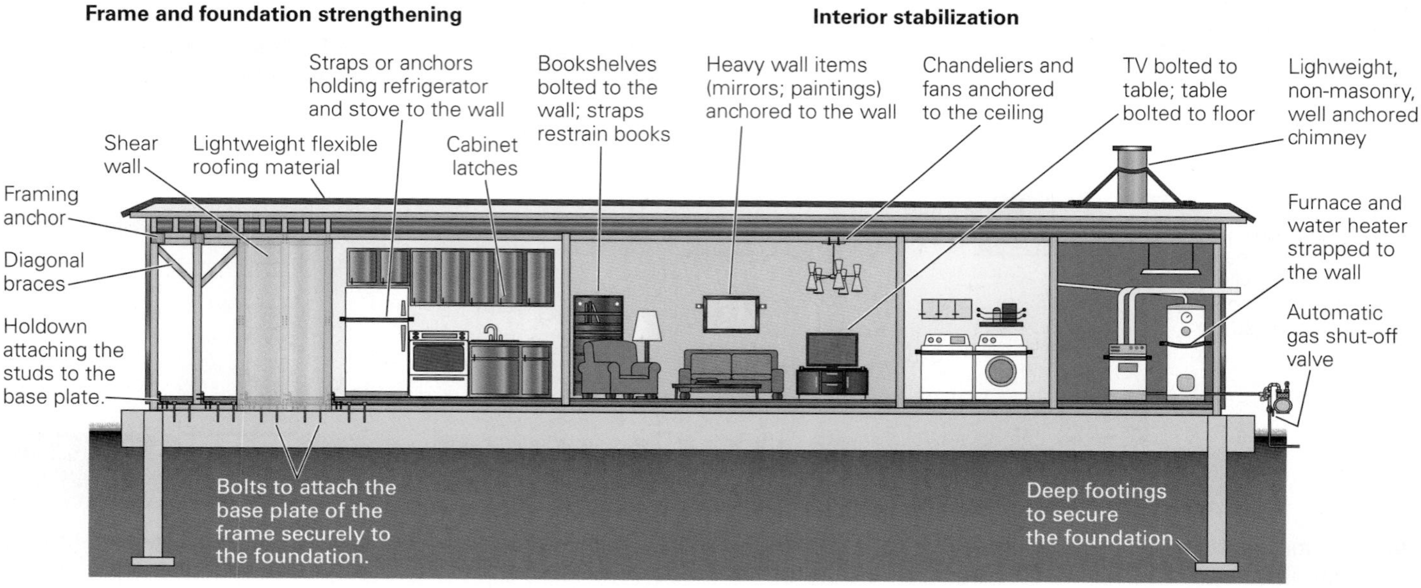

FIGURE 3.53 Seismic retrofitting.

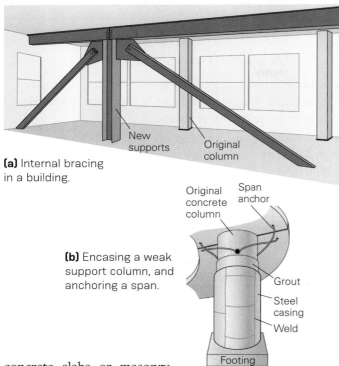

(a) Internal bracing in a building.

(b) Encasing a weak support column, and anchoring a span.

FIGURE 3.54 A map showing liquefaction hazard in the San Francisco Bay area.

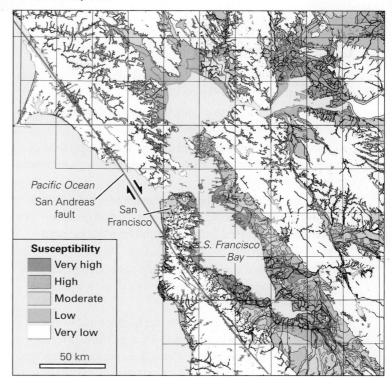

concrete slabs or masonry, which cracked, crumbled, and pancaked (see Fig. 3.49b), whereas many buildings in California were reinforced or had timber and steel frames, so they flexed, but very few fell down and crushed people. But in places where the hazard is lower, the choice is not so clear-cut. Money spent on improving the earthquake resistance of buildings, or on seismic retrofitting, alternatively might go to other worthy causes, such as health care or schools.

Prevention of Casualties

Even careful construction and planning can't prevent all earthquake casualties. Therefore, communities in seismic belts should draw up emergency plans to deal with disaster. They should put in place strategies to provide personnel, equipment, and supplies for rescue and recovery. Individuals living in regions of high seismic risk should also take personal responsibility for protecting themselves and their homes from earthquake damage. Simple precautions include securing the nonstructural components of a building, as we described earlier, as well as knowing how to shut off the gas and electricity, having a fire extinguisher handy, and knowing where to find family members. Schools, factories, and offices should hold earthquake preparedness drills, and individuals should know where to go to seek protection from falling objects, and how to find the building's exits (FIG. 3.55). People should also prepare for displacement by packing an emergency bag with necessary clothes, medicine, money, and key financial records. Likewise, people living in zones of potential earthquake-related tsunami inundation should know evacuation routes and areas of safety. (We'll discuss tsunamis, and how to stay safe from them, further in Chapter 6.) As long as the Earth's lithosphere plates continue to move, earthquakes will continue to shake us. But we can reduce the chances of damage or injury by being prepared.

Take-home message . . .

Earthquakes are a fact of life on this dynamic planet. People in regions facing high seismic risk should build on stable ground, avoid unstable slopes, and design structures that can survive shaking. Individuals should have a plan for what to do if an earthquake happens.

QUICK QUESTION What factors influence the degree of devastation during an earthquake?

FIGURE 3.55 A sturdy table can provide some protection from falling objects during an earthquake.

3.9 Prevention of Earthquake Damage and Casualties

Chapter 3 Review

Chapter Summary

- Earthquakes are episodes of ground shaking. They are generally a consequence of slip on a fault.
- Geologists distinguish among three types of faults based on the relative direction of slip on the fault.
- According to elastic-rebound theory, during slip, the rock rebounds to its original shape. Fracturing and rebound send out seismic waves.
- Some earthquakes are due to formation of new faults, but most happen when stress overcomes friction on a pre-existing fault and the fault slips. Faults typically exhibit stick-slip behavior.
- The place where a fault begins to slip is an earthquake's focus. The point on the ground surface directly above the focus is its epicenter.
- Earthquake energy travels in the form of seismic waves. Body waves include P-waves and S-waves. Surface waves pass along the surface of the Earth. Seismic waves undergo attenuation as they travel. When seismic waves pass into soft sediment, they may be amplified.
- A seismograph can detect and record seismic waves. Seismograms—records of earthquakes—demonstrate that different types of seismic waves travel at different velocities. Seismologists can pinpoint the epicenter of an earthquake.
- The Modified Mercalli Intensity scale measures the size of an earthquake at a locality by assessing damage and people's perception of ground shaking. Earthquake intensity decreases with distance from the epicenter.
- Magnitude scales, which characterize the amount of energy released at the focus of an earthquake, are based on the amount of ground motion, as indicated on a seismogram. There is one magnitude number for any given earthquake. These days, seismologists use the moment magnitude (M_W) scale.
- An M_W 8 earthquake yields about 10 times as much ground motion as an M_W 7 earthquake, and it releases about 32 times as much energy.
- Seismicity takes place mainly in seismic belts, the majority of which lie along plate boundaries. Intraplate earthquakes happen in the interior of plates.
- Earthquake damage results from ground rupture, ground shaking, landslides, sediment liquefaction, tsunamis, and fire. In some cases, buildings resonate during an earthquake until they collapse.
- Because of the destruction of infrastructure, epidemics of disease may follow a major earthquake.
- Providing an estimate of when and where earthquakes will happen is very difficult. Such assessment takes into account the history of seismic events in an area as well as other information.
- Overall, places that have had earthquakes in the past are more likely to have them in the future, so seismic belts represent higher risk than other areas.
- Seismic hazard can be represented by statements about the recurrence interval and the annual probability of earthquakes. In some cases, the presence of a seismic gap may indicate increased potential for seismicity.
- Short-term predictions of seismicity are not possible. But earthquake early-warning systems can alert cities before seismic waves arrive.
- Earthquake hazard maps outline areas where seismic risk is greatest. Officials use this information to design building codes and set insurance rates.
- Earthquakes can damage buildings in many ways. Ground motion can cause buildings to shear, crack, and collapse.
- Liquefaction causes buildings to sink, tip over, or crack.
- Earthquake engineering can save lives. Reinforcing existing buildings to better withstand earthquakes is called seismic retrofitting.
- Because certain locations have higher seismic risk, it makes sense for those communities to establish sensible building codes and zoning rules to decrease vulnerability.

Review Questions

Blue letters correspond to the chapter's learning objectives.

1. What is an earthquake, and what causes most earthquakes on the Earth? **(A)**
2. Distinguish among different types of faults. Which accommodate shortening? **(A)**
3. Explain how elastic-rebound theory accounts for the generation of earthquake energy. Why do faults, once formed, exhibit stick-slip behavior? **(A)**
4. Distinguish between the focus and the epicenter of an earthquake, and label them on the diagram. **(B)**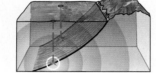
5. Distinguish among foreshocks, mainshocks, and aftershocks. Which are the largest? **(B)**
6. Describe the four different types of seismic waves. **(B)**
7. Why do seismic waves cause less shaking farther from the focus? How do they change as they enter soft sediment? **(B)**
8. How does a seismograph detect and record an earthquake? **(C)**
9. Explain the difference between earthquake intensity and earthquake magnitude. What do the numbers on these scales mean? **(D)**
10. How much more energy does an M_W 9.0 earthquake release than does an M_W 5.0 earthquake? **(D)**
11. What is a seismic belt, and how are seismic belts related to plate boundaries? Where do megathrust earthquakes take place, and why are they so dangerous? At which type of plate boundary do deep-focus earthquake take place? **(E)**
12. What are intraplate earthquakes, and why might they be dangerous? **(E)**
13. Describe some ways earthquakes can cause damage. Which type does the image show? **(F)**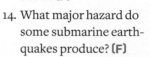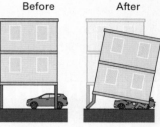
14. What major hazard do some submarine earthquakes produce? **(F)**
15. Why might buildings constructed on weak, wet sediment be particularly vulnerable to damage by ground shaking? **(F)**
16. What kinds of tertiary disasters may be triggered by a major earthquake? **(F)**
17. What is the primary basis for estimating seismic hazard? What are the recurrence interval and the annual probability of an earthquake? **(G)**
18. Can short-term predictions of earthquakes be trusted? What is an earthquake early warning? **(G)**
19. How can seismic risk be represented on a map? **(G)**
20. How can earthquake engineering help prevent injury and building damage during an earthquake? What features can make a building stronger? What features can diminish the amount of motion that a building endures? **(H)**
21. What policies can a community subject to significant seismic risk enact in order to make its members less vulnerable to earthquake hazards? **(H)**

On Further Thought

22. Is seismic risk greater on the east coast of South America, or on the west coast? **(G)**
23. Imagine that you are responsible for siting a hospital in a city that lies within a seismic belt. What characteristics of the area should you study to identify a site that is relatively safer than other sites? **(H)**

4

THE WRATH OF VULCAN
Volcanic Eruptions

By the end of the chapter you should be able to . . .

A. explain why melting happens inside the Earth, and how molten rock becomes igneous rock, both underground and at the surface.

B. recognize major types of igneous rocks, and occurrences of igneous rocks, and explain why the differences exist.

C. relate occurrences of igneous activity and volcanism to plate tectonics.

D. distinguish among the various kinds of volcanic eruptions and volcanic edifices.

E. characterize different types of lava flows and volcanic explosions, discuss how the volcanic explosivity index classifies explosions, and learn to identify eruptive styles.

F. describe hazards produced by volcanic eruptions, and discuss how these hazards can cause disasters locally and worldwide.

G. evaluate methods used to predict and mitigate volcanic hazards.

4.1 Introduction

For many years, the Roman town of Pompeii prospered at the foot of Mt. Vesuvius, near the western coast of the Italian Peninsula. Pompeiians thought that Vesuvius, which at the time towered 3 km (nearly 10,000 ft) above the sea, was just another scenic mountain. Sadly, they were wrong—very wrong! In fact, Mt. Vesuvius is a volcano. (The word **volcano** refers both to a *vent* or opening from which material emerges from below a planet's surface, and to a hill or mountain built from the materials that come out of a vent.) For several weeks in the spring of 79 C.E., earthquakes had jolted Pompeii with unnerving frequency, and Vesuvius grumbled like distant thunder. People browsing in the town's markets, however, paid little heed. Then, at 1:00 P.M. on August 24, Vesuvius suddenly roared, and a dark, mottled cloud churned out of its summit (**FIG. 4.1a**). This was no ordinary cloud. Instead of just water, it contained poisonous gases, *ash* (composed of tiny flakes of glass and pulverized rock), and marble-sized rock fragments. The cloud spread over Pompeii, turning day into night. Choking fumes filled the air, ash sifted down like heavy snow, and rock fragments fell like hail. Panic ensued as inhabitants rushed to escape, but sadly, for most it was too late. Vesuvius suddenly exploded, blasting out immensely more hot gases and debris. Much of this material swirled high into the atmosphere, but some rushed downslope in a fiery-hot avalanche that swept over Pompeii, wiping the town off the map.

The blanket of volcanic debris that buried Pompeii protected the town's ruins so well that 18 centuries later, archaeological excavations revealed an amazingly complete record of daily life in the Roman Empire (**FIG. 4.1b**). Occasionally, workers found enigmatic open spaces in the debris while digging. These remained a puzzle until someone thought to fill one with plaster before removing the surrounding debris. The spaces turned out to be molds of Pompeii's ill-fated inhabitants, twisted in agony or huddled in despair, at their moment of death (**FIG. 4.1c**).

On that fateful day in 79 C.E., Vesuvius had undergone a **volcanic eruption**, an event during which *melt* or *molten rock*—a sticky material derived by transformation of solid rock into liquid at very high temperatures—flows or sprays out of a vent, or solid debris blasts out of a vent. Ancient Romans thought that volcanic eruptions happened when Vulcan, the god of fire, fueled his subterranean forges to manufacture swords for other gods. No one believes that myth anymore, but Vulcan's name has been immortalized as the Latin root of the English word *volcano*. Modern research

◀ Lava erupting from Mt. Etna, Sicily, occasionally flows down the flanks of the volcano and overruns homes and farmland in communities established at its base.

FIGURE 4.1 The eruption of Vesuvius and destruction of Pompeii.

(a) A painter's visualization of the 79 C.E. eruption.

You can see ruts carved into streets by chariots.

(b) Excavation of Pompeii has provided an image of daily life almost 2,000 years ago.

(c) Plaster casts of victims that were buried in the ash.

demonstrates that volcanic eruptions are a manifestation of **igneous activity** (from the Latin word *ignis*, meaning fire), the general name that geologists use for the production of melt underground and its migration upward, followed by either the solidification of the melt underground or its eruption from a volcano and solidification at the ground surface. Geologists refer to molten rock underground as **magma**, and to molten rock that has erupted at the surface as **lava**.

Clearly, as Pompeiians would attest if they could, volcanoes are natural hazards, and volcanic eruptions can be devastating natural disasters. In the past two centuries alone, more than a quarter of a million people have died as a direct consequence of volcanic eruptions. Vastly more die when an eruption triggers famine, disease, and mass migration. Even a relatively small eruption can shut down air traffic on major routes, significantly disrupting global trade. To understand volcanic eruptions and their consequences, we begin this chapter by discussing why rocks melt inside the Earth to produce magma in the first place (in the context of plate tectonics theory) and by describing the products of melting. Next, we focus on volcanic eruptions themselves and distinguish among different kinds, or *styles*, of eruptions. With this background, we can describe the primary and secondary disasters resulting from eruptions and the ways in which they can affect human society. We conclude the chapter by discussing how society can mitigate volcanic hazards.

4.2 Melting Rock

Causes of Melting in the Earth

Popular media sometimes give the impression that the mantle beneath the Earth's crust consists entirely of molten rock, and that volcanoes simply erupt wherever a conduit connects this imaginary "underground magma sea" to the Earth's surface. In fact, our planet's crust and mantle both consist almost entirely of solid rock. Melting of pre-existing rock to produce magma takes place only in certain locations in the upper mantle and the crust, in response to local changes in temperature, pressure, or chemical composition. Let's look at each of these causes of melting individually.

- *Melting due to a decrease in pressure:* When hot rock from deep in the mantle (where pressures are higher) rises to shallower depths (where pressures are lower) without cooling significantly, it undergoes **decompression**

FIGURE 4.2 Decompression melting.

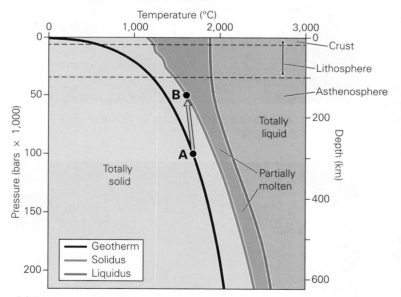

(a) Decompression melting takes place when the pressure acting on hot rock decreases. As this graph shows, when mantle rock rises from point A to point B, the pressure decreases a lot, but the rock cools only a little, so the rock begins to melt. The solidus line indicates conditions at which melt begins to form, and the liquidus line indicates conditions at which the rock melts completely. The geotherm indicates the change in temperature with depth in the Earth.

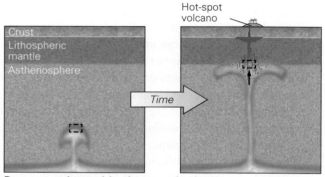

Decompression melting in a mantle plume

Decompression melting beneath a rift

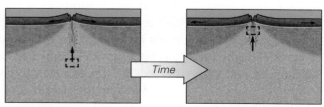

Decompression melting beneath a mid-ocean ridge

(b) The conditions leading to decompression melting occur in several geologic environments. In each case, a volume of hot mantle rock (outlined by dashed lines) rises to a shallower depth, and magma (red areas) forms.

melting because at lower pressures, bonds between atoms break more easily (**FIG. 4.2a**). Decompression melting can take place where a mantle plume rises at a hot spot, beneath a rift, or beneath a mid-ocean ridge (**FIG. 4.2b**).

- *Melting due to the addition of volatiles:* Geologists refer to compounds such as H_2O and CO_2 that can evaporate relatively easily as **volatiles**. When volatiles seep into very hot mantle rock, the rock begins to melt, a process called **flux melting** (**FIG. 4.3a**). Flux melting takes place in the mantle directly above a subducting plate, for as the oceanic crust gets carried down into the hot asthenosphere, it warms up and releases volatiles.

- *Melting due to heat transfer:* Very hot magma rising from the mantle into the crust can provide enough heat to melt surrounding crustal rocks, a process called **heat-transfer melting** (**FIG. 4.3b**). This can happen where magma rises into continental crust because the rocks in continental crust melt at a relatively low temperature. (To picture the process, imagine injecting hot fudge into a scoop of ice cream.)

FIGURE 4.3 Flux melting and heat-transfer melting.

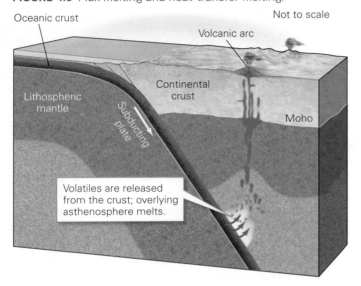

(a) Flux melting takes place at a convergent boundary as the downgoing plate sinks, heats up, and releases volatiles, which cause melting in the overlying asthenosphere.

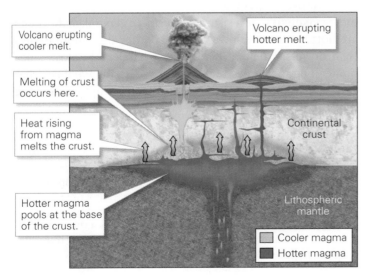

(b) Heat-transfer melting takes place where very hot magma rising from the mantle brings enough heat with it to melt continental crust.

What Are Magma and Lava Made of?

Unlike water, which contains a single compound (H_2O), molten rock contains many different compounds. Geologists represent these chemicals as *oxides*, meaning molecules consisting of metal atoms bonded to oxygen (**TABLE 4.1**). Different melts contain different proportions of these oxides. Geologists distinguish among four types of melts—*felsic*, *intermediate*, *mafic*, and *ultramafic*—by the proportion of silica relative to iron and magnesium oxide in the melt (**TABLE 4.2**). Felsic melts contain the greatest amount of silica, and ultramafic ones contain the least. The specific composition of a melt depends on many factors:

- *Source-rock composition:* The original composition of the source rock that starts to melt determines which chemicals are available to go into a melt. For example, if a source rock contains a lot of silica, the melt can be rich in silica.

- *Partial melting:* Not all minerals melt at the same temperature, so when a source rock begins to melt, some chemicals go into the melt and some remain in solid minerals that haven't yet melted. Therefore, if magma forms

TABLE 4.1 Principal Oxide Compounds in Magma

Oxide	Chemical formula	Informal name
Silicon oxide	SiO_2	Silica
Aluminum oxide	Al_2O_3	Alumina
Potassium oxide	K_2O	—
Iron oxide	FeO or Fe_2O_3	Total iron
Sodium oxide	Na_2O	Soda
Calcium oxide	CaO	Lime
Magnesium oxide	MgO	Magnesia
Titanium oxide	TiO_2	Titania

TABLE 4.2 The Four Major Types of Magma

Type	Ratio of silica to iron and magnesium oxides	% silica	Melt temperature	Viscosity
Felsic	High	66–76	700°C	High
Intermediate	↓	52–66	900°C	↑
Mafic		45–52	1,100°C	
Ultramafic	Low	38–45	1,300°C	Low

FIGURE 4.4 The cloud of material rising from this Alaskan volcano contains various gases.

by *partial melting*—meaning that only part of the source rock melts, and the melt seeps away before the source rock has melted entirely—its composition will be different from that of the source rock. Minerals containing more silica tend to melt at lower temperatures than do those containing more iron and magnesium oxides, so magma formed by partial melting is less mafic than the source rock from which it was extracted. For example, partial melting of ultramafic rock produces a mafic magma.

- *Fractional crystallization*: When magma starts to solidify, mineral crystals grow in a sequence: those containing more iron and magnesium oxides grow first, and those containing more silica grow later. Early-formed minerals, therefore, preferentially extract iron and magnesium oxides from magma. Due to such *fractional crystallization*, the composition of the remaining magma becomes progressively richer in silica as solidification takes place. Consequently, a rock formed by solidification of magma that has undergone fractional crystallization will be more felsic than one formed from magma that has not.

- *Assimilation of wall rock*: When very hot magma enters cracks or pores in solid rock, it may absorb or dissolve chemicals from the surrounding solid rock, or *wall rock*. Such *assimilation* can change the magma's composition before it eventually solidifies.

In addition to the oxides listed in Table 4.1, magma contains volatiles. Some of these, as we've seen, are added to the source rock during the process of melting. Others come from the source rock itself (for volatile compounds can be bonded to solid crystals and detach from those crystals when the rock melts), and still more can be absorbed from the wall rock that magma passes through. Under the high pressures at the depths where melting takes place, molecules of volatiles remain dissolved in magma, just as gas remains dissolved in an unopened can of cola. But when magma approaches to within about 5 km (3 mi) of the Earth's surface, pressure decreases sufficiently for volatiles to come out of solution and form bubbles, just like the bubbles that form when you open a can of cola. Some of these bubbles remain trapped in igneous rock that solidifies from a melt, whereas others escape through a volcano's vent (FIG. 4.4). Most gas in magma consists of H_2O. But some gases—including H_2O, CO_2, sulfur dioxide (SO_2), and hydrogen sulfide (H_2S)—pose a direct hazard to people that inhale them. In addition, when they get trapped in high-viscosity lava, they can contribute to making a volcano explosive.

Why discuss melt composition in a chapter on volcanic disasters? The silica and gas content of molten rock affects the temperature at which a melt remains liquid, and it affects a melt's **viscosity**, or resistance to flow. In general, mafic melts are hotter and have lower viscosity (they flow more easily) than do felsic melts. This contrast happens because silica builds long molecules that can't move easily. We'll see that this contrast controls the style of volcanic eruptions and the types of disasters that result.

Take-home message...
The Earth's crust and mantle remain solid except in specific locations where rock undergoes melting. Melting can be triggered by a decrease in pressure, addition of volatiles, or injection of hot magma from greater depth. Geologists classify magma by the proportion of silica that it contains. Magma composition varies for a variety of reasons. Magma can contain dissolved gases.

QUICK QUESTION How does the composition of magma affect its viscosity?

4.3 Forming Igneous Rocks

The Rise of Magma

When magma forms, it tends to rise. Why? First, magma is buoyant relative to the surrounding solid rock because rock expands and becomes less dense when it transforms into liquid. Second, the high pressure in the Earth's interior squeezes magma upward. Magma moves upward initially by seeping along grain boundaries and into cracks (**FIG. 4.5a**). As it rises, it may collect in and pass upward through larger conduits. Generally, rising magma accumulates, at least temporarily, in a **magma chamber** (**FIG. 4.5b**), an underground space containing a *crystal mush* composed of melt and dispersed mineral grains. Magma chambers typically develop at depths of 6–10 km (4–6 mi), though some may be as shallow as 3 km (2 mi) and others as deep as 50 km (31 mi). Sometimes magma solidifies within a magma chamber. In other cases, it rises to higher levels in the crust, or even to the Earth's surface, where it erupts as lava at a volcano.

How Do Melts Become Solid?

The Earth's temperature decreases toward the surface, so as magma rises, it enters a cooler environment. As it does so, it releases heat to its surroundings, for heat always flows from hotter to colder material. The loss of heat causes atoms or molecules in a melt to slow down, so that they can attach to one another to produce a solid. Solidification doesn't happen all at once, however, because not all components of a magma or lava solidify at the same temperature. (Technically, solidification begins at a temperature known as the *liquidus*, and becomes complete at a temperature called the *solidus*; see Fig. 4.2.) In fact, solidification of a melt may take place over a range of about 200°C. In contrast, water, since it only contains one chemical, solidifies into ice entirely at one temperature at a given pressure.

The specific temperature at which magma or lava starts to solidify depends on the chemical composition of the molten rock, as well as on the amount of water present and on the depth at which solidification takes place. Mafic magmas start to solidify at about 1,200°C (2,192°F), whereas felsic melts start to solidify at around 700°C (1,300°F). These temperatures are much higher than those in your home oven, which can reach a temperature of only 230°C (450°F).

FIGURE 4.5 Melt formation and migration into a magma chamber.

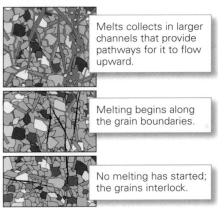

(a) Melt starts to form, and then rises, accumulating in channels as it does.

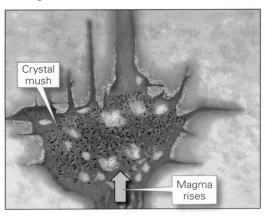

(b) Eventually, magma reaches a level where it accumulates, along with crystals, to form a crystal mush in a magma chamber. Magma in the chamber may interact with the wall rock.

The time it takes for a melt to solidify depends on how fast heat transfers from the melt into its surroundings. For example, magma that cools deep in the crust, surrounded by warm rock, cools slowly—like coffee in a thermos bottle—because the surrounding rock serves as a good insulator. It may take a million years for a large blob of such magma to solidify. In contrast, lava flowing onto the cold surface of the Earth may solidify in hours or days, since it's not well insulated. If lava comes into contact with water, it cools especially fast, because water removes heat very efficiently. Since a body of melt loses heat to its surroundings only at its surface, bodies with a large surface area per unit volume cool faster than do those with a smaller surface area per unit volume. Consequently, a thin layer or small drops of melt cool faster than a large blob of the same volume (**FIG. 4.6**).

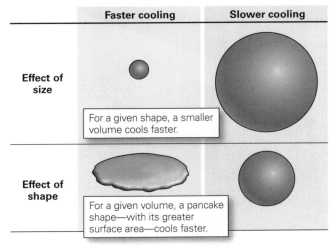

FIGURE 4.6 The cooling rate of a molten mass depends on its size and shape.

FIGURE 4.7 The difference between the intrusive and extrusive environments of igneous activity.

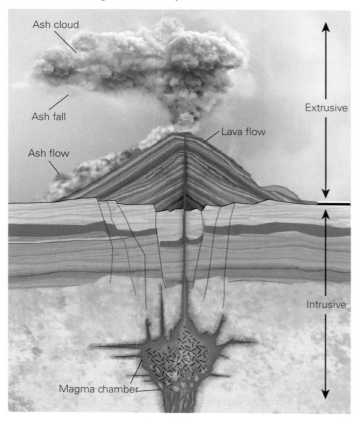

The Variety of Igneous Rocks

Though the picture of an erupting volcano may come to mind when someone mentions igneous activity, most igneous rock actually solidifies underground **(FIG. 4.7)**. Geologists refer to all igneous rock formed underground as **intrusive igneous rock**, because it intruded, or was injected into, pre-existing rock while still molten, and use different names for different shapes and sizes of intrusions. Blob-shaped intrusions are *plutons* **(FIG. 4.8a)**, wall-shaped intrusions (formed where magma solidifies in large cracks) are *dikes*, and table-top-shaped intrusions (formed where magma squeezes between pre-existing layers) are *sills* **(FIG. 4.8b)**. All igneous rock formed from melt that has come out of a volcano is called **extrusive igneous rock**. Extrusive igneous rock includes both **lava flows** (masses of lava that flow onto the Earth's surface) and **pyroclastic debris** (ash and other fragments blasted out of a volcano) **(FIG. 4.8c, d)**.

The texture of igneous rock depends, in part, on the rate at which it cools. Simplistically, melt that solidifies quickly, before the ions or molecules organize into an orderly arrangement, yields a **glassy texture**, whereas melt that solidifies more slowly, so that ions or molecules have time to arrange themselves into mineral crystals, yields a **crystalline texture**. The grains in rocks with a crystalline texture interlock like pieces in a jigsaw puzzle. The size of crystals generally reflects the cooling rate: rocks with

FIGURE 4.8 Examples of intrusive and extrusive igneous rocks.

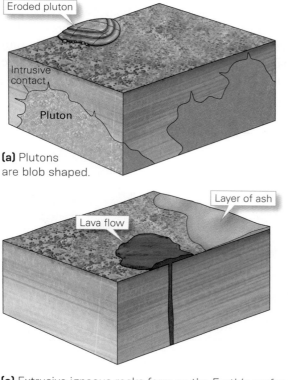

(a) Plutons are blob shaped.

(b) Dikes are wall shaped and cut across bedding. Silla are tabular and are parallel to bedding.

(c) Extrusive igneous rocks form on the Earth's surface.

(d) Extrusive rocks include lava flows and beds of pyroclastic debris.

finer grains have cooled more quickly than have those with coarser grains. Igneous rocks with a **fragmental texture** form either when pyroclastic debris welds together while still extremely hot, or when cool pyroclastic debris accumulates and later undergoes compaction and cementation. Note that the types of igneous rocks that form reflect the environment of cooling: intrusive igneous rocks are crystalline (with the coarsest grains generally found in plutons), whereas extrusive igneous rocks include crystalline, glassy, and fragmental textures.

Geologists classify igneous rocks by their composition and texture. To characterize igneous rock composition, we use the same adjectives that we used when describing magmas: felsic, intermediate, mafic, and ultramafic rocks. We characterize igneous rock texture with the adjectives *crystalline*, *glassy*, or *fragmental*. **Figure 4.9** illustrates and names six major types of crystalline igneous rocks. Note that color can help us identify these rocks: felsic rocks tend to have lighter colors, while mafic and ultramafic rocks tend to be darker. Glassy igneous rocks include *obsidian*, a mass of felsic glass; **pumice**, a rock composed of tiny *vesicles* (trapped gas bubbles) surrounded by screens of glass; and *scoria*, a mafic volcanic rock that contains abundant vesicles (**FIG. 4.10a**). Fragmental igneous rocks include

FIGURE 4.9 Examples of common crystalline igneous rocks. Grain size typically reflects cooling rate: fine-grained igneous rocks generally cool rapidly, while coarse-grained igneous rocks tend to cool more slowly.

FIGURE 4.10 Examples of glassy and fragmental igneous rocks.

(a) Glassy igneous rocks do not contain mineral crystals. Pumice may be light enough to float.

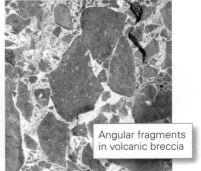

(b) Fragmental igneous rocks can contain fragments of various sizes. Tuff consists mostly of ash.

4.3 Forming Igneous Rocks • **129**

tuff, a fine-grained rock composed mostly of volcanic ash; *volcanic agglomerate*, composed mostly of rounded fragments; and *volcanic breccia*, composed of angular chunks (FIG. 4.10b).

> **Take-home message...**
>
> Magma rises because it is relatively buoyant and because it's under pressure. When it enters a cooler environment, it solidifies into igneous rock. The rate of cooling, which depends on the cooling environment and on the size and shape of the melt body, affects a rock's texture. Intrusive igneous rocks form underground, whereas extrusive igneous rocks form at the Earth's surface. Geologists classify igneous rocks by their composition and texture.
>
> **QUICK QUESTION** Distinguish among crystalline, glassy, and fragmental igneous rock textures.

4.4 Where Does Igneous Activity Take Place?

Igneous activity plays a critical role in the Earth System, for it serves as the only means of transferring material from the Earth's interior to the crust, hydrosphere, and atmosphere. Why does melting happen where it does? The theory of plate tectonics provides some answers (FIG. 4.11).

Igneous Activity at Divergent Boundaries

During seafloor spreading at a mid-ocean ridge, lithosphere plates move apart. Very hot *peridotite* (the coarse-grained ultramafic rock making up the mantle) rises from the asthenosphere to shallower depths, constantly replenishing the

FIGURE 4.11 A map of the Earth's volcanoes, and the geologic settings in which they form.

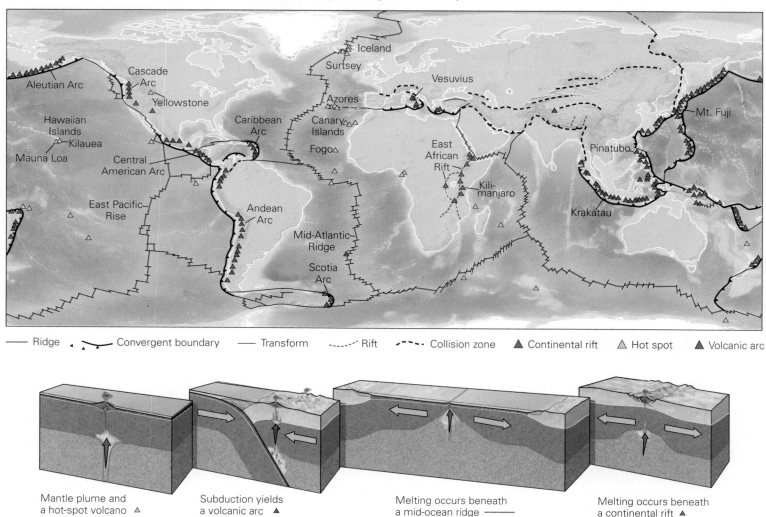

130 • CHAPTER 4 • The Wrath of Vulcan: Volcanic Eruptions

region beneath the ridge with new rock. (Recall from Chapter 2 that even though it's solid, this peridotite is soft enough to flow plastically.) When the peridotite rises above a depth of about 150 km (93 mi), it undergoes partial melting due to decompression. During this partial melting, some of the iron and magnesium oxides of the peridotite remains behind in solid minerals, so the rising magma has a mafic, rather than an ultramafic, composition. The mafic melt accumulates in a magma chamber beneath the ridge axis. Slow solidification along the margins of this chamber produces *gabbro* (a coarse-grained mafic rock), which forms the lower part of the oceanic crust. Some mafic magma continues its upward movement, filling vertical cracks and solidifying to form dikes, and some reaches the seafloor, where it erupts and cools rapidly to form piles of glass-encrusted blobs called *pillows* **(FIG. 4.12a)**. Rock that solidifies in dikes, or from lava that spills onto the seafloor, consists of *basalt* (a fine-grained mafic rock).

Volcanic eruptions at mid-ocean ridges take place at vents along open vertical cracks, or **fissures**, parallel to the ridge axis. Seawater can percolate down into the seafloor through these fissures. When this water approaches the magma chamber and the new dikes, it warms and rises. As this convective flow takes place, the hot water reacts with minerals in the crust—some of the ions in the crust dissolve in the hot water, and some of the water molecules become incorporated into the remaining minerals. After it passes through the crust, the hot and now mineralized water spurts out at *hydrothermal vents*. The emerging water is instantly cooled by mixing with the surrounding seawater. In this cooler water, tiny mineral grains precipitate, producing a dark cloud. Consequently, the hydrothermal vents are known as **black smokers (FIG. 4.12b)**.

Though most of the world's volcanism takes place along mid-ocean ridges—enough to have completely resurfaced 70% of the Earth's surface in less than 200 million years—we don't generally see the submarine volcanoes of mid-ocean ridges because most lie beneath 2 km (1 mi) of seawater. Consequently, this igneous activity doesn't pose much of a risk to society.

Igneous Activity at Convergent Boundaries

Most **subaerial volcanoes**—those that erupt into the air—occur along convergent boundaries. In fact, the convergent boundaries surrounding the Pacific Ocean host so many subaerial volcanoes that this ocean margin has come to be known as the **Ring of Fire**. The magma that feeds convergent-boundary volcanoes forms because the crust of the downgoing oceanic plate carries volatile compounds along with it as it sinks into the mantle. Some of these volatiles were trapped in sediment on the surface of the crust, and some were attached to minerals in basalt when the minerals were altered by interaction with hot seawater at mid-ocean ridges. At a depth of about 150 km, the surface of the subducted plate becomes hot enough for volatile molecules to separate from minerals and rise into the overlying asthenosphere, where they cause flux melting. Generally, the downgoing plate itself doesn't melt significantly.

As was the case with magma generated at mid-ocean ridges, only part of the peridotite in the mantle above a downgoing plate melts. Iron and magnesium oxides are left behind, so the resulting magma has a mafic composition. Some of this magma seeps up through the lithosphere of the overriding plate to the Earth's surface and erupts to form basalt. But some of it accumulates in magma chambers within the crust. Heat-transfer melting, along with fractional crystallization and assimilation, produces magma of a felsic to intermediate composition, especially where this igneous activity happens in continents. Melt that solidifies underground in these magma chambers forms plutons

FIGURE 4.12 Igneous activity along a mid-ocean ridge.

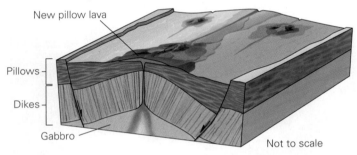

(a) At a mid-ocean ridge, some magma solidifies under the seafloor as gabbro or in dikes as basalt, and some erupts on the seafloor, forming pillows.

(b) A hydrothermal vent (black smoker) along a mid-ocean ridge.

composed of granite and similar rock. Melt that rises from these magma chambers to the Earth's surface erupts to form *andesite*.

Eruptions at a convergent boundary build a chain of volcanoes called a **volcanic arc**. If the chain grows on oceanic lithosphere, it becomes an *island arc* **(FIG. 4.13a)**, whereas if it grows on continental lithosphere, it becomes a *continental arc* **(FIG. 4.13b)**. Geologists use the word *arc* in this context because many such chains have a curving trace on a map.

Igneous Activity in Continental Rifts

During rifting, continental lithosphere stretches horizontally and thins vertically. Thinning of lithosphere causes decompression melting of the underlying asthenosphere, which produces mafic magma. Some of this magma rises straight to the Earth's surface and erupts to form basalt, but some incorporates melt produced by heat-transfer melting within the crust, or undergoes fractional crystallization and assimilation, and erupts to form lava (*rhyolite*) and felsic pyroclastic debris **(FIG. 4.14)**.

FIGURE 4.13 Igneous activity at convergent boundaries.

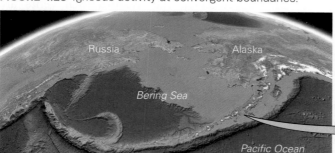

(a) An island arc, the Aleutian Arc, forms where the Pacific Plate subducts beneath Alaska.

(b) A continental arc, the Andean Arc, forms where the Pacific Plate subducts beneath western South America.

FIGURE 4.14 Igneous activity in a continental rift.

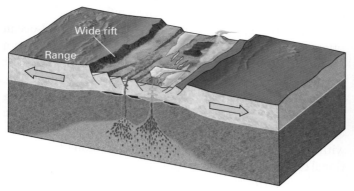

(a) Stretching of the lithosphere in a rift causes decompression melting. Both mafic and intermediate lava can be extruded.

(b) An aerial photo of the East African Rift shows several faults caused by stretching. Volcanoes rise in the distance.

Igneous Activity at Hot Spots

According to the mantle-plume hypothesis (see Chapter 2), hot spots form where a column of very hot peridotite rises from deeper in the mantle up to the base of the lithosphere. When it rises above a depth of about 150 km, this rock undergoes decompression melting to yield mafic magma. When a hot-spot volcano forms on oceanic lithosphere, it can build an oceanic island composed of mafic lava and mafic pyroclastic debris. The Big Island of Hawai'i grew this way (**FIG. 4.15a**). As plates move, a hot-spot volcano is eventually carried off the hot spot. When this happens, the volcano stops erupting, and a new volcano forms above the hot spot. This process repeats over time, producing a chain of former volcanoes called a *hot-spot track*. In the case of Hawai'i, the track includes the other Hawaiian islands as well as a chain of seamounts to the northwest (**FIG. 4.15b**). Hot spots beneath continents produce a greater variety of lava types because magma rising from the mantle changes in composition as it passes through, and melts, continental crust. For example, the Yellowstone hot

FIGURE 4.15 Igneous activity due to hot spots.

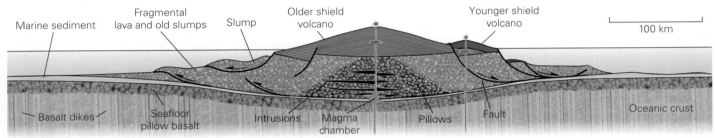

(a) The subaerial volcanoes of Hawai'i grew on top of older seafloor.

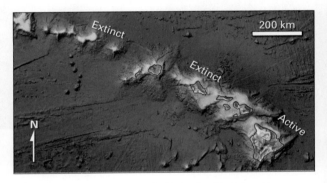

(b) Hawaiian hot-spot track. Only the Big Island has erupted recently. The other Hawaiian islands, and the seamounts to the northwest, have not erupted recently and will not erupt again, so they are extinct.

(c) Yellow tuff deposits at Yellowstone.

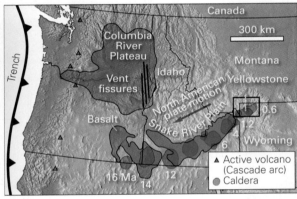

(d) The Yellowstone hot-spot track. Calderas of the Snake River Plain get older to the west (labels indicate age).

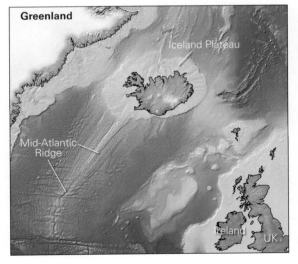

(e) Iceland straddles the Mid-Atlantic Ridge, forming a plateau.

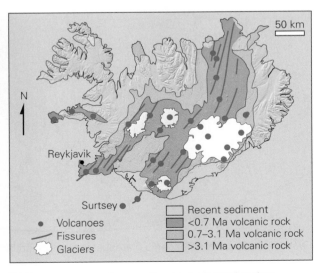

(f) Because spreading continues during volcanism, a rift traverses the island.

spot, which now underlies Yellowstone National Park, has produced not only mafic lava, but also thick layers of felsic tuff (the "yellow stone" of the park; **FIG. 4.15c**). Note that the Yellowstone hot-spot track is marked by a succession of former eruptive centers under what is now the Snake River Plain of southern Idaho (**FIG. 4.15d**).

The two examples of hot-spot volcanoes that we've mentioned (Hawai'i and Yellowstone) are in plate interiors, far from any plate boundary. But not all oceanic hot-spot volcanoes occur in the middle of a plate. Iceland, for example, formed over a hot spot under the axis of the Mid-Atlantic Ridge. Because of this hot spot, far more magma erupts in Iceland than at other places along the ridge, and these eruptions have built a broad plateau of basalt (**FIG. 4.15e**). Because Iceland straddles a divergent boundary, plate motion is actively stretching the island apart (**FIG. 4.15f**).

Large Igneous Provinces

In several locations around the world, particularly large volumes of mafic lava have erupted over a relatively short time and spread out in vast flows, some of which extend 500 km (311 mi) from the vent. Geologists refer to the rocks formed from these flows as **flood basalts** (**FIG. 4.16a**). Successive flood-basalt eruptions build a broad *basalt plateau* whose total volume may be immense (**FIG. 4.16b**). Such a region is known as a **large igneous province** (**LIP**) (**FIG. 4.16c**).

FIGURE 4.16 Flood basalts and large igneous provinces (LIPs).

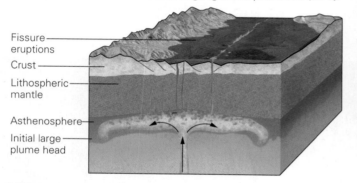

(a) Flood basalts may form when a rift overlies a mantle-plume head.

(b) Successive layers of flood basalt formed the Columbia River Plateau.

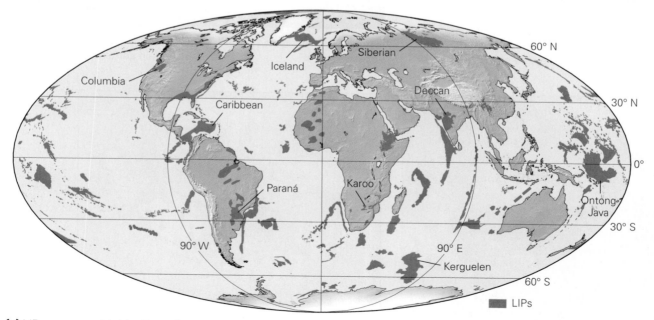

(c) LIPs occur worldwide. Some form oceanic plateaus; others are on land and are related to rifting.

What causes flood-basalt eruptions? According to one hypothesis, flood basalts form when a large mantle plume first arrives at the base of the lithosphere. Melting in the bulbous head of the plume produces a particularly large volume of magma. Faults formed by rifting in the crust over the plume provide avenues for the melt to reach the Earth's surface. No LIPs are erupting today—if one were, its eruptions would wreak havoc globally.

Take-home message . . .

Plate tectonics explains why igneous activity occurs where it does. Decompression melting takes place at hot spots, mid-ocean ridges, and rifts. Flux melting produces melts at convergent boundaries. Heat-transfer melting also takes place at rifts, hot spots, and convergent boundaries. At some locations, immense amounts of lava erupt to form large igneous provinces.

QUICK QUESTION Where do subaerial volcanoes occur—a convergent boundary or a divergent boundary?

4.5 Building a Volcano

Lava Flows, Lakes, and Fountains

In 1707, the town of Garachico, on Tenerife in the Canary Islands, was a prosperous port servicing ships from Europe and the Americas. That changed when the Trevajo volcano, only 8 km (5 mi) away, suddenly erupted. People startled by the rumble of the eruption turned toward the volcano and saw seven streams of bright red lava flowing toward the town. Everyone managed to escape alive, but much of the port was covered by the new igneous rock that formed when the lava finally stopped flowing and solidified.

Geologists use the term *lava flow* both for molten lava moving over the Earth's surface and for the layer of igneous rock formed when lava solidifies. The characteristics of a given flow reflect both the composition and the viscosity of the lava. For example, relatively cool lava with a felsic composition has such high viscosity that it builds into a bulbous mound called a **lava dome** that cannot flow very far from the vent (FIG. 4.17a). Intermediate lava extrudes as short, blocky flows or, on a steep slope, as a jumble of hot blocks that tumble downslope (FIG. 4.17b). Hot mafic lava has such low viscosity that on steep slopes near its source, it rushes along at speeds up to 30 km/h (19 mph) in rivers up to a couple of hundred meters across, 20 m (66 ft) thick, and many kilometers long (FIG. 4.17c). Cooling eventually causes the surface of a mafic flow to crust over, but the internal part of the flow can remain molten. This lava can continue moving through tunnel-like conduits called **lava tubes**, which have diameters ranging from 1 to 15 m (3–49 ft). Lava may eventually drain out of a lava tube, so that it remains as an empty tunnel after the flow has completely solidified.

FIGURE 4.17 The manner in which lava flows depends on the lava's viscosity.

(a) Felsic lava has high viscosity and may build into a lava dome.

(b) Intermediate lava flows slowly down a volcano's flanks. It may form glowing blocks that tumble downslope.

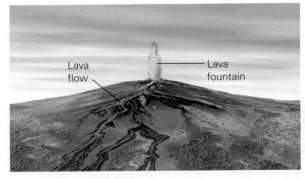

(c) Mafic lava has relatively low viscosity and can flow far and fast.

During eruptions of mafic melt, lava may initially collect around the vent to form a **lava lake**, a pool of lava tens to hundreds of meters deep (FIG. 4.18a). A skin-like crust of rock forms on the surface of the lava lake, but because of convective circulation in the lake, this skin stretches and cracks so that the red lava below peeks through. Also, slabs of the skin may sink back into the lava. In some cases, erupting mafic lava squirts upward into the air to produce a **lava fountain** (FIG. 4.18b). These fountains form when particularly high pressure develops in the lava, either due to expansion of gas bubbles or where lava rises through an upwardly narrowing conduit. (You can mimic the latter phenomenon by constricting the end of a water hose.)

The surface texture of a mafic lava flow depends on the volume and rate of the flow. Slower-moving, low-volume flows have soft, pasty surfaces and wrinkle into smooth, glassy, rope-like ridges—geologists refer to such flows by their Polynesian name, **pāhoehoe** (pronounced pa-hoy-hoy) (FIG. 4.19a, b). Flows containing a larger volume of faster-moving lava crust over while the flow is still moving, so the solidified crust breaks into a jumble of jagged fragments, forming a rubbly flow also known by its Polynesian name, **'a'ā** (pronounced ah-ah) (FIG. 4.19c, d). Mafic flows that erupt underwater look very different from those that erupt on land because the lava cools particularly quickly when in contact with water. Commonly, submarine mafic lava extrudes to form pillows (FIG. 4.20). Alternatively, the sudden cooling of lava may cause it to shatter into glassy chunks.

Pyroclastic Debris

For much of February 1943, small earthquakes had caused the ground to tremble frequently in a portion of the Central American volcanic arc west of Mexico City. As Dionisio Pulido, a local farmer, walked to his field to burn a small pile of debris, he came upon a fissure 2 m (6 ft) wide that looked like a tear in the ground. Suddenly, a strong earthquake jolted the ground, and the surface of the field around the fissure bulged upward. Sulfurous smoke, as well as fragments of rock, spewed into the air, and Dionisio fled. By the following morning, the field lay buried beneath a mound of gray-black debris 40 m (131 ft) high. Dionisio had witnessed the birth of a new volcano, named Paricutín. Within a week, the mound was 150 m (492 ft) high, and within 2 years, debris from Paricutín had nearly buried two nearby towns (FIG. 4.21).

As we've seen, geologists refer to fragments ejected during a volcanic eruption like that of Paricutín as *pyroclastic debris* (from the Greek word *pyro*, meaning fire; such material can also be called **tephra**). Pyroclastic debris can form by solidification of drops or clots of lava in air, by ejection of solidified or partially solidified lava from a volcano's vent, or by the breakup and ejection of pre-existing igneous rock making

FIGURE 4.18 Lava lakes and lava fountains.

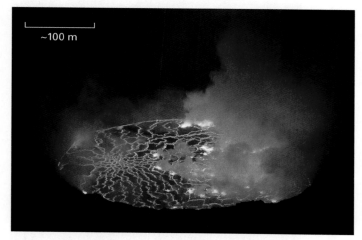

(a) A lava lake typically has a crust that is broken into slabs, with melt visible in between. Here, lava spurts upward in several locations.

(b) During the 2018 eruption on Hawai'i, a lava fountain rose out of a fissure cutting across a housing development and fed a lava flow.

FIGURE 4.19 Types of mafic lava surfaces.

(a) Pāhoehoe forms when a relatively small amount of lava flows slowly.

(b) The surface of pāhoehoe looks like coiled ropes.

(c) ʻAʻā forms when a voluminous flow continues to flow after its surface solidifies.

(d) The surface of ʻaʻā is very rough, and could slash the soles of your feet.

FIGURE 4.20 Pillow basalt forms when mafic lava slowly extrudes underwater.

FIGURE 4.21 Paricutín erupting in 1943.

4.5 Building a Volcano • **137**

up the volcano. Different names apply to different sizes of pyroclastic debris. **Ash** consists of tiny flakes or slivers of glass or pulverized rock **(FIG. 4.22a)**. **Lapilli** are marble- to golf-ball-sized pieces that may be composed of blasted-apart pumice and other rock **(FIG. 4.22b)**, or of ash that sticks together like little snowballs when mixed with water in clouds above a volcano. Lapilli formed from solidified clots in a lava fountain may be called **cinders (FIG. 4.22c)**. Larger, angular chunks are **blocks**, and blocks that were still soft during eruption, so that they became streamlined as they fell, are **volcanic bombs (FIG. 4.22d)**.

Volcanic Edifices

What's inside a volcano? Over time, solidified lava or pyroclastic debris may build up around the vent to produce a hill or mountain, a landform known as a **volcanic edifice**. The peak of the edifice is the volcano's *summit*, and the side slopes are its *flanks* **(FIG. 4.23)**.

A magma chamber may lie within or beneath the edifice. Magma and gases rise from the chamber through a conduit to the vent, where they erupt as lava or pyroclastic debris. In some volcanoes, the conduit through which lava and gases reach the Earth's surface has a chimney-like shape, and is topped by a circular depression, or **crater**, that resembles a bowl **(FIG. 4.24a)**. The top part of the chimney beneath the conduit is sometimes called the volcano's *throat*. Craters, which may be up to about 500 m (1,640 ft) across and about 200 m (656 ft) deep, develop in two ways: some form during eruptions when pyroclastic debris builds up around the vent, whereas others form after an eruption when the volcano's summit collapses into the drained conduit. Though eruptions commonly take place from a *summit vent*, not all do—some eruptions come from *flank vents*. Nor do all eruptions come

FIGURE 4.22 Examples of pyroclastic debris.

(a) Clouds of ash rising from an erupting volcano. The inset is a photomicrograph of ash.

(b) Lapilli made of pumice fragments.

(c) Cinders are lapilli formed from clots of lava.

(d) Blocks and bombs that have fallen on a bed of lapilli. The inset is a close-up of a volcanic bomb.

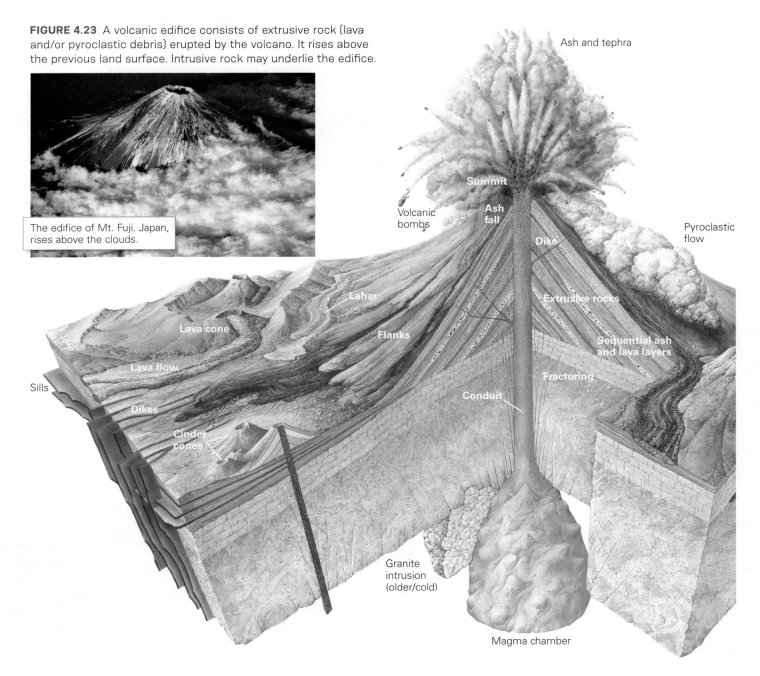

FIGURE 4.23 A volcanic edifice consists of extrusive rock (lava and/or pyroclastic debris) erupted by the volcano. It rises above the previous land surface. Intrusive rock may underlie the edifice.

The edifice of Mt. Fuji, Japan, rises above the clouds.

from a circular crater. As we noted earlier, during **fissure eruptions**, lava spews from a long crack, or *fissure*. Some fissure eruptions produce a *curtain* of lava, an elongate lava fountain that may extend for kilometers, whereas others take place at several unconnected vents aligned along the fissure. Some fissures radiate outward from the summit of a large volcano, whereas others align parallel to faults (FIG. 4.24b). When magma within a fissure solidifies, it becomes a dike. The overall shapes and sizes of volcanic edifices vary significantly, but geologists group them into three broad categories.

Shield volcanoes, so named because they resemble a soldier's shield lying on the ground, are broad, gentle domes (FIG. 4.25a). Volcanoes develop this shape when they erupt low-viscosity mafic lava that flows for a long distance in a thin sheet before solidifying. Far from the vent, the volcanic edifice consists almost exclusively of lava flows, layered one on top of the other (FIG. 4.25b), but closer to the vent, mafic pyroclastic debris may occur between the flows (FIG. 4.25c). A **caldera**—a large circular or elliptical depression, significantly bigger than a crater—may develop at the summit of a shield volcano. The caldera of Kilauea, a shield volcano on the Big Island of Hawai'i, is a few kilometers across and several hundred meters deep. A lava lake may occupy all or part of the caldera at times. Calderas on shield volcanoes develop due to collapse of the summit following withdrawal or eruption of magma from below (FIG. 4.25d).

4.5 Building a Volcano

FIGURE 4.24 Contrasting summit and fissure eruptions.

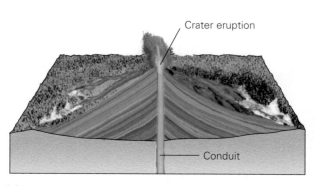

(a) During a summit eruption, lava and pyroclastic debris come out of a summit crater.

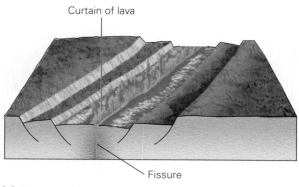

(b) During a fissure eruption, lava and pyroclastic debris come out of a long crack.

FIGURE 4.25 Architecture of a shield volcano.

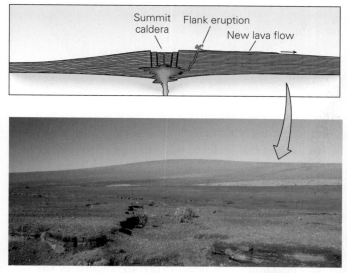

(a) A shield volcano has gentle slopes and consists mostly of mafic flows.

(b) Cross section of eroded shield volcano on Kaua'i. Note the thin lava layers.

(c) Layers of pyroclastic debris near the caldera of Kilauea.

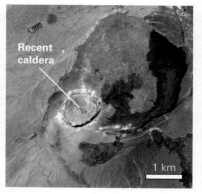
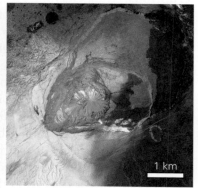

(d) The caldera of Kilauea, as viewed looking straight down from above in 2003 (left), and after the 2018 eruption when collapse caused it to grow wider (right).

FIGURE 4.26 Formation of a cinder cone.

(a) This cinder cone, on the flank of Mt. Etna, in Italy, is forming from lava clots erupted in a fountain.

(b) This cinder cone in Arizona is 1.2 km (3937 ft) wide. A lava flow covers the land surface in the distance. Note the road for scale.

Cinder cones are symmetrical, cone-shaped piles of tephra ejected by a lava fountain (**FIG. 4.26**). Cinder cones can form on the flanks of a much larger volcano or at a distance from a larger volcano. They may lie in a chain that defines the trace of a fissure, or may occur in a cluster.

Stratovolcanoes, also known as *composite volcanoes*, tend to be large (up to a few kilometers high and on the order of 25 km, or 16 mi, across) cone-shaped mountains made from alternating layers of lava and pyroclastic debris (**FIG. 4.27**). The prefix *strato–* emphasizes the stratified (layered) character of the igneous rock that makes up these volcanoes. Their shape, exemplified by Japan's Mt. Fuji (see inset of Fig. 4.23), serves as the classic image most people have of a volcano. In most examples, the upper, steeper part of the cone consists of erupted rock that has collected in place, whereas the lower, less steep part consists of debris formed when erupted rock tumbled downslope, later moved downslope in landslides, or was later picked up and redeposited as sediment by water. Stratovolcanoes commonly erupt intermediate-composition lava, so the lava doesn't flow far from the summit. Large explosions during eruptions may blast off the summit or side of a stratovolcano. When this happens, a new volcanic edifice grows within the remnants of its predecessor and eventually covers those remnants. In fact, a stratovolcano that we see today may be built on the remnants of several predecessors, each of which grew to mountain size, then exploded, providing a foundation for yet another edifice to grow. Particularly large explosions of stratovolcanoes produce calderas that are much larger than those of shield volcanoes, as we'll see shortly.

The three types of volcanic edifices differ significantly in size (**FIG. 4.28**). Mafic shield volcanoes are the largest, and can

FIGURE 4.27 Internal structure of a stratovolcano.

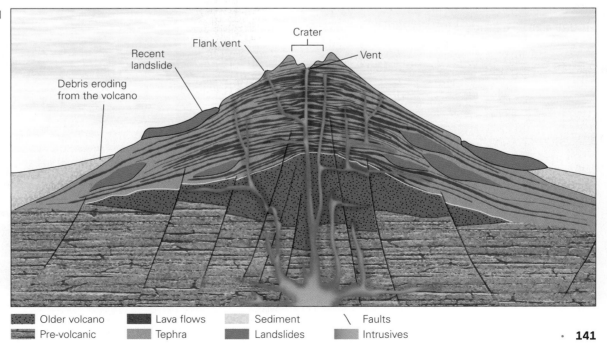

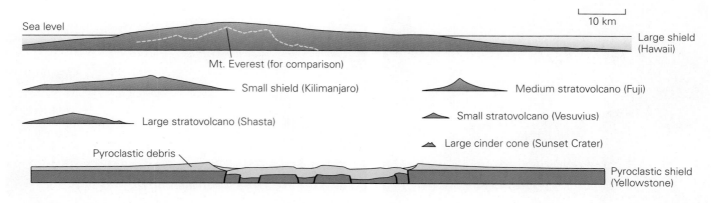

FIGURE 4.28 Comparing the sizes of volcanoes, in profile.

be up to 100 km (62 mi) across. Stratovolcanoes range from 19 to 50 km (12–31 mi) across, and cinder cones tend to be much smaller. As we'll see shortly, an immense explosion, known as a *supervolcanic eruption*, can produce a broad accumulation of pyroclastic debris known as a *pyroclastic shield volcano* up to 120 km (75 mi) across (see Fig. 4.28).

Take-home message . . .

Mafic lava has low viscosity and tends to spread over a large area, while felsic lava has high viscosity and may build a lava dome over the volcanic vent. Pyroclastic debris includes ash, lapilli, cinders, blocks, and bombs. Types of volcanic edifices include shield volcanoes (broad, gentle domes), cinder cones (composed of cone-shaped piles of lapilli), and stratovolcanoes (formed by alternating layers of lava and pyroclastic debris).

QUICK QUESTION What characteristic of lava causes shield volcanoes to differ in shape from stratovolcanoes?

4.6 Will It Flow or Will It Blow?

We've now learned that volcanoes can produce lava and pyroclastic debris. Which of these products dominates during an eruption serves as the basis for distinguishing *effusive eruptions* from *explosive eruptions*. Let's examine the nature of these eruption categories.

Effusive Eruptions

An **effusive eruption** is a lava-dominated eruption. Typically, such eruptions yield fountains and flows of mafic lava and relatively little pyroclastic debris. As we've noted, if the lava emerges in a crater or caldera, it may initially pool to form a lava lake around the vent. The mafic lava has relatively low viscosity, so on steep slopes it can travel relatively rapidly, and it can cover land tens of kilometers from the vent (**FIG. 4.29a**). Effusive eruptions occur at any

FIGURE 4.29 The contrast between effusive and explosive eruptions.

(a) A lava flow from an effusive eruption in Iceland in 2015.

(b) 1992 explosive eruption of Mt. Pinatubo, in the Philippines.

FIGURE 4.30 Volcanic explosions involving high-pressure steam.

(a) A phreatomagmatic eruption in Tonga. The eruptive cloud contains ash, lapilli formed from lava clots, volcanic gases, and lots of steam condensed from water vapor.

(b) A deadly phreatic eruption on White Island, New Zealand, in 2019.

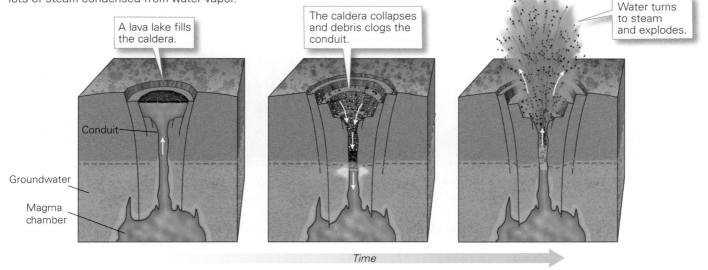

(c) Phreatic eruptions may involve both steam (from groundwater) and volcanic gases (from below). The gas pressure may blow debris from the throat of the volcano.

location where decompression or flux melting of the mantle has produced mafic magma. On land, such settings include rifts, hot spots, and island arcs. Underwater effusive eruptions take place along mid-ocean ridges and typically produce pillow basalt.

Explosive Eruptions

An **explosive eruption** involves energetic blasts that forcefully eject pyroclastic debris from a volcano (FIG. 4.29b). Such eruptions occur in a range of sizes. Why do explosive eruptions take place? We'll consider two of the most common reasons.

EXPLOSIONS DUE TO INTERACTION WITH WATER. If a volcanic edifice cracks, magma in the throat of the volcano or in a near-surface magma chamber may come in contact with water. When that happens, the surface of the magma suddenly cools to form a solid shell, which immediately contracts and cracks—a process that exposes more magma to more water. Because of this very rapid progressive cracking, a lot of magma can come in contact with water and can cool very quickly. The heat released by the magma instantly transforms the liquid water it touches into water vapor, a gas. Water expands dramatically when it becomes a gas, so the newly formed water vapor generates extreme outward pressure. This pressure blasts blobs of lava, fragments of pre-existing volcanic rock, and water vapor out of the vent. As the water vapor enters the atmosphere, it condenses into tiny droplets to form *steam*. Such an event—involving interaction of molten rock with water to produce an outward blast of lava and other pyroclastic debris, mixed with steam—is called a **phreatomagmatic eruption** (FIG. 4.30a).

Not all explosions of volcanoes caused by interactions of water and magma eject lava. In some cases, only pyroclastic debris bursts out of a vent. Such **phreatic eruptions** may be

FIGURE 4.31 The 1883 explosive eruption of Krakatau (Krakatoa), Indonesia.

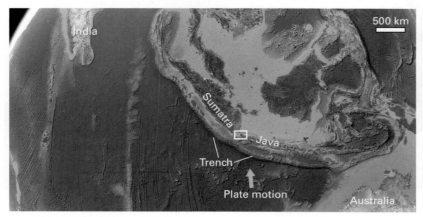

(a) Krakatau was an island volcano lying between Sumatra and Java, along the Java Trench (a convergent boundary). The white box indicates the location of the images of Krakatau in part (b).

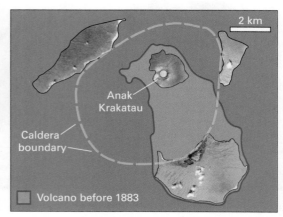

(b) A map of Krakatau showing the area of the volcano before the 1883 eruption and the caldera that formed after the eruption.

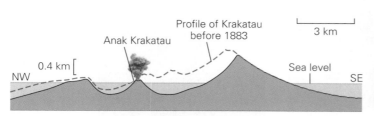

(c) A profile showing the caldera and Anak Krakatau.

(d) A painting of one of the amazing sunsets caused by ash erupted into the atmosphere by Krakatau.

relatively small, simply blasting *ballistic blocks* (large rocks ejected skyward like cannonballs) from a fissure. But some can be larger, blasting hot ash and debris up for a few kilometers (**FIG. 4.30b**), and some can send finer pyroclastic debris 10–15 km (6–9 mi) up into the atmosphere. For example, during the 2018 eruption of Kilauea, sinking of lava in the volcano's conduit allowed debris from the walls of the conduit to tumble in and clog the volcano's throat (**FIG. 4.30c**). Gas pressure, from both vaporizing groundwater and volcanic gases rising from below, became sufficient to blast the debris skyward in orangish-tan clouds, which rained ash, lapilli, and blocks on the surrounding landscape. In some cases, the blockage of a conduit that traps gases may be due to mineralization that produces a concrete-like seal. When the seal suddenly cracks, the pressure immediately releases, causing an eruption.

Explosions can also happen where a lava flow enters seawater. Though usually relatively small, these explosions can be strong enough to blast pyroclastic debris—including blocks—for hundreds of meters. A ballistic block blasted from the spot where lava from the 2018 eruption of Kilauea entered the sea went through the roof of a tourist boat and broke the arm of a passenger.

Where does the water for phreatomagmatic or phreatic eruptions come from? In the case of volcanoes on continents, such eruptions can occur when magma interacts with groundwater, downward-percolating rainwater, melting ice, or a lake. In the case of volcanoes erupting underwater or in island arcs, a huge explosion may take place if large volumes of seawater suddenly gain access to the magma chamber via cracks in the edifice. Such a phenomenon may explain the violent explosion of Krakatau (also known as Krakatoa), a stratovolcano in the Sunda Strait, between the Indonesian islands of Sumatra and Java, where the Indian Ocean floor subducts beneath Southeast Asia (**FIG. 4.31a**). The edifice of Krakatau had grown to become an island 9 km (6 mi) long, rising 800 m (2,625 ft) above the sea. On May 20, 1883, the volcano began to erupt. A series of large explosions produced ash that settled as far as 500 km away. Smaller explosions continued through June and July, and steam and ash rose from the island, forming a huge black cloud that rained ash into the surrounding straits. Krakatau's demise came at 10 A.M. on August 27. Researchers suggest that sudden flooding of seawater into the magma chamber triggered a phreatomagmatic blast 5,000 times more powerful than the Hiroshima atomic-bomb explosion. Tsunamis (see Chapter 6) generated by the explosion slammed into nearby coastal towns, killing over 36,000 people. When the air finally cleared, Krakatau was gone, replaced by a submarine caldera about 7 km (4 mi) wide and 300 m (984 ft) deep

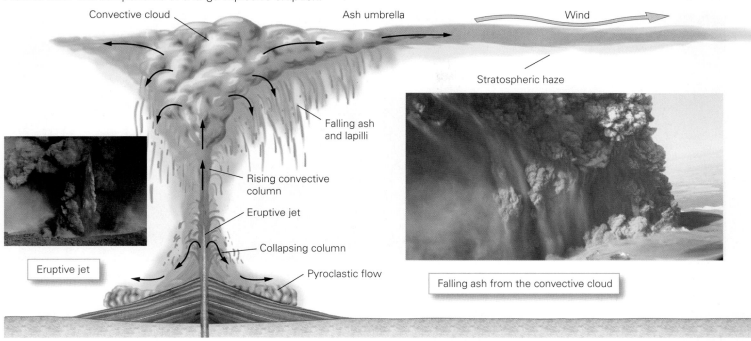

FIGURE 4.32 The components of a large explosive eruption.

(FIG. 4.31b, c). All told, the eruption sent nearly 20 km³ (5 mi³) of rock skyward—the ash that reached stratospheric altitudes caused spectacular sunsets during the next several years. Their strange light and colorful skies appeared in several notable paintings by European artists of the time (FIG. 4.31d).

EXPLOSIONS DUE TO TRAPPED GAS BUBBLES. Mafic magma has relatively low viscosity, so bubbles of magmatic gases that form within it can rise faster than the magma, and most escape at the volcano's vent. As a result, while pressure in mafic magma can become sufficient to drive a lava fountain, it tends not to cause huge explosions (although small to moderate phreatic explosions may happen during effusive eruptions). Felsic and intermediate magmas, in contrast, are so viscous that gas bubbles cannot escape. In addition, they tend to contain more volatiles than do mafic magmas. As a consequence, gas pressure in such lavas can become extreme, setting the stage for a large explosion.

What sets off the violent explosion of a volcano erupting felsic or intermediate lava? If a lava dome formed at the summit of the volcano suddenly cracks, or the flank of the volcano's edifice suddenly slumps away, the inward pressure acting on partially solidified, bubble-rich lava in the volcano's conduit abruptly decreases (BOX 4.1). Consequently, the bubbles suddenly try to expand even more, and the pressure within them becomes sufficient to break the glassy walls surrounding them. Instantly, the gases expand explosively, blasting the shattered bubble walls and adjacent rock out of the vent. Large explosions incorporate material from the volcanic edifice. The pressure of a large explosion drives an **eruptive jet**, composed of hot pyroclastic debris and some lava, up several hundred meters (FIG. 4.32). Hot ash in the jet mixes with and heats the surrounding air, producing a buoyant cloud, and this *convective cloud* continues to rise to stratospheric heights (FIG. 4.33a; see also Fig. 4.29b).

Eventually, debris that rises in the convective cloud falls back to the Earth. Lapilli, because of their size, tend to fall closer to the volcano, whereas ash can blanket regions extending far from the volcano. (Rock formed from such ash is called *air-fall tuff* or *ash-fall tuff*.) Not all of the material in the eruptive jet, however, rises in the convective cloud. Some collapses due to gravity and avalanches down the flanks of the volcano as a devastating **pyroclastic flow**, a turbulent cloud of fiery ash, pumice lapilli, and hot air that rushes at high velocity (FIG. 4.33b). When the turbulent flow comes to rest, compacts, and undergoes lithification, it becomes **ignimbrite**, or *ash-flow tuff*. If the ash is still very hot when it settles or stops moving, the glassy shards in it can meld together to form a fairly hard rock, known as a *welded tuff*.

Volcanic Explosivity Index

As our descriptions suggest, explosive volcanic eruptions come in a range of sizes, from ones that eject a few ballistic blocks, to ones that cover the surrounding landscape with lapilli, to ones that remove much of a volcanic edifice from the landscape and replace it with a large caldera. Researchers have determined that the largest explosive eruptions to have happened during the past few million years actually dwarf any that have ever been observed during historic time.

FIGURE 4.33 The difference between a convective cloud and a pyroclastic flow.

(a) The convective cloud above Mt. Redoubt, Alaska, in 1990.

(b) A small pyroclastic flow rushing down the side of a volcano.

To classify explosive eruptions by size, researchers have developed a logarithmic scale, called the **volcanic explosivity index** (**VEI**), based on the volume of erupted pyroclastic debris (tephra) as measured in cubic kilometers (**TABLE 4.3**) (**FIG. 4.34**). A VEI 0 eruption does not eject any tephra, though it may produce large quantities of lava. Therefore, an effusive eruption, such as the 2018 eruption of Kilauea, can serve as an example. The 79 C.E. eruption of Mt. Vesuvius and the 1980 eruption of Mt. St. Helens were VEI 5 events. The explosive

TABLE 4.3 Volcanic Explosivity Index (VEI)

VEI	Erupted volume*	Frequency†	Column height	Description‡
0	≤0.0001	Continuous	<100 m	Effusive
1	>0.0001	Daily	100 m–1 km	Gentle
2	>0.001	2 weeks	1–5 km	Explosive
3	>0.01	3 months	3–15 km	Catastrophic
4	>0.1	18 months	10–20 km	Cataclysmic
5	>1	12 years	10–20 km	Paroxysmic
6	>10	50–100 years	10–20 km	Colossal
7	>100	500–1000 years	>20 km	Super-colossal
8	≥1,000	>50,000 years	>20 km	Mega-colossal

*Volume of tephra ejected (km^3).
†Frequency refers to the average interval between events of this size somewhere on the Earth, not to their frequency at an individual volcano.
‡These terms are informal, and are not used uniformly.

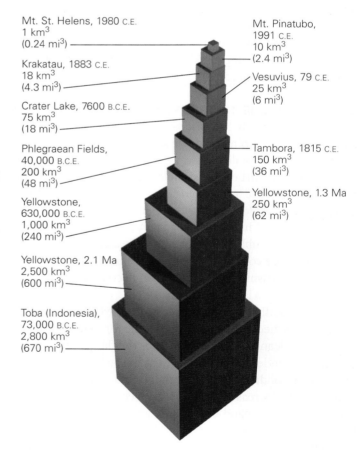

FIGURE 4.34 A comparison of the volumes of pyroclastic debris expelled in well-known explosive eruptions. The volume of erupted pyroclastic debris is the basis of the volcanic explosivity index (VEI).

FIGURE 4.35 Examples of calderas formed by large explosive eruptions.

(a) A panorama of Crater Lake, a caldera in the Cascade volcanic arc. Wizard Island is a new volcanic edifice built after the caldera formed.

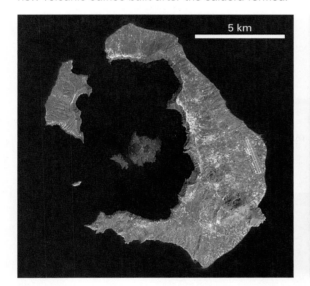

(b) The Santorini caldera as seen from space.

(c) The cliff face on the side of the caldera shows many layers of tuff from previous eruptions.

eruption of Mt. Pinatubo, a stratovolcano in the Philippines, in 1991 was a VEI 6 event, sending a convective cloud 35 km (22 mi) into the sky (see Fig. 4.29b) and massive pyroclastic flows down its side. During the past two millennia, four VEI 7 eruptions have taken place. The most recent of these happened at Tambora, a stratovolcano in the Indonesian volcanic arc, in 1815, and ejected 150 km³ (36 mi³) of pyroclastic debris—that affected climate globally.

Large Caldera Eruptions and Supervolcanoes

Eruptions that have a VEI greater than 6 eject so much lava and pyroclastic debris that much of the volcano itself collapses to produce a huge caldera, much larger than those that occur at the summit of a cinder cone or shield volcano. During the collapse, blocks of debris sink down into the drained magma chamber beneath the volcano. Initially, calderas produced by huge explosions have a flat floor covered by pyroclastic debris. But because their floors are so low, some fill with water. For example, Crater Lake in Oregon, 8 km (5 mi) wide and 655 m (2,149 ft) deep, formed 7,700 years ago when a large stratovolcano, Mt. Mazama, blew its top in a VEI 7 eruption (FIG. 4.35a). Another huge caldera, 12 km (7 mi) long by 7 km (4 mi) wide, formed about 1600 B.C.E. when Thera, an island arc volcano in the Mediterranean Sea south of mainland Greece, exploded. This VEI 6 or 7 eruption probably impacted civilizations around the eastern Mediterranean and may be alluded to in Greek myths and the Old Testament. The Greek island of Santorini borders the caldera. To visit Santorini today, tourists can

BOX 4.1 DISASTERS AND SOCIETY

The 1980 explosion of Mt. St. Helens

Mt. St. Helens was a beautifully symmetrical, snow-crested stratovolcano in the Cascade Range of Washington State, in the northwestern United States. The volcano had last erupted in 1857, so 20th-century residents of the surrounding area didn't think of it as a threat. But on March 20, 1980, an earthquake, as well as increased emissions of gases, announced that the volcano was awakening once again. In response, the US Geological Society (USGS) placed observers at various locations around the volcano to monitor such changes. Seismographs recorded earthquakes that increased in frequency over the next month, indicating the movement of magma within the volcano. Ash and gases began to rise from the summit, and tiltmeters indicated that the volcano was slowly inflating, like a giant balloon. In particular, the volcano's north side began to bulge. Volcano observers became so concerned that an eruption was imminent that they advised local authorities to evacuate people from the area. Fortunately, the area surrounding the volcano was a wilderness in which not many people lived. Most residents heeded the evacuation orders, but some refused to budge.

The main eruption came suddenly. At 8:32 A.M. on May 18, David Johnston, a USGS geologist monitoring the volcano from a station 10 km (6 mi) away, shouted over his two-way radio to

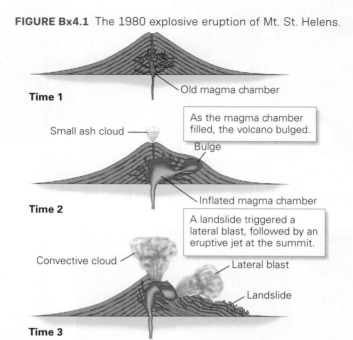

FIGURE Bx4.1 The 1980 explosive eruption of Mt. St. Helens.

(a) The eruption was triggered by decompression caused by a landslide.

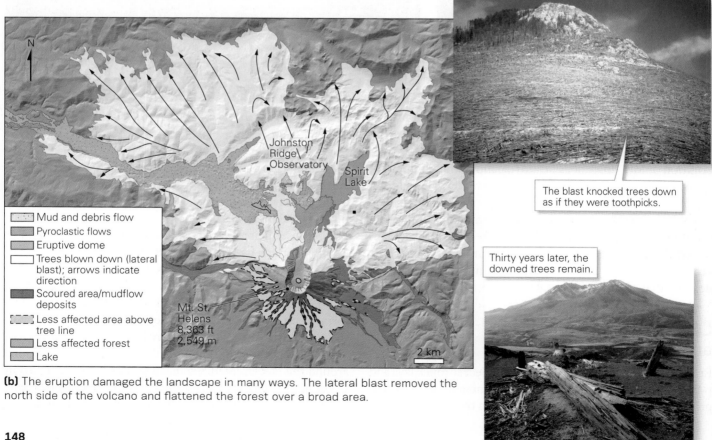

(b) The eruption damaged the landscape in many ways. The lateral blast removed the north side of the volcano and flattened the forest over a broad area.

(c) Ash from Mt. St. Helens fell on towns in the surrounding area. Residents had to blow it off their cars.

headquarters, "Vancouver, Vancouver, this is it!" An earthquake had caused 3 km³ (nearly a cubic mile) of the volcano's weakened north side to slip away, producing a high-velocity landslide that carried material as far as 18 km (11 mi) from the volcano. Removal of this material released pressure on the magma in the volcano, causing a sudden and violent expansion of gases that blew through the volcano's side as a giant *lateral blast* **(FIG. Bx4.1a)**. Rock, steam, and ash screamed northward. Close to the blast, forests shattered into splinters. Farther away, trees were ripped from the ground and laid out parallel to the blast direction. The land up to 15 km (9 mi) from the vent, covering an area of 600 km² (232 mi²), was devastated **(FIG. Bx4.1b)**. Tragically, Johnston, along with 60 other people, some of whom had ignored evacuation orders, vanished beneath the debris.

Following the lateral blast, a vertical eruptive jet rose from the summit and remained active for the next 9 hours. It fed a convective cloud that carried about 1 km³ (over 500 million tons) of ash skyward. Some of the ash reached stratospheric heights, where high-altitude winds transported it around the globe. But most fell closer to the volcano, producing an ash layer up to 12 cm (5 in) thick covering areas as far as 500 km from the volcano. Thinner ash deposits from the eruption could be detected on the ground as far as 2,000 km (1,243 mi) away. In nearby towns, the blizzard of ash reduced highway visibility to almost zero; buried fields, houses, roads, and vehicles; and insidiously coated even the interiors of houses and machinery **(FIG. Bx4.1c)**.

On the flanks of the volcano, the ash mixed with melting snow and rainwater to produce muddy slurries, or *lahars*, that flowed down the mountain, picking up boulders and logs along the way. The first lahar started flowing within minutes of the eruption, and several more followed. As they rushed along stream channels—at speeds of up to 55 km/h (34 mph)—the lahars, together with the heavy objects they carried, battered down trees, bridges, railroad trestles, and streamside homes, and covered the areas bordering the streams with meters of sediment.

When the eruption was finally over, the peak of Mt. St. Helens had disappeared, and the once snow-covered conical peak was a gray mound with a large gouge in one side. Spirit Lake, once a sparkling tree-lined fishing destination, was a gray mud puddle, and tens of thousands of trees lay on the ground. By some estimates, the eruption caused $1 billion in damage. The consequences for residents of surrounding areas were challenging, to say the least. Farmers, in particular, endured severe crop losses because the ash both coated leaves, preventing photosynthesis, and decreased soil permeability.

sail on a ship right into the caldera, whose floor lies over 100 m (328 ft) below sea level **(FIG. 4.35b, c)**.

Renewed injection of magma into a chamber beneath the caldera floor may cause it to bulge upward, forming a *resurgent lava dome*, and subsequent eruptions may build a new volcano that rises from the floor of the caldera. For example, Wizard Island in Crater Lake grew during eruptions after the caldera's formation (see Fig. 4.35a). About 50 years after the Krakatau eruption, a new volcanic island, Anak Krakatau (translated as Child of Krakatau), rose above sea level near the middle of the caldera (see Fig. 4.31). In December 2018, Anak Krakatau erupted and partially collapsed, leading to an undersea landslide, which in turn caused a tsunami that struck Sumatra and Java, killing hundreds and injuring over 14,000 people.

VEI 7 or 8 eruptions are so big that they have been called "super-colossal" and "mega-colossal," respectively, and can have global consequences. In fact, geologists now refer to volcanoes that produce VEI 8 eruptions as **supervolcanoes** to emphasize their size. No VEI 8 eruption has happened during recorded human history, but at least 60 have been identified in the geologic record. Three supervolcanic eruptions, the

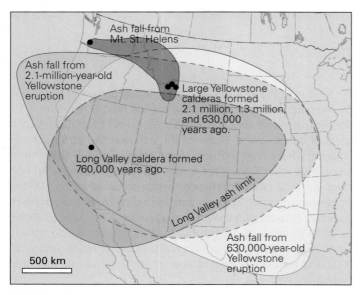

FIGURE 4.36 Areas covered by substantial quantities of ash from VEI 8 eruptions in the western United States during the past 2.1 million years. The ash fall from Mt. St. Helens is provided for comparison.

volume of these eruptions, they would dwarf any observed effusive eruptions in historic time. Some of the flows they produced were so hot, had such low viscosity, and were so voluminous that they traveled as far as 600 km (nearly 400 mi). The total volume of lava erupted to form the Columbia River Plateau was about 175,000 km^3 (about 42,000 mi^3). Over geologic time, even larger flood basalts covered areas of eastern Brazil, southern Africa, and Siberia. The Siberian eruptions were the largest, producing about 7 million km^3 (nearly 1.7 million mi^3) of lava.

> **Take-home message . . .**
>
> Geologists distinguish between effusive (lava-dominated) and explosive eruptions. Explosive eruptions can be caused by trapped magmatic gases and interaction of hot lava or magma with water. The largest volcanic eruptions occurred prior to historic time, but super-colossal eruptions with an explosivity of index VEI 7 have happened within human history.
>
> **QUICK QUESTION** How frequently do eruptions producing 1–10 km^3 of pyroclastic debris occur?

most recent at 630,000 B.C.E., left behind the gigantic caldera (70 km, or 43 mi, in diameter) that now underlies much of Yellowstone National Park. The largest of these eruptions sent over 2,500 km^3 (600 mi^3) of debris into the atmosphere, which fell to cover much of the United States, as did the ash from the Long Valley caldera, another supervolcano in eastern California (**FIG. 4.36**). The world's most recent VEI 8 eruption, at the Toba volcano on Sumatra in about 73,000 B.C.E., sent 2,800 km^3 (672 mi^3) of pyroclastic debris skyward, leaving a caldera that measures 83 km (52 mi) long by 30 km (19 mi) wide. Some researchers suggest that the ash from this eruption so disrupted the environment that it caused the deaths of nearly all humans on the Earth, but this proposal remains a subject of intense debate. The largest known supervolcano erupted about 27 million years ago, in southwestern Colorado, and ejected about 5,000 km^3 (1,200 mi^3) of pyroclastic debris. Supervolcanic eruptions such as those at Yellowstone generate immense ash falls as well as pyroclastic flows. It is difficult to imagine what the consequences of a VEI 8 eruption today would be, and whether our global society has the resilience to recover from such a cataclysm.

While the word *supervolcano* has generally been used in the context of explosive eruptions, some researchers suggest that it should also apply to eruptions of flood basalts. The most recent such eruption produced the Columbia River flood basalts of eastern Washington and Oregon between 17 and 14 million years ago (see Fig. 4.15d), before the earliest hominin species had even appeared, so no one has ever witnessed a flood-basalt eruption. But given the

4.7 The Style of Volcanic Eruptions

Geologists informally refer to the specific manner in which material comes out of a volcano as the **eruptive style** of the volcano. Below, we introduce the most common eruptive styles, roughly in order of their VEI. Note that the designation of an eruptive style applies to an individual eruption—sometimes, the same volcano hosts eruptions of different styles at different times.

Hawaiian Eruptions

Hawaiian eruptions are predominantly effusive and produce mafic lava flows that build over time into shield volcanoes. Eruptions of this style occasionally produce phreatic explosions, effectively clearing the volcano's throat of debris. Over the past millennia, dozens of Hawaiian eruptions have taken place on the Big Island of Hawai'i. Many have emanated from fissures radiating outward from the summit caldera of Mauna Loa (**FIG. 4.37**). More recent eruptions have come from Kilauea (**BOX 4.2**).

Surtseyan Eruptions

The name of the next eruptive style comes from Surtsey, an island that grew off the coast of Iceland in 1963. *Surtseyan eruptions* are substantial phreatomagmatic eruptions driven

FIGURE 4.37 Hawaiian eruptions on the Big Island of Hawai'i.

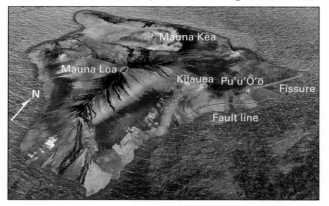

(a) On Hawai'i, lava flows of the past few centuries are obvious because they have not yet weathered and developed soils, so they are still dark colored. The East Rift Zone, where the fissure is labeled, is the site of 2018 eruptions.

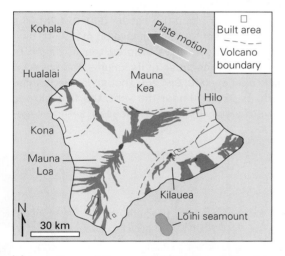

(b) Lava flows of the past 1,000 years. A new island (Lō'ihi) is growing underwater, to the southeast of Hawai'i.

by the interaction of magma with water, which causes the development of explosive pressure that can blast material out of the volcano in clouds of debris and steam, forming jagged fragments of glassy volcanic rock.

Surtsey announced its birth by rattling the seafloor with earthquakes that were recorded by seismographs in Iceland. On November 14, 1963, a fishing boat saw what looked like black smoke rising from the sea. Thinking it might be a boat on fire, the crew investigated and found instead a column of black ash and steam. For the next 10 days, the eruption blasted blocks, lapilli, and ash skyward (**FIG. 4.38**). By the end of this time, a pile of pyroclastic debris had risen above sea level. Phreatomagmatic eruption stopped when the island became so big that water could no longer seep downward and come in contact with magma. At that point, the style of the eruption became more Hawaiian, producing lava fountains and flows. While the erosive power of ocean waves would have been able to wash away an island made only of pyroclastic debris, the lava flows effectively armor-plated the volcano and allowed the island to become permanent. By 1965, Surtsey had an area of almost 3 km^2 (more than a square mile).

Strombolian and Vulcanian Eruptions

The eruption of Paricutín that Dionisio Pulido watched in 1943 serves as an example of a *Strombolian eruption*, named for a volcano on a small island off the north coast of Sicily. A Strombolian eruption differs from a Hawaiian eruption in that it does not produce large lava flows, but rather ejects burst-like lava fountains and eruptive jets of scoria bombs, lapilli, and some ash (**FIG. 4.39a**). Strombolian eruptions are more explosive than Hawaiian eruptions because Strombolian magma is more viscous, so gas bubbles can't escape

FIGURE 4.38 The formation of Surtsey, off Iceland.

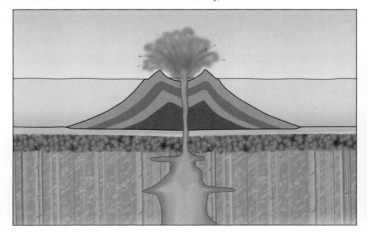

(a) A Surtsey grew up from the seafloor, it produced large phreatomagmatic eruptions.

(b) Surtsey, just after rising above sea level, was a cinder cone.

BOX 4.2 DISASTERS AND SOCIETY

The 2018 eruption of Kilauea

Kilauea, on the island of Hawai'i, has been erupting nearly continuously since 1983 at varying rates. Magma chambers feed lava lakes at the summit caldera of the volcano, as well as in a flank vent called Pu'u 'Ō'ō. Between 1983 and 2018, lava seeped out of the base of Pu'u 'Ō'ō to feed a stream that flowed south to the shore of the Pacific Ocean. By the time the lava reached the coast, it was flowing only through lava tubes, which would either empty onto the surface to form small pāhoehoe flows or spill into the sea. Visitors to Hawai'i Volcanoes National Park could walk out on the flows, poke molten lava with sticks, and stand near the cloud of steam rising where lava entered water. Such fun became impossible in the spring of 2018.

In late April, the summit lava lake in Kilauea's caldera overflowed, and molten rock spread out onto the floor of the caldera. Then, without warning, the level of the lava lake dropped, as did the level of the lava lake inside the crater of Pu'u 'Ō'ō, as if the lakes were being drained. In fact, that's precisely what was happening. The distribution of earthquake activity indicated that a northeast-trending fissure 40 km (25 mi) long, known locally as the East Rift Zone, had opened due to sudden slip along a fault at the boundary between the volcanic edifice and the seafloor sediment beneath it. Magma withdrawn from the magma chambers moved northeast along the fissure **(FIG. Bx4.2a)**. About 20 km (12 mi) northeast of Pu'u 'Ō'ō, a constriction in the fissure caused the magma to force its way up to the ground surface—right beneath a housing development! As this upward flow began, cracks opened at the ground surface. Steam from heated groundwater began to rise out of the cracks, followed by sulfurous gases. Occasional bursts of steam and gases ejected heavy ballistic blocks hundreds of meters into the air.

Then came the lava, first fountaining out of one vent, then another, and then another along the fissure **(FIG. Bx4.2b)**. The most energetic fountain sent a flood of lava 90 m (295 ft) skyward (see Fig. 4.18b). Enough lava erupted that it began to spread over the land, and because of its volume and speed, it produced a thick 'a'ā flow **(FIG. Bx4.2c)**. Though lava streamed quickly (20 km/h, or 12 mph) near the vent, the flow slowed near its toe. Nevertheless, it moved relentlessly, setting vege-

FIGURE Bx4.2 The 2018 Lower Puna eruption on Hawai'i.

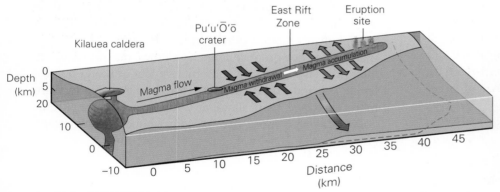

(a) Slip on a fault caused a fissure to open, so that magma from beneath the Kilauea caldera and Pu'u 'Ō'ō moved northeast and erupted.

(b) Lava erupted from fissures that cut through forests, homes, and roads.

(c) Lava flowed over and blocked roads.

(d) Lava burned down and then engulfed houses.

(e) Pyroclastic debris ejected by phreatic explosions at the summit of Kilauea.

tation and buildings on fire (**FIG. Bx4.2d**). When the edges of the flow solidified, lava was restricted to a channel up to a few hundred meters across and 10 m (33 ft) thick. Lava eventually reached the sea and spilled into the water, causing billows of acidic steam to rise skyward. Occasionally, a mass of lava exploded on contact with the water, ejecting blocks and lapilli of glassy, newborn rock. Eventually, lava built a delta of new rock extending half a kilometer into the ocean. Throughout May, June, and July, lava kept coming out of various vents along the fissure.

Draining of the magma chambers eventually caused the summit caldera of Kilauea to collapse. Debris from the caldera floor fell into the summit vent and clogged it. When the pressure beneath the blockage became large enough, sudden phreatic eruptions ejected clouds of ash and lapilli, as well as heavy blocks, skyward from the summit (**FIG. Bx4.2e**). Ash fell like snow in the tropical forests surrounding the summit, coating leaves so that the normally bright green forests looked as though their color had been drained. While the 2018 eruption produced only air-fall ash and lapilli, researchers studying deposits from ancient eruptions determined that in 1790, a particularly large eruption of pyroclastic debris from Kilauea's summit caldera produced a pyroclastic flow that killed over 400 people, some of whom were Native Hawaiian warriors retreating from a battle.

When the 2018 eruption—which came to be known as the Lower Puna eruption—finally ceased, lava had covered over 35 km^2 (nearly 14 mi^2) of land with a brand-new rock layer that was 15 m (50 ft) thick in places. Sadly, the flow of this lava burned and destroyed over 700 homes, displacing over 2,000 people. But because of timely evacuations, no one was killed.

from it as easily. Instead, they coalesce into large bubbles. When these large bubbles burst through a thin skin of frozen lava in the volcano's throat, lapilli and bombs fly into the air. Individual bursts during a Strombolian eruption tend to produce only small quantities of pyroclastic debris, so each eruption has a relatively low VEI.

Vulcanian eruptions, named after the Italian island of Vulcano, involve lava whose viscosity exceeds that of Strombolian lava (**FIG. 4.39b**). Eruptions of this style begin with the growth of a lava dome at the summit vent, which causes pressure to build in the vent. Over time, small explosions eject lapilli and blocks, weakening and thinning the lava dome until it can no longer hold back the pressure of the underlying lava, and a large eruption of ash takes place. Some of the ash reaches kilometers into the atmosphere.

Peléan and Plinian Eruptions (The Really Big Ones!)

In 1902, the town of Saint-Pierre was a quiet port on the island of Martinique, in the island arc formed by the subduction of Atlantic Ocean floor beneath the Caribbean Plate. Saint-Pierre lies about 6 km (4 mi) south of the summit of Mt. Pelée. The mountain was known as a dangerous volcano, for it had erupted over 30 times in the past 5,000 years, and

FIGURE 4.39 Strombolian and Vulcanian eruptions.

(a) Strombolian eruptions burst through a thin crust of lava in the volcano's throat.

(b) Vulcanian eruptions occur when pressure builds up in a vent blocked by a lava dome.

FIGURE 4.40 The 1902 eruption of Mt. Pelée, Martinique.

(a) The pyroclastic flow rushes downslope.

(b) The destruction of Saint-Pierre.

small eruptions had taken place in 1851. But no one fully appreciated how dangerous it could be. In late April of 1902, sulfurous gases began to billow from vents around the crater. A few days later, the ground shook, and a cloud of ash rose from the summit. Climbers went up to investigate and found a new cone of tephra around the vent. Soon, volcanic gases were wafting down to Saint-Pierre, making breathing difficult, and ash was contaminating the town's water supplies. By May 2, the eruptions were getting more frequent and violent, and thousands of people from farms around the volcano sought refuge in the town. Unfortunately, local authorities claimed that the town itself was safe, and didn't call for an evacuation. This decision was a disastrous mistake.

On May 8, the summit of the volcano exploded. The eruptive jet fed a convective cloud that headed skyward in the shape of a gigantic mushroom, and the collapse of the cloud fed a pyroclastic flow that raced down the volcano's flanks **(FIG. 4.40a)**. This flow consisted of a mixture of volcanic gases and ash at temperatures of about 1,000°C (1,832°F)—no wonder such flows were once called *nuées ardentes*, French for fiery clouds. The flow reached speeds of several hundred kilometers per hour, and when it reached Saint-Pierre, it spread over the town and swept out into the harbor. The force of the flow toppled buildings, and its heat incinerated everyone and everything over an area of about 20 km² (8 mi²) **(FIG. 4.40b)**. Only a handful of people survived, out of a population of about 30,000. Mt. Pelée gave its name to *Peléan eruptions*, ones characterized by pyroclastic flows.

Plinian eruptions are huge explosive eruptions, like the one that took place at Mt. Vesuvius in 79 C.E. They are named for the Roman scholar Pliny the Elder, who observed and described the destruction of Pompeii. They happen at stratovolcanoes erupting very viscous intermediate to felsic lava. These volcanoes erupt when gas bubbles account for about 75% of the magma volume beneath them, so that immense pressures develop as the magma rises. During an eruption, large quantities of pumice formed by the cooling of frothy lava in the volcano's conduit get blasted out, along with fragmented and pulverized pre-existing rocks of the volcanic edifice. During a large Plinian eruption, immense volumes of ash loft to stratospheric levels, sometimes as high as 45 km (28 mi) above the Earth **(FIG. 4.41)**, and catastrophic pyroclastic flows rush down the volcano's flanks. The difference between a Plinian eruption and other explosive eruptions is partly its size, but also the existence of a relatively long-lasting eruptive jet and a huge convective cloud.

Take-home message . . .

Differences in eruptive style depend on the volume and character of material that erupts. Some eruptions are lava-dominated, yielding large mafic flows. If an underwater volcano builds to sea level, it erupts fragmented lava and steam. Some volcanoes blast out blocks, bombs, and lapilli; others produce pyroclastic flows. The largest eruptions blast out huge volumes of ash in both convective clouds and pyroclastic flows.

QUICK QUESTION
How does the viscosity of lava affect eruptive style?

FIGURE 4.41 The 2015 Plinian eruption of Calbuco in Chile.

(a) During the day, the convective cloud can be tracked to the stratosphere.

Lightning forms ash clouds.

(b) At night, hot ash glows.

4.8 Beware! The Many Hazards of Volcanoes

From the descriptions of individual volcanic eruptions earlier in this chapter, it's clear that the products of eruptions—lava, pyroclastic debris, gases, and explosions—present hazards to society. Indeed, eruptions can trigger natural disasters affecting nearby towns, and in extreme cases, global society. Due to the rapid expansion of urban areas near volcanoes, far more people live in dangerous proximity to volcanoes today than ever before, so if anything, the hazards posed by volcanoes have gotten worse. Let's look at the different kinds of threats posed by volcanic eruptions and review additional examples of their consequences.

Hazards Due to Lava

When you think of a volcanic eruption, perhaps the first threat that comes to mind is red-hot lava. Mafic lava from effusive eruptions poses the greatest threat because it can flow relatively quickly and spread over a broad area. As a lava flow advances, its heat sets grass, trees, and shrubs on fire, so the edge of the lava flow is marked by a rim of smoke. As lava advances over roads, it blocks access to the affected area and may sever evacuation routes **(FIG. 4.42a)**. The heat from the lava also melts and vaporizes asphalt, producing toxic fumes. Vehicles trapped by the flow heat up and catch fire **(FIG. 4.42b)**. The plastics in vehicles and the rubber in tires release more toxic fumes when burned. In addition, the gas in the fuel tank may explode, scattering metal shrapnel. In rural communities, heat from lava can cause underground septic tanks and

FIGURE 4.42 Mafic lava flows can be a hazard to communities.

(a) Lava flowed into this Hawaiian neighborhood and blocked roads.

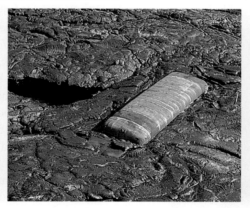

(b) Lava submerged this empty school bus on Hawai'i.

(c) Lava from Mt. Nyiragongo covered the streets of Goma.

household propane tanks to blow up. As lava flows over power lines or pipelines, it burns them and breaks connections. When lava approaches a building, the structure bursts into flames before the molten rock even touches it. In some cases, the lava flows around the building, but more often, it simply surges over the charred remnants. People usually have time to get out of the way of such flows, but not with all their possessions.

We've seen in Box 4.2 how lava threatened housing developments in Hawai'i. A similar lava flow, produced by the 2002 eruption of Mt. Nyiragongo, a volcano lying along the East African Rift, affected many more households because it entered the city of Goma, in the Democratic Republic of the Congo. This lava flow covered streets in the center of the city as well as a substantial portion of the airport (FIG. 4.42c). This event was particularly disastrous because temporary housing in Goma, a city of 1 million, was already filled with refugees escaping genocide and war, so the newly displaced people had nowhere to go.

Hazards Due to Ash Falls and Ash Clouds

Large explosive eruptions blast huge quantities of pyroclastic debris into the air, as illustrated by the Plinian eruption of Mt. Vesuvius (see Fig. 4.1). Close to the volcano, ballistic blocks or bombs that tumble from the sky can crush people and crash through the roofs of buildings or vehicles. Lapilli can fall for distances as far as 20 km (12 mi) from the vent. Ash can be carried over even great distances and thus can be a threat over a broad area. For example, in the Philippines, a typhoon spread ash from the 1991 eruption of Mt. Pinatubo over a 4,000 km² (1,544 mi²) area. Ash from the 1995 eruption of the Soufrière Hills volcano on the Caribbean island of Montserrat blanketed its port city in ghostly white and made the southern half of the island uninhabitable.

FIGURE 4.43 Air-fall ash can reduce visibility, bury streets and fields, foul machinery, and cause breathing problems.

Dense clouds of ash not only reduce visibility like a blizzard (FIG. 4.43), but can also cause a lot of damage. First, the sharp glass shards they contain can cause lung damage when inhaled, and in larger quantities can cause silicosis. Second, given that an ash fragment is about 2.5 times heavier than an equal-sized piece of ice, an accumulation of ash can crush roofs. Specifically, wet ash, on average, has a density of 2,000 kg/m³, so 0.5 m (1.6 ft) of wet ash spread over the roof of a small house (30 m², or 323 ft²) can weigh 30,000 kg (33 tons), enough weight to break support beams. Ash can also abrade paint, so removing it from vehicles can ruin the finish. And if ash gets into the moving parts of machinery, it can cause severe abrasion. Ash can damage or destroy crops and pastures. Heavy ash falls kill off crops by crushing them or by coating leaves and preventing photosynthesis. Even light ash falls are a problem because rainfall can leach acids and fluorine out of the ash, adversely affecting soil chemistry. Similarly, ash can be deadly to livestock by causing an array of gastrointestinal problems. Fortunately, these negative effects on agriculture are temporary, and after a few years, nutrients provided by the ash may actually be beneficial to soil.

Volcanic ash that rises to high altitudes can be a hazard to airplanes. Ash particles scratch windows and damage the fuselage. Ash sucked into a jet engine melts to produce a glassy coating that restricts air flow, decreasing the efficiency of the compressors that make the thin air at high altitudes dense enough for engine operation, so the plane's engines backfire and shut down. This happened to a British Airways 747 that flew through the ash cloud rising from a volcano on Java in 1982. The plane shimmered with an electric blue light due to the static caused by the flow of ash around it. Then, everything went silent. The pilot announced to the passengers, "Ladies and gentlemen, this is your captain speaking. We have a small problem. All four engines have stopped. We're doing our damnedest to get them going again." For 13 minutes, the plane dropped in a steep glide from 11.5 km (over 37,000 ft) as the crew frantically tried to restart the engines. Finally, at 3.7 km (12,000 ft), the air was dense enough, and enough ash had been blown out of them, for the engines to roar back to life. The plane headed to Jakarta, where, without functioning instruments and without being able to see through most of the windshield, the pilot managed to bring the 263 passengers and crew in for a safe landing. A similar event happened in 1989 when a KLM 747 flew through a cloud of ash produced by Mt. Redoubt (see Fig. 4.33a), in the Aleutian island arc. Fortunately, after falling more than 4.2 km (14,000 ft), the plane got its engines restarted and landed safely.

Because of the obvious risk that ash represents to air travel, all airspace in Europe was closed for about a week following the 2010 ash-rich eruption of Iceland's Eyjafjallajökull (FIG. 4.44). Though its name can be unpronounceable to

FIGURE 4.44 Ash rising from an eruption in Iceland in 2010 disrupted air traffic in Europe.

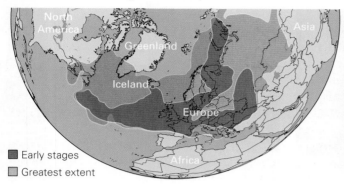

(a) The spread of ash (dark gray) from the eruption of Eyjafjallajökull (red dot) across Europe in late April 2010.

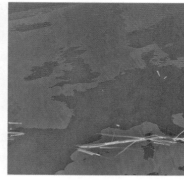

(b) Comparison of air traffic during the closure of airspace (left) and after airspace reopened (right) (not including France) gives a sense of the disruption caused by the eruption. White lines are flights.

non-Icelandic speakers (try saying Ay-uh-fyat-luh-yoe-kuut-le-uh), the VEI 4 eruption, which ejected about 0.1 km^3 (0.02 mi^3) of pyroclastic debris, garnered world headlines because the closure stranded millions of passengers and disrupted the global economy. Some 95,000 flights were grounded, severely impacting commerce, tourism, funerals, and weddings as well as cultural and sports events.

Hazards Due to Pyroclastic Flows

A pyroclastic flow racing down the flank of a volcano could mean instant death to anyone caught in its path (FIG. 4.45). It can travel so fast (over 100 km/h, or 60 mph) that a person is unlikely to outrun it, even in a speeding car. In effect, a pyroclastic flow is like an avalanche, except that it contains volcanic gases and ash at a temperature of up to 1,000°C (1,800°F). The lower, denser part of the flow, which hugs the ground, has so much momentum that it can go up the sides of hills and ridges in its path, and even shoot out over the surface of a water body. We've already seen how pyroclastic flows devastated Saint-Pierre after the 1902 eruption of Mt. Pelée, and Pompeii after the 79 C.E. eruption of Mt. Vesuvius.

Pyroclastic flows come in a great range of sizes. Eruptions of supervolcanoes, for example, have left behind thick ignimbrites that extend hundreds of kilometers from the eruption and can be meters to tens of meters thick. But even a flow that contains only a small amount of ash can prove deadly because of its temperature and its ability to cause suffocation, as illustrated by the deadly consequences of a 1991 volcanic eruption in Japan. A group of 41 people, which included some internationally known volcano researchers, were engulfed and killed by a pyroclastic flow, but when they were found, they were covered by less than 5 mm (0.2 in) of ash.

Hazards Due to Lahars

If ash from an explosive eruption mixes with water, it can start to flow down the side of a volcano as a muddy slurry resembling dilute concrete. The water can come from rain, melted snow, or ice. The viscous slurry, known as a **lahar**, is denser than water, so it packs more *dynamic pressure* (the push caused by a moving fluid) than a comparable flow of clear water. As a result, lahars can carry away nearly

FIGURE 4.45 A pyroclastic flow threatening people in Sumatra, Indonesia.

4.8 Beware! The Many Hazards of Volcanoes • 157

FIGURE 4.46 Lahars flow when ash mixes with water and becomes a slurry.

(a) A lahar inundating a floodplain.

(b) A lahar destroyed this village in Indonesia.

everything in their path, including bridges, boulders, and huge trees (FIG. 4.46).

We've already described the lahars that happened during the eruption of Mt. St. Helens. An even worse one accompanied the 1985 eruption of Nevado del Ruiz, a volcano in the Andean volcanic arc of Colombia. The lahar formed when pyroclastic flows emitted from the summit flowed over and melted glaciers. The ash mixed with the meltwater and surged down the mountainside into a valley, burying the sleeping town of Armero under a thick layer of mud and entombing over 20,000 residents. In New Zealand, numerous lahars have flowed down the slopes of Mt. Ruapehu. In 1953, one of these lahars moved a bridge pier along an express railroad line, causing the bridge to collapse when a train started to cross it minutes later, an accident that killed 151 people.

Hazards Due to Volcanic Gases

Volcanoes emit significant amounts of gases. These gases play an important role in the Earth System because they include H_2O, essential to life, and CO_2, a greenhouse gas that regulates the atmosphere's temperature over geologic time. On a few occasions, however, accumulations of CO_2 have proved deadly. This happened in Cameroon in 1986, when CO_2 seeping from a volcanic vent dissolved in the cold water at the bottom of Lake Nyos, which fills the volcano's crater (FIG. 4.47). A small landslide into the lake disturbed the cold water layer, causing it to rise, so that the gas came out of solution and bubbled up to the surface of the lake. An invisible cloud of CO_2, which is heavier than air, silently drifted down the flank of the volcano and suffocated over 1,700 people in villages downslope. Today, a fountain circulates the lake water in hopes of preventing a future CO_2 buildup.

Sulfur-bearing gases can be problematic during eruptions when they form choking clouds that make breathing difficult near the vent. Even at low concentrations, sulfurous gases such as SO_2 and H_2S can be dangerous, so people entering a volcanically active zone should wear a respirator to avoid inhaling them (FIG. 4.48a). These gases can also react with oxygen and water vapor in the presence of sunlight to produce acidic aerosols. (*Aerosols* consist of solid particles or liquid droplets that are so small that they can remain suspended in air for a long time.) If an *inversion* develops—a weather condition during which warmer air lies above cooler air so that air near the ground can't rise—these acidic aerosols accumulate to produce **vog** (from *v*olcanic sm*og*), polluted air that can cause health problems (FIG. 4.48b). Chemically, vog differs from urban smog in that it forms from sulfur-based gases, while smog forms from hydrocarbon emissions from vehicles. Following the 2018 eruption of Kilauea, prevailing winds blew gases emitted from the eruption to the leeward side of the island, where they accumulated to form vog that made the air of the coastal city of Kona extremely hazy.

Laze (from *la*va ha*ze*), another atmospheric hazard of eruptions, forms at locations where lava enters the sea. The mixing of 1,000°C lava with salty water (containing NaCl) produces a cloud of steam that contains significant amounts of hydrochloric acid as well as fine particles of ash (FIG. 4.48c). Laze, though less well known because it generally affects only local areas, can be very dangerous to breathe, and it kills off vegetation in areas that it blows over.

Hazards Due to Explosions

A major explosive eruption, like any explosion, sends shock waves, high-pressure winds, and ear-shattering sound outward and can cause devastation directly by flattening forests and buildings. Eruption-generated landslides and pyroclastic flows can drive hurricane-force winds in front of them, which add to the destruction. The 1883 explosion of Krakatau and the 1980 explosion of Mt. St. Helens illustrate these consequences. Krakatau's explosion has been called the loudest sound in history—it broke the eardrums of sailors on ships as far as 60 km (37 mi) from the volcano, and was heard by

FIGURE 4.47 CO_2 gas released from Lake Nyos caused over 1,700 deaths.

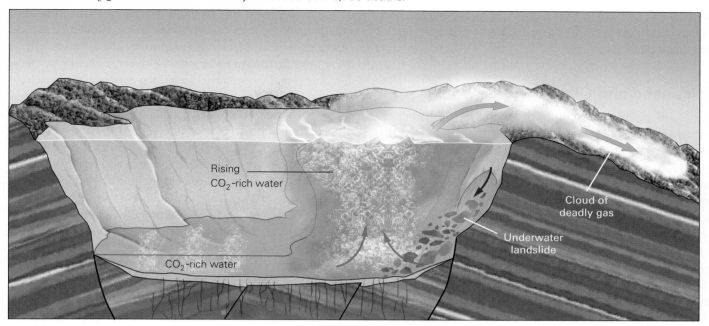

(a) CO_2 emitted by a volcanic vent dissolved in the cold water at the bottom of the lake. A landslide caused the CO_2 to bubble out all at once and flow downslope.

(b) Lake Nyos after the disaster.

(c) In addition to people, large numbers of cattle died.

FIGURE 4.48 Poisonous volcanic gases and associated hazards.

(a) Sulfurous gases emitted from volcanic vents are dangerous. A respirator should be worn if these gases are present.

(b) Vog surrounds the base of Mauna Loa during the 2018 eruptions at Kilauea.

(c) Laze develops where lava enters the sea, producing steam that contains hydrochloric acid.

4.8 Beware! The Many Hazards of Volcanoes • **159**

people over 4,800 km (2,983 mi) away. Sensitive gauges determined that the air-pressure pulse due to the explosion circled the globe at least four times.

Volcanoes and Climate

In 1783, Benjamin Franklin was serving as the American ambassador to France. He noticed that the summer of that year seemed to be unusually cool and hazy, and that the harvest in Europe was terrible. Franklin, an accomplished scientist as well as a statesman, couldn't resist seeking an explanation for this phenomenon. He learned that earlier in the year, a volcanic eruption had taken place in Iceland, and wondered if the "smoke" from the eruption had prevented sunlight from reaching the Earth. Franklin reported this idea at a scientific meeting, and by doing so, may have been the first scientist to suggest a link between volcanic eruptions and climate. In Iceland, gases and ash from the 1783 eruption killed off most crops and half the livestock, causing a famine that led to the deaths of about a quarter of the island's population.

Franklin's idea was confirmed in 1815, when the VEI 7 eruption of Tambora blew the top off a mountain 4.3 km (3 mi) high, replacing it with a caldera 7 km (4 mi) wide. This event sent devastating pyroclastic flows and surges down its flanks—some continuing over the sea surface for as much as 20 km. The eruption directly killed at least 10,000 people, and the tsunamis it generated killed thousands more. But because it occurred before transoceanic telegraph communication was possible, most of the world did not know that the eruption had happened. Soon, however, the whole world indeed felt its consequences, and the death toll related to the eruption climbed, by some estimates, into the millions.

The Tambora eruption produced immense amounts of fine ash and sulfur-bearing gases such as SO_2 and H_2S. At altitudes of 10–50 km (6–31 mi) in the atmosphere, ultraviolet light breaks apart the molecules of these gases, allowing chemical reactions to take place that produce acidic aerosols. Air at these altitudes doesn't mix much with the air below, so these aerosols remain suspended for a long time, and can encircle the Earth. Ash and acidic aerosols from Tambora blocked enough incoming solar radiation that temperatures worldwide dipped. In the northern hemisphere, the average temperature for 1816 dropped about 1°C (1.8°F), enough to cause frost and snow in the northeastern United States and eastern Canada during June. Europe suffered throughout the summer from this weather. In fact, it remained so cold that 1816 became known as "the year without a summer." The dreary weather confined the author Mary Shelley and her literary friends indoors, where they read ghost stories. One guest, Lord Byron, challenged the others to write their own ghost stories. Shelley took the task to heart, and penned *Frankenstein*.

The weather anomaly, and the associated acidification of rain, caused by the Tambora eruption had far-reaching effects. It caused crops to fail, leading to starvation and emigration. It caused the pattern of monsoons in Asia (see Chapter 8) to change, and probably contributed to severe flooding in China and India. The flooding, in turn, led to an increase in cholera, which become a global epidemic in 1817. Short-term climate changes in Europe may also have contributed to a typhus epidemic there. The Tambora eruption took place before modern scientific measurement techniques were available, but studies of atmospheric conditions following the 1991 eruption of Mt. Pinatubo in the Philippines provided clear documentation of the short-term effects that an eruption can have on global atmospheric temperature **(FIG. 4.49)**.

Over the course of Earth history, volcanic eruptions may have had profound effects on the Earth's climate, possibly even causing some of the global mass extinctions that delineate

FIGURE 4.49 The short-term climate consequences of Mt. Pinatubo's 1991 eruption.

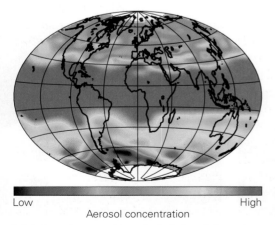

(a) Ash and aerosols encircled the globe, especially at low latitudes.

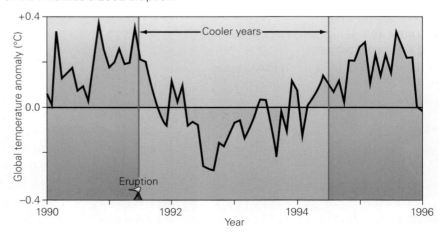

(b) During the years following the eruption, global temperatures were lower than expected. The horizontal line represents average temperature over many years, so a temperature anomaly is the change relative to that average.

boundaries between subdivisions of the geologic time scale. For example, at the boundary between the Paleozoic and Mesozoic Eras, 252 million years ago, a large percentage of the species living on the planet suddenly vanished. So much life disappeared that geologists informally refer to that time interval as "the Great Dying." Some geologists attribute this immense and sudden drop in biodiversity to the eruptions that formed the Siberian Traps, an LIP that covers much of central Siberia. These eruptions may have emitted so much CO_2 (a greenhouse gas) that the climate warmed significantly, after a period of cooling caused by ash and aerosols. The injection of basalt sills into coal beds underground may have led to the release of other toxic chemicals that destroyed the Earth's ozone layer. Ozone absorbs ultraviolet radiation, and without it, that radiation would have reached the Earth's surface. Recent studies suggest that this radiation made many plant species infertile and drove them to extinction.

> **Take-home message . . .**
> Volcanoes can be dangerous! Lava flows, ash clouds, pyroclastic flows, lahars, volcanic gases, explosions, and tsunamis produced during eruptions can destroy cities and farmland. Ash in the air can be a hazard to air traffic. Aerosols produced by reactions of volcanic gases high in the atmosphere reflect sunlight back to space and can cause a temporary global drop in temperature.
>
> **QUICK QUESTION** Why can a lahar do so much more damage than a flood of clear water?

4.9 Protecting Ourselves from Vulcan's Wrath

Clearly, volcanic eruptions are natural hazards (**VISUALIZING A DISASTER**, pp. 166–167). Can society do anything to predict eruptions and minimize the casualties and damage they cause? In this section, we examine evidence that can indicate whether a volcano has the potential to erupt, and steps that people can take to mitigate the consequences of an eruption.

Active, Dormant, or Extinct?

Most volcanoes do not erupt continuously. Rather, a volcano erupts for a period of time, and then ceases erupting and remains quiet for a while, and then erupts again—in some cases, the quiet times may last for centuries or millennia. A volcano doesn't last forever, though, because plate boundaries, mantle plumes, and other geologic features that cause igneous activity themselves have limited lifetimes. Eventually, a volcano dies completely and will never erupt again. The degree to which a volcano represents a natural hazard depends on which of the above situations characterizes its state. To make this distinction, geologists distinguish among active, dormant, and extinct volcanoes as follows:

- *Active volcano:* An **active volcano** is one that is erupting currently, has erupted in historic time, or displays signs that a magma chamber exists beneath it so that it will probably erupt in the future.

- *Dormant volcano:* A **dormant volcano** is an active volcano that hasn't erupted for decades to millennia and doesn't show current signs of erupting in the near future, but still has the potential to erupt because the geologic cause of igneous activity at the location still exists. Some dormant volcanoes may come to life after as much as 10,000 years of inactivity.

- *Extinct volcano:* An **extinct volcano** is one that has shut off entirely and will never erupt again because the geologic conditions leading to eruption no longer exist. For example, a volcano on an island that has been carried off a hot spot is extinct, because without the hot spot beneath it, there's no magma source.

How do geologists determine whether a volcano is active, dormant, or extinct? In the case of volcanoes near populated areas, written or oral histories provide clues to when eruptions took place. Generally, such histories go back only a few centuries, and at most, a few millennia. To extend the record as far back as 70,000 years, geologists use C^{14} dating (a type of radioisotopic dating applied to carbon in organic material) to determine the age of wood fragments buried by pyroclastic debris. This age indicates how long ago the tree died and the wood was buried. For older eruptions, dates from minerals in extrusive rocks produced by the volcano may provide insight. The landscape on a volcano's surface also provides clues to its activity. A volcano whose surface exposes lava or pyroclastic debris that has not been vegetated or eroded by streams must have erupted recently (**FIG. 4.50a**), whereas a volcano that still displays the shape of a volcanic edifice but whose surface has become vegetated and incised by streams has not erupted for a while, but probably was active during the past 10,000 years. A volcano that has completely eroded away, so that underlying intrusions crop out, has probably become extinct (**FIG. 4.50b**). Using these criteria, geologists recognize about 1,500 active or dormant volcanoes worldwide, of which about 10% are in the United States (**FIG. 4.51**).

Because an active volcano (including dormant ones) may be underlain by a hot magma chamber, groundwater beneath the volcano may be heated to high temperatures. Where this water rises to the surface and burbles out of the ground, **hot springs** (pools of hot water) and *geysers* (fountains of hot water and steam) may develop. Where a layer of ash overlies a

FIGURE 4.50 Progressive erosion of a volcano.

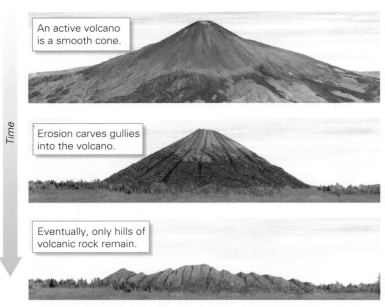

(a) As a volcano remains dormant, erosion makes its surface rougher, and vegetation grows on it.

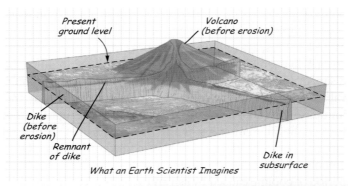

(b) Erosion of an extinct volcano produced Shiprock, New Mexico. Dikes radiate from an intrusion that was once within the volcano.

FIGURE 4.51 Active volcanoes and volcanic hazards in the United States are due to tectonic activity along subduction zones, in rifts, and at hot spots.

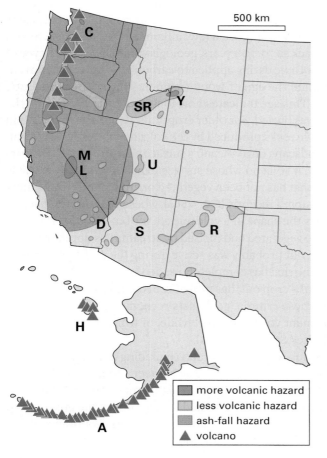

- *Cascade Volcanic Arc:* Subduction of the Juan de Fuca plate beneath North America has produced 18 stratovolcanoes in the Cascades (C), at least 7 of which erupted during the past 200 years. Future eruptions could be disastrous to Washington, Oregon, and California.

- *Basin-and-Range Rift:* Most of the rift has been volcanically inactive for the past million years. But volcanism happened in Death Valley (D) about 800 years ago and in Utah (U) as recently as 660 years ago. Huge explosive events at the boundary between the Basin and Range and the Sierra Nevada produced the Long Valley caldera (L) and Mono Lake (M). The Inyo Craters erupted 500 years ago. A magma chamber still underlies the area.

- *San Francisco Peaks:* Near the southern edge of the Colorado Plateau, eruptions during the past six million years produced a 3.8 km (2.4 mi) high stratovolcano. Numerous younger cinder cones, including Sunset Crater, have erupted during the last 1,000 years.

- *Rio Grande Rift:* During the past two million years, 700 eruptions occurred along this rift in New Mexico. A supervolcano eruption at 1.6 Ma produced the Valles Caldera. Volcanism occurred in the Valley of Fires about 5,200 years ago.

- *Yellowstone:* Hot-spot volcanism caused a VEI 8 explosion 640,000 years ago, and it continues to heat geysers and hot springs. Other eruptions happened along the Snake River Plain hot-spot track (SR), during the past few million years. Craters of the Moon lava erupted 2,000 years ago.

- *Aleutian Arc:* This 3,300 km (2,000 mi) long volcanic arc, resulting from subduction of the Pacific Plate beneath North America, hosts 50 active stratovolcanoes. Some erupt ash that is a hazard to airplanes.

- *Hawai'i:* The Big Island, at the southeast end of the Hawaiian hot-spot track, hosts four shield volcanoes that erupted during the past one million years. Kilauea and Mauna Loa erupted during the past one million years. Kilauea and Mauna Loa erupted multiple times during the past 1,000 years, and Lō'ihi is growing offshore.

FIGURE 4.52 Seismicity may indicate that an eruption is impending.

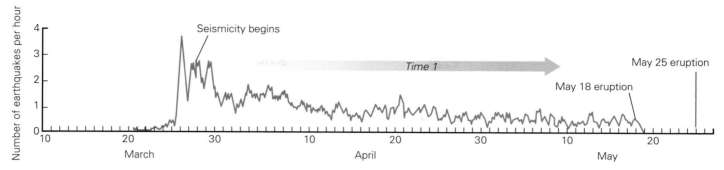

(a) The rate of earthquakes increased just before the 1980 eruption of Mt. St. Helens. Seismicity remained above normal until the May 18 explosive eruption.

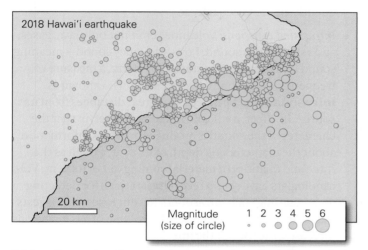

(b) Locations of earthquakes in southeastern Hawai'i recorded by USGS from January 1 to May 5, 2018. The earthquakes signaled volcanic unrest at Kilauea.

hot spring, bubbling *mudpots* may develop. And because magma under a volcano may release gases, an active volcano may also host **fumaroles**, vents emitting volcanic gases and steam. The deadly sulfurous gases coming out of fumaroles can precipitate mounds of sulfur crystals.

Predicting Eruptions

LONG-TERM PREDICTIONS. As is the case for earthquakes (see Chapter 3), the distribution of present-day volcanism serves as a reliable clue to where volcanic eruptions will happen in the future, for the positions of plate boundaries, rifts, and hot spots do not change significantly on a human time scale. New volcanoes occasionally do appear, as Paricutín did, but only within established regions of igneous activity.

Insight into the likelihood of an eruption on a time scale of decades to centuries may be gained by analyzing the **recurrence interval** of activity at a volcano, meaning the average time between eruptions. If, for example, the recurrence interval is 50 years, then a volcano will probably erupt a few times over the course of a couple of centuries. If the recurrence interval is 5,000 years, the likelihood of an eruption taking place in your lifetime is slim, though it's not impossible. Because the recurrence interval merely gives the average time between eruptions, it's not possible to predict the exact timing of an eruption years in the future.

SHORT-TERM PREDICTIONS. Fortunately, observers monitoring a volcano can detect signs that it may erupt within days to months. Such forecasting can lead to the evacuation of people from a hazardous region before an eruption. (In this regard, eruption prediction differs from earthquake prediction, in that no known technology can provide an accurate short-term forecast of an earthquake.) For example, when Mt. Pinatubo showed signs of activity in 1991, it had not erupted for 500 years, so many people weren't aware it was a hazard. But danger signs led geologists to convince officials to evacuate regions around the volcano, saving many lives.

To determine whether a volcano might be getting ready to erupt, observers look for signs that the volcano has entered a state of **volcanic unrest**, meaning that there has been a change from its normal state (the behavior that it has displayed over a period of time). Several kinds of changes can be clues to volcanic unrest:

- *Seismicity*: Seismic activity typically occurs when magma rises and accumulates in a magma chamber within or beneath a volcano, or when magma starts rising from the magma chamber into conduits within the volcano. Rocks surrounding the magma chamber break and slip, and small explosions may take place in the magma chamber. All this movement causes earthquakes, so in the days or weeks preceding an eruption, the region beneath a volcano becomes seismically active **(FIG. 4.52)**.

- *Changes in heat flow*: The intrusion of hot magma into a volcano increases local *heat flow*, the amount of heat passing through rock. In some cases, the increase in heat melts snow or ice on the volcano.

- *Increases in gases:* Even when magma remains below the surface, gases bubbling out of the magma percolate upward through cracks in the Earth and rise from volcanic vents, fissures, and fumaroles. So an increase in the volume of gas emission indicates that magma has entered the ground below. In the case of the 2018 eruption of Kilauea, the emission of SO_2 gas from a fissure signaled that magma was getting close to the surface. Because it may be extremely dangerous to approach the vent of a volcano about to erupt, geologists now employ drones to fly over vents to collect gas samples.

- *Changes in shape:* As a magma chamber inside a volcano fills, the magma pushes outward and can cause the surface of the volcano to bulge. The resulting shape changes can be measured in a number of ways. Traditionally, *tiltmeters* (devices that measure changes in slope) were installed on the flanks of volcanoes to detect such changes. More recently, satellite-based technologies have come into use. For example, very accurate GPS measurements can detect movements of a volcano's surface, and InSAR gives researchers the ability to detect elevation changes of as little as a few millimeters (**FIG. 4.53**).

- *Groundwater changes:* Prior to an eruption, changes in the permeability of rocks in the volcanic edifice and heating of groundwater may cause changes in groundwater levels and may produce new hot springs. Also, rising volcanic gases dissolving in groundwater may change the water's chemical makeup.

- *Computer models of volcanic systems:* High-speed computers have allowed researchers to make models depicting stresses around a volcano related to shape changes. These models can give insight into when and where faults and cracks might develop, weakening the volcano in ways that could trigger an eruption.

- *Movement of lava:* The appearance of lava at vents indicates that an eruption is starting. In some cases, the lava may be hidden by erupting gases and ash, but infrared imagery from satellites or drones may be able to "see" through the smoke and record the lava as a band of bright light. Similarly, changes in the level of lava lakes in vents may be an indication of lava movement and, consequently, of increased unrest.

- *Infrasound emissions:* Volcanoes emit not only lava, gases, and ash, but also sound. Low-frequency sound—meaning sound under 20 Hz (Hz stands for Hertz, a measure of frequency; 1 Hz is one cycle per second)—known as **infrasound**, can be detected at great distances from its source. In effect, a volcano is like a trumpet, in that the character of the sound it produces provides information about the diameter and depth of its vent, as well as the force and quantity of material coming out of its vent. Volcanologists have set up recording stations for detecting and recording such infrasound. Recent research suggests that as lava rises in a vent and the volume of gas emissions increases, the tone of the infrasound that a volcano produces changes.

FIGURE 4.53 An InSAR image of a volcano represents elevation change by rainbow-like bands. Each band represents 3 cm of uplift.

Interferomic synthetic aperture radar (InSAR) measures the elevation at different times and calculates the amount of movement between times.

FIGURE 4.54 Warnings from USGS volcano observatories.

(a) USGS field teams use drones to monitor volcanic gas emissions.

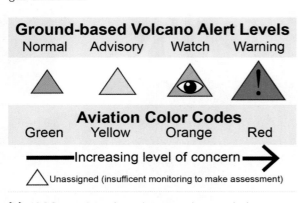

(b) USGS warnings for volcanoes that are being monitored.

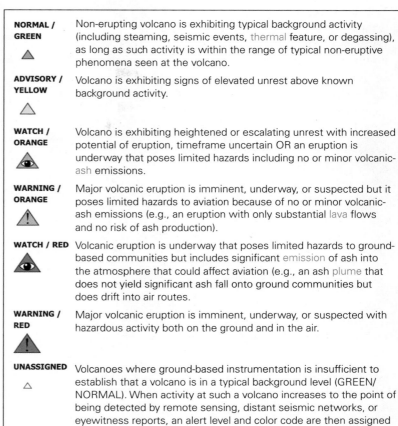

(c) Warnings used for alerting pilots to aviation hazards.

Volcano Observatories and Eruption Warnings

Because of the real threat that volcanoes pose to society, governments with sufficient resources have set up agencies or groups to monitor volcanic activity. In the United States, for example, the USGS oversees five **volcano observatories** (Alaska, California, Cascades, Hawai'i, and Yellowstone) where staff monitor volcanoes for seismicity, shape changes, gas emissions, and other signs of unrest. Each observatory has set up rapid-response teams that can land on a volcano and monitor an eruption in real time. Modern technology, of course, helps in this effort (FIG. 4.54a). During the 2018 eruption of Kilauea, geologists from the Hawai'i Volcano Observatory (together with colleagues flown in from other observatories) not only sent out field teams, but used helicopters, drones, and satellite imagery to map the path and volume of lava flows daily.

USGS volcano observatories also provide volcano alerts. For ground-based hazards, they issue a *Volcano Activity Notice* (*VAN*) that specifies one of four alert levels. If the VAN is Normal, there's no particular concern. If there's evidence of volcanic unrest, they issue an Advisory. When the likelihood of an eruption increases, they issue a Watch, and if an eruption is imminent or underway, they issue a Warning. The USGS also provides color-coded *Volcano Observatory Notifications for Aviation* (*VONA*), with green representing normal conditions, yellow indicating an advisory, orange indicating a watch, and red indicating that emission of ash is imminent or already underway (FIG. 4.54b, c).

Other countries have similar volcano monitoring schemes. For example, New Zealand, which faces significant volcanic hazards on its North Island—the city of Rotorura, for example, was built over a region of hot springs, and sulfurous steam sometimes pushes up through lawns and roads—has defined a series of volcanic alert levels issued by GeoNET (the national agency charged with assessing geologic hazards). In the GeoNET scheme, alert level 0 means there is no unrest, levels 1 and 2 indicate unrest, and levels 3–5 indicate that an eruption is taking place. Unfortunately, eruptions can happen suddenly and with little warning. In December 2019, the alert level for White Island (Whakaari), an island-arc volcano off the north

VISUALIZING A DISASTER
Volcanic Eruptions

Here, we see several types of volcanic hazards. (In the real world, all these hazards aren't likely to occur simultaneously.) The main Plinian eruption produces a convective plume, from which ash falls. Pyroclastic flows and lahars flow down the flank of the volcano and engulf towns at the base. Lightning sparks in the plume, and a jet encountering the ash glows with St. Elmo's fire. Note that the presently erupting summit vent occurs within the shell of an older, exploded volcano whose flank has eroded so it now hosts gullies. Lava from a flank eruption at a cinder cone flows across the landscapes and into a town. Fumeroles along a fissure produces sulfurous smoke. Where the lava enters the sea, laze forms. A lahar from an earlier lahar builds a delta.

- Vog
- Old cinder cone
- Fumaroles along a fissure
- Lava fountain
- Active cinder cone
- Old lava flow
- Active lava flow
- Burning buildings
- Lava spilling into the sea to produce laze
- Delta of pyroclastic debris

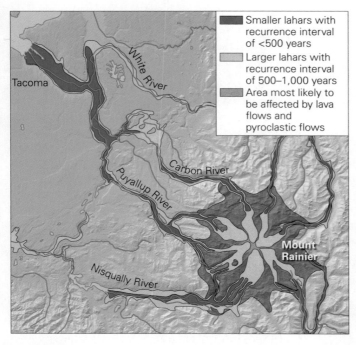

FIGURE 4.55 Volcanic hazard assessment map for Mt. Rainier, a volcano to the southeast of Tacoma, Washington.

shore of the North Island, had just been raised to a 2 because of rising emissions of SO$_2$ gas and increasing seismicity. At this level of unrest, cruise lines were still landing tourists on the island. Sadly, shortly after 50 tourists had been dropped off, the island underwent a VEI 2–3 phreatic eruption, blasting out tephra and hot gases, which caused several fatalities and severe injuries.

Evacuations in response to eruption warnings are often the best way to ensure civilian safety, but undertaking them is not a simple matter. Evacuations are expensive and can be dangerous in their own right. And although volcanic unrest tends to be fairly obvious, the exact moment of an eruption is generally a surprise, and the VEI of a particular eruption can't be predicted. Occasionally, unrest will increase dramatically and then decrease without an eruption. This possibility complicates the decision of whether to issue an evacuation order. For example, an evacuation of Naples, Italy, in response to unrest at Mt. Vesuvius could save tens of thousands of lives (BOX 4.3). But it could also cause mayhem and great expense, and if an eruption did not happen, would be likely to lead to interminable legal proceedings.

Mitigating Risk from Volcanic Hazards

There's no way to stop an eruption, so what can we do to prevent loss of life and property near an active volcano? As an important first step, geologists compile a **volcanic hazard assessment map** for the region (FIG. 4.55). This map delineates areas that lie in the path of potential lava flows, lahars, or pyroclastic flows. If an eruption is clearly imminent, people within these danger zones should evacuate. In the United States, the National Guard plays a significant role in overseeing evacuations. For example, during the 2018 Kilauea eruption, the National Guard coordinated its efforts with USGS geologists (FIG. 4.56). The geologists used gas detection meters to identify portions of the fissure that were about to erupt, so the Guard knew where to evacuate people from and could set up roadblocks to prevent people from entering the dangerous areas.

In some cases, lahars can travel down river valleys for distances significantly greater than might be expected for pyroclastic flows. So people living in the floodplains of river valleys that might host lahars should pay attention to warnings and should have evacuation routes planned. Lahar monitoring systems have been set up along some valleys to give a few minutes' warning to people downstream once a lahar has been detected.

In a very few cases, people have tried to divert or stop a lava flow. For example, during a 1669 eruption of Mt. Etna, a mafic lava flow approached the town of Catania. Fifty townspeople boldly hacked through the solidified side of the flow to create an opening through which lava could exit. They hoped to cut off the supply of lava feeding the end of the flow that was approaching their homes. Their strategy worked, but unfortunately, the diverted flow began to move toward the neighboring town of Paterno. Five hundred Paternoans chased away the Catanians. The hole was closed, and the flow headed back toward Catania, eventually overrunning part of the town. More recently, people have used high explosives to blast breaches in the flanks of lava flows and have employed bulldozers to build dams and channels to divert lava. Major

FIGURE 4.56 A USGS geologist tests for volcanic gas near Kilauea. A rise in gas production indicates where the next vent may erupt.

FIGURE 4.57 Efforts to stop the advance of lava flows.

(a) Officials used bulldozers to build a levee to redirect a lava flow from Mt. Etna.

(b) Residents of Heimaey sprayed a lava flow with seawater in an attempt to freeze the lava front.

(c) A reporter describing a lava flow to a television audience.

efforts to divert flows from 1983, 1992, and 2011 eruptions of Mt. Etna were successful (FIG. 4.57a). In 1935, in an effort to divert a flow threatening Hilo, Hawai'i, the US Army bombed lava tubes, hoping the tubes would collapse and block the flow. The bombing had no effect. During a 1973 eruption of lava from a fissure on the Eldfell volcano on the island of Heimaey, off Iceland, a lava flow began to move toward the port. Facing the likelihood that the lava would overwhelm the town, citizens thought of a novel way to stop the flow. They brought in fire hoses and, over the course of 5 months, continuously drenched the front of the lava flow with icy seawater (FIG. 4.57b). Their goal was to freeze the lava front to keep it from advancing, with the hope that the lava would find another path that did not threaten the town and its harbor.

The effort seemed to have worked, for while part of the town was covered, the harbor was spared. But whether the cold shower made a difference, or the flow stopped of its own accord, remains unknown.

An overview of volcanic risk mitigation suggests that, with few exceptions (such as the channeling of lava flows we have just described), evacuation provides the only confirmed approach to protecting people from a volcanic hazard. But because of the expense, and inherent risk, of evacuations, they can't be undertaken lightly. For that reason, volcanologists remain hard at work trying to refine their tools for interpreting how risk correlates with volcanic unrest, and they are working to improve lines of communications so that people potentially affected by an eruption can be made aware of the hazards and of their options for responding to them (FIG. 4.57c).

Take-home message . . .

Geologists distinguish among active, dormant, and extinct volcanoes, and they can provide short-term predictions of eruptions, allowing people to take precautions. There are many clues to volcanic unrest, such as increased seismicity, heat flow, and gas emissions; changes in the shape of volcanoes; and changes in groundwater. Volcanic hazard assessment maps help determine who should be evacuated. Rarely, it's possible to divert lava flows.

QUICK QUESTION Why can the decision to evacuate an area threatened with a volcanic eruption be so difficult?

BOX 4.3 DISASTERS AND SOCIETY

Naples, Italy: A city between two volcanoes

When Pompeii was destroyed by the eruption of Mt. Vesuvius in 79 C.E., it wasn't the only town to be wiped off the map. The nearby town of Herculaneum also disappeared beneath up to 20 m of pyroclastic debris, and a few other towns were severely affected. Nevertheless, archaeologists guess that the total population within the hazard zone of Mt. Vesuvius was probably less than 50,000. Today, in contrast, the city of Naples and several other towns, totaling about 4 million in population, lie within 20 km of the volcano's summit (FIG. Bx4.3a, b).

Vesuvius represents a hazard for the surrounding population because it remains an active volcano. While the eruption in 79 C.E. is the most famous, the volcano has erupted about 50 times since then. An explosive eruption in 1631 killed about 4,000 people, and a series of eruptions in 1774–1776 so fascinated William Hamilton, then the British ambassador to Naples, that he published a beautifully illustrated description of the eruptions. This work made a broader audience aware of volcanic activity and marked the founding of *volcanology*, the study of volcanoes. The most recent series of eruptions at Vesuvius took place between 1913 and 1944. These eruptions not only produced significant lava flows, but blew out immense clouds of pyroclastic debris. In 1944, American bombers based in the Naples area in preparation for World War II raids on Germany had to be swept free of ash before they could take off. Vesuvius still smokes and fumes today,

FIGURE Bx4.3 The city of Naples lies between Vesuvius and the Phlegraean Fields.

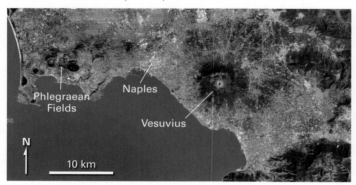

(a) A satellite image shows how urbanized the region in this volcanically active area has become. Most of the area shown was covered by a thick layer of ignimbrite from a prehistoric eruption in the Phlegraean Fields.

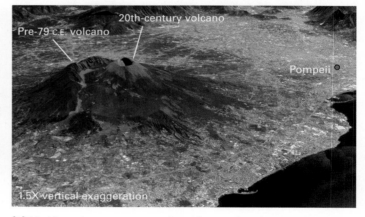

(b) Mt. Vesuvius towers over densely populated areas. A new volcano is growing within the remains of an older one.

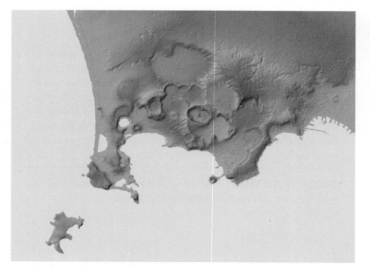

(c) A computer-generated shaded-relief map of the Phlegraean Fields emphasizes that this region is a nest of calderas.

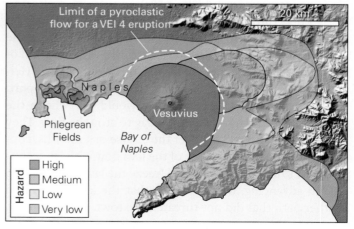

(d) A simplified volcanic hazard assessment map of the Naples region.

and it is being closely monitored to detect unrest. Despite the risk, neighborhoods are still being built on its flanks, with some homes sitting on igneous rocks born in 1944.

If the threat of Vesuvius weren't enough for Neapolitans to worry about, Naples has the dubious distinction of also sitting just to the southeast of the Phlegraean Fields (*Campi Flegrei*). The name, from the Greek word for flaming, is apt, for the 120 km² (46 mi²) area encompasses several calderas and today hosts numerous hot springs and fumaroles **(FIG. Bx4.3c)**. In some respects, the Phlegraean Fields may pose an even greater volcanic hazard than does Vesuvius, for geologic studies have revealed that it hosted a VEI 7 eruption about 40,000 years ago that ejected about 200 km³ (48 mi³) of debris. Some of this debris formed an ignimbrite 50 m (164 ft) thick that covered the entire area of present-day Naples. Another eruption 12,000 years ago produced a caldera 15 km (9 mi) in diameter, and a VEI 2 eruption less than 500 years ago produced a cinder cone now called *Monte Nuovo* (new mountain). Evidence of unrest in the area comes from studies of ground-surface movement. Surveys reveal that the land surface of the Phlegraean Fields moves up and down by up to a few meters over the course of several years. A rise of up to 1.8 m (6 ft) in 1984 led to the evacuation of 40,000 people. Fortunately, it turned out to be a false alarm. Movements in 2013 reached a rate of 3 cm (more than an inch) per month, and again raised alarm. Researchers believe the movements to be caused by either inflation and deflation of a magma chamber or the expansion of volcanic gases.

Given the proximity of Naples to two active volcanic systems, the volcanic hazard faced by the city should not be ignored. Officials have set up a volcanic observatory to monitor land-surface movements, gas emissions, and seismicity, and have developed volcanic hazard assessment maps to delineate areas that might be susceptible to ash falls and pyroclastic flows **(FIG. Bx4.3d)**. Evacuation plans are under development to accommodate a VEI 4 event. These plans assume that officials will have 2 weeks' advance warning of an eruption, for it would take that long to physically transport the large numbers of people living in the area to safer locations.

ANOTHER VIEW Volcanic eruptions—like this 2009 eruption of Santa María in Guatemala—can produce ash clouds that rise hundreds of meters to several dozen kilometers into the air, and affect atmospheric conditions.

Chapter 4 Review

Chapter Summary

- The term *volcano* can be used both for a vent from which gas, molten rock, and/or pyroclastic debris erupts, and for the edifice built from erupted material.
- Melt underground is called magma, whereas melt at the Earth's surface, erupting from a volcano, is called lava.
- Production of enough magma to cause igneous activity occurs where hot rock in the upper mantle or lower crust melts in response to decompression, addition of volatiles, or heat transfer.
- Magma composition depends on the composition of the source rock, the degree of partial melting, changes in magma composition as crystals form, and the extent of interaction with surrounding rock.
- Once formed, magma rises because of its buoyancy and because pressure squeezes it upward. Some magma solidifies underground to become intrusive igneous rock. Melt that reaches the Earth's surface erupts at a volcano and solidifies to become extrusive igneous rock.
- Igneous rocks are classified by their texture and composition. Crystalline igneous rocks contain interlocking crystals, whereas fragmental igneous rocks consist of pyroclastic debris. Grain size in crystalline igneous rocks depends on cooling rate.
- Plate tectonics theory can explain where igneous activity happens. Magma forms at convergent boundaries due to flux melting, and at divergent boundaries, rifts, and hot spots due to decompression melting. In continental crust, heat-transfer melting can also take place. At large igneous provinces (LIPs), immense amounts of lava have erupted.
- Lava viscosity depends on its composition. Viscous felsic lava flows tend to pile into domes at a volcano's vent, whereas less viscous mafic lava can flow great distances.
- Eruptions may come from a chimney-shaped conduit or from a long fissure. Some eruptions occur in a crater at a volcano's summit, and others on the volcano's flanks. Collapse into an emptied magma chamber produces a caldera.
- Shield volcanoes are broad, gentle domes formed by eruptions. Cinder cones are symmetrical hills. Stratovolcanoes can become large and consist of alternating layers of pyroclastic debris and lava.
- Effusive volcanic eruptions are dominated by mafic lava flows. During explosive eruptions, large quantities of pyroclastic debris are blasted skyward. Some rises in a convective cloud, but some rushes down in pyroclastic flows. Phreatic and phreatomagmatic eruptions involve interaction with water. Other explosive eruptions involve gas-rich, viscous felsic lavas.
- Geologists classify volcanic explosions using the volcanic explosivity index (VEI). Volcanoes that produce VEI 8 eruptions ($>$1,000 km^3 of debris) are called supervolcanoes. The most recent such eruption happened 73,000 years ago.
- The eruptive style of a volcano depends on the volume and character of igneous material produced, and on the explosivity of the eruption. Hawaiian eruptions are predominantly effusive, whereas Plinian eruptions involve large explosions.
- Volcanic eruptions pose many hazards: lava flows overrun roads and towns; ash falls from ash clouds crush buildings, damage fields and equipment, and can be a hazard to air traffic; pyroclastic flows incinerate everything in their path; lahars bury the land surface with muddy debris; and volcanic gases can suffocate people.
- Volcanic gases and ash from large explosive eruptions can affect the climate. Short-term climate change has happened in association with historic eruptions. Longer-term climate change may have resulted earlier from production of large igneous provinces.
- We can distinguish among active, dormant, and extinct volcanoes based on their eruption history and on physical clues.
- Indicators of volcanic unrest include seismic activity, changes in heat flow, emission of gases, edifice shape, groundwater, and infrasound emissions.
- Volcano observatories monitor volcanic unrest and issue warnings if eruptions are imminent or underway. Officials use these warnings to guide evacuation efforts on the ground and to issue warnings to air traffic.
- Volcanic hazard assessment maps indicate where lava flows, ash falls, pyroclastic flows, and lahars are most likely to occur.

Review Questions

Blue letters correspond to the chapter's learning objectives.

1. Describe the processes responsible for causing rock inside the Earth to melt. **(A)**
2. What factors determine the composition of magma or lava? Why does molten rock rise from depth to the surface of the Earth, and why does it solidify? **(A)**
3. Explain the difference between an intrusive and an extrusive igneous rock. Describe different types of intrusions. What factors control the cooling rate of molten rock? **(A, B)**
4. What factors control the viscosity of a melt, and how does viscosity affect the behavior of magma or lava? **(B)**
5. Describe the different kinds of material that can erupt from a volcano. Identify them on the figure. **(B)**
6. How are igneous rocks classified? Why do different crystalline igneous rocks have different grain sizes? **(B)**
7. Explain why igneous activity occurs where it does, in the context of plate tectonics theory. Identify the type of melting that takes place at each type of plate boundary, and at hot spots. **(C)**
8. Distinguish among shield volcanoes, stratovolcanoes, and cinder cones. What factors determine which type of volcanic edifice develops? **(D)**
9. What is the difference between effusive and explosive eruptions, in general? **(D)**
10. Explain the various causes of volcanic explosions. What is the volcanic explosivity index? **(E)**
11. How do large calderas form? Have there been any supervolcanoes during human history? **(E)**
12. How do Hawaiian, Surtseyan, and Plinian eruptive styles contrast with one another? **(E)**
13. How can lava flows affect people? How can volcanic ash affect people? (In your answer, distinguish between the effects of an ash cloud and the effects of a pyroclastic flow.) Which feature does the photo show? **(E)**
14. What is the difference between a pyroclastic flow and a lahar? Which moves faster—a lava flow or a pyroclastic flow? Why can volcanic ash be a hazard to air travel? **(F)**
15. How does CO_2 pose a hazard to people living near a volcano? What's the difference between laze and vog, and how do they affect people? Can volcanic gases cause climate change? **(F)**
16. What are the differences among active, dormant, and extinct volcanoes? What evidence can provide insight into which of these categories a particular volcano belongs in? Did the volcano in the image erupt recently? **(F)**
17. Why is there an association between seismicity and volcanic unrest? How can changes in the shape of a volcano be detected? What other clues may signal that an eruption could happen soon at a given volcano? **(G)**
18. Does the recurrence interval of eruptions provide insight into the likelihood of an eruption at a particular volcano during, say, the next 50 years? **(G)**
19. What types of warnings do volcano observatories provide? What features appear on a volcanic hazard assessment map? **(G)**
20. What steps can be taken to protect people from the effects of eruptions? Can people stop an eruption? Can they prevent lava from overrunning a town? **(G)**

On Further Thought

21. Look at a map showing the Earth's plate boundaries and explain why the Andean volcanic arc is so much longer than the Cascade volcanic arc. **(C)**
22. Do people living near the volcanoes of Hawai'i face the same kinds of volcanic hazards as do people living near Mt. Rainier, a volcano in the northwestern United States? **(E)**

5

UNSTABLE SLOPES
The Danger of Mass Wasting

By the end of the chapter you should be able to . . .

A. define mass wasting, and relate it to the more familiar concept of a landslide.

B. explain why topographic relief and slopes exist on the Earth.

C. evaluate factors that determine whether a slope is stable or unstable.

D. describe the characteristics and consequences of different types of mass wasting.

E. discuss the variety of events that can trigger a mass-wasting event.

F. explain how mass-wasting hazards can be identified, evaluated, and possibly prevented.

5.1 Introduction

Sunday, May 31, 1970, was a market day, and thousands of people were shopping in Yungay, a rural town along the edge of the Rio Santa valley of Peru. Suddenly, seismic waves from an M_W 7.9 earthquake jolted the town, and its buildings swayed and cracked. The seismic waves had been generated a few minutes earlier, about 120 km (75 mi) to the west, when a fault along South America's convergent boundary suddenly slipped. The waves had lost some of their energy when they reached Yungay, so destruction due to ground shaking in the town wasn't as extreme as it had been nearer the epicenter. Unfortunately for Yungay itself, a secondary disaster was about to unfold, with catastrophic consequences.

Near the peak of Nevado Huascarán, a 6.6 km (4 mi) high mountain just 14 km (9 mi) east of Yungay, ground shaking caused an 800 m (nearly half a mile) wide ice slab to break off a glacier. As the ice tumbled downslope, it disintegrated into blocks that accelerated to over 300 km/h (186 mph). Due to frictional heating, the ice melted and became a torrent of water, which, as it ripped up and incorporated sediment, evolved into a wet slurry viscous enough to carry house-sized boulders. When the slurry reached the mountain's base, 3.5 km (2 mi) below the summit, some of it rushed down a canyon and spilled into the Rio Santa Valley, where it came to rest. The rest flew up over a low ridge bordering the canyon, became airborne, and descended onto Yungay. Moments later, 4 minutes after the earthquake, debris had buried the town, along with almost 20,000 people **(FIG. 5.1)**. Today, the site is a grassy memorial to the victims entombed below.

On our dynamic planet, we can't always assume that land is terra firma, a solid foundation, as the disaster at Yungay illustrates. In fact, many locations on the Earth's surface host *unstable slopes*, meaning tilted ground that might start moving in response to gravity. Geologists and engineers refer to the gravity-driven transport of any material—bedrock that has broken free, **regolith** (unconsolidated sediment, rock fragments, and soil), or ice and snow—down a slope or escarpment as **mass wasting**, or *mass movement* **(FIG. 5.2)**. Landslides and avalanches, to use the common English terms you may recognize, are types of mass wasting. Geologists also distinguish several other types that we will discuss later in this chapter. Rock and regolith can become susceptible to mass wasting because fracturing and weathering at or near the Earth's surface weaken what was once strong, solid rock. Mass wasting plays an important role in the Earth System, for it contributes to the erosion of landscapes, and it serves as a primary source of sediment. But it also represents a danger to society, for it can cause disasters.

◄ A mass of moving mud and debris flowed over this highway after heavy rains, sadly, burying several vehicles and their occupants. Rescuers are trying to reach victims. Where there are slopes, instability can cause a disaster.

FIGURE 5.1 The May 1970 landslide disaster in Yungay, Peru.

This chapter introduces the hazards posed by unstable slopes. Specifically, we examine the various types of mass wasting as well as their causes and consequences. We also describe the precautions society can take to protect people and property from the threat posed by mass wasting.

5.2 Setting the Stage for Mass Wasting: Slopes

Uplift, Subsidence, and Slopes

If the Earth's surface were flat, mass wasting wouldn't take place. But our planet's surface isn't flat. Why? As we discussed in Chapter 2, movements and interactions of lithosphere

FIGURE 5.2 Both bedrock and regolith can undergo mass wasting on slopes.

(a) Red sandstone bedrock has broken free of cliffs and has fallen down to form aprons on slopes of gray shale.

(b) A layer of regolith in Death Valley has started to crumble. Loose clasts are rolling down a small slope.

plates can cause **uplift**, upward movement of the land surface, as well as **subsidence**, downward movement of the land surface. Such movements generate **relief**, differences in elevation between locations, and therefore, **slopes**, formally defined as tilted ground between locations at different elevations (FIG. 5.3a). When tectonic uplift or subsidence generates regional relief—as happens, for example, during mountain building—other components of the Earth System kick into action and modify the relief locally. Specifically, moving water, ice, and air drive **erosion**, the grinding away and removal of material at the Earth's surface, which can carve valleys and sculpt mountains, thereby producing local slopes.

Slope Steepness and Shape

You can easily walk up some slopes, but you'll need specialized climbing equipment to scale others. We can describe this difference, in everyday English, by using vague adjectives to characterize *slope steepness*, the angle that a slope makes relative to the horizontal (see Fig. 5.3a). Roughly speaking, *gentle slopes* have angles of less than 5°, *moderate slopes* range from 5° to around 35°, and *steep slopes* exceed about 35° (FIG. 5.3b, c). A *vertical slope* makes an angle of 90° relative to the horizontal, and an *overhang* exists where a line from a point on the slope straight down intersects the ground beyond the base of the slope (FIG. 5.4a). Technically, a *cliff* is a nearly vertical or overhanging slope, but in practice, you might consider any slope greater than about 60° to be a cliff, since you might tumble down it if you're not holding on. Highway engineers generally specify slope on a road as a percentage, called the road's *grade*. A 7% grade, the steepest that a superhighway can be, means that the surface of the road rises vertically by 7 m for every 100 m of horizontal distance.

A 7% grade represents a slope of 4°. A 30% grade (~17°) is about as steep as any public road gets (FIG. 5.4b).

Slopes come in a variety of shapes, including *convex up*, *planar*, and *concave up* (FIG. 5.5a). Some slopes display a *stair-step profile* on which cliffs alternate with gentler slopes (FIG. 5.5b). Slope shape depends on three key factors:

- *Strength*: The **strength** of a material characterizes its ability to avoid **failure** (breaking, flowing, or collapsing). A material's strength depends on whether it is *coherent* (a continuous solid) or *unconsolidated* (composed of separate pieces), and on whether it contains *planes of weakness* (discrete surfaces across which material is not strongly attached). Stronger materials can sustain steeper slopes. Therefore, steep cliffs are commonly composed of strong rock, and stair-step cliffs form where strong layers alternate with weak layers.

- *Climate*: Climate (meaning a region's average weather conditions and range of conditions; see Chapter 15) affects the character of slopes, for it determines whether running water, flowing ice, or wind is the main agent of erosion or deposition, and how fast these agents operate. Climate also controls vegetation growth and soil development on slopes. Consequently, a hill that has gentle slopes underlain by thick soil in a temperate climate might instead be a steep cliff of bedrock in a desert climate.

- *Process of slope formation*: Slopes formed by bedrock erosion look different from slopes formed by deposition of debris. Furthermore, different types of erosion form different types of slopes. For example, glacially carved valleys have a U shape in profile, whereas river-carved valleys have a V shape in profile (FIG. 5.5c, d).

FIGURE 5.3 Examples of relief and slopes.

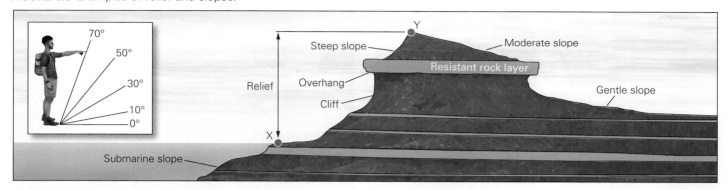

(a) Relief is the difference in elevation between two locations (X and Y). A slope is the tilt of the ground between locations at different elevations. The inset shows several slope angles, which vary from gentle to steep. You would need to scramble to get up a 50° slope, and a 70° slope is too steep to climb without ropes.

(b) Gentle slopes near the west coast of England. Such slopes are less susceptible to mass wasting.

(c) The steep slopes of a mountain range are the products of erosion acting on uplifted land. Such steep slopes are more susceptible to mass wasting.

FIGURE 5.4 You may come across slopes while hiking or driving.

(a) A cliff, with overhangs, near Sedona, Arizona.

(b) A sign warning of a steep grade along a road in England.

FIGURE 5.5 Various shapes of slopes.

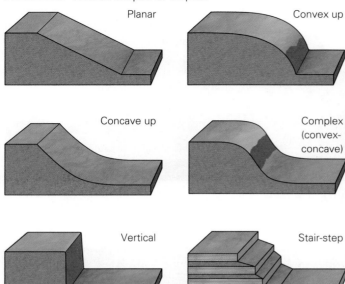

(a) Slopes can be classified by their overall shape. Most slopes are complex—different parts of a slope have different shapes.

(b) The walls of the Grand Canyon display a stair-step profile due to the alternation of strong and weak beds.

(c) A U-shaped glacially carved valley in Utah.

(d) A V-shaped stream-carved valley in Utah.

Take-home message . . .

Slopes exist because of uplift or subsidence. They are modified locally by erosion. Slope steepness varies with location and depends on the strength of the material forming the slope, as well as on climate and the process of slope formation.

QUICK QUESTION Why do stair-step slopes form?

5.3 Slope Stability

The Safety Factor

Let's do a simple thought experiment. Imagine placing a brick-shaped block of rock on a horizontal surface. Gravitational force pulls the block straight downward, because gravity pulls objects toward the center of the Earth. We can represent the force of gravity as a vector that points straight

FIGURE 5.6 The stresses acting on a block sitting on a slope change as the slope angle changes. If the downslope stress exceeds the resistance stress (which depends on friction and cohesion), the block slides.

σ_n represents the normal stress across the contact surface due to the pull of gravity. This stress is the component of gravity that squeezes the block against the surface of the slope.

σ_d represents the downslope shear stress. This stress is the component of gravity that pulls the block down, parallel to the slope.

σ_r represents the resistance stress (also called shear strength) across the contact surface. It keeps the block from moving and, as we'll see, it depends on friction and cohesion. Notably, decreasing σ_n decreases σ_r, because when surfaces are squeezed together less tightly, friction decreases.

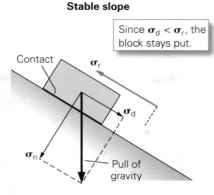

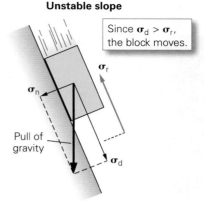

(a) On a slope, gravity resolves into a normal stress across the contact, and into a downslope shear stress parallel to the contact. The resistance stress (shear strength) can be symbolized by an arrow that opposes the downslope shear stress.

(b) Relative to case (a), resistance stress on a steeper slope decreases and downslope shear stress increases. In this example, σ_d exceeds σ_r, so sliding takes place.

down. Recall that a **vector** is a quantity that has both magnitude and direction and can be represented by an arrow. When we are describing a force, the length of the arrow represents the magnitude of the force, and the arrow's orientation represents the direction of the force. What happens to the block? If the material beneath the block were composed of water, the block would sink through the water surface. But if the block were placed on strong, solid bedrock, it would just stay on the surface, without moving. The block, due to gravity, applies force, a push, to the Earth's surface. Since this push spreads over the area of contact between the block and the ground, we refer to it as **stress**, defined as force divided by the area over which the force acts (see Chapter 2). Written as an equation, stress = force ÷ area, or in symbols, $\sigma = F \div A$, where the Greek letter sigma represents stress. The stress is perpendicular to the surface, so we refer to it as a **normal stress**.

Now, imagine placing the block on a gentle slope **(FIG. 5.6a)**. The vector representing the pull of gravity still points straight down. But now we can separate that vector into two components: a normal stress perpendicular to the slope, and a downslope vector, or **shear stress**, parallel to the slope. What happens to the block? The block doesn't necessarily start moving down the slope because a *resistance stress* (also known as *shear strength*), symbolized by another vector pointing upslope, also acts on the block. If the upslope resistance stress equals or exceeds the downslope shear stress, the block stays in place. The resistance stress depends on phenomena such as friction and cohesion, which we discuss later.

What happens if we put the block on a steeper slope? Again, we separate the vector into a normal stress component perpendicular to the slope and a shear stress component parallel to the slope. Note that the relative magnitudes of these

FIGURE 5.7 Regolith tumbles downslope when the downslope shear stress exceeds friction and cohesion between grains.

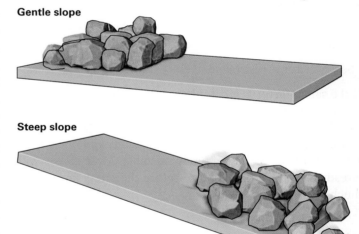

stresses are different than they were on the gentler slope—the normal stress has become smaller, and the downslope shear stress has become bigger. When the downslope shear stress becomes larger than the upslope resistance stress, the block moves downslope **(FIG. 5.6b)**.

Let's repeat the above thought experiment one last time by replacing the block of solid rock with a mound of dry gravel. On a horizontal surface, the gravel doesn't move, and on the gentle slope, it still doesn't move. But on the steep slope, it starts sliding downslope **(FIG. 5.7)**. Because the gravel isn't a coherent block, its downslope movement involves the slipping and rolling of individual fragments past one another.

From this thought experiment, we can draw a general conclusion about movement on slopes: on a **stable slope**, slope

FIGURE 5.8 The Earth's surface is weak enough to fail on slopes because rock near the surface contains joints, and because regolith is unconsolidated.

(a) Joints break sandstone layers into blocks, as seen in the walls of the Grand Canyon.

(b) Blocks of rock have broken away along joints and tumbled down this cliff in Utah.

(c) Regolith on this excavated slope in central China includes unconsolidated sediment and soil.

materials tend to stay in place because the resistance stress exceeds the downslope shear stress, whereas on an **unstable slope**, slope materials tend to move downslope because the downslope shear stress exceeds the resistance stress. When material starts moving on an unstable slope, we say that **slope failure** has occurred.

Engineers represent the stability of a slope with a number called the **safety factor**, defined by the following equation:

Safety factor = resistance stress ÷ downslope shear stress

If the safety factor is greater than 1, then the slope is stable, but if it's less than 1, then the slope is unstable. For example, a slope with a safety factor of 1.2 is stable. A slope with a safety factor of 1 is on the verge of being unstable, and one with a safety factor of 0.8 is definitely unstable. Most slopes will already have undergone failure before the safety factor gets much below 1.

What Provides Resistance to Mass Wasting?

If the Earth's surface were covered by unweathered and unbroken rock, mass wasting would be of little concern, for such *intact rock* has great strength, provided by the chemical bonds that hold minerals together. But the rock of the Earth's upper crust isn't intact. Most bedrock near the Earth's surface has been fractured by natural cracks in rocks, called **joints**, and blocks of rock can break free of bedrock along those joints (FIG. 5.8a, b). In addition, much of the Earth's surface has a cover of regolith resulting from the weathering of rock and the production of soil (FIG. 5.8c). Therefore, the question of what produces resistance to mass wasting on a slope has three parts: (1) What can cause a joint to grow until it completely separates solid rock from bedrock? (2) What prevents a block of rock from slipping on its substrate? (3) What prevents unattached grains in regolith from moving relative to their neighbors?

In the case of solid bedrock on a cliff face, resistance comes from chemical bonds that hold mineral crystals together. Breaking takes place when downslope stress becomes great enough to break these bonds. Such large stresses can develop at the tip of an already existing crack—to demonstrate this, make a small cut in a sheet of paper, and then pull on the sides of the paper (FIG. 5.9). Without having to pull very hard, the tear grows quickly and separates the paper in two. Similarly, pre-existing imperfections in rock can grow into joints due to stresses generated when bedrock from deep underground gets closer to the Earth's surface due to the erosion and removal of overburden (overlying material). This process, which takes millions of years, causes the rock to cool and

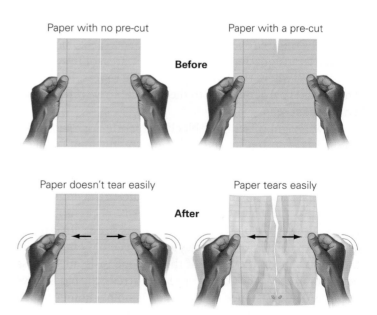

FIGURE 5.9 Joint growth resembles growth of a tear in a sheet of paper. When you pull on an intact sheet of paper, it does not tear easily. When you pull on a sheet of paper that has been pre-cut, the paper tears easily. Similarly, once a joint has started to form, it can propagate relatively easily.

FIGURE 5.10 Joints provide planes of weakness that allow rock to break away from bedrock.

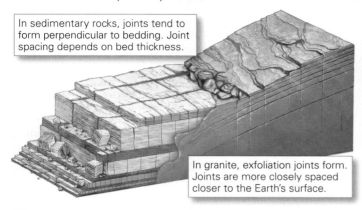

In sedimentary rocks, joints tend to form perpendicular to bedding. Joint spacing depends on bed thickness.

In granite, exfoliation joints form. Joints are more closely spaced closer to the Earth's surface.

(a) Joints in sedimentary beds break the beds into rectangular blocks, whereas exfoliation joints are parallel to the Earth's surface.

(b) Sedimentary rocks in Bryce Canyon in Utah contain vertical joints

(c) Exfoliation in granite of the Sierra Nevada in California produces slope-parallel joints.

contract, and decreases the pressure acting on it, causing the rock to expand upward. Stress produced by such expansion and contraction can crack rock. Once the rock reaches the surface, joints may grow further due to local stresses caused by the freezing and thawing of ice, or the growth of tree roots, or perhaps even daily heating and cooling by the Sun.

Notably, the character of joints varies among rock types **(FIG. 5.10)**. For example, in sedimentary rocks, joints tend to form perpendicular to bedding, and they are more closely spaced in thin beds than in thick ones. In horizontal beds, these joints tend to be vertical, so they break the beds into rectangular blocks. In igneous rocks such as granite, *exfoliation joints*, which are parallel to the land surface, form in addition to joints perpendicular to the land surface.

FIGURE 5.11 Various phenomena cause resistance to sliding.

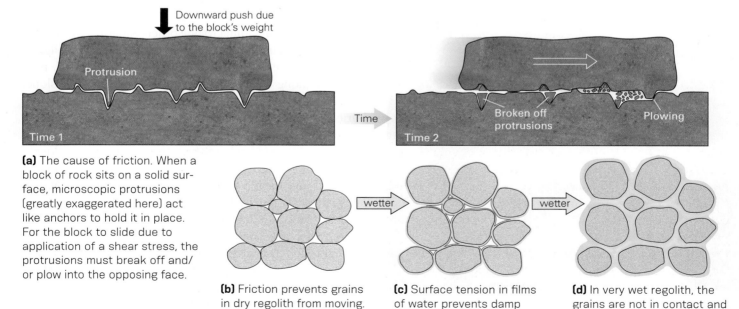

(a) The cause of friction. When a block of rock sits on a solid surface, microscopic protrusions (greatly exaggerated here) act like anchors to hold it in place. For the block to slide due to application of a shear stress, the protrusions must break off and/or plow into the opposing face.

(b) Friction prevents grains in dry regolith from moving.

(c) Surface tension in films of water prevents damp grains from moving.

(d) In very wet regolith, the grains are not in contact and can flow with the water.

5.3 Slope Stability • **181**

When a joint finally separates a block from bedrock completely, the block can fall or slide unless friction prevents its movement. **Friction** is the resistance to sliding that exists because natural microscopic protrusions on one surface dig into the opposing surface and act like little anchors (**FIG. 5.11a**). Slip can take place only when the downslope shear stress on the block becomes large enough to break off these protrusions, or cause them to gouge a scratch into the surface below.

In the case of dry unconsolidated material, friction between individual grains typically provides resistance to sliding (**FIG. 5.11b**). In some materials, however, additional resistance comes from **cohesion**, the attraction between grains caused by weak electric charges that attract one grain to another. In the case of damp unconsolidated material, meaning material in which microscopic films of water coat grain surfaces, the *surface tension* of water (caused by the electrical attraction of water molecules to one another) provides cohesion (**FIG. 5.11c**)—that's why you can make a steep-walled sand castle out of damp sand, but not out of dry sand. But when unconsolidated material becomes *saturated*, meaning that liquid water completely fills the *pores* (open spaces) between grains and separates them, surface tension can no longer prevent grains from moving (**FIG. 5.11d**).

The resistance stress acting on a dry granular material, such as sand or gravel, causes the material to form a cone-shaped pile when poured on a flat surface. The steepest slope angle such a cone can attain without collapsing is called the material's **angle of repose** (**FIG. 5.12a**). Most dry, unconsolidated materials (such as dry sand) have an angle of repose between 30° and 37°. The specific angle depends partly on the shape and size of the grains. For example, slopes composed of large, irregularly shaped grains have steeper angles of repose (up to 45°). Blocks that fall from cliffs in dry climates build into **talus piles** whose surface slopes represent their angle of repose (**FIG. 5.12b**). When sand gets damp, its angle of repose increases because of the cohesion that surface tension provides.

FIGURE 5.12 Angles of repose for granular materials.

(a) The angle of repose is different for different materials.

(b) Quartzite blocks have accumulated to form a talus pile at a steep angle of repose in the Uinta Mountains, Utah.

Failure Surfaces

On August 17, 1959, vibrations from a strong earthquake jarred the region of Madison Canyon, Montana, just west of Yellowstone National Park. The bedrock beneath the walls of the canyon includes schist. This metamorphic rock contains lots of *mica*, a mineral that consists of thin sheets. In schist, mica sheets are aligned parallel to one other—in other words, this rock displays *foliation*—and the bonds between the aligned sheets are quite weak. So when the ground vibrated, rock detached along a foliation plane and tumbled downslope. Unfortunately, 28 campers sleeping on the valley floor were buried under a mound of debris 45 m (148 ft) thick. This debris also dammed the Madison River, causing a lake to fill rapidly upstream. To avoid having the lake breach the dam suddenly and cause flooding downstream, the Army Corps of Engineers frantically dug a spillway to lower the lake level (**FIG. 5.13**). The lake, called Earthquake Lake, still exists today.

The Madison Canyon Slide illustrates how an underground *plane of weakness* (meaning a surface with less strength than the overall shear strength of the material above it) can control the resistance to sliding at a location. If downslope movement begins on the plane of weakness, we say that the plane has become a **failure surface**. At Madison Canyon, a layer of schist served as the failure surface. Geologists

FIGURE 5.13 The Madison Canyon Slide immediately after it happened in 1959. The debris blocked a river, causing Earthquake Lake to fill with water. The inset shows the lake today.

recognize several different kinds of features that can become failure surfaces **(FIG. 5.14a–c)**, including layers of wet clay or wet sand; joints; bedding planes (shale and salt beds are particularly weak); and metamorphic foliation planes. (Notably, a clay layer that failed in the past is more likely to fail again in the future, because slip causes clay flakes to align with one another, making them even more slippery.) The orientation of a plane of weakness relative to a slope determines the likelihood of its becoming a failure surface **(FIG. 5.14d)**. Specifically, planes of weakness oriented parallel to the slope of the land surface are more likely to fail than are planes perpendicular to the slope.

Take-home message . . .

On a stable slope, material (rock or regolith) remains in place, whereas on an unstable slope, it is likely to move downslope. Stability depends on the balance between upslope resistance stress and downslope shear stress. The ratio of these numbers gives the slope's safety factor. The source of resistance (chemical bonds, friction, surface tension) depends on the slope composition and on the presence of failure surfaces.

QUICK QUESTION What is the angle of repose, and what does it depend on?

FIGURE 5.14 Planes of weakness in different kinds of rock.

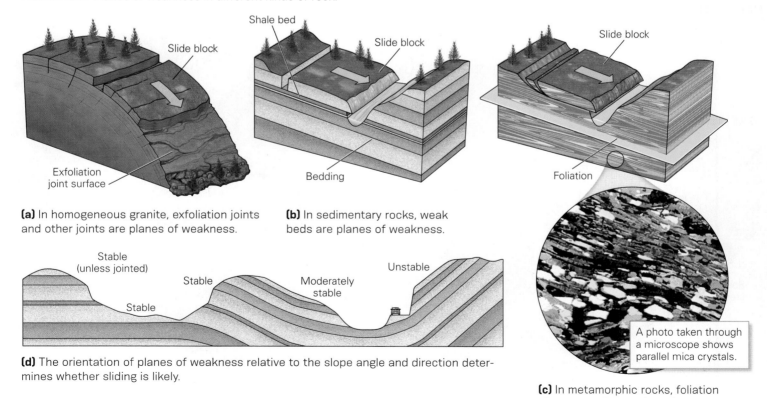

(a) In homogeneous granite, exfoliation joints and other joints are planes of weakness.

(b) In sedimentary rocks, weak beds are planes of weakness.

(c) In metamorphic rocks, foliation planes are planes of weakness.

(d) The orientation of planes of weakness relative to the slope angle and direction determines whether sliding is likely.

5.4 Types of Mass Wasting

Geologists and engineers find it useful to distinguish among different types of mass wasting on the basis of four features: (1) the type of material involved (rock, regolith, mud, ice, or snow); (2) the velocity of movement (slow, intermediate, or fast); (3) the character of the moving mass (coherent, chaotic, or cloudlike); and (4) the environment in which the movement takes place (subaerial or submarine). In this section, we examine the various types of mass wasting that occur on land, roughly in order from slow to fast (FIG. 5.15). You'll see that each type has a name—in everyday English, the term **landslide** informally substitutes for slumps, mudflows, debris flows, and sometimes rockfalls.

Creep and Solifluction

Creep involves the slow, gradual downslope movement of regolith. It happens when regolith alternately expands and contracts due to freezing and thawing (because water expands when it freezes and contracts when it thaws); wetting and drying (because some types of clay, a common mineral in regolith, expand when they absorb water and contract when they dry out); or warming and cooling (because regolith expands when it warms and contracts when it cools). To exemplify how creep works, let's consider the consequences of seasonal freezing and thawing.

In winter, when water in the pores between grains in a layer of regolith freezes and, therefore, expands, the layer thickens overall, and particles in the regolith move outward, in a direction perpendicular to the slope. During the spring thaw, when the water becomes liquid again, gravity makes the particles sink vertically. Consequently, the particles migrate downslope slightly (FIG. 5.16a–c). The presence of liquid water allows the grains to slide past one another during this process. You can't see creep by staring at a hillslope because it occurs too slowly, but over a period of years, creep causes trees, fences, gravestones, walls, and foundations built on a hillside to move downslope. Since the amount of creep varies with location and depth, structures don't move uniformly downslope, and consequently they may break apart. Notably, regolith closer to the ground surface moves faster than regolith deeper down, so tombstones and walls embedded in the regolith rotate and tilt downslope (FIG. 5.16d). Similarly, trees with deep roots tilt because shallower roots (and the tree above them) move downslope while the deeper roots don't. Because trees grow vertically, they develop a pronounced curvature at their base. Creep generally does not cause casualties, but it can destroy property and infrastructure.

In polar or high-elevation regions, which remain very cold most of the year, regolith can freeze to become **permafrost** (permanently frozen ground). During summer, however, the uppermost 0.5–3 m (1.6–10 ft) of the permafrost may thaw. Because meltwater can't sink into the underlying, still-frozen

FIGURE 5.15 Various types of mass wasting on land, roughly in order from slower (left) to faster (right).

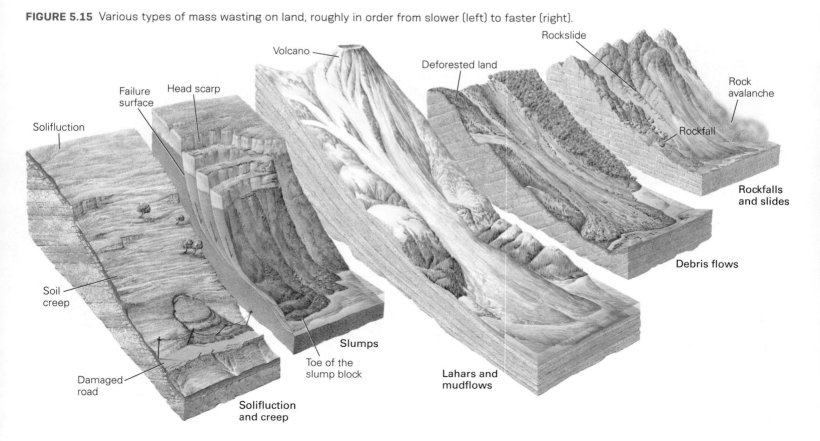

FIGURE 5.16 The process of creep, a slow type of mass wasting.

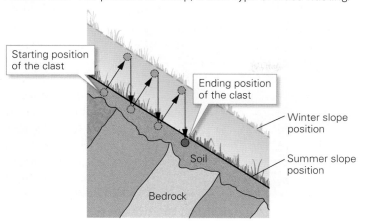

(a) When regolith expands, particles rise perpendicular to the ground. When regolith shrinks, particles move vertically downward. After several cycles, a particle ends up far downslope.

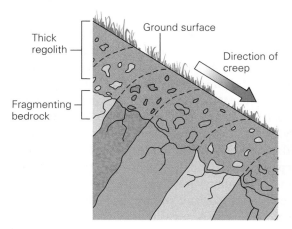

(b) As rock layers weather and break up, the resulting regolith creeps downslope.

(c) Creep is visible in this quarry in China. The quarry exposes vertical rock layers. Near the ground surface, these layers tilt downslope, as indicated by the dashed line in the inset.

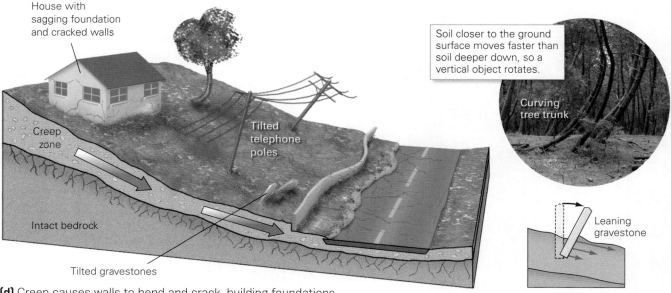

(d) Creep causes walls to bend and crack, building foundations to sink, and power poles and gravestones to tilt.

5.4 Types of Mass Wasting • **185**

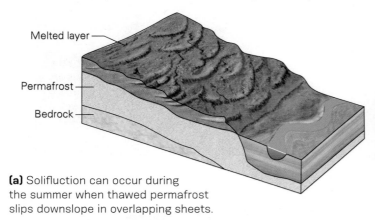

FIGURE 5.17 The process of solifluction in regions with permafrost.

(a) Solifluction can occur during the summer when thawed permafrost slips downslope in overlapping sheets.

(b) An example of solifluction on a hillslope.

permafrost, the melted layer becomes soggy and weak and flows slowly downslope in overlapping sheets. Geologists refer to this kind of creep as **solifluction** (FIG. 5.17). It has proved to be a hazard to people building structures, such as houses and pipelines, in cold regions.

Slumps

The majestic Holbeck Hall Hotel perched on a 60 m (197 ft) cliff overlooking the eastern coast of England from 1879 until June 5, 1993. On that day, a block of land extending inland 70 m (230 ft) from the cliff face slipped downslope toward the sea, taking half the hotel with it (FIG. 5.18). Fortunately, telltale cracks in the ground weeks before, and a smaller collapse the day before, had led management to evacuate the hotel, so no one was hurt. But the landmark, along with the substrate beneath it, eventually slid out onto the beach, where the pounding surf carried it away.

A relatively slow-moving mass-wasting event like the one that destroyed the Holbeck Hall Hotel, during which rock or regolith stays mostly coherent as it moves a relatively short distance down a distinct failure surface, is called a **slump**, and the moving mass itself is a *slump block*. We distinguish two types of slumps by the shape of the failure surface. A *translational slump* moves down a planar failure surface (FIG. 5.19a), whereas a *rotational slump* moves down a concave-up, spoon-shaped failure surface (FIG. 5.19b). During a rotational slump, the slump block rotates around a horizontal axis.

A variety of landscape features develop during slumping. The exposed upslope edge of the failure surface is a cliff called a **head scarp** (FIG. 5.19c), and the downslope end of the slump block becomes the slump's *toe*. Typically, the toe moves up and over the pre-existing land surface, producing a curving mound. But along seacoasts or riverbanks, the toe ends up in the water, where currents or waves erode it away (FIG. 5.19d). The slump block may have a **hummocky** surface, meaning that it's bumpy and irregular, with many small depressions and hills. In some cases, the upslope and downslope ends of a slump block break into a series of discrete slices, each separated from its neighbor by a small sliding surface. These slices produce *transverse ridges* (perpendicular to movement direction) near the toe, and *transverse cracks* and *transverse scarps* near the head scarp. Rotational slump blocks tend to spread out laterally at the toe, so *radial cracks* (oriented parallel to the overall movement direction) may also develop in the slump block (FIG. 5.19e).

FIGURE 5.18 Collapse of the Holbeck Hall Hotel, in Scarborough, England, in 1993. The inset shows an enlargement of the damaged building.

FIGURE 5.19 Slumps are relatively slow mass wasting events. Slip can occur on a curved or a planar failure surface.

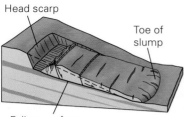

(a) A translational slump. The moving mass slides on a planar failure surface.

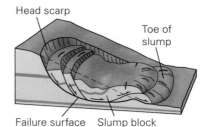

(b) A rotational slump. Note the curving transverse scarps near the head scarp, and the small transverse ridges near the toe.

(c) A developing slump along a highway.

(d) The river has washed away the toe of this small slump.

(e) A large slump in Romania. Note that the slump block broke up.

Slumps come in all sizes, from only a few meters across to tens of kilometers across. They move at speeds from millimeters per day to tens of meters per minute. Even slow-moving slumps like the one at Portuguese Bend, California **(BOX 5.1)**, can cause significant damage because movement and deformation within the slump block break foundations, roads, pipes, and utility lines.

Mudflows, Debris Flows, and Lahars

When a mass of regolith moving downslope contains sufficient water, it becomes too weak to maintain coherence and begins to behave like a viscous fluid. Geologists refer to a moving slurry of mud as a **mudflow** (or *mudslide*, or *earthflow*) **(FIG. 5.20a)**, and a moving mixture of mud and larger rock fragments as a **debris flow** (or *debris slide*) **(FIG. 5.20b)**. Because

FIGURE 5.20 Examples of mudflows and debris flows.

(a) After a heavy rain in 2014, a slump quickly evolved into a mudflow, which buried part of a village in Afghanistan, causing hundreds of casualties.

(b) Debris flowed down a valley following heavy rain in the Uinta Mountains, Utah. Note that the debris flow consists of a chaotic mixture of rock chunks and mud.

BOX 5.1 DISASTERS AND SOCIETY

The Portuguese Bend slump

Along the coast of southern California, in the Los Angeles area, housing covers much of the landscape. An exception can be seen along a 3 by 2 km (2 by 1 mi) stretch of coast called Portuguese Bend. Here, relatively few houses stand, and the land has a hummocky surface **(FIG. Bx5.1)**. Why?

The substrate beneath the land surface at Portuguese Bend consists of volcanic ash that has weathered to form weak clay. The clay layer overlies shale bedrock. At times in the past, a plane of weakness, about 60 m (197 ft) long, at the boundary between the weak clay and the underlying bedrock had become a failure surface, allowing the land above to slowly slip seaward. In other words, Portuguese Bend was a large translational slump that transported regolith downslope, where it was eroded away by ocean waves. Due to this erosion, there was nothing solid at the toe to hold back further movement, so only friction on the failure surface provided resistance stress.

In the early 1950s, a time when California's population was booming and new housing was being constructed at a rapid pace, the Portuguese Bend slump wasn't moving. Developers built houses on its slopes, easily attracting buyers who loved the spectacular ocean views. In retrospect, construction of a housing development on an old slump wasn't the wisest choice, because it added weight to the slump block and therefore increased the downslope shear stress. The weight came not only from the houses (each of which weighed about 100 tons) and roads (which cumulatively weighed over 30,000 tons), but also from water that infiltrated the ground from septic tanks and lawn watering. The added water also caused the clay minerals to expand, making the failure surface weaker. The increased downslope stress and the weakening of the failure surface caused the safety factor to drop below 1, and downslope movement began again in 1956, at rates on the order of 1–2 cm (0.4–0.8 in) per day. Consequently, foundations and walls cracked, floors began to tilt, water mains sprang leaks, and power cables snapped. Eventually, 150 houses were destroyed. Because the slump's rate of movement was so slow, it caused no casualties, but it led to millions of dollars in damage, and to most of the houses being abandoned. Some homeowners who couldn't afford to leave tried to deal with the disaster by repairing pipes and by re-leveling their homes using hydraulic jacks.

Could the Portuguese Bend disaster have been avoided? Probably. Careful surveying and geologic analysis could have revealed the potential for slumping. But in the case of Portuguese Bend, such studies were carried out only after the fact, in association with lawsuits. Resolution of the liability issues took years: Should the developers pay for undertaking construction without adequate risk assessment? Should homeowners suck up the costs because they didn't think about risk and exacerbated the problem by watering lawns and flushing toilets? In the end, it was an entity with money—Los Angeles County—that covered part of the cost. The county was fined $10 million for its role in adding weight to the slump block by building roads.

FIGURE Bx5.1 The Portuguese Bend slump on the coast of California.

mudflows and debris flows have greater density than clear water, they can buoy up and carry large chunks of rock as well as houses and cars (BOX 5.2). When a flow moves down a slope, it may spread out to form a broad layer at its toe. If a flow occurs at the head of a valley or along the side of a valley, it may continue to flow for a distance down the valley.

The speed at which the mud or debris in a flow moves depends on the slope angle and on its water content. Flows move faster on steeper slopes, and they move faster if they contain more water, which makes them less viscous. Therefore, on a gentle slope, a drier mudflow moves relatively slowly, but on a steep slope, very wet mud may move at rates of up to 100 km/h (62 mph).

Slower mudflows and debris flows generally aren't deadly, but nevertheless they can cause expensive destruction. For example, in 1983, after a winter of heavy snow and a spring of heavy rain, the mountain slope bordering the town of Thistle, along the Spanish Fork River in Utah, became a slow debris flow because water saturation had increased the weight of regolith. No one saw the slope start moving, so people didn't realize what was happening until railroad tracks at the base of the slope warped and the adjacent highway buckled. After 3 days of movement, the railroad and the highway had to be closed. The town itself did not lie in the path of the debris flow, but was just upstream. Unfortunately, the debris spread out over the river, building a dam 67 m (220 ft) tall, and backing up a lake that submerged Thistle (FIG. 5.21). No one was injured, but due to the loss of the railroad, highways, and town, the debris flow caused over $200 million in damage.

Rapid mudflows and debris flows move so quickly that people can't get out of their way in time. For this reason, such flows can cause not only devastating damage, but also loss of life. The greatest loss of life associated with debris flows occurred in December 1999, along the mountainous Caribbean coast of Vargas, a state in northern Venezuela. This loss, which came to be known as the Vargas tragedy, was due not to a single flow, but to several over a 2-day period when an intense storm drenched the landscape with 91 cm (36 in) of rain. This storm happened only a couple of weeks after a previous one, so the regolith underlying the steep slopes of the Vargas region was already saturated. Mudflows and debris flows had happened many times before in the region. In fact, the largest towns were built on fans of debris from past, long-forgotten flows, because the fans provided flat ground on which construction was easier. During the 1999 storm, thousands of small landslides carried regolith down steep slopes into flooding stream valleys, in which the debris mixed with water to become a fast-moving slurry that flowed out of canyons onto the densely populated fans. The muddy debris overwhelmed houses and apartment buildings and buried streets. When the skies finally cleared, up to 30,000 people had died and more than 75,000 had been displaced. Unfortunately, rescue operations were slow, and relatively little financial assistance was provided to the afflicted population. More than a decade later, thousands of people remained homeless.

In some cases, the material in a debris flow comes from piles of mine waste rather than from natural regolith. The failure of a waste pile caused the 1966 Aberfan disaster in Wales. Aberfan had been a coal-mining town for almost a century when, in 1958, workers began to dump piles of waste rock and mud from the mining operation onto a hillslope overlooking the town. One of these waste piles buried several groundwater springs. After weeks of heavy rainfall, not only had the waste pile become saturated, but the flow emerging from the springs beneath it had increased. The additional weight in the debris pile increased downslope stress, and the greater flow of the springs effectively reduced friction at the contact between the pile and the original slope, making the original slope behave like a failure surface. Consequently, the safety factor of the pile dropped below 1, and the entire pile became unstable and began to flow toward the town at speeds of up to 30 km/h (19 mph). It overwhelmed farms and houses in its path, and at its toe, buried a school in a 10 m (33 ft) thick layer. Horrifically, 112 children and 28 adults died beneath the debris. The loss of so many children, from an event that might have been avoided, made the Aberfan disaster a national tragedy for the United Kingdom and led to the enactment of new safety regulations.

FIGURE 5.21 The Thistle, Utah, debris flow.

5.4 Types of Mass Wasting • 189

BOX 5.2 DISASTERS AND SOCIETY

Mudflows in Rio de Janeiro

In recent decades, the population of Rio de Janeiro, Brazil, has grown so much that densely populated communities of makeshift shacks cover many of the slopes rising toward the sugarloaf mountains that surround the city **(FIG. Bx5.2a)**. These shantytown communities, called favelas, were built on the thick regolith produced by long-term weathering of igneous and metamorphic bedrock in Brazil's tropical climate. That bedrock, which once consisted of interlocking crystals of quartz, feldspar, and other silicate minerals, transformed over millions of years into *saprolite*, a highly weathered, soft, porous rock composed mostly of clay and quartz. Saprolite is much more vulnerable to mass wasting than the rock from which it was derived.

Because the favelas are built independently of any urban government, they have inadequate storm drains and sewage systems. And because the homes are constructed informally, they lack adequate foundations. As a result, when particularly heavy rains saturate the regolith, it transforms into a slurry of mud, resembling wet concrete, that flows downslope, sometimes detaching at a failure surface between regolith and solid bedrock. Whole sections of favelas can disappear in a matter of minutes, replaced by a hummocky muddle of mud and debris. And at the bases of the slopes, the flow of mud and debris knocks over and buries buildings of all sizes **(FIG. Bx5.2b)**. Similar mass wasting in hilly rural areas rips away forests **(FIG. Bx5.2c)**.

The mudflow disasters affecting Rio's favelas emphasize a correlation between poverty and mudflow risk. Wealthier families can afford to pay the higher cost of living on more stable ground. Poorer communities end up on unstable slopes because, due to the known hazard, land there costs very little. An estimated 8 million people face mudflow or flood risk in Brazil. Addressing these risks has proved to be very difficult because the affected communities tend to be isolated from social services. Even if there were better communication between favelas and civil authorities in Rio, funding to move the communities, or to install drainage and other services that might lessen the risk of mudflows, unfortunately does not exist.

FIGURE Bx5.2 Examples of mudflows in Brazil. Note that the soil that forms in Brazil's tropical climate has a red color due to the iron oxide it contains.

(a) Favelas are built on saprolite on slopes rising toward the bases of unweathered bedrock sugarloafs. The arrow points to a favela on a steep slope.

(b) A mudflow can destroy houses built on a slope or at the slope's base.

(c) Mudflows can also destroy forests.

Rapidly moving debris flow disasters have struck La Conchita, California, twice in the past few decades. This community lies along the coast of California, where the interaction between the Pacific Plate and the North American Plate over millions of years has broken up the bedrock of the region. The resulting cracks allow water to enter the bedrock, so that it undergoes pervasive weathering to form weak clay. Along the coast, waves erode this weak land and redistribute sediment to produce flat *platforms*. While this sedimentation is taking place, plate interaction causes gradual uplift, so platforms become small plateaus, or *terraces*, bounded by a steep escarpment, or *bluff*. One such terrace lies at an elevation of 180 m (591 ft) above sea level, about 500 m (1,640 ft) east of the present-day beach at La Conchita (**FIG. 5.22a**). Rain that falls on the face of the bluff drains away quickly via a network of temporary streams that have carved numerous deep gullies in the bluff, but rain falling on the terrace infiltrates the ground and saturates subsurface clay. Therefore, the bluff occasionally becomes unstable, so a mass of mud and debris flows down the face of the escarpment. If La Conchita were uninhabited, such

FIGURE 5.22 The 2005 La Conchita mudflow along the coast of California.

(a) La Conchita lies along an escarpment bounding a terrace. In some places, the escarpment is gullied; in other places, it has been carved into benches to stabilize it. In still others, it has slumped and flowed.

(b) Rescuers searching for victims beneath the mud.

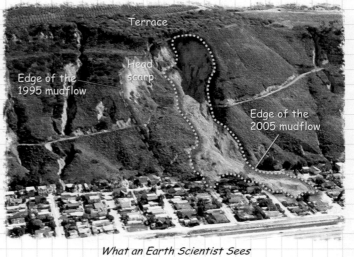

What an Earth Scientist Sees

(c) Two mudflows, in 1995 and in 2005, have happened since houses were built.

5.4 Types of Mass Wasting • **191**

mass wasting would simply be part of the natural process of landscape evolution. But when downslope movements take place at La Conchita, they make headlines, because developers built a community housing 350 people on the modern bench between the shore and the base of the bluff. In 1995, a mud and debris flow overwhelmed 9 houses at the base of the bluff. An even more devastating flow happened in 2005, burying 13 houses, damaging 23, and killing 10 people (FIG. 5.22b, c).

Particularly devastating mudflows spill down river valleys bordering volcanoes during and after eruptions. As we saw in Chapter 4, these mudflows, known as **lahars**, consist of a mixture of volcanic ash and water that can travel tens of kilometers or more from the volcano at speeds up to 60 km/h (37 mph) (FIG. 5.23a). In fact, following the 1877 eruption of a volcano in the Andes of Ecuador, a lahar reached the Pacific Ocean, 320 km (199 mi) away. Because lahars can be activated not only by melting of snow and ice due to an eruption, but also by heavy rains, they can be hazards for years after a large eruption of pyroclastic debris. In fact, devastating lahars incorporating ash from the 1991 eruption of Mt. Pinatubo spilled down valleys into the surrounding lowlands in the Philippines for each of the next four rainy seasons after the eruption (FIG. 5.23b). Fortunately, casualties caused by Pinatubo's lahars were minimized because of timely evacuations of people in their path.

As noted in Chapter 4, lahars following the 1980 Mt. St. Helens eruption destroyed homes along stream valleys at the base of the volcano. An even greater lahar disaster took place on November 13, 1985, in Colombia when a major eruption of pyroclastic flows melted the thick snowcap and glaciers of the Nevado del Ruiz volcano, producing scalding lahars that rushed down river valleys. A lahar swept over the town of Armero, 60 km (37 mi) away, burying it in places to a depth of over 5 m (16 ft)—near the town, the lahar locally was 10 times that thickness. Of the 29,000 residents, over 20,000 perished. Though a hazard potential map showing the susceptibility of Armero to lahars had been recently prepared, most of the people in the town had not seen it, or did not know how to interpret it. And when the eruption happened, officials discussed evacuation, but never ordered one.

FIGURE 5.23 Lahars develop when volcanic ash mixes with water from rain or melting snow and ice.

(a) A lahar that flowed down the side of Mt. St. Helens, Washington.

(b) A lahar following the 1991 eruption of Mt. Pinatubo in the Philippines submerged towns tens of kilometers from the volcano.

Lateral Spreading

During the 1964 Good Friday earthquake in southern Alaska, the Turnagain Heights neighborhood of Anchorage, along with about 10 m of underlying ground, broke up and slipped toward the shore along a nearly horizontal failure surface in weak clay (see Fig. 3.36). Geologists refer to mass wasting that involves displacement on a horizontal failure surface as **lateral spreading**. It occurs when an area overlying a plane of weakness is unsupported on one side, so that there is space for the moving mass to move into. In the case of Turnagain Heights, the neighborhood was built on a terrace bordered on the shore side by a steep bluff 20 m (66 ft) high.

FIGURE 5.24 During lateral spreading, land above a horizontal plane of weakness collapses, most commonly due to liquefaction of a weak sediment layer, causing blocks of land to tilt and break apart.

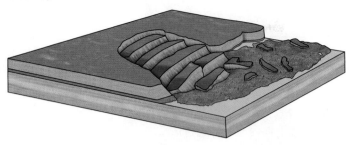

(a) A lateral spread. Liquefied sediment from the weak layer flows beyond the end of the blocks.

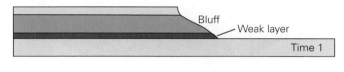
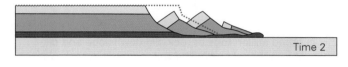
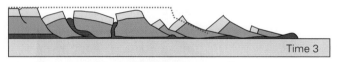

(b) The progression of lateral spreading. A succession of head scarps forms, each new one farther inland (upslope) than the one before.

(c) A lateral spread in New Zealand, as seen from the air.

Lateral spreading may be triggered by earthquakes or by drenching rains. It starts when a plane of weakness suddenly becomes weaker, commonly because of *sediment liquefaction*, a process by which a wet sedimentary layer changes and behaves like a viscous fluid, as we'll see later in the chapter. During lateral spreading, the boundary separating the land that has already started to move from the land that has not moved shifts progressively away from the bluff, as new slices of land become part of the moving mass and new head scarps form (**FIG. 5.24**). The liquefied sediment tends to seep out of the toe of the moving mass, and tends to squeeze upward into gaps that form between the blocks of broken-apart land.

Rock and Debris Slides and Falls

In the early 1960s, engineers built a dam across a deep but narrow gorge that emptied into the broad Piave Valley of northern Italy. The structure, known as the Vaiont Dam, was an engineering marvel, a concrete wall rising 260 m (853 ft), as high as an 85-story skyscraper, above the valley floor (**FIG. 5.25a**). Its purpose was to trap water in a reservoir to drive a hydroelectric power station. Unfortunately, the dam's builders failed to recognize the mass-wasting hazard posed by nearby Monte Toc. The side of the mountain that faced the reservoir was underlain with limestone beds interlayered with weak shale beds whose bedding planes were parallel to the slope of the mountain and curved under the reservoir. As the reservoir filled, the flank of the mountain cracked and shook. Local residents began to call Monte Toc *la montagna che cammina* (the mountain that walks).

After several days of rain, Monte Toc began to rumble so much that, on October 9, 1963, the engineers lowered the water level in the reservoir. They thought the wet ground might slump a little into the reservoir, with minor consequences, so no one suggested evacuating the nearby town of Longarone, in the Piave Valley, or other nearby communities. Unfortunately, the engineers had underestimated the problem. At 10:30 that evening, 600 million tons of rock detached from the mountain and rapidly slid down into the reservoir, displacing most of the water it held (**FIG. 5.25b**). Some of the water in the reservoir moved as a tsunami that rushed away from the dam and washed away villages upstream of the reservoir, killing 1,000 people. But most of the displaced water splashed over the top of the dam and rushed down to the Piave Valley below as a fast-moving flood. When the flood had passed, nothing of Longarone or its 1,500 inhabitants remained (**FIG. 5.25c**). Though the dam itself still stands, it holds back only debris and has never provided any electricity (**FIG. 5.25d**).

Geologists refer to a sudden movement of rock and debris down a failure surface on a nonvertical slope as a **rockslide** if the moving mass contains mostly rock, or as a *debris slide* if it consists mostly of regolith. (In the case of the Vaiont Dam disaster, a weak shale bed served as the failure surface.) In the process, the moving mass breaks up into a chaotic jumble, and the debris may continue tumbling down a slope beyond the end of the failure surface. The slide

FIGURE 5.25 The Vaiont Dam disaster was caused by a catastrophic rockslide that displaced the water in a reservoir with rock debris.

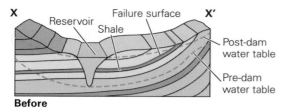
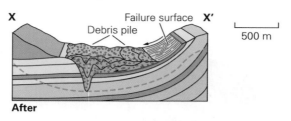

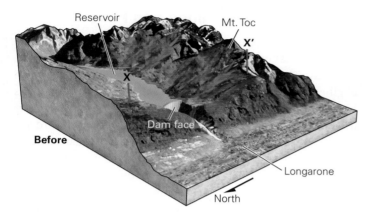
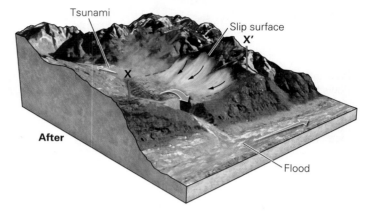

(a) Before the rockslide, Monte Toc was forested. When the reservoir filled, its slope became unstable. A weak shale bed a few hundred meters below the ground surface became a failure surface.

(b) Six hundred million tons of rock slid downslope and displaced water in the reservoir. The water surged over the dam and swept away towns in the valley below. A tsunami washed away towns at the upstream end of the reservoir.

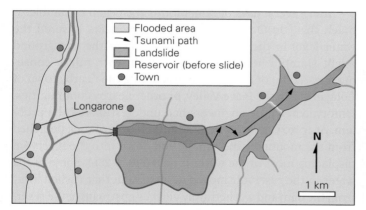

(c) A map shows the area affected by the rockslide, flood, and tsunami.

(d) Today, a forest grows over the rockslide debris. The top of the dam is visible at the bottom of the photo.

leaves a scar on the slope and forms a debris pile at the base of the slope **(FIG. 5.26)**. Slides, like slumps, occur at a variety of scales. The Vaiont Dam disaster serves as an example of a large one. Small slides happen where excavations for tunnels or mines intersect planes of weakness that tilt toward the excavation. If the excavation removes rock or regolith supporting upslope regions, a slide may result.

Rockfalls and *debris falls*, as their names suggest, occur when a mass free-falls for part of its journey down a cliff or steep slope. The resulting rock fragments may shatter or bounce when they land on the lower slope **(FIG. 5.27a)**. Some fragments effectively explode into dust. Some rockfalls happen when a slab of rock separates from a cliff face along a steep plane of weakness (a joint, a foliation plane, or a bedding plane). A column or wall of rock may drop straight downward if there is no support beneath it. Alternatively, rock masses may tilt away from the cliff face and tip over, a phenomenon called *toppling* **(FIG. 5.27b)**. Rockfalls and debris falls happen fairly frequently in areas with steep cliffs and jointed rock **(FIG. 5.27c–e)**.

Rockfalls happen not only along natural cliffs, but also along human-made escarpments. For example, they are fairly

FIGURE 5.26 Examples of rockslides.

(a) Rock detached from a cliff and then slid down a grassy slope in Peru.

(b) This rockslide buried a forest bordering a small lake in Utah.

FIGURE 5.27 Examples of rockfalls.

(a) A rockfall just starting to tumble down a mountain in the Swiss Alps. 700,000 m³ are in motion at the instant the picture was taken.

(b) Rockfalls can take place in different ways.

(c) This column of rock will soon topple.

(d) Failure occurred along bedding and joints at this locality in Bryce Canyon.

(e) Fresh rock is lighter in color than weathered rock, as seen here in the Grand Canyon. Where rock fell away, the cliff face shows as lighter.

(f) A rockfall along a highway.

5.4 Types of Mass Wasting • **195**

FIGURE 5.28 Rockfalls can travel far from their origin.

(a) The aftermath of the 1881 Elm rockfall.

(b) Fallen rocks litter the base of this cliff.

common along steep highway road cuts, **(FIG. 5.27f)**, leading to the posting of Falling Rock Zone signs. Such rockfalls commonly take place soon after construction because blasting and excavation leave loose rocks on the slope above the road. But rockfalls may continue to take place long after construction, as mechanical and chemical weathering weaken the slope over time. In temperate climates, road-cut rockfalls are common in spring, when winter ice that has pushed open joints melts. Large rockfalls can close whole segments of highways and have struck vehicles, causing casualties.

Rockfalls also happen as a consequence of quarrying. A tragic example took place in the Swiss town of Elm in 1881. At the time, the building of new schools to improve public education had created a demand for slate chalkboards. Elm lay at the foot of a mountain of slate, so workers from the town excavated a cut 180 m (591 ft) long and 65 m (213 ft) deep into the mountain to extract slate. But the workers left the cut unsupported, and by doing so, produced a dangerous overhang. In 1876, a fissure up to 3 m across formed in the mountaintop above the overhang, hinting that failure was beginning. But quarrying continued until 1881, when small rockfalls began to happen with increasing frequency. The quarry and a few houses were evacuated, but that was not enough. A massive rockfall later that year dropped 10 million m³ (353 million ft³) of rock down the cliff face. The rock disintegrated as it fell and became a cloud of debris that spread outward from the cliff base, flew into the air, and landed on Elm, burying the town and killing 115 people **(FIG. 5.28a)**.

Friction and collision with other rocks may bring falling blocks of rock to a halt before they reach the bottom of the slope; these blocks pile up as talus, as we have seen (see Fig. 5.12b). Alternatively, some larger blocks may continue to tumble, bounce, and roll for a long distance, even beyond the base of the slope **(FIG. 5.28b)**. Rockfalls that travel all the way to the base of a large cliff may keep moving away from the cliff for a long distance, occasionally riding on a cushion of air. In some cases, very large rockfalls push the air in front of them, creating a short blast of hurricane-like wind—the wind in front of a 1996 rockfall in Yosemite National Park flattened 2,000 trees.

Avalanches

In the winter of 1999, an unusual weather system passed over the Austrian Alps. First it snowed. Then the temperature warmed, and the snow began to melt. But then the weather turned cold again, and the melted snow froze into a hard, icy crust. This cold snap ushered in a blizzard that blanketed the ice crust with half a meter (1.6 ft) of new snow. With the frozen snow layer underneath acting as a failure surface, 200,000 tons of new snow began to slide down the mountain. As it accelerated, the mass transformed into a **snow avalanche**, a chaotic jumble of snow surging downslope. At the bottom of the slope, the avalanche overran a ski resort, crushing and carrying away buildings, cars, and trees, and killing over 30 people **(FIG. 5.29a)**.

Depending on the temperature, snow avalanches can be wet or dry: *wet* avalanches move as a slurry of solid and liquid water, whereas *dry* avalanches transform into a cloud of powder mixed with air **(FIG. 5.29b, c)**. What triggers snow avalanches? Some happen when a *cornice*, a large drift of snow that builds up on the lee (downwind) side of a windy mountain summit, suddenly gives way and falls onto slopes below, where it knocks free additional snow. Others happen when a broad slab of snow on a moderate slope detaches from its substrate along an icy failure surface **(FIG. 5.29d)**. Commonly, avalanches follow the same paths, called **avalanche chutes**, time after time due to the shape of the land surface **(FIG. 5.29e)**. These chutes may be visible in summer because the avalanches prevent trees from growing.

FIGURE 5.29 Snow avalanches.

(a) Aftermath of a deadly avalanche in Austria.

(b) A wet avalanche.

(c) A dry avalanche.

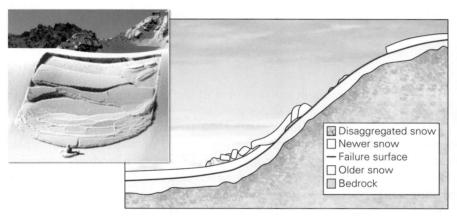

(d) A slab avalanche may detach from a slope along a frozen or granular failure surface.

(e) Avalanche chutes in the Canadian Rockies.

Though in everyday English, the word *avalanche* tends to apply only to mass wasting of snow, the same term has also been applied to debris falls during which significant quantities of air mix with the debris, so that the mass becomes a churning cloud. *Debris avalanches* can accelerate to high speeds (300 km/h, or 186 mph), especially if they trap compressed air beneath them, and can travel long distances from their origin.

Complex Landslides

If a given mass-wasting event involves more than one type of movement, we can refer to it as a **complex landslide**. As an example, a slope may start to fail by displacement of a coherent block on a failure surface to form a slump. But once the slump starts moving, it may break up, liquefy, and transform into a debris flow. Similarly, a mass-wasting event may start out as a rockfall, but as the rock bounces downward, it shatters, and the debris mixes with air to become a debris avalanche.

Take-home message . . .

Mass-wasting events differ from one another in their speed of movement and in the coherence and the composition of the moving material. Creep, solifluction, and slumps are slow mass-wasting events. Mudflows, lahars, and debris flows move faster, and avalanches, rockslides, and rockfalls move the fastest. Translational slumps take place on planar failure surfaces, and rotational slumps on curved ones. Many mass-wasting events are complex, in that they involve more than one style of movement.

QUICK QUESTION What is lateral spreading, and how does it differ from a slump?

FIGURE 5.30 Submarine mass wasting.

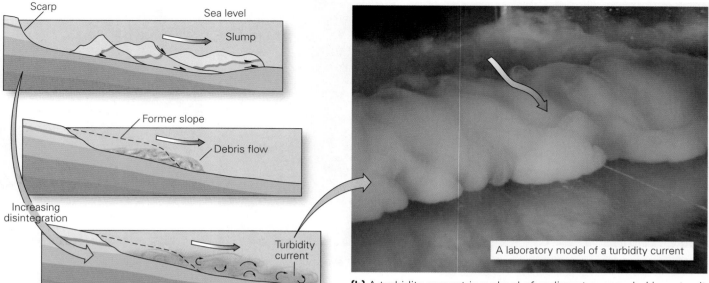

(a) Geologists distinguish three different types of submarine mass wasting.

(b) A turbidity current is a cloud of sediment suspended in water. It flows near the seafloor because it is denser than clear water.

5.5 Submarine Mass Wasting

So far, we've focused on mass wasting that occurs *subaerially* (in the air, above water). Twentieth-century research revealed that substantial mass wasting also happens underwater. Geologists distinguish three types of such **submarine mass wasting**, or *submarine slides*, by whether the slide remains coherent or disintegrates as it moves (**FIG. 5.30a**). In *submarine slumps*, semicoherent blocks comprising layers of sediment slide downslope on planes of weakness. In some cases, the layers become contorted as they move, like a tablecloth that has slid off a table. In *submarine debris flows*, the moving layers break apart to form a slurry containing larger fragments suspended in a mud matrix. And in *turbidity currents*, sediment disperses completely to form a turbulent cloud of sediment, suspended in water, that rushes downslope like an underwater avalanche (**FIG. 5.30b**).

In recent years, geologists have used satellite-based technologies as well as sonar to map out the extent of submarine slides on the seafloor. The shapes of head scarps and sediment deposits stand out on modern high-resolution bathymetric maps (**FIG. 5.31a**). Geologists have found that submarine slopes bordering both hot-spot volcanoes and active margins are carved by many immense head scarps, and at the bases of these scarps, large areas of the seafloor have lumpy, irregular surfaces indicating that they are covered by debris. Clearly, tectonic activity frequently jars these areas with earthquakes, which set masses of sediment and rock in motion. For example, numerous large submarine slides have happened during the past several million years around the Hawaiian Islands (**FIG. 5.31b, c**). Some of these slides represent the collapse of a substantial portion of an adjacent island, and have produced immense *debris fans* that extend as far as 200 km (124 mi) from the islands. Because of such slides, some of the islands are small remnants of what were once much larger shield volcanoes.

Significantly, passive margins are not immune to sliding. Most smaller slides happen without our knowledge. But researchers know exactly when an earthquake-triggered submarine slump flowed down the continental slope of North America's east coast, about 300 km (186 mi) south of Newfoundland. On the evening of November 18, 1929, the flow abruptly broke 12 trans-Atlantic telegraph cables, and telegraph transmissions between the United States and Europe suddenly stopped. (The event also produced a deadly tsunami, as we'll see in Chapter 6.) One of the largest known passive-margin slumps, known as the Storegga slide, happened along the coast of Norway about 8,000 years ago. Modern bathymetric maps reveal the 300 by 200 km (186 by 124 mi) head scarp where 5,000 km³ (1,200 mi³) of material broke away from the edge of the continental shelf. Computer models suggest that it took half an hour for the debris to move down the continental slope to the abyssal plain, a distance of 700 km (435 mi). As we'll see in Chapter 6, this slide produced a tsunami that may have destroyed Stone Age villages along the coasts of the North Sea.

FIGURE 5.31 Bathymetric maps reveal large submarine slides.

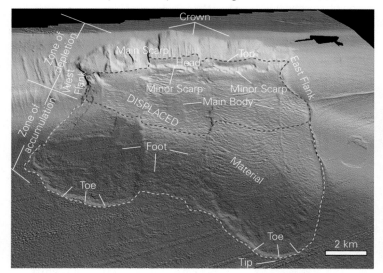

(a) A bathymetric map of a slump along the coast of California.

Take-home message...
Mass wasting occurs under the sea in the form of slumps, debris flows, and turbidity currents. Huge slumps have developed along active and passive margins and along hot-spot oceanic islands.

QUICK QUESTION Why does submarine slumping happen more frequently along active margins than it does along passive margins?

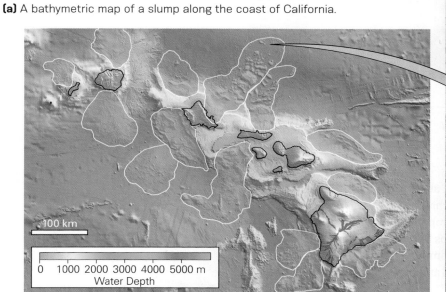
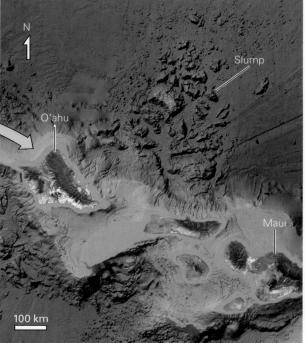

(b) A bathymetric map of the area around the Hawaiian Islands shows several slumps. The enlargement shows the detailed bathymetry of the Nu'uanu slump, which reduced the size of O'ahu.

(c) An oblique satellite image shows the giant slump scarp on the northeast side of O'ahu.

FIGURE 5.32 The town of Qushan, China, was destroyed by a combination of seismic shaking and landslides in 2008. Note the landslide scars on the hills and the debris flow that covers the ground in the town.

Sediment Liquefaction

Seismic waves and other sources of vibration can cause **sediment liquefaction**, a process by which a wet, but solid, layer of sediment turns into a slurry that can flow like a liquid. Sediment liquefaction can also be caused by heavy rains or changes in the level of the *water table* (the boundary below which pores in rock or regolith are completely filled with water and above which pores contain mostly air). Liquefaction in sand happens when the sand is saturated. Vibrations cause grains to shift slightly and therefore to pack together more tightly. This movement causes the pressure in the water between grains to increase until it starts to push the grains apart. When this happens, friction or cohesion between the grains no longer exists, so the entire mass can flow. Liquefied sand can squirt upward through cracks in the ground to form *sand volcanoes*. In some cases, erupted sand

5.6 What Factors Trigger Mass Wasting?

What phenomena can trigger a mass-wasting event at a given locality? Or, in technical terms, what factors can change the balance between downslope shear stress and upslope resistance stress so that the safety factor of a slope drops below 1, causing the slope to fail and mass wasting to commence? Here, we review various phenomena—some natural and some human-caused—that trigger slope failure.

Shocks and Vibrations

Landslides are responsible for many of the casualties that happen during an earthquake, as we saw in Chapter 3 (**FIG. 5.32**). Earthquake tremors can cause a mass that was on the verge of moving to start moving. Vibrations caused by strong wind (that causes trees to sway), a clap of thunder, the passing of a large truck, or blasting at a construction site can have the same effect. Back-and-forth or up-and-down movements can break bonds that hold a mass in place, causing cracks or planes of weakness to grow and become failure surfaces. They can also cause a mass to accelerate upward slightly and separate from its substrate, leading to a decrease in friction and allowing movement to begin even without the growth of new planes of weakness. Finally, they can cause sediment liquefaction, as we describe in the next section.

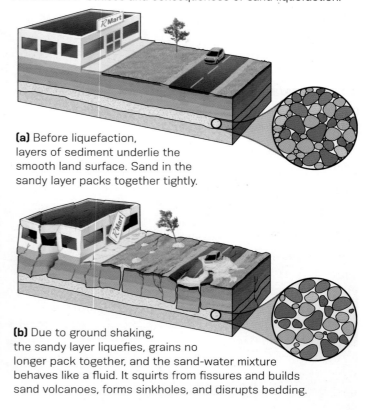

FIGURE 5.33 Causes and consequences of sand liquefaction.

(a) Before liquefaction, layers of sediment underlie the smooth land surface. Sand in the sandy layer packs together tightly.

(b) Due to ground shaking, the sandy layer liquefies, grains no longer pack together, and the sand-water mixture behaves like a fluid. It squirts from fissures and builds sand volcanoes, forms sinkholes, and disrupts bedding.

FIGURE 5.34 Causes and consequences of quick clay liquefaction.

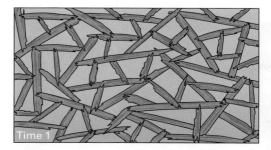

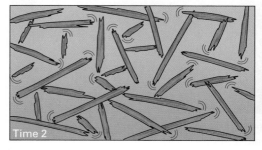

(a) At Time 1, the clay flakes in quick clay are held together by weak electrostatic bonds. At Time 2, after shaking or saturation, the bonds break and the clay flakes can move freely.

(b) After heavy rains, liquefaction of quick clay carried these lakeside houses and surrounding trees into the water.

spreads over the land surface. Liquefaction of sand underground allows overlying layers to break up and collapse to form *disrupted bedding* **(FIG. 5.33)**. Most importantly, a layer of liquefied sand underground can serve as a failure surface for lateral spreading or other types of landslides.

A similar phenomenon happens in certain types of clay known as *quick clay* or *sensitive clay*. Such clay behaves as a stiff solid when undisturbed because surface tension or electrostatic charges hold clay flakes together. But when it is shaken, these bonds break, and the clay becomes a mud-like fluid. Quick clay typically forms from glacial mud deposited in saltwater. As this mud compacts, the positive ions in salt play a role in binding together negatively charged and randomly oriented clay flakes. Later, when the glacier melts away and the land surface is uplifted, the clay layer ends up above sea level. This situation allows fresh groundwater to infiltrate the clay and remove the salt ions and salt crystals that held the clay in its open framework. As a result, the structure of the clay becomes very unstable. Shaking causes the wet clay grains to separate entirely from one another, producing a wet, muddy slurry with no strength **(FIG. 5.34a)**. As quick clay flows, the clay flakes within it tend to align. Because of the ease with which quick clay undergoes liquefaction, a layer of quick clay can easily become a failure surface, leading to lateral spreading or to slumping **(FIG. 5.34b)**.

A tragic example of a large lateral spreading event caused by sediment liquefaction happened during a major earthquake that struck Palu, Indonesia, in 2018 **(FIG. 5.35)**. Ground shaking triggered liquefaction of sand underground. Due to the weakness of the liquefied layer, the land above, at several locations, started to undergo lateral spreading and then broke apart. Videos show that, while moving, the land looked like it was flowing, carrying along houses as if they were floating rafts. At the toe of the slide, some of the material of the slide block, along with extruded liquefied sediment, transformed into a slurry, which continued moving well beyond the toe of the slide as a debris flow. What had once been a landscape of streets and houses became a chaotic jumble of debris engulfing over 1,000 homes.

Changing Downslope Stress or Resistance Stress

If the weight of a slope increases due to construction of buildings on top, or by saturation of regolith with water from heavy rains, leaking pipes, or septic systems, the downslope stress increases and may exceed the resistance stress. The La Conchita debris flows described earlier in this chapter illustrate this phenomenon. Another example occurred near Jackson Hole, Wyoming, in 1925. This mass-wasting event, known as the Gros Ventre Slide, remains the largest observed natural rock and debris slide on a hillslope in US history. During this slide, which took place on the flank of Sheep Mountain near Jackson Hole, 40 million m³ (1.4 billion ft³) of rock, as well as the overlying soil and forest, detached from the side of the mountain

FIGURE 5.35 Widespread sand liquefaction during a 2018 earthquake caused lateral spreading that led to the destruction of whole neighborhoods in Palu, Indonesia.

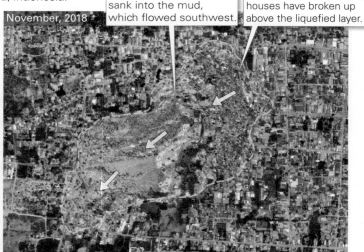

(a) Before the liquefaction event, a densely populated neighborhood covered the land.

(b) Liquefaction destroyed all homes within the dashed yellow line, an area of about 1 km³.

and slid 600 m (about 2,000 ft) downslope, filling a valley and forming a natural dam 75 m (246 ft) high across the Gros Ventre River **(FIG. 5.36)**. The slide followed heavy rains, which saturated the sandstone bedrock above a shale failure surface. The addition of this water, which replaced air in the pore spaces of the sandstone, added weight to the slope, increasing downslope stress, and also weakened the failure surface.

Removal of support at the base of a slope triggers many slope failures by effectively decreasing the resistance stress. For example, **undercutting** of a cliff—by stream erosion, wave erosion, or excavation—can remove the support from

FIGURE 5.36 Stages leading to the 1925 Gros Ventre Slide in Wyoming.

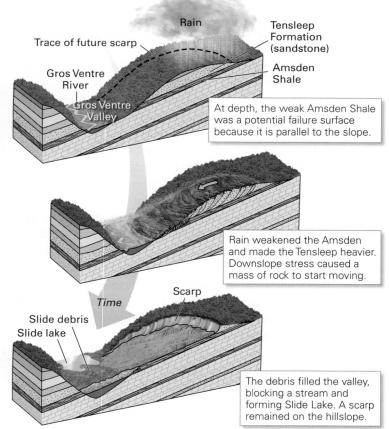

202 • CHAPTER 5 • Unstable Slopes

FIGURE 5.37 Removal of support at the base of a slope leads to collapse.

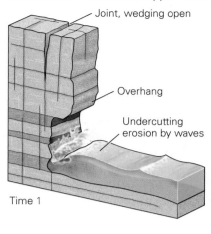

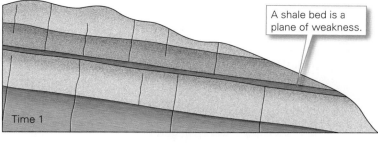

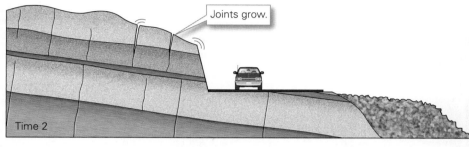

(a) Undercutting of a cliff by erosion can trigger a rockfall.

(b) Excavation of a road cut can remove support for rock above a plane of weakness so that a rockslide happens.

beneath a mass of rock, triggering a rockfall (FIG. 5.37a), as was the case at Elm, Switzerland. Even on gentler slopes, the material at the base of a slope acts like a wall holding back the material farther up the slope. When it's removed, either by natural erosion or by human excavation, the resistance stress decreases significantly, and slumps and other types of mass wasting may take place (FIG. 5.37b).

Changing Slope Angle

Steepening a slope can set the stage for failure because it increases the downslope shear stress. Steepening can be caused either by natural processes or by human activities. For example, when a river erodes a deeper valley, the slopes on the side of the valley become steeper, and can become unstable and fail (FIG. 5.38a). And when contractors cut a flat bench into a slope, they effectively remove a wedge-shaped block of the slope (FIG. 5.38b). If they then dump the removed material over the edge of the bench, this becomes a wedge of weak debris. The consequence of this approach is to create two segments of a steeper slope than existed before, and these steeper segments may be unstable.

When miners dig an *open-pit mine* (a basin- or trough-shaped mine cut to expose rock), they are producing a slope where one didn't exist before. Typically, they cut a series of benches into the slope, which provide surfaces on which to place equipment, catch falling debris before it can reach the bottom of the mine, and decrease the overall downslope stress on the slope. But miners try to make the overall slope inside an open-pit mine as steep as possible to avoid having to remove large quantities of waste rock, a cost that might make the mining operation unprofitable. If the overall slope becomes too steep, it becomes unstable and can collapse. This was the cause of a massive rock avalanche in the Bingham Canyon copper mine of Utah, one of the largest open-pit mines in the world (FIG. 5.38c). On April 10, 2013, two slumps evolved into debris slides that carried away 70 million m³ (2.5 billion ft³) of debris from one wall. The volume of material moved was almost twice that of the Gros Ventre Slide.

FIGURE 5.38 Processes that can steepen slopes and make them unstable.

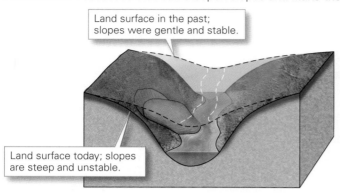

(a) Downcutting by a river can produce a deeper valley, with steeper walls that become susceptible to mass wasting.

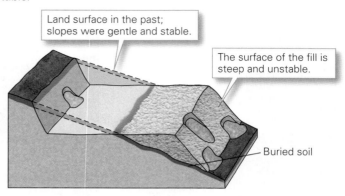

(b) Excavation and filling can produce slopes that are steeper than the angle of repose.

(c) A huge debris flow in the Bingham Canyon copper mine, Utah.

Changing Slope Strength

If the material underlying a slope weakens over time, the slope may become unstable. Three factors influence the strength of slopes:

- *Weathering*: Over time, chemical weathering produces weaker minerals, and physical weathering breaks rocks apart. Consequently, weathering transforms formerly intact rock composed of strong minerals into weaker rock or regolith that could fail.
- *Vegetation cover*: In the case of a slope underlain by regolith, vegetation tends to strengthen the slope because the roots hold otherwise unconsolidated grains together. Also, plants absorb water from the ground, keeping it from turning into slippery mud, and prevent water from infiltrating the ground. The removal of vegetation has the net result of making slopes more susceptible to mass wasting. For this reason, continuing deforestation and the increase in the size and number of forest fires in recent years have had the effect of increasing the number of mass-wasting events (FIG. 5.39a). In California, areas burned by wildfires have become susceptible to debris flows. In 2018, for example, creekside communities in the town of Montecito were inundated by debris flows 5 m (16 ft) high and moving at 30 km/h (19 mph) (FIG. 5.39b). The flows destroyed over 100 homes, cut off power to 20,000 people, and closed roads. In the immediate area, 30,000 people had to be evacuated, and a state of emergency was declared.
- *Water content*: As we've seen, water affects the strength of materials constituting slopes in many ways. For example, surface tension in damp regolith may help hold it together. But if the water content of the regolith increases, water pressure may push its grains apart so that it liquefies and can begin to flow. Water infiltration may make planes of weakness underground more slippery, or may push surfaces apart and decrease friction.

FIGURE 5.39 Removal of vegetation increases the susceptibility of slopes to mass wasting.

(a) A slump on a deforested hillslope in Brazil.

(b) Mudflows devastated Montecito, California, after drenching rains saturated wildfire-devastated slopes upstream.

Take-home message . . .

A variety of geologic phenomena can trigger a mass-wasting event. Examples include shocks and vibrations, liquefaction, changing slope angles, removal of support from the base of a slope, and modification of a slope's strength by weathering, deforestation, or an increase in the water content of the slope.

QUICK QUESTION Why is water content such an important factor in slope instability?

5.7 How Can We Protect against Mass-Wasting Disasters?

Identifying Regions at Risk

Clearly, mass wasting is a natural hazard we cannot ignore (**VISUALIZING A DISASTER**, pp. 206–207). Too many of us live in regions where landslides, mudflows, or slumps have the potential to injure people and destroy property. In many cases, the best solution is avoidance: don't build, live, or work at a location where mass wasting has a high probability of happening. But avoidance is possible only if we know where the hazards are.

To pinpoint dangerous regions, geologists look for landforms known to result from mass wasting, for where mass wasting has happened in the past, it could happen again in the future. Features such as head scarps, swaths of forest in which trees have been tilted, piles of loose debris at the bases of hills, and hummocky land surfaces all indicate recent mass wasting. But even if there's no evidence of recent movement, a danger may still exist—just because a steep slope hasn't collapsed in the recent past doesn't mean it won't in the future.

In recent years, geologists have begun to identify such potential hazards by using computer programs to evaluate factors that trigger mass wasting. Examples of such factors that we've discussed in this chapter include: slope steepness; strength of the substrate; degree of water saturation; orientation of bedding, joints, or foliation relative to the slope; the nature of vegetation cover; the potential for heavy rains; the potential for undercutting to occur; and the likelihood of earthquakes. Surveys may also be used to identify areas of a slope that lie between previous slumps, which may represent unstable land that has not yet slumped (**FIG. 5.40**). From such hazard assessment studies, geologists

VISUALIZING A DISASTER
Mass Wasting

FIGURE 5.40 Slumps along a cliff on the Pacific coast of California. Some of the housing developments at the top of the cliff are perched on land between slumps, in an area where the base of the cliff is being undercut.

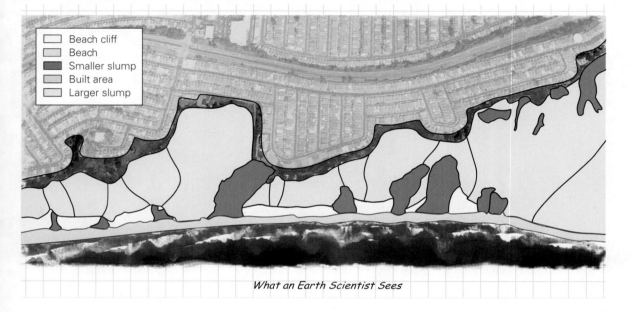

- Beach cliff
- Beach
- Smaller slump
- Built area
- Larger slump

What an Earth Scientist Sees

compile **landslide potential maps**, which rank regions according to the likelihood that mass wasting will occur **(FIG. 5.41)**, and *liquefaction potential maps*, which identify areas overlying sediments that might liquefy. Common sense suggests that building on or below particularly dangerous slide-prone slopes should be avoided, as should construction on substrates that could liquefy.

Notably, NASA has developed a computer system that predicts landslide potential for an area based on the amount of rainfall it receives, since rainfall plays a major role in triggering landslides. The system uses satellite-based technology that provides data on precipitation every half hour **(FIG. 5.42)**. Monthly maps show how landslide potential varies globally due to seasonal changes in precipitation.

In many cases, the potential for mass wasting is only clear in retrospect. The 2014 Oso landslide in Washington State serves as an example **(FIG. 5.43)**. Erosion by the Stillaguamish River had cut a canyon into relatively weak glacial sediments. As the river eroded the base of the canyon's slopes, several slumps and mudslides occurred along the walls of the canyon. Unfortunately, a housing development was built across the river from an area that had slumped previously and was recognized as unstable. On March 22, 2014, mass wasting suddenly happened again. A small upper slump brought part of the canyon wall down and removed support from the wall behind it. This led to a second, even larger, slump. The moving material transformed into a debris flow as it slid downslope, spread across the river, and covered 49 houses in the development, killing 43 people. A state of emergency was declared. Military search-and-rescue teams, along with local officials and 160 volunteers, sifted through the debris. In the aftermath, the county bought the property because the site was deemed too dangerous to rebuild on, and $60 million in settlements were paid out.

FIGURE 5.41 Landslide potential maps.

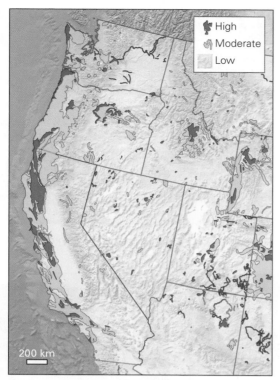

(a) General landslide potential map of the western United States.

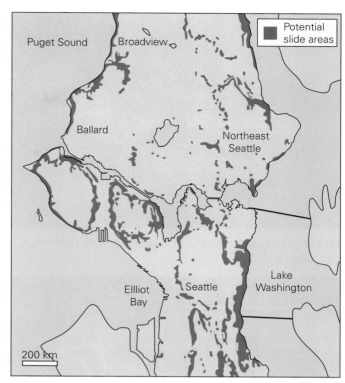

(b) Detailed landslide potential map of the Seattle, Washington, area.

FIGURE 5.42 Satellite measurements of rainfall can help predict mass wasting.

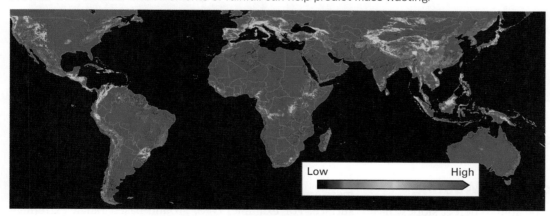

(a) A NASA map showing landslide potential, compiled from data on both landscape features and precipitation amounts. New maps are produced frequently, to reflect variations in precipitation amounts. The scale shows the likelihood of a landslide.

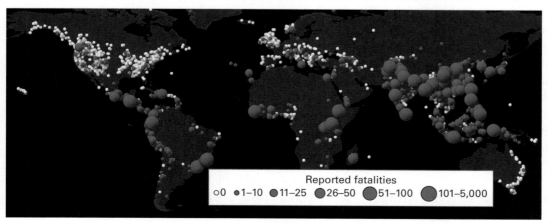

(b) A NASA map showing reported landslides and fatalities between 2007 and 2017. About 20% of the reported landslides caused fatalities. White dots represent those that caused no fatalities. The sizes of the purple dots indicate the relative number of fatalities. Comparing this map with the one above shows how rainfall can serve as a predictor of landslides.

5.7 How Can We Protect against Mass-Wasting Disasters? • **209**

FIGURE 5.43 The Oso landslide.

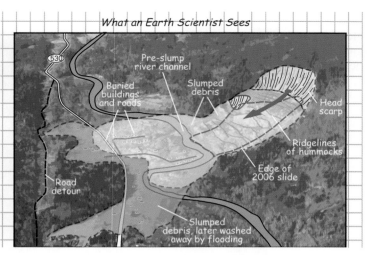

(a) The slide surged over the Stillaguamish River, damming it temporarily, and tragically buried a small community on the opposite bank.

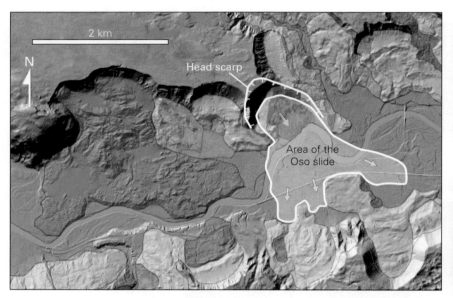

(b) A digital elevation model, looking vertically down, of the Oso area before the slide. Colored areas are various landslides that happened in the past 2,000 years. Arrows indicate movement of the Oso slide.

(c) Photos looking west along the Stillaguamish valley in 2012 and 2020. Note that an older scarp is visible in 2012.

Detecting the Start of Mass Wasting

In some cases, geologists may be able to detect slopes that are *beginning to move*. For example, roads, buildings, and pipes may begin to crack over unstable slopes. Power lines may become too tight or too loose because the poles to which they are attached move together or apart (FIG. 5.44). Visible cracks may form on the ground at the potential head of a slump, and the ground may bulge upward at the potential toe. In some cases, subsurface cracks may drain the water from an area and kill off vegetation, whereas in other areas land may sink and form a swamp. And, as in the case of creep, slow movements may cause trees to develop pronounced curves at their base.

More recently, new, extremely precise surveying technologies have permitted geologists to detect the beginnings of mass wasting that may not yet have visibly affected the land surface. For example, GPS, ground-based radar, extremely accurate surveys using lasers, and satellite-based radar (InSAR) can detect the slight movements that may be precursors to a slope failure. InSAR data compiled over time can be used to calculate rates of movement. Careful surveying, for example, detected movements prior to the massive debris slide in the Bingham Canyon mine (see Fig. 5.38c), allowing mine operators to evacuate the mine. When the mine wall did fail catastrophically, no miners were injured, though expensive equipment that couldn't be moved in time was destroyed. Efforts are now underway to develop acoustic sensors that will detect the weak, high-frequency vibrations produced by moving grains along a failure surface just before a slide begins to move.

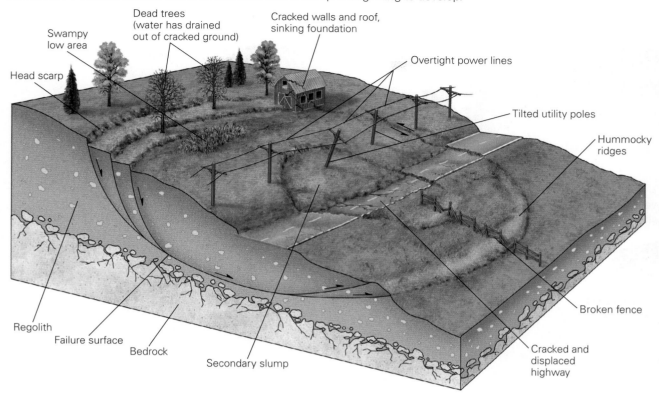

FIGURE 5.44 Various surface features can warn that a slump is beginning to develop.

Preventing and Mitigating Mass Wasting

In areas where mass wasting is a hazard, people can take certain steps to help stabilize a slope or to mitigate the consequences if it fails:

- *Revegetation:* The stability of a slope can be improved by planting trees, shrubs, or grasses with deep roots (FIG. 5.45a). The roots bind regolith together and absorb water. Both processes strengthen the regolith and prevent potential failure surfaces from activating.

- *Lowering the water level in reservoirs or lakes:* When the water level in a lake or reservoir at the base of a slope is lowered, the water table beneath the slope drops (FIG. 5.45b). This change decreases the weight of the slope and also makes potential failure surfaces less slippery.

- *Preventing erosion at the base of a slope:* A stream flowing at the base of a slope erodes the slope by removing support. To prevent instability, engineers can reposition the stream channel so that it no longer comes in contact with the base of the slope (FIG. 5.45c), or they can line the stream channel with stone or concrete to strengthen the channel walls. In places where waves are eroding the base of a slope, engineers can install breakwaters composed of *riprap* (loose blocks of rock or concrete) or anchored blocks of concrete to absorb wave energy and decrease erosion rates (FIG. 5.45d).

- *Excavating benches:* Removing material to cut a bench can decrease the overall downslope stress, so that the slope becomes less likely to fail. If debris does fall, it may be trapped on a bench before it tumbles all the way to the bottom of the slope, where it might cause damage (FIG. 5.46a).

- *Draining slopes:* Because groundwater and soil water not only weaken the material beneath a slope, but also add substantial weight to that material (increasing downslope stress), slope stability can be improved by preventing water from entering a slope or by removing water already underground. As we've seen, vegetation can help extract water from slopes. Other preventative measures include installing drainage culverts at the top of a slope and down its face (to remove water before it infiltrates), or inserting perforated pipes into the slope, positioned so that they drain groundwater and soil water from beneath the slope into a drainage culvert at its base. Draining groundwater lowers the water table, thereby decreasing the weight of the regolith (FIG. 5.46b).

- *Construction of mitigation structures:* In some cases, the best way to prevent damage from mass wasting is to

FIGURE 5.45 Steps that can help mitigate the threat of landslides.

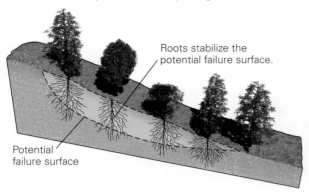

(a) Revegetating a slope can strengthen the substrate and can prevent a potential failure surface from slipping.

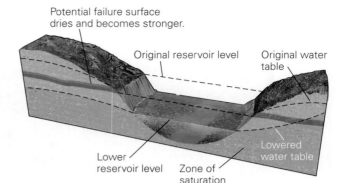

(b) Lowering the water table can decrease downslope stress and can make a failure surface less likely to slip.

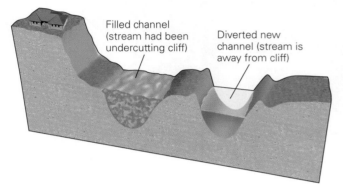

(c) Relocating a river channel can prevent undercutting that could destabilize a slope.

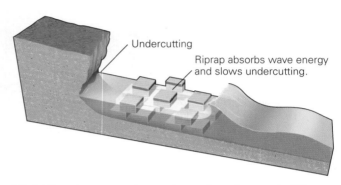

(d) Adding riprap can slow undercutting of coastal cliffs.

build a structure that will protect a region downslope from debris if mass wasting does occur, or keep material from moving in the first place. Such goals can be achieved by several methods. For example, a wall built in front of a slope's base can catch tumbling debris before it reaches a road, railroad, or town (FIG. 5.46c); a shed built out over a road can divert debris from a rockslide or avalanche, so that the debris flows over the shed's roof without affecting people or vehicles below (FIG. 5.46d); *shotcrete* (concrete sprayed onto a surface through a nozzle) can cover a slope underlain by weathered rock with a protective shell that holds the material underneath in place (FIG. 5.46e); a screen of chain link fencing draped over the face of a slope can hold loose material in place (FIG. 5.46f); *rock bolts* screwed into the rock face can attach potentially loose slabs of rock that are breaking off along joints to the intact bedrock behind to prevent the slabs from breaking free and tumbling (FIG. 5.46g); and a rigid *retaining wall* of steel, stone, or concrete placed directly against a slope can provide resistance to downslope stress (FIG. 5.46h).

In this chapter, we've seen how gravity, by pulling rock or regolith downward, can cause unstable slopes to fail. Significantly, if these movements take place on slopes next to or under bodies of water (lakes, reservoirs, or the ocean), they can trigger tsunamis. Chapter 6 describes these waves—which can also result from earthquakes, explosive volcanic eruptions, and meteorite impacts—in greater detail. Chapter 7 addresses another aspect of gravity's role in loss of land—namely, land subsidence and sinkhole formation.

Take-home message . . .

Various features of the landscape may help geologists to identify unstable slopes and predict sites of future mass wasting. This information leads to the production of landslide potential maps. New surveying technologies can help identify land that has started to move. Where mass-wasting hazards exist, a number of techniques can be used to stabilize slopes.

QUICK QUESTION What clues indicate that land is starting to move?

FIGURE 5.46 Engineering solutions to improve slope stability.

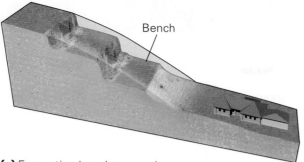

(a) Excavating benches can decrease the slope's weight and can catch debris before it tumbles to the base of the slope.

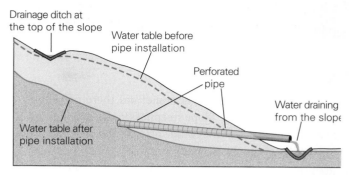

(b) A slope can be drained by digging culverts and installing perforated pipes.

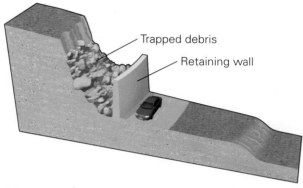

(c) A strong retaining wall can hold back debris.

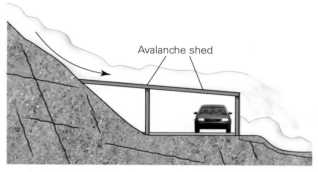

(d) Constructing a shed over a road can keep rockfalls or avalanches from landing on the road.

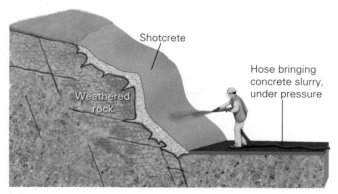

(e) Spraying shotcrete onto weathered rock can keep the rock in place.

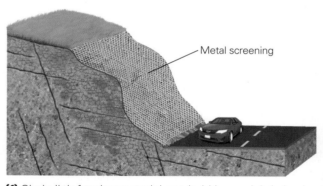

(f) Chain link fencing material can hold loose debris in place.

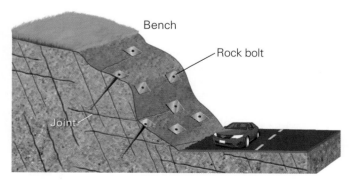

(g) Rock bolts attach fractured material at the surface of a rock face to solid bedrock behind.

(h) This concrete retaining wall holds back sediment that would otherwise have to be graded to a gentle slope in order to remain stable.

5.7 How Can We Protect against Mass-Wasting Disasters?

Chapter 5 Review

Chapter Summary

- Uplift and subsidence produce relief and, therefore, slopes on the Earth's surface. Slope steepness varies depending on many factors, such as the composition of the material forming the slope, and the climate.
- Rock or regolith has the potential to move downslope due to gravity. Such movement is called mass wasting. In the Earth System, mass wasting plays an important role in landscape evolution and in sediment production.
- A slope is unstable when the downslope shear stress exceeds the resistance stress that holds material in place. Put another way, sliding occurs when the safety factor becomes less than 1.
- The steepest angle at which a pile of unconsolidated material can remain without collapsing is its angle of repose. Talus slopes reach an angle of repose.
- In solid rock, chemical bonds provide resistance to sliding. Friction prevents blocks and dry regolith from sliding. In wet regolith, resistance comes from surface tension in water.
- Joints, bedding planes, weak sediments, and foliation planes can serve as failure surfaces during mass wasting, especially if the plane of weakness tilts downslope.
- Slow mass wasting caused by the freezing and thawing of regolith is called creep. In places where slopes are underlain by permafrost, solifluction causes a melted layer of regolith to flow. Creep and solifluction can result in significant property damage by causing structures built on slopes to tilt and crack.
- A slump is a semicoherent mass of material that moves down a failure surface. Beneath a translational slump, the failure surface is planar, whereas beneath a rotational slump, it's spoon-shaped. A slump can not only carry away structures (and their inhabitants) on the surface of a slope, but can also bury communities at the base.
- Mudflows and debris flows occur where regolith has become saturated with water and moves downslope as a slurry. They can move rapidly, so they can cause many casualties.
- During lateral spreading, a horizontal failure surface becomes so weak that a layer of land above breaks up and moves horizontally.
- Rockslides and debris slides move very rapidly down a slope. In a rockfall or debris fall, the material free-falls down a vertical cliff. During snow avalanches, snow moves as a slurry or turbulent cloud. Rockslides, debris falls, and avalanches can destroy and cover everything in their path.
- Mass wasting can take place on underwater slopes. Submarine slumps are movements of semicoherent blocks, debris flows are slurries, and turbidity currents are avalanche-like.
- Downslope movement can be triggered by shocks and vibrations, changes in the steepness of a slope, removal of support from the base of a slope, changes in slope strength, or changes in slope weight. Water plays a major role in mass wasting by affecting the strength of planes of weakness as well as the weight of a slope. Liquefied sediment can become a particularly weak failure surface.
- Geologists can produce landslide potential and liquefaction potential maps to identify areas susceptible to mass wasting. Various clues in the landscape may indicate that slope failure is starting. New survey techniques may detect the initiation of movement.
- Engineers can help prevent mass wasting by using a variety of techniques to stabilize slopes.

Review Questions

Blue letters correspond to the chapter's learning objectives.

1. What is mass wasting, and why does it happen? **(A)**
2. How does solid rock differ from regolith? Which tends to be stronger? **(A)**
3. Why do slopes develop on the Earth? What factors can control the steepness of a slope? **(B)**
4. Explain the difference between a stable and an unstable slope. What does the safety factor of a slope represent? Is a slope with a safety factor of 1.2 stable or unstable? **(C)**
5. What phenomena provide materials with resistance to failure and slip on a slope? **(C)**
6. Why does the cliff in this photo have a stair-step profile? **(B)**

7. Approximately what is the value of the angle of repose for most dry, unconsolidated materials? How does the presence of water affect the angle of repose? **(C)**
8. What features in rock are likely to serve as failure surfaces? Does their orientation relative to a slope affect their likelihood of failure? **(C)**
9. Discuss the phenomena that can cause a stable slope to become unstable and fail. **(C)**
10. What factors do geologists use to distinguish among various types of mass wasting? **(D)**
11. Explain how creep operates. Why do gravestones tip over in response to creep? How does creep differ from solifluction? **(D)**

12. Identify the key differences between a slump, a debris flow, a lahar, an avalanche, a rockslide, and a rockfall. Does the diagram show a translational slump or a rotational slump? What is lateral spreading? **(D)**

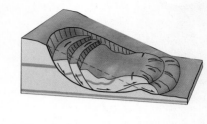

13. Of the various types of mass-wasting events, which have been responsible for thousands of casualties, and which have not? What aspects of a mass-wasting event determine whether it will cause casualties? **(D)**
14. Explain why sediment liquefaction occurs and what phenomena can trigger it. What is quick clay? **(E)**
15. Discuss the role of vegetation and water in slope stability. Why can fires and deforestation lead to slope failure? **(E)**
16. What factors do geologists take into account when producing a landslide potential map? **(F)**
17. How can people avoid landslide disasters? How does the process illustrated in the diagram help improve the stability of the slope? **(F)**

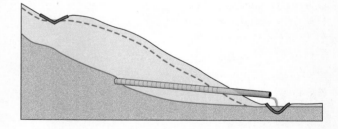

On Further Thought

18. Imagine that you have been asked by the World Bank to determine whether it makes sense to build a dam in a steep-sided, east-west-trending valley in a small central Asian nation. Construction would provide power, irrigation water, and jobs. The rock of the valley wall consists of schist containing strong foliation that is parallel to the slope of the valley wall. Outcrop studies reveal that abundant small faults and moderate earthquakes have rattled the region in the past. How would you advise the bank? Explain the hazards present and what might happen if the reservoir were filled. **(F)**

19. Due to rapid population growth, construction of new homes and apartment buildings has been taking place in a town built on a broad uplifted terrace overlooking the sea. In some places, the terrace is underlain by granite that has not been substantially weathered, whereas in others, the terrace is underlain by layers of clay and sand that were deposited before the terrace was uplifted by tectonic processes. If you were on the town planning board, where would you suggest that construction be avoided? **(F)**

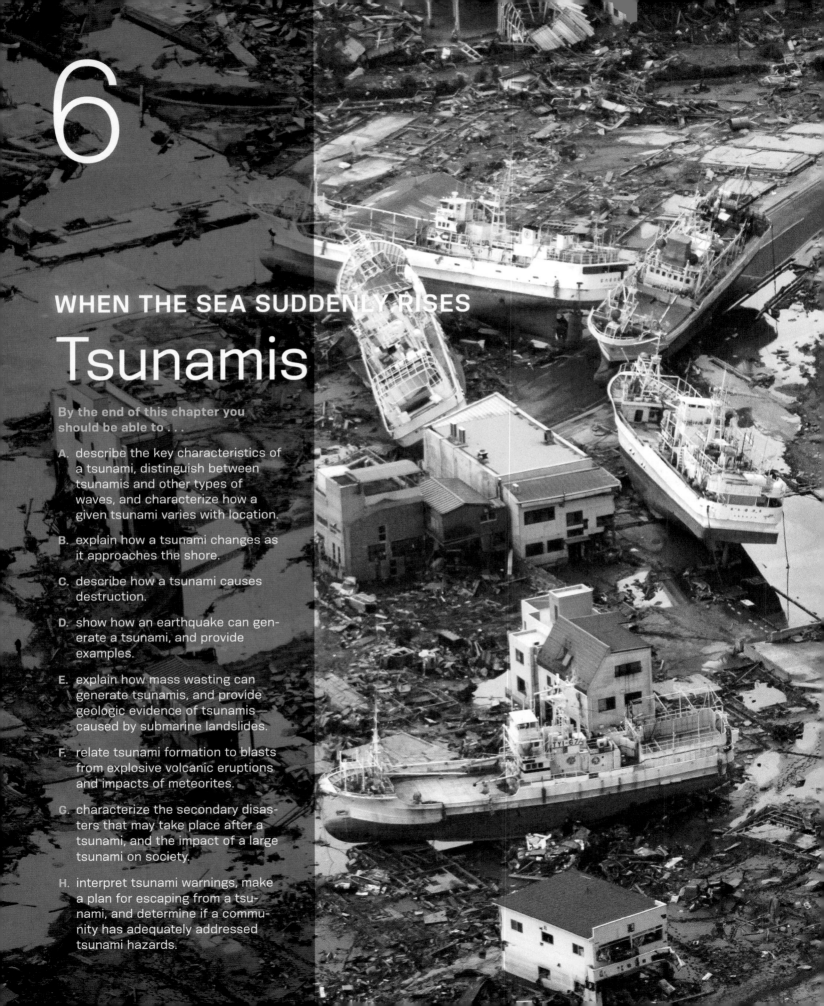

6

WHEN THE SEA SUDDENLY RISES
Tsunamis

By the end of this chapter you should be able to . . .

A. describe the key characteristics of a tsunami, distinguish between tsunamis and other types of waves, and characterize how a given tsunami varies with location.

B. explain how a tsunami changes as it approaches the shore.

C. describe how a tsunami causes destruction.

D. show how an earthquake can generate a tsunami, and provide examples.

E. explain how mass wasting can generate tsunamis, and provide geologic evidence of tsunamis caused by submarine landslides.

F. relate tsunami formation to blasts from explosive volcanic eruptions and impacts of meteorites.

G. characterize the secondary disasters that may take place after a tsunami, and the impact of a large tsunami on society.

H. interpret tsunami warnings, make a plan for escaping from a tsunami, and determine if a community has adequately addressed tsunami hazards.

6.1 Introduction

It was 7:00 A.M. on April Fools' Day, 1946. The streets of downtown Hilo, on the eastern shore of Hawai'i, were just beginning to fill with people heading to work or school. Everything seemed normal... until it wasn't. Suddenly, the water in Hilo Bay began to drain, for no obvious reason—there hadn't been an earthquake or volcanic eruption on the Big Island that day. Moments later, boats that had been peacefully anchored in the harbor settled into newly exposed mud. Then the sea began to rise, as if the tide were coming in, but much faster. The rising water didn't stop where the tide normally did, or even where storm waves normally do, but just kept rising, lifting boats and tearing them from their anchors. Frothing water incorporated mud and coral as it swirled landward, crossed the beach, and began to flood the downtown streets. People who had been standing and staring now turned and sprinted inland, but unfortunately, many couldn't reach high ground fast enough (FIG. 6.1a). The water slammed into buildings, crushing the walls of some, toppling others, and floating many off their foundations (FIG. 6.1b). Groves of trees were uprooted, bridges were carried off their footings, and cars bobbed and rolled in the water before being crumpled against other objects. The moving water became a viscous jumble of wreckage that traveled inland for as far as 1.5 km (1 mi) before stopping. Then it began to surge back seaward, carrying debris and victims out into Hilo Bay. Sadly, the devastation wasn't over. Over the course of the next few hours, several more waves arrived and resubmerged parts of the city before receding. The highest of these waves rose 12 m (39 ft) above sea level, covering several blocks of the downtown (FIG. 6.1c). When the devastation was over, 159 people had died, and property worth tens of millions of dollars had been destroyed.

What happened in Hilo? The city, and much of the rest of Hawai'i's shoreline, had been inundated by several tsunamis. Unlike the more familiar storm waves produced by wind blowing across the sea surface (see Chapter 13), a *tsunami* forms when a mass suddenly moves and displaces water. The tsunamis that struck Hawai'i in 1946 were generated when slip on a fault during an M_W 8.6 earthquake near Unimak, in the Aleutian island arc southwest of Alaska's mainland, caused the seafloor to push upward against the overlying water. At Unimak itself, a tsunami 40 m (130 ft) high—much higher than the ones that reached Hilo—washed away the five-story-high Scotch Cap Lighthouse and its occupants. Tsunamis can cross the widest oceans very quickly. In fact, those that reached Hilo had traveled 3,850 km (2,392 mi) across the Pacific in just under 5 hours.

◂ Vessels lie on the ground of a devastated area in Kesennuma, northern Japan, after a powerful earthquake-triggered tsunami hit the country's east coast in March 2011.

FIGURE 6.1 In 1946, several tsunamis traveled across the Pacific from Alaska and washed over part of downtown Hilo, Hawai'i.

(a) People ran from the approaching tsunami.

(b) Buildings were toppled and crushed.

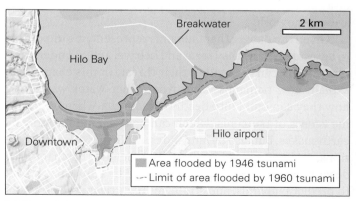

(c) Part of downtown Hilo was inundated.

FIGURE 6.2 The 2004 tsunami approaching the shore of Indonesia.

(a) First, the sea withdrew, exposing mud and coral. Then the wave roared in.

(b) The wave blasted onto the shore.

Globally, noticeable tsunamis happen about twice a year, but fortunately, most of them don't result in major casualties or damage. Large tsunamis—the ones that cause devastation and make headlines—have a recurrence interval of about 15–20 years. These events have produced some of the most devastating natural disasters of historic time. The tsunamis produced by the 2004 Sumatra earthquake (see Chapter 3), for example, killed over 230,000 people along the coasts of the Indian Ocean (FIG. 6.2).

Though people have been aware of tsunamis for millennia, modern study of these waves did not begin until 1896, when a large one struck Japan. The word *tsunami*, which comes from the Japanese word for harbor wave, came into use centuries ago because most coastal inhabitants live near harbors, so most of the impact that tsunamis have on society happens in harbor communities. *Tsunami* first entered the English language after the 1896 event. Global public awareness of these waves increased dramatically after the 2004 Indian Ocean disaster, as digital images and videos of the waves were seen around the world.

Given the danger that tsunamis represent, it's important to be aware of how they form and behave, and to understand how society can respond to their threat. In this chapter, we explain how tsunamis differ from other water waves and how they can build into deadly behemoths. Then we examine causes of tsunamis, which can be triggered not only by earthquakes, but also by landslides, explosive volcanic eruptions, and meteorite impacts. We conclude the chapter by describing secondary disasters caused by tsunamis, efforts to predict tsunamis, and strategies to mitigate the disasters that they can cause.

6.2 The Origin and Movement of Tsunamis

What Is a Tsunami?

On our dynamic planet, most surface water moves constantly, for reasons that may already be familiar to you:

- *Currents* circulate water around the global ocean.
- *Tides* cause the sea surface to rise and fall, generally twice daily, by as much as 16 m (52 ft) at a given location.
- *Wind-driven waves* form when moving air shears the water surface; during intense storms, such waves can rise as high as 29 m (95 ft) above normal sea level.

You may be less aware of tsunamis because they are relatively rare, and because most are triggered by phenomena that people don't see directly. We can define a **tsunami** as a water wave generated by the sudden movement of a mass against water. The mass can be the seafloor (when it lurches up or down during an earthquake), a submarine landslide, a subaerial landslide that falls into a body of water, a pyroclastic flow or air blast from an explosively erupting volcano, or a meteorite. Note that tsunamis are defined by the mechanism of their origin, not by their size.

Tsunamis were poorly understood until relatively recently, as reflected by the terminology and imagery used to represent them in popular media. For example, tsunamis were once referred to as "tidal waves" because they commonly affect a broad area, and because the front of an arriving

FIGURE 6.3 Tsunami-generating forces. Each view shows the instant after the mass has displaced water.

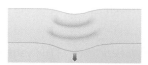

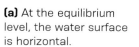

(a) At the equilibrium level, the water surface is horizontal.

(b) A downward-moving mass indents the water surface and sends waves to the sides.

(c) If the seafloor suddenly rises, it pushes up the water surface.

(d) A submerged mass sliding down a slope pushes water in front of it and draws water down behind it.

(e) A drop in the seafloor pulls the water surface downward.

tsunami may resemble the frothy step at the front of a rising tide. They have also been portrayed in films, incorrectly, as particularly large storm waves. As we will see, however, tsunamis differ from real tides and real storm waves in many important ways.

Generation of Tsunamis

To picture how the sudden movement of a mass against water produces a tsunami, imagine a stationary tub of water. The downward pull of gravity acts on the water in the tub, so the surface of the water forms a horizontal plane, called the *equilibrium level* (**FIG. 6.3a**). If a sudden movement of a mass pushes water upward from below, the water surface rises above this equilibrium level, and if the mass pushes water downward from above or pulls it downward from below, the water surface sinks below this equilibrium level (**FIG. 6.3b–e**). Regardless of the cause, as soon as the movement of the mass ceases, gravity is the water surface to start returning to its equilibrium level. In other words, gravity is a *restoring force*. Consequently, the surface of an upward bulge moves down, and the surface of a downward depression moves up (**FIG. 6.4**). Because of *inertia* (the tendency of an object in motion to remain in motion), however, the water surface doesn't stop moving when it reaches the equilibrium level, but keeps going past it. In fact, the initial input of energy from the moving solid ultimately causes the water surface to bounce up and down several times, until it eventually returns to the equilibrium level. The up-and-down movement generates several tsunamis that propagate outward. Researchers refer to a wave whose formation involves gravity as a restoring force as a *gravity wave*. So, technically speaking, a

FIGURE 6.4 Progressive formation of a tsunami wave train. Gravity is the restoring force, so a tsunami is a type of gravity wave.

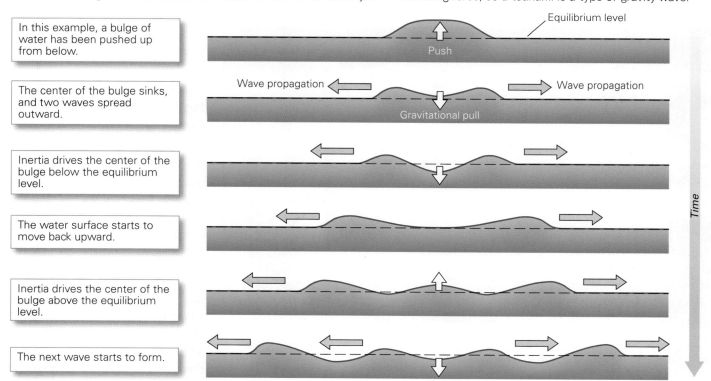

6.2 The Origin and Movement of Tsunamis • **219**

TABLE 6.1 Wave Train Terminology

Trough	The line along which water depth is lowest
Crest	The line along which water depth is highest
Wave height	The vertical distance between crest and trough
Wave amplitude	The vertical distance between the equilibrium level and a crest (= half the wave height)
Wavelength	The horizontal distance between successive crests (or successive troughs)
Wave period	The time between the passage of successive crests (or successive troughs)
Wave velocity	The horizontal speed at which a crest (or trough) moves

FIGURE 6.5 Tsunami wave trains.

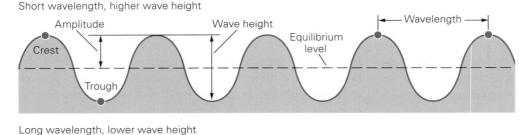

(a) The basic terminology for describing a wave train.

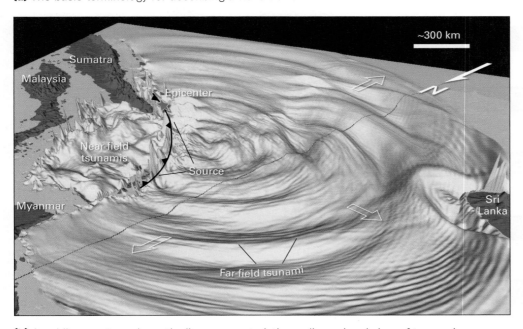

(b) An oblique, extremely vertically exaggerated, three-dimensional view of tsunami wave trains generated by the 2004 Sumatra earthquake, as viewed looking southeast. Note that these real tsunami wave trains are quite complex, and that overlapping waves can occur.

tsunami is a type of gravity wave generated when sudden movement of a mass disturbs the water surface, and gravity then works to restore it to its equilibrium level. Note that a single push by a mass against a body of water produces several waves. We refer to this group of waves as a **tsunami wave train**, and use the terms listed in **Table 6.1** (similar to those we used to describe seismic waves in Chapter 3) to describe those waves (**FIG. 6.5a**).

The characteristics of a tsunami wave train depend on the area over which the generating mass moves, on the amount of movement, and on the velocity of movement. For example, a tsunami wave train caused by the push of a broad area of seafloor against the base of the deep ocean during an earthquake may have wavelengths of 200–500 km (124–311 mi, 40–100 times the ocean's average depth) and periods ranging from 10 minutes to 2 hours. In contrast, a tsunami produced by a local landslide may have a wavelength of a few hundred meters and a period of a few minutes. Because the depth of water affected by a wave is roughly equal to half the wavelength, the entire depth of the ocean is affected by the passage of a long-wavelength tsunami wave train. Note that the velocity at which the solid pushes or pulls the water also plays a role in tsunami generation: very slow movements do not generate tsunamis, because water has time to flow around the moving object without the equilibrium level being disturbed, but fast movements can generate tsunamis.

Why Does a Tsunami's Height Vary in the Open Ocean?

The height of a tsunami at a location depends in part on the distance between the tsunami source and the location, for the energy carried by a wave spreads out along a greater crest length the farther the wave travels. Therefore, **near-field tsunamis** (also known as *local tsunamis*), those that reach shore close to their source, tend to be higher than **far-field tsunamis** (also known as *distant tsunamis* or *teletsunamis*), those that cross an ocean and reach shore far from their source (**FIG. 6.5b**). The heights of far-field tsunamis in the open ocean aren't the same everywhere. Initially, this variation happens because the amount of push in a given direction depends on the shape of the mass pushing the water, and different degrees of push will produce different wave heights. In addition, by the time a tsunami wave train has traveled a significant distance from the source, it will have become quite complex, with numerous smaller waves traveling in a variety of directions in the wake of the main waves (**FIG. 6.6a**). This complexity develops because tsunamis undergo *refraction* (bending) and *reflection* (see Chapter 3) when they encounter seamounts and oceanic islands. For example, waves coming around opposite sides of an island bend toward each other, so on the side of the island farther from the source, they not only come ashore, but also interfere (overlap) with one another. Positive interference (overlapping of crests) produces higher waves, whereas negative interference (overlapping of a crest and a trough) reduces wave height (**FIG. 6.6b**). Importantly, the highest tsunamis caused by an elongate (long and narrow) source, such as a fault, occur in a relatively narrow band oriented perpendicular to the source. This production of a narrow band of higher tsunamis is known as *tsunami beaming* (**FIG. 6.6c**).

Take-home message...

A tsunami is a wave produced when sudden movement of a mass displaces water. Gravity acts to restore a bulge or depression on the surface of the water to equilibrium level, causing the surface to move up and down repeatedly, generating a tsunami wave train. Tsunamis radiate outward from a source. Wave height can vary with direction depending on the shape of the source, and decreases overall with distance from the source.

QUICK QUESTION What's the difference between a far-field and a near-field tsunami?

FIGURE 6.6 The heights of waves in a tsunami wave train vary in the open ocean.

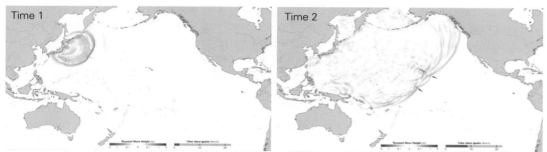

(a) Satellite imagery tracked the 2011 Tōhoku tsunami wave train as it moved across the Pacific. Initially, the waves were fairly simple. But as they interacted with seamounts and oceanic islands, wave refraction, reflection, and interference caused them to become quite complicated.

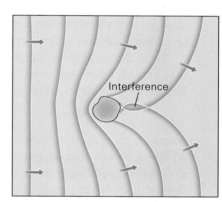

(b) When waves interact with an oceanic island, they undergo refraction and interfere with one another on the far side of the island.

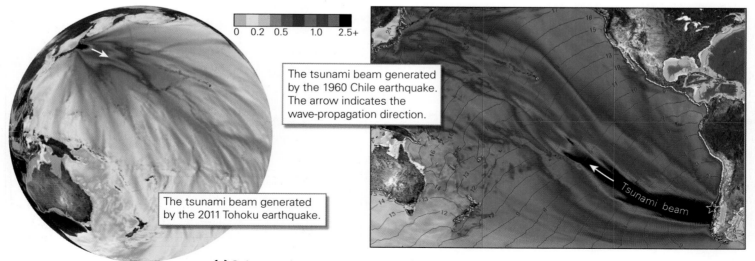

(c) Colors on these maps represent the maximum height of tsunamis, determined after they have crossed the ocean. Note that the highest tsunamis (black to purple) occur in a relatively narrow "beam" perpendicular to the fault that generated the tsunami. The width of the beam is comparable to the length of the slipped portion of the fault. Both examples are due to slip along a convergent plate boundary.

6.3 When a Tsunami Comes Ashore

How Does a Tsunami Change as It Enters Shallow Water?

Over the centuries, countless ships have foundered far out in the open ocean when they've encountered towering storm waves (see Chapter 13). But no ship has ever been lost far from shore due to a tsunami. That's because even a tsunami that towers tens of meters high and causes terrible damage when it washes ashore is no higher than a couple of meters (and generally less than 30 cm, or 1 ft) while crossing the open ocean. Also, tsunamis are so broad (have such long wavelengths) that while one wave is passing, the crest of the next wave may still be over the horizon. In fact, sailors on a ship crossing a tsunami out in the open ocean won't even notice the wave! The giant tsunamis that destroy coastal towns are a product of changes that the waves undergo as they approach the shore and enter shallower water—a process called *shoaling*.

To understand how tsunamis change as they shoal, let's first consider how wave velocity depends on water depth. Researchers define this relation by the equation

$$\text{Velocity} = \sqrt{g \times \text{depth}}$$

where g represents the acceleration of gravity. Put simply, this equation means that as waves enter shallower water, they slow down. Specifically, a tsunami wave train moving over the abyssal plains of the ocean, where water depths range between 4 and 5 km (2–3 mi), will be traveling at about 700 km/h (435 mph), almost the speed of a jet plane! When these waves start to pass into shallower water nearer to land, however, they slow down. For example, over the continental shelf, where water depth is about 200 m (656 ft), the waves slow to about 160 km/h (99 mph), and when the waves approach a beach, they slow to 40 km/h (25 mph) or less. As the waves flow over land, they slow even more because of friction between water and ground, and because turbulence within a wave saps its energy.

The relationship between wave velocity and water depth causes the height of an individual tsunami to change as it approaches the shore. Specifically, because a tsunami is so wide (as measured perpendicular to the wave crest), the rear of the wave, meaning the part farther offshore, is still traveling fast in deep water when the front of the wave enters shallow water and starts to slow down. Therefore, water near the rear of the wave starts to catch up with water toward the front, so the wave grows taller **(FIG. 6.7a)**. This change can be dramatic. For example, a tsunami that was only 30 cm high as it moved over an abyssal plain may build into a monster that rises 14–40 m (46–130 ft) above normal sea level when it reaches the shore.

The wavelength of a tsunami wave train similarly decreases as the train approaches shore, because waves toward the rear catch up with waves toward the front, while the period of the waves remains the same **(FIG. 6.7b)**. For example, a tsunami wave train that has a wavelength of 200 km (124 mi) over the abyssal plain may have a wavelength of 50 km (31 mi) over the continental shelf, and when this train arrives at the shore, its wavelength may be 10 km (6 mi). As the wavelength decreases, the width of elevated water in a single tsunami also decreases. So, when the tsunami arrives at the shore, it may have a width of about 5 km (3 mi). But even though its width near shore is much less than it was out in the deep ocean, the mound of elevated water in a single tsunami is still vastly wider than a storm wave, and it is much wider than a beach.

FIGURE 6.7 How a tsunami changes as it enters shallow water.

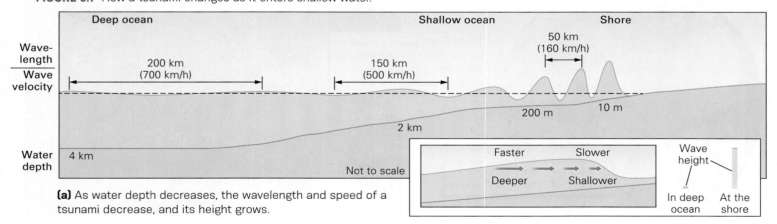

(a) As water depth decreases, the wavelength and speed of a tsunami decrease, and its height grows.

(b) Because the deeper part of the wave travels faster, the back of the wave catches up to the front, and the wave grows taller to accommodate the water volume.

Describing a Tsunami That Reaches the Shore

Researchers have developed a specialized vocabulary to describe the components of an arriving tsunami and its surroundings (FIG. 6.8). The *normal shoreline* refers to the intersection of sea level with the land before the tsunami arrives; its position depends on the tide at the time of arrival. If the trough of a tsunami arrives first, observers will see an initial **drawback**, or *drawdown*, that lowers the sea surface below sea level and may expose an area of seafloor that is normally submerged even at low tide. When the crest of the incoming wave arrives, the sea surface rises above sea level. The greatest vertical distance between the crest of the tsunami and sea level as it reaches shore is called the **tsunami elevation**. As the tsunami advances over land that slopes gently up from the shore, the vertical distance between the ground at any point and the water surface defines the *inundation depth* at that point. Eventually, the tsunami advances as far as it will go. The line on land at which the water stops is called the *inundation limit*. The horizontal distance between the normal shoreline and the inundation limit is the **inundation distance**, and the vertical distance between the normal shoreline and the inundation limit is the **run-up elevation**. Run-up elevation may be less than tsunami elevation if friction slows the advancing water to a halt before the crest of the tsunami arrives. Alternatively, run-up elevation may be greater than tsunami elevation if inertia carries the water up the slope.

The distance that water in a tsunami travels inland depends not only on the wave's height, but also on the slope of the coast (FIG. 6.9a). Specifically, where a steep cliff defines the coast, the arrival of a tsunami will cause the water level at the face of the cliff to rise. In some cases, the wave front may splash up the cliff and toss debris and boulders onto the ground inland from the cliff. But the water in a tsunami arriving at a steep cliff will not flow inland unless the wave's height exceeds that of the cliff. In contrast, where the coast has a gently sloping land surface, the tsunami height may be equal to land height many kilometers in from shore. Therefore, because tsunamis are so wide, water in the wave can flow many kilometers inland.

Significantly, bathymetry just offshore, as well as the shape of the shoreline, can influence wave height. For example, a tsunami that crosses an offshore *sand bar* (an underwater ridge of sand) before reaching the shore will become higher than a tsunami that reaches the shore where there is no bar because, as we've seen, waves get narrower and higher when they cross shallow water (FIG. 6.9b). And where a wave enters a funnel-shaped (in map view) bay or estuary, water in the wave gets focused into a progressively smaller area, so the wave height becomes greater than it would be in a place where the wave washes onto a straight shore (FIG. 6.9c) (see also Fig. 3.39d).

Because of all the complicated factors that can affect the heights of waves, the relative heights of individual waves in a tsunami wave train vary, as does the time between arrival of successive waves. *Tidal gauges* (devices that record the rise and fall of the sea surface over the course of a day) show that a tsunami wave train typically includes 6–25 noticeable waves that arrive over the course of 3–6 hours (FIG. 6.10). Earlier waves are, overall, higher than later ones, but the first wave to arrive isn't necessarily the highest. Notably, in cases where an arriving tsunami spreads out over a broad, nearly horizontal coastal plain, a new wave may arrive before the water of the previous wave has completely drawn back to the sea. Interaction of the water flowing seaward with the water flowing landward may cause the arriving tsunami to have a frothing, steep-walled face.

How Does a Tsunami Differ from a Wind-Driven Wave?

Visit a beach during a storm, and you'll see the drama of towering wind-driven waves crashing onto the sand (see Chapter 13). When they reach the shore, these waves can be as high as, or higher than, a large tsunami, but they behave very differently. Understanding the difference can help you understand the unique threat that tsunamis pose.

FIGURE 6.8 Terminology for describing aspects of tsunami inundation.

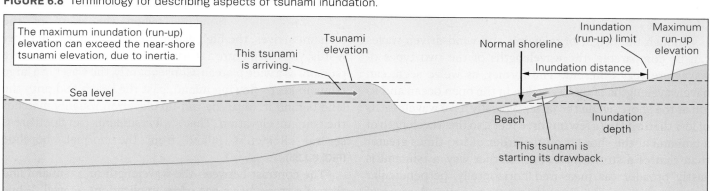

FIGURE 6.9 Interaction of a tsunami with a coast.

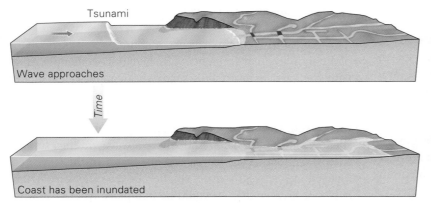

(a) When a tsunami arrives on shore, water rises along cliffs and spreads over lowlands.

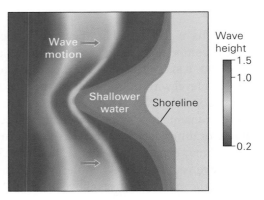

(b) If the seafloor near shore varies in depth, the wave builds higher over shallower water.

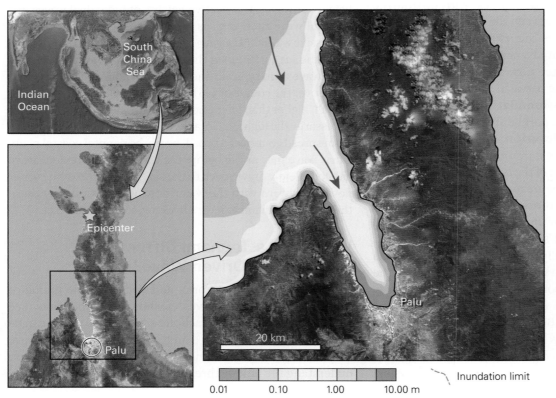

(c) The 2018 Palu tsunami was amplified as it moved up a funnel-shaped bay.

Though the height of tsunamis and wind-driven waves can be comparable, the wavelengths of the two types of waves are very different. The former, as we've seen, can have wavelengths of up to 200 km in the open ocean and 10 km at the shore, while the latter typically have wavelengths of less than 0.1 km (a few hundred feet), so the wavelength of a tsunami at the shore is on the order of 100 times greater than that of a storm wave. Put another way, a tsunami is vastly broader (as measured horizontally, perpendicular to its movement direction) than a storm wave. Because of this difference, the volume of water carried by a tsunami greatly exceeds that in a storm wave. So, when a storm-wave trough arrives at a beach, water draws back to the seaward edge of the beach, but not much farther than that, and when a storm-wave crest crashes down on the shore, the water rushes up to the top of the beach, but not much beyond that **(FIG. 6.11a)**. (In fact, the portion of the shore washed over by storm waves determines the width of the beach.) In contrast, when the trough of a large tsunami arrives, water draws back much farther than it does during even the lowest tide, and when the crest of the tsunami arrives, the high water keeps on coming. In other words, you can picture a storm wave as a narrow ridge, and a tsunami as a wide plateau. Consequently, the water in a large tsunami keeps flowing inland, past the beach and onto the land behind it, until the run-up elevation roughly matches the tsunami elevation. That's why tsunamis can flood areas several kilometers inland from the normal shoreline **(FIG. 6.11b)**.

The contrast between the wavelength of a tsunami and that of a storm wave has other implications as well. While

FIGURE 6.10 Tidal records showing the elevations of far-field tsunami wave trains generated by the M_W 9.5 Valdivia earthquake in Chile on May 23, 1960.

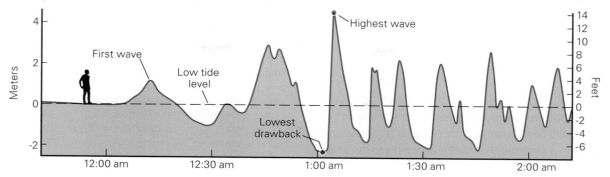

(a) A partial tidal record from Hilo, Hawai'i. At least a dozen tsunamis were detected over a 3-hour period. The dashed line represents low tide. The person is for scale.

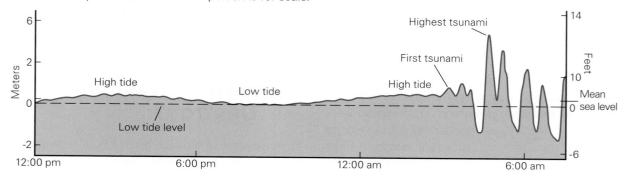

(b) A tidal record from coastal Japan. This record shows the tides of the 14 hours before the arrival of the first tsunami. Note that during this event, the tsunamis were much higher than high tide.

FIGURE 6.11 The difference between wind-driven waves and tsunamis.

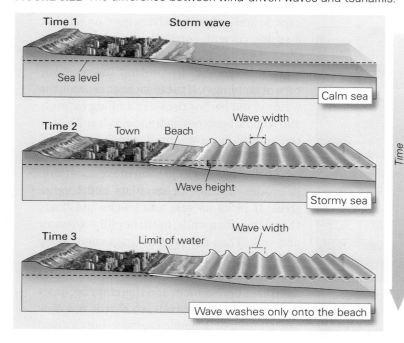

(a) Wind-driven waves are narrow, as measured perpendicular to the crest, so the volume of water in a single wave is relatively small, even if the wave is high. The wave runs out of water when it reaches the top of a beach.

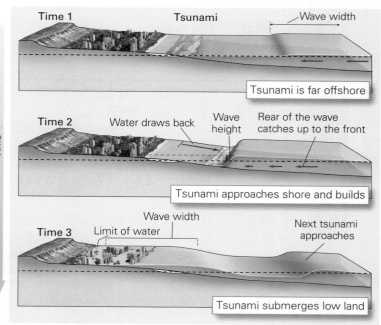

(b) Tsunamis are much wider, so a high one doesn't stop at the top of a beach, but keeps going, and can inundate land up to a few kilometers inland from the shore.

6.3 When a Tsunami Comes Ashore • **225**

successive crests of storm waves arrive at intervals of 5–20 seconds, those of a tsunami wave train arrive at intervals of between 10 minutes and 2 hours. As we'll see in Chapter 13, the *wave base*, meaning the depth down to which a passing wave affects the movement of water, equals half of the wavelength. So, because of its short wavelength, a storm wave is a near-surface phenomenon—even the largest storm waves affect water down to a depth of only about 50 m (164 ft)—whereas, as we noted earlier, a tsunami affects the entire depth of the ocean. The motion of water in a storm wave also differs from that in a tsunami **(FIG. 6.12)**. As we'll see in Chapter 13, the water in a storm wave near shore follows an elliptical path. In contrast, the water throughout a shoaling tsunami flows horizontally, roughly toward the shore—in this regard, its movement resembles an advancing tide. The velocity of water in a tsunami as it reaches the shore also differs from that of a storm wave. A storm wave in the open ocean can move at 30–50 km/h (19–31 mph), but as it interacts with the beach, it slows to about 6–7 km/h (4 mph). A tsunami arriving at the same beach may slow to 30–40 km/h (19–25 mph). The push of a moving fluid—its **dynamic pressure**—increases with the fluid's viscosity (or density) and the square of its velocity. This means that if the water in a tsunami is moving 6 times faster than the water in a storm wave when it reaches the shore, it applies about 13 times more push against the objects that it strikes. Therefore, a tsunami may push over walls that a storm wave of similar height can't.

FIGURE 6.12 The motion of water in a tsunami differs from that in a storm wave.

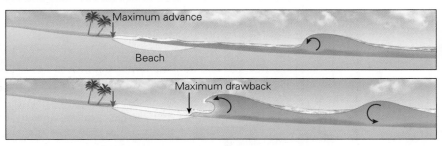

(a) As the positions of the vertical arrows indicate, even tall breakers advance and retreat only over the width of the beach. Water in these waves follows a circular to elliptical path.

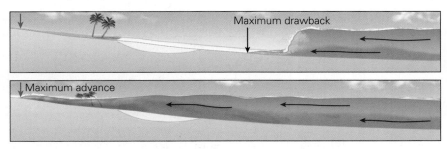

(b) A tsunami draws back much farther, exposing muddy seafloor or coral, and advances much farther. Water in the wave does not follow a circular or elliptical path.

Take-home message . . .

Tsunamis can travel as fast as a jet plane when crossing the deep water of an ocean. When a tsunami enters shallower water near shore, the back of the wave starts to catch up with the front, so the wave height increases. Also, the wavelength of a tsunami wave train decreases as it moves into shallower water. Tsunamis have such long wavelengths that, unlike wind-driven waves, they don't run out of water at the top of a beach, but can keep surging inland until the run-up elevation is comparable to the tsunami elevation at the shore.

QUICK QUESTION Which travels faster when it reaches the shore—a wind-driven wave or a tsunami? How does this difference affect the dynamic pressure (push) that a tsunami applies to objects that it strikes?

6.4 How Do Tsunamis Destroy?

The character of damage caused by a tsunami depends on location. Let's consider the consequences of a tsunami entering a town built on flat land next to the shore. Boats anchored offshore will be lifted by the rising water, and if the water rises high enough, their anchors may pull up, or their anchor chains may snap, so that the boats are carried toward shore. (If their anchors hold, the boats may capsize and sink.) The advancing tsunami, like a strong current, carries boats inland, where they may be crushed beneath bridges or stranded high and dry on land (even on the tops of buildings). If the town has a seawall, the water in the harbor rises like a filling bathtub, and when it becomes high enough, it spills over the seawall like a waterfall, which flows until the water on both sides attains the same depth.

As the water of a tsunami moves onshore, it erodes beaches or mudflats and topples trees. It picks up cars and trucks and floats them along for a while, until they fill with water and sink. Then the current drags them along and crushes them against one another, or against buildings or trees. The moving water also picks up sediment and therefore becomes denser and more viscous, so that it not only applies greater dynamic pressure, but can also more easily buoy up large objects, which can act as battering rams, destroying vehicles, bridges, fuel tanks, or other objects in their path. The wave can crush buildings **(FIG. 6.13a)**,

FIGURE 6.13 Some of the ways in which tsunamis cause damage.

(a) The dynamic pressure of fast-moving muddy water can crush buildings in its path.

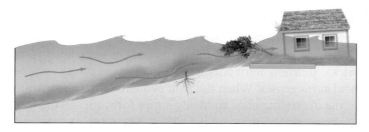

(b) Some buildings are buoyed off their foundations and float away.

(c) The dynamic pressure of the advancing Tōhoku tsunami crushed buildings.

but even buildings that withstand its dynamic pressure are filled with water up to the wave height and their interiors ruined. Buildings may also be destroyed by being pushed or buoyed off their foundations (FIG. 6.13b).

When the wave reaches its inundation limit, the water slows to a stop. But its destruction, unfortunately, hasn't finished. As drawback begins, the water, along with everything that it carries, starts rushing back to the sea (FIG. 6.14). Mobile, but no longer floating, debris is rolled and crushed, and can further damage still-standing structures. The moving water may scour into the ground and erode more sediment—in fact, the retreating tsunami may strip a beach of its sand. Retreating water, and all the sediment and debris that it carries, flows out to sea, where it may swirl in large whirlpools. Some of the debris ends up kilometers offshore—in fact, after the Tōhoku tsunami of 2011, a house was found floating 3 km (2 mi) offshore. Offshore debris may eventually be picked up by ocean currents to be carried far from its origin. When the sea returns to normal, a tsunami-battered coast may look like it has been blasted by a nuclear bomb.

What happens to people and animals caught in the path of a tsunami? Sadly, many are knocked over, submerged, and tumbled in the water, and consequently can be crushed and drowned. Some individuals may be picked up and carried with the water. These victims may be battered by objects, crushed against buildings, or dragged into flooding rooms. If they're still conscious, some victims may be able to grab onto or climb on top of the debris, float along with it, and possibly escape by climbing back onto dry land, or by climbing standing trees and bridges. There are records of people who have been carried a kilometer or more inland by a tsunami and lived to tell of it. A few victims have even been rescued after being carried offshore by the drawback.

FIGURE 6.14 The drawback of water from a tsunami carries debris out to sea. The arrow indicates the direction of the flow.

> **Take-home message...**
>
> Tsunamis do damage both while advancing and while retreating. The dynamic pressure of the advancing wave can crush buildings and people. The waves pick up sediment and become more viscous, which allows them to do even more damage. Some buildings and vehicles are buoyed along with the wave, but most are crushed and destroyed. During drawback, debris and sediment are carried out to sea.
>
> **QUICK QUESTION** What may happen to boats in a harbor when a tsunami arrives?

6.5 Tsunamis Related to Earthquakes

When a fault slips beneath the seafloor, not only can it trigger an earthquake, but it can also displace the seafloor, sending a pressure pulse to the sea surface that causes the surface to rise or fall relative to the equilibrium level. Such displacements are most likely to happen at convergent boundaries. During normal faulting, which can happen where oceanic lithosphere bends and stretches as it starts to descend into a subduction zone **(FIG. 6.15a, b)**, the hanging wall slides downward and suddenly produces a depression. During thrust faulting, which can happen along the contact between the downgoing plate and the base of the accretionary prism, or on faults within the accretionary prism itself, the hanging wall moves upward and pushes the overlying water up **(FIG. 6.15c)**. Vertical motion does not generally happen due to strike-slip faulting, so motions on transform faults typically do not generate tsunamis. (A surprising exception happened during the 2018 Sulawesi earthquake in Indonesia, when vertical movement, perhaps along bends in a strike-slip fault, was locally significant enough to trigger a tsunami that caused many fatalities in Palu.)

On occasion, very large areas of the thrust fault between a downgoing and an overriding plate slip, resulting in huge *megathrust earthquakes* (see Chapter 3). As the plates slowly move toward each other, the margin of the overriding plate sticks, shortens horizontally, and thickens vertically, so the land surface rises. When the fault suddenly slips, the margin of the overriding plate suddenly jumps forward, and the face of the accretionary prism is uplifted **(FIG. 6.16)**. At the same time, land that had been slowly rising farther from the trench suddenly subsides. The initial movement produces an N-shaped wave at the sea surface. Gravity then immediately causes the uplifted sea surface to sink, and the down-dropped sea surface to rise, in order to return to the equilibrium level. The initial wave separates into a far-field tsunami heading off into the open ocean over the downgoing plate, and a near-field tsunami that heads toward the nearby shore of the overriding plate. Researchers have reason to worry that the northwestern United States is vulnerable to near-field megathrust tsunamis **(BOX 6.1)** similar to the megathrust-generated tsunamis we describe below.

THE 1960 CHILE TSUNAMI. Devastating tsunamis were produced by the 1960 Valdivia megathrust earthquake in Chile, the largest earthquake ever recorded (M_W 9.5). Near-field tsunamis from this event reached heights of 25 m (82 ft), and far-field tsunamis propagated across the Pacific. The tsunamis that reached Hilo 14.8 hours after the earthquake ravaged the coast and cost the lives of 60 people. When the waves reached Japan, 22 hours later, they were still 3 m (10 ft) high (see Fig. 6.6c and Fig. 6.10).

FIGURE 6.15 Sudden displacement on a submarine fault can produce tsunamis.

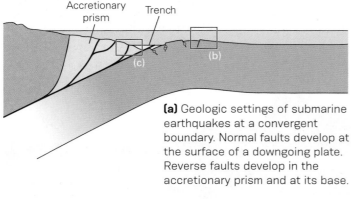

(a) Geologic settings of submarine earthquakes at a convergent boundary. Normal faults develop at the surface of a downgoing plate. Reverse faults develop in the accretionary prism and at its base.

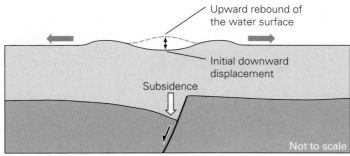

(b) Downward displacement on a normal fault causes subsidence of the seafloor, pulling the water surface down.

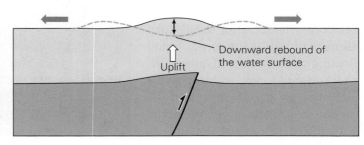

(c) Upward displacement on a thrust fault causes uplifting of the seafloor, pushing the water surface up.

FIGURE 6.16 Formation of megathrust-related tsunamis.

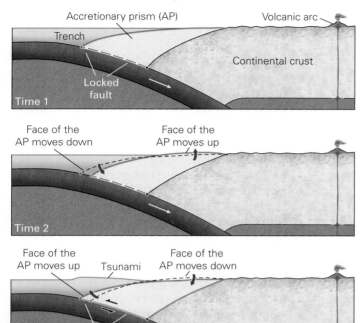

(a) Stages of stick-slip behavior at a subduction-related thrust fault.

(b) In detail, slip on the thrust fault causes uplift of the accretionary prism nearer the trench and subsidence of the part nearer the volcanic arc. Both movements affect the shape and size of the tsunami.

THE 1964 ALASKA TSUNAMI. The M_W 9.2 Good Friday megathrust earthquake that struck Alaska in 1964 (see Chapter 3) caused displacement of an area about the size of California. Notably, a band of the crust between the trench and the coast rose by as much as 12 m (39 ft), while a parallel band from the coast inland subsided by as much as 2.5 m (8 ft) (FIG. 6.17a). The movement generated near-field tsunamis that caused nearly 120 fatalities. In aerial photos of coastal towns, the inundation limit is clearly visible because the tsunamis washed away snow seaward of that line (FIG. 6.17b). Far-field tsunamis from this event caused significant damage along the west coast of the United States and Canada.

FIGURE 6.17 The tsunami associated with the 1964 Good Friday earthquake in Alaska.

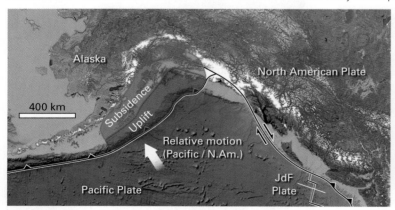

(a) A megathrust earthquake along the Pacific Plate–North American Plate boundary caused uplift and subsidence of southern Alaska (on the overriding plate).

(b) In Valdez, shown here, the inundation limit stands out because the waves washed away snow cover.

BOX 6.1 DISASTERS AND SOCIETY

Ghost forests of Cascadia

Along several tidal marshes and swamps of coastal Washington, in the Cascadia region, dry and weathered trunks of dead cedar trees rise above the grass **(FIG. Bx6.1a)**. The trees are rooted in a soil that now lies beneath a meter of peat, the deposits of coastal wetlands. When first recognized, the origin of these *ghost forests* was a mystery. What phenomenon could cause the land to subside below sea level, only to rise again above sea level a few hundred years later? A trench cut into the substrate of coastal areas in nearby Oregon provided clues. Layers of sediment visible in the trench walls revealed that the land on which Native American communities had lived had suddenly been covered by a layer of sand that contains sedimentary structures (such as cross beds) indicative of deposition in a strong current, before being covered by seawater in which a tidal marsh grew **(FIG. Bx6.1b)**. Radiocarbon dating of charcoal in Native American fire pits buried by the sand and tidal-marsh sediments, along with tree-ring dating of the dead trees in ghost forests, revealed that the event happened between 1695 and 1720 c.e., but because there was no written record of the event, the exact date remained unknown.

Geologists eventually concluded that the sand layer was deposited by a near-field tsunami produced by a megathrust earthquake along the Cascadia subduction zone, where the Juan de Fuca plate slides underneath North America (see Chapter 3). With this concept in mind, the origin of the ghost forests and the sand layer became clear. As stress builds due to plate convergence, the continental margin of the Pacific Northwest shortens horizontally and rises vertically. On the resulting dry land, forests grew. When the megathrust earthquake happened, slip on the plate boundary allowed the continental margin to thrust westward. As a result, the land

FIGURE Bx6.1 Formation of the ghost forests of Cascadia.

(a) A ghost forest as it appears today.

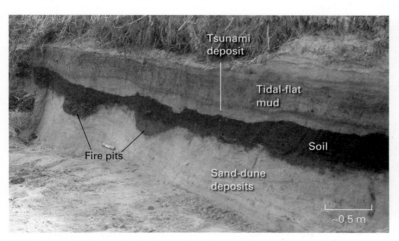

(b) The sand layer buried a layer of soil containing the remains of Native American fire pits. Mud then buried the sand.

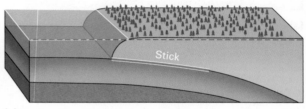

(c) During the sticking phase of stick-slip motion at the Cascadia subduction zone, the toe of the accretionary prism moved landward, the coastal area rose above sea level, and forests grew.

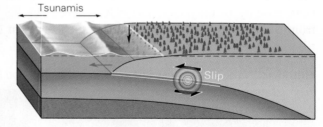

(d) When the thrust fault slipped, the toe of the accretionary prism moved abruptly seaward. The coastal area, and the forests on it, became submerged. Large tsunamis washed ashore.

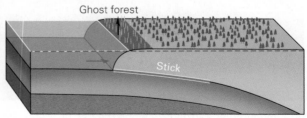

(e) During the next sticking phase, the coastal area rose above sea level again, along with the now-dead forests.

surface at the coast suddenly subsided while the accretionary prism offshore underwent sudden uplift. (The same phenomenon had been directly observed in association with the 1964 Good Friday earthquake in southern Alaska.) The double whammy of sudden uplift offshore and subsidence nearer the shore generated a wave train of large tsunamis. These waves deposited the sand layer. In areas that had subsided to become salt marshes, the forests, flooded by saltwater, died. Over time, peat produced by the decay of the marsh grass buried the trees. Then, during the next couple of centuries, while the thrust fault defining the plate boundary remained stuck, shortening again led to the uplift of the land containing the ghost forests (FIG. Bx6.1c).

Scientists recognized that if a megathrust earthquake produced near-field tsunamis in Cascadia, it likely produced far-field tsunamis that reached Japan. Starting in 684 C.E., historians in Japan started making a record of every tsunami to have reached the nation's coast. Study of this record revealed that an *orphan tsunami*, one that arrives in an area that has not felt an earthquake, indeed struck Japan in 1700—in the time window of the Cascadia megathrust earthquake. Based on the arrival time of the orphan tsunami, researchers calculated that the Cascadia earthquake happened at 9 A.M. (Pacific time) on January 26, 1700. And because of the size of the tsunami, researchers determined it was caused by a $> M_w$ 8.6 earthquake.

In the past few centuries, the population of coastal Oregon and Washington has increased substantially. If (when?) another megathrust earthquake occurs along the Cascadia fault zone, the resulting tsunamis could submerge coastal communities in which an estimated 70,000 people live, as well as harbor and industrial areas of Seattle and Tacoma. Because of the realization that another megathrust earthquake, generating a devastating tsunami, could occur, emergency officials in the Pacific Northwest have started to take planning for tsunami evacuation very seriously. Maps showing inundation levels for both near-field and far-field tsunamis are being prepared for much of the coast.

FIGURE 6.18 The tsunami associated with the 2004 M_w 9.3 earthquake off Sumatra.

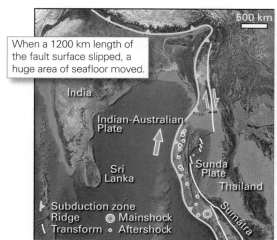

(a) Plate configuration and location of epicenters.

(b) Part of the accretionary prism uplifted, and part subsided.

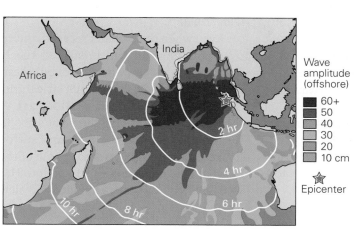

(c) Tsunami beaming sent the highest waves to the west and southwest. Yellow areas indicate areas of coast that were significantly damaged.

THE 2004 INDIAN OCEAN TSUNAMI. The azure waters and palm-fringed islands of the Indian Ocean's eastern coast hide the Sunda Trench, a seismically active convergent boundary. Just before 8:00 A.M. on December 26, 2004, an area of the thrust fault 1,300 km (808 mi) long by more than 100 km (62 mi) wide beneath the accretionary prism of this boundary slipped, and the overriding plate lurched westward by as much as 15 m (49 ft). This slip triggered the M_w 9.3 Sumatra megathrust earthquake (FIG. 6.18a). Closer to the trench, the seafloor rose by as much as a few meters, and closer to shore, it subsided by as much as a meter (FIG. 6.18b), generating tsunamis that spread across the Indian Ocean (FIG. 6.18c).

Within minutes, the near-field waves reached Sumatra. At beach resorts near Banda Aceh, the sea receded much farther than anyone had ever seen. Unsuspecting tourists walked out onto the exposed seafloor in wonder. When the first tsunami crest arrived, it resubmerged the exposed seafloor and then spread over the entire beach and beyond, floating away deck chairs and swamping the beachside bars. As that wave retreated, the next one

FIGURE 6.19 The 2004 tsunami reached heights of 15–30 m at Banda Aceh, and it washed away towns and fields.

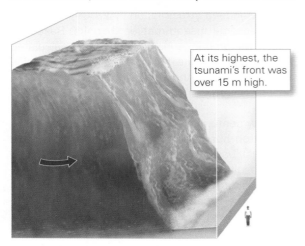

(a) The wave was many times higher than a person.

At its highest, the tsunami's front was over 15 m high.

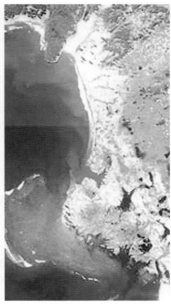

(b) These before and after satellite images show how the waves stripped the landscape.

(c) Only a few of the most strongly built houses remained standing.

approached, appearing in the distance as a wall of frothing water that dwarfed the yachts anchored offshore (see Fig. 6.2a). In places, the wave front reached heights of 15–30 m (49–98 ft) (FIG. 6.19a). With a rumble that grew to a roar, the tsunami rushed over the beach and kept going (see Fig. 6.2b), in places flooding land as far as 7 km (4 mi) inland (FIG. 6.19b). When gravity finally pulled the water back seaward, it left the landscape almost stripped of buildings and covered in rubble (FIG. 6.19c). Sadly, the horror of Banda Aceh was just a preamble to the devastation that would soon visit other stretches of Indian Ocean coast. Far-field tsunamis crossed the Indian Ocean and struck Sri Lanka and India 2–3 hours after the earthquake, and the coast of Africa, on the west side of the Indian Ocean, 8–10 hours after the earthquake. More than 230,000 people died that day.

THE 2011 TŌHOKU TSUNAMI. During the 2011 M_W 9.0 Tōhoku earthquake, a 60,000 km² (23,166 mi²) area of the fault separating Japan from the Pacific Plate slipped by up to 40 m. This movement produced tsunamis that ravaged the northeastern coast of Japan, as seen worldwide in high-definition video. The 10 m (33 ft) high seawalls that had been built to protect towns along the coast were no match for the advance of the wave, which locally rose to heights of 30 m at the shore (FIG. 6.20a). The sea swept over beaches and moved inland (see Fig. 3.39c), and it overtopped seawalls and spilled into towns (see Fig. 1.1b). As the waves spread over farm fields, they picked up dirt and debris and evolved into a viscous slurry with the consistency of wet mud (FIG. 6.20b). When they swept through towns bordering estuaries, they left behind a debris-strewn wasteland (see Fig. 6.9c). Some buildings that were strong enough to survive the wave were capped by ships carried in by waves and left behind on roofs (FIG. 6.20c).

Even when the tsunami receded, the catastrophe was not over. The tsunami had inundated the Fukushima Daiichi nuclear power plant, where it not only destroyed power lines, cutting the plant off from the electrical grid, but also drowned the backup diesel generators (FIG. 6.21). Without electricity, the water pumps driving the plant's cooling system stopped functioning, and the cooling water surrounding the plant's hot reactor cores boiled away. Some of the water became so hot that H_2O molecules separated into H_2 and O_2 gas, which, when ignited by a spark, exploded and blew the tops off three

FIGURE 6.20 The 2011 Tōhoku tsunami, triggered by a megathrust earthquake east of Japan, caused immense damage. (See Figure 3.39c, d.)

(a) The rising water of the tsunami filled harbors and spilled over seawalls.

(b) The waves picked up mud and debris and became a viscous slurry.

of the plant's four containment buildings, contaminating the surroundings with radioactivity. A decade after the event, officials have not yet found a safe way to dispose of radioactive water still trapped in the plant.

> **Take-home message . . .**
> Both normal faulting and thrust faulting can cause sudden vertical displacement of the seafloor and generate tsunamis. Strike-slip faults generally do not produce tsunamis. Large tsunamis develop in response to megathrust earthquakes, which displace vast areas of the seafloor.
>
> **QUICK QUESTION** Along which type of plate boundary is slip most likely to produce tsunamis?

(c) Ships that floated inland on the wave were left stranded when the water retreated.

FIGURE 6.21 The 2011 Tōhoku tsunami knocked out power to the Fukushima Daiichi nuclear power plant.

(a) The power plant before the tsunami. Due to loss of cooling water, a partial meltdown of the reactor cores took place.

(b) Heat caused water to disassociate into H_2 and O_2. Explosions of hydrogen gas blew the tops off reactor containment buildings.

6.6 Landslide-Generated Tsunamis

Landslides—including slumps, rock and debris slides, and rockfalls—can happen either on land or underwater (see Chapter 5). Subaerial landslides that fall into the sea push down on its surface, producing a large depression, which generates a tsunami as the sea surface bounces up and down on its way back to achieving equilibrium. Submarine landslides generate tsunamis partly because the downward motion of a solid mass underwater pulls the surface of the ocean down and partly because the toe of the landslide pushes the water above it up. You can simulate this phenomenon by placing your hand just below the surface of water in a tub, and then moving it suddenly down. In this section, we consider a few examples of such landslide-generated tsunamis.

Subaerial Landslides

On the night of July 9, 1958, an M_W 7.8 earthquake took place along a strike-slip fault that forms part of the boundary between the Pacific and North American Plates in southeastern Alaska. The fault cuts across the head of an 11 km (7 mi) long *fjord* (a water-filled, glacially carved valley), known as Lituya Bay, that opens into the Pacific Ocean. Displacement on the fault itself did not generate a tsunami, because the strike-slip motion did not cause the seafloor to go up or down. But the ground shaking destabilized the steep mountain slopes at the head of the bay and triggered sudden mass wasting—30 million m³ (1 billion ft³) of rock and debris fell hundreds of meters into the bay.

When it hit the water of Lituya Bay, the rockfall was traveling close to 300 km/h, so the impact shoved all the water below it out of the way. Because the debris landed next to a cliff, the water had nowhere to go except outward into the fjord. Some water sprayed upward in a giant splash, and some started to move outward from the impact site in a tsunami wave train. As the splash water fell, the lead tsunami emerged from beneath the spray and barreled forward (FIG. 6.22a). A kilometer from the rockfall, a ridge jutted

FIGURE 6.22 The Lituya Bay tsunami was generated by a rockfall shaken loose during an earthquake along the Fairweather fault.

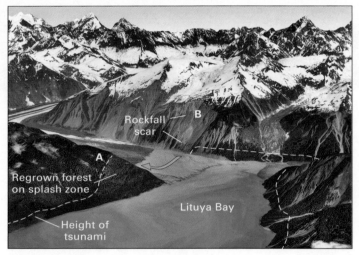
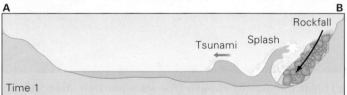
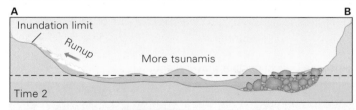
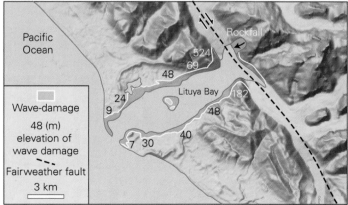

(a) The rock plunged down a cliff and into the bay, where it produced a huge splash and a tsunami.

(b) The tsunami scoured trees away from the banks of the fjord, exposing bare rock.

partway out into the fjord, into the path of the tsunami. The end of the tsunami that rammed into the ridge had so much inertia that its run-up shot up the side of the ridge, ripping away trees up to 524 m (1,719 ft—as high as a 104-story building) above sea level (see Point A in the figure), the highest inundation elevation ever recorded for a tsunami! As the rest of the wave headed farther down the axis of the bay, its height decreased, but even 3 km (2 mi) from the site of the rockfall, it washed away the forest 200 m above sea level (FIG. 6.22b). When the wave exited the mouth of the bay and entered the Gulf of Alaska, 11 km (7 mi) from the rockfall, it was still 30 m high. Due to its extreme height, the Lituya Bay wave has been called a **megatsunami** (defined as a tsunami over 100 m, or 328 ft, high). There are no communities along the bay, so fortunately this wave did not cause a disaster. But of the three fishing boats anchored in the bay, two were lost and the third was carried over a kilometer out to sea.

As you can tell from the description above, the Lituya Bay megatsunami differed in many ways from the earthquake-related tsunamis that we described earlier. It began as a much higher wave, but because the total volume of water displaced at the site of impact was relatively small (compared with the volume displaced during a megathrust earthquake), the wave didn't propagate very far into the open ocean, so there were no far-field effects. Similar megatsunamis have also happened in lakes and reservoirs. For example, the Vaiont Dam rockfall produced a tsunami in a reservoir that took many lives (see Chapter 5).

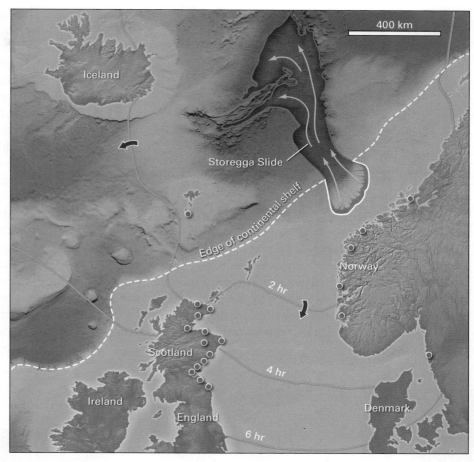

FIGURE 6.23 About 8,000 years ago, the Storegga Slide produced tsunamis (solid orange lines) that destroyed Stone Age villages (red dots) along the coast of the North Sea.

Submarine Slumping along Continental Margins

Geologists studying sediment layers along the coast of the North Sea observed a 10,000-year-old sand layer containing seashells and marine plankton lying 4 m (13 ft) above sea level at many locations. The origin of this layer remained a mystery until 1983, when bathymetric surveys revealed that a large chunk (300 km, or 186 mi, across, as measured parallel to the shore) was missing from the edge of the continental shelf off the coast of Norway (see Chapter 5). The missing chunk is the scar left when as much as 5,000 km³ (1,200 mi³) of debris slid down the continental slope, either all at once or in a succession of events. About 8,000 years ago, one of the events, the Storegga slide, generated a tsunami responsible for the mystery sand layer (FIG. 6.23). This wave may have swept away most Stone Age villages along the North Sea coast.

The existence of the Storegga slide emphasizes that submarine landslides can occur along passive margins. Such margins rim almost the entire edge of the Atlantic Ocean, so while the Atlantic coast has far fewer earthquakes than active margins do, it still faces the risk of tsunamis. While these events are clearly very infrequent, examples have happened in historic time. Most notably, when the Grand Banks earthquake of November 1929 that triggered a submarine slump and turbidity currents that broke trans-Atlantic telegraph cables (see Chapter 5), it also produced a tsunami 3–7 m (10–23 ft) high that, 2 hours after the earthquake, struck the Burin Peninsula of Newfoundland, where it flooded 40 communities, killing 28 people (FIG. 6.24). The tsunami not only destroyed houses and ships, but also washed away the season's catch of cod, which was being salted in sheds next to the piers.

FIGURE 6.24 A submarine slump caused by an earthquake, and associated turbidity currents, broke undersea telegraph cables on the continental slope. It also created a tsunami that inundated the coastal towns of Newfoundland.

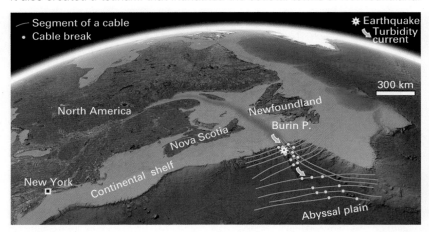

Major submarine slumps happen along active continental margins as well, probably even more frequently than along passive margins, because earthquakes triggering motions happen more frequently along active margins. Researchers have recently realized that some tsunamis attributed to fault-related displacement are actually due to earthquake-triggered landslides (BOX 6.2).

Flank Collapse of Volcanic Islands

The Big Island of Hawai'i, which consists of overlapping shield volcanoes (see Chapter 4), has a rounded shape overall. But on its southeastern side, about 10 km (6 mi) from the summit of the Kilauea volcano, a large northeast-trending scarp has formed (FIG. 6.25a). To the southeast of this scarp, the land surface has dropped downward by 100–500 m (328–1,640 ft). This scarp delineates the Hilina fault zone, the upslope edge of the Hilina slump, a piece of Hawai'i 70 km (43 mi) long by 10 km (6 mi) wide that is subsiding at a rate of about 10 cm (4 in) per year. Studies of the subsurface structure of Hawai'i hint that some of the faults in the Hilina fault zone connect to a *detachment*, a roughly horizontal fault along the base of the volcanic edifice. The detachment probably lies in the weak layer of clay-rich oceanic sediment onto which the earliest lavas that built Hawai'i were erupted (FIG. 6.25b). Could there be a sudden, catastrophic movement that removes the entire flank of Hawai'i on the seaward side of the Hilina fault zone and abruptly drops it into the ocean in a giant submarine

FIGURE 6.25 Submarine slumps around the Hawaiian Islands may have caused tsunamis in the past.

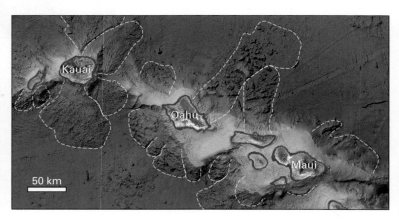

(a) Topography of the southern half of the Big Island of Hawai'i shows the Hilina fault system and the Ka'oki fault system.

(c) Numerous submarine slumps surround the Hawaiian Islands. If they moved fast enough, the slumps could have caused tsunamis.

(b) A schematic cross section of the Big Island of Hawai'i from Mauna Loa to the coast. Note that a detachment, which lies in the sediment layer under the volcanic edifice, underlies part of the island. Parts of Mauna Loa and Kilauea are slowly slipping seaward.

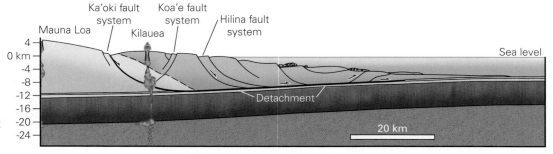

BOX 6.2 DISASTERS AND SOCIETY

The enigmatic Papua New Guinea tsunami of 1998

On July 17, 1998, an M_W 7.0 earthquake occurred along the northern coast of Papua New Guinea (FIG. BX6.2a). Though the earthquake caused some damage, it wasn't catastrophic, but in the villages along a 100 km (62 mi) stretch of the coast, what followed was. Minutes after the shaking ceased, a large wave became visible offshore. People tried to run inland, or climb trees, but most were caught in the roiling water of the 10–15 m (33–49 ft) wave and were tumbled over the sand and coral or struck by debris. Over 2,000 people died, 1,000 more were injured, and 10,000 were displaced (FIG. BX6.2b).

When researchers began to study this disaster, they suspected that the waves were near-field tsunamis triggered by displacement on a nearby fault. That assumption, however, proved wrong, for the timing of the wave's arrival didn't match the researchers' predictions. Eventually, they concluded that the wave wasn't due to the fault displacement itself, but rather to a submarine landslide triggered by the earthquake (FIG. BX6.2c, d). Bathymetric studies indicated that the slump block involved was about 5 km (3 mi) in diameter, and that its head scarp formed at a depth of about 1 km (0.6 mi).

The tsunami was devastating, not only because of the immediate loss of life, but also because of the health challenges that followed. The affected region did not have easy access to modern medical facilities, so injured victims were cared for in makeshift shelters without proper hygiene. Several victims succumbed to infections because of the lack of antibiotics. The event was so distressing that witnesses suffered post-traumatic stress syndrome and became distraught when discussing the event, even years later. To prevent a recurrence of this devastating disaster, communities moved inland where they could, and a video about tsunamis, explaining why they occur and what to do in the event one happens, was shown to residents. Before watching the video, residents did not know the warning signs that a tsunami might happen (a significant earthquake) and might be imminent (drawback of the sea), and had no plans to protect themselves. Now they do.

FIGURE Bx6.2 The 1998 Papua New Guinea tsunami was caused by a submarine landslide following an earthquake.

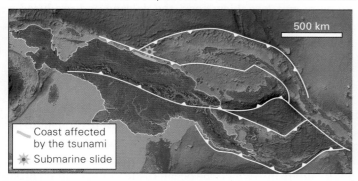

(a) Papua New Guinea borders several microplates, so it endures many earthquakes.

(b) The force of the wave flattened palm trees on a barrier island between the sea (behind the photographer) and a lagoon (in the background).

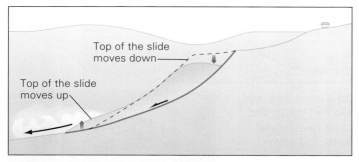

(c) The landslide produced tsunamis even though it took place entirely underwater. The top of the slide sank, so it pulled water down. The toe of the slide pushed water up.

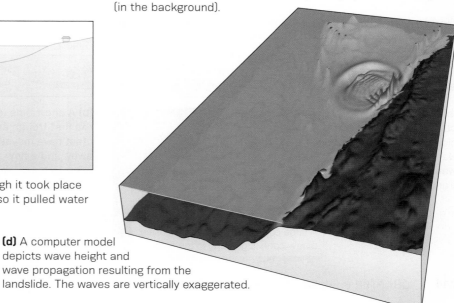

(d) A computer model depicts wave height and wave propagation resulting from the landslide. The waves are vertically exaggerated.

landslide? If so, could such a sudden **flank collapse** generate a large tsunami? This question remains a subject of debate: most researchers have concluded that a collapse won't be happening any time soon, and that the current movement rates are fortunately too slow to trigger a tsunami.

Seafloor bathymetry around the Hawaiian Islands shows that immense submarine slumps in the past have spread debris on the abyssal plain as far as 200 km (124 mi) from the islands **(FIG. 6.25c)** (see Chapter 5). The distance that the slump blocks have moved implies that they were traveling fairly quickly, and therefore, that some of the mass-wasting events on and around the islands have caused tsunamis in the past. Gravels composed of 100,000-year-old coral fragments, now found at an elevation of over 300 m (984 ft) above sea level on the island of Lanai, 125 km (78 mi) northwest of the Big Island, are interpreted as the deposits of such a wave.

Recent studies in the Cape Verde Islands, volcanic edifices formed over a hot spot 700 km (435 mi) west of Senegal, confirm that flank collapse of volcanic islands can produce megatsunamis. On one of the islands, Santiago, giant blocks of rock—up to 12 m (39 ft) across—were plucked from a cliff now at 170 m (558 ft) above sea level, and then were swept inland by as much as 650 m (2,133 ft) to locations now up to 220 m (722 ft) above sea level **(FIG. 6.26)**. Only the run-up of a megatsunami could have moved such large blocks so far. Researchers attribute the megatsunami to a flank collapse on Fogo, one of the most active volcanoes in the Atlantic, located just 60 km (37 mi) to the west of Santiago. During the collapse, which took place 73,000 years ago, when sea level was 50 m lower than it is today, about 160 km³ (38 mi³) of debris rushed down Fogo's submarine slope. This slide triggered a megatsunami that was probably over 250 m (820 ft) tall at the source. Deposits found on other islands in the region contain records of older megatsunamis. Similar events could happen in the Canary Islands, another group of hot spot–related islands 1,500 km (932 mi) north of the Cape Verde Islands.

> **Take-home message . . .**
> Landslides that dump large quantities of rapidly moving debris into the ocean can produce tsunamis by pushing the water surface abruptly down. Submarine landslides can also generate tsunamis, as the sliding mass pulls the water surface down, and the motion of the toe pushes the water surface up. While mass wasting capable of generating tsunamis happens most frequently on active margins and by flank collapse of oceanic islands created by hot-spot volcanism, it can also happen along passive margins.
>
> **QUICK QUESTION** Which affects a broader area—a megatsunami caused by a local rockfall, or a tsunami caused by a megathrust earthquake? Why?

6.7 Other Causes of Tsunamis

Tsunamis due to Explosive Volcanic Eruptions

The 1883 eruption of Krakatau, in the Sunda Strait between Sumatra and Java, blew apart the volcano and produced a large caldera (see Chapter 4). The eruption itself, and the pyroclastic debris it produced, caused many casualties. But by far the greatest loss of life—an estimated 36,000 people—happened when tsunamis triggered by the eruption reached land **(FIG. 6.27a)**.

How explosive eruptions generate tsunamis remains a subject of research. Originally, the waves were thought to be primarily due either to the collapse of the volcanic edifice, which would resemble the sudden drop of a piston, producing a caldera into which the ocean rushes, or to a lateral blast that pushes water away. While these phenomena may play a

FIGURE 6.26 Computer model of tsunamis generated by a flank collapse on Fogo, in the Cape Verde Islands.

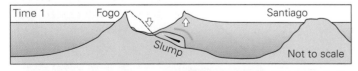

(a) The initial movement pulls the water surface close to the island downward, and pushes the water surface farther from the island upward.

A large part of the volcano is missing; a new cone is growing in the scar.

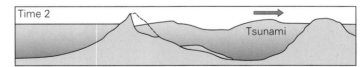

(b) A short time later, the first trough and crest have moved away, and a new crest forms close to the island.

A huge boulder thought to have been tossed by the 240 m-high tsunami.

FIGURE 6.27 Formation of the tsunamis associated with Krakatau's 1883 explosion.

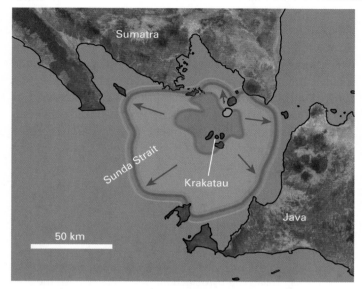

(a) Krakatau lies between Sumatra and Java. The waves propagated outward and struck the nearby coasts.

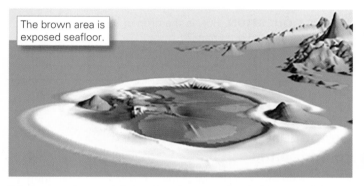

(b) A computer model showing that pyroclastic flows could have pushed away the sea over a broad area.

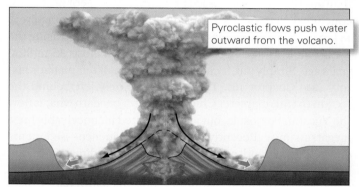

(c) Alternative models proposed to explain the formation of the tsunamis.

role in tsunami generation, researchers have concluded more recently that much of the energy that produces these tsunamis comes from the collapse of the debris cloud to produce gigantic pyroclastic flows. In the pyroclastic flows following Krakatau's eruption, over 18 km³ (4 mi³) of debris collapsed downward and slammed into the sea surface. Models suggest that when the debris hit the Sunda Strait, it pushed all the water away from the volcano for a distance of up to 10 km (6 mi) from the island, briefly laying bare the seafloor (FIG. 6.27b, c). The resulting tsunamis rose to heights of 15–35 m (50–115 ft) and washed away over 160 coastal communities in Sumatra and Java.

A similar event happened in the Mediterranean Sea during the eruption of Thera (also known as Santorini) in about 1600 B.C.E. The explosive eruption yielded immense pyroclastic flows—even larger than those of Krakatau—that generated tsunamis. The waves swept across the eastern Mediterranean, disrupting sediment on the seafloor, and struck the coastline of the sea and of many islands (FIG. 6.28). Historians speculate that the eruption and the accompanying tsunamis affected the course of Western civilization by initiating the decline of the Minoan civilization on Crete and leading to political turmoil in Egypt.

Tsunamis due to Meteorite Impacts

You can produce mini-tsunamis by dropping a pebble into a puddle (FIG. 6.29a). Imagine that, instead of a relatively slow-moving pebble, the falling object is a giant meteorite—such as an asteroid or comet—traveling at 72,000 km/h (44,700 mph), and instead of a puddle, the water body being struck is the ocean. The result will be a tsunami larger by far

FIGURE 6.28 The tsunamis associated with the eruption of Thera at about 1600 B.C.E. A study of coastal sediments has shown that the eruption sent huge waves across the eastern Mediterranean as far as modern-day Israel. The waves would have reached the coast of Israel and Lebanon about 1.5 hours after the eruption.

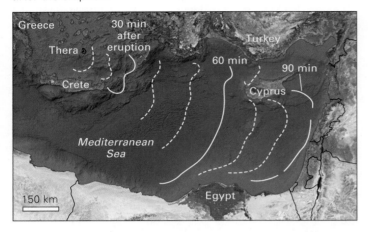

than any that we've discussed previously. Most geologists have concluded that such a collision—the impact of an asteroid 14 km (9 mi) in diameter—played a major role in bringing the Mesozoic Era—the Age of Dinosaurs—to a close 66 million years ago. Simply put, it changed the course of our planet's history (see Chapter 16). The impact not only left behind the 150 km (93 mi) diameter Chicxulub crater beneath what is now the Yucatán Peninsula, but also generated inconceivably huge megatsunamis. Recently discovered evidence—sediments deposited on land as well as disrupted sediments buried beneath the deep seafloor—demonstrate that these tsunamis raked across the deep seafloor around the world **(FIG. 6.29b)**. The tsunami may have swamped land along the Gulf coast of North America as far inland as 100 km.

Researchers have developed computer models to simulate the impact of a huge object like the one that formed the Chicxulub crater. The object would have instantly slammed through the entire depth of the ocean, producing a lateral air blast that would have laid bare the ocean floor and generated a megatsunami 1.5 km (1 mi) high. Huge chunks of debris thrown up by the impact plummeting down again as far as 200 km from the crater could each have generated other tsunamis, and waves would also be generated by the sea rushing back into the crater. Fortunately, no such event has ever happened during human history.

Take-home message . . .

When a volcanic island erupts explosively, it may produce a lateral blast, its edifice may collapse to form a caldera, and its pyroclastic flows may collapse onto the sea surface. All of these phenomena contribute to generating tsunamis. Collapsing pyroclastic flows probably have the greatest effect. A meteorite impact like the one that occurred at the end of the Mesozoic Era can produce towering megatsunamis.

QUICK QUESTION How is the drop of a pebble into a puddle similar to a tsunami?

6.8 Tsunamis and Society

Secondary Disasters Related to Tsunamis

So far we've focused on the causes and major consequences of tsunamis (**VISUALIZING A DISASTER**, pp. 242–243). The destructive consequences of tsunamis that we have described

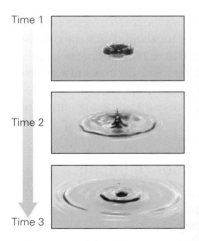

(a) The impact of a pebble on the surface of a puddle produces mini-tsunamis.

FIGURE 6.29 The asteroid impact that formed the Chicxulub crater and marked the end of the Mesozoic Era must have also produced megatsunamis.

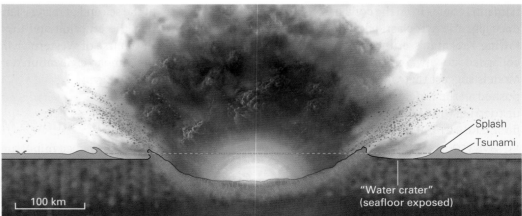

(b) The seafloor would have been exposed by the blast from the impact.

240 • CHAPTER 6 • When the Sea Suddenly Rises: Tsunamis

FIGURE 6.30 Secondary disasters associated with tsunamis.

(a) Debris floating on water can catch fire. Blocked streets prevent firefighters from reaching fires, or victims from escaping.

(b) When seawater evaporates, it leaves salt in the soil. The salt prevents crops from growing.

Some locations may be abandoned to prevent future inundation.

(c) Destruction of housing can leave communities homeless long after the waves recede.

throughout this chapter can be amplified by secondary disasters. For example, because tsunamis rip away or bury roads, bridges, power lines, water lines, gas lines, and sewage lines, they cut off transportation routes and prevent essential services from reaching victims. Fires that ignite in debris, like those that follow earthquakes, can't be contained by fire crews, who can't access equipment and can't reach the fires (FIG. 6.30a). Tsunamis, however, not only produce debris, but also drench it in water, so that it soon begins to mold and rot. And tsunamis not only drench agricultural fields, but they also saturate them with salt, destroying existing crops and preventing new ones from being planted (FIG. 6.30b). Tragically, loss of life due to a tsunami can be so staggering that the community, even with support from outside, can't cope, so corpses and sewage contaminate drinking water supplies, leading to epidemics of cholera and dysentery. Puddles and soggy ground remain even after the waves recede, allowing mosquitoes to breed and cause many cases of malaria.

Over the long term, displacement of populations, loss of housing, and loss of employers leads to a variety of health and social issues, particularly in poorer and less resilient communities (FIG. 6.30c). The destruction of harbors, for example, can cause failure of fishing and shipping industries that may have been the main source of revenue and employment for the community. Loss of transportation and flooding of fields interrupts food supplies, and loss of factories, stores, and offices eliminates jobs. Communities may take years to recover, and in some cases, are simply abandoned.

Tsunami Relief and Recovery

When the sea returns to equilibrium, relief efforts begin. Unfortunately, rescue workers seeking to find survivors and free them from debris face a daunting task because of the blockage of roads and loss of communications systems. To overcome this challenge after the 2004 Indian Ocean tsunami, some rescuers arrived by hovercraft, and ships became floating hospitals (FIG. 6.31a). In Japan, after the 2011 Tōhoku event, bulldozers had to plow debris off key access routes before emergency and relief vehicles could gain access to the devastated areas. Once the immediate needs of affected communities have been addressed and displaced people have food, water, temporary shelter, and health care, communities still face the question of what to do with literal mountains of debris. It takes years to clear some areas (FIG. 6.31b). Then, the process of rebuilding, and of mitigation for the future, begins.

VISUALIZING A DISASTER
Tsunamis

- Whirlpools in the backwash of the previous wave
- Exposed reef
- Wave front becoming a breaker
- Impact
- Tsunami damage and saltwater puddles
- Brownish water containing suspended sediment
- Debris fire
- Seawall
- Wave spilling over a seawall
- Wave moving up a channel
- Ships carried inland
- Nearshore structures have washed away
- Highlands remain dry

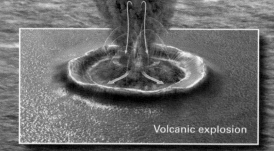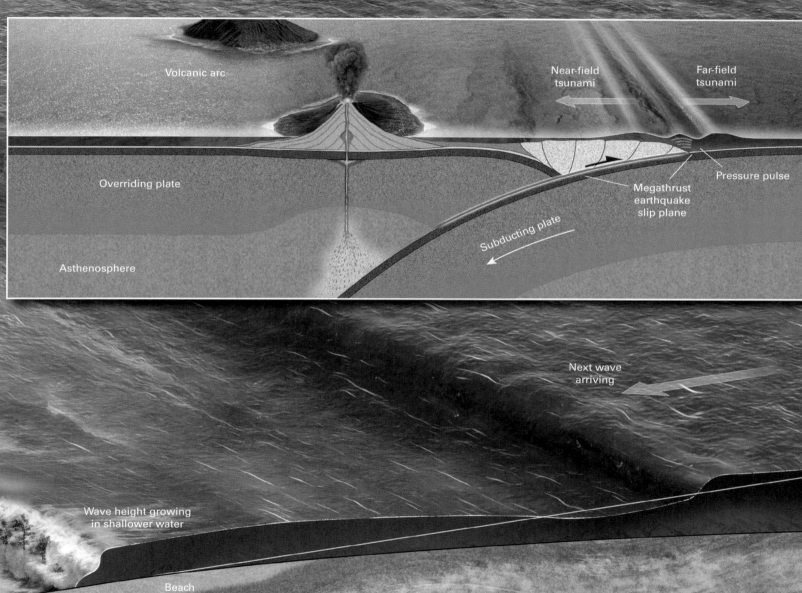

Tsunamis are waves produced when a mass suddenly pushes against water. The mass can be the seafloor (pushing up or dropping down during an earthquake), a submarine landslide, a volcanic explosion, a subaerial landslide that drops into water, or a meteorite impact. Tsunamis, unlike wind-driven waves, have large wavelengths, so the waves are more like broad plateaus of water rather than narrow ridges. The main image here depicts large tsunamis arriving at a coast. The waves slow as they enter shallow water, so wave height increases and wavelength decreases. But arriving waves are still very wide, and they carry so much water that they can wash over the beach and cover low-lying land as far as several kilometers inland. Tsunamis occur in trains that include several waves. Here, we see the second wave arriving while the first wave is retreating. The arriving wave has grown to the height of trees, so water is spilling over seawalls and will shortly cover the debris left by the previous waves. Another wave will arrive soon. The insets show the causes of tsunami production.

FIGURE 6.31 Relief and recovery after tsunamis.

(a) A hovercraft brings in supplies to Indonesia after the 2004 Indian Ocean tsunami.

(b) After victims have been tended to, communities face the daunting task of removing debris.

Tsunami Prediction and Mitigation

Is it possible to predict tsunamis? Earthquakes are the most common cause of tsunamis, so research on tsunami prediction has focused on those produced by earthquakes. Seismic waves travel about 20 times faster than open-ocean tsunamis, so seismographs can detect a potentially tsunami-generating seismic event and identify its location long before far-field tsunamis might reach a distant shore. Therefore, when a large earthquake happens beneath the sea, seismologists send out an alert to say that tsunamis might have formed. Unfortunately, the science of predicting whether a given seismic event actually will produce a tsunami remains imperfect—some do, but some don't. To understand why, researchers are developing computer models to simulate ocean-floor displacements associated with earthquakes.

Since measurements with seismographs cannot confirm that a tsunami wave train has been produced, NOAA has deployed **DART buoys** (for *D*eep-Ocean *A*ssessment and *R*eporting of *T*sunamis) (FIG. 6.32). They have put in place about 60 DART buoys around the Pacific Ocean. New networks of buoys are now being installed in other oceans. Each buoy overlies a pressure sensor on the seafloor. When the sensor detects the pressure pulse associated with a tsunami, it sends an acoustic signal (a sound pulse) through the water to the buoy. A computer in the buoy then sends a radio alert, which travels at the speed of light to a receiver on land. Officials monitoring the receiver can then take action to warn the public.

An alert can save lives only if officials have clear procedures to follow. To design procedures and coordinate efforts to warn coastal inhabitants of impending tsunamis in the Pacific Ocean, NOAA operates the *Pacific Tsunami Warning Center* in Hawai'i and the *National Tsunami Warning Center* in Alaska. If a large earthquake happens somewhere beneath the Pacific, staff at the centers calculate potential tsunami arrival times and then broadcast a **tsunami watch**, stating that conditions exist that may have led to tsunami formation. If a DART buoy then detects a tsunami, the center issues a **tsunami warning**, stating that tsunamis are approaching. In communities that might be affected, sirens blare, mobile phones vibrate, and radio and TV stations broadcast alerts. Unfortunately, the warning may come too late for communities at risk from near-field tsunamis, which may arrive only minutes after the earthquake. In cases where tsunamis are triggered by submarine slumps, there may also be little warning. But for far-field tsunamis, NOAA's alerts may provide hours of advance warning.

Tsunami warnings can save lives only if communities have a way to receive them, and if members of those communities know how to respond. The loss of life during the 2004 Indian Ocean tsunami emphasizes the unfortunate consequences of not having communication and response plans. When the 2004 Sumatra earthquake happened, seismologists knew that a tsunami was a possibility. But there was no network available that they could use to notify officials, and even if officials had been notified, the officials did not know how to communicate a warning to people in the areas at risk, and even if those people had received such a warning, they wouldn't necessarily have known where to go.

Public awareness is key to fostering appropriate public response to tsunamis. Centuries ago, local officials in Japan erected tsunami stones indicating the level to which water had risen in past tsunamis (FIG. 6.33a). The inscription on one near the village of Aneyoshi reads, "High dwellings ensure the peace and happiness of our descendants.... Remember the calamity of the great tsunamis. Do not build any homes below this point." Indeed, the run-up elevation of the 2011 Tōhoku

FIGURE 6.32 DART buoys are deployed to detect tsunamis.

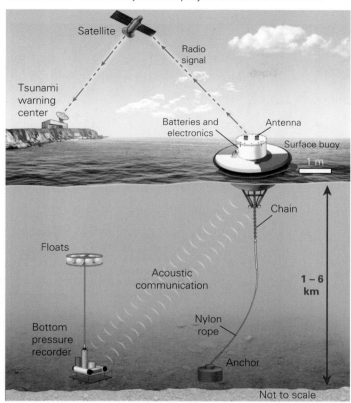

(a) A DART buoy connects to a pressure sensor on the seafloor. If the sensor reports a tsunami, the buoy sends a radio alert to staff.

tsunami was just a few hundred feet from the stone. Today, these informal markers have been replaced by *tsunami inundation maps*, which highlight areas that could be covered in water **(FIG. 6.33b, c)**. Worldwide, many coastal communities have posted signs labeling evacuation routes (in some cases, with yellow arrows) that will take residents out of an inundation area. Efforts continue to educate the public about what tsunami warnings mean and how to find and follow those evacuation routes **(FIG. 6.34)**. Ideally, people should head for a hill that's over 40 m tall, but if no such hill exists, they should head to a location far enough inland to be out of the inundation zone. If it's too late to get out of the inundation zone, they should head to the top of a strong building that's over five stories tall. Because a tsunami wave train includes many waves, evacuees should stay elevated for at least a few hours, or until an all-clear signal has been given.

Individuals can improve their survival chances by knowing the signs of an impending tsunami (such as drawback, or the appearance of a high wave that doesn't look like a normal breaker offshore), by understanding the meaning of their community's inundation map, and by having a personal evacuation plan that will get them to a safe location. People living in tsunami-susceptible areas need to keep in mind that the first tsunami to arrive most likely won't be the last. More than a dozen waves might arrive over several hours, and the first wave often isn't the highest. When the sirens blow, it's time to go . . . fast! Tsunamis can travel a lot faster than a person can

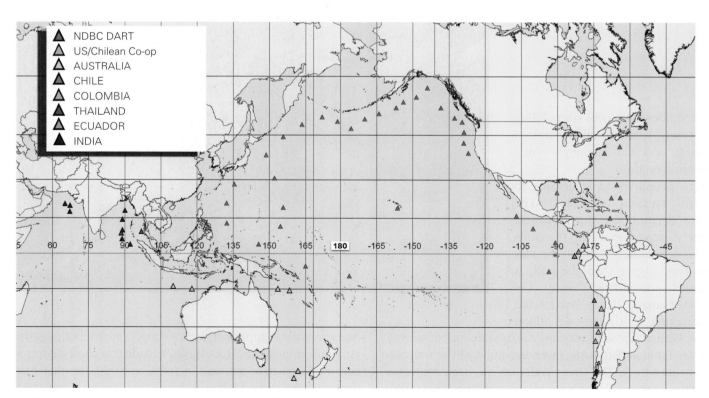

(b) DART buoys have been deployed around the Pacific. Similar buoys are now being installed in other oceans.

6.8 Tsunamis and Society • **245**

FIGURE 6.33 Coastal cities have developed tsunami inundation maps, highlighting areas that must be evacuated if a tsunami approaches.

(a) A tsunami stone on a hillside in Japan.

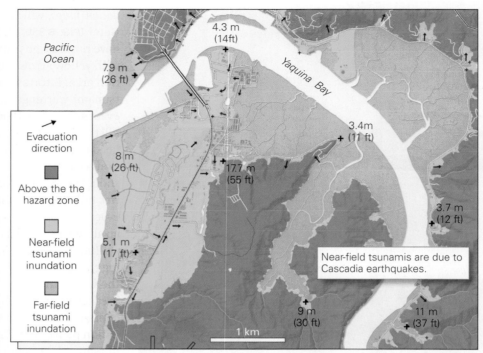

(b) A simplified tsunami-inundation map for Newport, Oregon. Near-field tsunamis are expected to be higher than far-field ones.

(c) Modern technology makes inundation maps easier to understand.

run, and once a tsunami starts filling city streets, water circulates around buildings and can seemingly come from any direction.

Can people build structures to prevent tsunami disasters? The answer depends on the tsunami's height. A seawall can stop tsunamis up to a certain height, but if the tsunami rises higher than the wall, it becomes ineffective. Since the Tōhoku tsunami, Japan has started construction of a nearly $7 billion "Great Wall of Japan," a 400 km (249 mi) network of seawalls 14 m (46 ft) high **(FIG. 6.35)**. Over 200 miles of the wall have already been built. The Great Wall plan has been controversial, because the wall won't necessary protect against all tsunamis, may disrupt ecosystems, and will block ocean views.

An alternative approach is to simply avoid building in tsunami-prone areas. Hilo, Hawai'i, made this choice. After a second destructive tsunami in 1960, coastal areas that had flooded were turned into parklands that can be submerged without significant damage or injury to the community.

FIGURE 6.34 Tsunami evacuation signs from Lima, Peru.

Take-home message . . .
Many secondary disasters follow in the wake of a tsunami, including fire, epidemics, and the loss of housing, infrastructure, industry, and jobs. Debris removal alone can take years. To prevent loss of life, planners produce tsunami inundation maps to predict the extent of flooding in coastal communities and to guide evacuations to safer land when necessary.

QUICK QUESTION Can tsunamis be predicted, and if so, how?

FIGURE 6.35 The "Great Wall of Japan," a new, higher seawall constructed after the 2011 Tōhoku tsunami, dwarfs its predecessor.

Chapter 6 Review

Chapter Summary

- A tsunami is a water wave produced by the sudden movement of a solid material against water. The displacement of water causes the water surface to move up or down relative to its equilibrium level; gravity then acts as a restoring force. Tsunamis differ from tides, wind-driven waves, and currents in many ways.
- Tsunamis occur in a wave train because the water surface at the site of displacement rises and falls several times before returning to the equilibrium level.
- A near-field tsunami reaches the shore close to the source of displacement, whereas a far-field tsunami crosses an ocean before washing ashore.
- In the open ocean, a tsunami travels as fast as a jet plane. When it enters shallower water, it slows down, and the back of the wave catches up to the front, so the wave can become very tall.
- In a given tsunami wave train, the highest wave isn't necessarily the first.
- Tsunamis have much longer wavelengths than do wind-driven waves. As a result, when a tsunami arrives on shore, it's like a plateau of water instead of a narrow ridge of water. Consequently, a large tsunami can cross the beach and go much farther than a wind-driven wave, which runs out of water at the top of a beach.
- Tsunamis cause damage in many ways. Because the water travels relatively fast, and because it becomes more viscous as it picks up debris, it exerts intense dynamic pressure against anything it hits. It can crush buildings, and can also buoy along cars and buildings so that they become battering rams. When drawback takes place, the water returns to the sea, carrying debris with it.
- Earthquakes cause tsunamis when the fault motion that generates the earthquake suddenly displaces the seafloor. Megathrust earthquakes produce the largest tsunamis. The 2004 Indian Ocean tsunami and the 2011 Tōhoku tsunami were associated with megathrust earthquakes.
- A large subaerial landslide along a water body can drop large volumes of debris, traveling at a high velocity, onto the surface of the sea. The impact creates a splash and tsunamis.
- The head of a submarine landslide pulls the water surface down, and the toe of the landslide pushes it up. These movements can produce a tsunami. Slumping can take place on both active and passive margins, and during flank collapse of volcanic islands.
- The explosive eruption of a volcano may generate tsunamis, particularly due to the downward collapse of huge pyroclastic flows from the convective cloud.
- The impact of a large meteorite striking the ocean would generate huge tsunamis. Such an event is believed to have happened at the end of the Mesozoic Era, but nothing like it has ever happened during human history.
- Large tsunamis are among the most calamitous of natural disasters. They can disrupt infrastructure, trigger epidemics, and destroy the economy of a region. It may take communities many years to recover.
- To prevent tsunami disasters in the future, agencies have set up tsunami warning systems that alert coastal areas to the possibility that a tsunami-generating event has taken place. DART buoys can signal that a tsunami is passing when pressure changes in ocean water are detected, allowing agencies to broadcast tsunami warnings.
- To prevent loss of life during tsunami inundation, agencies have produced tsunami inundation maps, showing areas that should be evacuated if a tsunami is detected. Residents in coastal communities should know how to read these maps, know where safe areas lie, and know how to follow evacuation routes to reach those safe areas quickly.
- To prevent tsunami-caused destruction, some communities have erected seawalls. But these walls work only if the wave height does not exceed the wall height. If possible, it's best to avoid building in tsunami-prone locations.

Review Questions

Blue letters correspond to the chapter's learning objectives.

1. What is a tsunami, what are its principal characteristics, and how do tsunamis differ from wind-driven storm waves? **(A)**
2. Once a tsunami has arrived and has retreated, is it safe to go back to the shore? **(A)**
3. What is a typical wavelength and period for a tsunami wave train in the deep ocean? How high is a large tsunami out in the deep ocean? How fast does a tsunami travel in the deep ocean? **(A)**
4. Explain how a tsunami changes as it passes from the deep ocean into shallower water. **(B)**
5. What is the difference between a near-field and a far-field tsunami? Which is likely to be higher, and why? **(B)**
6. What factors influence the height of a tsunami when it arrives at the shore? **(B)**
7. Why were tsunamis once called "tidal waves" in English? In what way do tsunamis resemble tides? What is the origin of the word *tsunami*? **(B)**
8. Explain the variety of ways that a tsunami can cause damage along a coast. **(C)**
9. Why is the dynamic pressure (the push) of water in a tsunami so much greater than that of the water in a wind-driven wave of similar height? **(C)**
10. What is the drawback of a tsunami, and how does it affect coastal areas? Distinguish between tsunami height and inundation limit. **(C)**
11. How does an earthquake generate a tsunami? Do all submarine earthquakes produce tsunamis? (Explain your answer.) Why do megathrust earthquakes produce large tsunamis? **(D)**
12. On the adjacent diagram, show where a slip takes place to produce a megathrust-related tsunami. **(D)**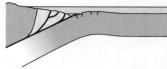
13. How does a subaerial landslide produce a tsunami? Will such tsunamis affect very distant shores? **(E)**
14. How does a submarine landslide produce a tsunami? Do such landslides occur only along active margins, or can they also occur along passive margins? **(E)**
15. What is the evidence that island volcanoes undergo flank collapse, and what is the evidence that such landslides generate tsunamis? **(E)**
16. How can the explosive eruption of an island volcano generate a tsunami? **(F)**
17. What process generates tsunamis during the impact of a large asteroid? Has any such impact happened during human history? **(F)**
18. Describe the range of challenges that communities in coastal regions endure after a large tsunami. **(G)**
19. On what basis do agencies send out tsunami warnings? **(H)**
20. How can communities decrease the loss of life and property should tsunamis strike in the future? **(H)**

On Further Thought

21. A tsunami-producing earthquake happened on the thrust fault at the base of the accretionary prism adjacent to the Aleutian Trench. The trace of where the slipped fault intersects the Earth's surface is marked by the white toothed line. The darkness of the blue color represents seafloor bathymetry—the lighter blue shows shallower locations. Exact depths are shown at the black points.

    Are tsunamis generated by this earthquake higher at location A or B? Explain your reasoning. Will the wave arrive at location A or location C first? (Locations that are similarly distanced from the epicenter.) Explain your reasoning. About how long will it take the first wave of a tsunami wave train to reach Sitka Sound at point S, 1,200 km (746 mi) from the epicenter?

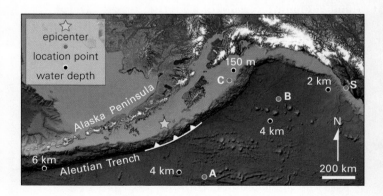

7

LOST GROUND
The Water Crisis, Sinkholes, Soils, and Subsidence

By the end of the chapter you should be able to . . .

A. trace water as it passes through reservoirs of the hydrologic cycle, and discuss how human intervention in the cycle has led to the global water crisis, a stealth disaster.

B. explain the challenges society faces due to the overuse, contamination, and salinification of surface fresh-water supplies.

C. define groundwater and the water table, and explain how overuse of groundwater lowers the water table and causes problems for society and the environment.

D. describe the evolution of karst terrains, how sinkholes form, and how they can cause local disasters.

E. describe the processes of land subsidence, and outline phenomena that can disrupt sediment near the ground surface.

F. distinguish between soil and other types of sediment, and explain how natural and human-caused phenomena can destroy soil.

FIGURE 7.1 The water crisis in Cape Town, South Africa.

(a) The mud on the floor of a reservoir cracked when it dried out.

(b) People in Cape Town lining up to fill water jugs.

7.1 Introduction

Day Zero is coming! This scary-sounding phrase headlined the news in Cape Town, South Africa, when the mayor declared that the city was facing such a severe water shortage that she might have to shut off the city's water supply. On Day Zero, initially set for April 22, 2018, zero water would come out of home faucets. Residents would then need to trek to distribution locations, where tanker trucks would supply a daily water ration of 25 L (6 gal) per person, about one-third the amount of water used during an average shower.

Why did Cape Town face Day Zero? The city depends on *reservoirs*—here meaning artificial lakes held back by dams—for its water supply. Beginning in 2015, a multi-year drought, together with increasing rates of water consumption, caused water levels in South Africa's reservoirs to begin dropping. When the water volume in the reservoirs decreased to 30% of the reservoirs' capacity, officials started worrying (FIG. 7.1a). The Day Zero warning went out when volume decreased to 15% of capacity, and at that point, people began lining up to fill water jugs (FIG. 7.1b). Day Zero itself would be the day when reservoir volume became less than 13.5% of capacity, so that water could no longer be pumped out (FIG. 7.1c). Fortunately, the dreaded event was avoided, this time, for two reasons. First, Cape Town implemented severe water-use restrictions, prohibiting the use of water for irrigation, filling pools, or washing vehicles, and allocating to each person a daily ration of 87.5 L (23 gal). This policy successfully cut water consumption by about half. Second, rains fell, and the reservoirs began to refill.

◄ To grow food in dry lands, farmers pump water out of the ground to irrigate fields (the circular areas), potentially leading to groundwater depletion. Loss of water and land at the Earth's surface represents a stealth disaster.

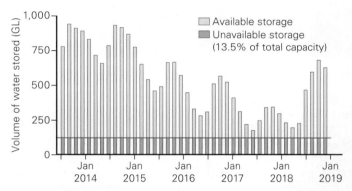

(c) Cape Town's water supply (shown here in gigaliters) decreased for 5 years in a row, and reservoirs almost dropped below 13.5% of their capacity (the red line), at which point they would have become unusable.

After being postponed for a few months, Day Zero was finally put off indefinitely. Unfortunately, similar disasters have already arrived elsewhere, as we'll see later in this chapter.

Cape Town's experience symbolizes the *global water crisis* that society faces—increasingly, there isn't enough clean **freshwater** (water that contains less than 0.05% salt) in places where people need it. Society's use of freshwater has increased dramatically over the past century, so we're now depleting supplies held at the Earth's surface and underground more quickly than they can be replenished by nature.

Running out of freshwater isn't the only threat to the **critical zone**, the surface and near-surface realm of the Earth System that hosts most life and most of the materials necessary for sustaining life. We are also losing ground, literally, due to a variety of phenomena:

- *sinkhole formation*: the collapse of the land into distinct circular depressions;
- *land subsidence*: the gradual sinking of the land surface over a broad area;

FIGURE 7.2 Our Blue Planet hosts a hydrologic cycle.

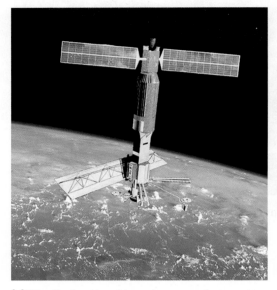

(a) The Earth glows blue when seen from space.

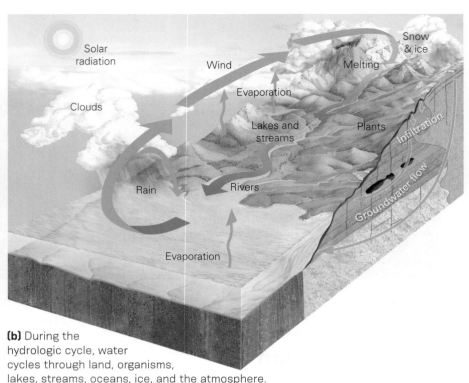

(b) During the hydrologic cycle, water cycles through land, organisms, lakes, streams, oceans, ice, and the atmosphere.

(c) A freshwater stream flows toward the ocean.

(d) A lake provides temporary storage for freshwater.

- *soil destruction:* the depletion of nutrients and carbon from *soil* (the layer of modified sediment at the Earth's surface that serves as the substrate for plants), the contamination of soil, and the erosion of soil by wind and running water.

In this chapter, we examine threats to society due to loss of freshwater, land, and soil. We can think of these threats as **stealth disasters**, situations that develop slowly enough that we might not recognize them until it's too late (see Chapter 1). We focus first on the water crisis. Then, we turn to the phenomena that damage land in ways that can result in casualties or financial loss. (The related issue of slope failure was already covered in Chapter 5.) Notably, some of the land-loss problems we'll examine have been exacerbated by the water crisis.

7.2 A Most Precious Resource: Water

The Hydrologic Cycle and Our Planet's Water Supply

Science writers often refer to the Earth as the "Blue Planet" because the oceans that cover much of our globe glow blue when seen from space (FIG. 7.2a). These oceans are the largest of several water reservoirs—here meaning portions of the hydrosphere—in the Earth System (see Chapter 2), and they are filled with **saltwater**, which, on average, contains 3.5% dissolved salt (TABLE 7.1). Water moves from reservoir to reservoir through the **hydrologic cycle** (FIG. 7.2b). The cycle

TABLE 7.1 Major Reservoirs Involved in the Hydrologic Cycle

Reservoir	Volume (km³)	% of total water	% of freshwater
Oceans and seas	1,338,000,000	96.53	—
Glaciers, ice caps, snowfields	24,064,000	1.74	68.7
Saline groundwater	12,870,000	0.93	—
Fresh groundwater	10,530,000	0.76	30.1
Permafrost	300,000	0.022	0.86
Freshwater lakes	91,000	0.007	0.26
Salt lakes	85,400	0.006	—
Soil moisture	16,500	0.001	0.05
Atmosphere	12,900	0.001	0.04
Wetlands	11,470	0.0008	0.03
Rivers and streams	2,120	0.0002	0.006
Organisms	1,120	0.0001	0.003

begins when water molecules evaporate from the oceans and drift into the atmosphere, leaving salt behind. Atmospheric water vapor eventually returns to the Earth's surface as rain or snow. About three-quarters of this freshwater precipitation falls back into the ocean, and the remainder falls on land. Some of the water falling on land evaporates directly back into the air, and some gets absorbed by organisms. Water in organisms soon returns to the air as a result of metabolic processes, or when the organisms die. The rest of that water becomes either **surface water** (as a liquid in lakes, streams, and wetlands, or as a solid in snowfields and glaciers) **(FIG. 7.2c, d)**; **soil moisture** (water that clings to minerals in soil or gets absorbed by organic matter in soil); or **groundwater** (water that fills small open spaces in rock or sediment underground). On a time frame of days to years, most surface water, soil water, and shallow groundwater ends up back in the oceans or evaporates back into the atmosphere.

As Table 7.1 emphasizes, most of the surface water and groundwater on our planet consists of undrinkable saltwater **(FIG. 7.3)**. While industrial-scale *desalination* (salt removal) can produce freshwater from saltwater, the process is expensive, and disposal of the salt can damage the environment. As a result, the freshwater reservoirs of the hydrologic cycle are society's primary water resource. By some estimates, readily accessible freshwater represents only 0.014% of our planet's total water volume **(FIG. 7.4)**. As we'll see next, this precious supply is now at risk.

The Global Water Crisis

The exponential growth of the human population in recent centuries, and the improvement of human living standards over that time, have dramatically increased freshwater consumption. Of this water, 70% goes to agriculture, 19% to industry, and 11% to municipalities and home use. Demand will continue to grow, perhaps by another 50%, during the lifetimes of most people living today **(FIG. 7.5)**. Because of this growth, coupled with the loss of some freshwater supplies due to contamination, the supply of freshwater can no longer meet demand for water in many regions. Consequently, society is on the brink of this **global water crisis**, a stealth disaster that is finally being recognized. For example, after the driest summer in half a century, and record-high temperatures, water shortages sparked protests in Iran. Demonstrators chanted the simple but alarming phrase, "I am thirsty!"

Because climate varies with location, and because the efficiency and fairness of water distribution systems depend on political and economic factors, water supplies vary across the globe, and some parts of the world face more dire shortages than others. Lack of sufficient clean water can lead to illness due to dehydration and to waterborne diseases carried by unclean water supplies. Lack of water also causes crops and industries to fail. History shows that the net decline in living standards caused by a water shortage can trigger wars and destabilize society, and can force multitudes of people to migrate.

FIGURE 7.3 The ocean is the largest water reservoir in the hydrologic cycle, but we can't use it directly because the water is salty.

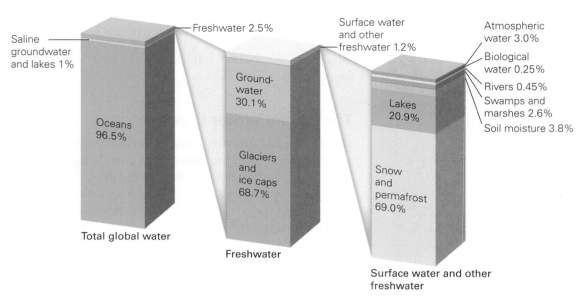

FIGURE 7.4 Relative sizes of reservoirs of the hydrologic cycle.

(a) Most of the water on the Earth is salty, and most freshwater is either frozen or underground.

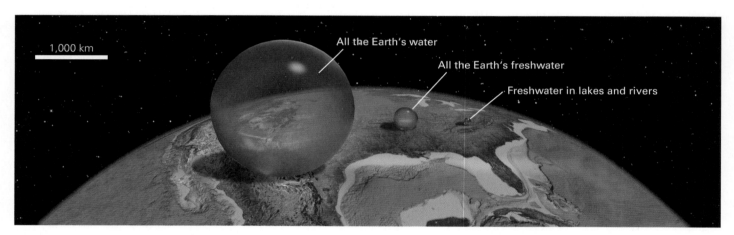

(b) All the water in the world could be collected in a sphere with a diameter of 1,300 km (808 mi). The freshwater portion would fill a sphere with a diameter of only 275 km (170 mi).

FIGURE 7.5 Society's water use worldwide.

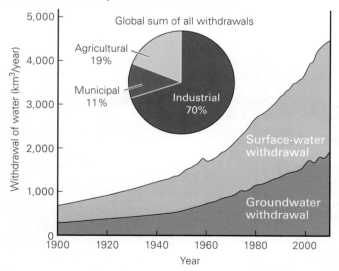

(a) A graph of water withdrawal, meaning the amount extracted from surface or subsurface water supplies, shows exponential growth.

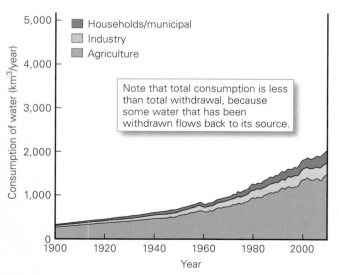

(b) A graph of water consumption, meaning the amount that does not return directly to the source after withdrawal, but instead evaporates or becomes incorporated in organisms or materials, shows that most water is used in agriculture.

Take-home message...

Water moves through many reservoirs in the hydrologic cycle. Some of the rain and snow that falls on land fills freshwater lakes, streams, and wetlands at the Earth's surface, and some sinks into the subsurface to become groundwater. Only a small proportion of the Earth's freshwater is accessible to people. Society faces a global water crisis because supplies of freshwater no longer meet demand in many parts of the world.

QUICK QUESTION Can desalination of ocean water currently satisfy global water demand?

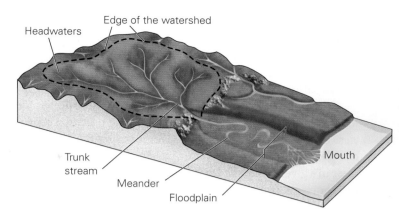

FIGURE 7.6 A watershed is an area of land that feeds water from rainfall and snowmelt to creeks, streams, and rivers, and then to outflow points such as reservoirs, bays, and the ocean.

7.3 Surface-Water Challenges

Surface Freshwater Depletion

The surface freshwater that society relies on comes from rivers, lakes, and reservoirs. Each water body receives its water supply from the rain and snow that falls over a defined area, called a **watershed**, and from groundwater seeping back to the surface within the watershed **(FIG. 7.6)**. **Freshwater depletion** takes place when the amount of water leaving a water body (by evaporation, sinking into the substrate, or diversion into pipes or canals) exceeds the amount being supplied by the watershed and by groundwater seepage. Such depletion may be due to insufficient precipitation or to overuse by society, as we will now see.

The fate of the Colorado River in the southwestern United States serves as an example of how freshwater depletion can become problematic. This river drains a watershed that spans parts of seven states **(FIG. 7.7a)**. Most of its water comes from snow and rain that falls in the highlands of eastern Utah and western Colorado. Over the past several million years, the

FIGURE 7.7 The Colorado River is being depleted of freshwater.

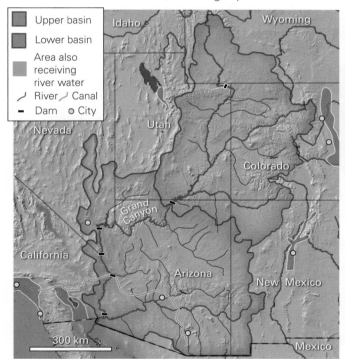

(a) The Colorado River watershed provides water to areas both within and outside the watershed.

(b) In the Grand Canyon, the river still has a substantial volume of water.

(d) Hoover Dam holds back Lake Mead, near Las Vegas. The arrow points downstream.

(c) Near its mouth in Mexico, the river is barely a trickle.

(e) The Central Arizona Project Canal carries water from the river to Phoenix and Tucson.

river's mighty currents carved the Grand Canyon and sustained a substantial flow all the way to the river's mouth in the Gulf of California (FIG. 7.7b). Today, however, the river is little more than a saline trickle near its mouth (FIG. 7.7c). How did this situation come to be?

To understand the depletion of the Colorado River, we must consider the river's history during the past two centuries. Until the 20th century, the river was free-flowing, so its *discharge* (the amount of water passing a cross section of a river in a given time; see Chapter 12) varied substantially over the course of a year, and the river often flooded. Beginning in the 20th century, however, construction of several major dams trapped its water in reservoirs (FIG. 7.7d), and canals and pipelines were built to transport water from these reservoirs to agricultural and urban areas of the arid Southwest. For example, about 90% of the water used in southern Nevada, and over half of the water used in southern California, comes from the Colorado River. And about 20% of Arizona's water supply comes from a single canal, the 540 km (335 mi) long Central Arizona Project Canal, which carries Colorado River water across the Sonoran Desert (FIG. 7.7e).

When the population of the Southwest began to expand rapidly, it became clear that areas within the region would compete for the river's water. So, to avoid perpetual arguments among states, officials in the 1920s established the *Law of the River*, specifying how much water each state can divert from the river. Unfortunately, instead of allocating percentages of the river's discharge, the law allocates specific volumes of water, and that strategy now causes problems. The river's overall discharge today is significantly less than it was in the 1920s, so the amount of discharge that has been allocated may be greater than the actual overall discharge in a given year. To accommodate the specified allocations, therefore, water in the river and its reservoirs has been depleted. The situation worsened in 2021 due to droughts and heat waves, causing the US Interior Department to declare a water shortage, for the first time, requiring states to cut off water supplies to certain areas. If water levels continue to drop, Lake Mead may become a "dead pool" that no longer provides water to the river downstream, or generates hydroelectric power.

In many cases, freshwater depletion has caused standing bodies of water to dry up. For example, the construction of the Los Angeles Aqueduct in 1913 led to the complete draining of Owens Lake, a 450 km² (174 mi²) body of water that once glimmered on the eastern side of the Sierra Nevada. Today, all that remains of the lake is a *salt flat*, a plain covered by clay, silt, and salt deposits. A similar fate beset the Aral Sea, a large inland lake in Kazakhstan. Rivers that once carried water to the Aral Sea were diverted into canals that provide water for irrigation. Consequently, the area of the lake has shrunk to a small fraction of what it once was, and the fishing fleets that once plied its waters are now rusting hulks sitting in the salty dust of the now-exposed lake bed (FIG. 7.8). Not only has this change been a socioeconomic disaster for the millions of people who live along the lake's coast, but because dust blowing from the exposed lake bed contains the residue of toxic pesticides washed in from the surrounding agricultural lands, it has also caused serious health problems.

FIGURE 7.8 Water was diverted from the rivers that fed the Aral Sea in order to supply irrigation projects. Consequently, the sea (actually, an inland lake) has almost completely dried up.

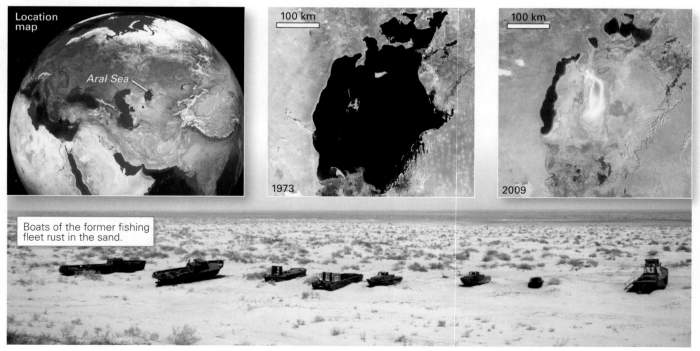

Surface Freshwater Contamination and Salinification

Society dumps sewage (containing dangerous bacteria, protozoans, and viruses), garbage, oil and gasoline, chemicals, fertilizer, pesticides, and animal waste into surface water in staggering quantities (FIG. 7.9a). Such **water pollution**—the contamination of water with materials in quantities that exceed those that would normally occur, and which natural systems cannot quickly remove—has been a problem for millennia. It not only makes water unsightly, but can also make it toxic, and can disrupt ecosystems. For example, fertilizer runoff and seepage from septic systems leads to **eutrophication**, algal blooms in surface-water bodies that result from overenrichment with nutrients. While the algae live, they turn the water green (FIG. 7.9b). Death and decay of the algae removes so much dissolved oxygen from the water that other organisms, such as fish, cannot survive.

By making water unusable, pollution effectively depletes water supplies, not only for humans, but also for other living components of the Earth System. With the advent of modern sewage treatment and efforts to stem the practice of dumping garbage into water bodies, the problem has decreased (though not entirely disappeared) in developed countries. But in some developing parts of the world, the problem has worsened as populations have grown.

Changes in salt concentration in freshwater can also decrease its usability. Recall that natural freshwater still contains a tiny amount of salt ($< 0.05\%$)—it's called "fresh" because its salinity is much lower than that of seawater (3.5%). When water evaporates, H_2O molecules rise into the atmosphere, but the ions of salt stay behind in the water. So, if the rate of H_2O lost from a river due to evaporation exceeds the rate of H_2O added from rain, other streams, or groundwater springs, then the *salinity* (salt concentration) of water in the river increases. The Colorado River is now undergoing **salinification**, an increase in salinity, for three reasons: (1) the rate of evaporation exceeds the rate of water addition, for when the river water sits in reservoirs behind dams, it has time to evaporate; (2) water input from dammed rivers that previously fed into the Colorado has been cut off; and (3) irrigation water that flows over agricultural areas picks up salt from soil and carries it into the river. Salinity of the Colorado River has more than doubled since the early 20th century, when the first dams were built. At times, the river's water in Mexico has reached a concentration of almost 0.3%, which makes it too salty to drink.

Salinification indirectly led to the devastating water crisis that struck Flint, Michigan, beginning in 2014, when the city stopped using water from Lake Huron and the Detroit River and began using water from the Flint River. The new supply contained high levels of chloride (Cl^-) because the city spreads halite (NaCl) on its streets during the winter to melt snow and ice, and meltwater carries the dissolved salt into the river. Chloride-contaminated Flint River water is almost 20 times as corrosive as Lake Huron or Detroit River water, so when it flowed through the city's century-old lead pipes, it extracted lead and other chemicals from the pipes. Consequently, the city's water supply not only became discolored and foul tasting, but it also contained dangerous concentrations of toxic lead. Over 100,000 people were exposed to this water, so the state and federal governments declared an emergency to make funds available to mitigate the situation. This tragedy led to legal suits and criminal charges.

> **Take-home message . . .**
>
> People use surface freshwater at unsustainable rates. Overallocation has drained the water from some rivers so that barely a trickle reaches the river's mouth, and lakes have dried up. Society also loses access to freshwater due to contamination and salinification.
>
> **QUICK QUESTION** Why do some water bodies undergo eutrophication?

FIGURE 7.9 Pollution and an oversupply of nutrients can damage freshwater supplies.

(a) A polluted river in Kathmandu, Nepal.

(b) Eutrophication in a Massachusetts pond.

7.4 Groundwater Challenges

Introducing Groundwater

The upper part of the Earth's crust behaves, in effect, like a giant sponge that can soak up water infiltrating down from the surface (**FIG. 7.10a**). Some infiltrating water descends only into the soil, where it adheres to the surfaces of mineral grains and organic fragments to become soil moisture. The rest sinks deeper into sediment or bedrock, where, along with water trapped at the time bedrock formed, it becomes groundwater (**FIG. 7.10b**). A region where water enters the ground, descending to become groundwater, is called a *recharge area*. Most groundwater within sediment or rock flows slowly and eventually returns to the Earth's surface at a place called a *discharge area* over a time frame of years to millennia. Some groundwater, however, has remained trapped in rocks for millions of years.

Groundwater resides in **pores**, the relatively small open spaces between solid mineral grains in sediment or rock, and in cracks of various sizes. We can use the term **porosity** for the total volume of open space (including pores and cracks) within a material, specified as a percentage of the material's total volume. For example, if we say that a block of rock has 20% porosity, then 20% of the volume of the rock consists of space that could be filled with groundwater (or air, or oil).

The degree to which pores or cracks within a material are connected to one another determines the material's **permeability**: groundwater flows easily through *permeable material* (**FIG. 7.11a**), flows slowly through *low-permeability material*, and cannot flow at all through *impermeable material* (**FIG. 7.11b**). With the concepts of porosity and permeability in mind, geologists distinguish between an **aquifer**, sediment or rock with high permeability and porosity, and an **aquitard**, sediment or rock with low permeability regardless of its porosity.

Typically, in the upper part of the subsurface, a realm called the *unsaturated zone*, pores contain just air, or both air and water. Deeper down, in a realm called the *saturated zone*, water completely fills the pores (**FIG. 7.12a**). The underground boundary between the

FIGURE 7.10 Groundwater infiltrates downward from the Earth's surface.

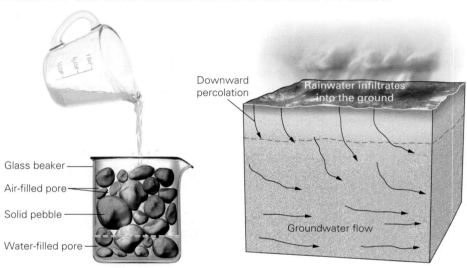

(a) You can observe infiltration by pouring water into a beaker full of gravel. The water fills open spaces between clasts.

(b) Water that falls on the Earth's surface can infiltrate downward to become groundwater.

FIGURE 7.11 The contrast between permeable and impermeable materials.

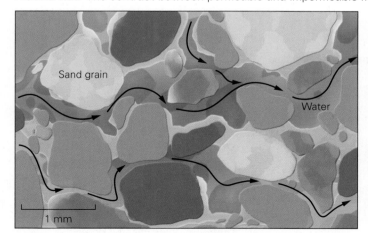

(a) In this permeable sandstone, there are open conduits between grains, so water can flow from pore to pore.

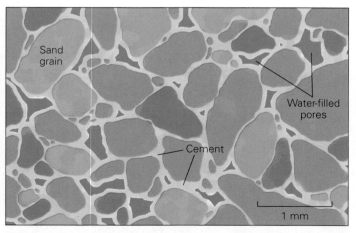

(b) In this impermeable sandstone, cement fills much of the space between grains and blocks conduits.

FIGURE 7.12 The concept of the water table, and how it can change.

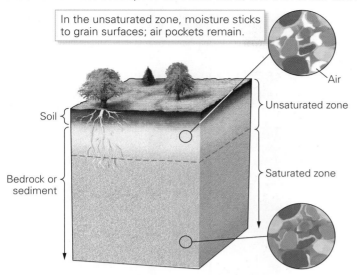

(a) The water table defines the boundary between the unsaturated zone and the saturated zone.

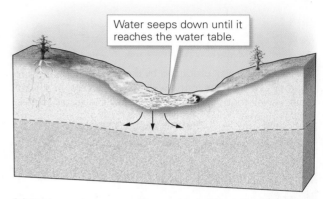

(b) During a wet season, the water table rises. Here, the water table intersects the surface of a pond.

(c) During a dry season, the water table sinks, and the pond dries up.

unsaturated and saturated zones defines the **water table**. Notably, the water table tends to follow topography, so it lies at a higher elevation beneath hills and at a lower elevation beneath valleys. Its depth beneath the ground surface varies greatly with location. For example, in arid regions, the water table may lie hundreds of meters down, whereas in tropical or temperate regions, it's generally less than several tens of meters below the ground surface. Where wetlands, lakes, and rivers occur in temperate and tropical regions, the water table corresponds to the surface of the water body and, therefore, sits at or above the solid surface of the

FIGURE 7.13 Natural springs.

(a) A spring coming out of the wall of the Grand Canyon.

Earth (**FIG. 7.12b**). Because the sources of new water in the groundwater supply are infiltrating rain or melting snow, the water table tends to rise during a wet season and drop during a dry season. If the water table drops below the bed of a water body, the water body can dry up (**FIG. 7.12c**).

In prehistoric time, people could obtain groundwater only from **springs**, natural outlets where groundwater seeps out of the ground (**FIG. 7.13a**). Springs, which occur in discharge areas, form for a variety of reasons (**FIG. 7.13b**), but they don't exist everywhere. Consequently, to obtain groundwater,

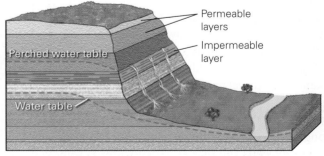

(b) Springs form in many different geologic settings. In this example, groundwater "perches" above an impermeable layer of rock, and water seeps out of cracks where the perched water table intersects a cliff face.

7.4 Groundwater Challenges

FIGURE 7.14 Examples of wells used to extract groundwater.

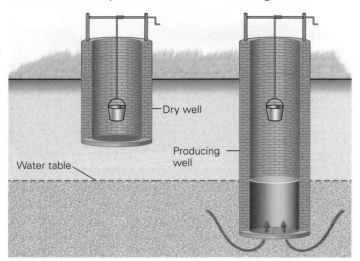

(a) An ordinary well produced by digging is simply a deep, steep-sided pit. Water rises in the well up to the level of the water table. A well that doesn't extend below the water table is dry.

(b) Women bringing up water from a traditional ordinary well.

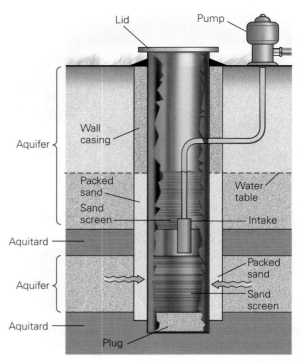

(c) In a modern ordinary well, a pump is used to suck up the water.

people eventually figured out how to dig or drill **wells**, pits or holes into which groundwater seeps (FIG. 7.14a). A *producing well*—one that provides water—must penetrate an aquifer below the water table; those that don't are called *dry wells*. In *artesian wells*, water rises under its own pressure, but in *ordinary wells*, people must pull the water up in a bucket or extract it with a pump (FIG. 7.14b, c). Where people lack access to sufficient surface water, groundwater becomes an essential source, or even the primary source, of water. Therefore, factors that decrease groundwater supplies, or render them unusable, are a threat to society.

Groundwater Depletion

As long as the rate at which groundwater seeps into a well equals or exceeds the rate at which water is removed, the level of the water table around the well remains the same. But pumping water out of a well faster than it can be replenished causes **drawdown**, the sinking of the water table.

Initially, drawdown produces just a downward-pointing *cone of depression* around the well (FIG. 7.15a). But if the well continues extracting water too fast over a long time, the water table over a wider area may become lower (FIG. 7.15b). Pumping from many wells throughout a broad region, for irrigation (FIG. 7.16a) or for a municipal water supply, can

FIGURE 7.15 Drawdown of the water table.

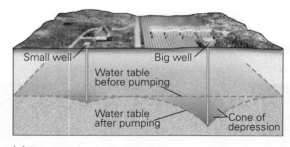

(a) Extraction of water from a well may draw down the water table, forming a cone of depression.

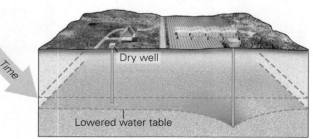

(b) If a well continues to extract groundwater at a rate faster than it can be replenished, the surrounding water table drops, and shallower wells run dry.

260 • CHAPTER 7 • Lost Ground: The Water Crisis, Sinkholes, Soils, and Subsidence

FIGURE 7.16 Intensive pumping of groundwater and its consequences.

(a) Irrigation of fields in Utah permits farming of otherwise arid land, but it requires large amounts of groundwater.

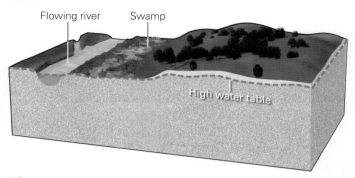

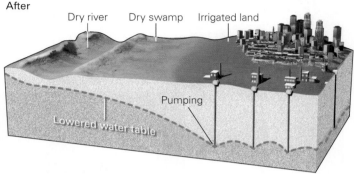

(b) Depletion of groundwater due to regional pumping can cause rivers and wetlands to dry up.

result in extensive **groundwater depletion** (FIG. 7.16b). Groundwater depletion can also happen when recharge areas are damaged by clearing of vegetation, compaction of the ground by grazing animals, or covering the ground with impermeable materials, all of which make it easier for water to flow away across the land surface before it has a chance to infiltrate the ground. *Recharge*, or refilling, of a depleted groundwater reserve to raise the water table back to its former level can take years to centuries in temperate or tropical climates, and centuries to millennia in arid climates. So, in the time frame of a human generation (years to decades), we must view groundwater in much of the world as a *nonrenewable resource*, in that nature cannot replace it as fast as people are extracting it.

A significant drop in a region's water table causes shallow wells, lakes, rivers, reservoirs, and wetlands to dry up and, as we'll see later in this chapter, allows pores in sediments and weakly cemented sedimentary rocks to collapse, causing overlying land to sink. Continued access to groundwater in a region where the water table has dropped requires wells to be drilled deeper. But there's a limit to how far down we can drill, for deeper groundwater is older and has had more time to dissolve minerals from the rocks it resides in—and these minerals give the water unpleasant characteristics. Deeper groundwater also tends to contain higher concentrations of salts. In fact, most very deep groundwater is too saline to be usable. Freshwater is less dense than saltwater, so fresh groundwater forms a layer above saline groundwater. Therefore, excessive pumping from a well may suck up saline water from below—a problem called *saltwater intrusion*.

Significant groundwater depletion has already happened in many locations. For example, extraction of groundwater from the Ogallala aquifer (an underground layer of porous and permeable sandstone that serves as a major aquifer for the High Plains region of the United States) is causing its water table to drop by about 0.6 m (2 ft) per year, so that in some places, the water table now lies more than 45 m (148 ft) deeper than it once did (FIG. 7.17a). An even greater drop in the water table has occurred beneath the cities of Phoenix and Tucson, Arizona, which pump groundwater from desert basins—in some places, the water table now lies 150 m (492 ft) or more below the surface. California, which suffered a severe drought in 2012–2016 (see Chapter 14), has endured significant water table drops. Between fall 2011 and fall 2012 alone, the water table in the southern part of the Central Valley dropped by as much as 10 m (33 ft) (FIG. 7.17b). Even recharge during a succession of wet years has not caused significant recovery of the water table (FIG. 7.17c).

Contaminated Groundwater

In most places, you can drink groundwater right from a spring or a well, for rocks and sediments serve as natural filters capable of trapping suspended solids and absorbing ions. Indeed, commercial distribution of bottled "spring water" has become a major business worldwide. Unfortunately, groundwater isn't directly usable everywhere. For example, groundwater containing calcium carbonate (known as *hard water*) can foul pipes with mineral precipitates known as *scale*, and groundwater containing iron oxide can cause staining; such groundwater needs to be treated before use.

FIGURE 7.17 Examples of groundwater depletion.

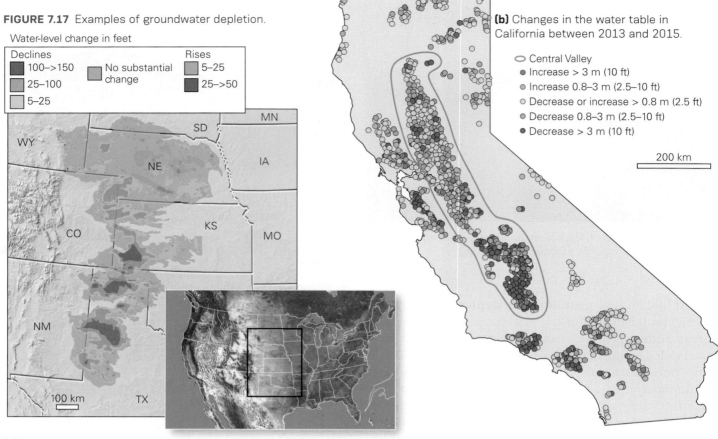

(a) Drawdown of the Ogallala aquifer, a major source of water in the High Plains region, east of the Rocky Mountains.

(b) Changes in the water table in California between 2013 and 2015.

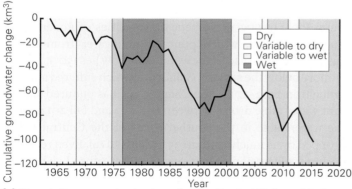

(c) Cumulative groundwater loss for the Central Valley of California between 1962 and 2015. Colored bands indicate whether overall conditions were wet or dry.

In addition, groundwater isn't safe everywhere—it may be contaminated by natural toxic chemicals leached (extracted) from surrounding sediment or rock, or by anthropogenic pollutants, and may need to be purified before use.

The occurrence of **arsenic contamination** in groundwater serves as an example of how natural toxic chemicals can become problematic to society. Trace amounts of arsenic occur in pyrite (FeS_2), a shiny mineral informally known as fool's gold, which occurs locally in sediments and rocks. Groundwater can leach arsenic, and in some regions, such as the Ganges River delta of southern Asia, the groundwater may eventually contain unsafe concentrations. Prior to the 1970s, arsenic wasn't a problem for people living on the delta, because they drank surface river water. But the river became so polluted that waterborne illness was becoming a major societal problem. So, beginning in the 1970s, governments and aid organizations began drilling wells into the delta to provide access to unpolluted groundwater. While the switch succeeded in diminishing waterborne diseases, it sickened people instead with arsenic poisoning. Fortunately, arsenic can be removed from groundwater in purification facilities, but not enough of these facilities have yet been built to solve the problem entirely.

In the past century, anthropogenic contamination of groundwater has become a global problem. Injection of effluents down wells, surface chemical or sewage spills, and leaking underground oil tanks all introduce pollutants into groundwater. A resulting *contaminant plume* moves outward from the source of contamination in the direction of groundwater flow **(FIG. 7.18)**. Contaminated groundwater can't be used, so contamination effectively decreases groundwater

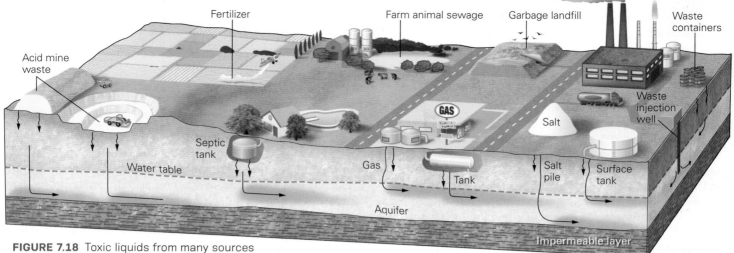

FIGURE 7.18 Toxic liquids from many sources can percolate below the water table and produce contaminant plumes.

supplies. *Groundwater remediation*, the removal of contaminants from groundwater, isn't easy. In some cases, natural processes can help clean the water. For example, some contaminants bind to clay, some react with oxygen and transform into less dangerous chemicals, and some get consumed and broken apart by natural bacteria. But when such natural processes can't remove contaminants, interventions of various types may be necessary. For example, engineers can pump steam into the ground at *injection wells* to drive contaminants to *extraction wells*, where they are pumped out and sent to a treatment plant. Or they can install *reactive barriers*, which are underground bands of materials, such as iron filings, that chemically react with contaminants, causing them to precipitate as solids that remain trapped underground. Another option is to inject oxygen and nutrients down wells into a contaminant plume to foster growth of bacteria that can break down the contaminants. Needless to say, such groundwater remediation methods are very expensive. In the United States, some locations where groundwater contamination is so severe that only long-term federal funding can cover the cost of cleanup have been designated as *Superfund sites*.

> **Take-home message . . .**
> Below the water table, groundwater completely fills pores and cracks in rocks and sediments. Aquifers are rocks or sediments with high porosity and permeability that contain significant volumes of groundwater. Removing groundwater from a well at a rate faster than it can be recharged causes groundwater depletion and, therefore, lowers the water table. Anthropogenic contamination of groundwater also leads to diminished water supplies.
>
> **QUICK QUESTION** Why should groundwater, in many places, be viewed as a nonrenewable resource?

7.5 Responding to the Water Crises

The *water footprint*, meaning the amount of water used per capita, of countries around the world has been increasing rapidly due to increases in industrialization, in the amount of land being irrigated, and in standards of living. Such increases, together with the growth of the global population, mean that *total water withdrawal*—a term used for the combined amount of groundwater and surface water extracted for human use—has increased dramatically worldwide, and has outstripped the water supply in many places. The 2021 water crisis led residents in the drought-afflicted western US to stockpile water trucked in from elsewhere, and to monitor fire hydrants for water theft. Some towns refused to issue new building permits, in efforts to limit further growth and consequent water demand. Day Zero—the shutdown of the city water system that Cape Town narrowly avoided—has already arrived in some cities **(BOX 7.1)**, and researchers predict that similar scenarios may become all too common during the next few decades. Consequently, many countries are facing **water stress**, meaning a lack of *water security* (access to safe and affordable water) **(FIG. 7.19a)**. By one estimate, one quarter of the Earth's human population now faces water stress. Water stress in many regions represents a *physical scarcity*—that is, the supply of available water on or in the ground isn't enough for the number of people who rely on it. But it can also represent an *economic scarcity*, in that the region contains enough water, but for economic and political reasons, the water is not being purified, or the region does not have a reliable or equitable water distribution system **(FIG. 7.19b)**.

Addressing water stress starts with *conservation* (decreasing the amount of water consumption), which can slow withdrawal rates. Conservation not only includes common-sense measures, such as shutting off faucets and repairing

BOX 7.1 DISASTERS AND SOCIETY

The 2019 water crisis in India

India's population is approaching 1.4 billion on a land area that's only about 30% the size of the United States. Given its rapid population growth and industrialization, demand for water in India has increased dramatically—and many regions in the country already face a water crisis.

India's water crisis has grown more severe due to many factors: agricultural and urban use has increased so much that groundwater has undergone severe depletion and the water table has dropped significantly; surface water has become severely polluted because only about a third of the sewage being produced undergoes treatment before being dumped into lakes and rivers; groundwater recharge areas have been deforested or covered over with impermeable concrete, so less infiltration takes place; wetlands have been filled to make room for housing; efforts to harvest rainwater or prevent water waste have been inadequate; and oversight of water purification and water distribution facilities has been insufficient and inefficient. To make matters worse, the volume of water that precipitates as seasonal rains has, on average, decreased—the average for the period of 2000–2020 is about 6% less than for the period of 1920–1960. Consequently, droughts have plagued parts of the country.

The net result is that municipal water supplies have been shut off in some communities, causing severe hardship. For example, when the four reservoirs supplying the city of Chennai, the country's sixth largest city, ran dry in 2019, the city reached Day Zero and the water supply was mostly shut off. Hundreds of thousands of people had to line up for water brought in by tanker trucks **(FIG. Bx7.1a)**. The daily allotment provided only enough water for drinking and cooking, so dishes had to be washed in contaminated water. Unfortunately, competition for water led to violence between neighbors, and to increased criminal activity. Illegal tanker trucks, filled with water pumped from illegal wells, became the main source of water in some communities, and the operators of these water sources began to charge exorbitant prices for the water. To avoid the extra cost, some people dug wells in the muddy floors of empty reservoirs to access remaining groundwater beneath the reservoirs **(FIG. Bx7.1b)**.

As yet, a national solution to the problem of water scarcity has not been implemented, and audits suggest that government response to the water crisis so far has been disorganized and, in places, corrupt. Part of the problem stems from the fact that about 60% of the urban population lives in makeshift communities that are not necessarily connected to municipal water supplies or sewers. The national government has promised to connect all communities to piped water by 2024, but it's not clear if this goal can be achieved, especially in the wake of the COVID-19 pandemic. Staff associated with various NGOs have gone into the countryside to help villagers manage local water supplies by increasing the efficiency of irrigation methods and redeveloping water harvesting systems. It is a race against time to see if water supplies can be conserved and improved before they run out on a national basis.

FIGURE Bx7.1 The Chennai water crisis.

(a) People line up to receive water allocations from a tanker truck.

(b) Some Chennai residents dug holes into the bed of a dried-out reservoir to reach the groundwater below.

FIGURE 7.19 Global water stress and its implications.

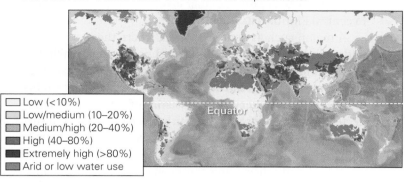

(a) Water stress by region. Colors indicate the ratio of a region's total annual water withdrawal to its total annual renewable water supply. Red areas use much more water than nature resupplies.

Low (<10%)
Low/medium (10–20%)
Medium/high (20–40%)
High (40–80%)
Extremely high (>80%)
Arid or low water use

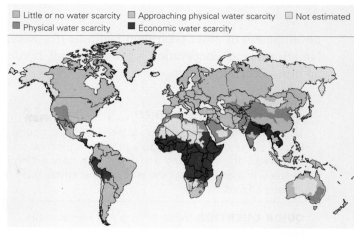

Little or no water scarcity   Approaching physical water scarcity   Not estimated
Physical water scarcity   Economic water scarcity

(b) Not all water scarcity is due to physical lack of water. In some places, there's enough water, but for political and economic reasons, it's not clean, nor is it distributed efficiently or equitably.

leaky pipes, but may also require lifestyle and resource management changes, such as switching to climate-appropriate landscaping, removing thirsty crops or herds from places with insufficient water supply, using more efficient irrigation methods where crops grow **(FIG. 7.20a)**, building recharge basins to capture rainfall and storm runoff so that it can sink into the ground **(FIG. 7.20b)**, increasing the efficiency of manufacturing and energy production processes that use water, and developing ways to reclaim and reuse water **(FIG. 7.20c-d)**. Communities could also decrease demand for water by removing subsidies that create artificially low prices. Regions undertaking these steps can see dramatic results. Australia, for example, cut its water consumption by about 40% during the past decade without slowing its economic growth. Decreasing water stress will also require appropriate regulation and monitoring,

FIGURE 7.20 Techniques used to conserve and reclaim water can lower rates of water withdrawal.

(a) Drip irrigation provides water slowly, directly to each plant, so much less water evaporates.

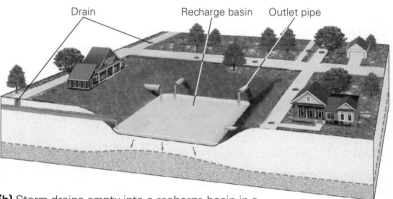

(b) Storm drains empty into a recharge basin in a neighborhood, which holds the water until it can sink into the ground.

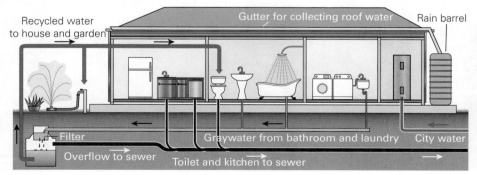

(c) Individual households can reclaim graywater and collect rainwater for household uses that don't require drinkable water.

(d) A municipal water treatment plant.

development of water treatment facilities (for water reclamation), and cooperation across national borders to address allocation challenges.

> **Take-home message...**
> The water crisis is a stealth disaster of global concern. Efforts are underway to encourage water conservation, improved water treatment, and water reclamation. Unfortunately, in many parts of the world, water stress has already arrived.
>
> **QUICK QUESTION** What actions can communities undertake to conserve water?

7.6 Sinkholes

When Mae Owens looked out her window on May 8, 1981, she saw that the large tree in the backyard of her Winter Park, Florida, home had suddenly disappeared. Going outside to investigate, she found that her whole backyard had become a deep, gaping pit! The pit continued to grow for a few days, eventually swallowing Owens's house and six other buildings, as well as part of a municipal swimming pool, a road, and several Porsches in a car dealer's lot (FIG. 7.21a). What had happened? It turns out that the bedrock beneath Winter Park consists of limestone, rock composed of calcite ($CaCO_3$), a mineral that dissolves in slightly acidic groundwater. Over time, the limestone beneath Owens's backyard had dissolved enough to produce a **cavern**, a sizable open space underground that never receives direct sunlight. On May 8, the roof of the cavern collapsed to form a circular depression or pit called a **sinkhole**. Water gradually filled the sinkhole to create a circular lake, similar to many others in central Florida (FIG. 7.21b).

A region such as central Florida, where the landscape is underlain by limestone bedrock and hosts many sinkholes, and in some cases, narrow ridges and spire-like pinnacles of limestone, is called a **karst terrain** (FIG. 7.22). Such terrains form where a **cave network**, consisting of many caverns connected by narrow *passages*, lies fairly close to the ground surface, so that the collapse of the cave network controls landscape evolution.

After briefly discussing the evolution of karst terrains, we'll focus on hazards associated with sinkhole formation in these landscapes. We will also touch on sinkholes that form in other geologic settings.

Origin of Karst Landscapes

When acidic groundwater comes in contact with calcite, a chemical reaction takes place that yields Ca^+ and CO_3^{2-} ions, which can be carried away with the groundwater as it slowly flows through the rock. Over time, this process yields a cave network. Although in a few localities the acidity of groundwater develops as the result of reactions between water and sulfur-containing minerals in surrounding rock, most groundwater acidity develops because rainwater percolating down through soil reacts with organic matter before reaching the water table. Most cave networks form at or just below the water table, for two reasons. First, the acidity of groundwater tends to be greatest at the water table, where downward-percolating acidic rainwater has not yet been diluted by groundwater from below. Second, groundwater

FIGURE 7.21 Sinkholes of central Florida.

(a) The Winter Park sinkhole as viewed from above.

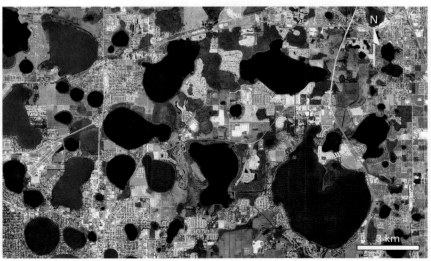

(b) A satellite view showing many sinkholes in central Florida. Note that some are in the midst of built areas.

FIGURE 7.22 The nature of karst terrains.

(a) Karst terrains form above cave networks, which include caverns and passages. Speleothems decorate caverns when they are empty of water.

(b) A karst landscape covered with sinkholes.

(c) Limestone pinnacles in southern China.

typically flows fastest near the water table, so it can carry away more dissolved ions near that level than it can deeper down, where it flows more slowly.

During the development of a cave network, caverns and passages are filled with water (FIG. 7.23). This water helps support the weight of overlying rock, because water can't be compressed to occupy a smaller volume, so sinkholes don't form. But if the water table drops, either because of groundwater depletion or because downcutting by streams deepens nearby valleys and allows groundwater to drain, caverns and passages fill with air. Calcite precipitating from water dripping through cracks in the ceiling of a cavern can build formations called *speleothems*, such as the icicle-shaped stalactites that delight tourists. Karst terrains develop when the rock overlying a cave network starts to dissolve and collapse to produce many sinkholes. Eventually, all that remains of the thick limestone layer that contained the cave network are the walls between the sinkholes. These walls eventually collapse, too, leaving behind pinnacles (narrow towers) of limestone.

Formation of Sinkholes

Let's look more closely at the different ways in which sinkholes form in karst terrains. *Dissolution sinkholes* develop where water draining from the ground surface into underground cracks or voids dissolves the rock at the surface and produces a depression. *Cover-subsidence sinkholes* form where loose sediment overlies caverns or passages and, over time, rainfall or surface drainage washes that sediment down through cracks into underground spaces, leaving a depression at the land surface. The formation of **collapse sinkholes** begins when weak weathered rock in the roof of a cavern breaks off and falls down into the cavern (FIG. 7.24a, b). Eventually, the bridge of rock that remains above the open cavern becomes too thin and weak to support itself, and it suddenly collapses. Because they can form so suddenly, as illustrated by the one that grew in Mae Owens's backyard, collapse sinkholes are the most dangerous type.

Commonly, the trigger that causes a collapse sinkhole to form is a natural phenomenon. For example, a heavy rain

FIGURE 7.23 The formation of karst landscapes as time passes.

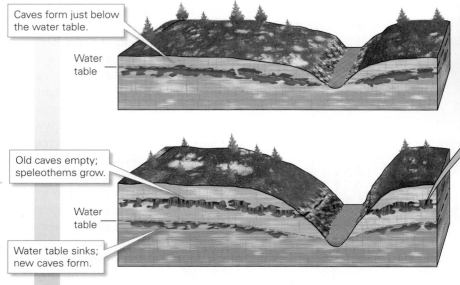
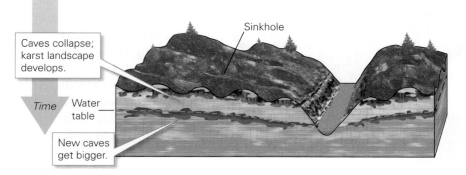

Speleothems in a cavern

Although we have focused so far on karst sinkholes, dangerous collapses of ground can also result from subsurface sediment liquefaction or subsurface dissolution of evaporites (BOX 7.2).

Dealing with the Risk of Sinkhole Formation

There are steps that we can take to avoid local disasters accompanying formation of sinkholes. Geologists start by determining whether a community lies within a karst terrain, or whether it overlies wet sediment or evaporite deposits (see Box 7.2). If so, they look for other sinkholes nearby, because new sinkholes may form near existing ones. New methods of surveying landscapes employ lidar (*l*ight *d*etection *a*nd *r*anging), a technique that uses lasers to produce high-resolution digital elevation models of the ground surface. These models can reveal sinkholes even when they are partially hidden by forests or fields (FIG. 7.25a).

Generally, determining whether a cavern actually underlies a particular piece of property isn't easy, but it may be possible. For example, a *resistivity survey* measures how easily electricity moves through the subsurface, so it can reveal the presence of air-filled voids, for air has relatively low conductivity. A *gravity survey* characterizes the local strength of the gravitational field. Because air has low density, the presence of a large, near-surface open space underground may cause a drop in local gravitational pull. And *ground-penetrating radar* (which can recognize a cavern roof and floor by the way they reflect radar beams) and local-scale *seismic-reflection profiling* (which bounces artificially generated seismic waves off subsurface layers) may provide images of the subsurface that could reveal underground openings (FIG. 7.25b). Unfortunately, none of these techniques is 100% reliable.

may saturate the ground above a cavern, and the added weight may cause the roof of the cavern to fail. But the activities of society in karst terrains can contribute to the risk. For example, constructing a building over a cavern creates an additional load on its roof. And in the case of a water-filled cave network, lowering the water table by pumping out groundwater can remove support from a weak cavern roof, allowing it to collapse.

For people directly affected by sinkhole formation, the consequences to life and property can be dire, so sinkhole formation can become a local disaster. Sinkholes can develop directly beneath buildings, so that the building breaks up and collapses. For example, in February 2014, a sinkhole suddenly grew beneath the showroom floor of the National Corvette Museum in Kentucky and swallowed eight vintage cars as it grew to a diameter of 15 m (49 ft). And in 2013, a portion of a residential building in Orlando, Florida, sank into a sinkhole 20 m (66 ft) wide (FIG. 7.24c). Fortunately, in these two cases, there were no casualties. Sadly, a sinkhole that engulfed a house in Seffner, Florida, in 2013 did claim the life of a resident. As communities grow in karst regions, the potential exists for more sinkhole disasters.

FIGURE 7.24 The formation of collapse sinkholes.

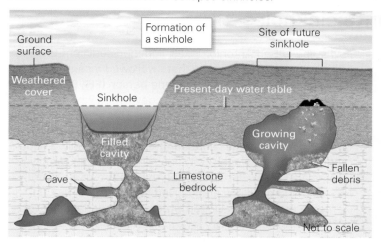

(a) Collapse sinkholes develop when a cavern grows upward and its roof thins, then breaks.

(b) Sinkholes forming in a field.

(c) An apartment building collapsing into a sinkhole.

FIGURE 7.25 Detecting potential for sinkhole formation.

(a) It's hard to see sinkholes in the satellite photograph of a location in Virginia (left) because of vegetation cover. But the sinkholes stand out clearly in a lidar image of the same location (right).

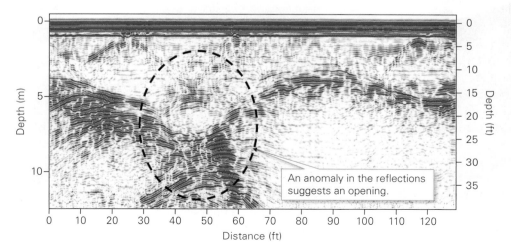

(b) Ground-penetrating radar reveals a depression in reflective horizons (colored bands) underground, suggesting the presence of a sinkhole. Drilling could confirm the existence of the sinkhole.

(c) Ground cracks beginning to develop. Note that the trees are starting to tilt.

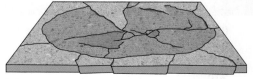

(d) Cracks and a depression forming in the concrete floor of a garage.

BOX 7.2 A CLOSER LOOK...

Liquefaction sinkholes and evaporite sinkholes

News stories about sinkhole development don't describe only sinkholes in karst terrain—they may on occasion refer to sinkholes that form in other ways. We've already discussed *sediment liquefaction*—the transformation of seemingly solid, wet sand or clay into a liquid-like slurry—in the context of earthquakes and mass wasting (see Chapters 3 and 5). Liquefaction can be triggered not only by shaking, but also by a sudden influx of water, due either to heavy rains or to the rupture of underground pipes such as water mains or sewer lines. Liquefaction caused by pipe ruptures has caused substantial property damage and has also resulted in casualties. For example, in 2016, a ruptured sewer line in San Antonio, Texas, liquefied sediment beneath a road, producing a sinkhole that swallowed a segment of road along with two cars, killing one person **(FIG. Bx7.2a)**. In 1957, a huge liquefaction sinkhole developed when a sewer line ruptured beneath a Seattle street. This sinkhole was so large that it could have engulfed a dozen homes, had it opened beneath homes instead of the street. Sadly, a liquefaction sinkhole 90 m (295 ft) deep that formed suddenly in Guatemala City in 2010, following the rupture of a sewer line, engulfed buildings and caused fatalities **(FIG. Bx7.2b)**.

Large areas of the United States are underlain by *evaporite* deposits, sedimentary layers composed of various salts, including halite, gypsum, and anhydrite. These salts are soluble in water, so rapid groundwater flow may dissolve the sediment to form caverns. Occasionally, the roof of a salt cavern collapses and a sinkhole develops. Because salts are hundreds to thousands of times more soluble than calcite, caverns can develop in an evaporite deposit in just days to years. Generally, geologically formed sinkholes aren't a major problem: they collapsed long ago, and while they can be distinct features in a landscape, as they are throughout the southwestern United States, they aren't particularly hazardous. Evaporite sinkholes are a hazard when human activity causes dissolution of salt underground. For example, collapse of a salt mine in Kansas in 1974 produced a sinkhole 100 m (328 ft) wide in just a few hours, leaving railroad tracks that had traversed the area before the collapse suspended in air **(FIG. Bx7.2c)**. Sinkhole formation can also be a hazard following oil drilling. If workers do not seal drill holes properly, water can seep in and dissolve the surrounding salt.

FIGURE Bx7.2 Examples of liquefaction and evaporite sinkholes.

(a) A liquefaction sinkhole that formed in San Antonio, Texas, in 2019, due to the release of water from an underground sewer line, caused a road to collapse.

(b) A liquefaction sinkhole in Guatemala City destroyed buildings in 2010.

(c) An evaporite sinkhole in Hutchinson, Kansas, formed directly beneath railroad tracks in 1974.

Government agencies do not provide property inspections to assess the likelihood of sinkhole development, nor do governments take the responsibility of filling sinkholes. For example, the Florida Department of Environmental Protection states, "There is a certain degree of risk in living on karst. However, most people accept the risk as one price to pay for living in the sunshine state." Given that sinkhole risk can't be known with certainty, homeowners should pay close attention to the wording of their insurance policies and should understand their coverage for "catastrophic ground-cover collapse."

If you live in a community at risk for sinkhole formation, you can avoid being a casualty by watching for telltale signs that a sinkhole is starting to form:

- Cracks in sidewalks, foundations, and walls
- A growing depression in the ground or in a concrete floor (FIG. 7.25c, d)
- Doors and windows that no longer fit in their frames
- Cloudiness in well water (due to sediment falling into caverns)
- A drop in the level of water in wells
- Tilting of trees or telephone poles
- Sudden changes in your water bill, due to leaking of pipes underground

Take-home message . . .

In places underlain by limestone bedrock, dissolution by acidic groundwater at or near the water table can produce cave networks. The collapse of such networks produces karst terrains. Sudden formation of sinkholes in karst terrains can be a local disaster that causes loss of life and property. Lowering of the water table may increase the risk of sinkhole formation in a region. Sinkholes can also form due to soil liquefaction or to dissolution of evaporite deposits.

QUICK QUESTION What clues can help you determine that a sinkhole may be developing?

7.7 Land Subsidence

Geologists refer to the slow downward movement or sinking of the land surface over a broad area as **subsidence**. Over the course of millions of years, *tectonic subsidence* (due to cooling and sinking of the lithosphere, or to slip on faults) produces *sedimentary basins*, accumulations of sedimentary beds that may be kilometers thick. Subsidence can also happen on a time scale of thousands of years in response to the growth of continental glaciers during an ice age, for the immense weight of such glaciers bends the lithosphere. While sedimentary basin formation and glacial loading are important geologic processes, they do not affect society significantly because they happen so slowly. In contrast, subsidence over shorter time scales (years to decades), due to human-caused phenomena or to changes in water and sediment supply, can represent a stealth disaster. Here, we examine the causes and consequences of land subsidence in a human time frame.

Subsidence due to Pore Collapse

In unconsolidated sediment, or in weakly cemented sedimentary rocks, water in pores helps hold the grains apart because, unlike a gas, water can't be compressed. Extraction of groundwater from the sediment or rocks eliminates the support holding pores open, because the air that replaces the water can be compressed into a much smaller volume (FIG. 7.26a, b). As a result, when air replaces water in pores, the weight of overburden causes grains to pack more closely together. Such **pore collapse** decreases the porosity of a rock or sediment, sometimes permanently.

When pore collapse takes place in a bed of sediment or sedimentary rock, the thickness of the bed decreases, so the overlying land surface subsides. Because the amount of pore collapse varies from place to place, the amount of subsidence tends not to be uniform. Varying degrees of subsidence may cause the land surface to tilt, and in some places it may cause fissures (open cracks or gashes) to open up. Such fissures can grow to be hundreds of meters long, a few meters across, and several meters deep (FIG. 7.26c, d). Where fissuring happens, building foundations and highways may crack; railroad tracks may bend; neighborhoods may become susceptible to flooding; wellheads may pop out of the ground; and the slopes of streams, drainage pipes, irrigation canals, sewer lines, and water mains may change, thereby disrupting gravity-driven flow.

Land subsidence due to groundwater depletion has been a particularly challenging problem in the western United States, where groundwater serves as a major freshwater resource. Some localities in the region consume groundwater at 50 times the rate of recharge. The resulting loss of groundwater has led to significant drops in the elevation of the land surface. For example, subsidence due to groundwater depletion has lowered the land surface of California's Central Valley—a major agricultural region—by over 15 m (50 ft) in some places (FIG. 7.27a, b). In the Houston, Texas, area, the water table has dropped by as much as 100 m (328 ft), leading to land-surface subsidence of up to 4 m (13 ft) (FIG. 7.27c). In some cases, this change has caused neighborhoods to become depressions that flood with water

FIGURE 7.26 Pore collapse and its consequences.

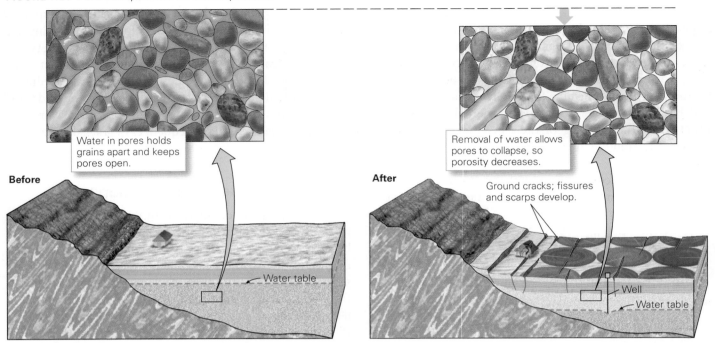

(a) Before collapse, groundwater holds the grains apart.

(b) After collapse, the grains fit more tightly together.

(c) The areas of the Sonoran Desert in southern Arizona that are undergoing significant subsidence are shown in yellow.

(d) Large fissures are developing in the subsiding areas.

during heavy rains. Beneath the desert cities of Arizona, the water table has dropped by over 150 m (492 ft), leading to subsidence of over 4 m (13 ft). In fact, over 8,800 km² (3,398 mi²) of Arizona has undergone subsidence (see Fig. 7.26c). Mexico City, which was built on an ancient lake bed, has so far subsided by almost 9 m (30 ft), mainly due to groundwater depletion (FIG. 7.27d). New techniques using satellite data have permitted researchers to detect subsidence in areas where ground survey data are not available. Satellite data, for example, have revealed the immense extent of subsidence in northwestern India (FIG. 7.27e).

Groundwater is not the only material that occupies pores underground. Oil and gas fill pores in regions where these materials have been trapped, so pumping oil and gas out of the ground can also lead to pore collapse. Their extraction from the subsurface of coastal California, for example, has caused land subsidence of up to 10 m (33 ft). To counter this problem, workers pump saltwater back into the ground under high pressure in order to refill the pores. Saltwater injection has reversed some, but not all, of the subsidence associated with oil and gas extraction.

FIGURE 7.27 Subsidence affects many locations.

(a) A record of subsidence in central California based on land surveys, from 1965 and 2016.

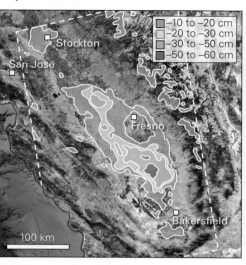

(b) A map of subsidence in central California, from 2015 to 2016.

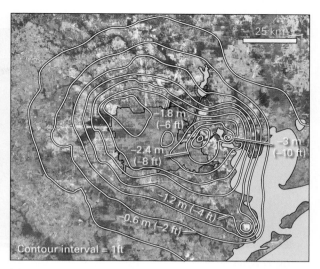

(c) A contour map showing subsidence of the Houston, Texas, region, between 1906 and 2016. The city limits of Houston are shaded.

(d) Subsidence in Mexico City has warped many buildings.

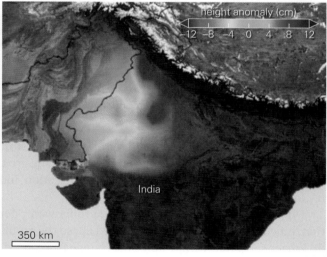

(e) Subsidence in northwestern India, based on satellite data.

Subsidence in Wetlands

Anthropogenic changes to surface-water flow in a region can have major impacts on wetlands. An example is the fate of the vast wetlands that once occupied a gentle depression, rimmed by a subtle ridge of limestone that rises only about 3 m (10 ft) above sea level, along the centerline of the Florida peninsula from Orlando southward. This depression extended across Lake Okeechobee and Everglades National Park to the southern tip of the peninsula. Prior to the 20th century, shallow water flowed slowly southward through the wetlands until it emptied into Florida Bay (**FIG. 7.28a**). Today, only a fraction of the wetlands remains—the largest remainder is now Everglades National Park.

For millennia, grasses and other vegetation lived and died in the central Florida wetlands. The dead organic matter accumulated to build a layer of *peat* (compacted and partially decayed organic matter) up to 4 m (13 ft) thick. When modern agriculture and rapid urbanization came to Florida in the 20th century, the wetlands changed forever. A long dike was constructed to block the water that once spilled over the southern shore of Lake Okeechobee, to prevent it from flowing southward, so that the region around the lake's southern shore could be used as farmland. Farther south, construction of a levee separated the wetlands on the west from agricultural and urban areas to the east. Now, the only water entering the wetlands comes from rain or canals (**FIG. 7.28b, c**). While some

FIGURE 7.28 Changes in central Florida wetlands over time, and the subsidence that has resulted.

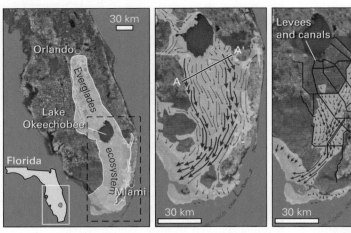

(a) The distribution of wetlands before the 20th century.

(b) Water flow in the wetlands before 1900.

(c) The canal network and consequent flow around 1990.

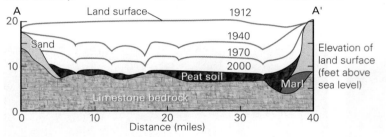

(d) A cross section showing how ground level subsided between 1912 and 2000.

of this water can flow southward across the land surface, much of it infiltrates into the subsurface to become groundwater, which cities and farms pump out for irrigation, industry, and household use.

Due to the loss of water supply and groundwater depletion, much of the wetland area has dried up, so the ground is no longer saturated with water. This change has caused the thick peat layer to dry and to undergo *compaction* (squeezing into a thinner layer due to the weight of overburden), *shrinkage* (inward collapse, like a wet sponge when it dries), and *oxidation* (reactions with oxygen that transform organic material into CO_2 and H_2O gases, which seep away). In addition, wildfires occasionally burn the peat. All of these processes have caused the volume of the peat layer to decrease, so the land surface over the peat layer subsides at a rate of about 25 mm (1 in) per year **(FIG. 7.28d)**. This subsidence, which has been going on for decades, has led to a decrease of land elevation by 1.5–3 m (5–10 ft). Given that the land surface was only about 6 m (20 ft) above sea level to start with, this subsidence means that restoring natural surface-water flow through the central Florida wetlands may become impossible—there is no longer a sufficient topographic slope to drive water flow to the sea.

Subsidence in Delta Plains, Floodplains, and Coastal Plains

Many cities and agricultural lands have been developed on *deltas* (wedges of sediment that a river deposits when it empties into standing water), *floodplains* (flat lands bordering a river that become submerged when the river overtops its banks), and *coastal plains* (broad, low-lying lands that occur along some continental margins) because such geologic settings provide large expanses of flat, fertile land. Unfortunately, these places can also be hard hit by subsidence. Before modern development, they were already undergoing natural subsidence because the thick layer of relatively young sediment beneath them had not yet lithified, so the weight of overburden was actively compacting the sediment. (In some cases, they were also undergoing tectonic subsidence.) But stream flooding, on average, brought in enough new sediment to fill in low areas and keep the land-surface elevation roughly constant.

Since the beginning of the 20th century, these areas have started to subside for several reasons. First, construction of dams and levees has prevented flooding and has therefore cut off the supply of new sediment. Second, the lack of flooding has allowed near-surface organic-rich sediments to dry out and undergo shrinkage and oxidation. And third, groundwater, oil, and gas extraction has caused pore collapse in sedimentary layers underground. Let's consider a few examples illustrating these processes.

- *Delta plain and floodplain subsidence in Louisiana:* The combination of factors we've just described has led to 2 m (7 ft) of subsidence under and around New Orleans and on the Mississippi Delta. New Orleans continues to subside at about 8 mm (0.3 in) per year, and now much of the city lies as much as 5 m (16 ft) below sea level. Only a system of levees, dikes, and pumps keeps the city from flooding. (When the dikes failed during Hurricane Katrina in 2005, the result was catastrophic, flooding vast areas of the city, as described in Chapter 11.) But subsidence isn't a problem only in New Orleans. The entire Mississippi Delta is undergoing subsidence, so much so that Louisiana has lost about 5,200 km² (2,008 mi²) of land in the past century. The effects of this subsidence on the shape of the region's coastline have been astounding **(FIG. 7.29a)**. About a third to a half of the land loss is due to pore collapse caused by extraction of oil and gas. The remainder is due to lack of sediment replenishment and compaction of organic-rich sediments. Though subsidence rates have slowed, the state still loses about 12–20 km² (4.6–8 mi²) of land per year along the coastline.

FIGURE 7.29 Subsidence of deltas.

(a) Change in the coastline of Louisiana between 1932 and 2011 due to subsidence of the Mississippi Delta.

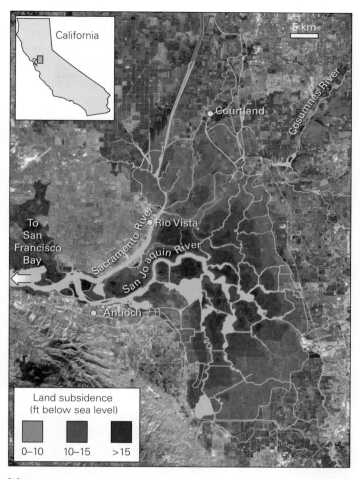

(c) Subsidence in the San Joaquin delta between the 1930s and the 1990s.

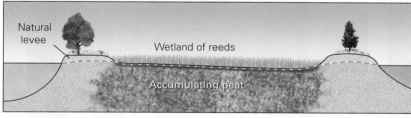

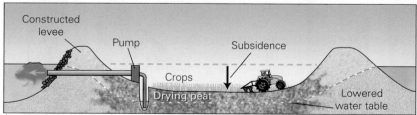

(b) Before modern agriculture, wetlands occupied areas between stream channels in the San Joaquin delta. Thick layers of peat accumulated in these wetlands. Since the wetlands were drained to create farmland, the peat has decayed and compacted, so the land has sunk below sea level.

(d) From the air, the levees that prevent subsided farmland from flooding stand out.

- *Delta plain subsidence in California:* Land subsidence in the San Joaquin–Sacramento River delta of central California, a major agricultural area, has caused the ground surface to sink by more than 8 m (26 ft) in the last century. This subsidence primarily reflects the drying out of peat that takes place when wetlands have been drained for agriculture, but it is also due to groundwater depletion (FIG. 7.29b–d).

- *Coastal plain subsidence in Venice:* Ironically, Venice, Italy, has become one of the most famous tourist destinations in the world because of subsidence. Venice, where canals serve as streets and boats serve as taxis, was built on a network of small islands that rise from a coastal lagoon (FIG. 7.30a, b). It's a long way down to bedrock, and the sediment underlying Venice has not been fully compacted. Construction of heavy buildings on the islands has accelerated natural compaction. Subsidence greatly accelerated in the 1930s due to groundwater extraction, causing the city to sink by 15 cm (6 in). The extraction was stopped in 1966, after a year of terrible flooding, and subsidence has been slower ever since. But new research using satellite data has revealed that the city continues to sink at a rate of 1–2 mm (0.04–0.08 in) per year, due to the continued compaction of sediment and to tectonic subsidence related to the subduction of the Adriatic Plate beneath Italy.

Venice's subsidence, coupled with rising sea level, means that particularly high tides—known as king tides or nuisance tides in English (see Chapter 13) and as *aqua alta* in Italian—have swamped the famed piazzas of the city. The year 2019 was the worst on record, not only for the number of *aqua alta*, but also for their height. Local authorities consider an *aqua alta* that rises 1.4 m (4.6 ft) above mean sea level to be "exceptionally high," and when these tides begin, sirens go off. Generally, the city accommodates such flooding by building elevated platforms for tourists to walk on (FIG. 7.30c), but a flood in November 2019 put 80% of the city underwater, and the water was so deep in places that the elevated platforms floated away. In order to decrease such flooding, the Italian government is investing billions of dollars in a network of floodgates to prevent tides from entering Venice's lagoon. In the meantime, many Venetians are simply moving away, and tourist bookings had dropped by a third since the flood, even before the COVID-19 pandemic in 2020.

FIGURE 7.30 The subsidence of Venice.

(a) Venice was built on an island in a coastal lagoon.

(b) Many of Venice's streets are canals.

(c) During very high tides, even the piazzas are flooded.

Subsidence due to Mine Collapse

People dig mines underground in search of many different types of resources. Geologists informally refer to mines dug in strong crystalline rocks, such as granite, as *hard-rock mines*, and to mines dug in weak sedimentary strata, such as coal and evaporites, as *soft-rock mines*. Both hard-rock and soft-rock mines can collapse, causing subsidence of the overlying land, but soft-rock mines are particularly susceptible because the weak rock they penetrate can crack and fail easily. We'll consider examples where large areas of land subsidence have accompanied collapses of two types of soft-rock mines: coal mines and salt mines.

Underground coal mines can be excavated in two ways, and both may result in land subsidence. In *longwall-advance mining*, large machines beneath an elongate enclosure grind coal and dump it on a conveyer belt that transports it out of the mine. As the grinder and enclosure move forward, the space behind the enclosure is allowed to collapse, and the overlying land sinks **(FIG. 7.31)**. In *room-and-pillar mining*, miners excavate coal in a grid of tunnels, leaving behind columns of coal to hold up the mine's roof. After mining has ceased, the relatively weak columns may fail and the mine may collapse, causing the ground surface above to subside. Land subsidence may also occur as a consequence of *coal-bed fires*. These fires, which burn underground, are almost impossible to put out and may last for years or even centuries. A fire that started near the town of Centralia, Pennsylvania, has been burning since 1962. In addition to causing land subsidence and cracking, it produces toxic fumes that rise through cracks in the ground, so the town was abandoned **(FIG. 7.32)**.

In locations where shallow seas or saline lakes existed over a long time in the geologic past, thick layers of halite (NaCl), gypsum ($CaSO_4 \cdot 2H_2O$), and other salts have accumulated. When subsequently buried, these evaporite deposits end up underground. In some places, evaporite deposits may be meters to tens of meters thick, and because the minerals in

FIGURE 7.31 Longwall-advance mining can lead to land subsidence.

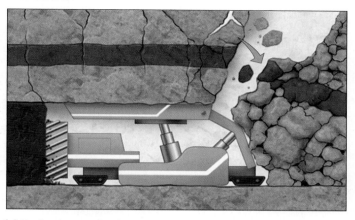

(a) During longwall-advance mining, a grinder moves horizontally, back and forth, along the coal seam. Hydraulic lifts support the roof. As the grinder and the roof supports advance, the roof behind them is allowed to collapse.

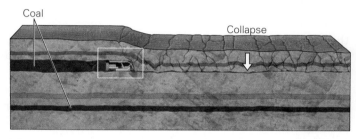

(b) A cross section drawn parallel to the advance direction shows the roof collapse. The rectangle shows the location of (a).

FIGURE 7.32 Now-abandoned roads near Centralia, PA, are hummocky because of subsidence that has happened as underground coal burns. Toxic gases, from the burning coal below, vent out of cracks.

FIGURE 7.33 The collapse of a salt mine, due to flooding by groundwater, led to the subsidence of land near Geneseo, New York.

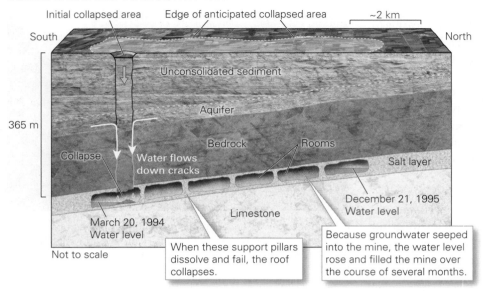

these layers are so weak that they can flow, the layers may change shape over time to become bulbous underground mounds called *salt domes*.

People mine salt for many uses. Highway departments, for example, spread halite (rock salt) on roads to melt winter snow. Salt mines pose a hazard, however, because salt dissolves in water so easily—we've already seen how such dissolution can cause sinkholes to form. If flowing groundwater enters a large salt mine, it may dissolve the salt and weaken support columns, leading to progressive collapse of the mine. A particularly destructive collapse began in 1994 near Geneseo, New York, when a portion of a room-and-pillar halite mine collapsed and groundwater from an aquifer above the mine began to pour into the mine. This groundwater dissolved more salt and caused further collapse **(FIG. 7.33)**. During the first 2 years following the initial mine collapse, the overlying land subsided by as much as 3 m (10 ft). Subsidence in the region is likely to continue until the entire mine area (40 km² or, 15 mi²) has collapsed.

Take-home message . . .

Land subsidence, the lowering or sinking of the land surface, can happen on a human time scale for several reasons. Most subsidence results from pore collapse due to pumping of groundwater, oil, or gas out of the ground. Land subsidence can result in the development of fissures, and can damage building foundations and infrastructure. Subsidence can also happen due to loss of sediment supplies or the draining of surface water.

QUICK QUESTION Which types of mining can lead to land subsidence, and why?

7.8 Soil Hazards

If you've ever had the chance to dig in a garden, field, or forest floor, you've seen firsthand that the soil in which flowers, crops, and trees grow looks and feels different from beach sand, potter's clay, or gravel. That's because **soil** consists of sediment that has been modified over time by physical and chemical interactions with rainwater, air, organisms, and decaying organic matter **(FIG. 7.34)**. Soil can form either from weathered rock or from a pre-existing deposit of sediment. Though rarely more than a meter (3 ft) thick, and commonly less than 10 cm (4 in) thick, soil represents one of our planet's most valuable resources. It is the heart of the critical zone, the amazing portion of the Earth System that provides the water and nutrients essential to life. Without it, the Earth would have no forests, grasslands, or crops, and therefore, no civilization.

Soil forms so slowly—it can take 100–400 years to produce 1 cm (0.4 in) of soil—that researchers consider it to be a nonrenewable resource on a human time scale. Shockingly, soil is being destroyed at a rate that far exceeds the rate at which new soil is being produced. This loss represents a stealth disaster, for it diminishes the world's capacity to produce food and could lead to famine and conflict. To provide a

FIGURE 7.34 Soil forms a thin layer at the Earth's surface.

(a) A cross section of soil beneath a pine woods.

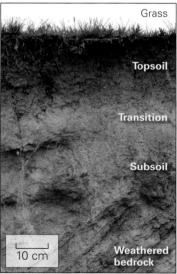

(b) A cross section of soil in a tropical climate.

FIGURE 7.35 The formation of soils.

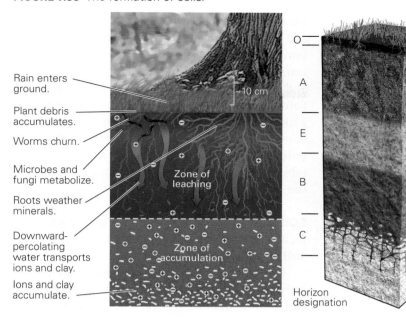

(a) Soils develop as water percolates downward, carrying ions and clay with it, and as organisms interact with mineral grains.

(b) Distinct layers or horizons develop in soil, each with a characteristic composition and texture.

Because different soil-forming processes operate at different depths, the character of soil changes with depth (**FIG. 7.35b**). The uppermost layer, the **topsoil**, has been modified by leaching. It contains most of the organic matter and *nutrients* that promote plant growth (chemicals such as nitrogen, phosphorus, potassium, and calcium that are key components of chemicals in living cells). The topsoil overlies **subsoil**, which forms in the zone of accumulation and does not contain much organic carbon—most plants don't grow as well in subsoil. Subsoil grades downward into weathered rock or sediment, which in turn grades downward into unweathered rock or sediment. Scientists refer to the distinct layers as *soil horizons*, and designate each with a letter.

As farmers, foresters, and ranchers know well, soil in one locality differs greatly from soil in another in composition, thickness, and texture. This diversity exists because soil makeup depends on several factors: climate, source-material composition, slope steepness, vegetation, and time. Climate controls the amount of water passing through soil and the types of organisms that live in it; source-material composition determines which minerals and grain sizes occur in the soil; slope steepness controls how thick the soil can become before it undergoes mass wasting; vegetation determines root density; and time determines how mature and how thick the soil will be.

context for understanding threats to soil, we first discuss the origin of soil. Next, we examine various processes, such as soil depletion and soil erosion, that destroy it. We also consider phenomena that change the volume and character of soil, thereby disrupting the land surface.

How Does Soil Form?

Three key processes contribute to soil formation. First, chemical and physical weathering breaks up a pre-existing rock or sediment to form loose grains and dissolved ions. Second, rainwater falling on the accumulation of grains sinks in and slowly percolates downward. Nearer the surface, in the *zone of leaching*, this water picks up ions and clay flakes, and then transports them deeper (**FIG. 7.35a**). Farther underground, in the *zone of accumulation*, new mineral crystals precipitate from the downward-percolating ion-rich water, and clay stops moving. The linked process of leaching and accumulation, therefore, redistributes chemicals in the soil. Third, organisms interact with loose mineral grains in many ways: microbes, fungi, plants, and animals alter grains by absorbing or releasing ions; root growth and burrowing animals (such as insects, worms, and gophers) move particles; animals leave behind organic debris as a by-product of their metabolism, and when they die, their remains decompose and mix into the mineral grains; and at the ground surface, leaves and other vegetation decompose and mix into the grains. As a result, soil consists not only of inorganic minerals, but also of carbon in various forms, as well as complex microbial communities.

Soil Degradation

We've seen that society faces many serious health disasters involving losses of water and land (**VISUALIZING A DISASTER**, pp. 286–287). Unfortunately, we need to add one more item to this list—**soil degradation**, loss of soil.

Researchers use this term to describe the overall consequences of the phenomena that contribute to decreasing the amount of soil available, or to making soil less capable of supporting vegetation. The rate at which soil degradation takes place has accelerated in recent decades, and it is a global problem (**FIG. 7.36**). Let's consider some examples of these phenomena.

SOIL DEPLETION. Growing crops in soil generally results in the removal of nutrients (primarily nitrogen, phosphorous, and potassium), as well as carbon, from the soil. Such

FIGURE 7.36 Soil degradation and destruction.

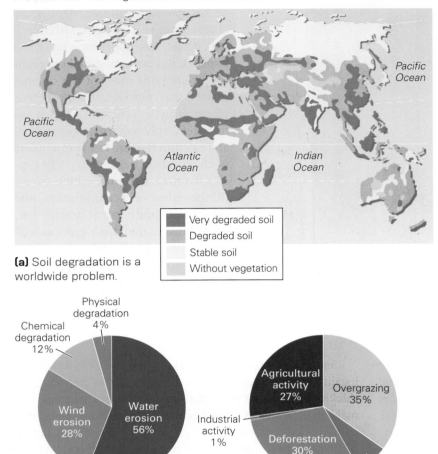

(a) Soil degradation is a worldwide problem.

(b) The major types of processes that lead to soil degradation.

(c) The underlying causes of soil degradation.

SOIL CONTAMINATION. Human activity can add toxic substances to soils. Sources of contaminants include chemical spills, overfertilization, oil spills, landfills, spoil piles at mines, and nuclear waste repositories, among others. Salinification of soils can result when poor irrigation techniques allow large volumes of water to evaporate on the soil, to leave salt behind. All **soil contamination** can make the soil inhospitable to plant life, or can make crops grown in the soil inedible.

SOIL EROSION. When wind or running water picks up soil particles and carries them away, **soil erosion** takes place.

Rates of soil erosion have increased dramatically in the past century due to deforestation, agriculture, and overgrazing (**FIG. 7.38**). Deforestation accelerates erosion because it removes both the protective cover that prevents rain from impacting the soil surface and the root networks that bind soil together. Agriculture accelerates erosion because the replacement of native plants with rows of crops not only decreases the density of roots binding soil, but also effectively transforms a field into a desert for much of the year. For example, in a typical cornfield, herbicides or hand weeding remove all plants except the corn stalks, so even during the growing season, the fields are mostly bare soil! And during the many months between harvest and when the removal, known as **soil depletion**, decreases soil fertility. Unless nutrients are replaced (by fertilization—leaving stalks of crops in the field to decay, or by planting nitrogen-replacing crops), the soil may no longer be able to sustain growth. Soil depletion can also take place in the wake of deforestation, especially in tropical environments, where loss of tree cover allows organic debris to decompose and oxidize so quickly that the carbon and nutrients in the debris are soon lost.

SOIL TEXTURE MODIFICATION. Good soils have just the right texture (determined by porosity, permeability, and stickiness) so that plant roots can grow and can access the appropriate amounts of water and air. Intense plowing, motion of heavy equipment, tromping by animals, and lack of vegetation input can ruin soil texture, making the soil impermeable. In some cases, soil becomes so hard that roots can't grow through it.

FIGURE 7.37 Slash-and-burn agriculture ultimately destroys rainforest soils.

FIGURE 7.38 Rates of soil erosion worldwide. The most rapid rates of erosion are occurring in southern Asia, the Midwestern US, east Africa, and southeastern Brazil; all places with large populations that rely on crops and herds for food.

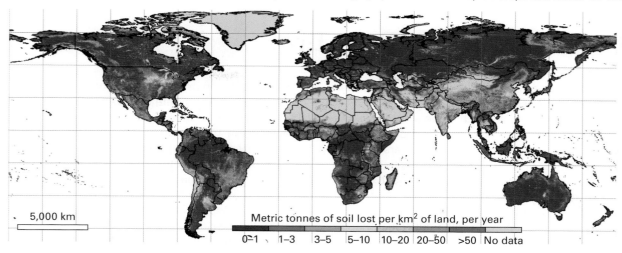

next crop has grown, topsoil lies completely exposed to wind and rain **(FIG. 7.39a)**.

Rain can cause soil erosion because raindrops striking the ground send soil particles bouncing upward to become suspended in *sheetwash* (a very thin layer of water that flows downslope). Sheetwash carries the particles into *rills* (small rivulets) that drain into deeper gullies, which in turn empty into streams that carry the particles away **(FIG. 7.39b–c)**. A single heavy rain can result in the loss of 82,000 kg (90 tons) of soil from an acre (about 4,000 m²) of farmland. Wind causes

FIGURE 7.39 Soil erosion.

(a) This farm field in Illinois, before a new crop grows, is as barren of vegetation as a desert, and is susceptible to erosion.

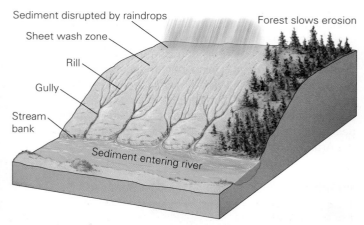

(b) Soil erosion by water carves channels into slopes.

(c) Close-up of rills. Note the sediment that flowed down the rills at the base of this small slope.

(d) As a tractor drives across a field, it breaks up soil and raises a cloud of dust.

BOX 7.3 DISASTERS AND SOCIETY

Dust storms and the Dust Bowl

When particularly strong winds suddenly blow across arid landscapes, **dust storms** (also known as *haboobs*) billow up, lofting soil and sediment particles high in the air and, in some cases, transporting them great distances. In fact, researchers have determined that dust blown off the Sahara, and the arid farmlands of the Sahel to its south, traverses the Atlantic and falls on South America and the Caribbean **(FIG. Bx7.3a)**.

Dust storms, many of which are due to the very strong winds or gust fronts that blow outward from beneath large thunderstorms or in association with cold fronts (see Chapter 8), typically begin abruptly. An approaching dust storm resembles a huge, roiling, opaque wall of dust, sometimes rising a mile high, dwarfing tall buildings **(FIG. Bx7.3b, c)**. During a severe dust storm, visibility may be reduced so much that airports and highways have to be closed. When a dust storm has passed, fine sediment coats the ground, plants, vehicles, and buildings. Dust also sifts into mechanical equipment and into building interiors.

Dust storms not only contribute to the stealth disaster of soil erosion, but they can also be a direct threat to human health. The fine particles in dust can become stuck in a person's lungs, so repeated exposure can lead to a lung disease called *silicosis*. Dust storms can also transport fungal spores, bacteria, and viruses, which can trigger diseases such as meningitis and pneumonia. And in places where storms incorporate dust from lake beds exposed by water depletion, the dust may contain carcinogenic agricultural chemicals and pesticide residue, as has happened due to the drying up of the Aral Sea (see Fig. 7.8).

Water depletion and agricultural practices have left large areas of formerly temperate land barren of plants. During the 1930s, many severe dust storms raked across the southern Great Plains in the United States. The affected region had been a highly productive frontier of agriculture in the first few decades of the 20th century, for federal farm policies supported a land boom that converted 5.2 million acres of native prairie grassland from Nebraska to Texas into fields of grain. The prairie grasses, with their deep, extensive root systems, had provided a wind-resistant surface on the land—but cornfields and wheat fields did not. In addition, farmers during this period practiced *dry-land farming*, which meant that after each rain, they would plow water into the soil to prevent it from evaporating. Unfortunately, dry-land farming destroyed soil texture—it pulverized the soil, converting what had been pea-sized clumps into very fine particles. When drought began in the early 1930s, the soil dried out, so that the winds that swept across the plains with wintertime cold fronts easily lofted it into walls of dust. These dust storms led to extreme soil erosion in the southern Great Plains **(FIG. Bx7.3d)**. In fact, researchers estimate that about 850 million tons of topsoil blew away in 1935 alone. Due to the dust storms and their aftermath, the affected region came to be known as the **Dust Bowl (FIG. Bx7.3e, f)**, becoming so inhospitable that countless families left the region and migrated westward to California, as described in Chapter 14.

To help prevent another Dust Bowl from happening, a Great Plains Committee was established in 1936 by the federal government to determine the causes of the catastrophe

FIGURE Bx7.3 Dust storms and the Dust Bowl.

(a) Dust from the Sahara blows over the Atlantic.

(b) A dust storm 1.5 km (1 mi) high approaches Phoenix, Arizona, in 2018.

and suggest preventive measures. The committee recommended that zoning regulations be developed to prevent lands that are particularly susceptible to erosion from being intensively farmed or grazed; that farmers change their plowing methods to keep furrows at right angles to wind direction; that post-harvest stubble remain in place; that windbreaks of tall trees be planted between fields to slow wind speeds; and that cover crops be planted on fields with high exposure to winds. Today, computer-based mapping of fields, instrumentation that monitors soil characteristics, and methods of no-till farming have decreased the Great Plains' risk of another Dust Bowl catastrophe. But increasing severity and frequency of droughts could overwhelm these preventive methods, so dust storms remain a concern.

FIGURE Bx7.3 Dust storms and the Dust Bowl (continued).

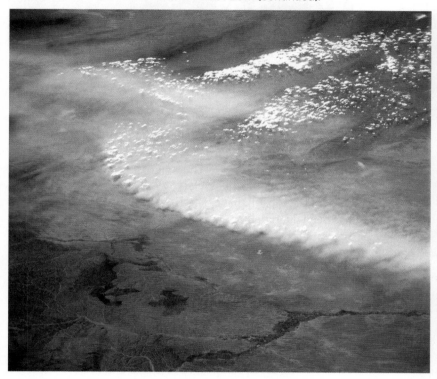

(c) A dust storm over Afghanistan, as viewed from the Space Shuttle.

(d) The area affected by the Dust Bowl dust storms.

(e) A family walking through a dust storm in the Dust Bowl during the 1930s.

(f) Equipment buried by a dust storm in the 1930s.

FIGURE 7.40 The damaging consequences of soil expansion for small buildings. Differential movement cracks a house's foundation and walls. When soil dries, the house sinks, and its foundation cracks even more.

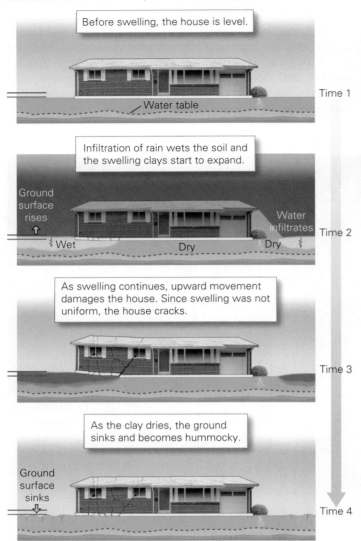

(a) Stages in the development of damage due to expansive soil.

(b) When exposed at ground surface, swelling clay has a popcorn-like appearance. Note the pencil for scale.

soil erosion by lofting fine-grained components into the air, especially when equipment drives across a bare field and breaks the weak bonds that hold soil particles together (FIG. 7.39d). Strong winds can carry soil for long distances, and particularly strong winds tear away huge amounts of soil and produce immense dust storms (BOX 7.3).

CONSTRUCTION OF IMPERVIOUS SURFACES. Significant amounts of soil have been lost simply because soil has been covered by impermeable materials, such as the asphalt and concrete that come with urban and suburban sprawl. Once such *impervious surfaces* cover soil, the soil degrades because soil-forming processes cease and microbial communities die off. Less obvious, but similarly problematic, are changes that happen when animals trample soil or farmers till it intensely. These processes alter soil texture.

SOIL CONSERVATION. The stealth disaster of soil destruction is creeping up on society. By some estimates, about a third of the topsoil that existed before the dawn of civilization has been degraded or destroyed, and if countries do not start reversing the trend now, much of the world's remaining soil could become unproductive within the next 60 years. Clearly, international efforts will be needed to slow soil destruction before the loss translates into famine and ecosystem failure. Governments and nongovernmental agencies, therefore, are increasing efforts to encourage **soil conservation**, a combination of practices intended to diminish soil erosion and soil depletion. In many cases, modifying techniques of plowing and rotating crops can prevent erosion and restore depleted nutrients and carbon. For example, planting grasses or winter crops in a field after harvest, instead of leaving it barren, decreases soil erosion rates, as does *no-till farming*, a technique of planting and harvesting that doesn't require turning over the soil. (Traditional deep plowing can disrupt soil texture, kill off important microbial communities in soil, and make soil susceptible to erosion.)

Expansive Soil

Imagine that the soil beneath a house expands during the rainy season, causing the house and its foundation to rise by a few centimeters. In this situation, the soil may shrink during the dry season, so the house and foundation would sink. If the amount of soil expansion and shrinkage varies along the length of the house, so that only part of the house and foundation goes up when the soil gets wet, and only that part goes down when the soil dries, the foundation, concrete floors, and the walls of the house could eventually become distorted and crack. This variable movement, in addition, would cause window frames and door frames to warp, windows to break, and the roof to develop leaks. Soon, the owners of the house would be facing a very expensive problem (FIG. 7.40).

The damage that we've just described can happen to houses built on **expansive soil**, soil that expands and shrinks due to wetting and drying, respectively. A soil can display this characteristic if it contains more than 5% of a clay mineral called smectite. Tiny smectite flakes can absorb water molecules into their crystal structure and expand—the way a book thickens when its pages get wet—and the flakes expel water molecules when they dry. Overall, adding water to an expansive soil can make the soil's volume increase by up to 10%. This expansion applies an upward push to buildings standing on the soil. When the soil dries out and shrinks, gravity causes structures built on it to sink. In places, shrinkage also causes soil to develop deep fissures called *desiccation cracks*.

Expansive soils occur widely, but they aren't always a problem. For example, in humid climates, where soils remain hydrated most of the time, they don't undergo alternating episodes of expansion and shrinkage. In regions with drier climates, where expansive soils dry out between rains, the soils can cause substantial damage. Because soils aren't homogeneous, and because the distribution of vegetation or impervious covers varies, the amounts of swelling and shrinkage can vary significantly over distances as small as a few meters.

As we've already noted, houses built on soils that undergo variable expansion and shrinkage can be severely damaged. Such damage primarily affects small buildings, such as single-family homes, because the upward-directed stress applied by an expansive soil can exceed the downward push due to the building's weight. In an open field, variable swelling and contraction of expansive soil produces a *hummocky land surface*, riddled with bumps and dimples. Where roads cross expansive soils, their surfaces can become very bumpy and they may undergo cracking. Similarly, variable swelling and contraction of expansive soil can damage fences, power lines, and pipelines.

An individual damaged building is a disaster only to the building's occupants and owners. But the sum of all property damage due to expansive soils in a given year can be comparable to the damage caused by more dramatic disasters. For example, by some estimates, expansive soils cause an average of about $13 billion in damage annually in the United States, a figure comparable to the value of average annual insured losses due to tornadoes (about $13 billion)!

What can be done to prevent this damage? Soil expansion occurs only down to a depth of a few meters, for below this depth, the moisture content of the ground doesn't vary seasonally. So, setting foundations on piles that extend below the affected depth will decrease building movement. If that is not feasible, improving drainage around a house can decrease the amount of water available for the soil to absorb. This can be done by installing drainage pipes underground, sloping the ground surface so that it tilts away from the house, and planting water-absorbing shrubs.

> **Take-home message...**
> Soil is the altered upper layer of sediment on the Earth's surface. Unfortunately, it is being destroyed by depletion, erosion, contamination, and urban sprawl faster than it's being produced. Some soil blows away in dramatic dust storms. Soil conservation efforts can counter soil destruction.
>
> **QUICK QUESTION** What is expansive soil, and how does it cause damage?

ANOTHER VIEW A karst terrain in Viet Nam. Here, most of the land surface that once existed has collapsed, leaving towers of limestone. The rise of sea level, relative to land, has submerged the low-lying areas.

Society can't flourish without adequate supplies of clean freshwater! Sadly, our planet faces a global water crisis because we are depleting supplies of both surface water and groundwater in places where people need water. The loss happens because climate change has diminished rainfall and snowfall, water from rivers has been diverted for use by farms and cities, and over-pumping has caused drawdown of the water table, causing fissuring. Loss of water has already caused some cities to temporarily turn off their municipal water supplies. Slowly, desertification transforms regions that were once temperate into arid regions, causing crops to be ruined and herds to die. Exposed land can be susceptible to immense dust storms. Both wind erosion and water erosion, as well as overuse, can cause soil degradation and loss. Water loss can also contribute to sinkhole formation and land subsidence. Regional land subsidence has resulted in loss of coastal land, increased flooding threats, and damage to buildings.

Chapter 7 Review

Summary

- Water in the surface and near-surface regions of the Earth passes through various reservoirs of the hydrologic cycle over time. The oceans represent the largest reservoir, accounting for about 97% of the water involved in the cycle.
- Accessible freshwater accounts for only 0.014% of our planet's water. It occurs as surface water, soil moisture, and groundwater. Groundwater occupies pores and cracks in sediment and rock underground.
- Due to population growth and improvement of living standards, water use has increased dramatically in recent centuries. The supply of freshwater is inadequate in many locations because of lack of supply and lack of efficient distribution. Groundwater depletion and surface-water overuse contribute to this global water crisis.
- If less water enters a river system than is removed (by evaporation or diversion), the river can run out of water and become a saline trickle. Surface-water depletion can also cause lakes to dry up.
- Loss of surface-water resources also happens when society pollutes surface-water supplies, which sometimes causes water to undergo eutrophication, or when surface freshwater undergoes salinification.
- Groundwater fills pores and cracks in rock and sediment below the water table. Accessible groundwater comes from springs or wells.
- Wells must penetrate an aquifer (rock or sediment with high porosity and permeability) below the water table in order to provide water.
- Groundwater depletion occurs when water is extracted from the ground at a rate greater than the recharge rate. Depletion around a well initially produces a cone of depression. Over time, excessive pumping can lower the water table over a broad region. When this happens, producing wells must be drilled deeper. Wells that go too deep may encounter saline water.
- Groundwater contamination can destroy a water supply and can be very hard to remediate.
- Many countries are facing water stress, partly due to physical scarcity of water (lack of enough surface water or groundwater to meet demand), and partly due to economic scarcity (lack of adequate water purification or distribution systems). Conservation and regulation can help rectify the problem.
- Sinkholes form where the ground collapses or subsides to form a circular depression. Most sinkholes form in karst terrains due to dissolution of limestone in acidic groundwater, but some form due to sediment liquefaction or evaporite dissolution.
- Karst sinkholes form when the land surface over a cave network starts to collapse. They can form under buildings or infrastructure, and thus represent a local hazard. Though some caverns can be detected by instruments, these techniques are not always reliable. Changes to the ground surface may serve as warnings that a sinkhole is forming.
- Land subsidence, the sinking of the land surface over a broad area, can occur on a human time scale for many reasons, including pore collapse (due to groundwater, oil, or gas extraction), shrinkage and compaction of organic material, loss of sediment supply, and mine collapse.
- Land subsidence can damage buildings and infrastructure and make areas more susceptible to flooding.
- Soil, formed by the interaction of sediment or weathered rock with rainwater, air, and organisms, is a thin layer, but without it, there would be no fields or forests, and civilization would not exist. Unfortunately, soil is a nonrenewable resource that is being destroyed faster than it forms.
- Soil develops very slowly. Rainwater percolating through it removes chemicals and clay from the zone of leaching and carries them down into the zone of accumulation. Organisms rework soil, and organic debris mixes into it, providing carbon.
- Processes that cause soil degradation include soil depletion (loss of nutrients and carbon), soil erosion (removal by wind and water), and soil contamination. Soil can also be lost when it is covered with impervious materials.
- Soil erosion increases when farming techniques remove native plants that protect the soil and leave the ground surface unprotected for much of the year.
- Soil conservation efforts can help preserve soil before too much has been lost.
- Expansive soils contain a significant proportion of an expanding clay mineral called smectite, which absorbs water. Expansion and contraction of these soils causes billions of dollars in damage to building foundations and infrastructure.

Review Questions

Blue letters correspond to the chapter's learning objectives.

1. Define the hydrologic cycle, list its various reservoirs, and describe how water passes through the cycle. **(A)**
2. In which reservoir does most water in the hydrologic cycle reside? **(A)**
3. How does the composition of freshwater differ from that of saltwater? **(A)**
4. Why is society facing a global water crisis, and why can the water crisis be thought of as a stealth disaster? **(B)**
5. What phenomena can lead to freshwater depletion, and how is freshwater depletion manifested in rivers and lakes? **(B)**
6. What is water pollution, and what aspect of it causes eutrophication? What processes can lead to the salinification of surface freshwater? **(B)**
7. Where does the water in groundwater come from, and where does it reside? **(C)**
8. Explain the difference between porosity and permeability, and between an aquifer and an aquitard. **(C)**
9. Describe the spatial relationships among the saturated zone, the unsaturated zone, and the water table. Label them on the figure. **(C)**
10. How does a well differ from a spring? Distinguish between a producing well and a dry well, and explain the relation of each to the water table. **(C)**
11. Under what conditions does the pumping of water from a well cause drawdown? When does pumping lead to a regional drop in the water table? **(C)**
12. Describe the consequences of groundwater depletion. **(C)**
13. Can natural chemicals adversely affect the quality of groundwater? If so, how? **(C)**
14. What is a contaminant plume, and what procedures might be followed to remediate groundwater contamination? **(C)**
15. Explain the concept of water stress as it affects society. Distinguish between physical and economic scarcity of water. How can water stress be diminished? **(C)**
16. How do karst landscapes develop? **(D)**
17. Describe the three ways in which sinkholes can form in karst terrains. Which type of sinkhole formation can result in a local disaster? **(D)**
18. Would a sinkhole be as likely to form if the water table in the adjacent figure were above the top of the cavern? Why? **(D)**
19. Under what conditions do sinkholes form outside of karst terrains? **(D)**
20. Is there a way to determine if a location might be susceptible to sinkhole formation, and is there any way to recognize the initiation of a sinkhole? **(D)**
21. Define land subsidence, and describe various processes that can lead to the development of subsidence. **(E)**
22. Why do coal mines and salt mines occasionally collapse, and how does their collapse affect the land surface? **(E)**
23. What is soil, how does it differ from sediment, and how does it form? How thick is a typical soil, and at what rate does it form? **(F)**
24. What phenomena are responsible for soil depletion and soil contamination? **(F)**
25. How does soil erosion occur, and what phenomena accelerate the rate at which it occurs? **(F)**
26. Why is soil degradation considered to be a stealth disaster, and what measures can be taken to prevent it? **(F)**
27. What is expansive soil, and how does the existence of such soil cause property damage? **(F)**

On Further Thought

28. Explain how groundwater depletion, surface-water depletion, and land subsidence can be linked to one another. **(E)**
29. A government is contemplating the construction of a large dam along a major river to provide water for irrigation. What impacts might this dam have on the river, its coastal plain, and its delta plain downstream? **(E)**

8

IN CONSTANT MOTION
The Earth's Atmosphere

By the end of the chapter you should be able to . . .

A. list the primary and trace gases that you inhale in a breath of air.

B. describe atmospheric properties that meteorologists use to describe the weather.

C. interpret a drawing that displays atmospheric layers.

D. explain what clouds are made of, and how they produce precipitation.

E. discuss why our planet's atmosphere constantly flows on a global scale.

F. explain why monsoons and ENSO events are so important to society.

G. detail the relationships among weather, fronts, and mid-latitude cyclones.

FIGURE 8.1 Disastrous consequences of storms in the atmosphere.

(a) An ice storm rips power poles from the ground.

(b) A coastline decimated by a hurricane.

(c) A blizzard brings city traffic to a standstill.

(d) Debris from homes destroyed by a tornado.

8.1 Introduction

Air—though we breathe it in and out all day, every day, we're usually unaware of its presence, because air is normally invisible, and unless it's moving, we don't sense it around us. But on some days, air flows at terrifying speed and immense quantities of water descend from the skies. During these episodes, air becomes a focus of fear, for it can obliterate homes, build towering waves, feed floodwaters, or bury the landscape in snow (see Chapters 9–12). At other times, air becomes so hot and dry that it threatens society with droughts that wither crops and trigger famines, and wildfires that consume forests and towns (see Chapter 14).

« Moisture in the air can condense to form clouds that can dump rain, snow, and ice on the landscape.

Formally defined, **air** is the unique mixture of gases that makes up the Earth's **atmosphere**, the layer of gas that surrounds our planet. The atmosphere makes life as we know it possible, because air includes carbon dioxide (CO_2) for photosynthesis in plants, oxygen (O_2) for metabolism in animals, and various other gases that protect our planet's surface from dangerous ultraviolet light coming from the Sun. Air also serves as the medium in which birds, insects, and airplanes fly, and through which our voices carry. Fortunately, the atmosphere constantly flows and stirs, producing *wind* (horizontal flows of air) as well as updrafts and downdrafts (vertical flows of air). This movement carries water from where it evaporates to where it falls as *precipitation* (rain, snow, and hail), which sustains lives and livelihoods. But during *storms*, wind and precipitation become dangerously intense, and we must protect ourselves from the atmosphere (FIG. 8.1). Though the atmosphere is very thin relative to the Earth's radius—99.9% of the gas molecules in air lie below an elevation of 50 km (31 mi)—it's a precious layer. Without it, the Earth would be as barren as the Moon.

Why do storms form, and why can they become violent, producing catastrophic winds? Why do rains sometimes flood the land, and why does lightning spark in the sky? Why do ice and snow sometimes overwhelm our cities and shut down our transportation networks? Why do rains sometimes fail to come, so that lands become parched and dry? How will society be affected by the overall warming trends in the atmosphere? To set the stage for exploring these questions, which address the *atmospheric hazards* (atmospheric conditions with the potential to cause natural disasters) that are discussed in the remainder of this text, this chapter provides a general introduction to the atmosphere. We begin by exploring the characteristics and behavior of the atmosphere, and how they vary with altitude. Next, we turn our attention to clouds, precipitation, and wind, some of the features that define the *weather* at a given place and time. We also examine the forces that drive atmospheric motion and cause weather to vary. Finally, we describe regional- to global-scale atmospheric circulations, which vary across the globe and play a role in development of hazardous weather.

8.2 Atmospheric Composition

The Recipe for Air

What are the chemicals in air? An average sample of *dry air*, meaning air from which all water has been removed, consists mainly of two gases: 78% (by volume) molecular nitrogen (N_2) and 21% molecular oxygen (O_2) (FIG. 8.2). The remaining 1% includes argon (Ar) and many other gases, such as carbon dioxide (CO_2), ozone (O_3), and sulfur dioxide (SO_2). Of the two major gases, molecular nitrogen is inert, meaning that it doesn't react with materials and doesn't play a role in sustaining life. Oxygen is not inert; it reacts with other chemicals in organisms to provide the energy of life, and with rocks and other materials in the Earth System. Some of the atmosphere's other gases, although they occur in such minuscule quantities that researchers refer to them as *trace gases*, play key roles in regulating atmospheric temperature (see Chapter 15).

Water (H_2O) is generally considered separately in descriptions of air composition, because the amount of water in air varies greatly from place to place and from time to time. We can't specify a single percentage that characterizes the average proportion of H_2O in any specific region of the atmosphere. Water in the atmosphere occurs in three forms: as an invisible gas called **water vapor** (consisting of isolated H_2O molecules) (FIG. 8.3a), as a liquid (in tiny *droplets* suspended in air, or in larger *drops* that fall as rain) (FIG. 8.3b), and as a solid (in the form of ice crystals, snowflakes, or hailstones) (FIG. 8.3c). The sizes of water particles in air vary greatly (TABLE 8.1). Looking up at the sky, we often see *clouds*—these consist of countless tiny water droplets or ice crystals.

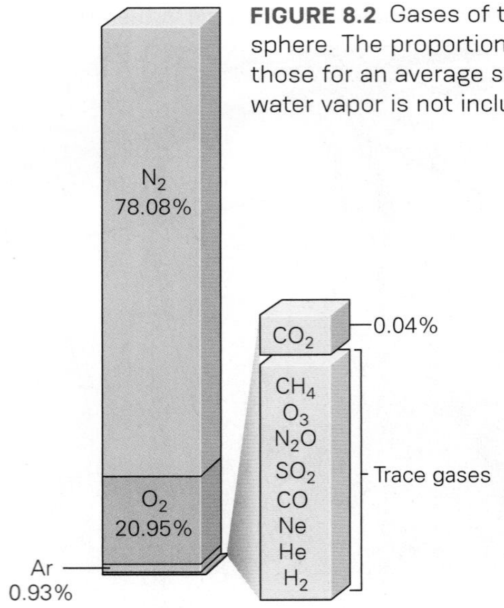

In addition to the gases we described earlier, air contains **aerosols**, liquid or solid particles that are far smaller than cloud droplets (see Table 8.1) (FIG. 8.4a). Natural inorganic aerosols include specks of mineral dust lofted by winds blowing over soil, salts carried into the air from sea spray, fine ash injected into the atmosphere by volcanic eruptions, and carbon particles (soot) billowing from forest fires (FIG. 8.4b). Some natural inorganic aerosols form by chemical reactions of acidic gases released from volcanoes with other components of air. Natural organic aerosols include pollen, bacteria, molds, and viruses, as well as detritus from decaying organisms. When air contains sufficient moisture, certain aerosols capture water molecules in the air and dissolve in the water to form microscopic droplets, which produce **haze** (FIG. 8.4c).

FIGURE 8.2 Gases of the Earth's atmosphere. The proportions shown here are those for an average sample of dry air; water vapor is not included.

TABLE 8.1 Sizes of Water Particles in Air

Name	Size range (mm)
Water molecule	0.0000003
Aerosol	0.0004–0.01
Cloud droplet	0.01–0.1
Drizzle	0.1–0.5
Small raindrop	0.5–1.0
Large raindrop	1.0–4.0
Very large raindrop	4.0–8.0

FIGURE 8.3 Water in the atmosphere occurs in three forms.

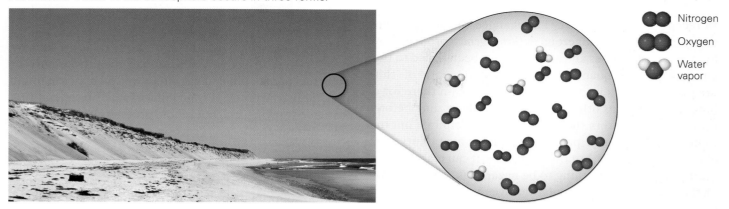

(a) Air contains invisible water vapor, even on a cloudless day. Water molecules in vapor are mixed in with other gas molecules.

(b) These clouds in the warm sky over Brazil consist of liquid water droplets.

(c) These high clouds consist of tiny ice crystals.

FIGURE 8.4 Atmospheric aerosols.

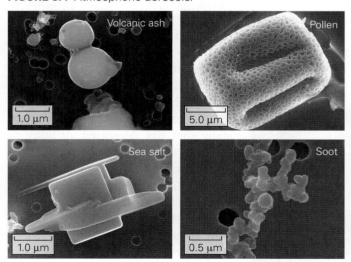

(a) Aerosols viewed through a scanning electron microscope (1 μm = 0.001 mm).

(b) Smoke rising from a forest fire carries soot particles into the atmosphere.

(c) Haze over the Blue Ridge Mountains, part of the Appalachian mountain range.

Air Pollution: The Danger of Dirty Air

In recent centuries, society has changed the recipe of air in profound ways. Dangerous *anthropogenic* (produced by human activity) gases and aerosols have been added to the air in immense quantities. Some components of this **air pollution** mix into the air and can remain there for a long time, whereas others remain for a short time before they are washed from the air by rain or precipitate out on their own. In isolated parts of the globe, air pollution is almost undetectable, but over big cities, it can make the air murky and cause breathing to be hazardous. The majority of the world's human population lives in cities, so about 90% of the world's population breathes polluted air to some degree. Air pollution endangers the health and well-being of people and other organisms, and it can disrupt the environment. In fact, the World Health Organization (WHO) estimates that air pollution leads to about 7 million premature deaths per year, in that it increases susceptibility to such illnesses as heart attack, stroke, lung cancer, pneumonia, and asthma.

Where does air pollution come from? *Primary pollutants*, those emitted directly from a source, come from burning fossil fuels (coal, oil, and gas) in power plants or automobiles, industrial processes and mineral smelting operations, pesticide spraying, and household products **(FIG. 8.5a)**. Such pollutants include soot (carbon aerosols), carbon monoxide (CO), a variety of sulfur-bearing gases, a variety of nitrogen-oxide compounds, toxic metals, and carcinogenic organic compounds. *Secondary pollutants* form by chemical reactions of primary pollutants with one another and with air in the presence of sunlight. For example, reactions between nitrogen-oxide compounds and organic gases produce ozone (O_3), which can irritate your lungs. Reactions of sulfur-bearing gases with water produce sulfuric acid aerosols. When incorporated into raindrops, they cause *acid rain*, which can kill trees and fish and can damage some structures.

The composition of air pollution varies from place to place, and has changed over time as the effluent of society's activities has changed. For example, dank, dark *smog* engulfed London and other industrial cities in the 19th century and into the mid-20th century. This pollution developed when steam engines in factories and on trains, together with home stoves, belched primary pollutants consisting of soot, carbon monoxide, and sulfur-bearing gases, into the air, where it mixed with a suspension of water droplets at ground level. Smog sometimes got so bad that it led to health disasters. For example, during the London smog event of December 5–9, 1952, over 6,000 people died. The *photochemical smog* of today's cities develops when exhaust from cars and trucks reacts with various gaseous compounds in the presence of sunlight to produce secondary pollutants such as ozone and nitric acid aerosols **(FIG. 8.5b)**.

In the Ganges River valley of India, rapid industrialization, coupled with continued use of wood- or dung-burning stoves in homes and routine burning of nearby fields, has led to photochemical smog that has decreased visibility to such an extent that it's rare to see the horizon. The same is true in large, highly industrialized, traffic-heavy cities across southern Asia. Many residents navigated through this dirty air daily without realizing how bad visibility had become—until the COVID-19 pandemic hit in 2020. That spring, due to shelter-at-home ordinances, traffic and industrial activity came nearly to a

FIGURE 8.5 Air pollution.

(a) Industrial smokestacks spew primary pollution into the atmosphere.

(b) Thick photochemical smog over Los Angeles, California, forms from auto exhaust and other pollutants.

halt. Air pollution decreased substantially, and the horizon became visible for the first time in anyone's memory.

Because of the hazard of air pollution, government agencies monitor it and issue health-hazard warnings as necessary. Often, these warnings focus on air containing *fine particulate matter*, defined as solid or liquid aerosols smaller than 2.5 microns (0.0025 mm) in diameter. (By comparison, a human hair has a diameter of 70 microns.) Fine particulate matter, also known by the abbreviation $PM_{2.5}$, can become embedded deep in people's lungs where it can't be dislodged. In the United States, the *Air Quality Index* ranks air on the basis of its concentrations of certain pollutants (fine particulate matter, coarser particulate matter, O_3, CO, SO_2, ammonia (NH_3), and lead). The US Clean Air Act, and comparable policies in other countries, have helped encourage society to diminish air pollution. In many locations, pollution has decreased due to engineering solutions such as scrubbers on smokestacks and catalytic converters on car exhausts.

> **Take-home message . . .**
> Dry air consists of nitrogen (78%), oxygen (21%), and trace gases (1%). The amount of water in the atmosphere varies considerably over time and space, and that water can occur as liquid, solid, or gas. The atmosphere also contains tiny particles called aerosols. Anthropogenic gases and aerosols cause air pollution, which significantly affects human health.
>
> **QUICK QUESTION** Distinguish between London's smog in the 1950s and photochemical smog in Los Angeles today.

8.3 Describing Atmospheric Properties

Everywhere in the world, people complain about, marvel at, or just simply button up and prepare for the **weather**, the specific atmospheric conditions that they find outdoors at a given time and location. Weather conditions vary substantially among different parts of the world. The average, range, and seasonality of weather conditions that occur over the course of decades or more in a specific region determine that region's **climate** (see Chapter 15). The atmospheric conditions that **meteorologists** (scientists who study and predict the weather) commonly use to describe weather include the following:

- *Temperature*: a measure of hotness or coldness.
- *Atmospheric pressure*: a measure of the weight of a column of air above a location.
- *Relative humidity*: a measure of the amount of water vapor in the air.
- *Wind direction*: a measure of the compass direction from which air moves horizontally.
- *Wind speed*: a measure of how fast the air moves horizontally.
- *Visibility*: a measure of how far one can see through the air.
- *Cloud cover*: the proportion of the sky that is covered by clouds.
- *Precipitation*: a measure of the amount of water falling from the air and reaching the ground during a time interval.

Let's look at each of these properties in more detail.

Air Temperature

Everyone pays attention to the air temperature—if it's hot, you might stroll about in shorts, but if it's cold, you need to put on a coat. But what exactly does *temperature* measure? In air, gas molecules move rapidly in random directions, continually colliding with one another. **Air temperature** represents the average speed at which air molecules move: the faster the average speed of the molecules, the higher the temperature. Therefore, temperature represents the *kinetic energy*, the energy of motion, of the molecules, which can be expressed by the equation

$$\text{Kinetic energy} = \text{mass} \times \text{velocity}^2$$

Note that *temperature* is not synonymous with the *thermal energy* contained in a material—thermal energy is the sum of all the kinetic energy due to vibrations and movements of all atoms or molecules in the material. According to these definitions, air within a container holding a few very fast-moving molecules may have a higher temperature than air within a container holding many slower-moving molecules, but the former can contain much less thermal energy. The more familiar term **heat** refers to the thermal energy transferred from one body to another. Heat always flows from hotter to cooler materials.

We measure temperature with a **thermometer**, calibrated in degrees according to a standard scale (**FIG. 8.6**). Most thermometers in the United States use the *Fahrenheit scale*, while the rest of the world uses the *centigrade scale*, also known as the *Celsius scale*. Each scale uses a *degree* (°) as its unit of temperature. At sea level, water boils at 212°F and freezes at 32°F on the Fahrenheit scale, and boils at 100°C and freezes at 0°C on the centigrade scale. Note that a degree represents a greater temperature change on the centigrade scale than on the Fahrenheit scale. Scientists also use another temperature scale, the Kelvin scale. At *absolute zero*

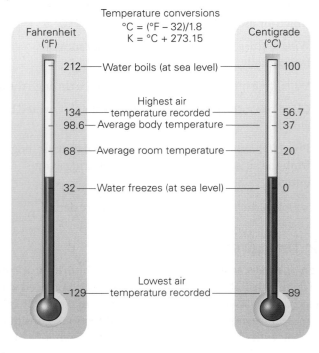

FIGURE 8.6 Temperature scales and the relationship between centigrade (Celsius) and Fahrenheit values for common temperatures.

(0 K, or −273.15°C), the lowest temperature possible, molecules stop moving or vibrating. Measurements on the Kelvin scale do not use a degree symbol.

At any particular location, air temperature fluctuates daily as the Sun rises and sets, and changes seasonally because the number of daylight hours and the Sun's position in the sky vary with latitude and time of year. Temperature at a location also changes when volumes of cooler or warmer air move in from elsewhere. The highest air temperature recorded on land was measured in Death Valley, California, where the thermometer reached 56.7°C (134°F), and the lowest air temperature was recorded at Vostok Station, Antarctica, where the thermometer dropped to −89.2°C (−128.6°F).

Atmospheric Pressure

Atmospheric pressure (or *air pressure*) refers to the force applied by air on a surface of a specified area. Atmospheric pressure exists because air, like any substance, has mass, so in the Earth's gravitational field, it has weight. You can picture atmospheric pressure as the weight of a column of air pushing down on the area at the base of the column. For example, using the English system of measurement, a column of air that is 1 inch by 1 inch at its base and extends from the ground at sea level all the way through the atmosphere to the edge of space weighs, on average, 14.7 pounds. Therefore, we can say that the average atmospheric pressure at sea level is 14.7 pounds per square inch. Several different units are used to specify atmospheric pressure (TABLE 8.2). These days, meteorologists prefer to use either *millibars* (mb) or the numerically equivalent *hectopascals* (hPa).

Pressure is a special type of stress, where *stress* is defined as the force applied per unit area. When an object is subjected to air pressure in an atmosphere with no wind, the stress acting on it is the same in all directions. (The same condition exists in any non-moving fluid.) Put another way, you won't get pushed over by air pressure in still air, because the air pressure pushing on your front is the same as that pushing on your back. And an empty can doesn't collapse, because the air pressure pushing on its inside walls equals that pushing on its outside walls. We'll see that when the wind blows or water flows, the pressure on one side of the fluid no longer equals that on the other.

Atmospheric pressure affects the density of air. As pressure increases, air molecules squeeze closer together, and as pressure decreases, they move farther apart. The density of a gas also depends on temperature—when air gets hotter, the air molecules move faster and spread out, so the air expands and becomes less dense, and when air cools, the air molecules slow down and move closer to one another, so the air contracts and becomes denser.

Atmospheric pressure, and therefore atmospheric density, always decreases as altitude increases, because the higher you go, less gas is pressing down from above you. Roughly speaking, atmospheric pressure decreases by about half for every 5.6 km (3.5 mi) of altitude gain (FIG. 8.7). As a result, about 50% of the atmosphere's gas lies below an altitude of 5.6 km, and 75% lies below an altitude of 11.2 km

TABLE 8.2 Common Units Used for Measuring Atmospheric Pressure

Unit	System	Average atmospheric pressure at sea level
Pounds per square inch	English	14.7 lb/in²
Hectopascals (hPa)	Metric	1,013.25 hPa
Atmospheres (atm)	Convention*	1.000 atm (= 1.013 bar)
Bars (= 1,000 hPa)	Metric equivalent	1.01325 bar (= 1 atm)
Millibars (mb) (= 0.001 bar)	Metric equivalent	1,013.25 mb (= 1,013.25 hPa)
Inches of mercury†	Traditional	29.92 in

*Defined by international agreement.
†Inches of mercury refers to the height to which atmospheric pressure pushes mercury up a vertical glass vacuum tube.

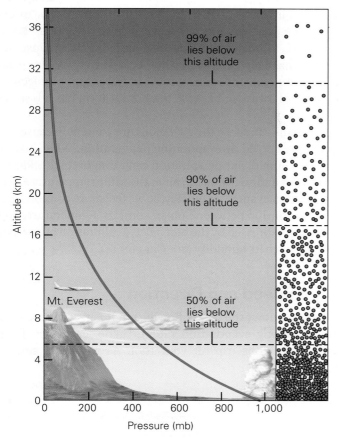

FIGURE 8.7 Atmospheric pressure and air density decrease with increasing altitude.

typhoon (a Pacific hurricane), where the pressure dropped to 870 mb. Regions of higher pressure develop beneath colder, denser air, while regions of lower pressure occur beneath warmer, less dense air. As we'll see later, horizontal variations in atmospheric pressure control the speed and direction of the wind; particularly large variations drive the violent winds of the Earth's hazardous storms.

Moisture and Relative Humidity

Water molecules enter the atmosphere by evaporation from the Earth's oceans, lakes, rivers, wetlands, and soil, and by transpiration from plants and forests. As we've seen, water in the atmosphere exists as water vapor, an invisible gas consisting of freely moving water molecules. The portion of total atmospheric pressure exerted by water vapor alone is called **vapor pressure**. At a given temperature, dry air has a lower vapor pressure than does humid air.

The atmosphere has a limited capacity for holding water vapor. When the atmosphere contains as much water vapor as it can hold, it becomes *saturated*—any additional water vapor added to saturated air will condense to form liquid droplets or solid ice crystals. Meteorologists refer to the vapor pressure at which the atmosphere reaches its holding capacity as the **saturation vapor pressure**. The atmosphere's saturation vapor pressure depends strongly on temperature: warm air can hold more water vapor than can cold air **(FIG. 8.8)**.

(7 mi), the altitude at which commercial jets fly. People can't survive long at altitudes above about 6 km (4 mi) because a breath can't bring in enough oxygen molecules. As a result, jet cabins must be pressurized by pumping in air. Though 99.9% of gas in the atmosphere lies below 50 km, there's still enough air above 50 km that meteoroids arriving from space will heat up and vaporize when they pass below an altitude of 70–120 km (43–75 mi). The "top" of the atmosphere, where the density of the atmosphere equals that of interplanetary space, varies between 350 and 800 km (217–500 mi) in altitude. To avoid slowing down due to friction with air, the International Space Station orbits at an altitude of about 400 km (250 mi).

The numbers given for average atmospheric pressure at sea level in Table 8.2 are just that—averages. The specific value of sea-level pressure varies over time at a given location, and varies with location at a given time. In fact, rarely will a direct measurement of atmospheric pressure at sea level yield a number exactly equal to the average sea-level pressure stated in the table. Corrected for elevation, the highest sea-level pressure ever recorded was 1,085 mb in winter in Siberia, whereas the lowest ever recorded occurred in the center of a

FIGURE 8.8 Saturation vapor pressure (SVP) decreases rapidly as air temperature (T) decreases.

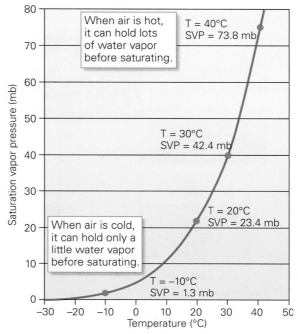

Significantly, it's not the absolute amount of water vapor in the air, as indicated by the vapor pressure, that makes air feel humid, but rather how closely air approaches saturation. To convey a sense of how humid air feels, meteorologists characterize the air's moisture content by specifying its **relative humidity** (**RH**). This measure is defined as the amount of water vapor in the air (the vapor pressure) divided by the air's capacity for holding water vapor (the saturation vapor pressure). We can express this quotient as a percentage:

$$RH = \frac{\text{vapor pressure}}{\text{saturation vapor pressure}} \times 100\%$$

Clearly, RH depends on both the amount of water vapor in the air and the air's holding capacity. For example, if two volumes of air contain the same number of water vapor molecules, the warmer one will have a lower RH (**FIG. 8.9a**). Therefore, a cool location may feel more humid because the air there has a high RH, while a hot location may feel dry, even if its air contains the same number of water vapor molecules as the cool location, because the hot location has a low RH (**FIG. 8.9b**).

Humans can sense the air's moisture content because our bodies are cooled by evaporating perspiration. When hot air becomes humid, our perspiration can't evaporate quickly, so we feel hot and sticky. In dry air at the same temperature, our perspiration evaporates quickly, so we feel more comfortable. The relationship between human comfort, temperature, and RH can be expressed by the *heat index*, a calibration of how hot air feels given its humidity (see Chapter 14).

Wind Speed and Direction

Everyone knows the feel of flowing air. It can cool your skin (by increasing evaporation rates), rustle leaves, or if it's especially strong, rip off your roof. As we noted earlier,

FIGURE 8.9 Relative humidity.

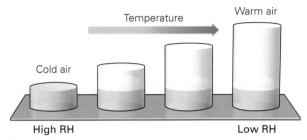

(a) The relative humidity is the ratio of the amount of water vapor in the atmosphere (represented by the liquid in the container) to the atmosphere's holding capacity (represented by the size of the container). The holding capacity increases as temperature increases, so for the same amount of water vapor, the relative humidity will decrease as the temperature increases.

(b) In the real world, both vapor pressure and air temperature vary from place to place. The size of each beaker represents the saturation vapor pressure. The ratio of the amount of water in the beaker to the beaker's capacity represents the relative humidity (RH).

atmospheric scientists define **wind** as the horizontal movement of air. Of course, air in nature doesn't move only horizontally, but can also flow upward in an *updraft* or downward in a *downdraft*. While up-and-down air movements have real consequences, such as triggering storms or causing discomfort for airline passengers, vertical air movements are not represented by a standard wind measurement.

Why do we feel the wind, and why can it cause so much damage? We mentioned earlier that when an object, such as a house, is subjected to air pressure in still air, the push acting on it is the same in all directions. When the wind blows, however, air molecules are moving, so molecules strike the side of the house facing the wind with greater force than the side away from the wind. Consequently, the side facing the wind feels greater stress. If this stress exceeds the strength of the walls or of the bolts that hold the roof on, then the roof tears away and the walls collapse. The stress caused by a moving fluid, such as air, is sometimes called *dynamic pressure*.

To describe wind, we use two familiar quantities: wind direction and wind speed. By convention, the **wind direction** refers to the direction from which the wind blows. For example, in the northern hemisphere, a north wind brings cold air from the north toward the south. **Wind speed** indicates the horizontal rate of air movement. Meteorologists display wind speed and direction on a weather map by placing *wind barbs* on it **(FIG. 8.10a)**. We can also display winds by drawing streamlines parallel to the wind direction **(FIG. 8.10b)**.

Visibility, Cloud Cover, and Precipitation

Meteorologists describe air transparency by specifying **visibility**, the distance from which a person with normal vision looking through air can identify objects. Pure air is transparent, so on a clear day, visibility may be "unlimited," in that you can see to the horizon and beyond. Smoke, haze, and fog all reduce visibility because aerosols, water droplets, or ice particles reflect, absorb, and scatter light.

While visibility pertains to what you see when looking in front of you, cloud cover refers to what you see when looking up into the sky. The water droplets and ice crystals that make up clouds scatter light that comes in from the Sun in random directions. The degree of light scattering depends on how thick and dense the clouds are. The amount of light reaching the ground depends on **cloud cover**, meaning the proportion of the sky that is hidden by clouds. A weather report routinely characterizes cloud cover using familiar words, such as partly cloudy or overcast.

In discussing the atmosphere, **precipitation** refers to water falling from clouds in the form of liquid (rain) or solid (snow or hail). It includes brief showers, heavy downpours, hailstorms, and blizzards. To characterize precipitation, meteorologists specify both the *precipitation rate*—a measure of how fast rain or snow is falling, specified in inches or millimeters per hour, and the *total precipitation*—the cumulative amount that falls during a given time period. Total rainfall

FIGURE 8.10 Displaying wind speed and direction.

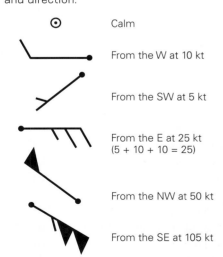

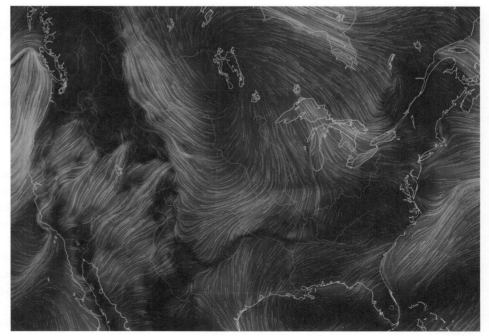

(a) Key to wind barbs used on a weather map. Meteorologists typically measure wind speed in knots (kt), where 1 kt = 1 nautical mile per hour, which converts to 1.85 km/h, or 1.15 mph.

(b) On this wind map, winds are represented by green streamlines. The lines, which are parallel to the wind direction, are thicker and more closely spaced where the wind is faster.

BOX 8.1 A CLOSER LOOK...

Weather radar

Every year, floods, tornadoes, and other types of hazardous weather cause loss of life and property damage. Meteorologists can warn the public of impending hazards by using weather radar **(FIG. Bx8.1a)**. Radar systems transmit microwaves, much like the waves in a microwave oven. But unlike your oven, a radar transmitter sends out microwaves in pulses that last only about 1 millionth of a second. When the microwaves encounter raindrops and hailstones, some of the waves scatter back to the radar system's antenna. The antenna collects this energy, called the *radar echo*, and measures the time it took the microwave pulse to travel out to the precipitation and back. Because microwaves travel at the speed of light, we can calculate the distance to the precipitation from the following equation: distance = velocity × time. By knowing the angle at which the antenna points to the sky, we can determine the precise altitude of the precipitation within the atmosphere.

The character of precipitation affects the strength of the returned signal, or the *radar reflectivity*. For example, the greater the size and number of particles the pulse intercepts, the greater the radar reflectivity will be. High radar reflectivity values, depicted with red and orange colors on a display, indicate heavy rain or hail. In contrast, low values, depicted with blues and greens, indicate light drizzle or nonprecipitating clouds **(FIG. Bx8.1b)**. By examining radar reflectivity measurements over time, meteorologists can estimate the total amount of rain that fell during the period of observation. These data help them predict hazards such as flash floods.

Radar systems transmit microwave energy at specific frequencies. When reflected energy returns to the antenna, the frequency of the returned energy has typically shifted slightly as a result of the movement of the raindrops or ice particles along the radar beam. **Doppler radar** systems measure this shift and use it to determine the speed of the wind in the direction parallel to the beam. Not only can Doppler radar identify strong straight-line winds, it can recognize rotation in the air flow to help spot potential tornadoes.

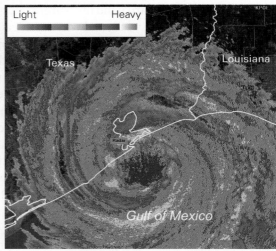

FIGURE Bx8.1 Weather radar and satellites.

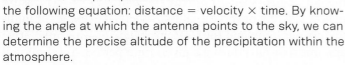

(a) A radar antenna used in research.

(b) A radar reflectivity image of Hurricane Ike as it made landfall near Galveston, Texas. Colors indicate reflectivity, which depends on rainfall intensity.

---

can be stated in inches or millimeters that fell during a single storm, a month, a season, or a whole year, depending on context. Meteorologists describe total snowfall either as an *accumulated depth*, meaning the thickness of the snow layer, or as a *water equivalent*, the thickness of a water layer that would be formed if the snow melted completely.

Measuring Atmospheric Conditions

To keep track of the weather and to make weather predictions, government agencies have set up **weather stations** at thousands of locations around the world. At each weather station, *thermometers* measure air temperature, *barometers* measure air pressure, *anemometers* measure wind speed and direction, and *hygrometers* measure relative humidity. These instruments are calibrated to specific standards, so that measurements from one station can be compared with those from another.

The weather that affects the Earth's surface develops as a result of atmospheric conditions and motions up to altitudes as high as 20 km. Therefore, to gain a more complete understanding of weather at a given time, meteorologists send up *rawinsondes*, balloon-lofted instruments that measure temperature, moisture, and wind at high altitudes. They also use radar to detect the intensity of precipitation and air movements above the ground **(BOX 8.1)**. Satellites can

FIGURE 8.11 Weather satellites provide images of cloud cover, moisture distribution, and other characteristics.

(a) This image shows cloud cover as revealed in the wavelengths of visible light.

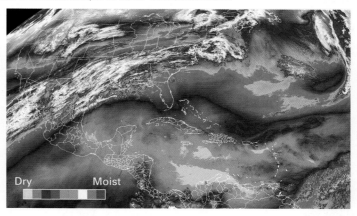

(b) By sensing certain wavelengths of infrared light, satellite images reveal variations in atmospheric moisture content.

acquire images of the atmosphere over a wide region. These images allow us to see cloud cover from above, and to detect regional variations in moisture content (**FIG. 8.11**). Time-lapse satellite images allow us to watch how individual storms develop and evolve.

> **Take-home message . . .**
> To characterize weather, meteorologists measure temperature, atmospheric pressure, relative humidity, wind speed and direction, visibility, cloud cover, and precipitation. These properties of the atmosphere vary with location at a given time and with time and altitude at a given location.
>
> **QUICK QUESTION** Why is total snowfall measured differently than total rainfall?

8.4 Vertical Structure and Heat Sources in the Atmosphere

Earlier in this chapter, we saw that air pressure decreases progressively as altitude increases. Temperature, too, changes with altitude. In some layers of the atmosphere, temperature decreases as altitude increases, whereas in others, temperature increases as altitude increases. Altitudes at which the direction of temperature change reverses (from decreasing to increasing, or vice versa) delineate boundaries that separate four distinct atmospheric layers (**FIG. 8.12**). Let's examine these layers in sequence, beginning at the Earth's surface. We will also consider how these layers are heated.

Because solar energy, mostly in the form of shortwave radiation (visible light and ultraviolet light), reaches the atmosphere from outside the Earth, it's a common misperception that the air around you becomes warm because it's "broiled from above." In fact, air is transparent to most

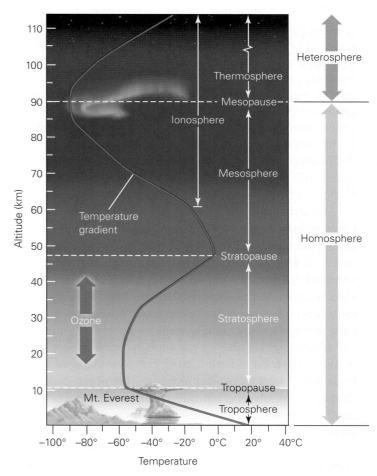

FIGURE 8.12 Layers of the atmosphere.

incoming sunlight—that's why you can see through clear air—so most solar energy passes through the atmosphere and reaches the Earth's surface, where it gets absorbed by molecules in rock, soil, or water. These surface materials then reradiate the energy back into the air as longwave infrared radiation, some of which gets trapped by molecules of certain trace gases, such as H_2O, CO_2, and methane (CH_4), causing the air to warm up. (We'll discuss this *greenhouse effect* in more detail in Chapter 15.) Put another way, the Earth's surface acts as a heater that warms the air at the base of the atmosphere—in effect, air is "baked from below."

Moving upward in the atmosphere, as distance from the Earth's surface—and thus from the atmosphere's heat source—increases, the air gets cooler. For this reason, if you climb a tall mountain, you'll find that as your elevation increases, the air temperature decreases. This decrease in temperature with increasing altitude continues up to about the cruising altitude of commercial jets, where air temperature ranges between −40°C and −60°C (−40°F and −76°F). Meteorologists refer to this lowest layer of the atmosphere, where temperature decreases with altitude, as the **troposphere**. The change of temperature with altitude in the troposphere, a quantity known as the **environmental lapse rate**, varies substantially from place to place and from time to time, but on average, it has a value of 6.5°C/km (3.5°F/1,000 ft). The thickness of the troposphere depends on the amount of solar energy arriving at the surface of the Earth, so it varies with latitude and time of year. For example, at the hot equator, the troposphere can be as thick as 20 km (12.5 mi), whereas during winter above the frigid poles, it's only about 7 km (4.3 mi) thick. All of the Earth's storms occur within the troposphere, for it is in this layer that the vertical air currents that create clouds and storms are possible. (We'll discuss the sources of these vertical air currents in the coming chapters.)

Above the *tropopause*, the top of the troposphere, air temperature increases as altitude increases. The layer where this takes place, the **stratosphere**, reaches up to a height of about 50 km. Temperature increases with altitude in the stratosphere because ozone (O_3) molecules in that layer absorb ultraviolet radiation from the Sun. This absorbed radiation causes air molecules to move faster, which makes their average kinetic energy—their temperature—increase. Above about 50 km, there's not enough ozone to heat the air by absorbing solar radiation, so air temperature again begins to decrease with increasing altitude. This point, called the *stratopause*, marks the bottom of the **mesosphere**, which continues upward to the *mesopause*, at an altitude of about 90 km (56 mi). Above the mesopause lies the outermost layer of the atmosphere. In this layer, the **thermosphere**, temperature increases with increasing altitude, reaching a maximum of 1,500°C (2,732°F) as the few remaining air molecules absorb high-energy radiation from the Sun.

Atmospheric scientists sometimes use two other characteristics to divide the atmosphere into layers: chemical composition and electric charge. In the *homosphere*, which encompasses the troposphere, stratosphere, and mesosphere, air circulation mixes the atmosphere so thoroughly that its chemical composition is more or less the same from place to place. In the *heterosphere*, which coincides with the thermosphere, gases sort gravitationally according to their molecular weight, forming layers with lighter gases above and heavier ones below. Meteorologists recognize another layer, the *ionosphere*, that is distinguished by its high concentration of ions. We'll revisit the higher layers of the atmosphere in Chapter 16, when we examine potential hazards from space and from the Sun.

> **Take-home message . . .**
> The atmosphere can be divided into four layers, the troposphere, stratosphere, mesosphere, and thermosphere, delineated by boundaries where the direction of temperature change with altitude reverses. All weather occurs in the troposphere.
>
> **QUICK QUESTION** Why is it so cold at the altitude where commercial jets fly?

8.5 Clouds and Precipitation

From the vantage point of an astronaut in the International Space Station, one of the most striking features of the Earth is the pattern of its clouds (**FIG. 8.13**). What do clouds consist of? As we've seen, they are not composed of pure water vapor, a transparent gas. Rather, a **cloud** is a visible collection of countless water droplets and/or tiny ice crystals suspended in the sky, and these particles reflect, scatter, and absorb light. Let's examine how clouds form and what conditions lead certain clouds to produce rain, snow, or hail. With an understanding of clouds, we can then explore storm formation.

Cloud Formation

We have already seen that air can hold only a certain amount of water vapor before it becomes saturated, and that the amount it can hold depends on its temperature: warm air can hold more water vapor than cold air. Thus, if saturated air cools, it becomes *supersaturated*, and when this happens, water molecules begin to attach to the surfaces of aerosols. If the temperature is above freezing, an aerosol can serve as a *condensation nucleus*, allowing a droplet of liquid water to

FIGURE 8.11 Weather satellites provide images of cloud cover, moisture distribution, and other characteristics.

(a) This image shows cloud cover as revealed in the wavelengths of visible light.

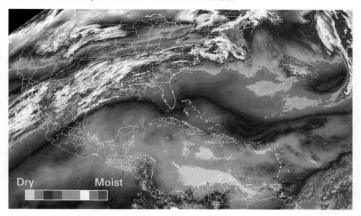

(b) By sensing certain wavelengths of infrared light, satellite images reveal variations in atmospheric moisture content.

acquire images of the atmosphere over a wide region. These images allow us to see cloud cover from above, and to detect regional variations in moisture content (FIG. 8.11). Time-lapse satellite images allow us to watch how individual storms develop and evolve.

> **Take-home message . . .**
> To characterize weather, meteorologists measure temperature, atmospheric pressure, relative humidity, wind speed and direction, visibility, cloud cover, and precipitation. These properties of the atmosphere vary with location at a given time and with time and altitude at a given location.
>
> **QUICK QUESTION** Why is total snowfall measured differently than total rainfall?

8.4 Vertical Structure and Heat Sources in the Atmosphere

Earlier in this chapter, we saw that air pressure decreases progressively as altitude increases. Temperature, too, changes with altitude. In some layers of the atmosphere, temperature decreases as altitude increases, whereas in others, temperature increases as altitude increases. Altitudes at which the direction of temperature change reverses (from decreasing to increasing, or vice versa) delineate boundaries that separate four distinct atmospheric layers (FIG. 8.12). Let's examine these layers in sequence, beginning at the Earth's surface. We will also consider how these layers are heated.

Because solar energy, mostly in the form of shortwave radiation (visible light and ultraviolet light), reaches the atmosphere from outside the Earth, it's a common misperception that the air around you becomes warm because it's "broiled from above." In fact, air is transparent to most

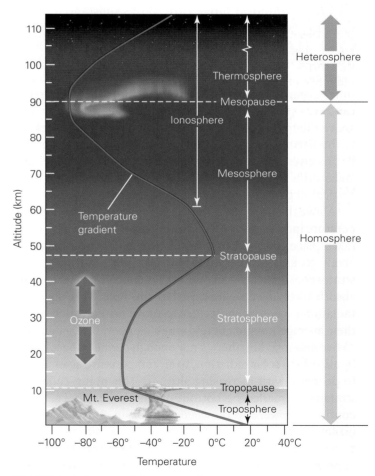

FIGURE 8.12 Layers of the atmosphere.

incoming sunlight—that's why you can see through clear air—so most solar energy passes through the atmosphere and reaches the Earth's surface, where it gets absorbed by molecules in rock, soil, or water. These surface materials then reradiate the energy back into the air as longwave infrared radiation, some of which gets trapped by molecules of certain trace gases, such as H_2O, CO_2, and methane (CH_4), causing the air to warm up. (We'll discuss this *greenhouse effect* in more detail in Chapter 15.) Put another way, the Earth's surface acts as a heater that warms the air at the base of the atmosphere—in effect, air is "baked from below."

Moving upward in the atmosphere, as distance from the Earth's surface—and thus from the atmosphere's heat source—increases, the air gets cooler. For this reason, if you climb a tall mountain, you'll find that as your elevation increases, the air temperature decreases. This decrease in temperature with increasing altitude continues up to about the cruising altitude of commercial jets, where air temperature ranges between −40°C and −60°C (−40°F and −76°F). Meteorologists refer to this lowest layer of the atmosphere, where temperature decreases with altitude, as the **troposphere**. The change of temperature with altitude in the troposphere, a quantity known as the **environmental lapse rate**, varies substantially from place to place and from time to time, but on average, it has a value of 6.5°C/km (3.5°F/1,000 ft). The thickness of the troposphere depends on the amount of solar energy arriving at the surface of the Earth, so it varies with latitude and time of year. For example, at the hot equator, the troposphere can be as thick as 20 km (12.5 mi), whereas during winter above the frigid poles, it's only about 7 km (4.3 mi) thick. All of the Earth's storms occur within the troposphere, for it is in this layer that the vertical air currents that create clouds and storms are possible. (We'll discuss the sources of these vertical air currents in the coming chapters.)

Above the *tropopause*, the top of the troposphere, air temperature increases as altitude increases. The layer where this takes place, the **stratosphere**, reaches up to a height of about 50 km. Temperature increases with altitude in the stratosphere because ozone (O_3) molecules in that layer absorb ultraviolet radiation from the Sun. This absorbed radiation causes air molecules to move faster, which makes their average kinetic energy—their temperature—increase. Above about 50 km, there's not enough ozone to heat the air by absorbing solar radiation, so air temperature again begins to decrease with increasing altitude. This point, called the *stratopause*, marks the bottom of the **mesosphere**, which continues upward to the *mesopause*, at an altitude of about 90 km (56 mi). Above the mesopause lies the outermost layer of the atmosphere. In this layer, the **thermosphere**, temperature increases with increasing altitude, reaching a maximum of 1,500°C (2,732°F) as the few remaining air molecules absorb high-energy radiation from the Sun.

Atmospheric scientists sometimes use two other characteristics to divide the atmosphere into layers: chemical composition and electric charge. In the *homosphere*, which encompasses the troposphere, stratosphere, and mesosphere, air circulation mixes the atmosphere so thoroughly that its chemical composition is more or less the same from place to place. In the *heterosphere*, which coincides with the thermosphere, gases sort gravitationally according to their molecular weight, forming layers with lighter gases above and heavier ones below. Meteorologists recognize another layer, the *ionosphere*, that is distinguished by its high concentration of ions. We'll revisit the higher layers of the atmosphere in Chapter 16, when we examine potential hazards from space and from the Sun.

Take-home message . . .
The atmosphere can be divided into four layers, the troposphere, stratosphere, mesosphere, and thermosphere, delineated by boundaries where the direction of temperature change with altitude reverses. All weather occurs in the troposphere.

QUICK QUESTION Why is it so cold at the altitude where commercial jets fly?

8.5 Clouds and Precipitation

From the vantage point of an astronaut in the International Space Station, one of the most striking features of the Earth is the pattern of its clouds (**FIG. 8.13**). What do clouds consist of? As we've seen, they are not composed of pure water vapor, a transparent gas. Rather, a **cloud** is a visible collection of countless water droplets and/or tiny ice crystals suspended in the sky, and these particles reflect, scatter, and absorb light. Let's examine how clouds form and what conditions lead certain clouds to produce rain, snow, or hail. With an understanding of clouds, we can then explore storm formation.

Cloud Formation

We have already seen that air can hold only a certain amount of water vapor before it becomes saturated, and that the amount it can hold depends on its temperature: warm air can hold more water vapor than cold air. Thus, if saturated air cools, it becomes *supersaturated*, and when this happens, water molecules begin to attach to the surfaces of aerosols. If the temperature is above freezing, an aerosol can serve as a *condensation nucleus*, allowing a droplet of liquid water to

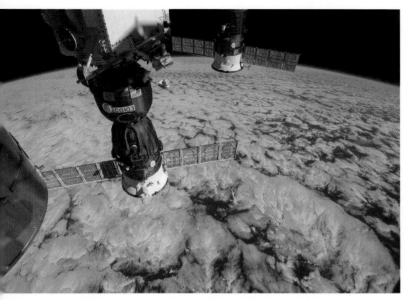

FIGURE 8.13 A view of the Earth and its clouds from the International Space Station.

FIGURE 8.14 Several mechanisms can cause air to rise.

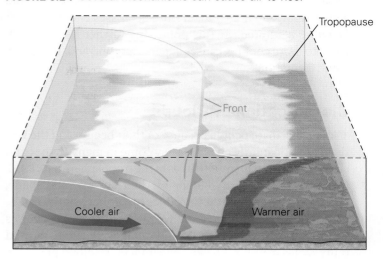

(a) Warm air flows up and over a mass of cooler, denser air.

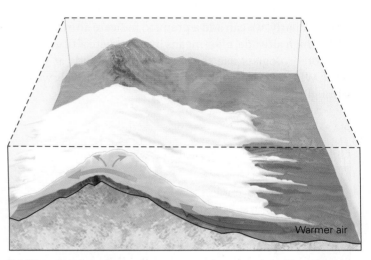

(b) When flowing air reaches a mountain range, it is forced to rise.

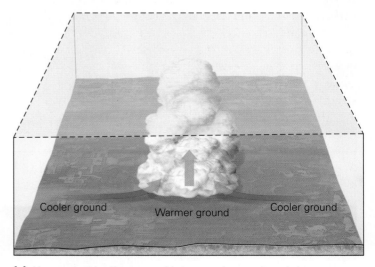

(c) Air warmed by the ground becomes buoyant and rises. Cooler air then flows in to replace it.

grow as more and more water molecules attach. If the temperature is below freezing, some aerosols can serve as *ice nuclei*, to which water molecules attach to produce crystals of solid ice. Water vapor is invisible, but water droplets and ice particles are visible—so, simply put, clouds form when water molecules in supersaturated air collect on aerosols to form tiny liquid droplets and ice particles. Every raindrop and snowflake you see formed initially on an aerosol. If there were no aerosols, there would be no rain, and life on land would never have evolved.

How does air become locally supersaturated? Generally, supersaturation happens where air rises. Air rises when it undergoes *lifting* by any of several processes that force air upward. Lifting can happen, for example, when less dense, warm air encounters and flows up and over a volume of denser, cooler air **(FIG. 8.14a)**. A similar phenomenon happens along a shoreline when cool air moves onshore as a sea breeze and lifts warm air. Air can also rise as it encounters and flows over a mountain range **(FIG. 8.14b)**. And air can rise when it becomes *buoyant*, meaning that it becomes less dense than its surroundings **(FIG. 8.14c)**. Buoyancy can develop, for example, when heat rising from the ground below warms air near the ground surface. The warmed air becomes less dense than the surrounding, cooler air, and therefore it rises. You've seen an example of this process if you've ever seen a hot-air balloon head skyward.

Atmospheric pressure decreases with altitude, so rising air expands as it moves upward. In order to expand, the air must push against surrounding air molecules, a process that takes energy. So the rising air loses energy and becomes cooler. Cooling, in turn, reduces the air's capacity to hold water vapor, and if its temperature decreases enough, the air becomes saturated. At this point, any further rising, which leads to more cooling, causes supersaturation and, therefore, cloud formation.

Classification of Clouds

Clouds fascinate people because of their beauty and the way that they constantly change. They come in many shapes and sizes, and form at many altitudes in the troposphere (FIG. 8.15). Meteorologists classify clouds by their altitude and shape, and by whether or not they produce precipitation. For example, *cirrus clouds* are high and wispy, *stratus clouds* occur as sheets that cover broad areas, and *cumulus clouds* are tall and puffy with a cauliflower-like appearance. If a cloud produces precipitation, we add the suffix *–nimbus* to its name, so a *cumulonimbus* cloud is a towering puffy cloud that produces rain. Similarly, we can add a prefix to indicate altitude: *cirro–* means high altitude, *alto–* means intermediate altitude.

Meteorologists use the term **fog** for clouds at ground level. When you walk through fog, you're effectively walking through a cloud.

Precipitation and Its Causes

Unless you live in a bone-dry desert, you've probably experienced precipitation (rain, hail, or snow) many times, and you're probably familiar with the common terms (drizzle, downpour, flurry, light, heavy) used to describe precipitation. Let's consider the conditions that lead to precipitation.

WARM-CLOUD PRECIPITATION. Meteorologists define a *warm cloud* as one in which the temperature stays above 0°C throughout, so that the cloud consists only of liquid water droplets (see Table 8.1). If the droplets in such a cloud remain small, they stay suspended in air and the cloud doesn't precipitate. Rather, the cloud survives until its droplets evaporate and become invisible gas again. If, however, cloud droplets start consolidating, they can grow into **raindrops**, spheres of water greater than 0.5 mm in diameter, heavy enough to overcome air resistance and fall toward the ground. Raindrops continue growing as they fall by combining with other drops (FIG. 8.16a). Generally, they can't get bigger than about 4 mm across, because once they do, colliding with other droplets causes them to break apart. On a few occasions, very large raindrops reaching a diameter of 8 mm have been recorded. **Rain** consists of a vast collection of

FIGURE 8.15 The most common cloud types. Clouds come in a variety of sizes and shapes.

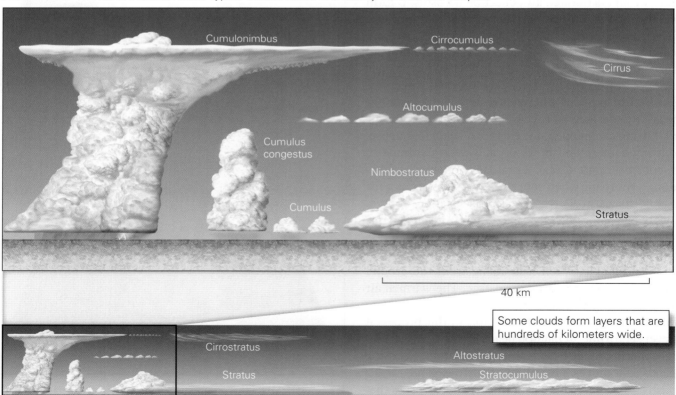

Some clouds form layers that are hundreds of kilometers wide.

FIGURE 8.16 Formation of precipitation.

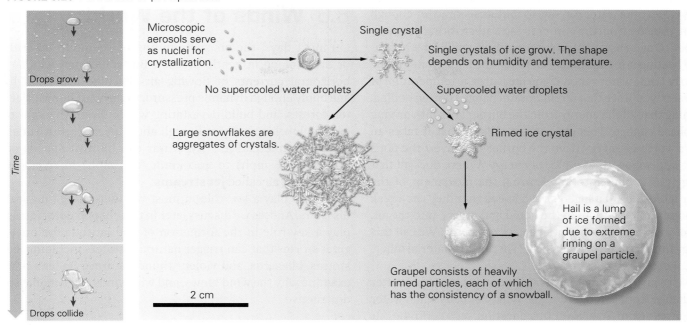

(a) Rain formation.   (b) Ice particle growth in a cold cloud.

raindrops—during a downpour, trillions of raindrops can fall on a single square kilometer of the Earth's surface.

COLD-CLOUD PRECIPITATION. Many clouds extend up to altitudes where atmospheric temperatures fall below 0°C. In such *cold clouds*, ice particles grow in one of two ways, depending on whether or not clouds contain *supercooled droplets*, meaning droplets that remain in a liquid state even though their temperature has dropped well below 0°C (FIG. 8.16b).

During direct growth of ice crystals, water molecules attach to solid ice nuclei and immediately become part of the solid. This process dominates only when there are sufficient ice nuclei of appropriate composition in the cloud. (Aerosols of appropriate composition for ice formation are those, such as salts or minerals, whose crystal structure resembles water ice, because such aerosols provide good attachment sites for water molecules.) As more and more water molecules attach, an ice nucleus grows into a hexagonal ice crystal. Ice crystals can grow into a wide variety of beautiful shapes, depending on the temperature and humidity. When they collide with neighboring ice crystals, they latch onto one another and clump together. The resulting composite of numerous ice crystals becomes a **snowflake**, which can be heavy enough to fall to the ground.

If sufficient ice nuclei do not exist in a cold cloud, water in the cloud remains in the form of supercooled water droplets. When an ice crystal does happen to form on an ice nucleus, and starts to fall through the cloud, the supercooled droplets colliding with it instantly freeze onto the ice to form *rime*, a coating of frozen droplets, on the crystal. Continued accumulation of rime produces soft spheres of ice called *graupel*. A graupel ball that continues to grow and harden becomes a **hailstone**. As we'll see in Chapter 9, hail can cause substantial damage.

Clouds and Energy

Recall that atmospheric water can exist in three states: gas (vapor), liquid, and solid (ice). A **phase change** happens when water in one state converts to another (TABLE 8.3). Phase changes happen when air containing water cools or warms, or dry air mixes into a cloud along its boundaries. Because phase changes don't take place instantly, and because temperature varies with altitude, a single cloud can contain all three phases at once.

Phase changes absorb or release energy. To understand the relationship between a phase change and energy, imagine a simple experiment. Put a pot full of water in the refrigerator

TABLE 8.3 Terminology for Phase Changes

Term	Nature of the change	Energy transfer
Melting	Solid → liquid	Absorbs heat
Freezing	Liquid → solid	Releases heat
Evaporation	Liquid → gas	Absorbs heat
Condensation	Gas → liquid	Releases heat
Sublimation	Solid → gas	Absorbs heat
Deposition	Gas → solid	Releases heat

to cool it, then take it out and place it on the red-hot burner of a stove. Heat the water for, say, 10 minutes—time for enough heat to transfer from the burner to the water to bring the water to a boil. Once the water has come to a boil, you'll have to continue heating it for another hour for all the liquid water in the pot to become water vapor. If you measure the temperature of the water, you'll see that it remains unchanged during boiling, meaning that all the energy from the burner goes into driving the phase change. This experiment tells us that it takes an hour's worth of energy from the burner to convert the pot's water into vapor—that's a lot of energy! Where does all that energy go? It goes into accelerating the movement of the water molecules to a high enough speed that they can break the bonds holding them in the liquid and escape into the air. Therefore, water vapor contains a lot of energy. We call this energy **latent heat** ("hidden heat") because a material inherits the heat from a phase change.

We've just seen how water vapor contains latent heat after the change from a liquid to a gas. What happens when water condenses again? To answer this question, let's continue our experiment. Imagine that you could magically force all the water vapor molecules that had boiled out of the hot pot back into the pot instantly so that they all became part of a liquid again. The latent heat in the water vapor (representing all the heat that had been added to the water from the burner) would be suddenly released all at once, and the pot (and probably the kitchen) would explode!

The experiment we've just described may seem odd, but the processes it describes occur in the atmosphere all the time. Water in the oceans absorbs solar energy, evaporates, and enters the atmosphere as vapor. This vapor stores the solar energy as latent heat. When clouds form, water vapor condenses to liquid, thereby releasing its latent heat. Its release, in turn, warms the air. Similarly, since it's necessary to add heat to ice to get it to melt, liquid water stores latent heat that will be released when the liquid water freezes. When clouds evaporate, they extract heat from the atmosphere to cause the phase change. We'll see in the next two chapters that the energy from these phase changes drives the development of the Earth's violent storms.

Take-home message . . .
Clouds form when water molecules attach to aerosols and form either liquid droplets or ice crystals. Droplets and crystals grow both by absorbing water molecules from the air and by collecting and aggregating together. When they become large enough, they fall as precipitation. Processes involved in cloud formation and precipitation absorb or release large amounts of energy.

QUICK QUESTION How does ice formation in cold clouds differ from water droplet formation in warm clouds?

8.6 Winds of the World

On most days, the air surrounding us seems to be still, or drifts along in gentle breezes. But sometimes winds—horizontal currents of flowing air—become fast enough, and apply enough dynamic pressure, to rip apart homes, flatten forests, and build devastating waves. Even when air at ground level is nearly still, high above us, at the altitudes where commercial jets fly, winds can blow at speeds of 90 km/h (56 mph) to 400 km/h (248 mph) in rivers of fast-moving air called **jet streams**.

Winds play a key role in most weather-related natural disasters. And, as we discuss later in this chapter, jet streams play a key role in the formation of mid-latitude cyclones, great storms that can trigger natural disasters including ice storms, blizzards, and violent thunderstorms. So let's first examine why the wind blows, and why winds can become so destructive.

Forces Acting on Air

Why does the wind blow? Three forces act to influence the direction and speed of the wind at a location—the pressure-gradient force, the frictional force, and the Coriolis force—as we now see.

PRESSURE-GRADIENT FORCE. The air pressure at one location in the atmosphere may differ from that at another location at the same altitude. These differences can develop, for example, due to variations in heat input into the base of the atmosphere—solar radiation provides more heat at the equator than at the poles, and land and sea absorb and release heat at different rates—and the resulting variations in air temperature can cause variations in air density. A difference in pressure at a given altitude causes a **pressure gradient**, defined as the change in pressure divided by the distance over which that change occurs. Where a pressure gradient exists, air will start to flow from the location of higher pressure to the location of lower pressure because more collisions happen on the side of an air molecule where the pressure is greater, so all the molecules are driven in the direction of lower pressure. (You can simulate such flow by opening the end of a long balloon full of air—the air moves from inside the balloon, where the pressure is higher, to outside the balloon, where the pressure is lower.) Because the velocity of air changes as this flow takes place, a pressure gradient causes air to accelerate, and the force causing this acceleration is the *pressure-gradient force*.

We can represent pressure gradients at a given altitude on a map by plotting contour lines, or **isobars**, along which all points have the same air pressure. The wind's speed depends on the magnitude of the pressure gradient, so where isobars lie closer together, meaning that the pressure gradient is larger (or steeper), faster winds blow **(FIG. 8.17a)**, and where isobars lie farther apart, meaning that the pressure gradient is smaller (or gentler), slower winds blow **(FIG. 8.17b)**.

FRICTIONAL FORCE. When you picture a *frictional force*, you might think of the force that slows a book that you've slid across a table, or that stops slip on a fault. Friction occurs in the atmosphere, too, either when a volume of faster-moving air shears against a volume of slower-moving air, or when a volume of air shears against the Earth's surface, or against trees and buildings rising from the Earth's surface. The frictional force always acts in the direction opposite to an object's motion, so friction reduces the speed of air flow. This force increases near the Earth's surface, both because moving air interacts with the surface and because air density increases toward the base of the atmosphere, so there are more molecules close to the surface to interact with it. The rougher the surface, and the more variable its elevation, the greater the frictional force. Because more friction exists between moving air and a city of tall buildings than between moving air and calm ocean water, winds tend to be stronger, overall, out over the ocean (see Fig. 8.10), than over adjacent coastal land. Atmospheric scientists refer to the layer of air adjacent to the Earth's surface, where friction significantly affects air movement, as the **boundary layer**. Over a smooth surface, the boundary layer is less than a half a kilometer thick, but over rough topography or over a city, it can be a few kilometers thick.

CORIOLIS FORCE. The Earth rotates on its axis once a day, and because of this rotation, a point on the equator moves at a speed of 1,670 km/h (1,038 mph), relative to an observer outside of the Earth. This speed decreases toward the poles, where it reaches zero, because a point at a higher latitude (closer to the poles) doesn't have as far to go in a day's

FIGURE 8.17 Isobars are contour lines used to represent air pressure at a particular altitude on a map.

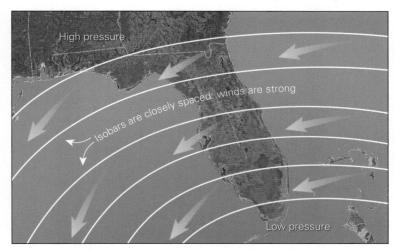

(a) Closely spaced isobars depict a steep pressure gradient and strong winds.

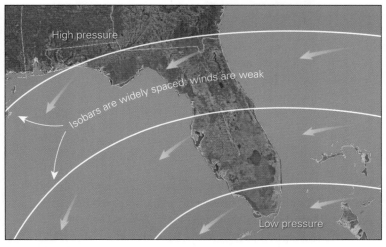

(b) Widely spaced isobars depict a gentle pressure gradient and weak winds.

FIGURE 8.18 A simplified explanation of the Coriolis effect.

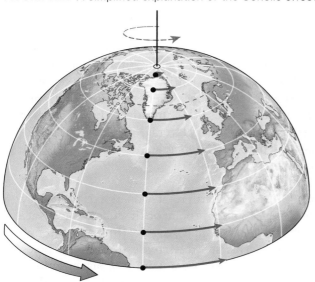

(a) Although the entire Earth rotates once a day, the velocity of any point on its surface depends on its latitude.

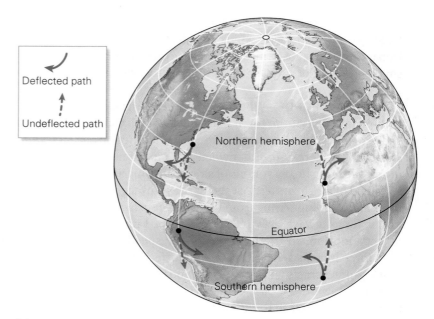

(b) In the northern hemisphere, the Coriolis effect causes objects initially moving north, from a lower to a higher latitude, to deflect to the right (east). Objects initially moving to the south, from a higher to a lower latitude, also deflect to the right (this time, west). In the southern hemisphere, objects moving north deflect to the left (west) and objects moving south also deflect to the left (this time, east).

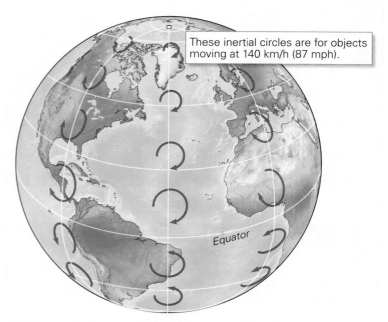

(c) Due to the Coriolis effect, an object that keeps moving on the Earth's surface follows a circular path called an inertial circle. The diameter of the circle depends on the object's latitude and speed.

rotation as a point at a lower latitude (**FIG. 8.18a**). Because of the Earth's rotation, the instant that air starts to move, it deflects away from its initial straight-line path and starts to travel in a different direction, following a curved path. The deflection of any moving object, including air, relative to the surface of a spinning object beneath is called the **Coriolis effect**, named for a French physicist.

You can picture the consequences of the Coriolis effect on the Earth simplistically. Imagine that you launch a projectile (such as a cannonball) from a higher latitude in the northern hemisphere in a due-south direction toward a lower latitude (**FIG. 8.18b**). The instant that the projectile starts on its journey, it's not only moving southward, but it's also moving around the Earth's axis at the same speed as the Earth at its launch point. As the projectile heads south, it moves eastward more slowly than the Earth below it, so from the perspective of an observer on the Earth below, it appears to deflect to the right (west) (see Fig. 8.18c). If, instead, you launch the projectile northward, from a lower latitude to a higher latitude in the northern hemisphere, the projectile not only has northward motion, but also has eastward motion. Therefore, as soon as it has moved a distance to the north, it's moving eastward at a rate faster than the Earth below it. This means that, relative to a point on the Earth below, it deflects to the right (east). If you were to repeat these experiments in the southern hemisphere, an object initially heading south would deflect to the left (east), and an object initially heading north would deflect to the left (west).

All moving objects have **inertia**, defined as the tendency of an object to remain in motion at the same velocity unless acted on by an outside force. If there were no friction, so that the object did not slow down, it would soon start moving in a circle, known as an *inertial circle*, whose diameter would depend on latitude and on the object's velocity (**FIG. 8.18c**).

Note that specification of the **velocity** of an object indicates both the speed of its motion and the direction of its motion. (We can represent velocity with an arrow, where the length of the arrow gives the speed and the orientation of the arrow gives the direction.) So, from the perspective of the projectile we described in the previous paragraph, the Coriolis effect seems to be due to a push by some invisible force, for only an external force can cause a moving object to deflect from its initial straight path and to change its direction of motion. The apparent force causing the deflection is called the **Coriolis force**, and it has the following properties: (1) it causes a deflection to the right in the northern hemisphere and a deflection to the left in the southern hemisphere; (2) it affects the direction in which an object moves across the Earth's surface, but not its speed; (3) its magnitude increases as the speed of the object increases; and (4) it has a value of zero on the equator and a maximum at the poles.

While our description above provides a useful visual image of the Coriolis effect, it doesn't provide a technically complete explanation of it. A more complete physics-based explanation reveals that the Coriolis effect results from the imbalance between changing gravitational and centrifugal forces acting on an object as the relative magnitudes of these forces change with latitude. Such details are not required to understand natural disasters as presented in this book.

The Effect of Heating and Cooling on Surface Air Pressure

As we've noted, the Earth's surface absorbs incoming solar energy, then reradiates it as infrared energy, which heats air at the base of the atmosphere. Consequently, that near-surface air expands and decreases in density. This change produces low atmospheric pressure at the Earth's surface. To see why, imagine a situation (at Time 1) in which temperature and, therefore, pressure at any altitude is uniform. In a diagram representing this condition, an *isobaric surface* (an imaginary surface where all points experience the same air pressure) at any altitude is horizontal, so pressure decreases uniformly with altitude (FIG. 8.19a). Now imagine a situation (at Time 2) in which infrared radiation rising from the Earth's surface heats a region of the atmosphere (FIG. 8.19b). Expansion of air in the heated region initially causes the isobaric surfaces at the top of the column of air to warp upward. This situation results in an outward-directed pressure-gradient force in the

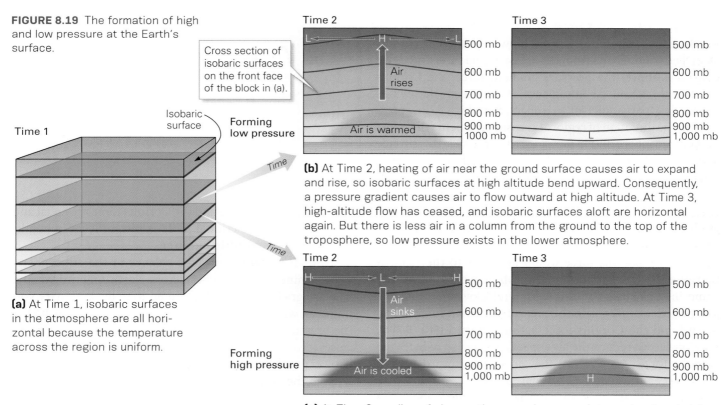

FIGURE 8.19 The formation of high and low pressure at the Earth's surface.

(a) At Time 1, isobaric surfaces in the atmosphere are all horizontal because the temperature across the region is uniform.

(b) At Time 2, heating of air near the ground surface causes air to expand and rise, so isobaric surfaces at high altitude bend upward. Consequently, a pressure gradient causes air to flow outward at high altitude. At Time 3, high-altitude flow has ceased, and isobaric surfaces aloft are horizontal again. But there is less air in a column from the ground to the top of the troposphere, so low pressure exists in the lower atmosphere.

(c) At Time 2, cooling of air near the ground causes air to contract and sink, so isobaric surfaces at high altitude bend downward. Consequently, a pressure gradient causes air to flow inward at high altitude. At Time 3, high-altitude flow has ceased, and isobaric surfaces aloft are horizontal again. But there is more air in a column from the ground to the top of the troposphere, so high pressure exists in the lower atmosphere.

upper troposphere, and air starts to flow out of the column. As a result, the total mass of air within the column decreases. Because air pressure at the Earth's surface represents the weight of the overlying column of air, the expansion associated with heating ends up decreasing the air pressure at the surface, as shown for Time 3.

The opposite effect happens when air at the surface cools (FIG. 8.19c). Once again, imagine air at Time 1 with a uniform temperature at any given altitude. Large-scale cooling—for example, over Canada or Siberia in winter—causes near-surface air to contract and increase in density. In the upper troposphere, this causes isobaric surfaces to warp downward, producing an inward-directed pressure-gradient force, so that air flows into the column (Time 2). As a result, the total weight of the air column increases, and high pressure develops at the Earth's surface (Time 3).

The creation of high pressure due to heating, or low pressure due to cooling, generates pressure gradients in the atmosphere, and therefore causes the wind to blow. Heating strongly influences air circulation in the tropics. Cooling strongly influences flows in the Earth's mid- and high latitudes, and can also affect flows in the tropics, as we will see.

What Controls Wind Direction?

In what direction does the wind blow near the Earth's surface? The wind results from the action of the three forces we discussed above: the pressure-gradient force, the Coriolis force, and friction. At first glance, it may seem that winds should flow perpendicular to isobars, from high to low pressure, in the direction of regional pressure gradients that exist in the atmosphere. But in fact, winds near the Earth's surface spiral inward around a low-pressure center; they move counterclockwise in the northern hemisphere. To understand why, suppose air was initially not moving in the vicinity of a low-pressure system. Under the influence of the pressure-gradient force, air would start moving toward the low-pressure center (FIG. 8.20a). However, once the air started moving, the Coriolis force would deflect it to the right. If there were no frictional force, the flow would eventually come into balance, rotating counterclockwise around the low-pressure center, with the pressure-gradient force acting inward and the Coriolis force acting outward (FIG. 8.20b). Friction acts to reduce the air's speed, which also reduces the Coriolis force. As a result, the air spirals inward toward the low-pressure center (FIG. 8.20c).

In contrast, winds higher in the atmosphere, above the boundary layer, where friction is no longer an important force, generally flow nearly parallel to isobars. That's because the winds at higher altitudes are controlled only by the pressure-gradient and Coriolis forces. In fact, winds will flow exactly parallel to isobars when the horizontal

FIGURE 8.20 Why air flows counterclockwise around a northern hemisphere low-pressure system.

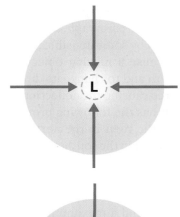

(a) If the Earth didn't spin, air would flow directly toward the low-pressure center, and would converge and rise.

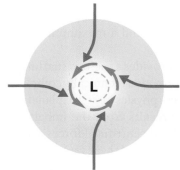

(b) Because the Earth does spin, the Coriolis force deflects air to the right, producing an overall counterclockwise rotation of wind.

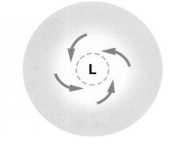

(c) In the real world, there is friction between moving air and the Earth's surface. Consequently, the overall wind follows a spiral path.

pressure-gradient force and the Coriolis force are equal and opposite and the frictional force is insignificant, a condition called **geostrophic balance**.

To understand geostrophic balance in the atmosphere, let's examine how air moves across North America when a pressure gradient exists. Picture an initially stationary volume of air within the region of a pressure gradient (FIG. 8.21a). The pressure-gradient force initially accelerates the air and causes it to move from a region of higher pressure toward one of lower pressure. However, as soon as the volume of air starts to move, the Coriolis force deflects it, so it starts to turn to the right. As a result, the air doesn't follow a path perpendicular to the isobars, but rather starts flowing at an angle relative to the isobars. The Coriolis force, and the resulting deflection, increase as the air volume accelerates to a critical speed, at which point the Coriolis force and the pressure-gradient force become equal and opposite. When this happens, the air has

FIGURE 8.21 Geostrophic wind.

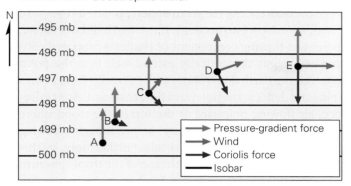

(a) A volume of air at point A starts from rest. As it moves from point A to point E and accelerates, the Coriolis force turns its path to the right until it flows parallel to the isobars.

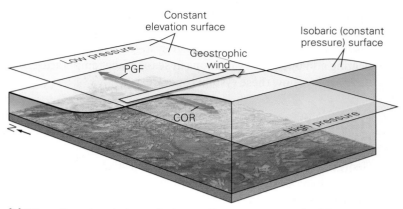

(b) When there is a balance between the pressure-gradient force (PGF) and the Coriolis force (COR), the resulting wind is called the geostrophic wind.

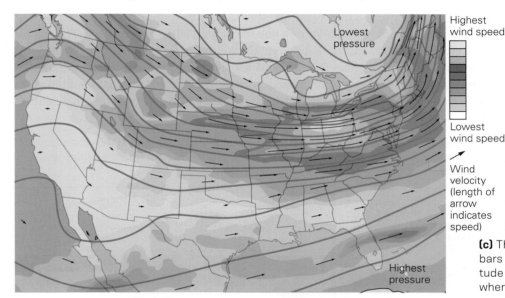

(c) The wind blows roughly parallel to the isobars at 10 km (6 mi) above sea level, the altitude of the jet stream. Winds blow faster where the pressure gradient is greater.

attained geostrophic balance (**FIG. 8.21b**), and the resulting wind, called the **geostrophic wind**, flows parallel to the isobars (**FIG. 8.21c**). Over much of the Earth, air is nearly in geostrophic balance at altitudes above the boundary layer. We'll see later in this chapter how deviations from geostrophic balance contribute to the formation of large storms in the Earth's mid-latitudes.

> **Take-home message . . .**
>
> Three forces (the pressure-gradient, frictional, and Coriolis forces) determine the speed and direction of the wind. Heating or cooling at the base of the atmosphere lowers or raises air pressure at the Earth's surface. Above the boundary layer, the wind is nearly in geostrophic balance.
>
> **QUICK QUESTION** Why is air deflected by the Coriolis force in opposite directions in the northern and southern hemispheres?

8.7 The Atmosphere of the Tropics and Subtropics

What image comes to mind when you hear the word *tropics*—warm sunny breezes and swaying palm trees? Indeed, the **tropics**—traditionally defined as the region between the Tropic of Cancer (at a latitude of about 23°N) and the Tropic of Capricorn (at about 23°S)—span the equator and are warm. They host the dense rainforests of the Amazon and Congo, but they also include a great variety of other landscapes, including deserts. As we'll see in Chapter 11, the warm waters of the tropics provide the fuel for devastating tropical cyclones (hurricanes and typhoons). The **subtropics**, a zone that traditionally includes regions between a latitude of 23° and a latitude of 35°, are also warm, but they tend to be drier and include some of the great *steppes* (grasslands) and deserts of the world. Weather in the tropics and subtropics, overall, results from global air circulation systems driven by solar heating and by related variations in sea-surface temperature. Let's explore how these systems operate.

Hadley Cells and the Intertropical Convergence Zone

George Hadley, a British lawyer who had a strong interest in meteorology, often wondered why winds blow. In 1735, he published an article in which he suggested that warming of air at the equator—where the Sun heats the Earth most intensely—would cause air to rise from low altitudes to the top of the atmosphere, where, unable to rise higher, it would start to flow toward the poles. As the poleward-moving air in the upper troposphere cooled and became denser, it would sink back toward the surface and, nearer the ground, would flow back toward the equator. The density-driven circulation that Hadley described is an example of convection; such a circulation pattern is called a **convective cell**.

Hadley's model was only partially correct, for he wasn't aware of the existence of the stratosphere, and he did not take the Earth's rotation into account. If the Earth did not rotate, the poleward-flowing component of Hadley's proposed convective cell might theoretically extend well into the polar regions. But the Earth does rotate, so the Coriolis force comes into play and deflects the poleward flow eastward. As a consequence, air flowing poleward at the top of the troposphere reaches a latitude of only about 25° before it begins flowing roughly parallel to that line of latitude. Furthermore, by this latitude, all of the air that rose to the top of the troposphere in a convective cell near the equator has begun sinking toward the Earth's surface. Similarly, as the sinking air reaches a low altitude and flows back toward the equator, it's deflected by the Coriolis force. Therefore, in the northern hemisphere, it curves southwestward, and as it approaches the equator, it flows almost due west, parallel to the equator. The mirror image happens in the southern hemisphere. This tropical surface flow constitutes the **trade winds**, steady winds that were so named because they drove merchant sailing ships westward across the oceans **(FIG. 8.22)**. Atmospheric scientists

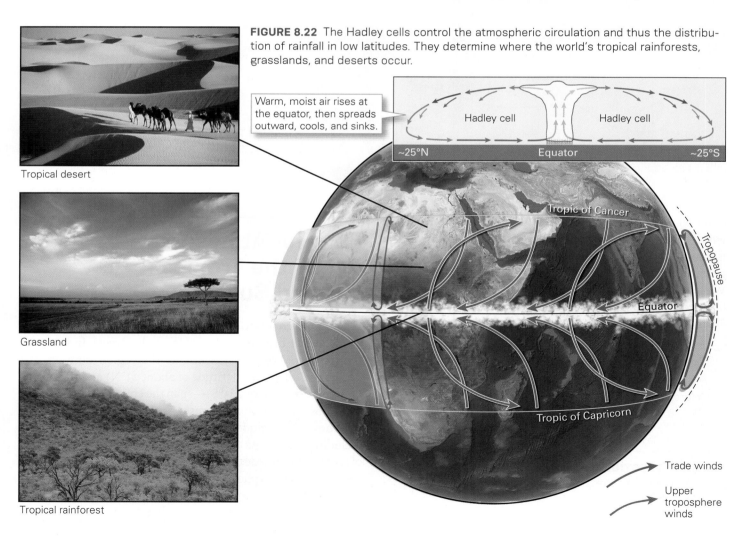

FIGURE 8.22 The Hadley cells control the atmospheric circulation and thus the distribution of rainfall in low latitudes. They determine where the world's tropical rainforests, grasslands, and deserts occur.

Tropical desert

Grassland

Tropical rainforest

312 • CHAPTER 8 • In Constant Motion: The Earth's Atmosphere

refer to this global convective cell, which extends from the equator to a latitude of about 25°, as a **Hadley cell**, in honor of George Hadley. A Hadley cell exists in each hemisphere.

Near the Earth's surface, the southwest-blowing trade winds of the northern hemisphere and the northwest-blowing trade winds of the southern hemisphere converge, producing the **intertropical convergence zone**, or **ITCZ**. The exact position of the ITCZ approximately follows the line along which the Sun's heat is most intense. Because of the Earth's tilt, the position of this line changes over the course of a year. In general, the ITCZ lies north of the equator during the northern hemisphere's summer and south of the equator during the northern hemisphere's winter. The exact position of the ITCZ also depends on the temperature of the Earth's surface below, so it is affected by ocean currents, land cover, and topography (FIG. 8.23a).

Air in the ITCZ has nowhere to go but up, so it forms the rising part of the Hadley cell. This air, because it's so warm, absorbs water evaporating from the ocean surface below. For reasons we'll discuss in Chapter 9, rising *moist air*, air with high relative humidity, can produce **thunderstorms**, localized areas of strong winds and heavy precipitation accompanied by lightning. On a satellite image, clusters of thunderstorms can be seen ringing the globe along the ITCZ (FIG. 8.23b). These storms produce the heavy rainfall that waters tropical rainforests.

Where the air of the Hadley cell descends from the cold upper troposphere to the middle and lower troposphere, it contains very little water, for most of its moisture has rained out close to the ITCZ. As this dry air descends, it compresses and warms, so its relative humidity decreases even more. Skies near the edge of the tropics and into the subtropics,

FIGURE 8.23 The intertropical convergence zone (ITCZ) forms at the junction of the Hadley cells in the two hemispheres.

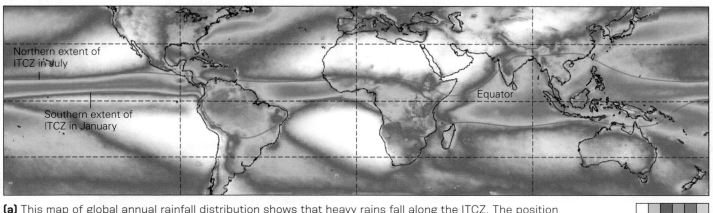

(a) This map of global annual rainfall distribution shows that heavy rains fall along the ITCZ. The position of the ITCZ varies over the course of a year.

(b) This satellite image shows thunderstorms along the ITCZ in the eastern Pacific. The inset represents the winds by streamlines. Note the trade winds curving into the ITCZ.

therefore, tend to host clear, hot, and dry weather. The steppes, for example, have a long dry season and a short rainy season—the rainy season occurs during the summer, when the ITCZ moves overhead. The great deserts, such as the Sahara, occur beneath the descending flow, where the air is driest (see Fig. 8.22).

The Southern Oscillation and El Niño

Often, when a major weather event happens, articles in the media blames El Niño or La Niña. What are these phenomena, and how do they affect the world's hazardous weather? To address the first question, we must first introduce the *Walker circulation*, a cell of flowing air that spans the Pacific Ocean along the equator. We'll defer answering the second question until later chapters.

The **Walker circulation**, named after a British physicist, extends vertically between the Earth's surface and the tropopause, and horizontally from South America's western coast to Australia and Indonesia, a distance of more than 16,000 km (9,942 mi). This circulation, which appears as a distinctive feature only within about 5° of the equator, can be thought of as a secondary, east-west-flowing convective cell within the Hadley cells. Due to the Walker circulation, air flows westward in the lower troposphere, rises over the western equatorial Pacific, returns eastward in the upper troposphere, and sinks over the eastern equatorial Pacific (**FIG. 8.24a**).

Why does the Walker circulation exist? The western Pacific Ocean tends to be warmer than the eastern Pacific because of the configuration of surface currents in the ocean. As a consequence, air over the western Pacific basin becomes warmer than that over the other parts of the Pacific, generating a region of low pressure at the Earth's surface. Meanwhile, a region of high pressure develops at the base of the atmosphere in the eastern Pacific, where the ocean water is cooler. Because the Coriolis force along the equator is very weak (it's zero at the equator), the pressure-gradient force drives air flow from the high-pressure zone of the eastern Pacific toward the low-pressure zone of the western Pacific.

When the temperature contrast between the eastern and western Pacific becomes large, so that a large pressure gradient develops, strong westward-blowing trade winds push surface water westward. This flow causes seawater to pile up so that sea level rises regionally by 10–20 cm (4–8 in) in the western Pacific relative to the eastern Pacific. Because surface water moves away from the eastern Pacific, cool, deep water wells up along the coast of South America to replace the surface water (**FIG. 8.24b**). This water brings nutrients up from the seafloor to nourish the plankton that fish eat, so fish populations thrive, and it reinforces the

FIGURE 8.24 The Walker circulation and the Southern Oscillation.

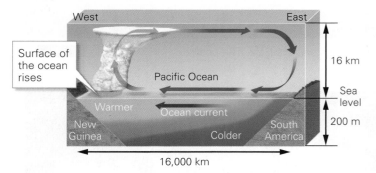

(a) Normal Walker circulation. During La Niña, this circulation strengthens.

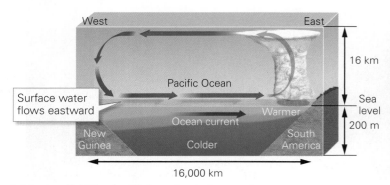

(c) During El Niño, the Walker circulation weakens.

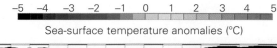

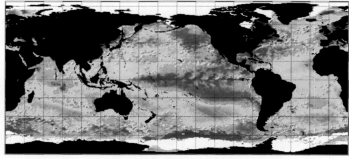

(b) Sea-surface temperature anomalies during La Niña.

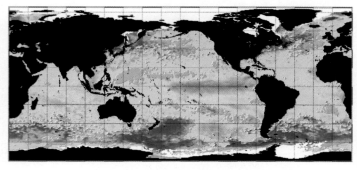

(d) Sea-surface temperature anomalies during El Niño.

formation of high pressure in the air over the eastern equatorial Pacific.

Every few years, the intensity of the Walker circulation weakens. This weakening reduces the speed of the trade winds blowing west across the equatorial Pacific. When this happens, the tilted sea surface flattens, and warm ocean water from the western Pacific spreads eastward along the equator all the way to the South American coast (FIG. 8.24c, d). As a result, the temperature of the ocean water becomes more uniform across the Pacific, so the pressure-gradient force in the air above the ocean surface decreases. In fact, surface water in the eastern Pacific may become warmer than that in the western Pacific, leading to a reversal in the pressure gradient, which in turn causes a brief reversal in the direction of the equatorial trade winds. Since surface water isn't being pushed westward by the winds, upwelling of deep water along the coast of South America stops. This situation is temporary; over the course of a year, the Walker circulation slowly strengthens, low pressure again develops in the western Pacific and high pressure in the eastern Pacific, and the trade winds again begin to blow westward.

Atmospheric scientists use the term **Southern Oscillation** to refer to the east-west seesaw in surface air pressure across the equatorial Pacific that accompanies these changes in the Walker circulation. The warming of ocean temperatures in the eastern equatorial Pacific when winds stop moving water westward has come to be known as **El Niño**, a term originally coined by Peruvian fishermen because the onset of the warming often coincides with the Christmas season (*El Niño* is Spanish for little boy, meaning the Christ Child). El Niño conditions are bad for fishermen, because without upwelling to nourish plankton, the fish population diminishes. Atmospheric scientists added the term **La Niña** (Spanish for little girl) to describe times when the Walker circulation becomes very strong and the upwelling of cold water along coastal South America increases. The acronym **ENSO** (pronounced enn-soh), which stands for El Niño/Southern Oscillation, is used for the overall seesaw pattern and its consequences. During the past 100 years, more than two dozen El Niño events have taken place—one of the strongest happened in 2015.

If it were just a local phenomenon, El Niño would not have become so famous globally. But the increase in sea-surface temperatures in the eastern Pacific affects many parts of the planet. Specifically, such warming shifts the center of tropical thunderstorm activity eastward, and these thunderstorms influence the strength of both the Hadley cells and associated weather. Therefore, ENSO modifies atmospheric circulation around the globe, even in the mid-latitudes, influencing weather worldwide. For example, El Niño triggers droughts in Australia and Indonesia, particularly rainy weather in the eastern Pacific, and flooding along South America's northwestern coast (FIG. 8.25a). In North America, El Niño can cause storms that normally move from the Pacific into Washington and Oregon to move northward toward the Gulf of Alaska, or cause storms that normally migrate to the south of the United States to move north instead, triggering floods in California and winter storms farther east. Strong La Niña conditions can similarly affect weather worldwide (FIG. 8.25b).

FIGURE 8.25 The Southern Oscillation affects global weather.

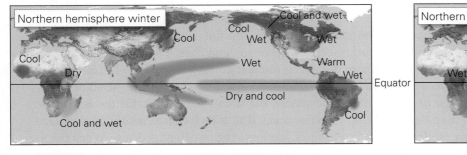

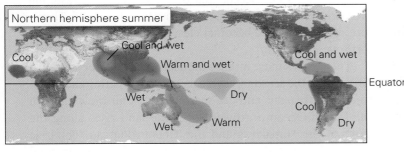

(a) Effects of El Niño.

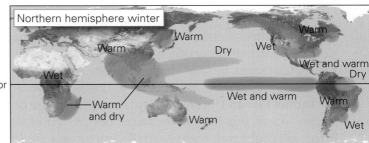

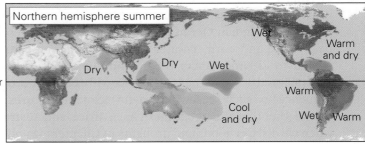

(b) Effects of La Niña.

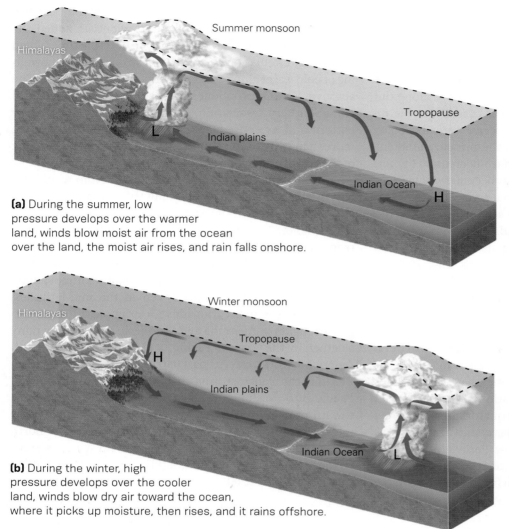

FIGURE 8.26 Formation of the monsoon in southern Asia.

(a) During the summer, low pressure develops over the warmer land, winds blow moist air from the ocean over the land, the moist air rises, and rain falls onshore.

(b) During the winter, high pressure develops over the cooler land, winds blow dry air toward the ocean, where it picks up moisture, then rises, and it rains offshore.

Monsoons

Think of the word *monsoon* and you probably picture heavy rain and overflowing rivers. In fact, the word has a more general meaning. Technically, a **monsoon** is a seasonal change in air circulation directions that results in seasonal changes in rainfall. Monsoons exist where regional winds blow over both sea and land.

Specifically, in regions that experience monsoons, summer winds blow from the ocean toward the land, and winter winds blow from the land toward the ocean. Why? During the summer months, solar energy heats the land more rapidly than it heats the ocean (**FIG. 8.26a**), for ocean water cannot gain or lose heat as quickly as rock or soil. (You may have noticed this contrast on a hot, sunny summer day—you'll burn your feet standing barefoot on concrete, but not if you step into a pool of water.) As a result, air over land becomes warmer than air over the adjacent ocean. The warm air over land expands and rises, so air pressure over land at the Earth's surface becomes less than that over the ocean. The resulting pressure gradient causes moist air to flow from the ocean toward and over the land. Once over the hot land surface, the moist air warms, becomes buoyant, and rises. The rising moist air produces drenching thunderstorms (as we'll see in Chapter 12), causing the monsoon flooding that makes headline news. During the winter, the situation reverses, for the land cools faster than the ocean (**FIG. 8.26b**). Consequently, air over land cools and sinks, forming a region of high pressure. When the air pressure over land exceeds that over the ocean, the resulting pressure gradient generates a wind that blows seaward. Therefore, air over the land becomes clear and dry, while rain clouds form over the ocean.

The world's most intense monsoon occurs over southern Asia, where the seasonal reversals in wind direction cause distinct rainy seasons and dry seasons. Moist air blowing from the Indian Ocean northward over the land rises along the front of the Himalayas and the Tibetan Plateau. This rise spawns strong thunderstorms that produce torrential

Does a phenomenon like the Walker circulation occur near the equator in other ocean basins? The answer is yes for the Indian Ocean, but not so much for the Atlantic Ocean. Meteorologists refer to the version in the Indian Ocean as the **Indian Ocean Dipole**. During the positive phase of the Indian Ocean Dipole, warmer water lies off the coast of Africa and cooler water near Australia, in a pattern similar to La Niña in the Pacific. Air consequently rises over the ocean near Africa and descends near Australia. During the negative phase of the Indian Ocean Dipole, cooler water is present near Africa and warmer water near Australia, in a pattern similar to El Niño. Under these conditions, air rises near Australia and descends near Africa. We will examine the consequences of such seesawing sea-surface temperatures on the weather that affects southern Asia, Africa, and Australia when we discuss floods and droughts in Chapters 12 and 14, respectively.

FIGURE 8.27 The southern Asian monsoon.

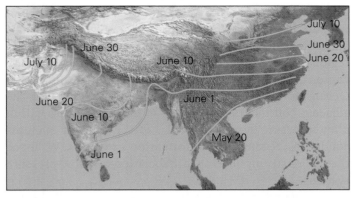

(a) The typical advance of monsoon rains in the summer season.

(b) Monsoon thunderstorms bring heavy floods.

(c) When the monsoon fails to bring rain, major droughts occur.

rainfall. Over a 2-month period, thunderstorms spread northward across Asia **(FIG. 8.27a)**, and the water they supply causes rivers to rise and flood large areas **(FIG. 8.27b)**. Monsoon rains determine the success or failure of agriculture in southern Asia—abnormally wet or dry monsoon years can destroy a region's crops and livestock and lead to economic disaster **(FIG. 8.27c)**. Monsoons also occur in other tropical and subtropical regions of South and North America, Australia, and Africa.

Take-home message...

Atmospheric circulations in the tropics are associated with the formation of large convective cells in the atmosphere. Air warmed by strong solar heating near the equator rises to the tropopause along the ITCZ and flows poleward, forming two Hadley cells, one north and one south of the equator. The surface flows of these convective cells are the trade winds. Other circulations caused by differential temperatures at the Earth's surface drive the Southern Oscillation and monsoons.

QUICK QUESTION What's the difference between El Niño and La Niña?

8.8 The Atmosphere of the Mid- and High Latitudes

People living at tropical and subtropical latitudes typically experience weeks or months of the same weather—dry during the dry season and wet during the wet season—and do not endure any days of cold and ice. Weather in the **mid-latitudes**, the regions between latitudes 30° and 60°, is quite different, for these regions host a **temperate climate**, characterized by weather that changes not only as seasons progress, but even during the course of a single day. At high latitudes, between about latitude 60° and the poles at 90°, the Earth has a **polar climate**, in which temperatures stay near or below freezing for much of the year.

The Earth's mid-latitudes serve as a battleground where the cold air of polar regions collides with the warmer air of tropical and subtropical regions, and their interaction generates immense storm systems known as *mid-latitude cyclones*. Let's explore what happens in this complex zone of air circulation. To begin this discussion, we need to describe the participants in these atmospheric battles—namely, air masses and fronts.

Air Masses and Fronts

An **air mass** is a broad body of air within which temperature and humidity are relatively uniform. Some air masses are local in scale, but at any given time, the atmosphere contains many distinct *major air masses*, each of which is several kilometers thick and a thousand or more kilometers across. Each can cover a quarter to a half of a continent or ocean. Over a period of days, they may remain nearly stationary, move slowly, or sweep across an entire continent or ocean. Meteorologists characterize air masses on the basis of their source region (continental or maritime) and their temperature. Air masses characterized by bitter cold temperatures are classified as arctic, those with cold temperatures (or cool temperatures in summer) as polar, and those with warm or hot temperatures as tropical. When an air mass moves out of its source region, a new, similar air mass is regenerates over that region. This regeneration can be caused by upward flow of heat into the air from warm ground, or by absorption of heat from the air by cold ground. In the case of maritime air masses, evaporation of water from the underlying ocean provides moisture. Seven major air masses typically lie over and around North America and influence its weather (FIG. 8.28). In winter, for example, frigid continental arctic air masses form in the Canadian Arctic and sweep southward from Canada into the United States, bringing extremely cold weather.

Recall that when air cools and sinks, high pressure develops in the air at the base of the atmosphere (see Fig. 8.19c). Therefore, cold air masses are associated with broad **high-pressure systems**. In winter, when large continental land areas such as Canada and Siberia become cooler than the surrounding oceans, semipermanent high-pressure systems develop beneath continental polar air masses generated in those regions. In summer, when the oceans tend to be cooler than the land, semipermanent high-pressure systems develop beneath maritime air masses. Examples of such high-pressure systems include the Bermuda High, over the North Atlantic, and the Pacific High, over the North Pacific. As we'll see, air in high-pressure systems tends to descend from high altitudes toward the Earth's surface. This flow causes regions beneath high-pressure systems to have clear skies.

FIGURE 8.28 Major air masses affecting North America. Maritime air masses are moist, while continental air masses are dry. Tropical air masses are warm, while arctic and polar air masses are cool.

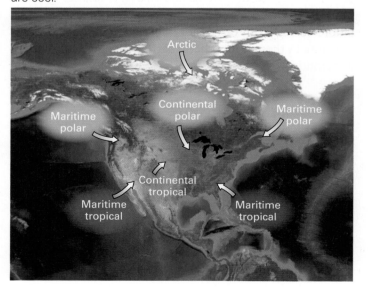

When air warms and rises, low pressure develops near the Earth's surface, so warm air masses are associated with broad **low-pressure systems** (see Fig. 8.19b). Semipermanent low-pressure systems develop over desert regions, such as the American Southwest, in summer. In winter, they form over the far northern Atlantic and Pacific Oceans, for ocean water remains warm compared with the adjacent continents. As air rises to higher altitudes within a low-pressure system, it cools, so the water it contains condenses, leading to the formation of clouds. We'll see that major storms develop in association with low-pressure systems.

The boundary between two air masses typically occurs along a narrow zone called a **front**. Fronts are sites where abrupt contrasts of temperature and, commonly, relative humidity occur. Most stormy weather in mid-latitudes, particularly during cooler seasons, develops along fronts. As moist air on the warm side of a front flows up and over the denser air on the cold side—a process called *frontal lifting*—it rises and cools, and the moisture within it condenses into precipitation.

Meteorologists classify fronts by the direction in which the colder air mass moves. At a **cold front**, a cold air mass advances under a warm air mass, forcing the warm air to rise (FIG. 8.29a). Commonly, the leading edge of the cold air mass develops a dome-like shape. Often, a line of thunderstorms develops along a cold front, and some of the storms may produce tornadoes. A **warm front** exists where a cold air mass retreats and a warm air mass advances (FIG. 8.29b). Typically, the retreating cold-air dome slopes more gradually ahead of a warm front than it does behind a cold front. An **occluded front** forms when an advancing cold front overtakes a warm front, so that the cold front lifts the cool air from behind the warm front, and the warm front no longer intersects the ground surface (FIG. 8.29c). In winter, ice storms or blizzards can develop along both warm and occluded fronts. A **stationary front** is one whose position does not move. Air on the cold side of the front flows nearly parallel to the front, but air on the warm side typically rises over the cold air (FIG. 8.29d). Along stationary fronts, the same weather may remain over a location for hours to days, a situation that can lead to disastrous floods, devastating ice storms, or heavy snowfall.

The Polar Front and the Jet Stream

We've just described cold and warm fronts as boundaries between cold and warm air masses, and we've seen that the differences between the types of fronts are based on whether the cold air is advancing or retreating. If we step back for a moment to take a global view, we can see that all these fronts are part of a much larger air mass boundary.

This boundary, which meteorologists call the **polar front**, separates warm air that originates in the tropics and sweeps across subtropical and temperate regions from cold air that originates in polar and subpolar regions. As we'll see, major storm systems called mid-latitude cyclones develop along this complex boundary.

The contrast in temperature across the polar front exerts a major influence on wind velocity at the top of the troposphere. Why? Warm air occupies more volume than cold air, so the vertical distance between isobaric surfaces is greater on the warm side of the front than on the cold side. Because isobaric surfaces are lower on the cold side of the front than on the warm side, they slope downward at the front from the warm side toward the cold side. Their slopes become progressively steeper at higher altitudes, and they are steepest near the tropopause (**FIG. 8.30**). As a result, the horizontal pressure-gradient force near the tropopause over the polar front is very strong—much stronger than it is at lower altitudes or far away from the front. Because pressure gradients, as we have seen, drive the wind, this very steep pressure gradient generates very fast high-altitude winds. The river of fast winds that flows over the polar front, at an altitude of about 10 km (6 mi), is called the **polar-front jet stream**. This jet stream does not parallel a line of latitude, but rather displays broad curves or waves in map view, so that in places it flows well into temperate latitudes. As we'll see, the positions of these wave-like undulations, which are influenced by the positions of air masses and fronts, change over time and strongly influence weather. For example, over North America, the polar-front jet stream

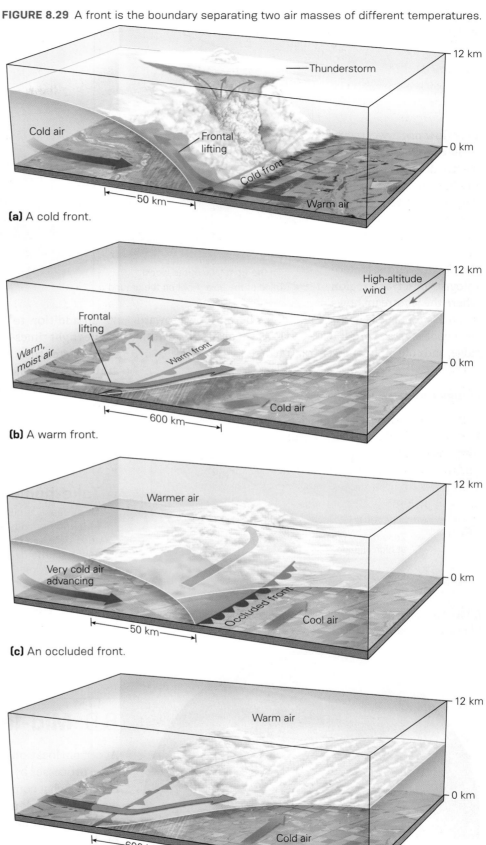

FIGURE 8.29 A front is the boundary separating two air masses of different temperatures.

(a) A cold front.

(b) A warm front.

(c) An occluded front.

(d) A stationary front.

8.8 The Atmosphere of the Mid- and High Latitudes • **319**

FIGURE 8.30 The relationship between the polar front and the polar-front jet stream. The strongest winds are at the center of the jet stream.

may flow northwestward over the Pacific Northwest, down along the Rockies into the southeastern United States, turn northeastward, and exit the United States flowing northward toward the North Atlantic. Meteorologists describe a region where the jet stream bows toward the poles as a **ridge** and one where it bows toward the equator as a **trough** (FIG. 8.31). Warm air lies beneath a ridge, and cold air lies beneath a trough.

FIGURE 8.31 Ridges and troughs in the upper troposphere.

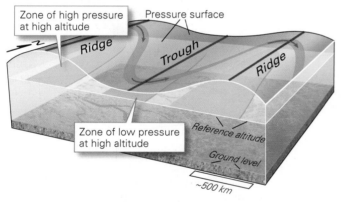

(a) An isobaric surface in the upper troposphere bows up beneath a ridge, and down beneath a trough.

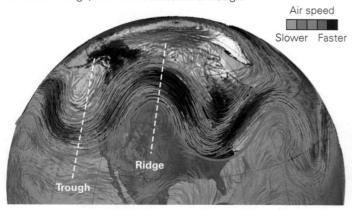

(b) Because of ridges and troughs, the jet stream has a wavy trace in map view.

In addition to the weather, the jet stream affects airplane flight times—flights moving with the jet stream travel much faster than those fighting against it. In fact, a 747 jet traveling with the jet stream made the transit from New York to London, which normally takes 7 hours, in a record 4 hours and 56 minutes—though the plane's air speed didn't exceed about 925 km/h (575 mph), at times its ground speed reached 1,287 km/h (800 mph).

Take-home message . . .

An air mass is a broad body of air within which temperature and humidity are relatively uniform. Fronts are boundaries between air masses. Meteorologists classify fronts based on the direction in which air on the cold side of the front is moving. The polar-front jet stream flows over the polar front in the upper troposphere and influences the development of mid-latitude cyclones.

QUICK QUESTION How can you identify whether a front is a cold, warm, or stationary front?

8.9 Mid-latitude Cyclones

At certain locations within the jet stream, air flow speeds up. (The processes by which air speed changes in the jet stream are somewhat complicated and not critical to understanding weather-related natural disasters, so we won't provide details.) Where air speeds up, more air goes out of a segment of the jet stream than enters that segment (FIG. 8.32). This process, called **divergence**, produces an air deficit at the top of the troposphere, and this deficit, in turn, causes the

FIGURE 8.32 The relationship between the jet stream and the formation of a mid-latitude cyclone. Low pressure develops at the Earth's surface below a zone of divergence within the jet stream, which flows near the tropopause. The colors denote the jet stream, and the white areas denote storm clouds.

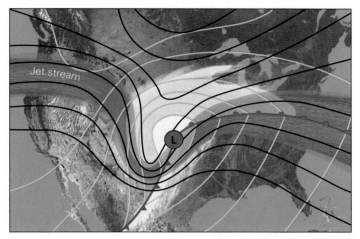

Upper-atmosphere isobars ———
Air pressure at sea level ———
Low-pressure center (L)

FIGURE 8.33 This winter mid-latitude cyclone covers the northern midwestern and southeastern United States.

weight of the underlying air column to decrease. As a result, air pressure decreases at the base of the troposphere, forming a low-pressure system, within which the position of lowest pressure defines a **low-pressure center**. The pressure gradient that develops around this low-pressure center, together with the Coriolis and frictional forces, causes air to flow counterclockwise around and spiral inward toward the low-pressure center. To replace the deficit of air at jet-stream altitude, air from lower in the troposphere rises in and around the low-pressure center. The vast weather system that forms within this spiraling region of air, along with the fronts and storms associated with it, is called a **mid-latitude cyclone** (also known as an *extra-tropical cyclone*, to contrast them with tropical cyclones). These systems are responsible for much of the devastating weather that afflicts regions with temperate climates. Mid-latitude cyclones can be as wide as 2,000 km (1,243 mi) across **(FIG. 8.33)**. To better understand these important weather systems, let's look more closely at how they evolve.

Initially, in a mid-latitude cyclone, a warm front typically extends eastward of the low-pressure center, while a cold front extends from the center toward the south or southwest **(FIG. 8.34a)**. At this stage, warm air south of the warm front flows up and over the warm front producing a widespread area of clouds and rain or snow. Meanwhile, warm air to the east or southeast of the cold front undergoes lifting, producing showers and thunderstorms as the cold air advances toward the southeast. As low-altitude air spirals counterclockwise around the low-pressure center, the cold front advances rapidly **(FIG. 8.34b)**. At the same time, the warm front migrates northward, but only slowly. In many cyclones, therefore, the cold front progressively wraps around the low-pressure center and catches up with the warm front **(FIG. 8.34c)**. When this happens, the cool air north of the warm front undergoes lifting along the face of the cold front, yielding an occluded front **(FIG. 8.34d)**. Once the occluded front forms, cold air from what is now the south side of the cold front contacts cold air from the north side of the warm front, and the warm air that once lay to the south of the warm front becomes a layer that overlies both cold air masses.

A typical mid-latitude cyclone develops and intensifies rapidly, generally reaching its maximum intensity (the lowest air pressure at the low-pressure center) within

FIGURE 8.34 Evolution of a mid-latitude cyclone.

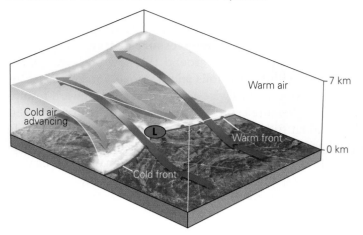

(a) Air initially flows upward over the fronts.

(b) The cold front advances southward and eastward around the center of low pressure.

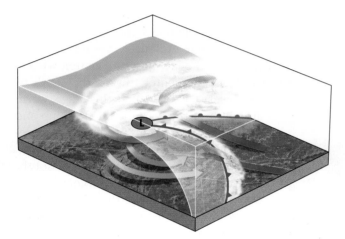

(c) Eventually, the cold front reaches the warm front.

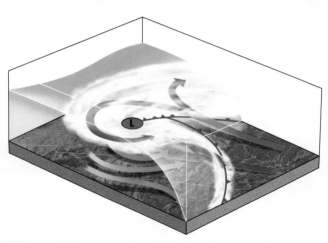

(d) The cold front overtakes the warm front, creating an occluded front.

36–48 hours of initiation. The storm can remain at maximum intensity for another day or two, but then it begins to weaken. After the occluded front forms, the cyclone will begin to dissipate, but it can continue to spin down over a period lasting for days to over a week.

The strong pressure gradient that develops in a mid-latitude cyclone generates strong winds. At ground level, these winds flow counterclockwise (in the northern hemisphere) (FIG. 8.35a). Lifting along the system's fronts causes low-altitude moist air to rise, producing broad areas of cloud cover and precipitation (FIG. 8.35b). The presence of the fronts and a low-pressure center in a mid-latitude cyclone means that, when seen from space, the distribution of clouds resembles the shape of a giant comma. The head of the comma, and the top part of the comma's curve, lie along and over the cyclone's warm and occluded fronts. In winter, this area is the potential home of blizzards and ice storms (FIG. 8.35c). The long tail of the comma corresponds to the trace of the cold front, and it typically consists of a line of precipitating clouds. Along the comma tail, thunderstorms, devastating tornadoes, and windstorms are possible (FIG. 8.35d,e).

Take-home message . . .

Mid-latitude cyclones are large, counterclockwise-rotating systems that can result in hazardous weather across the mid-latitudes. They form along the polar-front jet stream in association with low-pressure centers. The most dangerous weather occurs along fronts, which move as the result of air flow within the cyclone.

QUICK QUESTION Why does the weather in the mid-latitudes change so frequently?

FIGURE 8.35 Features of a large mid-latitude cyclone.

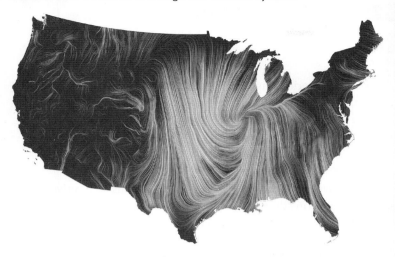

(a) A wind map shows the circulation of air in a mid-latitude cyclone.

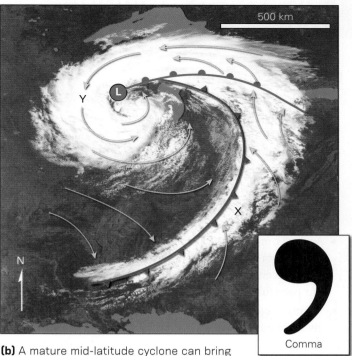

(b) A mature mid-latitude cyclone can bring different types of hazardous weather over broad geographic regions.

(c) A blizzard occurs at point Y on map (b).

(d) When mid-latitude cyclones affect coastal areas of New England, they are known as nor'easters, because at the height of the storm, winds come from the northeast.

(e) A thunderstorm occurs at Point X on map (b).

Chapter 8 Review

Chapter Summary

- Dry air consists mostly of molecular nitrogen (78%), molecular oxygen (21%), and trace gases such as argon, carbon dioxide, ozone, and sulfur dioxide (1%). Although present in small quantities, trace gases are important in many atmospheric processes.
- Water is found in three forms in the atmosphere: as an invisible gas called water vapor, as a liquid in clouds and raindrops, and as a solid in the form of ice, snow, or hail.
- Aerosols, microscopic particles in the atmosphere, are introduced from both natural sources and human activities. In high concentrations, they contribute to air pollution, both as fine particulate matter and as photochemical smog.
- Weather refers to the specific atmospheric conditions at a given time and location. We describe weather using properties such as temperature, pressure, relative humidity, wind direction and speed, visibility, cloud cover, and precipitation.
- Meteorologists monitor and warn of impending weather hazards using radar systems and satellite images. To measure conditions high in the atmosphere, meteorologists launch balloons carrying rawinsondes.
- The atmosphere is divided into four layers, delineated by boundaries at which the direction of temperature change with altitude reverses: the troposphere, stratosphere, mesosphere, and thermosphere. All weather occurs in the troposphere.
- Clouds form when air is lifted or when it becomes buoyant and rises.
- Clouds form when water molecules in supersaturated air collect on aerosols to form liquid droplets and/or ice particles.
- Cloud droplets grow into raindrops by coalescing with neighboring droplets. Ice particles grow to precipitation size by first incorporating water molecules, and then by either linking together to form snowflakes or by collecting supercooled water droplets to form rimed ice particles, graupel, or hail.
- Water in the atmosphere constantly undergoes phase changes, releasing or absorbing latent heat in the process.
- Three forces—the pressure-gradient force, the frictional force, and the Coriolis force—together generate and control the direction and speed of winds in the atmosphere.
- Atmospheric circulations in the tropics are associated with the formation of large convective cells in the atmosphere. These Hadley cells form when air warmed by strong solar heating near the equator rises to the tropopause along the ITCZ and flows poleward, then descends and returns toward the equator as the trade winds.
- The Walker circulation is an east-west-flowing convective cell covering the equatorial Pacific. An El Niño event occurs when the Walker circulation weakens, and a La Niña event occurs when it strengthens. The resulting pattern of changes in surface air pressure, called the Southern Oscillation or ENSO (El Niño-Southern Oscillation), can affect weather around the world.
- A monsoon is a seasonally changing air circulation. The largest monsoon is located over southern Asia, where it brings heavy rains in summer and dry weather in winter.
- A major air mass is a body of air with relatively uniform temperature and humidity that covers a large region of the Earth.
- Major air masses associated with semipermanent high-pressure systems develop over regions where air is cooled, such as Canada in winter.
- Major air masses associated with semipermanent low-pressure systems develop over regions where air is warmed, such as the oceans south of Iceland and the Aleutian Islands in winter.
- The boundaries between air masses are called fronts. Meteorologists distinguish among cold fronts, warm fronts, and stationary fronts by the direction in which the colder air mass is moving. If a warm front lifts over a cold front, an occluded front forms.
- The polar-front jet stream is a narrow band of strong winds that encircles the Earth in the upper troposphere and lies over the polar front, the boundary between tropical and polar air masses within the mid-latitudes.
- Mid-latitude cyclones form in the mid-latitudes when a low-pressure center develops at ground level beneath a zone of divergence in the jet stream.
- From above, the clouds of a mid-latitude cyclone have the shape of a comma. The comma tail follows a cold front, and along it, thunderstorms often form. The comma head overlies warm and occluded fronts, where rain, snow, or freezing rain may fall. Blizzards and ice storms can occur under the clouds of the comma head.

Review Questions

Blue letters correspond to the chapter's learning objectives.

1. While in a large city, you step outside in the afternoon and take a breath of air. What gases and other substances are contained in that breath? **(A)**
2. Explain why aerosols are needed for life on the Earth to exist, but why high concentrations can threaten human health. **(A)**
3. Explain why a person may feel uncomfortable and sticky in Florida when the temperature is 90°F, but feel quite comfortable in Arizona when the temperature is also 90°F. **(B)**
4. Describe the weather at your location today. Which properties of the atmosphere discussed in this chapter did you use in your report? **(B)**
5. The atmosphere consists of four distinct layers. List these layers in order, from the bottom to the top of the atmosphere. In which layer does all weather occur? In which layer is a large concentration of ozone found? **(C)**
6. Would you expect the tropopause to be higher or lower in the tropics compared to the polar regions? Explain your answer. **(C)**
7. Why do clouds normally form over cold fronts, such as the front shown here? **(D)**
8. What does a cloud consist of? **(D)**
9. When water vapor condenses into liquid droplets to form a cloud, is heat released to the atmosphere, or absorbed from the atmosphere? How about when liquid droplets freeze to become ice crystals? **(D)**
10. When a cloud evaporates, does the air containing the cloud become cooler or warmer? **(D)**
11. What forces act on air and cause it to flow? Which of these forces can cause the wind to increase in speed? **(E)**
12. Of the three forces discussed in the chapter, one causes air to speed up, one to slow down, and one to change direction. Which force causes which effect? **(E)**
13. What is a Hadley cell? Where are the Hadley cells located on the Earth, and how do they vary with the seasons? **(F)**
14. If you lived in a region that experiences monsoons, how would the weather you experience vary throughout the year? **(F)**
15. Suppose you book a vacation on an island in the western Pacific near the equator. If you want sunny days, would you be better off booking during an El Niño or a La Niña? **(F)**
16. Does this figure represent conditions during La Niña or El Niño conditions? **(F)**
17. What is an air mass, and what are the major source regions for air masses in North America? **(G)**
18. What is a boundary between two air masses called, and how can we distinguish among different types of these boundaries? **(G)**
19. What types of weather hazards are associated with stationary fronts? **(G)**
20. What is a mid-latitude cyclone, why does it form, and how is it related to the polar-front jet stream? What kinds of weather form in association with mid-latitude cyclones? **(G)**
21. What is the shape of a mid-latitude cyclone? Which types of weather typically occur in different parts of the cyclone? **(G)**

On Further Thought

22. You decide to climb a mountain and take along a thermometer and a barometer, a device that measures atmospheric pressure. During the climb, you take readings. At the end, you plot graphs of the results. What will the graphs look like? **(C)**
23. Outside, it is cool and rainy, and the wind is from the east. Two hours later, the rain has ended, the temperature has increased, and the air now feels humid. The wind is from the south. What type of front passed your location? **(G)**

9

LOCAL TEMPESTS
Thunderstorms and Tornadoes

By the end of the chapter you should be able to . . .

A. explain the conditions that lead to the formation of a thunderstorm, and describe the difference between an ordinary and a supercell thunderstorm.

B. provide an explanation for the destructive winds in thunderstorms, and for the origin of supercell thunderstorm rotation.

C. illustrate the steps in a lightning stroke, and identify safe practices to avoid being struck by lightning during a thunderstorm.

D. discuss the formation of hail, and describe the damage that hailstorms can cause.

E. describe how tornadoes form, how they are classified, and how they cause destruction.

F. explain how to protect yourself against thunderstorm hazards, including tornadoes.

9.1 Introduction

In the late morning of Tuesday, July 13, 2004, strong thunderstorms began to develop over Illinois. Meteorologists at the Storm Prediction Center of the National Weather Service (NWS) were concerned that such storms could eventually produce tornadoes, so they issued a *tornado watch* for the region, a formal statement that conditions are favorable for tornado development. Over the next few hours, even stronger thunderstorms developed, and at 2:29 P.M., the NWS issued a *severe thunderstorm warning*. A warning means that a specific storm has been observed that is capable of producing dangerously strong winds, as well as hail, and possibly a tornado.

At the Parsons manufacturing plant in Roanoke, Illinois, the NWS warning interrupted normal broadcasting on the radio in the front office, causing the factory's designated emergency response team leader to look outside. The view was terrifying—a tornado was approaching! He grabbed a microphone and bellowed the words no one wanted to hear: "This is not a drill. Move immediately to designated tornado shelters ... now!" Thanks to a training exercise three months earlier, everyone knew exactly where to go, and they moved quickly.

The plant's owner, when constructing the plant, had insisted that restroom walls and ceilings be made of reinforced concrete 20 cm (8 in) thick so that they could serve as tornado shelters. All 140 employees in the building were already on their way into these shelters when the NWS broadcast a *tornado warning*, meaning that a tornado had been sighted. This fearsome funnel of destruction, with winds blasting to nearly 290 km/h (180 mph), was less than a mile away when the employees bolted the shelter doors closed. Two minutes later, there was a deafening roar, and the tornado struck. Cars in the parking lot tumbled like toys, the roof was ripped away, and the walls of the plant collapsed. For a terrifying minute, winds howled and debris clattered and crashed. Then the sound and motion stopped—the tornado had passed. Employees emerged from shelters to find a chaotic jumble of twisted beams, smashed machines, and crushed vehicles (FIG. 9.1). But not one person was seriously injured—warnings and preparations had saved everyone's lives.

Every year, we hear of devastating thunderstorm and tornado damage, often with results much more tragic than what we've just described. Thunderstorms and tornadoes are examples of weather systems that affect a relatively small region—between a few square kilometers and several hundred square kilometers in area—but can destroy businesses and homes and leave a trail of death. Although thunderstorms and tornadoes can take place in isolation, they can also occur in clusters or lines that affect a broader area, and they are often part of much larger weather systems, such as mid-latitude cyclones and hurricanes. In this chapter, we focus on thunderstorms and related hazards, including lightning, hail, and tornadoes. We discuss how they form and evolve, and how to avoid being injured when these storms strike.

FIGURE 9.1 The Parsons plant, before and after the tornado that destroyed it.

◂ A violent tornado crosses the Great Plains.

FIGURE 9.2 Lightning flashes during an intense thunderstorm in Nebraska.

9.2 Conditions That Produce Thunderstorms

What Is a Thunderstorm?

Imagine you are sitting outside on a warm, calm summer day. Suddenly, thunder rumbles in the distance. Soon dark, turbulent clouds hide the Sun. Then gusts of cool wind begin to blow, followed by drenching rain, stronger winds, and possibly hail. As the sky crackles with lightning flashes, you hear the booms of thunder. A thunderstorm has arrived (FIG. 9.2).

A **thunderstorm**, formally defined, is a towering cloud that produces lightning and thunder. Thunderstorms are typically accompanied by strong winds, heavy rains, and sometimes hail. The towering clouds in such storms typically extend from an altitude of less than 2,000 m (6,500 ft), all the way up to the top of the troposphere, which lies at an altitude of 12–20 km (7–12 mi), depending on latitude. As we will see, a single thunderstorm contains a strong **updraft** (a vertical upward flow of air) and evolves to have strong **downdrafts** (vertical downward flows of air).

In general, a single thunderstorm is less than 30 km (19 mi) in diameter, and may move across the ground at rates of up to 80 km/h (50 mph), driven by regional winds. Therefore, the storm may pass over a particular location in under 30 minutes. Commonly, however, a location experiences a cluster or line of thunderstorms, so it may endure stormy weather for a few hours. Thunderstorms can happen at any time of day. During the daytime, the thick clouds may block so much daylight that the sky becomes almost as dark as night, while during the nighttime, lightning can momentarily fill the sky with flashing displays. Thunderstorms vary in size and intensity, but all pose danger because of the risk of lightning. The NWS defines a **severe thunderstorm** as one that produces either winds exceeding 93 km/h (58 mph), hail greater than 2.5 cm (1 in) in diameter, a tornado, or any combination of these phenomena.

Thunderstorms take place frequently over some regions of the Earth, but rarely or never over others (FIG. 9.3). Most occur in the tropics, where they provide the rain that sustains rainforests. In the temperate climates of mid-latitudes, they occur fairly frequently during warm seasons. They take place only rarely at high latitudes, over subtropical deserts, or over most of the world's oceans. In the United States, thunderstorms occasionally develop over western mountainous

FIGURE 9.3 The global distribution of thunderstorms, as estimated from lightning flashes. Note that thunderstorms concentrate in specific regions, and that most occur over land.

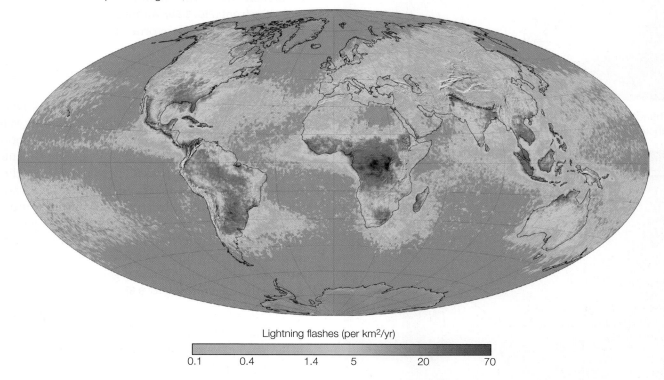

areas, but they are more common and intense over the interior plains east of the Rocky Mountains and are particularly frequent over the southeastern states. In fact, in the eastern two-thirds of the United States, any given location will experience 30–50 *thunderstorm days* (days on which at least one thunderstorm occurs) per year (FIG. 9.4). Note that a day can qualify as a thunderstorm day even if less than an hour of the day is actually stormy. Florida holds the nation's record for thunderstorm days: parts of the state endure 65–80 thunderstorm days per year.

Thunderstorms do not occur everywhere all the time, because three specific conditions must be met in order for them to form: (1) the lower atmosphere must hold *moist air*; (2) a *lifting mechanism* must be present to initiate an updraft that can transport the moist air to a higher altitude; and (3) *atmospheric instability* must be present to allow that air to rise buoyantly to the top of the troposphere. Let's consider each of these conditions in turn.

The Source of Moist Air

Amazingly, a typical thunderstorm contains nearly 4 million tons of water (equivalent to the amount of water in about 1,400 Olympic swimming pools) at any given time and, over its lifetime, it processes many times that amount. Therefore, in order for thunderstorms to develop, air in the lower atmosphere must have high relative humidity. Where does this moisture come from? Most evaporates from the oceans. For example, the water vapor fueling thunderstorms in the central and eastern United States typically comes from the Gulf of Mexico and the Atlantic Ocean—this moisture is carried over the land by regional winds.

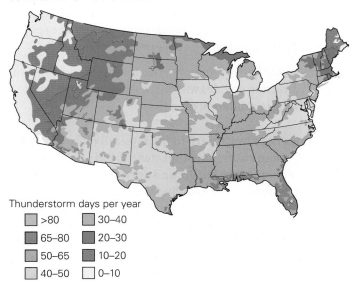

FIGURE 9.4 Number of thunderstorm days per year across the continental United States.

9.2 Conditions That Produce Thunderstorms

FIGURE 9.5 Thunderstorms may form as a result of frontal lifting.

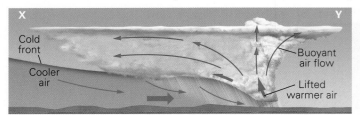

(a) Thunderstorms form as warm air is lifted ahead of an advancing cold front.

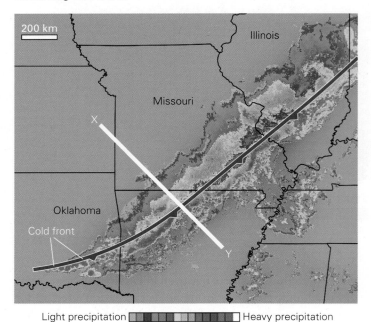

(b) A radar reflectivity image showing thunderstorms along an advancing cold front in the central United States. The line X–Y shows the location of the cross section in part (a).

Water that enters thunderstorms can also come from lakes and wetlands, damp soils, and transpiration by plants.

Lifting Mechanisms

To initiate the growth of a thunderstorm, moist air must begin to rise from the lower troposphere to a higher altitude. A process that forces air to rise is called a **lifting mechanism**. As you'll see, lifting mechanisms depend on the fact that cool air is denser than warm air (see Chapter 8). Therefore, cool air will remain near the ground and will flow underneath and lift a body of warm air. Atmospheric scientists distinguish among several types of lifting mechanisms:

1. *Lifting due to movement of a front*: In Chapter 8, we learned that a *front* is a boundary between two large air masses. The movement of advancing cool, dense air behind a cold front can physically lift warm, moist air ahead of the front (FIG. 9.5a). Lifting by an advancing cold front may produce several thunderstorms, at roughly the same time, in a line along the front (FIG. 9.5b). Although it's less common, lifting that triggers thunderstorms can also take place along a warm front.

2. *Lifting due to gust fronts*: As we'll see later in this chapter, when rain starts to fall within a thunderstorm, a downdraft of cool air rushes to the base of the storm. When this downdraft reaches the solid ground, it spreads outward to the sides of the storm to produce a dome-shaped pool of dense air, called a *cold pool*, which spreads outward into areas beyond the storm's base. The leading edge of the advancing cold pool is called a **gust front**. As it advances, the gust front acts like a local cold front, lifting warm air ahead of it, and this lifting can trigger new thunderstorms. At places where gust fronts from different initial storms collide, or where a gust front and a regional cold front collide, focused lifting may generate particularly intense thunderstorms (FIG. 9.6).

3. *Lifting along less distinct boundaries*: Lifting can take place not only at fronts, but also along less distinct *boundaries* in the lower troposphere, where two local air masses of different temperature come in contact. A difference in temperature between local air masses might arise, for example, where the character of the Earth's surface within a given region varies, so that the overlying air warms by different amounts in different locations. For example, air above a fallow field of dark soil may heat up more than air over a forest, and air above an island may heat up more than air above the surrounding ocean. At such boundaries, air in the cooler, denser air mass can flow under the warmer air mass and lift it.

4. *Lifting due to high relief*: When wind reaches hills or mountains, air can no longer travel horizontally, and instead, must flow upslope. Such **orographic lifting** can carry moist air to higher altitudes and set the stage for thunderstorm formation above the mountains.

5. *Convergence lifting*: Lifting can occur when air flowing in a given direction collides with slower-moving air, or with air flowing in a different direction.

Air does not always have to be physically lifted in order to rise. In some cases, moist air near the ground starts to rise simply because the hot ground warms it enough to make it buoyant relative to surrounding air, like the air in a hot-air balloon. This process typically happens during the summer, when intense sunlight strikes the ground. Such warming can take place above plains as well as along mountain ranges. For example, on a hot afternoon, the slopes of steep mountains warm air adjacent to the slopes. This causes the

FIGURE 9.6 Development of thunderstorms along colliding gust fronts.

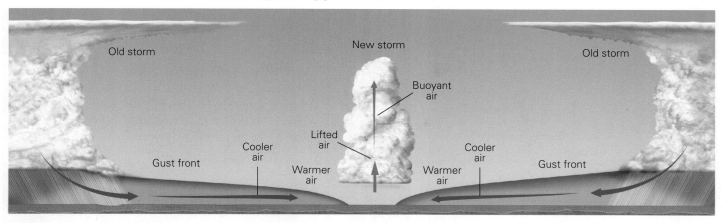

(a) Each storm sends out a gust front from its base. Where the gust fronts collide, air rises and a new storm forms.

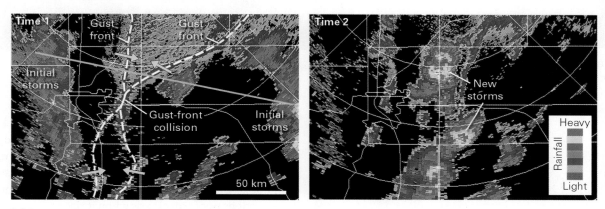

(b) Radar reflectivity images show new thunderstorms forming where two gust fronts collide. The yellow line shows the location of the cross section in part (a). The arrows give wind direction.

air to rise, eventually triggering late afternoon thunderstorms (**FIG. 9.7a**). People living near the Front Range of the Rocky Mountains witness this process nearly every day during the summer (**FIG. 9.7b**).

Development of Atmospheric Instability

Lifting of air alone, by any of the mechanisms described above, will not produce a thunderstorm. The location where the lifting takes place must also have the potential to develop **atmospheric instability**. This means that it must be possible for the lifted air to become less dense than the air that surrounds it, so that it can continue to rise buoyantly as an updraft to even higher altitudes. For a volume of lifted air to become *unstable*, meaning that it's less dense than the air surrounding it, it must become warmer than the surrounding air. If a volume of air does not become warmer than its surroundings after lifting, we say that the air is *stable*, and the lifted air will stop rising or sink back down. Thunderstorms can develop only when air becomes unstable.

The explanation of why atmospheric instability does or doesn't develop at a location is rather complicated. To make the process easier to follow, let's examine what can happen to an *air parcel*, a specified volume of air, that has just undergone lifting. As the air rises to progressively higher altitudes, keep in mind that in the surrounding air (the air parcel's *environment*), the air pressure, and in most cases the temperature, decreases. As a result of the decrease in pressure, the rising air parcel will expand, just as a balloon would expand as it rises to higher altitudes.

If there is no exchange of heat or mass between the air parcel and its surroundings during its expansion—that is, if it undergoes what physicists call **adiabatic expansion**—the temperature of the parcel decreases as it rises. Why? In order for the parcel to expand, air molecules within it must push aside the surrounding air, and this pushing requires energy (**FIG. 9.8**). Because the energy that does this work comes from heat energy inside the parcel, the parcel must give up some heat energy as it expands, causing the air in the parcel to cool.

Whether lifted air is stable or unstable and, therefore, whether an air parcel stops rising or becomes buoyant and continues rising, depends on two factors. First, stability depends on the **environmental lapse rate**, the rate at which the temperature of the air in the environment surrounding

FIGURE 9.7 Thunderstorms commonly develop over mountains in midafternoon during the summer.

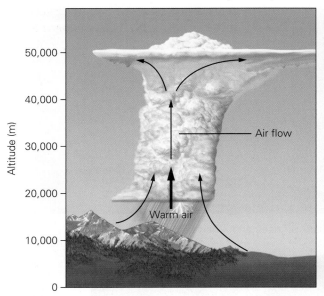

(a) During a hot afternoon, the slopes of steep mountains warm the adjacent air, which may begin to rise.

(b) An example of thunderstorms brewing over mountains.

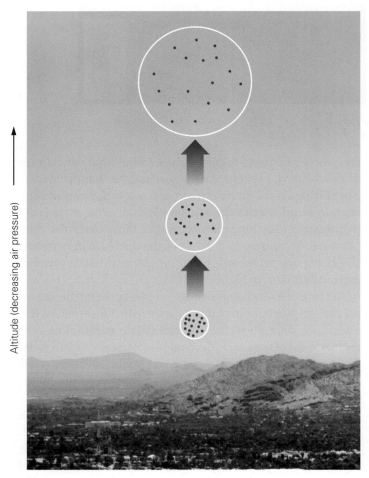

FIGURE 9.8 A parcel of air expands as it rises. Note, however, that the number of air molecules in the parcel doesn't change as the parcel rises.

the rising parcel decreases with altitude. Typically, environmental lapse rates range between 4°C and 9°C per kilometer (12–26°F/mi), depending on location. Second, it depends on the moisture content of the rising air parcel. *Saturated air*—air that is saturated with water vapor—forms clouds as it rises and cools. *Unsaturated air*—which may contain some water vapor—does not form clouds when it rises.

The reason that cloud formation within a rising air parcel plays such an important role in determining its stability is that the rate at which air cools during adiabatic expansion depends on whether or not the water vapor within it condenses into cloud droplets. Let's see why.

When unsaturated air rises, it undergoes *dry adiabatic expansion*. As long as the air remains unsaturated, even though the air's relative humidity increases as it cools, cloud droplets will not form. Dry adiabatic expansion causes the temperature of the air parcel to decrease by about 10°C for each kilometer (29°F/mi) that the parcel rises. This rate, known as the **dry adiabatic lapse rate**, does not depend on altitude: an air parcel undergoes the same temperature change rising between altitudes of 1 and 2 km (0.6–1.2 mi) as it does rising between 9 and 10 km (5.6–6.2 mi). Now here's a key point: if the dry adiabatic lapse rate is greater than the environmental lapse rate, a rising unsaturated air parcel eventually becomes colder, and therefore denser, than the surrounding air. Therefore, the parcel becomes stable and will not rise further on its own.

In contrast, rising saturated air cools at a rate of only 6°C per kilometer (17°F/mi) of altitude. This rate, known as the **moist adiabatic lapse rate**, is less than the adiabatic lapse rate for unsaturated air because cloud droplets condense out of saturated air as the air rises and cools. As we saw in Chapter 8, conversion of water vapor to liquid water releases latent heat, and adding that heat to the rising air parcel slows the rate of cooling during the parcel's ascent. Because the moist adiabatic lapse rate can be less than the environmental lapse rate, a saturated, cloudy air parcel has the potential to become buoyant, and it can remain buoyant as it continues to rise. Meteorologists refer to such air as *conditionally unstable*.

Pulling all these ideas together, we see that whether or not air parcels become buoyant and rise to the tropopause, leading to thunderstorm formation, depends on their environment and their moisture content. If air parcels are unsaturated, they cool more quickly than surrounding air when the parcels have undergone lifting. Consequently, they become denser than the surrounding air and don't rise further, so thunderstorms don't develop (FIG. 9.9a). If lifted air parcels are saturated, water in them begins to condense as the parcels rise. Such moist air cools more slowly than does unsaturated air, and can become conditionally unstable (FIG. 9.9b). What happens next depends on the lapse rate of the air in the environment surrounding the parcels. If the lifted, moist parcels do not become warmer and less dense than the surrounding air, they become cloudy, but remain stable and won't rise to the tropopause, so thunderstorms don't develop. But if the lifted air parcels become warmer than surrounding air in their environment, they become unstable and will rise buoyantly to the tropopause, setting the stage for thunderstorm formation.

In unstable air, an *updraft* develops. Within the updraft, air blows vertically upward. If the updraft is strong enough, and carries enough moisture to high elevations, towers of billowing cumulonimbus clouds form, and a thunderstorm develops. Moisture drawn into the base of the storm effectively provides the "fuel" to drive the storm. Conditions leading to atmospheric instability and thunderstorm formation exist only when air in the lower troposphere is warm and moist while air in the upper troposphere is cool. These conditions can happen throughout the year in the tropics, but in mid-latitudes they occur primarily during late spring through early fall.

Take-home message . . .

Three conditions are necessary for a thunderstorm to form: moist air near the Earth's surface, a lifting mechanism, and local atmospheric instability. Once lifted, a saturated air parcel can become buoyant when its moisture condenses into cloud droplets because the release of latent heat makes the parcel warmer and less dense than its surroundings. Such atmospheric instability can produce an updraft that carries air up to the tropopause, producing large, billowing clouds that develop into a thunderstorm.

QUICK QUESTION What mechanisms can lift air to trigger thunderstorm development?

9.3 Types of Thunderstorms

Towering dark clouds, flashes of lightning, claps and rumbles of thunder, torrential rains, and sometimes the clatter of hail—these are the signatures of thunderstorms. But even though all thunderstorms share these basic characteristics, not all are alike. Here, we examine the different ways that thunderstorms organize.

Ordinary Thunderstorms

Meteorologists call an isolated thunderstorm that does not rotate—meaning that the air in the storm does not spin around a vertical axis—an **ordinary thunderstorm**. Such storms are *single-cell thunderstorms* because each has one updraft and one downdraft. Another term used for these storms is *air-mass thunderstorms*, because they tend to form in the middle of an air mass, rather than along a front. Ordinary thunderstorms form where, at the time of storm formation, regional winds do not change substantially in direction or speed with altitude.

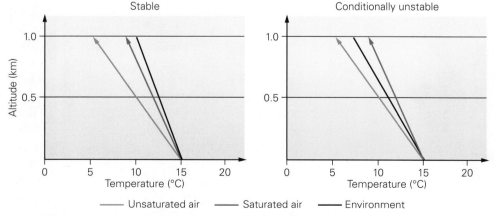

FIGURE 9.9 Atmospheric stability under different environmental conditions.

(a) Under these conditions, both saturated and unsaturated air parcels, when lifted, would cool more rapidly with altitude than would the surrounding air, and would return to their original altitude.

(b) Under these conditions, an unsaturated air parcel, when lifted, would return to its original altitude, but a saturated air parcel would cool more slowly than the surrounding air, and would rise buoyantly.

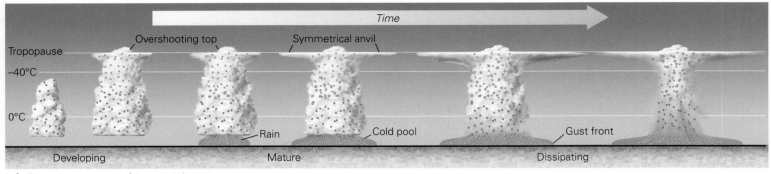

FIGURE 9.10 Stages in the development of an ordinary thunderstorm.

This situation can happen when air undergoes orographic lifting or gets warmed by the ground. In general, ordinary thunderstorms develop during hot afternoons and last an hour or so before they dissipate. Only rarely do they produce severe weather such as hail, strong winds, or tornadoes. Worldwide, at any given time, as many as 2,000 ordinary thunderstorms are active.

The life cycle of an ordinary thunderstorm can be divided into three stages (FIG. 9.10). During the *developing stage*, warm, moist air is lifted to an altitude at which it becomes buoyant and unstable. Recall that this happens because the air expands and cools, the water vapor within it condenses into liquid droplets, or at higher altitudes, solidifies into tiny ice crystals, thereby releasing enough latent heat to maintain the air's buoyancy. The air surges upward in an updraft, producing towering cauliflower-like cumulus clouds that will rise to the tropopause. Air cannot continue to rise far into the stratosphere, because in that atmospheric layer the environmental air temperature increases with altitude, while rising parcels encountering the stratosphere continue to cool, so an updraft pushing into the stratosphere no longer remains buoyant. At the top of the thunderstorm, clouds spread laterally into a symmetrical sheet just below the tropopause. A storm cloud with this shape is called an **anvil cloud**, so named because of its resemblance to an anvil used by a blacksmith (FIG. 9.11). If, however, the updraft is strong

FIGURE 9.11 When a thunderstorm rises to the tropopause, the top of the cloud spreads out into a flat layer, making the cloud overall resemble a classic blacksmith's anvil.

enough, it surges briefly into the stratosphere, producing a bulge of clouds called an **overshooting top**. These clouds then sink back into the anvil.

With the formation of the anvil cloud, the thunderstorm enters the *mature stage* of its life cycle. During this stage, ice particles aloft grow into snowflakes and graupel as they cascade downward through the cloud, then fall into the warmer lower atmosphere, where they melt into raindrops (see Fig. 9.10). As this precipitation falls, a downdraft develops. Falling rain initiates a downdraft because as it falls, it encounters drier air mixing in from the sides of the cloud, and some of the raindrops evaporate. This evaporation cools the air, making it denser, so it sinks, intensifying the downdraft. Falling rain also pushes air downward ahead of it, contributing to the intensity of the downdraft.

The updraft that triggers formation of an ordinary thunderstorm is nearly vertical because, as we've noted, the regional wind velocity where ordinary thunderstorms form does not change significantly with altitude. Therefore, when the downdraft forms during the mature stage of the storm, it opposes the updraft. Eventually, as the downdraft intensifies, it suppresses the updraft entirely and shuts off the source of the moisture that feeds the storm. Without this "fuel," the thunderstorm weakens and enters its *dissipating stage* (see Fig. 9.10). The storm's clouds begin to dissipate by evaporating and dispersing. Even as the storm is dissipating, however, the cold pool, bordered by a gust front, can be spreading out near the ground, lifting warm air ahead of it, and triggering the formation of new thunderstorms.

Squall-Line Thunderstorms

Squall-line thunderstorms are organized in a distinct line—a *squall line*—along either a gust front or a cold front. Meteorologists sometimes refer to these storms as *multicell storms*, since many thunderstorm updrafts occur along a line. As the line of storms moves along, the rain they produce may cover a broad area. Indeed, squall-line thunderstorms are responsible for much of the summer rainfall over the interior plains of the United States.

When a squall line approaches, you'll probably first see a dark and ominous planar cloud called a *shelf cloud*, which forms over the front (**FIG. 9.12**). Typically, only a small amount of precipitation falls from the shelf cloud as it passes. Behind it come **straight-line winds**, winds that blow in a single direction rather than in a spiral as in a tornado. Strong straight-line winds can damage buildings, topple trees, and pose a hazard for low-flying airplanes.

Once the shelf cloud has passed overhead, you will see a line of thunderstorms approaching—as you face the squall line, the storms will extend into the distance to your right and left. As the storms pass overhead, lightning streaks the sky, heavy rains fall, and the winds become *gusty* (variable in velocity and direction) and occasionally very strong. After the heavy rain passes, lighter rain may continue for hours, with an occasional flash of lightning and crack of thunder.

How do squall-line thunderstorms form, and how can they last so long? The answer depends on the season. Squall lines form along gust fronts during the warm season (late spring through early fall) in regions with temperate

FIGURE 9.12 A shelf cloud develops when warm air is lifted over an approaching gust front rushing out from beneath a thunderstorm.

climates. The process begins during the late afternoon or early evening when ordinary thunderstorms form. Initially, a disorganized cluster of several ordinary thunderstorms grows (**FIG. 9.13a**). As time progresses, downdrafts from these storms coalesce to form a single large cold pool at the ground. This cold pool spreads outward, and its leading edge forms a strong gust front. As the initial thunderstorms go through their life cycle, the cold pool continues to expand outward, and lifting along the gust front triggers the formation of new thunderstorms. These new storms align with the gust front, forming an organized line of storms—a squall line (**FIG. 9.13b**). As the cold pool advances forward into warmer air, the squall line widens, updrafts carry moisture over the top of the cold pool, and rainfall becomes more widespread. Eventually, the cold pool spreads out so much that the squall line weakens. Lighter rain may continue for several more hours as the storms eventually dissipate.

In cooler seasons (late fall through early spring), lifting becomes focused along strong cold fronts. Very long lines of thunderstorms can develop along such fronts. From a regional perspective, such **frontal squall lines** lie along the "tail" of a comma-shaped mid-latitude cyclone (**FIG. 9.14**). Typically, frontal squall lines move with the front as it crosses a continent. With relatively long lives (many hours to more than a day), they can produce severe weather.

In some squall-line events, severe straight-line winds can extend over a large geographic region. Such a widespread thunderstorm-generated windstorm is called a **derecho** (from the Spanish, meaning direct, or straight). One of the worst derechos on record occurred June 29–30, 2012 (**FIG. 9.15a**). The winds raced eastward from western Indiana along the Ohio River Valley, over the central Appalachians, and into the mid-Atlantic states. It caused 22 fatalities, most due to falling trees, and damage totaling $2.9 billion (**FIG. 9.15b**).

Supercell Thunderstorms

Between April 25 and 28, 2011, many violent thunderstorms developed and continually moved across the southeastern United States. An estimated 360 tornadoes emerged from these storms, producing $10.2 billion in damage and breaking

FIGURE 9.13 A squall line develops when ordinary thunderstorms coalesce and a large cold pool forms.

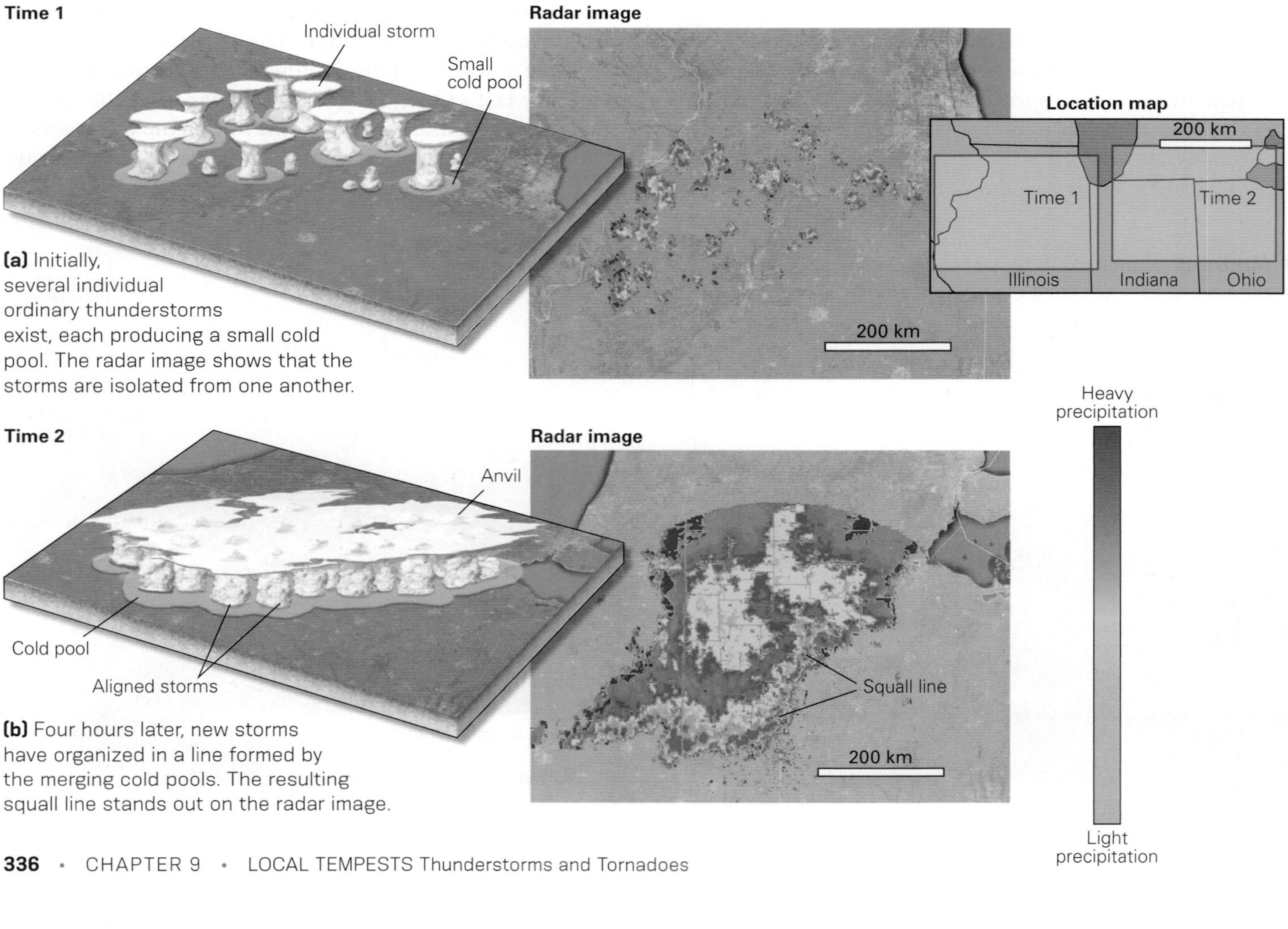

(a) Initially, several individual ordinary thunderstorms exist, each producing a small cold pool. The radar image shows that the storms are isolated from one another.

(b) Four hours later, new storms have organized in a line formed by the merging cold pools. The resulting squall line stands out on the radar image.

FIGURE 9.14 Radar image of a frontal squall line along a cold front within a mid-latitude cyclone.

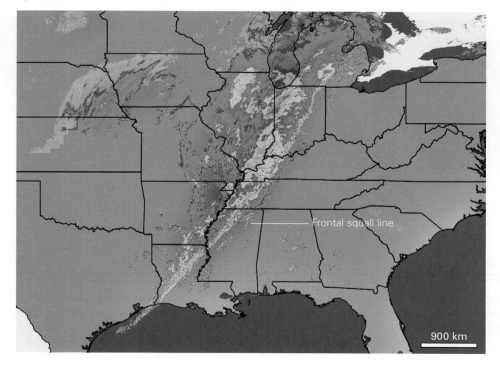

the record for tornadoes in a 4-day period and for the month of April. Despite advance warnings, 346 people were killed. Weather reports called these terribly destructive storms supercell thunderstorms. Why?

Formally defined, a **supercell thunderstorm** (or *supercell*) is a thunderstorm containing a particularly strong, long-lasting, rotating updraft. In contrast, ordinary thunderstorms, as we noted earlier, do not rotate. Because of their distinct structure, supercell thunderstorms often become severe and can produce strong tornadoes. Supercells occur much less frequently than ordinary thunderstorms, and they are typically isolated from neighboring thunderstorms. Supercells typically have an overall base diameter of around 50 km (31 mi), larger than that of ordinary thunderstorms. The most dangerous portion of the storm, which lies beneath the storm's updraft, has a typical diameter of 5–10 km (3–6.2 mi).

What conditions lead to supercell formation? Supercells develop only when strong *vertical wind shear*—a condition in which wind near the ground differs in speed and direction from wind at a higher altitude—exists in the lower part of the troposphere, and strong winds are present in the upper troposphere. In the central United States, vertical wind shear typically develops when slow winds near the ground blow from the south-southeast, bringing in warm, moist air from the Gulf of Mexico, while faster winds 1–2 km (1 mi) above the ground blow from the west-southwest.

As a result of wind shear, air in the lower troposphere tends to swirl around a horizontal axis. The shape of this movement can be visualized by considering a rolling horizontal cardboard tube (**FIG. 9.16a**). When a strong, thunderstorm-producing updraft develops, it draws up the horizontally rotating air. To get a sense of this phenomenon, imagine that the cardboard tube shown in **Figure 9.16a** gets pulled

FIGURE 9.15 The June 29–30, 2012, derecho.

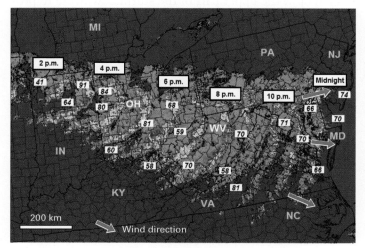

(a) A composite of successive superimposed radar images as the derecho moved from western Indiana to the US east coast in 10 hours. The numbers in boxes are wind speeds (in mph).

(b) Tree damage like that seen here occurred all along the path of the derecho.

9.3 Types of Thunderstorms • 337

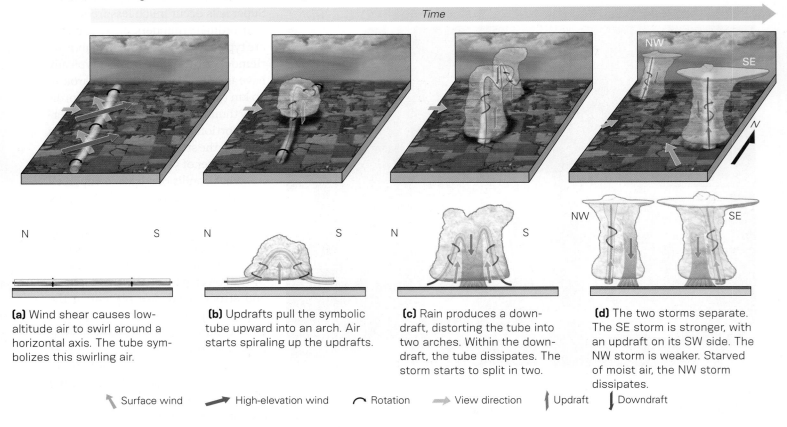

FIGURE 9.16 Stages in the development of a supercell thunderstorm.

(a) Wind shear causes low-altitude air to swirl around a horizontal axis. The tube symbolizes this swirling air.

(b) Updrafts pull the symbolic tube upward into an arch. Air starts spiraling up the updrafts.

(c) Rain produces a downdraft, distorting the tube into two arches. Within the downdraft, the tube dissipates. The storm starts to split in two.

(d) The two storms separate. The SE storm is stronger, with an updraft on its SW side. The NW storm is weaker. Starved of moist air, the NW storm dissipates.

upward in the middle and becomes an arch. Because the updraft also triggers cloud formation, most of the rotating air cylinder lies hidden within a growing cumulus cloud (**FIG. 9.16b**). Due to the updraft, air on the steeply tilted sides of the arch spirals upward. As soon as precipitation begins and the cloud becomes a thunderstorm, a downdraft forms, driving the center of the arch downward. This process separates the initial arch into two columns of counter-rotating air. Very quickly, however, turbulence causes the rotation within the central downdraft to weaken, and the original storm starts to split in two (**FIG. 9.16c**).

As the storm splits, two distinct thunderstorms exist. In the northwestern storm, air rotates clockwise as it spirals upward, whereas in the southeastern storm, air rotates counterclockwise. Because the warm, moist air that feeds the storms comes from the south-southeast, the southeastern storm draws in moisture and grows, while the northwestern storm, starved of warm, moist air, dies out (**FIG. 9.16d**). Therefore, supercells that pose a hazard in the central United States are always rotating counterclockwise.

Meteorologists refer to the rotating updrafts at the heart of these storms as **mesocyclones**, meaning medium-sized (much smaller than a hurricane or a mid-latitude cyclone), counterclockwise-spinning winds. Within the updrafts of supercells, vertical wind velocity can reach 160 km/h (100 mph). Because of the strong winds in the middle and upper troposphere, the updrafts of supercell thunderstorms tilt downwind, so that the axis of the updraft within the storm is not exactly vertical.

When the updraft of a supercell thunderstorm nears the tropopause, the high-altitude winds of the polar-front jet stream blow the upper part of the storm downwind to produce a large, asymmetrical anvil cloud. Therefore, supercell thunderstorms have a distinctive structure characterized by an asymmetrical anvil at the top and a strong mesocyclone within (**FIG. 9.17a**). The updraft in the mesocyclone can be so strong that it forms a particularly large overshooting top (**FIG. 9.17b**). At the base of the updraft, a lowering cloud—called a **wall cloud**—typically develops (**FIG. 9.17c**). It is from this feature that a tornado, if it occurs, will form.

Meteorologists refer to the northeastern side of a northeast-moving supercell, such as those of the central United States, as the *forward flank*, and to the southwestern side as the *rear flank*. Heavy rain on the forward flank produces an intense downward-directed air flow, called the *forward-flank downdraft* (FFD), which descends to the ground and spreads out. Meanwhile, high-altitude winds are bringing dry air from the southwest into the high clouds

FIGURE 9.17 Supercell thunderstorms have a dramatic appearance that includes several distinct features.

(a) A composite image of a supercell viewed from the southwest.

(b) Again from the southwest, but from a high altitude, this view of a supercell thunderstorm shows the anvil and the overshooting top.

on the rear flank of the storm. Consequently, droplets in these clouds evaporate, a process that cools air. The resulting cooled air sinks rapidly, producing a strong *rear-flank downdraft* (RFD). Because of the storm's northeastward tilt, the FFD lies to the northeast of the updraft. Therefore, unlike the updraft of an ordinary thunderstorm, which tends to be destroyed by the downdraft, the updraft of a supercell thunderstorm is not suppressed by the downdraft. As a result, the supercell's updraft can survive for a relatively long time (a few hours), and the storm can maintain its identity, as it moves hundreds of kilometers across the land.

A radar image of a supercell highlights the storm's key features (**FIG. 9.18**). (Recall from Chapter 8 that the colors on a radar image are related to the intensity of precipitation.) A distinctly curving *hook echo* outlines the southwestern side of the updraft. The updraft itself appears as an echo-free region within the hook because no rain falls in the updraft. The large echo to the northeast of the updraft indicates the presence of heavy rain in the FFD. Hail, if it forms, falls primarily on the northeastern side of the updraft, and a tornado, if it develops, occurs at the tip of the hook echo.

(c) A close-up view of the base of a supercell thunderstorm, showing the wall cloud and a tornado.

Take-home message...

An ordinary thunderstorm has a vertical, nonrotating updraft and tends to dissipate fairly quickly. Squall-line thunderstorms form in a line, either along the gust fronts of ordinary thunderstorms or along a regional cold front. The most violent thunderstorms, supercells, form where vertical wind shear exists. A supercell has a rotating updraft (called a mesocyclone), a broad asymmetric anvil, and a large overshooting top.

QUICK QUESTION Why do supercell thunderstorms tend to survive for a relatively long time?

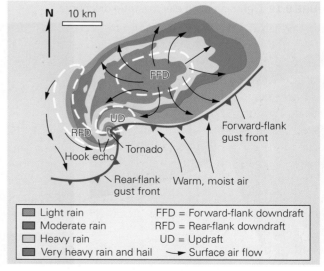

FIGURE 9.18 A diagram of a radar image of a supercell, showing the wind flow and other key features.

9.3 Types of Thunderstorms • **339**

9.4 Hazards of Thunderstorms

Lightning

Thunderstorms owe their name to the clap, crash, and rumble of thunder that accompanies them. Thunder is a consequence of lightning. It's the lightning in a storm that is hazardous—thunder, while frightening, doesn't cause casualties or damage. Here we examine why lightning happens.

FORMATION OF LIGHTNING. If you rub your shoes on a carpet and then touch a metal doorknob, you'll see—and feel—an electrical spark (FIG. 9.19a). This spark appears because electrons (negatively charged particles) from the carpet's atoms flow into your body, making your skin negatively charged relative to the doorknob. Put another way, a **charge separation** develops between the doorknob and your finger. When your finger gets close enough to the doorknob, electrons suddenly flow from your skin to the doorknob, and this short-lived current generates a spark that arcs across the charge separation and tingles your nerves. Physicists refer to such a spark as an *electrostatic discharge*. Simply put, **lightning** is a huge electrostatic discharge, caused when a large charge separation develops in a thunderstorm's clouds (FIG. 9.19b).

The spark that jumps from your finger to a doorknob is about 1 mm (0.04 in) long, with a diameter less than 0.1 mm (0.004 in), and it transmits a current of about 0.0001 amps. (An *amp*—short for *ampere*—is a measure of the quantity of electrons flowing in a current during a 1-second interval.) In contrast, a typical **lightning stroke**, or *lightning bolt*, the jagged spark produced by a thunderstorm, can be more than 5 km (3 mi) long and 2–3 cm (about an inch) in diameter, and it can carry a current of 15,000 to 30,000 amps. A single lightning stroke produces about the same amount of energy as is required to power a typical house for 2 months.

Why does the charge separation that triggers lightning develop? As particles collide in a cloud, electrons jump from tiny ice crystals to larger ice particles, such as hailstones. As a consequence, the tiny ice crystals become positively charged, because they have a deficit of electrons, while the larger particles become negatively charged, because they have an excess of electrons. Updrafts in the storm sweep the tiny, positively charged ice crystals upward to high altitudes in the cloud, while the larger, heavier, negatively charged particles fall toward the lower part of the cloud, producing the charge separation. As it develops, charge separation can be maintained in the cloud because air is an *electrical insulator*, meaning that it prevents electrons from flowing easily between negatively and positively charged regions.

As charge separation develops, the *electrostatic potential*, meaning the potential energy that could be available to drive electrons from the region of negative charge to the region of positive charge, grows. A lightning bolt forms when the electrostatic potential in a thunderstorm becomes great enough to drive electrons across a large expanse of insulating air.

CLOUD-TO-CLOUD LIGHTNING. Most lightning occurs when electrons from a negatively charged part of a storm cloud jump to a positively charged part. Such **cloud-to-cloud lightning** may hit an airplane flying through a storm. In fact, jet planes are struck, on average, once or twice a year. Fortunately, electricity can be conducted along the metal skin of a plane without penetrating its interior, so lightning typically causes minimal damage, if any, to airplanes.

FIGURE 9.19 Lightning is an electrostatic discharge.

(a) An electrostatic discharge from a finger to a doorknob resembles a tiny lightning stroke.

(b) Some lightning strokes go from cloud to cloud, or stay within a cloud. Some go from cloud to ground.

CLOUD-TO-GROUND LIGHTNING. Because humans live on the ground, **cloud-to-ground lightning**, which strikes buildings, trees, people, or other objects, poses the greatest hazard to us. Cloud-to-ground lightning requires development of charge separation between the cloud and the ground. Such separation develops when negative charges accumulate toward the base of a cloud. Then, just as the negative end of a magnet repels the negative ends of other magnets, the negatively charged cloud base drives away negative charges on the ground, leaving a region of the ground with a net positive charge.

A cloud-to-ground lightning stroke begins when electrons surge first toward the cloud base and then toward the ground in a series of steps, producing a *stepped leader* **(FIG. 9.20a)**. Each step is 50–100 m (164–328 ft) long and takes only a few millionths of a second to form. Typically, electrons take many different paths downward, so the stepped leader has several branches. The branch that happens to approach the ground surface first attracts positively charged particles. These particles jump upward from an object on the ground, producing a feature called a *positive streamer*. Such streamers typically rise from a narrow high point, such as a tower, a tree, a chimney, or an unfortunate person. When the positive streamer connects with the stepped leader, a channel for electron flow becomes established. The channel provides a pathway for electron flow because its charged particles make it behave like an electrical conductor, unlike the air around it. The instant that the positive streamer connects with the stepped leader, a powerful *return stroke*—the main electrostatic discharge—flashes **(FIG. 9.20b)**. The stroke runs from the cloud to the point where

FIGURE 9.20 The steps leading to the production of a cloud-to-ground lightning bolt.

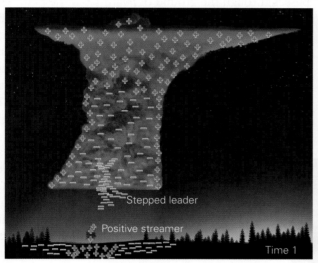

(a) A stepped leader descends from a cloud, while a positive streamer rises from the ground below.

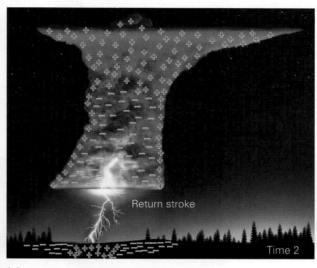

(b) A return stroke makes a brilliant flash.

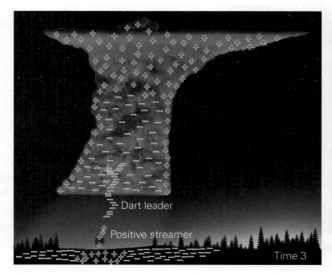

(c) A dart leader descends from the cloud, while a positive streamer flows upward.

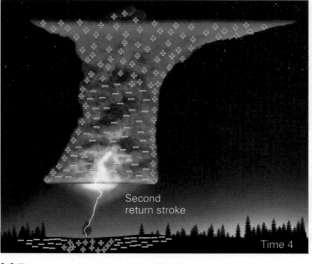

(d) The second return stroke flashes.

the positive streamer started. The name *return stroke* seems curious, because nothing actually returns to the cloud when it flashes. This name remains from a time before researchers understood why lightning initiates.

In the return stroke, vast numbers of electrons stream from the cloud to the ground. The process of forming a leader, positive streamer, and return stroke typically repeats multiple times, generally within a second or two **(FIG. 9.20c, d)**. Subsequent leaders, after the first stepped leader, are called *dart leaders*. They follow the previously established channel through the air, since it has become an electrical conductor. The whole process happens so fast that you may see it as a single bolt that flashes like a strobe light.

Meteorologists refer to the type of lightning stroke that we've just described as a *negative polarity cloud-to-ground stroke* **(FIG. 9.21)**. As we've seen, it occurs when the negative charge in the base of a storm jumps to the positive charge on the ground. A rarer *positive polarity cloud-to-ground stroke* occurs when electrons in the negatively charged ground jump all the way to the positively charged anvil of the storm cloud. Because the distance between the anvil and the ground is greater than that between the base of the storm and the ground, a large electrostatic potential must develop for such a stroke to take place, so these strokes can release an immense amount of energy. And because they occur beneath the anvil, they may happen in advance of the arrival or after the departure of the storm's rain, when some blue sky may still be visible. For this reason, they are sometimes called "bolts from the blue."

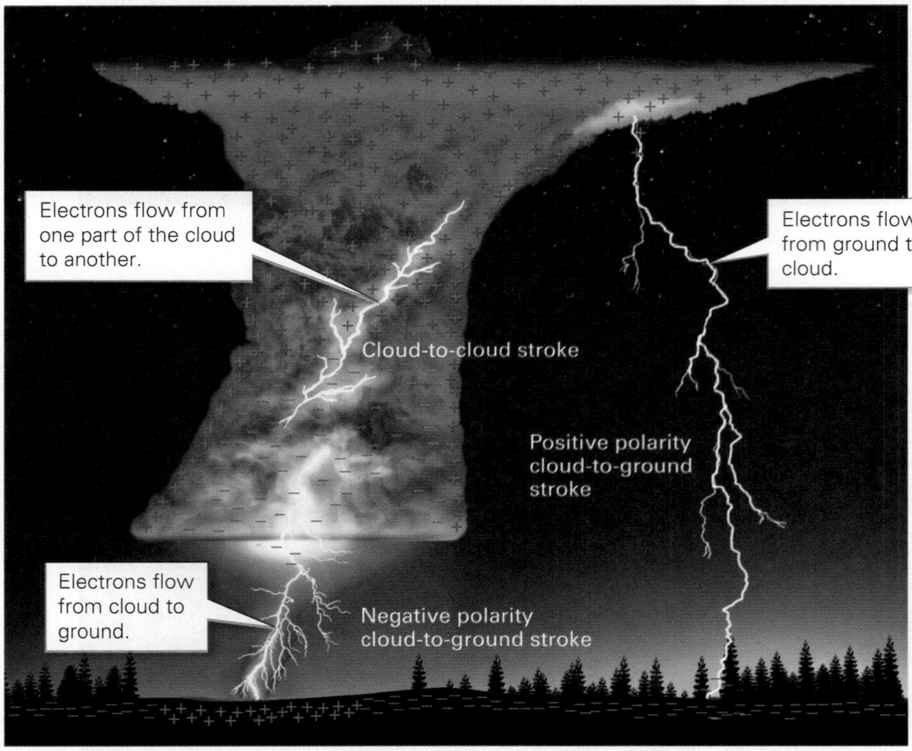

HAZARDS OF CLOUD-TO-GROUND LIGHTNING.

Needless to say, lightning is dangerous. For example, if lightning strikes a tree, it can cause the sap in the tree to vaporize instantly, so that the tree explodes, and it can set the tree on fire **(FIG. 9.22a)**. Many devastating forest fires have been ignited by lightning strikes (see Chapter 14). When lightning strikes a house, it can pass into the frame of the house and ignite flammable materials, setting the house on fire. And if lightning strikes power lines, it can blow out transformers and cause a blackout.

Tragically, when lightning strikes, or even strikes near, a person, death or injury can result. Many human casualties are caused not by a direct lightning strike, but by the current that passes through the ground outward from the point where a lightning bolt strikes **(FIG. 9.22b)**. Deadly currents can flow as far as 20 m (66 ft) from a lightning strike. Lightning can also jump from an object that has been struck to a nearby person—since people are essentially made of water, and water has low electrical resistance, the current will take the easiest path, through the person. People may also be injured not by the electricity, but by the debris produced when lightning blasts apart a tree or other object.

A lightning strike, like any strong electrical shock, can produce several different kinds of injuries. It can leave severe wounds and burns and can cause paralysis of limbs and organs, heart failure, or breathing failure. The expansion of air and water heated by lightning can cause internal damage to organs. Fortunately, only 10% of people who are struck are killed, but unfortunately, 70% of lightning strike victims suffer serious injuries, such as burns or neurological damage **(FIG. 9.22c)**. Globally, the number of people struck ranges between 2,000 and 6,000 annually, depending on the source of information. In the United States, lightning strokes kill, on average, about 50 people per year.

How can you avoid being a victim of lightning? First and foremost, go indoors when a thunderstorm approaches. Never stand in the middle of an open area or near isolated tall objects, and avoid being on a boat in open water where you would be the highest object—lightning tends to strike tall objects that can serve as the

FIGURE 9.21 Lightning discharges can occur in three ways: as negative polarity cloud-to-ground strokes, positive polarity cloud-to-ground strokes, and cloud-to-cloud strokes.

FIGURE 9.22 Hazards of lightning.

(a) A tree split by lightning.

(b) Streaks on a golf course show how lightning can cross the ground from its strike point.

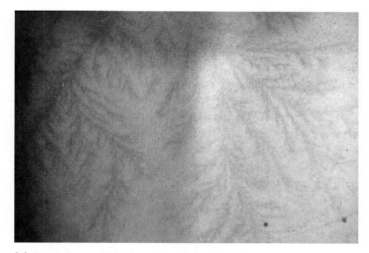

(c) Scars from a lightning strike.

THE BOOM OF THUNDER. Regardless of the path a lightning stroke takes, it generates thunder, a booming, cracking, or rumbling sound that travels through air, like any sound, in the form of waves that cause air to alternately expand and contract. How does the electrical spark of lightning produce sound? The temperature in a lightning-stroke channel can reach 30,000°C (54,032°F)—five times hotter than the

FIGURE 9.23 When you touch a Van de Graf generator (the metal ball in the photo) it sends electrons into your body, including your hair. Since like charges repel, your negatively charged hairs repel each other and stand on end. A similar phenomenon may happen beneath a thunderstorm. The negative charge at the base of the storm repels the electrons in your hair, so your hairs become positively charged, repel each other, and stand on end. If this happens to you when you're near a thunderstorm, you are in danger—lightning may strike at any second!

source of a positive streamer. When inside, stay away from appliances, plumbing, and corded telephones that could conduct electricity. If you are outdoors and feel your hair standing on end (FIG. 9.23), take cover immediately or get into a car. While lightning can strike cars, it rarely harms the occupants; the metal skins of cars, like those of airplanes, act as a cage that conducts the electricity and keeps it from entering the vehicle. If there is no cover available, assume a stooping position, crouching down as close to the ground as possible with your feet together, and minimize contact with the ground by staying on your toes or heels. Do not lie down!

Lightning often strikes buildings, especially tall ones. It's a good idea to protect your home with *lightning rods*, pointed metal rods anchored to rooftops and connected by wires or cables directly to the ground outside a building (FIG. 9.24). If lightning strikes a building that has lightning rods, it will typically hit a rod, so the electricity of the stroke will travel down the wire, outside the building, and into the ground, and stay out of the building's interior.

surface of the Sun—so the electrical energy of the stroke instantly heats the air around it. This heating causes the air to expand explosively. Then, when the stroke ceases, the air contracts violently. The sudden expansion and contraction of air generates shock waves that travel through air—these shock waves are what we hear as **thunder**.

Sound travels at approximately 340 meters per second (1,236 km/h, or 768 mph), so it moves about 1 km in 3 seconds (1 mile in 5 seconds). Light travels so fast (300,000,000 m/s, 186,000 mi/s) that it arrives at our eyes in an instant. You can estimate the distance to a stroke of lightning by counting the seconds that pass between the flash and the thunder, then dividing by 3 (to get the distance in kilometers) or 5 (to get the distance in miles).

Downbursts

As we noted earlier, very strong downdrafts develop within thunderstorms when it rains, for two reasons: cooling of air due to evaporation and the downward push of the falling rain. When a downdraft reaches the ground, the air must turn and flow outward. The outrushing wind emerging from an intense downdraft is called a *downburst*. A downburst flows radially outward from the base of the downdraft, but at any given location, the wind of a downburst flows straight and parallel to the ground. These straight-line winds may travel a few kilometers in front of the storm, generally at speeds of 30–100 km/h (19–62 mph). In severe downbursts, they may be as fast as 100–130 km/h (62–81 mph).

A small, intense downburst, known as a *microburst*, can develop in association with rain falling from a tall cumulonimbus cloud, even if the cloud hasn't become a thunderstorm.

FIGURE 9.24 To protect your home (or barn), you can install lightning rods on its rooftop.

FIGURE 9.25 Microbursts can be a serious hazard in the vicinity of an airport.

(a) A microburst over an airport in Denver, Colorado.

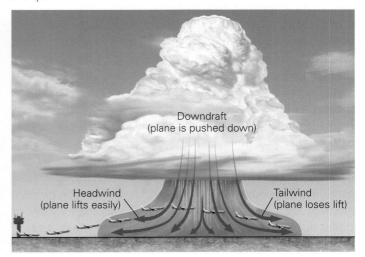

(b) As an airplane approaches a microburst, it encounters strong headwinds, then strong tailwinds after it passes the microburst center.

FIGURE 9.26 Hail occurs over a relatively small area, but it can cause significant damage.

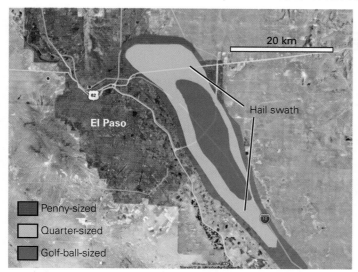

(a) Map of a hail swath near El Paso, Texas.

(b) Hail damaged this car's back window.

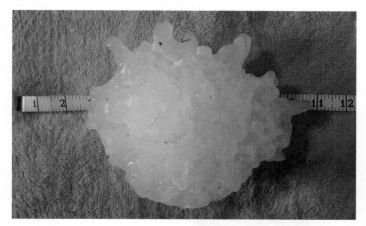

(c) A hailstone that fell in Vivian, South Dakota, on July 23, 2010, holds the US record for diameter (20.3 cm, or 8 in) and weight (0.879 kg, or 1.94 lb).

Microbursts affect an area only about 5 km² (2 mi²) in diameter, but even so, they can be dangerous to aircraft approaching or departing airports (**FIG. 9.25a**). In news stories describing air traffic hazards, microbursts are commonly referred to as wind shear, because wind speed and direction change rapidly across a microburst. If an airplane takes off into approaching winds from a microburst, it lifts into the sky easily. But when the plane crosses into the downdraft itself, it gets pushed down, and when it enters the portion of the microburst blowing in the same direction that it's flying, it loses lift (**FIG. 9.25b**).

Since 1950, at least 10 airline disasters have been attributed to microbursts. To decrease the threat of downburst-related calamities, many airports have now installed special Doppler radar systems and wind sensors that detect downbursts and automatically alert pilots. After these systems were installed in the late 1990s, no commercial aircraft have been lost to microbursts in the United States. Other countries have adopted similar technology.

Hail

In nearly all thunderstorms, **hail**, which consists of spherical or irregularly shaped lumps of solid ice (each called a *hailstone*), forms at high altitudes. The process begins when updrafts carry liquid cloud droplets to altitudes where temperatures are well below freezing. These droplets become *supercooled*, meaning that they remain liquid even though their temperature lies below 0°C. When ice particles falling from farther aloft collide with the supercooled droplets, the droplets bond to the ice particles and immediately freeze, so the ice particles quickly grow into graupel and then into hailstones (see Fig. 8.16). In an updraft, upward air flow keeps hailstones aloft in the cloud, where they can keep growing. Eventually, however, a hailstone drifts out of the updraft, or its weight overcomes the force of the updraft, and it starts to fall. It collects even more supercooled droplets on the way down, continuing to grow until it encounters above-freezing temperatures in the lower part of the cloud and starts to melt.

Hailstones reach the ground only if they grow large enough to avoid melting away entirely as they fall. This happens when a thunderstorm contains particularly strong updrafts, which is why the most damaging hail tends to develop in supercell thunderstorms. Conditions that cause a **hailstorm**, a period during a thunderstorm when hail reaches the ground, may last for a few minutes, or even over an hour. During this time, the thunderstorm moves, so hail falls on the ground along a path called a *hail swath* or a *hail streak* (**FIG. 9.26a**). About 4,000 damaging hailstorms strike the United States in given year—far less than the number of thunderstorms, but still a significant number.

BOX 9.1 DISASTERS AND SOCIETY

Hailstorms of Mendoza, Argentina

South of the equator, in west central Argentina (in the vicinity of the Sierras de Córdoba mountains and the Andes foothills, near the city of Mendoza), some of the most intense thunderstorms in the world threaten a major agricultural area **(FIG. Bx9.1a)**. The moisture fueling these storms blows southward into Argentina from the Amazon rainforest, along the eastern face of the Andes. The storms that form near Mendoza, like those of the Great Plains in the United States, are supercell thunderstorms, except that they rotate clockwise because of the direction of the Coriolis force in the southern hemisphere. In addition to the hail threat, the storms regularly produce tornadoes, severe winds, and flash floods. Little about the intensity of these thunderstorms was known until NASA satellites discovered that these storms produced the most frequent lightning strikes anywhere on the planet. Studies then confirmed that the Argentina thunderstorms were also some of the worst with respect to the frequency of large hail.

In the general vicinity of Mendoza, these thunderstorms threaten a major wine-producing agricultural area. The hailstorms can be highly localized, damaging one field of vines, but not a neighboring field. Farmers have spread out their vineyards to hedge against losing all their crops in one location. Depending on the time of the growing season when a hailstorm occurs, and on the size and intensity of the hail, damage to the leaf canopy can occur. This damage inhibits the vines' ability to capture sunlight and pass energy and nutrients to the grapes, which reduces yields and allows for the spread of rot. The Agrelo wine district, for example, lost an estimated 15% of its grapes as one large storm in 2007 spread hail over 50 km^2 (more than 12,000 acres) of land. The frequency and intensity of the hailstorms is so great, and they produce so much damaging hail annually, that farmers routinely place hail-catching nets across their fields **(FIG. Bx9.1b)**. Without hail nets, hail would destroy crops so frequently that farming would become unprofitable.

FIGURE Bx9.1 Hailstorms in Argentina.

(a) Hail-producing supercells over Argentina, as seen from space.

(b) Hail nets cover crops near Mendoza, Argentina.

Hailstorms cause serious damage. Hailstones are often pea-sized or smaller, but they can cover the ground, making a summertime landscape look like it's been buried by snow. Generally, accumulations of hail are less than 2.5 cm thick, but in 1959, a layer of hail 43 cm (17 in) thick was recorded in Kansas. Even relatively small hail can flatten crops and destroy gardens **(BOX 9.1)**. Larger hail can cause direct damage by killing livestock, denting cars, shattering windows, damaging building exteriors and roofs, and taking down power lines **(FIG. 9.26b)**. The heaviest hailstone ever measured in the United States, which fell during a 2010 storm in South Dakota, weighed 0.879 kg (1.94 lb) **(FIG. 9.26c)**. Fortunately, hail rarely kills people, who usually have time to take shelter. But in 1888, a hailstorm over an agricultural area in India killed 246 people and injured many more. Crop losses to hail in the United States are estimated at $1.3 billion annually, representing between 1% and 2% of the annual crop value. Property losses caused by hail have been increasing over time, to about the same as crop losses—nearly $1 billion.

Take-home message . . .

Thunderstorms produce several dangerous phenomena. When the charge separation in a thunderstorm becomes great enough, a lightning stroke flashes from one part of a cloud to another or from the cloud to the ground. Strong downdrafts that accompany heavy rains can cause damaging straight-line winds. Strong updrafts in thunderstorms can produce large hail.

QUICK QUESTION How does a cloud-to-ground lightning stroke develop?

FIGURE 9.27 Examples of tornadoes of different sizes.

(a) A small tornado in Kansas.

(b) A large tornado (1 km wide) in Oklahoma.

9.5 Tornadoes

Perhaps no other weather phenomenon can be as terrifying as a **tornado**, a violently rotating funnel, or *vortex*, of air that extends from the base of a severe thunderstorm down to the ground (FIG. 9.27). The strongest tornadoes can damage or destroy steel-reinforced concrete structures, throw trucks long distances, sweep houses completely away, and tear off the surface of a highway. At its base (the end touching the ground), a tornado typically ranges in width from about 50 m to 1 km (164 ft–0.6 mi), but some are much larger. The largest tornado ever recorded, which hit El Reno, Oklahoma, on May 31, 2013, had a width of 4.2 km (2.6 mi) at its base. Tornado winds generally blow between 100 and 320 km/h (62–199 mph), but the fastest tornado wind speed ever recorded by radar, also in the El Reno tornado, clocked in at 486 km/h (302 mph). A tornado moves across the countryside along with its host thunderstorm, so movement at its base can vary from almost none to a forward speed of close to 110 km/h (68 mph). In the United States, a northeast-trending swath running from northern Texas across the midwestern states, where warm, moist air from the Gulf of Mexico borders cold air from Canada along strong fronts in spring and early summer, hosts the world's largest number of tornadoes, giving this region the nickname *Tornado Alley* (FIG. 9.28). But tornadoes are also common across the southeastern United States, including Florida, and they can occur elsewhere in the world. The greatest number of deaths in a single tornado, nearly 1,300 people, occurred in Bangladesh in 1989.

Most tornadoes form over land. (When tornadoes develop over water, they are called *waterspouts*; these tornadoes tend to be fairly weak.) As a tornado moves, it leaves a swath of destruction, called a **tornado track**, on the ground surface (FIG. 9.29a). Most tornadoes are short-lived, traveling only a few kilometers from where they touch down to where they lift and dissipate. But occasionally, a tornado can survive for over an hour and produce a very long track. The longest tornado track on record was left by the great Tri-State Tornado of 1925, which tore a swath of devastation 352 km (219 mi) long across Missouri, Illinois, and Indiana and killed 747 people. In some cases, numerous tornadoes develop during a set of related storms within a relatively short time. The world's largest such **tornado outbreak** was the one that took place in April 2011, which we mentioned earlier in this chapter (FIG. 9.29b).

Most tornadoes develop within supercell thunderstorms, but some form along squall lines or within hurricanes. Let's now look at how and where tornadoes form within supercells.

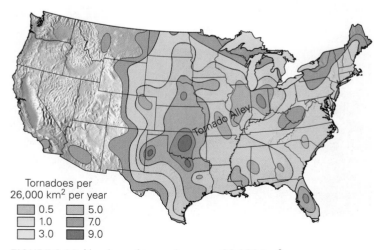

FIGURE 9.28 Number of tornadoes per 26,000 km² per year across the continental United States, averaged over a 27-year period.

FIGURE 9.29 Tornado tracks.

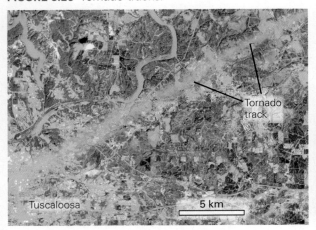

(a) This false-color satellite image shows the track of a tornado near Tuscaloosa, Alabama. Red areas are forested—the tornado ripped out the trees along its track.

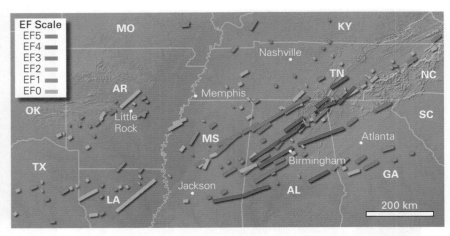

(b) A series of tornado tracks from an outbreak of tornadoes in April 2011.

Supercell Tornadoes

To understand tornado formation in supercell thunderstorms, let's focus on how a tornado develops in a supercell moving from southwest to northeast across the midwestern United States (FIG. 9.30a). Recall that a supercell thunderstorm contains a strong rotating updraft, called a mesocyclone, and that strong downdrafts form on both the forward (northeastern) flank and the rear (southwestern) flank of the mesocyclone (FIG. 9.30b). Tornado development occurs in association with the rear-flank downdraft. As the rapidly downward-flowing air of this downdraft approaches the ground, it spreads outward and produces a roll of rotating air near the ground along the downdraft's spreading outer boundary (FIG. 9.31a). The air in this roll spins around a horizontal axis. A tornado forms when the part of the roll that passes under the base of the mesocyclone gets pulled upward into the mesocyclone (FIG. 9.31b). Initially, the roll warps into a tight arch, but the clockwise-rotating arm, rotating in the opposite direction of the larger mesocyclone, quickly weakens and dissipates, while the counterclockwise-rotating arm, the one rotating in the same direction as the mesocyclone, gets carried into the updraft, where it stretches and narrows to become a spiral or vortex of spinning air. If the spinning vortex is in contact with the ground, it forms a tornado (FIG. 9.31c). The immense wind speed of the tornado

FIGURE 9.30 Setting the stage for tornado formation in a supercell thunderstorm.

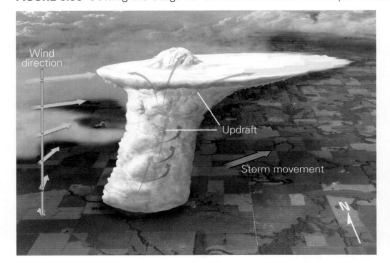

(a) A supercell migrates northeastward across the plains.

(b) Strong winds aloft cause the updraft (red arrows) to tilt. Both forward-flank and rear-flank downdrafts develop. When the downdrafts reach the ground, they spread outward.

FIGURE 9.31 A close-up of the rear-flank downdraft of a supercell, showing the formation of a tornado.

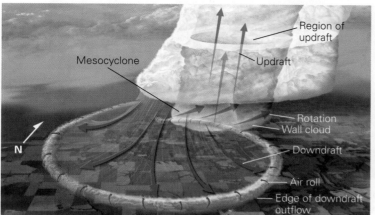

(a) The boundary of the rear-flank downdraft forms a horizontal roll of rotating air close to the ground.

(b) The part of the roll under the updraft gets drawn upward and tilted into the updraft. As it stretches and thins, it spins faster.

develops because as the vortex stretches vertically and narrows, the air within it rotates faster and faster, similar to the way a figure skater spins faster when she pulls in her arms. Note that although the process by which a tornado forms resembles the way a mesocyclone forms, a tornado and a mesocyclone are different phenomena, and they have different dimensions. A tornado extends downward from within a wall cloud, and it is much narrower than the wall cloud.

If a tornado were composed only of air, it would be invisible. Some tornadoes appear only as a dust swirl below the wall cloud of a supercell. Most tornadoes become visible because the rapidly rotating air within them creates very low air pressure. As a result, as moist air from around the tornado is drawn into it, the moisture condenses and forms a spinning cloud—the visible *funnel cloud* that we see. When a tornado moves across the ground, dirt and debris it rips from the ground become incorporated into the funnel, making it even more visible and causing its dark appearance.

As we've noted, most tornadoes are small and short-lived, but some grow to be over 1 km wide. These very large tornadoes may include several small vortices (known as *suction vortices*) spinning around the main vortex (FIG. 9.32). How can a tornado become particularly wide, and how do these wide ones evolve into *multiple-vortex tornadoes*? Tornadoes, as we've seen, consist of rapidly rotating, rising air. In some tornadoes, the extremely low pressure that develops at the base of the vortex can draw air downward within the tornado core. In other words, a downdraft forms aloft within the tornado circulation and descends, inside the tornado, toward the ground. This process pushes the sides of the tornado outward, so that the tornado becomes wider. When the downdraft reaches the ground, the difference in wind direction between the downdraft outflow and the rotating winds of the tornado

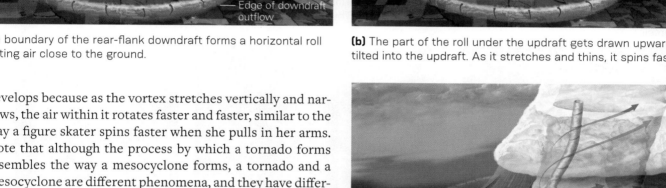

(c) The vortex tightens into a tornado and protrudes from the base of the wall cloud.

causes the main tornado to break down into several smaller vortices moving around the main tornado. The most violent winds in the tornado occur in the suction vortices, because the speed of the wind in the main tornado is enhanced by the rotational speed of the suction vortex. The most destructive and intense tornadoes are nearly always multiple-vortex tornadoes, which have long lifetimes and produce particularly wide swaths of devastation. The 2013 El Reno, Oklahoma, tornado, for example, which set records for size and wind speed, was a multiple-vortex tornado. It was on the ground for 40 minutes, tracking over 26 km (16 mi). The tornado caused 8 fatalities and 151 injuries, but only $40 million in damage because it did not cross a major metropolitan area. The Xenia, Ohio, tornado in 1974, the first multiple-vortex tornado to be filmed, killed 32 people and destroyed a significant portion of the town.

FIGURE 9.32 Formation of a multiple-vortex tornado.

(a) A small tornado, consisting of a single vortex, forms beneath the wall cloud at the base of a supercell.

(b) A downdraft forms in the middle of the tornado and pushes the sides of the tornado outward. The right figure indicates air flow.

(c) The central downdraft reaches the ground. The whole tornado is now almost as wide as the wall cloud.

(d) Within the overall huge tornado, a number of short-lived but very intense suction vortices form. These vortices contain the most intense winds.

Non-supercell Tornadoes

The largest tornadoes occur in association with supercells, but not all tornadoes form in such storms. Tornadoes can also develop along squall lines and in hurricanes. Such non-supercell tornadoes develop where significant *horizontal wind shear* has developed. This means that air on one side of a front is moving horizontally in a direction opposite to the air on the other side.

In the case of a squall line formed along a northeast-southwest-aligned cold front, for example, the wind on the warm side of the front may be blowing toward the northwest, whereas the wind on the cold side may be blowing toward the east (FIG. 9.33a). If it is strong enough, shear between these two winds causes air to start rotating around a vertical axis, producing small vertical vortices (FIG. 9.33b). The updraft of an overlying thunderstorm can pull one of these vortices upward, causing it to stretch vertically and spin faster to become a short-lived tornado. This process can take place simultaneously at many locations along a front—in rare cases, producing a line of tornadoes (FIG. 9.33c). The winds within these tornadoes are generally weaker than those of supercell tornadoes, but they can cause local damage.

As we'll see in Chapter 11, the outer regions of hurricanes consist of large lines of thunderstorms that spiral outward from the central core, or eyewall, of the hurricane. These lines of storms, called rainbands, produce heavy rain closer to the hurricane center and lighter rain in the outer margins of the hurricane. If the air along either side of a rainband is moving horizontally at different velocities, this horizontal shear can produce vertical vortices very similar to those formed along a squall line. Because they are cloaked by heavy rain, the tornadoes within hurricanes aren't easy to see, but, like weak tornadoes in squall lines, they can cause local damage.

Defining the Intensity of a Tornado

The damage caused by a tornado depends on wind velocity, so it would be convenient if we could classify tornadoes based on their wind speed. However, instruments that can measure wind speed are rarely available near the path of a tornado, and

FIGURE 9.33 Formation of non-supercell tornadoes along a squall line.

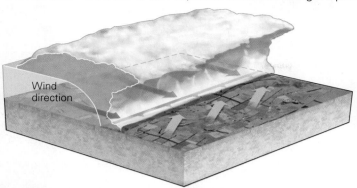

(a) Horizontal wind shear along a front.

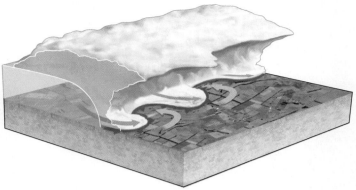

(b) Horizontal wind shear forms small vortices that may be pulled upward by thunderstorms.

they would be at risk of destruction anyway if a tornado struck them. To address this classification challenge, Ted Fujita, a professor at the University of Chicago, proposed classifying the intensities of tornadoes on the basis of the damage they caused. His system, published in 1971, came to be known as the *Fujita scale*. In 2007, a more accurate version of the Fujita scale, called the **Enhanced Fujita (EF) scale**, became available. The EF scale takes into account laboratory studies that better constrain the relationship between wind speed and observed damage, and it is this scale that we use today (TABLE 9.1).

The EF scale divides tornadoes into six categories, labeled EF0 through EF5. An EF0 tornado is very weak, whereas an EF5 tornado is catastrophic, as its wind speeds exceed 322 km/h (200 mph) at the time EF5 damage occurs. An EF5 tornado can sweep even the sturdiest house away and can rip up the ground like a monstrous bulldozer (BOX 9.2). But even an EF1 tornado can be dangerous to people who are not in adequate shelters.

Use of the EF scale still presents some challenges. First, the EF rating does not characterize the size of a tornado, but in general, more intense tornadoes are larger. Second, as a

(c) This process can form vortices at several locations, as in this row of tornadoes.

tornado evolves, its EF rating may change. For example, it may start out by causing EF1 damage, accelerate to causing EF5 devastation, and then weaken to an EF1 before dissipating entirely. Its EF rating in the record books is based on the worst damage the tornado causes along its track (**VISUALIZING A DISASTER**, pp. 352–353). Finally, not all tornadoes cross

TABLE 9.1 The Enhanced Fujita Scale

EF rating	Wind velocity (km/h)	Wind velocity (mph)	Typical damage
EF0	105–137	65–85	*Minor damage:* Peels off some shingles; breaks branches; topples weak trees
EF1	138–177	86–110	*Moderate damage:* Strips roofs of shingles; overturns mobile homes; breaks windows
EF2	178–217	111–135	*Considerable damage:* Tears off roofs; shifts houses off foundations; destroys mobile homes; uproots large trees; flings debris and lifts cars
EF3	218–266	136–165	*Severe damage:* Destroys upper stories of well-built houses; severely damages large buildings; overturns trains; throws cars; debarks trees
EF4	267–322	166–200	*Extreme damage:* Completely levels well-built houses; throws cars and trucks
EF5	>322	>200	*Catastrophic damage:* Collapses tall buildings; severely damages structures made of reinforced concrete; carries cars and trucks more than a kilometer

VISUALIZING A DISASTER
Life Cycle of a Large Tornado

The tornado touches down.

EF0

The tornado strengthens.

EF1

EF2 and EF3

The tornado first reaches its peak strength.

Each of the photos below illustrates characteristic damage at the specified intensity.

EF0

EF1

EF2

BOX 9.2 DISASTERS AND SOCIETY

The threat to urban centers

A tornado evolves over time: it starts small, grows larger, and then eventually dissipates. The swath of destruction that a tornado produces as it passes through an area typically corresponds to the shape and size of the tornado at various times during its life cycle. Most violent tornadoes start as a narrow funnel, then widen into a giant vortex, often with multiple suction vortices. The tornado maintains this size and intensity for minutes to an hour or more, but eventually narrows and stretches into a rope-like shape before it dissipates.

One of the deadliest tornadoes to follow this pattern was the Joplin, Missouri, tornado of May 22, 2011 (FIG. Bx9.2a).

FIGURE Bx9.2 The Joplin, Missouri, EF5 tornado.

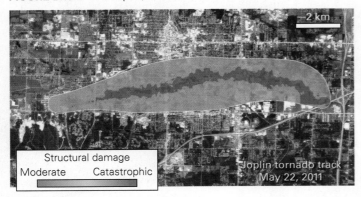

(a) The track of the Joplin tornado, showing EF ratings for damage in different parts of the city.

(b) Aerial view of the area of EF5 damage, including St. John's Regional Medical Center.

(c) Helicopter blown off the roof of the medical center.

(d) Complete destruction of a neighborhood.

Leaving 158 people dead, it was the single deadliest tornado since 1947, and the seventh-deadliest tornado in US history. Estimates of damage approached $3 billion, and over 16,000 insurance claims were filed. The tornado reached a maximum width of nearly 1.6 km (1 mi) during its rampage through the southern part of Joplin. The St. John's Regional Medical Center endured so much damage to its foundation and structural system that the building had to be demolished (FIG. Bx9.2b, c). Homes across Joplin were wiped clean off their foundations (FIG. Bx9.2d).

How does a city like Joplin prepare for a tornado? The first step is to identify vulnerable facilities and train facility staff about procedures to move people to safety. Locations where people gather—schools, churches, park districts, hospitals, shopping malls, theaters, sports and entertainment venues, transportation centers, and care facilities—all must have emergency action plans. Hospitals must have medical personnel available to respond to an emergency, even one that strikes at a distance from the hospital. When the hospital itself is threatened, staff must safely move patients to the building interior, where they will be better protected. This task is a challenging one. One hospital, which usually has about 350 patients located near exterior walls, estimated that at least 15–20 minutes would be required to move all patients to its interior—more than the typical lead time of NWS tornado warnings. Schools, similarly, must quickly move students to safe locations in building interiors. Sports venues, such as baseball stadiums, can host thousands of people in a small open-air space. Imagine trying to move everyone in a full stadium to safety quickly without creating panic. Each type of facility has unique vulnerabilities, requirements for emergency actions, and lead times required to execute them.

Some issues are hard to manage, even with the best preparation. For example, trains transport hazardous materials right through many city centers. A tornado strike on such a train would set in motion a particular hazardous event that would require a particular emergency response: evacuating people from the plume of hazardous material emerging from destroyed train cars.

To prepare for the range of possible scenarios, emergency managers, first responders, and NWS personnel hold regular planning sessions and exercises to anticipate potential emergency response scenarios. They meet with individuals in charge of emergency response for vulnerable facilities to understand their needs and requirements. Expert training, good use of technology, and clearly established lines of communication are all part of society's efforts to reduce casualties when disasters like the Joplin tornado strike.

through areas containing buildings that may be damaged. For example, the El Reno multiple-vortex tornado, whose record wind speeds could have caused EF5 damage, was classified as an EF3 because the worst damage found along its path was EF3 damage. Tornadoes, even large ones that remain over open fields, are classified as EF0 or EF1 because they do little damage.

Tornado Destruction, Detection, and Safety

As we've seen, tornadoes destroy property and landscapes, and they often kill and injure (FIG. 9.34). On average, about 60 people per year die in tornadoes in the United States, but in some years, death tolls are much higher. For example, in 2011, the year of the Joplin tornado, 553 people lost their lives due to tornadoes. Tornado outbreaks, when many events occur in association with a particularly strong front in a mid-latitude cyclone, can leave a high death toll in their wake. Most injuries and fatalities happen when people are struck by flying debris. Casualties also happen when vehicles carrying people are picked up and tossed in the air, or when roofs, walls, and trees collapse onto people.

Given the devastation wrought by tornadoes, accurate prediction of their arrival is a key to survival. Today, with Doppler radar (see Box 8.1), tornado signatures appear clearly on a radar screen. For example, flying debris scatters radar energy, so the *tornado debris signature* (TDS) generated by a tornado appears as a bright spot on the screen at the tip of a hook echo (FIG. 9.35). Doppler radar can also detect rotating winds because the part of a vortex moving toward the observer looks different from the part moving away. Therefore, when a meteorologist sees a supercell thunderstorm with a hook echo on the southwestern side, rotation within the hook, and a debris signature at the tip of the hook, it's a sure sign that a tornado is on the ground and that a warning should go out.

FIGURE 9.34 During a strong tornado, trees snap off and even brick walls fall.

FIGURE 9.35 Radar image of a tornado that struck Alabama in 2013. The bright spot at the tip of the hook echo comes from debris thrown upward by the tornado.

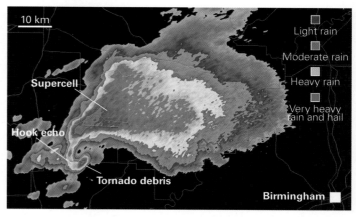

The NWS Storm Prediction Center issues a tornado watch for specific areas when weather conditions favor tornado formation. A local NWS office issues a tornado warning either when observers have spotted a tornado visually or when one has been detected by radar. NWS warnings trigger weather alerts on radio, TV, and cell phones; in some communities, emergency managers activate warning sirens.

The NWS recommends that families, schools, and businesses prepare safety plans and hold regular tornado drills. In general, when a tornado warning goes out, you should move to a predetermined safe location, preferably a basement. If an underground space is not available, the safest choice is an interior room or a hallway on the building's lowest floor. In an interior bathroom, plumbing provides additional wall support. Take shelter with as many walls as possible between you and the outdoors, and avoid windows. Mobile homes are dangerous, as they may be crushed or rolled. Abandon your car, especially in an urban or congested area, and seek a sturdy shelter. If you are caught outdoors, move away from potential sources of airborne debris, and lie flat in the lowest spot available. In all cases, when severe weather threatens, early action can be the key to safety and survival.

Take-home message . . .
Tornadoes are rapidly rotating funnels of air that extend from the ground to the base of a severe thunderstorm. They form when a rotating cylinder of air gets drawn into the storm's updraft, which causes it to stretch, narrow, and spin ever faster. The largest tornadoes form within supercell thunderstorms, but tornadoes can also develop along squall lines and in hurricanes. Tornado intensity is classified using the Enhanced Fujita scale, which is based on the damage caused.

QUICK QUESTION What creates the visible funnel of a tornado?

Chapter 9 Review

Chapter Summary

- For a thunderstorm to form, three conditions must exist: abundant moisture in the lower atmosphere, a lifting mechanism, and atmospheric instability.
- Lifting occurs when air is forced upward by a front, gust front, or high relief. Air can also rise when it is warmed from below by the Earth's surface.
- Because condensation of water vapor releases latent heat, rising saturated air becomes buoyant relative to its surroundings when atmospheric instability is present, and it may flow upward in an updraft that extends all the way to the tropopause. In stable atmospheric conditions, lifted air sinks back down or will not rise beyond where it is lifted.
- Ordinary thunderstorms form where winds don't vary much with altitude, so these storms do not rotate. They evolve through three stages: the development stage, when warm, moist air is lifted and becomes buoyant; the mature stage, when precipitation produces a downdraft; and the dissipation stage, when the updraft ceases and the clouds dissipate.
- In squall-line thunderstorms, several thunderstorms develop along a gust front or cold front. The combined outflow of cool air from such storms can trigger formation of new thunderstorms and produce strong straight-line winds.
- Supercell thunderstorms develop in the presence of vertical wind shear, which results in a rotating updraft. Supercell updrafts are so strong that a large overshooting top rises above the asymmetrical anvil cloud of the storm.
- The winds in the upper troposphere cause a supercell to tilt downwind, so that rain falls downwind from the updraft. As a result, downdrafts associated with falling rain don't counteract the updraft, and the supercell can survive for several hours.
- Lightning is a form of electrostatic discharge. It is caused by charge separation in a thunderstorm: the base of the cloud accumulates negative charge, and the upper parts of the cloud accumulate positive charge.
- Because of charge separation, a spark—lightning—can jump across a cloud, or between a cloud and the ground.
- If lightning strikes a person, instant death or permanent harm, such as burns and neurological damage, can result.
- Heat within the lightning-stroke channel causes air to expand suddenly and then contract violently. The resulting shock wave is heard as thunder.
- Hailstones form when supercooled water droplets freeze onto ice particles in clouds. The largest hailstones form in supercell thunderstorms.
- Hail can cause major agricultural losses by flattening crops and other economic losses by damaging property.
- Tornadoes typically occur on the rear flank of a supercell thunderstorm. They form when a roll of rotating air formed near the ground along the boundary of the rear-flank downdraft gets drawn into the storm's updraft.
- Tornado intensities are determined from the damage they cause. The Enhanced Fujita (EF) scale classifies tornado intensities from 0 to 5. An EF0 tornado produces minor damage, while an EF5 tornado produces catastrophic damage.
- To avoid injury during a tornado, seek shelter in a basement or strong building. Put as many walls as possible between yourself and the tornado. Abandon cars and mobile homes. If outside, lie in the lowest spot available.

ANOTHER VIEW A shelf cloud marks the arrival of strong winds ahead of a squall line in Oklahoma.

Review Questions

Blue letters correspond to the chapter's learning objectives.

1. What distinguishes a thunderstorm from other kinds of rainstorms? What characteristics does a thunderstorm have if it is classified as "severe"? **(A)**
2. The supercell thunderstorm is viewed from which direction on the figure? **(A)**
3. List the three conditions that must exist for a thunderstorm to develop. **(A)**

4. Describe the various lifting mechanisms that can trigger initiation of a thunderstorm. **(A)**
5. Explain the difference between stable and unstable atmospheric conditions. **(A)**
6. Are strong surface straight-line winds associated with a thunderstorm updraft or downdraft? Explain your response. **(B)**
7. When looking at a radar image of a supercell, how would you identify the location of the updraft, forward-flank downdraft, and rear-flank downdraft? **(B)**
8. Explain how a supercell thunderstorm forms and how it is different from an ordinary thunderstorm. Why do these storms typically have broad, asymmetrical anvils and overshooting tops? What do they look like on a radar screen? **(B)**
9. Relative to a supercell thunderstorm, what is a hook echo, and where is a tornado typically located relative to the hook? **(B)**
10. What is a tornado debris signature? **(B)**
11. Identify the positive polarity and negative polarity cloud-to-ground strokes on the figure. **(C)**

12. When charge separation occurs in a thunderstorm, in which part of the clouds do the positive and negative charges accumulate? **(C)**
13. During a lightning storm, are you safer inside or outside of a car? **(C)**
14. What is a "bolt from the blue"? **(C)**
15. What is supercooled water, and what is its role in hail formation? **(D)**
16. What is a hail swath? **(D)**
17. Why is hail more common in supercell thunderstorms than in other types of thunderstorms? **(D)**
18. What sector of the economy is particularly vulnerable to hail damage? **(D)**
19. Explain the basic process of tornado formation. Why do tornadoes typically develop on the rear flank of a supercell thunderstorm? **(E)**
20. Is it possible for a giant tornado with 250 mph winds to be classified as an EF1? Explain your answer. **(E)**
21. On what basis do meteorologists classify tornadoes? **(E)**
22. How is a tornado watch different from a tornado warning? What should you do if a tornado is coming your way? **(F)**
23. Suppose you are caught outside in an open area during a thunderstorm. If your hair stands on end, what is the best position to take? If you are about to be struck by a tornado, what is the best position to take? Are these the same? **(F)**
24. What is a microburst, and how can it pose a hazard to aircraft? **(F)**

On Further Thought

25. You are the emergency manager at a sports stadium filled with people, and a thunderstorm approaches. What dangerous conditions are possible, and what measures should you take to keep people in the stadium safe? **(F)**
26. Emergency managers grapple with the question of how much warning time the public should receive for the best safety response in advance of an approaching tornado. Discuss potential pros and cons of providing warnings centered on 5, 10, 20, or 30 minutes prior to a tornado strike on a town. **(F)**

10

COLD AND ICY
Hazardous Winter Weather

By the end of the chapter you should be able to . . .

A. distinguish between the air temperature and the wind chill temperature.

B. identify actions to take to avoid the dangers of whiteout conditions and hypothermia during blizzards.

C. describe how ice storms form, and why they can result in power outages.

D. explain where different types of winter weather typically occur within a mid-latitude cyclone.

E. contrast the hazards posed by winter mid-latitude cyclones over continents and along coasts.

F. explain why communities east and south of the Great Lakes experience heavy snowfall in winter.

G. give reasons why mountain snowstorms are both necessary and dangerous.

H. list safety precautions to take in winter, and the many ways that winter weather affects human health and the economy.

10.1 Introduction

The Atlanta, Georgia, metropolitan area is home to 5.5 million people spread across many separate municipalities. The sprawl of this metro area requires children to take school buses or be driven to school, and workers to commute long distances on crowded highways. Though generally thought of as a warm-weather city, Atlanta actually lies on the southern fringe of the region of North America that is subject to *hazardous winter weather*: atmospheric conditions during which cold, snow accumulation, ice accumulation, and sometimes wind can disrupt traffic and commerce and possibly cause casualties and damage. Because Atlanta endures such weather so rarely, officials and residents aren't accustomed to planning for it, and the metro area doesn't have sufficient equipment to address its consequences. The events of January 2014 demonstrate how this situation led to a local disaster that came to be known in the media as the "snowpocalypse."

The story of the snowpocalypse began at 12:07 P.M. on Sunday, January 26, when the US National Weather Service (NWS) issued a *winter weather advisory* **(TABLE 10.1)** for the Atlanta region, announcing that snow would fall across the region sometime on Tuesday, January 28, as a cold front moved across the southern United States **(FIG. 10.1a)**. At 3:12 P.M., the NWS issued a more formal *winter storm watch*, and a day later, it upgraded the watch to a *winter storm warning*, stating that 2.5–5.1 cm (1–2 in) of snow accumulation would begin near midday and would last into the night. The NWS emphasized that reduced visibility would make travel dangerous, so it urged people not to travel except in an emergency. As expected, snow began to fall on Tuesday at 12:30 P.M.

Despite the NWS warning, schools, businesses, and government offices opened as usual on Tuesday morning. But because of the storm, most decided to release students and employees early in the afternoon. Consequently, 2 hours after the snow started falling, the highways began to clog with drivers trying to get home or trying to pick up children from school. Many of these drivers had little experience with controlling vehicles on icy roads, so there were over 1,200 traffic accidents (causing several deaths) in Georgia, some of which blocked highways. By 3:00 P.M., traffic across the Atlanta metro area stood at a standstill, and tens of thousands of people were stranded on highways **(FIG. 10.1b, c)**. The traffic jams trapped almost 100 school buses, so over 2,000 children had to spend the night in buses or at police stations. Many more children were stuck for the night at school because parents couldn't reach them. Eventually, some motorists sought shelter in supermarkets or restaurants, but many shivered in their cars for as long as a day, and emergency vehicles couldn't reach people in need. One truck driver was stopped for 30 straight hours, and at least one baby was born in a trapped car.

◀ Cars disappear as they move down the road into the blinding snowfall of a blizzard.

In addition to the disruption of road travel, businesses, and schools, thousands of flights into and out of Atlanta's airport were canceled, because runways couldn't be kept clear, planes couldn't be de-iced, and reduced visibility slowed flight operations. Because the airport—the world's busiest—serves as a major travel hub, commerce, tourism, and shipping were disrupted internationally.

TABLE 10.1 Terminology for Discussing Hazardous Winter Weather

Winter weather advisory	There is an anticipation that freezing rain and/or snow will accumulate sufficiently to cause inconvenience and hazards.
Winter storm watch	There is a possibility of hazardous winter weather in the next 48 hours; the weather can be a combination of snow, sleet, ice accumulation, and wind, significant enough to cause damage and/or life-threatening situations.
Winter storm warning	The conditions stated for a winter storm watch are occurring or are imminent.
Blizzard warning	Conditions expected to occur during the next 12–18 hours include snow and/or blowing snow, reducing visibility to ≤ 0.40 km (≤ 0.25 mi) for 3 hours or longer, and sustained winds of ≥ 56 km/h (≥ 35 mph).
Ice storm warning	A hazardous amount of ice (generally meaning ≥ 0.64 cm, or ≥ 0.25 in) will accumulate on roads, trees, power lines, and other structures.
Freeze watch	Temperatures are expected to fall below 0°C (32°F) within the next 1–1.5 days during the growing season, damaging plants and crops.
Freeze warning	Temperatures will fall below 0°C (32°F) during the growing season, damaging plants and crops.
Frost advisory	A potential exists that temperatures may fall below 0°C (32°F) during the growing season, damaging plants and crops.
Wind chill advisory	The combination of cool temperatures and wind could lead to frostbite for exposed skin or to hypothermia.
Wind chill warning	The combination of cool temperatures and wind will quickly lead to frostbite for exposed skin or to hypothermia.

Note: The temperatures, amounts of snowfall, or amounts of ice accumulation that trigger advisories, watches, and warnings vary with location because conditions that represent a hazard in a region that rarely receives winter weather are different from those in a region that commonly experiences such weather.

FIGURE 10.1 A snowstorm can paralyze a city.

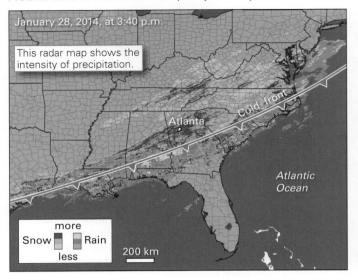

(a) Precipitation along a cold front brought winter weather to Atlanta in 2014.

(b) Traffic was at a standstill for days as accidents clogged highways.

(c) Highways in the Atlanta metro area that experienced standstill traffic are marked in red, and those with slow traffic are marked in yellow.

How, in this age of information, when the NWS had issued a warning well in advance, could such a disaster happen? There are many reasons. Reliance on commuting in private cars, due to the lack of widespread public transportation, set the stage for the disaster. Managers at businesses and administrators at schools didn't think to send people home before the storm, or to spread their departures over a range of times to avoid overcrowding the roads in foul weather. Drivers inexperienced in handling slippery roads caused accidents, which became the tipping point for entirely halting traffic flow. Finally, the distribution of authority across so many municipalities made coordinating emergency responses to the storm nearly impossible.

Clearly, significant casualties, as well as costly disruptions of commerce, transportation, and infrastructure, can turn hazardous winter weather into a disaster! This chapter characterizes such weather and the dangers it can present. We begin by examining the consequences of cold itself. We then turn our attention to the characteristics of winter storms that make them hazardous. Using this background, we describe the specific weather conditions that produce such hazards, most notably mid-latitude cyclones. Finally, we describe the impacts of winter weather on society and consider how society can mitigate these impacts.

10.2 The Dangers of Cold

Cold has long been a cause of human misery, and indeed, it has altered human history. Napoleon's Grand Armée of France, several hundred thousand soldiers strong, marched against Russia in 1812, but they fell victim to the Russian winter as troops by the thousands perished in the cold. During World War II, Hitler's army was weakened on its eastern front by the coldest winter in a century. Frostbite, in epidemic proportions on the German side, turned the course of the war to benefit the well-prepared Russian army (FIG. 10.2). In nearly every war in wintry climates, battles have been decided or advances halted by unrelenting cold and snow.

FIGURE 10.2 Winter has changed the outcome of wars, including World War II.

Cold is a hazard not only to armies, but to economies and ordinary people. Early arrival of cold weather can wipe out crops. In past centuries, before worldwide transportation of food, early frosts portended famine and death as food supplies dwindled over the long winter. On a more personal scale, cold can lead to deadly hypothermia and frostbite.

Hypothermia, Frostbite, and Wind Chill

Cold is a relative term—what you perceive as cold depends on where you live. A resident of Fairbanks, Alaska, for example, may consider a January outdoor temperature of −23°C (−10°F) warm, while a resident of southern Florida would consider 4.5°C (40°F) bone chilling. Nevertheless, regardless of where we live, we humans are warm blooded, and we need to maintain our *core body temperature* (that of the brain, heart, lungs, and internal organs) within a very limited range centered on 37°C (98.6°F). If it drops below that range, our normal muscular and brain functions will be impaired. We accommodate to cold by wearing clothing that acts as *insulation* (slows the outward flow of heat) and by burning fuel to heat the buildings we occupy.

What happens to our bodies when we can't maintain our core temperature? Exposure to extreme cold for a short time, or moderate cold for a prolonged time, can drive a person's core temperature below the safe range. If this happens, **hypothermia** sets in. The initial symptoms of hypothermia, minor impairment of motor functions and involuntary *shivering* (a reflex that causes muscles to make small movements to generate warmth), begin when the body's core temperature falls to 37°C–35.5°C (98.6°F–96°F). If the core temperature continues to fall, dropping to 35°C–33.8°C (95°F–93°F), violent shivering, accompanied by irrational behavior and loss of coordination, ensues. Life-threatening hypothermia begins when the core temperature drops below 33.3°C (92°F). At this stage, shivering stops, the body shuts down all nonessential functions in an effort to preserve its remaining heat, and the victim cannot stand upright. Without immediate action to warm the victim, death soon follows.

Hypothermia can result from prolonged exposure to cold, but above-freezing, temperatures. **Frostbite**, the actual freezing of body tissues, can happen only when temperatures drop below freezing. A human body tries to preserve heat in its core, so when subjected to freezing temperatures, it restricts the flow of blood and heat to the extremities. As a result, frostbite initially affects the ears, nose, fingers, and toes. While you can recover from mild frostbite with few ill effects, severe frostbite kills skin deeper down, and in the worst cases, it can result in death of muscle or even of bone tissue. Dead tissue cannot fight off infection and therefore can suffer gangrene. If frostbite treatment is not prompt, such problems may lead to surgical removal of tissue, or even amputation of fingers, toes, or limbs.

While walking outside on a bitterly cold day, you feel a chill on your exposed skin. The chill becomes more intense if the wind is blowing, and especially if your exposed skin is wet. What is happening? On a dry, windless cold day, your body transfers heat to air right next to your exposed skin. In still air, a thin layer of warm air remains next to your skin. Because the rate of heat transfer from a warmer material to a cooler material depends on the temperature difference between the two, the presence of this warm layer prevents your body from losing heat quickly. Wind strips away that warm layer, replacing it with colder air, which extracts heat from your body more quickly. As the wind strengthens, it becomes harder to maintain a warm layer. As a result, stronger wind causes your body to lose heat at a faster rate. Skin cools even faster when it's wet, for the process of turning liquid to vapor during evaporation extracts additional heat from your body (see Chapter 8).

To communicate the combined effects of cold and wind to the public, meteorologists use the **wind chill temperature**, or *wind chill factor*, to represent the heat loss that dry human skin experiences at a particular air temperature and wind speed. **Table 10.2** shows how the wind chill temperature varies with air temperature and wind speed, and it depicts the duration of exposure that will lead to frostbite. Progressively darker shaded areas of the table show the temperatures and wind speeds at which frostbite will begin to affect exposed skin in less than 30, 10, and 5 minutes, respectively.

TABLE 10.2 Wind Chill Temperatures

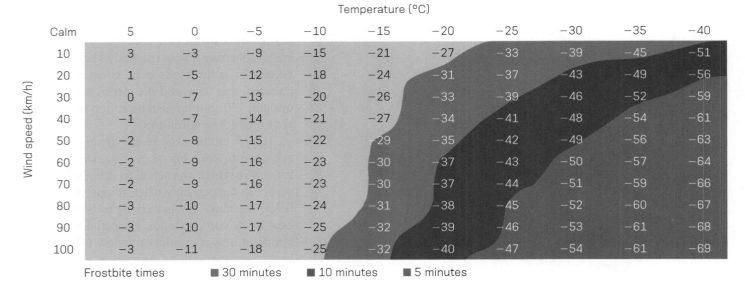

Wind speed (km/h) \ Temperature (°C)	5	0	−5	−10	−15	−20	−25	−30	−35	−40
Calm										
10	3	−3	−9	−15	−21	−27	−33	−39	−45	−51
20	1	−5	−12	−18	−24	−31	−37	−43	−49	−56
30	0	−7	−13	−20	−26	−33	−39	−46	−52	−59
40	−1	−7	−14	−21	−27	−34	−41	−48	−54	−61
50	−2	−8	−15	−22	−29	−35	−42	−49	−56	−63
60	−2	−9	−16	−23	−30	−37	−43	−50	−57	−64
70	−2	−9	−16	−23	−30	−37	−44	−51	−59	−66
80	−3	−10	−17	−24	−31	−38	−45	−52	−60	−67
90	−3	−10	−17	−25	−32	−39	−46	−53	−61	−68
100	−3	−11	−18	−25	−32	−40	−47	−54	−61	−69

Frostbite times ■ 30 minutes ■ 10 minutes ■ 5 minutes

Cold Waves and the Polar Vortex

Meteorologists refer to a relatively rapid fall in temperature that requires society to take precautions against cold as a **cold wave**. The criteria for a cold wave are regional—no precise range of temperatures defines a cold wave, as the temperatures considered to be cold in Florida and in northern Minnesota, for example, will naturally be different.

To understand the cause of cold waves, we must first focus our attention on the nature of atmospheric circulation at high latitudes in the middle and upper troposphere. Recall from Chapter 8 that the *polar front*, the boundary between cold polar air and warmer temperate and subtropical air, stretches around the globe. As the polar-front jet stream travels counterclockwise in the upper troposphere above the polar front, the cold air residing to the north of it (in the northern hemisphere) also circulates counterclockwise around the North Pole. Since any flow of a fluid circulating around an axis is called a *vortex*, this overall flow of cold air (a fluid) circulating around the North Pole (an axis) is called the **polar vortex** (FIG. 10.3a). In map view, the polar-front jet stream effectively traces out the southern edge of the polar vortex.

Because cold air, and therefore the polar front, may extend farther south at some longitudes than at others, the polar-front jet stream, in map view, follows a somewhat wavy path, more or less within latitudes of 40°–60° N (see Fig. 10.3a), close to the Canada-US border. In some winters, the waviness of the flow becomes more extreme, particularly after a mid-latitude cyclone has passed across North America. The polar front, under these conditions, can extend far to the south and may pass over the Gulf Coast and into central Florida. When this happens, very cold air can blanket a large area of the continent. During such cold air invasions, the polar vortex distorts so that a large wave appears in the jet stream along and over the polar front (FIG. 10.3b). Once established, this configuration of the polar vortex can persist for days or even weeks, often with arctic air reinforcing the cold by continuously sweeping southward over the continent. Notably, North American popular media now associate the term *polar vortex* with cold waves, because the distortion of the polar vortex brings cold weather so far south that it becomes a hazard for people who are not used to such weather, since they live in normally temperate climates.

Cold waves were quite common throughout most of the 20th century. Beginning in the 1990s, and continuing into the 21st century, they have become less frequent. North America's first major cold wave of the 21st century occurred in January 2014. Temperatures fell to record lows at locations all across the United States. Green Bay, Wisconsin, endured −28°C (−18°F) conditions, and Houston, Texas, had a morning low of −6°C (21°F). In fact, on January 7, at least 49 records for low temperatures were set across the United States. Economists estimate that this cold wave contributed to a 2.9% drop in the gross domestic product (GDP) of the country for the month and caused economic losses of $5 billion. These losses reflected disruption of transportation (20,000 flights were canceled, freight and passenger trains were stranded, and roads were impassable), power outages, and destruction of crops such as citrus, primarily in Florida. Over a dozen deaths due to highway accidents and hypothermia were associated with the cold wave. We'll discuss the 2021 cold wave later in this chapter.

FIGURE 10.3 Changes in the flow of the polar vortex can bring cold air to normally temperate regions.

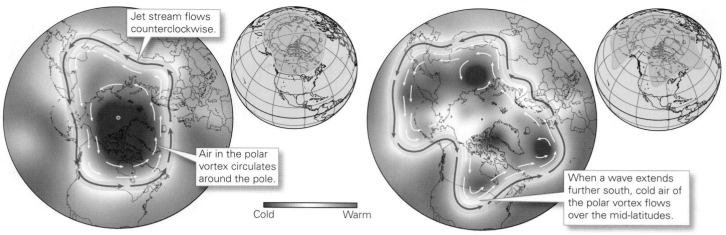

(a) The polar vortex, under normal conditions, flows in a roughly circular pattern around the North Pole.

(b) During cold waves, the polar vortex develops a wavy structure as cold air advances southward out of the Arctic.

Take-home message . . .

Human responses to exposure to extreme cold include frostbite and hypothermia, which progresses to death if not treated. The NWS reports the wind chill temperature to communicate the combined effects of cold and wind on exposed skin. Cold waves are rapid drops in temperature associated with changes in the shape of the polar vortex.

QUICK QUESTION Why is there no universal definition of a cold wave?

10.3 Hazardous Conditions during Winter Storms

Snowfall

Let's turn our attention from cold as a hazard of winter weather to the precipitation that occurs during winter. When the air temperature is cold enough, ice crystals grow in clouds and aggregate to form flakes, which descend to the ground as snowfall. Rates of snowfall vary dramatically. In the United States, classification of *snowfall intensity* depends on visibility through the falling snow: when visibility is greater than 1 km (0.6 mi), the precipitation is termed *light snow*; when it's between 1 km and 0.5 km (0.3 mi), *moderate snow*; and when it's less than 0.5 km, *heavy snow*. Accumulation of snow during a snowstorm depends on how long the snow falls as well as on its intensity. Heavy snow typically produces a layer about 2.5 cm (1 in) thick in an hour. The record for the most snow accumulated in a single 24-hour period was 1.9 m (~6 ft) at a Colorado mountain location. Needless to say, a major snowfall halts commerce and travel (FIG. 10.4a, b). But even a thin layer of snow can have major consequences if it occurs in a location that is not accustomed to snowfall, as we saw in the case of Atlanta at the start of this chapter.

Over the course of a winter, each successive snowfall buries the previous one, unless a warm spell has melted older snow away. Consequently, the **snowpack**, the composite layer of snow that accumulates in an area, can become quite thick, even accounting for compaction by the weight of overlying snow. The snowpack in the mountains of the western United States can exceed 13 m (~42 ft) locally. When this snowpack melts in the spring, it can cause flooding (see Chapter 12).

Blizzards

In everyday English, we tend to equate blizzards with snowstorms. The NWS, however, is more specific: a snowstorm is a **blizzard** only when its winds exceed 56 km/h (35 mph) and visibility is less than a quarter mile for at least 3 hours. Blowing snow during a blizzard sometimes produces *whiteout conditions*, in which visibility decreases to nearly zero and an observer can't distinguish the sky from the ground (FIG. 10.4c). Being in a whiteout can be disorienting and dangerous—people have walked away from a stranded vehicle, only to become lost in the whiteout and die of hypothermia. Blizzard winds can transport snow horizontally and pile it into mounds or ridges called **snowdrifts**. Even if the total accumulation of new snow isn't extreme, snowdrifts can grow to many meters high, and they can completely bury a road and the vehicles on it (FIG. 10.4d).

FIGURE 10.4 Hazards of snowstorms.

(a) A snowstorm creates hazardous driving conditions.

(b) Air traffic may be grounded during a snowstorm.

(c) A blizzard is characterized by reduced visibility.

(d) Snowdrifts may block highways.

Surprisingly, blizzard conditions are possible even in otherwise clear weather, if strong winds blowing across freshly fallen snow loft enough snow into the air. Such **ground blizzards** typically extend less than 10–15 m (33–49 ft) above the ground, but can persist for hours, long after snowfall has stopped. For a such an event to develop, the snow must be *dry*, meaning that snowflakes are completely frozen and don't stick together. Ground blizzards can make highways dangerous by creating and obscuring ice patches on the road surface. Regions in the Great Plains often experience ground blizzards because the absence of trees permits strong winds to blow across open fields at ground level.

Ice Storms

In December 2008, a winter storm crossed the northeastern portion of the United States and southeastern Canada. In the northern parts of the storm, heavy snow fell, and in the southern part, rains drenched the countryside. But in a relatively narrow band that crossed New England, precipitation reached the ground in liquid form, but instantly froze on cold surfaces. Quickly, the ground, tree branches, power lines, and highways in this region were coated with ice. Eventually, the ice formed a layer 19–25 mm (0.75–1.0 in) thick. Unlike fresh snow, which contains a lot of air and has low density overall, solid ice can add hundreds of kilograms of weight to a tree limb or power line. All through the night, people living in the affected area could hear the sharp snap of branches and tree trunks, followed by a crash as they fell to the ground. When the Sun rose in the morning, the area looked like a fairyland of sparkling ice, but roads were blocked by fallen branches and power lines. Some 1.7 million people were without electricity for a week or longer.

New England had experienced a devastating **ice storm**, an event during which the landscape and everything on it becomes covered with ice. Ice storms most commonly develop when a warm, moist air mass flows upward over a cold air mass along a warm front, forming a wedge of warm air aloft. Under these conditions, air with temperatures above freezing overrides a layer of cold air with temperatures below freezing. Clouds that develop in the moist, rising air above the warm front produce

FIGURE 10.5 Formation of freezing rain and sleet.

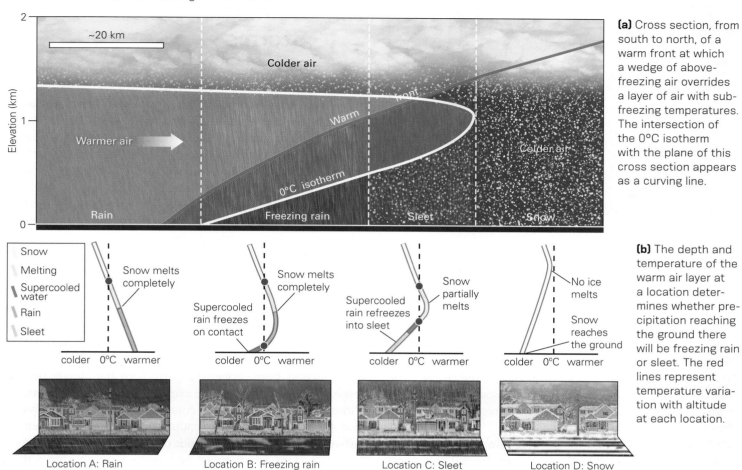

(a) Cross section, from south to north, of a warm front at which a wedge of above-freezing air overrides a layer of air with sub-freezing temperatures. The intersection of the 0°C isotherm with the plane of this cross section appears as a curving line.

(b) The depth and temperature of the warm air layer at a location determines whether precipitation reaching the ground there will be freezing rain or sleet. The red lines represent temperature variation with altitude at each location.

precipitation. In the winter, the *0°C isotherm*, the invisible boundary that separates warmer air (above freezing) from colder air (below freezing), envelops a wedge or tongue of warmer air that crosses the warm front **(FIG. 10.5a)**. The character of the precipitation that reaches the ground at a given location varies depending on the position of the location relative to this wedge. At high altitude, above the 0°C isotherm, all precipitation falls as snow. Above Location A **(FIG. 10.5b)**, the snow passes down across the 0°C isotherm at an elevation of 0.5–1.5 km (0.3–1 mi) and descends into warmer air, where it melts. The air stays above freezing all the way down to the ground, so the precipitation reaches the ground as rain. Above Location B, however, falling rain crosses the 0°C isotherm again, close to the ground. It then descends in colder air, so raindrops become *supercooled*, meaning that they remain in liquid form, even though they are below freezing. These supercooled drops freeze on contact with the ground, buildings, trees, or power lines. Ice formed from supercooled rain that freezes when it contacts a cold solid surface is called **freezing rain**. Above Location C, the wedge of warm air aloft is not as deep, and its temperature, although above freezing, is

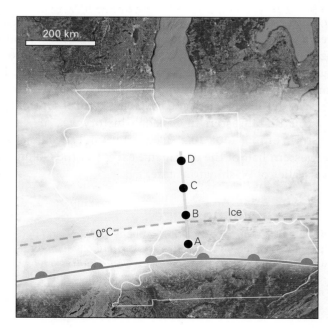

(c) The zone where ice accumulates (pink) is typically narrow and lies north of the warm front. The line from point A to point D shows the location of the cross section diagrammed in part (a).

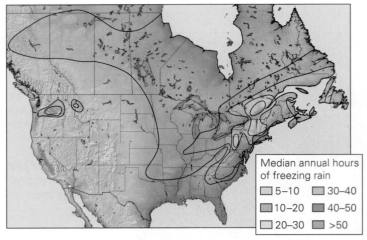

FIGURE 10.6 The average annual number of hours of freezing rain across North America.

not as warm. Snow falling though the warm air wedge at that location melts only partially, and as it falls into the cold air, it refreezes. These frozen raindrops, called **sleet** or *ice pellets*, simply pile up like snow on the ground and don't stick to anything. Finally, beyond the tip of the warm air wedge (Location D), snow falls through cold air all the way to the ground. The diagram in Fig. 10.5a represents a north-south cross section across a warm front (**FIG. 10.5c**). Only the region that receives freezing rain will have ice accumulation.

If a front associated with freezing rain remains stationary or moves very slowly, freezing rain can fall on a local area for a long time, and thick layers of ice can accumulate. If the front moves quickly, freezing rain may occur for a short time, but an ice storm will not develop. Nevertheless, even a small amount of freezing rain can become a deadly hazard on roads and sidewalks, causing vehicles to lose traction and slide or spin, and people to slip and fall. In North America, freezing rain occurs most commonly over the eastern United States and Canada (**FIG. 10.6**). One of the worst natural disasters in Canada's history was the Great Ice Storm of 1998 (**BOX 10.1**). Ice storms rarely occur across the western half of the continent, except in the Columbia River Valley of Washington State.

> ### Take-home message . . .
> During blizzards, blowing snow reduces visibility, and people may become disoriented and lost outdoors. Ice storms occur when freezing rain continues over a long enough time that trees and power lines accumulate ice and collapse. Even small amounts of freezing rain can pose hazards for ground transportation.
>
> **QUICK QUESTION** Why is it dangerous to walk away from your vehicle in a blizzard?

10.4 Causes of Winter Storms

Nearly all major winter storms that affect populated regions of North America and Europe occur within *mid-latitude cyclones*, the huge weather systems that develop in association with low-pressure centers and fronts (see Chapter 8). Here, we first examine the areas within these storms where blizzards and ice storms typically occur over the North American continent. Then we look at settings where local conditions amplify winter hazards within mid-latitude cyclones—these include coastlines, lake shores, and mountain ranges. We conclude by briefly examining causes of hazardous winter weather in Asia.

Continental Mid-latitude Cyclones

Mid-latitude cyclones are more frequent, more intense, and occur over a broader latitude range in winter than in summer. Their development is linked to the temperature contrast between the poles and the tropics, which is greatest in winter. When viewed from space, the pattern of the clouds that develop within a mid-latitude cyclone resembles a comma. Winter weather occurs most commonly across the comma head, the region of clouds extending north of the warm front and wrapping around the west side of the low-pressure center (**FIG. 10.7**). Here, warm, moist near-surface air flows up and over cold air north of the warm front. This warm rising air produces precipitating clouds, as well as the wedge of

FIGURE 10.7 Precipitation within a mid-latitude cyclone moving across the north-central United States in winter.

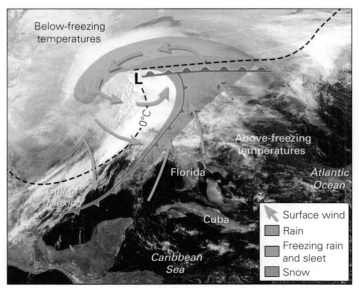

366 • CHAPTER 10 • Cold and Icy: Hazardous Winter Weather

BOX 10.1 DISASTERS AND SOCIETY

The Great Ice Storm of 1998

During the first week of January 1998, a succession of storms producing freezing rain moved across the northeastern United States and Canada. Ice coated extensive areas in the Canadian provinces of Ontario, Quebec, New Brunswick, and Nova Scotia, as well as the American states of New York, Vermont, New Hampshire, and Maine. The worst effects were on the Canadian side of the border, where ice accumulations exceeded 8 cm (3 in) **(FIG. Bx10.1a)**. Many millions of trees collapsed due to the weight of the ice, both within cities and in rural areas. Collapsing trees fell across roads, covered yards, and made parks look like battlefields **(FIG. Bx10.1b)**. Fruit trees and syrup-producing maple trees were decimated.

The thick ice effectively shut down two major cities, Montreal and Ottawa, because their roads and sidewalks were impassable. To make matters worse, chunks of ice falling from buildings and trees posed a deadly hazard for people below. The damage to Canada's infrastructure was immense. Not only were power lines snapped by the weight of the ice or pulled down by falling trees or tree limbs, but over a thousand steel transmission towers and 35,000 wooden transmission towers collapsed, cutting power to over 4 million people in Canada and half a million people in the United States, in many cases for as long as a month. In some places, the electrical grid had to be completely reconstructed **(FIG. Bx10.1c)**. Because this power outage occurred in the middle of winter, people struggled to stay warm, and about 30 deaths occurred due to hypothermia and carbon monoxide poisoning. Animals also suffered—because livestock could not be kept warm, untold numbers of farm animals died, and forest creatures starved because they couldn't reach food. The loss of power shut down industries throughout the region, thereby disrupting the economy.

The amount of ice accumulation in the 1998 storm was more severe than anticipated, and few preparations had been made in advance of the storm. Though emergency services in the regions were accustomed to dealing with hazardous winter weather, they were so overwhelmed by the extent of the disaster that the Canadian government deployed over 16,000 military personnel to help relief efforts. Unfortunately, these efforts were hampered by blocked highways. Total damage from the ice storm approached $6 billion in Canada and the United States combined. In terms of cost, the Great Ice Storm of 1998 ranks as the third worst natural disaster to occur in Canada (following floods in 2013 and wildfires in 2016).

FIGURE Bx10.1 The 1998 eastern Canada and New England ice storm.

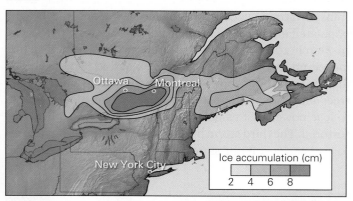

(a) Ice accumulation during the storm.

(b) Fallen trees and power lines blocked roads across eastern Canada.

(c) The weight of the ice collapsed electrical transmission towers.

above-freezing air aloft associated with freezing rain and sleet. Typically, in a mid-latitude cyclone, the ground-surface location of the 0°C (32°F) isotherm—the boundary between above-freezing and subfreezing temperatures—occurs somewhere north of the warm front. Freezing rain occurs just to the north of the 0°C isotherm in a zone normally no more than 20–100 km (12–62 mi) wide. The narrowness of this zone makes its precise position hard to pinpoint in forecasts. Farther to the north, precipitation falls as snow, heavier toward the south, lighter toward the north (see Fig. 10.5). In a strong mid-latitude cyclone with a very low central air pressure, strong pressure gradients across the region of snow lead to strong surface winds, producing blizzard conditions. In the region of ice accumulation, these same strong winds apply stress to ice-coated power lines and tree limbs, enhancing the likelihood that they will collapse.

Although it is not shown in **Figure 10.7**, snow can also occur in other parts of a mid-latitude cyclone. In cases where a mid-latitude cyclone has pulled down particularly frigid air from the polar vortex, snow may fall west of the cold front along the comma tail of the storm, reaching as far south as Florida, producing measurable accumulations over regions unaccustomed to winter weather.

Nor'easters and Oceanic Cyclones

Water has a high *heat capacity*, meaning that it can store a large quantity of heat energy, and it can therefore remain warmer than the overlying air during winter. Consequently, large bodies of water transfer large amounts of heat energy into the atmosphere at the interface between water and air. In addition, evaporation provides moisture that can be converted to precipitation in clouds. For these reasons, winter storms that form along coastlines can produce particularly hazardous winter weather.

When mid-latitude cyclones form along the east coast of North America, they can intensify rapidly as they draw in moisture from the Atlantic Ocean. These storms, traditionally called **nor'easters** because strong winds north of the low-pressure center blow from the northeast, produce dangerous waves at sea, crashing breakers at the shore, and intense blizzards and ice storms inland (**FIG. 10.8**). Nor'easters can have severe economic consequences, shutting down transportation, schools, and businesses along the heavily populated urban corridor of the northeastern United States. In March 2018, for example, four consecutive nor'easters moved up the east coast in a 3-week period (**FIG. 10.9**). The first storm left 15 people dead, 5 of whom were hit by falling trees and branches. The second caused up to 1 million people to lose power, just as power lost due to the first storm was restored. The third storm brought blizzard conditions and up to a meter (1–3 ft) of snow across New England. The final

FIGURE 10.8 Hazards unleashed during nor'easters.

(a) A storm surge associated with strong onshore winds.

(b) Heavy snowfall has buried cars in Boston.

(c) Streets clogged with wet snow in Manhattan.

FIGURE 10.9 Four nor'easters pounded the northeastern coast of the United States in March 2018.

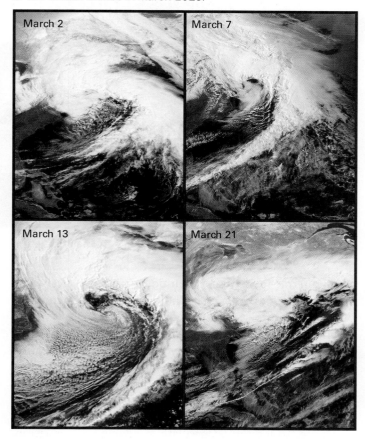

FIGURE 10.10 Oceanic mid-latitude cyclones.

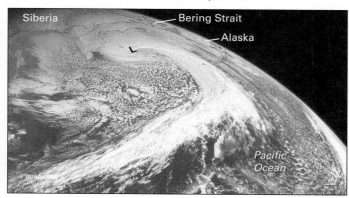

(a) An intense oceanic cyclone over the Bering Sea.

(b) A massive wave caused by an oceanic cyclone's winds.

storm dumped half a meter (19 in) of snow on Long Island, causing 4,400 flights to be canceled at airports along the east coast, and 88,000 electrical customers to lose power. The economic losses caused by nor'easters like these four storms run into billions of dollars.

Extremely strong wintertime mid-latitude cyclones can also develop in the northeastern Pacific Ocean, south of the Aleutian Islands, and in the Atlantic Ocean, south of Iceland **(FIG. 10.10a)**. These **oceanic cyclones** typically become more intense than their continental counterparts, both because of the expanse of water beneath them and because winds over the ocean are not slowed by topography, trees, or buildings. The huge waves caused by these winds endanger fishing fleets, transport ships **(FIG. 10.10b)**, and offshore drilling platforms. The spray from such storms can coat vessels with ice, making them top heavy and subject to capsizing.

Winter oceanic cyclones also cause havoc when they reach the shore. As they move eastward from the Pacific, such storms lash the west coast of North America with heavy rains and intense winds, and bury coastal mountain ranges in wet snow. Seattle, Washington, for example, has experienced sustained winds of 96–113 km/h (60–70 mph) during these storms, with gusts approaching 145 km/h (90 mph), which have damaged not only power lines, trees, and buildings, but also the floating bridges that span local waterways **(FIG. 10.11)**. Along the west coast of Europe, particularly strong winds, known as sting jets, develop at the tip of an

FIGURE 10.11 Strong winds batter a floating bridge in Seattle, Washington.

FIGURE 10.12 Oceanic cyclones and sting jets.

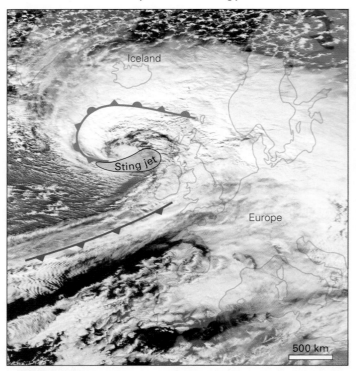

(a) Sting jets form near the back tip of the comma head.

(b) Wind damage in Birmingham, UK, resulting from a sting jet.

oceanic cyclone's comma head and can cause widespread destruction **(FIG. 10.12)**.

Cold Air Damming

The high frequency of ice storms in the eastern United States (see Fig. 10.6) is due not only to mid-latitude cyclones, but also to *cold air damming*. This phenomenon results from the trapping of a low-altitude cool air layer by a mountain range—in this case, the Appalachian Mountains **(FIG. 10.13)**. Cold air damming takes place after a cold front moves across the Appalachians, placing a layer of cold, dense air over the coastal plain that lies between the eastern side of the mountain range and the Atlantic shore. Effectively, the mountains behave like a dam, holding the cold air in place. When warm, moist air flows from the ocean toward the mountains north of the low-pressure center during a nor'easter, it rises over the trapped layer of cold air. Precipitation then falls through the warm layer as rain, but it becomes supercooled in the cold layer dammed against the mountains, and freezes on contact with the ground. Because the mountains are stationary, the cold air can stay trapped for a long time, so ice can accumulate, producing an ice storm. Because of cold air damming, the northeastern coasts of the United States and Canada experience the most days with freezing rain in North America. Notably, cold air damming can cause freezing rain in coastal states as far south as South Carolina and Georgia.

FIGURE 10.13 Cold air damming by the Appalachian Mountains.

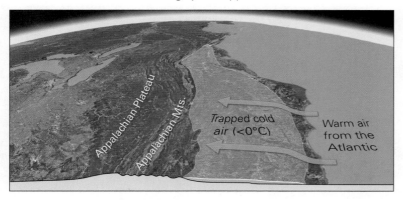

Lake-Effect Snowstorms

The heat and moisture that a large lake provides to the atmosphere can produce local weather systems called **lake-effect snowstorms** **(FIG. 10.14a)**. In North America, lake-effect snowstorms commonly occur over and adjacent to the Great Lakes when cold air from northern Canada flows southward over the lakes. As this cold air passes over each lake, heat and moisture rise from the warm lake surface into the boundary layer of the atmosphere, while the air farther aloft remains cold. The warmed air becomes buoyant and rises to form cumulus clouds, which eventually precipitate snow on the downwind shore of the lake **(FIG. 10.14b)**.

The most extreme lake-effect snowstorms in the Great Lakes region occur between late November and mid-January, when air moving across the lakes can be

FIGURE 10.14 Lake-effect snowstorms.

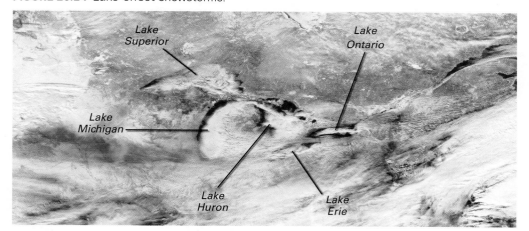

(a) The clouds of a lake-effect snowstorm as seen from space.

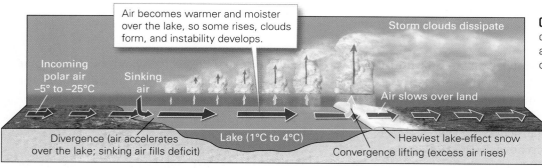

(b) Lake-effect snowstorms develop as cold air is warmed and moistened by its passage over the Great Lakes.

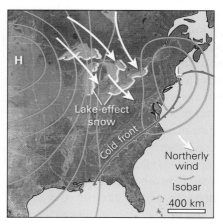

(c) The weather pattern that is common during lake-effect snowstorms.

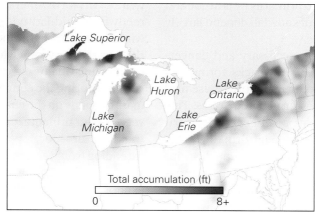

(d) Heavy snow occurs locally on the downwind shores of the lakes.

(e) The aftermath of a lake-effect snowstorm.

very cold, but the lakes remain relatively warm and ice free. The weather pattern that leads to the most intense lake-effect snowstorms develops after a mid-latitude cyclone's cold front has passed to the east of the lakes, so that cold air behind the front flows southeastward across the lakes (**FIG. 10.14c**). Regions downwind of the lakes then receive very heavy snowfalls, sometimes continuing for days at a rate of more than 2.5 cm (1 in) per hour. Up to 150 cm (60 in) may fall during a single lake-effect snowstorm (**FIG. 10.14d, e**). Virtually all of the snow falls between the lake shore and a distance of about 50–80 km (31–50 mi) inland, where most of the moisture the lake has supplied to the air has been removed as precipitation. Lake-effect snowstorms have major economic impacts across the Great Lakes region, mainly on large cities bordering the lakes, due to high snow-removal costs, hazardous driving conditions, and lost work and school days.

Mountain Snowstorms

Winds blowing toward a mountain range have nowhere to go but up, so on the upwind, or *windward*, side of the range, air undergoes orographic lifting. Such lifting produces

FIGURE 10.15 Mountain snows and avalanches.

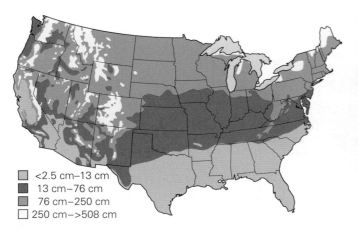

(a) Mean annual snowfall across the United States, 1961–1990.

<2.5 cm–13 cm
13 cm–76 cm
76 cm–250 cm
250 cm–>508 cm

(b) Snow avalanches pose a danger to drivers on some mountain roads.

clouds and precipitation that, in winter, result in **mountain snowstorms** on the windward flank of the range. These storms are among the heaviest and most persistent snowstorms in the world, and they can build immense snowpacks. The distribution of annual snowfall across the United States reveals the effect of mountains on snowfall (FIG. 10.15a). East of the Rockies, amounts of snowfall generally increase from south to north, as they are controlled primarily by latitude and temperature. In the west, amounts of snowfall depend largely on the locations of mountain ranges.

Mountain snowstorms are both necessary and dangerous. Seasonal melting of mountain snowpacks formed by these snowstorms provides water for many otherwise arid areas of the world, including the western United States. This water is important for hydroelectric power generation, irrigation of agricultural fields, and for supplying drinking water to residents of urban areas. The snowpack also enables winter sports. However, mountain snowstorms can also be hazardous, setting the stage for traffic accidents and prolonged highway closures. The major east-west interstate highways that cross mountain ranges in the western United States are especially vulnerable to these winter storms. Donner Pass, in the Sierra Nevada of California (named for a party of pioneers who were trapped there by winter storms in 1846–1847 and resorted to cannibalism to survive), receives frequent heavy snowfall that can close Interstate 80 for days. Such closures can result in enormous losses for commerce and tourism. Mountain snowstorms also set the stage for snow avalanches (FIG. 10.15b) (see Chapter 5). For example, the road that connects Salt Lake City, Utah, to ski resorts in Utah's Wasatch Mountains is only 14.5 km (9 mi) long, but it passes 42 avalanche chutes. The road must be closed many times each winter so that experts can trigger avalanches artificially. Without these safety measures, snow would tumble down the mountainsides unexpectedly, and could bury some of the thousands of cars that drive the road daily during the winter.

Hazardous Winter Weather in Asia

We've noted that nearly all significant hazardous winter weather in North America and Europe results directly or indirectly from mid-latitude cyclones. Asia also hosts hazardous winter weather, but the meteorological conditions that lead to this weather are somewhat different. Recall from Chapter 8 that distinct major air masses develop over land and water surfaces during the course of a year. The character of each air mass depends on the character of the region that it overlies. A very large continental polar air mass—an immense region of frigid, dry air—develops over northern Asia during the winter, extending from eastern Europe across Russia through Siberia and Mongolia. The land areas beneath this frigid air mass experience very harsh winters in which intense cold can become deadly. The large series of mountain ranges extending from the Alps and Carpathian Mountains in the east to the Tibetan Plateau in the west form the southern boundary of this frigid air mass. Under some conditions, its cold air can intrude into western Europe, triggering a cold wave, or can spill southward over the mountains, producing a strong, cold windstorm called a *bora*. When the Siberian air mass moves southeastward and around the east side of the high Tibetan Plateau, it can cause a cold wave in the temperate regions of China. Sometimes the cold air mass moves so far south that it encounters moist tropical air moving northeastward from the Indian Ocean. When this happens, the moist air rising over the front bounding the southern edge of the Siberian air mass generates hazardous winter storms that impact densely populated areas of eastern China (BOX 10.2).

BOX 10.2 DISASTERS AND SOCIETY

China's worst winter

China is home to nearly 1.5 billion people, most of whom live in large cities in the eastern part of the country. In just over 3 weeks in January and February 2008, four pulses of hazardous winter weather blanketed the region in snow and ice, causing the worst winter disaster in more than a century. This disaster was especially disruptive to society, given that it happened while the country was undergoing rapid urbanization and modernization.

The winter storms were triggered when the frigid Siberian air mass moved southward from Siberia and Mongolia, across central and eastern China, and down to the tropics, nearly reaching northern Vietnam at 20°N **(FIG. Bx10.2a)**. The snow and ice storms originated when moist air flowed northeastward, from warmer latitudes, over the cold air mass. The resulting series of slow-moving storms dropped wave after wave of heavy snow in the north and coated the southern and central parts of China with a layer of ice 5–16 cm (2–6 in) thick **(FIG. Bx10.2b)**.

Six eastern provinces were severely affected by these storms. The weight of snow and ice collapsed roofs, destroying 220,000 homes and damaging 850,000 others. Snow blocked highways and forced airports to close. Cold weather froze and burst water pipes, disrupting water supplies, and many people lost power, telecommunications, and internet service for weeks. In fact, over half of China's provinces experienced some power disruptions. The failure of the electrical grid triggered breakdowns of other electricity-dependent systems. Not only did cooking and heating become impossible for millions, but hospitals and factories closed, mines flooded (because pumps couldn't operate), and electrified trains stalled.

China's food supply during winter depends on the agricultural lands of the south. Ice destroyed over 40% of the country's winter crops and led to the deaths of 870,000 pigs, 450,000 sheep, and 85,000 head of cattle. China's aquaculture was also hard hit, as millions of tons of fish at hatcheries died. Also, 75,000 km² (18.6 million acres) of trees—amounting to 10% of the country's forests—and 28,000 km² (6.9 million acres) of bamboo collapsed under the weight of ice. Reports from more than 800 natural reserves noted that starvation and freezing killed hundreds of protected wild animals, including monkeys and sambars, and over 10,000 birds. The population density of butterflies decreased by 90%, and 60% of all butterfly species previously known in one reserve could not be found after the event.

Secondary disasters followed the winter storms. Damage to forests led to soil erosion and landslides. The massive numbers of dead trees resulted in insect infestations and tree disease, and large accumulations of dead material on forest floors fueled wildfires. In Hunan Province in March 2008, wildfires were 11 times more frequent than in the previous 10 years, and the area consumed by fire was 5 times greater than average.

According to statistics from China's Ministry of Civil Affairs, the direct economic loss from the storms was over $22.3 billion, and their indirect effects, such as loss of economic productivity, were even greater. Of course, these numbers pale in comparison to the cost of the COVID-19 pandemic within China, but they are greater than the cost of most natural disasters in the country.

FIGURE Bx10.2 The winter of 2008 was China's worst in a century.

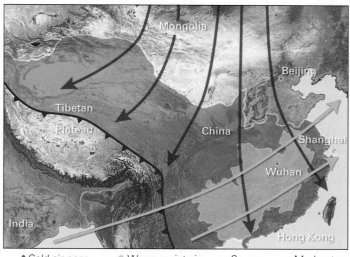

(a) An encounter between cold air moving southeastward from Mongolia and moist tropical air moving northeastward up and over the cold air triggered the event.

(b) Ice and snow made roads and streets impassable.

> **Take-home message . . .**
>
> Ice storms, blizzards, and snowstorms occur within mid-latitude cyclones, primarily within the comma head region of the storms. Oceanic cyclones such as nor'easters have stronger winds and generate more snow than their continental counterparts. Particularly heavy snowstorms occur on the upwind side of mountains, and downwind side of large lakes. Extreme cold air forms over Siberia in winter and can trigger cold waves in Europe and severe winter weather in East Asia.
>
> **QUICK QUESTION** Where on the Earth do the heaviest snowstorms occur?

10.5 Dealing with Hazardous Winter Weather

Hazards to Personal Safety

Winter can not only be inconvenient, it can be deadly (**VISUALIZING A DISASTER**, pp. 382–383). Causes of winter fatalities include traffic accidents, heart failure while shoveling snow, falls on slippery surfaces, infection from frostbite, and hypothermia. Attempts to keep warm during power outages can result in carbon monoxide poisoning from inadequately vented generators, heaters, or woodstoves. Use of these devices also carries the risk of fires, which are very difficult to fight in freezing weather.

Travel by car during hazardous winter weather should be avoided if possible. If you must travel, take a cell phone and a charger, a car battery charger, warm clothes, blankets, water, and food. Assume that you could go off the road and spend hours waiting for help. If you get stuck, remain with your vehicle, as rescuers look for stuck vehicles. By abandoning your vehicle, you risk hypothermia. A safely placed candle can keep the interior of a car reasonably comfortable until help arrives. If you use your engine and heater, avoid carbon monoxide poisoning by making sure the tailpipe is clear of snow. If you drive, turn on your lights, go slow, stay far behind the vehicle in front of you, and avoid sudden braking and swerving that could send you into a skid.

Global statistics on cold-related fatalities are not readily available, but in the United States, the Centers for Disease Control and Prevention (CDC) reports an average of about 1,300 deaths per year—roughly 70% of the number of deaths caused by Hurricane Katrina. Cold-related deaths occur disproportionately among the elderly and homeless. Casualties are also proportionally higher in regions where cold occurs less frequently, and where homes may not be well insulated or may not have functioning heating systems.

Hazards to Aircraft

We've seen that winter weather can close airports. Icing during flight in winter can also be a particular hazard to airplanes. When supercooled droplets in clouds come in contact with a surface whose temperature is below 0°C, such as the wing or fuselage of an airplane, they immediately freeze onto that surface. Therefore, as an airplane flies through a cloud containing supercooled water, ice may build up on its surfaces, disrupting airflow and making the plane difficult to control **(FIG. 10.16)**. Such was the case on October 31, 1994, when a commuter aircraft flew in a holding pattern over northern Indiana, waiting for clearance to land at Chicago's airport. The holding pattern kept the plane in a cloud containing supercooled water droplets, and the ice buildup on the plane became so extreme that the pilot lost control and the aircraft went down, killing 68 passengers and crew.

What can be done to mitigate winter flying hazards? Before planes take off, they must be de-iced with a heated solution of glycol and water. This solution, sprayed on a plane just before it departs from the gate, removes existing ice and can prevent new ice from forming during takeoff. To prevent icing while flying, modern commercial jets send heat from the engine through pipes to heat the wings, and larger commercial propeller planes have mechanisms in their wings that expand and contract, breaking off the ice. If a plane's design doesn't allow for in-flight de-icing, the

FIGURE 10.16 Ice on the wings can make an aircraft difficult to control.

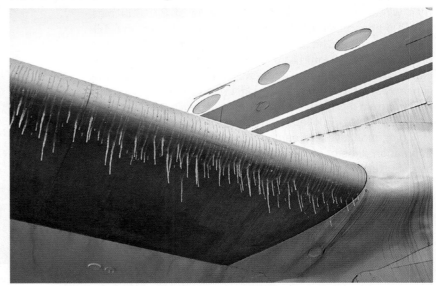

FIGURE 10.17 Cold weather has many impacts on the economy.

(a) Ice jams like this one on the Mississippi River near Hannibal, Missouri, shut down transportation on waterways.

aircraft must avoid flying in winter weather. The 1994 crash caused airlines to move their similar commercial planes to warmer climates, where ice buildup is unlikely, and led the NWS Aviation Weather Center to provide updated maps showing the predicted altitudes, locations, and potential severity of icing to assist pilots with flight planning.

Hazards to Property, Infrastructure, and Agriculture

PROPERTY DAMAGE. Property damage at all scales happens during winter cold waves. In homes, water freezes in pipes and causes them to burst, so that water floods interior rooms. More broadly, a cold wave can freeze city water mains, cutting off water to whole neighborhoods. In addition, as we have described, winter storms can disrupt the electrical grid over broad areas, leading to a cascade of disruptions affecting tens of millions of people (see Box 10.2). Highway, railroad, and airport closures—some due to ice and snow cover and associated traffic accidents, and some to loss of power—can shut down transportation networks. Even transport on rivers and canals may come to a halt because of *ice jams* (FIG.10.17a). Transportation shutdowns, in turn, impact all sectors of society. Supplies can't reach stores, causing some to close and others to be overcrowded. In some cases, shortage of supplies leads people to hoard items, thereby making the problem worse. Similarly, materials and workers can't reach factories or offices, slowing or halting manufacturing and business operations; because students and teachers can't commute,

(b) These citrus crops were ruined by cold.

schools close. All told, winter-weather damage and closures can end up costing communities millions of dollars.

AGRICULTURAL DAMAGE. The effects of hazardous winter weather on agriculture can also result in economic disaster, and in extreme cases, can trigger famines. For example, freezes that takes place after spring planting destroy root networks that send up grasses and grains, as well as buds on trees or bushes that would have grown into fruit. Freezes in the fall can destroy crops and fruit before harvest. During cold waves in 1983 and 1985, for example, Florida's citrus-growing industry incurred losses of $9.6 and $7.7 billion (in 2021 dollars), respectively. Similarly, a January 2007 cold wave in California destroyed more than $1 billion worth of citrus crops (FIG.10.17b), and an April 2007 freeze in California destroyed

VISUALIZING A DISASTER
Winter Storms

Closed airport

De-icing needs

Expensive plowing and salting

Huge drifts

Dead farm animals

Blocked roads

Snow-related accident

Buried cars

Traffic jam

Ice-coated power lines

In the wake of a winter mid-latitude cyclone, an unlucky location may first get a heavy coating of ice, which can collapse trees and disrupt the power grid, and may then endure a blizzard with strong wind and intense snow. Ice and snow can cause accidents, snarl traffic, shut down airports, and block rail lines. Cold and associated frostbite and hypothermia threaten people who are trapped outside, especially if their clothing is inadequate. Farm animals and wildlife can die, not only due to the cold, but also because snow hides their food supplies. Yet, despite these challenges, the beauty of a frozen landscape invites outdoor play.

$1 billion worth of peaches and apples. Cattle and hogs may die in large numbers in winter storms if snow and ice prevent them from reaching shelter, food, and water.

THE 2021 WINTER-WEATHER DISASTER. Many of the impacts described earlier in this chapter made headlines in February 2021 when a massive cold wave struck the central United States. That month, the jet stream migrated south, allowing north winds of the polar vortex to bring the frigid air of the Canadian continental polar air mass down to the Gulf Coast (FIG.10.18). Snow fell over a broad area; along the stationary front bounding the east edge of the cold air, ice storms glazed the ground, trees, and power lines. Southern states that normally had fairly mild winters faced the most hazardous winter conditions they'd seen in years. Sadly, slippery roads caused numerous accidents, including a fatal 135-vehicle pile-up on an interstate highway that trapped victims for hours beneath a horrific jumble of mangled trucks and cars. Nationwide, about 60 people died as a direct result of the cold wave. Many more were put at risk because distribution of COVID-19 vaccine was hampered, and because social distancing couldn't be maintained in shelters. The cold would have been routine for northern states, and indeed caused few problems there. But it was unusual for the south-central United States, and many states struggled. Texas, in particular, faced so many challenges that the US president declared the state to be a federal disaster area, clearing the way for FEMA and other entities to come in and help out.

What happened in Texas? The winter-weather disaster in the state was made worse, in large part, by a partial failure of the state's electric power grid. Texas depends on natural gas and other fossil fuels, and to a lesser extent on wind turbines and solar arrays, for power generation. But due to political decisions made in the 1990s, power production and delivery in the state is largely deregulated, so Texas' grid is managed by a network of competing companies. This decision was intended to spur competition and decrease costs, which it did in normal times. It also caused Texas' grid to become isolated from the national grid, so that the state's grid would not have to comply with federal regulations in place elsewhere. Such regulations require winterization of grids, and

FIGURE 10.18 A map showing the extent of cold air, snowfall (total accumulation), and power outages during the February 2021 winter weather event. The inset provides a snapshot of precipitation in the early afternoon of February 15th, to indicate the extent of precipitation at a given time. The blue shading is the colder air, and pink shading is warmer air. The two air masses are separated by a cold front.

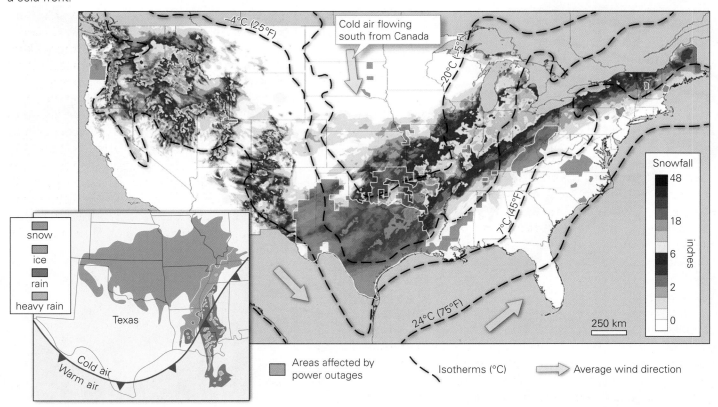

378 • CHAPTER 10 • Cold and Icy: Hazardous Winter Weather

establishment of *reserve capacity* (the ability to produce more electricity when demand increases). Consequently, when the unusually cold weather of 2021 arrived, pipelines were not warmed by heaters, generating equipment was outdoors and exposed to the cold, and wind turbines did not include de-icing equipment. As a result, equipment for both fossil-fuel generation and wind generation froze. In addition. As a result, insufficient reserve capacity was available.

When the cold wave hit, people began to turn on electric heaters, but just as the need for electricity increased, power plants went offline, one after the other. Without adequate power, companies that pumped gas out of the ground couldn't operate, so natural gas supplies dwindled, and those power plants that could still operate couldn't get enough gas, making the problem worse. Overall, the power supply dropped by about 40% statewide. (The grid came very close to shutting down entirely, and if it had, the state would have faced a catastrophe, for it can take weeks to months to restart an entire grid.) Since Texas' electrical grid isn't integrated with the national grid, the state was unable to tap into electricity supplies from out of state to make up for the lack of generating capacity. Without adequate power, utilities tried to ration electricity by instituting rolling blackouts. The intent was to cut off electricity for a given region for fifteen minutes or so, in order to provide enough electricity for other regions. But the plan didn't work, so some places went without electricity for hours to days. Some consumers had energy plans that pegged their energy costs to wholesale rates, allowing costs to go up and down. Normally, consumers with variable rates could save money, but with generating capacity crippled, wholesale costs rose dramatically—some customers received monthly bills of several thousand dollars, over 50 times what they would normally pay.

The shortage of power meant that many families could not stay warm or cook food. People who relied on power for life-saving medical equipment at home were put at great risk, and some died. Hospitals could not keep all facilities operating. Supermarkets could not store perishable food, and shoppers quickly purchased what food remained on shelves, so that food supplies dwindled, causing customers waiting to enter stores to wait in lines that wrapped around the block. Because many homes were not insulated for cold conditions, people were warned to protect their water pipes from freezing by allowing faucets to drip continuously, so water usage went up. But without power to drive pumps, water supplies couldn't be replenished, so municipal water supplies soon dried up. When bottled water started to run out, some people resorted to melting icicles and snow over open fires to obtain water. During the cold wave, some people died of hypothermia, and others who tried to heat houses with gas or kerosene stoves, or by running parked cars, succumbed to carbon-monoxide poisoning. The problem didn't end when temperatures warmed, for as burst pipes thawed, thousands of homes flooded. Not surprisingly, insurance companies were overwhelmed with claims. And, since water purification facilities had been shut down, cities had to issue boil-water notices. The 2021 winter disaster emphasizes that vulnerability to winter weather disasters tends to be significantly greater in locations that seldom experience, and are inadequately prepared for, winter weather.

PREPARING FOR WINTER-WEATHER DISASTERS. Vulnerability to winter weather is increasing due both to climate change (see Chapter 15) and to inadequate or decaying infrastructure. How can the danger, disruption, and cost of winter storms be mitigated? For municipalities and institutions, the answer is complicated, and requires careful cost-benefit analysis. Certainly, the purchase of plows, salt trucks, sand trucks, backup generators, and other equipment will decrease a community's vulnerability. So will protecting electrical and communication infrastructure by moving overhead wiring underground, pruning trees next to wires, and reinforcing cell-phone towers and utility poles. And, as the 2021 Texas disaster illustrates, power-generating and transmission equipment must be designed to handle excessive cold as well as excessive heat, and energy reserves need to be available.

Unfortunately, solutions to decrease winter-weather vulnerability are expensive, and they may deplete financial resources that could be used to meet other challenges. Communities and corporations should take into account the recurrence intervals of hazardous winter weather when making such decisions. People living in locations that endure blizzards and ice storms multiple times a year may find the cost to be worth it, while those living where the recurrence interval of such storms is 20 years (meaning an annual probability of 5%) may conclude that funds would be better spent to solve other problems. But, as the 2021 disaster emphasizes, low-probability events do happen. Regardless of financial investments in equipment and infrastructure, all communities and institutions should develop emergency action plans to prepare for extreme incidents.

> **Take-home message . . .**
> Hazards posed by winter weather tragically lead to loss of life and billion-dollar losses to the economy. Strategies to mitigate losses are generally expensive and often ignored.
>
> **QUICK QUESTION** If your car goes off the road in a blizzard, what should you do?

Chapter 10 Review

Chapter Summary

- The temperatures and amounts of snowfall or ice accumulation that trigger winter weather advisories, watches, and warnings vary with location because conditions representing winter hazards differ among climates.
- Hypothermia occurs when the body's core temperature decreases to a point where normal muscular and brain functions become impaired. Frostbite occurs when body tissues freeze.
- To communicate the effects of cold and wind, meteorologists report the wind chill temperature, the temperature that dry human skin feels due to heat loss caused by the combined effects of temperature and wind speed.
- A cold wave is a rapid fall in temperature that requires society to take precautions against cold.
- The polar vortex is the circulation of cold air in the middle and upper troposphere around the North Pole. Changes in the southern boundary of the polar vortex are associated with the onset of cold waves.
- Hazards of winter storms include heavy snowfall, blizzards, and ice storms.
- A blizzard is a snowstorm in which winds exceed 56 km/h (35 mph) and visibility is reduced to less than a quarter mile for a period of at least 3 hours.
- Freezing rain occurs when a warm, moist air mass overrides a cold air mass along a warm front, so that air with temperatures above freezing lies above air whose temperatures are below freezing. Ice storms occur when prolonged episodes of freezing rain produce large accumulations of ice on trees, power lines, roads, and other surfaces.
- Nearly all hazardous winter weather in North America and Europe occurs within mid-latitude cyclones. Winter precipitation typically occurs along and north of the warm front, under the clouds composing the comma head of the storm.
- Nor'easters along the east coast of North America produce heavy snow, blizzards, and high waves. Treacherous conditions for shipping and air traffic are a likely result.
- Oceanic mid-latitude cyclones produce high winds over a huge area, causing massive ocean waves to form and threaten ships.
- The east coast of North America has the continent's greatest number of winter days with freezing rain. This is partially due to cold air damming, in which cold air is trapped along the eastern side of the Appalachian Mountains.
- Lake-effect snowstorms can form over large lakes in winter as cold air passing over the water surface picks up moisture and heat. These storms bring heavy snowfall to downwind lake shores.
- On the windward side of a mountain range, heavy snows build a large snowpack and can set the stage for snow avalanches.
- An immense continental polar air mass develops over northern Asia in winter. Hazardous winter weather develops over China when this frigid, dry air moves over the country.
- Safety measures for winter travel by car include carrying items such as a cell phone and charger, blankets, water, food, and candles. If you go off the road, remain with your vehicle.
- Aircraft icing can occur when supercooled water droplets in clouds come in contact with the wings and other surfaces of an aircraft. In extreme cases, it can cause planes to lose control.
- Winter cold and storms affect the broader economy by shutting down transportation, causing schools and businesses to close, and decimating crops.

ANOTHER VIEW As the storm clouds dissipate, sunlight makes newly fallen snow on the slopes of the Wasatch Mountains, Utah, sparkle. While beautiful, snow in the mountains can become hazardous, for it may suddenly slide in avalanches that bury homes and roads.

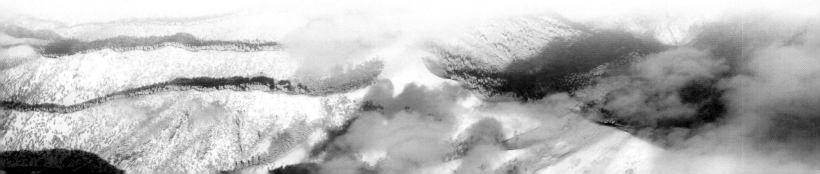

Review Questions

Blue letters correspond to the chapter's learning objectives.

1. How long would exposed skin take to freeze if the temperature outside was −5°F and the wind was 40 mph? (Hint: See Table 10.2.) **(A)**
2. Describe the physical changes that accompany the progressive stages of hypothermia when humans are exposed to extreme cold. **(B)**
3. What is frostbite, and what parts of the body are normally first to suffer? **(B)**
4. On the map, indicate the circulation direction of air in the polar vortex. Does the figure show a time when the midwestern United States is enduring a cold wave, or a time when the region is relatively warm? **(B)**

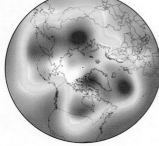

5. What is the difference between a blizzard and a ground blizzard? **(B)**
6. Suppose the front shown in the figure below is stationary. At what location will an ice storm occur? **(C)**

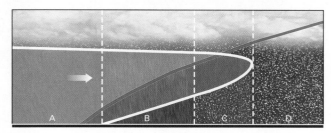

7. Is an ice storm likely if the atmosphere above a location has temperatures entirely below 0°C (32°F) from the Earth's surface to the tropopause? Why or why not? **(C)**
8. Why might it be dangerous to walk under trees immediately after an ice storm? **(C)**
9. When freezing rain occurs, is it likely that falling raindrops will supercool before striking the ground, trees, or power lines? **(C)**
10. Where in a mid-latitude cyclone are blizzard conditions most likely to occur? How about freezing rain? Label these on the figure. **(D)**

11. If you were taking a ship from London, England, to Boston, Massachusetts, and wanted to minimize your chances of getting seasick in rough conditions, what time of the year might be the best time to travel? (Hint: Consider all types of storms that occur over the ocean.) **(E)**
12. Why are the surface winds within oceanic mid-latitude cyclones stronger than those in continental mid-latitude cyclones? **(E)**
13. Explain why lake-effect snowstorms are typically more intense in December than in February. **(F)**
14. Why are winter snowstorms in mountainous regions considered to be both a hazard and a necessity to society? **(G)**
15. Describe all the items that travelers should place in a car for safety during winter. **(H)**
16. List ways that cold weather affects the economy. Which sectors of the economy are most threatened by cold-weather hazards? **(H)**
17. Southern regions of the United States, particularly Texas, endured a disaster during a cold wave in 2021. Why did this cold wave cause so many problems for residents? How could such problems be avoided in the future? How do communities balance the costs of preventative steps with the costs of other societal needs? **(H)**

On Further Thought

18. If a glass of water is placed outside on a bitterly cold, very windy day, will it cool down to the air temperature or to the wind chill temperature? **(A)**
19. If you were going to travel through the Great Lakes region in winter and wanted to experience a lake-effect snowstorm, would it be better to visit Chicago, IL, Detroit, MI, or Buffalo, NY? Explain your choice. **(F)**
20. You are building a house in a location where the temperature falls below 0°C (32°F) on average once every 10 years. If you take steps to protect your house against frozen pipes and other winter damage, including power outages, the construction costs will increase by 15%. Would you spend the money? Why or why not? **(H)**

11

THE DEADLIEST STORMS
Hurricanes

By the end of the chapter you should be able to . . .

A. understand how a tropical cyclone differs from other types of storms.

B. explain how wind speed serves as a basis for distinguishing among different categories of tropical cyclones.

C. characterize the main structural features of a hurricane, such as the eye, eye wall, and spiral rainbands.

D. explain where the energy of a hurricane comes from, and how hurricanes form.

E. show where most hurricanes track, and explain why hurricanes don't occur everywhere.

F. discuss ways that hurricanes cause damage, and why they are nature's most destructive storms.

G. describe how the landfall of tropical cyclones impacts different parts of the world.

H. interpret a prediction of a hurricane path, and explain the importance of hurricane damage mitigation and evacuation.

FIGURE 11.1 Destructive 2017 hurricanes.

(a) Flooding in Port Arthur, Texas, due to Hurricane Harvey.

(b) Buildings destroyed in the Virgin Islands by Hurricane Irma.

(c) Power lines downed by Hurricane Maria in Puerto Rico.

11.1 Introduction

On August 26, 2017, Hurricane Harvey slammed into the Texas coastline. The swirling mass of clouds, wind, and rain covered an area more than 400 km (249 mi) in diameter and caused widespread damage near the shore. Once over land, the storm drifted along the coastline until it stalled over Houston. There it remained for days, dumping up to 150 cm (60 in) of rain and setting a United States record for rainfall from a single storm (FIG. 11.1a). Water filled low areas and submerged over a quarter of the metropolitan area. Harvey killed 108 people and displaced thousands more. It caused damage to infrastructure and to commercial, public, and private property totaling $105 billion.

Unfortunately, the meteorological disasters of 2017 were far from over. As Harvey *dissipated* (weakened, lost its structure, and eventually broke up) over Louisiana, another storm grew on the eastern side of the Atlantic. As regional winds carried this storm westward, its *intensity* increased—meaning that its winds accelerated and its rainfall rates increased—and it became Hurricane Irma. This storm, one of the strongest hurricanes ever observed over the Atlantic, packed *sustained winds* (lasting for over a minute) as high as 300 km/h (186 mph). These winds, together with pounding waves and torrential rain, left a trail of destruction across several Caribbean islands (FIG. 11.1b). Irma crossed the Florida Keys and caused damage along Florida's west coast before dissipating. Immediately following Irma, yet another storm formed over the Atlantic, and grew to become Hurricane Maria. Maria slammed into Puerto Rico and Dominica, causing these islands' worst-ever natural disasters (FIG. 11.1c).

◄ Violent wind, torrential rain, and coastal flooding cause widespread destruction as a hurricane makes landfall.

As the deadly storms of 2017 made clear, hurricanes are the most destructive and costliest storms on the Earth. Meteorologists use the word **hurricane** (from the native Caribbean word *hurakán*, god of wind) for a large, spiral-shaped storm formed in the tropical latitudes of the Atlantic or eastern Pacific in which sustained winds exceed 119 km/h (74 mph). Over the northwestern Pacific, the same type of intense storm is called a **typhoon**, and over the Indian Ocean and throughout the southern hemisphere, it's called a **cyclone**. In this chapter, for simplicity, we'll refer to all such storms as hurricanes, unless we are discussing a specific region for which one of the other names applies. Hurricanes are the most intense type of **tropical cyclone**, a name used for any storm that originates between tropical latitudes of 5° and 30° and rotates around a low-pressure center.

In this chapter, we begin by introducing terminology used to classify tropical cyclones and to characterize the intensity of hurricanes. Then, we examine the structure of hurricanes, discuss the source of their energy, show how they form and evolve over time, and describe the damage that they cause. We next examine the natural disasters these violent storms have produced worldwide. Finally, we consider steps that communities and individuals can take to mitigate the consequences of these storms.

11.2 What Is a Hurricane?

Distinguishing among Tropical Cyclones by Wind Speed

Hurricanes develop over tropical ocean waters. The atmospheric feature that may eventually grow to become a hurricane begins as a **tropical disturbance**, defined as a cluster of thunderstorms that grows in the tropics and lasts for more than a day. If the disturbance forms or moves over tropical ocean waters, begins to circulate around a *low-pressure center* (an area in which the atmospheric pressure at sea level has dropped significantly below values in surrounding regions), and has maximum sustained winds with speeds between 37 and 61 km/h (23–38 mph), it becomes a **tropical depression**. Because of the Coriolis force, such storms circulate counterclockwise in the northern hemisphere, and clockwise in the southern hemisphere (see Chapter 8). If its winds increase to 63–118 km/h (39–73 mph), the depression evolves into a **tropical storm**, and if a tropical storm intensifies to the point that it has sustained winds of over 119 km/h (74 mph), it becomes a hurricane.

Meteorologists from the US National Weather Service (NWS) assign names to Atlantic and eastern Pacific tropical storms, which the storms retain if they become hurricanes. The names vary among English, Spanish, and French, the languages used in countries of the affected region, and alternate between traditionally male and female names. Each year, the name given to the first storm starts with the letter A, and each successive name starts with the next letter of the alphabet, skipping some letters not commonly used for names. The Japan Meteorological Agency assigns names to western Pacific typhoons, and the India Meteorological Department assigns names to cyclones in its region, the Indian Ocean.

To communicate the intensity of a particular storm at a particular time, meteorologists use the *Saffir-Simpson scale*, which assigns categories to hurricanes based on their maximum sustained wind speed **(TABLE 11.1)**. The intensity, and thus the category, of a given storm evolves over time, and a hurricane reaches its peak intensity for only a portion of its lifetime. Note that a hurricane's category provides no information on the storm's diameter, the atmospheric pressure at its center, the amount of rain it produces, or the damage it ultimately causes.

What's Inside a Hurricane?

When astronauts look at a hurricane from space, they see the top of the storm, a broad sheet of cirrus clouds at the level of the tropopause **(FIG. 11.2a)**. A nearly cloud-free

TABLE 11.1 The Saffir-Simpson Scale for Hurricane Intensity

Category	Sustained wind speed	Types of damage resulting from wind
1	119–153 km/h 74–95 mph	**Dangerous winds produce some damage:** Well-constructed frame homes could have damage to roof shingles, vinyl siding, and gutters. Large branches of trees snap, and shallowly rooted trees may topple. Some power outages occur.
2	154–177 km/h 96–110 mph	**Extremely dangerous winds cause extensive damage:** Well-constructed frame homes sustain major roof and siding damage. Many shallowly rooted trees are snapped or uprooted, blocking roads. Near-total power outages occur.
3	178–208 km/h 111–129 mph	**Devastating damage occurs:** Well-built frame homes may incur major damage or removal of roof decking. Large trees snap or are uprooted. Numerous roads become blocked. Electricity and water are unavailable for days to weeks.
4	209–251 km/h 130–156 mph	**Catastrophic damage occurs:** Well-built homes sustain severe damage. Most trees snap or are uprooted, and most power lines are downed. Debris isolates residential areas, and power outages last weeks to months. The area becomes temporarily uninhabitable.
5	252 km/h or higher 157 mph or higher	**Total catastrophic damage occurs:** Most homes are destroyed. Only strongly reinforced buildings remain standing. Debris isolates large areas, and power outages last for weeks to months. Most of the area remains uninhabitable for weeks or months.

Category 1   Category 2   Category 3   Category 4   Category 5

FIGURE 11.2 Structure of a hurricane.

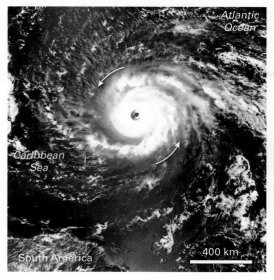

(a) Hurricane Irma (2017), as viewed from space, near its time of maximum intensity.

(b) The eye and eye wall of a hurricane, as viewed from an aircraft inside the eye.

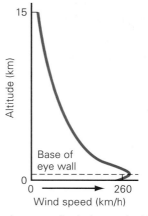

(c) Variation in eye wall wind speed with altitude.

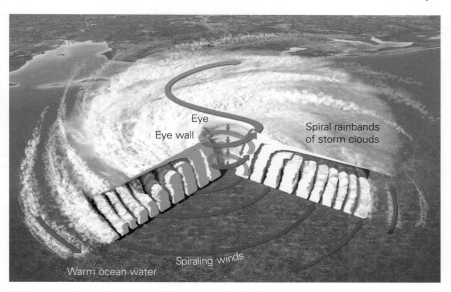

(d) Spiral rainbands surround the eye wall of a hurricane.

circular area, the top of the hurricane's **eye**, lies at the center of the storm. In three dimensions, a hurricane's eye tapers downward, extending from the top of the storm down to the low-level clouds located at an altitude of less than 1 km (0.6 mi) above sea level **(FIG. 11.2b)**. At its base, the eye ranges between 10 and 65 km (6–40 mi) in diameter—the top of the eye is roughly twice the diameter of the base. The eye is surrounded by a ring of dense, rapidly swirling clouds called the **eye wall**. The fastest winds within a hurricane flow around the base of the eye wall. Air spirals upward within the eye wall, slowing with altitude, so that high in the eye wall, the winds are not as fierce **(FIG. 11.2c, d)**. As very intense hurricanes evolve, the eye wall contracts in diameter, while a second eye wall forms around the first. The first eye wall then dissipates as the second one contracts, a process called *eye wall replacement*.

Because of the importance of pressure gradients in controlling wind velocity, a reading of sea-level air pressure in the eye of a hurricane reflects the intensity of the storm. The lowest air pressure at sea level ever measured (870 mb) occurred beneath the eye of Typhoon Tip, in 1979. The resulting gradient between pressure at the eye's center and pressure surrounding the storm caused Tip's sustained winds to reach an amazing 305 km/h (190 mph)—equivalent to those of an EF4 tornado, but covering a much broader area.

Spiral rainbands, nearly continuous arc-shaped cloud banks consisting of tall, aligned thunderstorms, curve outward from the eye to define a spiral shape when the storm is viewed from above (see Fig. 11.2d). Torrential downpours come from the base of the rainbands. The overall air flow in a rainband carries clouds and thunderstorms inward along the spiral, toward the eye wall. The heaviest rain falls from the eye

FIGURE 11.3 Images of spiral rainbands in a hurricane.

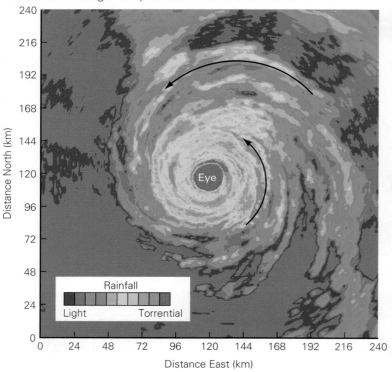

(a) A radar reflectivity map view of a hurricane as viewed from above. Greater reflectivity means more rain, so on this image, the most intense rain is in the red area along the north side of the eye.

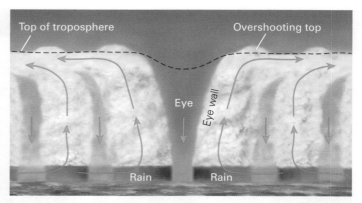

(b) A simplified cross section through the center of a hurricane shows the pattern of air flow within and between the spiral rainbands.

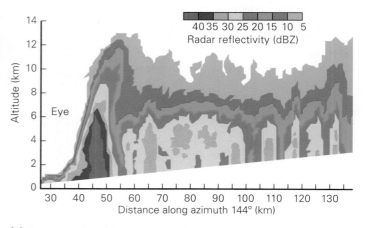

(c) A radar reflectivity cross section through Hurricane Andrew (1992) shows that rain occurs in narrow bands, separated by regions of less or no rain. The heaviest rain falls under the eye wall. Note that the eye is wider at the top than at the bottom.

wall; rainfall rates decrease outward in a hurricane, but very heavy rain can still be falling as far as a few hundred kilometers from the eye. In the gaps between rainbands, there's less rain or no rain, and no rain at all falls beneath the eye **(FIG. 11.3a)**. Air rises in updrafts within rainbands and slowly sinks back down in the regions between rainbands **(FIG. 11.3b)**. At any given time, a line drawn outward from the eye to the margin of a hurricane may cross several rainbands, depending on the size of the storm **(FIG. 11.3c)**.

A hurricane can be divided into two zones on the basis of air flow in and around the storm. Within the *inner core*, which extends from the center outward, air moving through the storm loops completely around the eye. This zone includes the eye wall **(FIG. 11.4a)**. In the *outer zone* of the hurricane, air does not make a complete loop, and the spacing between rainbands increases, resulting in broad areas between rainbands that produce little rainfall. Generally, the inner edge of the *principal rainband*, the widest and most continuous one, defines the boundary of the inner core. *Outer rainbands* lie outside the core. In most storms, the boundary of the inner core is not a perfect circle **(FIG. 11.4b)**.

Thunderstorms form, mature, and dissipate many times within rainbands over the lifetime of a hurricane. At any given time, many thunderstorms are active. Bulges of clouds appear at the top of the eye wall and the rainbands—these bulges are overshooting tops of thunderstorms, indicating places where updrafts in a cell have pushed upward through the tropopause **(FIG. 11.5)**.

Short-lived EF0 to EF2 tornadoes (see Chapter 9) sometimes appear at the bases of rainbands, particularly in the outer zone of a hurricane, once the hurricane starts moving over land. These tornadoes are hard to detect because they are embedded in heavy rain. How they form is not well understood, since none have yet been observed with advanced instrumentation. Such tornadoes can cause localized areas of extreme damage.

A hurricane drifts along with the regional winds in its environment. This *forward motion* typically takes place at a speed that averages 20–35 km/h (12–22 mph) (see Fig. 11.5), although speeds of forward motion exceeding 100 km/h (62 mph) have been recorded. Some hurricanes stall, causing long-lasting downpours leading to flooding beneath the storm. In 2019, Hurricane Dorian, one of the strongest hurricanes to develop

FIGURE 11.4 Zones within a hurricane.

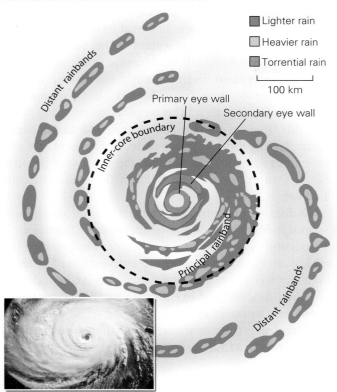

(a) As seen from above, the inner core of a hurricane is the region where wind circulates completely around the storm. Outside the inner core are spiral rainbands.

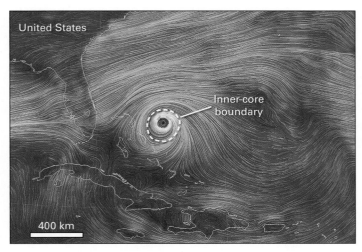

(b) A wind streamline map of Hurricane Isaias (2020). The streamlines are drawn parallel to the wind direction, and their brightness indicates relative wind speed. The redder areas are faster winds. The dashed white line is the edge of the inner core.

FIGURE 11.5 Hurricane Florence (2018), as viewed from the International Space Station. Overshooting tops of thunderstorms are visible above the high cirrus clouds. The forward speed of this hurricane was 10 km/h (6 mph), so it took 20 hours to move from the location where it was photographed to the location of the observer.

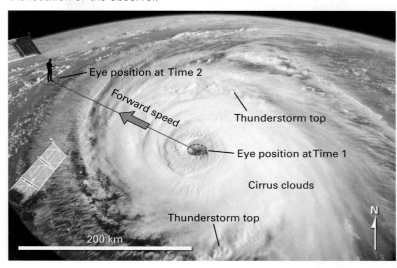

in the Atlantic, stalled near Grand Bahama, and its winds ravaged the island for 24 hours.

Because of the rotation of a hurricane, an observer in the northern hemisphere facing an approaching hurricane will first be buffeted by winds blowing from left to right (see Fig. 11.5). As the eye of the storm approaches, winds strengthen and rainfall gets heavier—as we've seen, the strongest winds and heaviest rains occur at the base of the eye wall. Once the eye arrives, wind speed suddenly drops, and the air clears—in fact, as the eye passes overhead, blue skies may be visible above. Then, as the other side of the eye wall reaches the observer, winds suddenly rise to full force, now blowing from right to left, and torrential rain pours down again. As the storm continues to move along its track, wind and rain gradually decrease in strength until the storm has passed entirely. The time it takes for a hurricane to pass depends both on the storm's diameter and on the speed of its forward motion.

Take-home message . . .

Tropical cyclones are spiral-shaped storms formed in tropical latitudes. The weakest, called tropical depressions, can intensify to become tropical storms and, eventually, hurricanes (also known as typhoons or cyclones, depending on location). The Saffir-Simpson scale classifies hurricanes by wind speed. Hurricanes consist of rainbands that spiral inward toward the eye wall, which surrounds a central eye. The most intense winds and heaviest rains occur at the base of the eye wall. Air within the eye itself is clear and calm.

QUICK QUESTION Does rain fall continuously as a hurricane passes?

11.3 Hurricane Energy and Evolution

Where Does the Energy in a Hurricane Come From?

Hurricanes develop only over tropical ocean waters with a temperature above 26°C (79°F), to a depth of at least 60 m (200 ft) (FIG. 11.6a). That's because it takes heat (thermal energy) to drive the air flow in a hurricane, and only warm ocean waters can provide enough heat to accelerate winds to hurricane speeds. Ocean waters are at their warmest during late summer and fall, which meteorologists refer to as **hurricane season**. In the Atlantic and eastern Pacific, hurricane season runs from June 1 to November 30 (FIG. 11.6b). About half of the tropical storms that form during this time develop into hurricanes (FIG. 11.6c). In the western Pacific, most hurricanes occur between July and October, but some may form throughout the year because the ocean waters retain heat even in the colder months (FIG. 11.6d).

How does warm ocean water transfer heat into the base of the atmosphere to provide energy that drives a hurricane? A small portion is conducted directly from the sea into the overlying air. But most comes from evaporation of water at the sea surface. This water vapor then rises and condenses into clouds. As discussed in Chapters 8 and 9, water vapor contains vast amounts of *latent heat*, and when water condenses, it releases this latent heat, which then warms the surrounding air (FIG. 11.7a). As air warms, its density decreases

FIGURE 11.6 Hurricane locations and seasons.

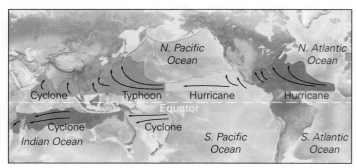

(a) Hurricanes develop only in the tropical oceans.

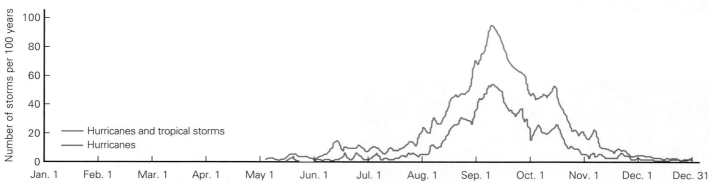

(b) Number of storms over the Atlantic Basin on each date over a period of 100 years. The greatest number of storms occurred in September.

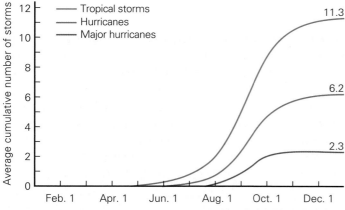

(c) Average cumulative number of tropical storms, hurricanes, and major hurricanes (Category 3 or stronger) over the Atlantic Basin during 1966–2009. Only about half of named tropical storms become hurricanes.

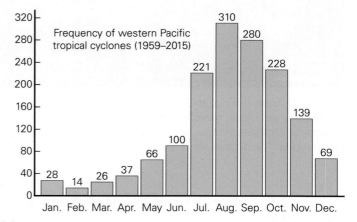

(d) Total number of tropical storms and typhoons in the western Pacific, by month, for a 56-year period (1959–2015). These storms can happen all year, but most occur between June and November.

FIGURE 11.7 The energy that fuels tropical cyclones comes from latent heat released when water vapor condenses into liquid droplets.

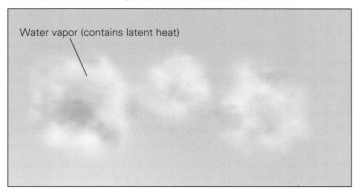
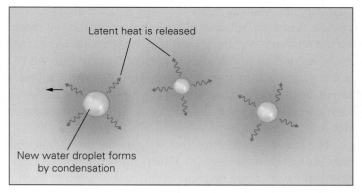

(a) Water vapor (left) contains latent heat. When the vapor condenses back to liquid water (right), latent heat is released and warms the surrounding air. As a result, the air becomes less dense.

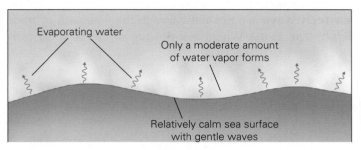
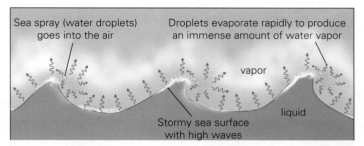

(b) When the ocean is calm but warm, seawater evaporates and produces water vapor. When the wind blows, so that waves build and sea spray enters the air, evaporation rates increase dramatically. Water evaporates from each droplet, so the surface area of evaporating water becomes much greater.

relative to its surroundings, so it becomes unstable and continues to rise, building towering storm clouds (see Chapter 9). Put another way, hurricanes feed on water vapor from evaporation of the ocean, and only warm water evaporates fast enough to provide sufficient fuel to drive these storms. Since cold ocean water does not provide much water vapor, it can't provide enough latent heat to power a hurricane. Because of this relationship, hurricanes strengthen when they drift over warmer ocean water and weaken when they move over colder water or over land.

Growth of a hurricane involves **positive feedbacks**. This means that once the storm has formed, it produces conditions that provide it with even more energy, so that it grows and intensifies. The first positive feedback comes into play due to the winds that the storm generates. Those winds build tall, frothy waves from which *sea spray*, which consists of tiny water droplets, enters the air. Once the droplets exist, evaporation takes place not only from the sea surface, but also from the surface of each droplet. The production of sea spray increases evaporation rates by a factor of 100 to 1,000, providing more latent heat to the storm (**FIG. 11.7b**). Just as adding gas to a running car engine by pressing the accelerator causes the car to speed up, adding latent heat to a storm by increasing evaporation rates causes the storm to intensify. A second positive feedback happens because the updrafts of early thunderstorms carry moisture into the middle and upper troposphere, areas which may otherwise be fairly dry. The resulting increase in humidity means that more of the moisture in subsequent thunderstorms can condense to form cloud droplets and then rain. Consequently, the amount of latent heat released by subsequent storms increases. The additional heat causes successive storms to grow progressively stronger.

How Do Hurricanes Form and Evolve?

LIFTING MECHANISMS THAT LEAD TO TROPICAL THUNDERSTORM INITIATION. We noted earlier that a hurricane begins with the development of thunderstorms. Recall from Chapter 9 that nearly all thunderstorms begin when a *lifting mechanism* causes air to rise at a location, and that if the lifted air becomes *unstable* (less dense than its surroundings), it will continue to rise to the tropopause. The lifting that produces nearly all thunderstorms that organize into tropical cyclones results from convergence of flowing air. **Convergence** happens either when faster-moving air merges with slower-moving air, or when air moving in one direction merges with air moving in another direction. The first process resembles the collision that happens when a faster car rear-ends a slower car on a highway. The second process resembles the collision that happens when two cars traveling in different directions collide. When two volumes of flowing air converge, the excess air has nowhere to go but up, so lifting is the result. If the lifted moist air rises to an altitude where it cools sufficiently for the water vapor within it to condense into clouds, it releases latent heat. This release warms the air so that it becomes unstable and rises still higher, causing thunderstorms to develop.

The convergence lifting that triggers the thunderstorms that ultimately organize into a hurricane occurs most commonly for one of two reasons. In the case of western Pacific typhoons and Indian Ocean cyclones, convergence lifting occurs along the *intertropical convergence zone* (ITCZ), where the southwest-flowing trade winds of the northern hemisphere converge with the northwest-flowing trade winds of the southern hemisphere (see Chapter 8). These trade winds represent the base of the Hadley cell in each hemisphere (FIG. 11.8a,b). Where the trade winds come together, excess air rises.

Modern computer programs use data produced by weather forecasting models to create *wind streamline maps*, visualizations that depict the orientation and velocity of surface winds. Such maps clearly display the development of tropical depressions along the ITCZ in the Pacific because of the spiraling winds associated with their rotation (FIG. 11.8c). Satellite images of the clouds within these depressions can show a succession of tropical storms and hurricanes developing over the ITCZ in the Pacific (FIG. 11.8d).

In the case of Atlantic and some Pacific hurricanes, convergence lifting takes place in easterly waves that develop within the southwest-flowing trade winds. In map view, an **easterly wave** is a distinct curve in the flow path of trade winds (FIG. 11.9a). When west-flowing air enters an easterly

FIGURE 11.8 Formation of tropical cyclones along the ITCZ in the tropical Pacific.

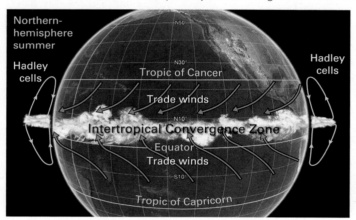

(a) At the ITCZ, the southwest-flowing trade winds of the northern hemisphere and the northwest-flowing trade winds of the southern hemisphere converge.

(b) In the western Pacific, convergence lifting happens over the ITCZ.

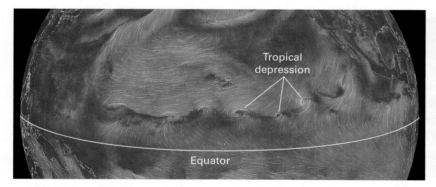

(c) A map of surface-wind streamlines (representing the direction and velocity of air flow) for August 15, 2020. (Blue areas have no wind or only gentle wind.) Note the several tropical depressions along the ITCZ, and that the ITCZ is more than 5° north of the equator.

(d) A satellite photograph of the Pacific Ocean at about the same time the wind streamline map was made. Note the clouds at the positions of the tropical depressions.

FIGURE 11.9 Convergence lifting initiating tropical cyclones in an easterly wave.

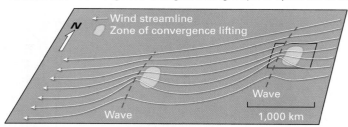

(a) Easterly waves develop in the trade winds, as seen here in map view. Convergence happens where air flow turns north on the eastern side of a wave.

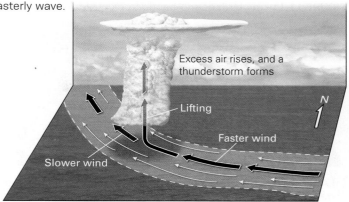

(b) As air flow in the trade winds turns north, it slows. Faster winds approaching from the east converge with these slower winds.

wave and turns to head northwest, forces act on the air to cause it to slow down, as explained in meteorology books. When faster-flowing air pushes into the slower-moving air in the easterly wave, convergence lifting takes place and thunderstorms develop (FIG. 11.9b). A satellite image shows the tropical storms forming in a succession of easterly waves over the Atlantic Ocean (FIG. 11.9c).

STAGES IN THE FORMATION AND EVOLUTION OF AN ATLANTIC HURRICANE. With our understanding of convergence lifting, we can now explore the life story of a typical Atlantic tropical cyclone from its birth, through its growth, to its demise, focusing our discussion on hurricanes that originate from thunderstorm clusters emerging from Africa. This discussion is keyed to the time markers in **Figure 11.10**, which show the successive positions of a single tropical disturbance as it moves from east to west.

- *Time 1:* Trade winds blowing over the Indian Ocean pick up moisture and carry it westward to Africa.

(c) A succession of hurricanes that formed in association with easterly waves across the Atlantic in 2017.

- *Time 2:* When the moist trade winds blow over northeastern Africa, their flow path is bent by interactions with mountains (particularly in the Ethiopian Highlands), forming an easterly wave. Thunderstorms may initially form over the highlands.

FIGURE 11.10 The life cycle of an Atlantic hurricane. Times during the storm's evolution (T1–T10) are described in the text. EH = Ethiopian Highlands; CVI = Cape Verde Islands.

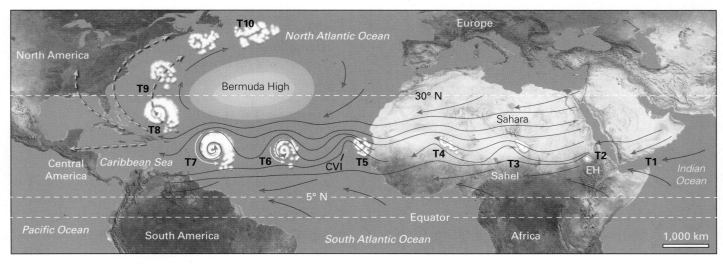

11.3 Hurricane Energy and Evolution • **391**

FIGURE 11.11 The birth of an Atlantic tropical storm.

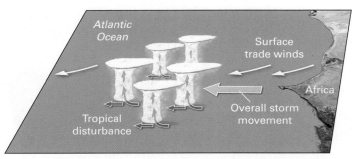

(a) A tropical disturbance moves out over the warm Atlantic and strengthens.

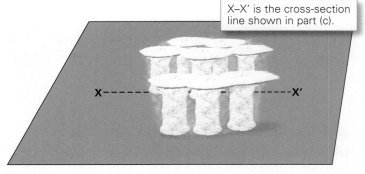

(b) Release of latent heat causes air in the growing thunderstorm cluster to expand and flow outward at high altitudes.

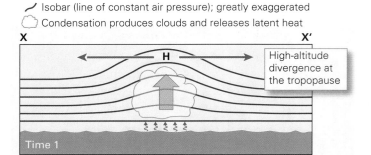

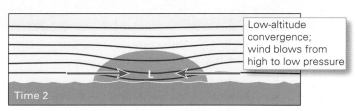

(c) High-altitude divergence causes a low-pressure center to form above the ocean surface.

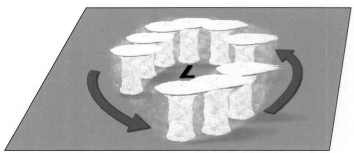

(d) As the thunderstorms start rotating around the low-pressure center, a spiral-shaped storm forms.

(e) Centrifuging of air in the rotating storm further lowers the central pressure, and an eye wall and spiral rainbands begin to form.

- *Time 3:* As winds curve north on the east side of the wave and slow, convergence lifting produces a tropical disturbance, a cluster of thunderstorms that persists as the wave exits the highlands and moves westward. Heating by the underlying land helps maintain the storms.

- *Time 4:* The easterly wave, with the associated tropical disturbance, continues moving westward across Africa, along the boundary between the Sahara and the Sahel. If the thunderstorms associated with the disturbance survive and move out over the Atlantic near the Cape Verde Islands, the seed of a hurricane may have been planted.

- *Time 5:* Once the tropical disturbance is over the warm waters of the ocean, an uninterrupted supply of water vapor is available to feed the storms within it, so those thunderstorms intensify, and additional thunderstorms form **(FIG. 11.11a)**. The release of latent heat within the storms adds thermal energy that causes air in the disturbance to expand laterally and move outward at high altitudes within the disturbance **(FIG. 11.11b)**. This *high-altitude divergence* removes air from a column that extends from the tropopause down to the sea surface in the interior of the tropical disturbance. The divergence therefore causes air pressure at sea level to decrease, producing a distinct low-pressure center at the center of the disturbance **(FIG. 11.11c)**.

Due to the Coriolis force, air rotates counterclockwise about the low-pressure center in a pattern called *cyclonic flow* **(FIG. 11.11d)**. Positive feedbacks accelerate the growth process as the rotating winds build waves and produce sea spray, which increases evaporation rates, causing the thunderstorms to increase in number and intensity. Each thunderstorm moistens the air around the low-pressure center. In turn, the increase in the number and intensity of thunderstorms causes more latent heat to be released, increasing the rate at which the low-pressure center intensifies. As the air pressure in the low-pressure center drops, the horizontal pressure gradient at the sea surface between air outside of the tropical disturbance and air within the disturbance increases in steepness, causing winds to accelerate, ocean waves to increase in height, and more sea spray to evaporate, so that more moisture is carried into the storm. This intensifying circulation gives birth to a tropical cyclone.

- *Time 6:* In the cluster of thunderstorms now rotating around a low-pressure center, winds eventually accelerate above 37 km/h, at which point the disturbance becomes a tropical depression, the weakest category of tropical cyclone. As the low-pressure center continues to intensify, the mass of thunderstorms begins to organize into spiral arms, and a central eye wall slowly takes shape **(FIG. 11.11e)**.

As the tropical depression rotates, air pressure at the base of the storm's center decreases even more. This effect of rotation also creates an outward-directed *centrifugal force* that causes air molecules to move outward, away from the low-pressure center, much as a ball on a string moves outward when you start swinging the ball around your head **(FIG. 11.12a)**. The resulting *low-altitude divergence* further reduces the weight of the air column at the storm's center, and therefore further decreases air pressure in the low-pressure center. This, in turn, steepens the pressure gradient and causes the winds to blow even faster. The progressive intensification of the low-pressure center due to *centrifuging* represents yet another example of positive feedback during the growth of a tropical cyclone. When winds circulating around the low-pressure center exceed 63 km/h, the tropical depression has become a tropical storm **(FIG. 11.12b)**.

- *Time 7:* When sustained winds exceed 118 km/h, the storm becomes a hurricane **(FIG. 11.12c)**. At this point, a distinct eye develops in the center of the storm. Formation of an eye requires the central pressure within the storm to be quite low, so the eye typically forms after a tropical storm has transitioned into a hurricane.

- *Time 8:* When a hurricane reaches the west side of the Atlantic, it can follow any of several different *tracks*, or paths, which may lead to landfall in Central America, on the shores of the Gulf of Mexico, or on the Atlantic coast. Here, we'll follow a hurricane that stays out over the Atlantic.

- *Time 9:* Regional air circulation first drives the hurricane north, then northeast, around the Bermuda High (a semi-permanent high-pressure system over the North Atlantic). The storm maintains hurricane-force winds for a while as it migrates north of the tropics, but eventually it moves over cooler ocean water, which starves it of "fuel" (evaporating water). Faster westerly winds associated with the polar-front jet stream also begin to tear it apart. As the storm weakens, it is *downgraded* to a tropical storm or tropical depression.

- *Time 10:* Eventually, the storm dissipates. The remnants of tropical cyclones can still produce heavy rains. Sometimes tropical cyclones evolve into, or combine with, mid-latitude cyclones and cause stormy weather along the east coast of North America or even in Europe.

FIGURE 11.12 Transformation from a tropical depression to a hurricane.

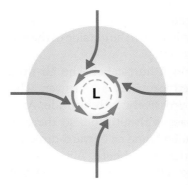

(a) Due to the Coriolis force, winds flowing toward the low-pressure center deflect to the right (in the northern hemisphere), so counterclockwise rotation takes place.

(b) Spiral arms forming in Tropical Storm Gaston (2004) due to the storm's rotation.

(c) Gaston has evolved into a hurricane. A distinct eye has developed, and spiral rainbands are clearly defined.

FACTORS THAT ALLOW A HURRICANE TO GROW AND LEAD TO ITS DEMISE. The evolution of a tropical disturbance into a hurricane can happen only if two criteria are met. First, the storm must have an uninterrupted supply of warm, evaporating seawater, the fuel that nourishes the storm. Second, the storm must develop in a region where high-altitude winds are not strong, for when winds increase rapidly with altitude, they apply *wind shear* to the storm and can tear apart the rotation within it.

Once a tropical cyclone has formed, a variety of factors controls its strength. If the storm moves over warmer seawater, evaporation rates increase, and the storm can intensify. If the tropical cyclone lies within a region where the troposphere is thick, as occurs in lower latitudes, there is more depth through which thunderstorms can rise. As a consequence, more condensation can occur, and more latent heat is released to power the storm, so the storm can intensify. If a tropical cyclone moves over rough terrain, such as a large mountainous island, friction slows its winds, and the absence of warm surface water reduces its intensity.

The demise of a hurricane happens for any of three reasons. First, if the storm moves to higher latitudes, where surface water is cooler and doesn't evaporate as rapidly, and where the troposphere is not as thick, the various positive feedbacks that intensify its winds are interrupted. Second, if a hurricane drifts over land, the supply of evaporating water decreases dramatically, and the wind slows as the flowing air interacts with landforms, trees, and buildings. Finally, if a hurricane drifts into a region of strong high-altitude winds, wind shear disrupts the cyclonic flow in the storm, effectively tearing the storm apart.

AIR FLOW IN A HURRICANE AND THE DEVELOPMENT OF AN EYE. The structure of a hurricane reflects the movement of air within the storm. Air near the Earth's surface spirals inward toward the eye of the hurricane **(FIG. 11.13)**. As it does so, it moves progressively faster, like spinning skaters who spin faster as they pull their arms inward. (This increase in speed is a manifestation of the law of conservation of *angular momentum*.) At the eye wall, the winds reach their greatest velocity. Because of high-altitude divergence, air rises in an ascending spiral within the eye wall until it reaches the tropopause (see Fig. 11.2d). At the tropopause, nearly all the rising air then spirals outward and slows, creating the layer of cirrus clouds at the top of the hurricane. Eventually, air sinks back toward the Earth's surface beyond the margins of the hurricane. Rainbands form as the thunderstorms outside the eye wall align with the wind and form narrow banks of clouds, separated by less cloudy regions. Air rises in the updrafts within rainbands and slowly descends between them. The atmospheric pressure at sea level decreases rapidly across the eye wall, with the lowest pressure at the center of the eye (see Fig.

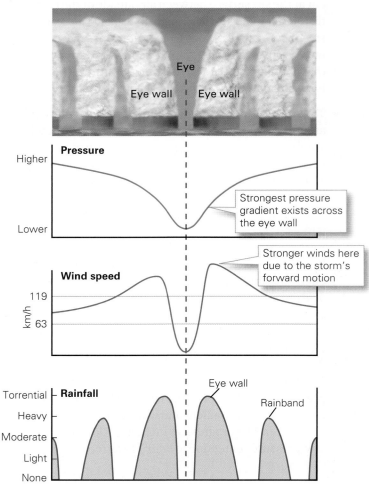

FIGURE 11.13 A cross section of a hurricane showing its air flow and variation in its physical characteristics. Air spirals into the eye along the rainbands, and then up the eye wall. It also rises within the rainbands. Physical conditions in the atmosphere beneath a hurricane reflect the structure of the storm.

11.13). The strongest winds occur under the eye wall because of the strong pressure gradient between the outside of the eye wall and the eye.

Why does an eye form in a hurricane? The air-flow pattern we've just described provides an answer. The density of air decreases progressively with altitude. As a result, pressure also decreases with altitude, so there is always a vertical pressure gradient, and an upward-directed pressure-gradient force, in the atmosphere. Air does not blow off into space because, almost everywhere on the Earth, this upward force is exactly balanced by the downward force of gravity **(FIG. 11.14a)**. One place where this is not true is in the eye of a hurricane. As winds in a tropical cyclone approach hurricane strength, latent heat release, together with centrifuging, causes the pressure in the center at the base of the storm to decrease enough that the downward force of gravity exceeds the upward

FIGURE 11.14 Formation of a hurricane's eye.

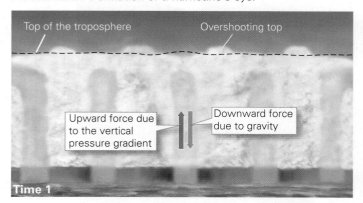

(a) Initially, the upward force due to the atmosphere's vertical pressure gradient equals the downward force of gravity in the cluster of thunderstorms.

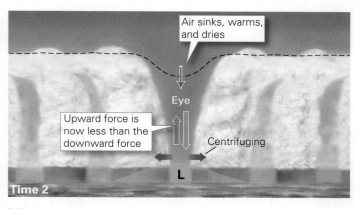

(b) When the wind speed gets high enough, centrifuging of air out of the center, together with latent heat release, makes air pressure in the low-pressure center low enough that the force of gravity exceeds the vertical pressure-gradient force. The air sinks, heats up, and dries, and clouds disappear.

pressure-gradient force. As a result, air sinks slowly downward in the storm's center. This descending air undergoes compression, and therefore warms and dries, so it become largely cloud free **(FIG. 11.14b)**.

Notably, as air in the eye of a hurricane sinks and warms, it also expands, and this expansion moves air molecules horizontally outward from the storm's center. The resulting low-altitude divergence further reduces the air pressure in the low-pressure center. This positive feedback causes the winds to become even stronger.

Hurricane Tracks and Dimensions

If a tropical disturbance develops close to the equator (at a latitude of 0°–5°), it may grow into a large thunderstorm cluster, but does not start rotating, and does not become a tropical cyclone. That's because close to the equator, the Coriolis force is too weak to initiate rotation (see Chapter 8). For that reason, hurricanes never form on, or cross, the equator **(FIG. 11.15a)**.

Over the course of its lifetime, a tropical cyclone moves across the Earth. The forward path that a hurricane's center follows over time, as plotted on a map, is known as a **hurricane track**. The prevailing winds that control a tropical cyclone's forward motion differ for each storm, so no two tracks are exactly the same, although most follow one of a few typical patterns. Because most tropical cyclones form either along the ITCZ or within the belt of trade winds, they first move westward with the trade winds during their early lifetimes. A tropical cyclone is sufficiently large that the strength of the Coriolis force varies from its equatorward side to its poleward side. This contrast causes the storm to drift poleward as it travels west.

Strong, semipermanent high-pressure systems are located over the Atlantic and Pacific Oceans north of 30° during the summer (see Chapter 8). For example, the Bermuda High typically resides over the North Atlantic during hurricane season (see Fig. 11.10). In the northern hemisphere, air flows clockwise around high-pressure centers. Consequently, as Atlantic hurricanes drift northwestward, most eventually come under the influence of the flow around the Bermuda High, and their tracks then curve north **(FIG. 11.15b)**. As they move north, they cross from the belt of easterly trade winds in the tropics into the belt of westerly winds in the mid-latitudes. Westerly winds steer a hurricane back toward the east on the north side of the high-pressure system. Some hurricanes do not move far enough north to be influenced by the Bermuda High. These storms can then strike the Central American coast or move over the Gulf of Mexico. Most hurricane tracks conform to one of the patterns described above, although some take highly erratic tracks that are more challenging to predict.

Along its track, the intensity of a tropical cyclone varies for the reasons that we've discussed. The storm begins as a tropical depression, strengthens to a tropical storm, then becomes a hurricane. The intensity of a hurricane also changes, sometimes increasing and decreasing more than once. Hurricane Andrew, the notorious 1992 Category 5 hurricane that destroyed communities along the east coast of Florida, displayed this behavior **(FIG. 11.15c)**.

A hurricane's dimensions can be specified using different measures, such as the diameter of the eye wall, the extent of the storm's hurricane-force winds, or the diameter of its cirrus-cloud shield as viewed from space. Hurricane-force winds can extend outward to about 40 km (25 mi) from the center of a small hurricane and to more than 250 km (155 mi) from the center of a large one. The area over which tropical storm–force winds occur is even greater, ranging as far as 500 km (311 mi) from the eye of a large hurricane **(FIG. 11.16a)**.

FIGURE 11.15 Hurricane tracks.

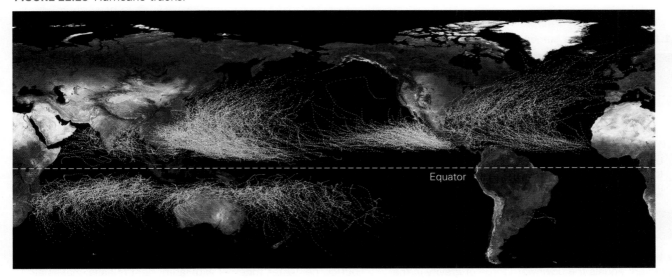

(a) Each line on this map represents a hurricane or tropical storm track. Light blue lines are weaker storms, and yellow lines are stronger storms. Note that none of these tracks crosses the equator.

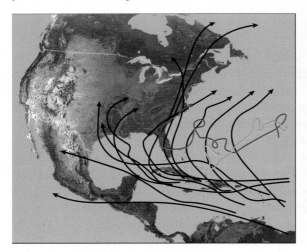

(b) Selected Atlantic hurricane tracks. Most storms (black lines) head west and then north. A few (colored lines) follow irregular paths.

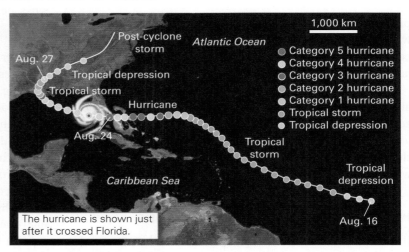

(c) The track of Hurricane Andrew (1992), which struck Florida as a Category 5 storm, and then Louisiana as a Category 1 storm. Note its evolution from tropical depression to Category 5, and then its progressive downgrading after it makes landfall.

FIGURE 11.16 The dimensions of hurricanes.

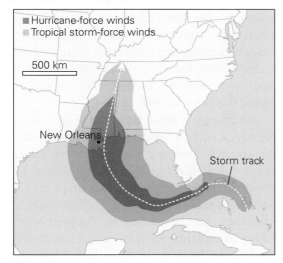

(a) Wind speeds along the track of Hurricane Katrina (2005). Hurricane-force winds extend outward from the eye. Tropical storm–force winds extend out even farther.

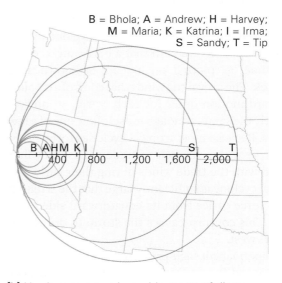

(b) Hurricanes come in a wide range of diameters. Typhoon Tip is the largest known example. But even smaller storms, such as Andrew, can reach Category 5 in intensity.

Hurricane diameters, as measured from space, range greatly, from the smallest at less than 50 km (31 mi) to the largest on record, Typhoon Tip, at close to 2,200 km (1,367 mi). The average hurricane has a diameter of about 500 km (311 mi) (FIG. 11.16b).

> **Take-home message . . .**
>
> Hurricanes form only over warm ocean waters, which provide an unlimited supply of rapidly evaporating water. The energy of a hurricane comes primarily from latent heat released when water vapor in the storm condenses. Hurricanes originate from clusters of thunderstorms in a tropical disturbance and grow into large spiraling storms. Atlantic hurricanes typically begin where convergence lifting takes place along easterly waves in the trade winds. Most western Pacific hurricanes form along the ITCZ. Air flow in a hurricane spirals inward near the ocean surface, with the strongest winds under the eye wall.
>
> **QUICK QUESTION** Why can't a hurricane form on the equator?

11.4 How Do Hurricanes Cause Damage?

Hurricanes cause damage and destruction by means of wind, storm surge, waves, flooding rains and landslides, and occasional tornadoes. Let's consider the way in which each of these hazards can contribute to a disaster (**VISUALIZING A DISASTER**, pp. 398–399).

Wind Damage

Sustained winds in a hurricane can range from 119 km/h (74 mph), for the weakest Category 1 storm, to greater than 252 km/h (157 mph) for a Category 5 storm. The strongest winds recorded during a hurricane occurred during Hurricane Patricia in 2015 as it crossed the eastern Pacific—they reached 345 km/h (214 mph). The extent of wind damage depends on both the intensity and duration of the winds (see Table 11.1). Almost no buildings can remain standing in the winds of a Category 5 hurricane, so in the wake of such a storm, coastal towns may be completely flattened.

How does wind cause damage? Wind, like any moving fluid, applies *dynamic pressure*, a directed push, to any object that it impacts (see Chapter 6). Engineers have calculated that a 250 km/h (155 mph) wind exerts over 1,200 times the pressure on a wall that a 1 km/h (0.6 mph) wind does, and that a 300 km/h (186 mph) wind exerts over 6,800 times the pressure. When wind diverts around a building, differences in air pressure at different locations effectively produce suction, like that of a vacuum cleaner, on some parts of the building (FIG. 11.17a). The push of dynamic pressure, and the suction forces caused by complex air-flow paths, can be strong enough to cause many types of damage (FIG. 11.17b). Weaker hurricane winds may rip loose siding or shingles off a building, while stronger winds may lift off roofs and blow down walls. Buildings may shear sideways so that walls bend and break, or slide off their foundations. Mobile homes can flip. Components of a building, such as windows or doors, may shatter. Windows or walls may be smashed by flying debris or toppled trees. Fallen power lines pose electrocution hazards and can also cause catastrophic power outages. When Hurricane Ida crossed Louisiana and Mississippi in 2021, it toppled about 30,000 utility towers, causing blackouts for over a million people. Loss of power hampered emergency services and left residents without clean water

FIGURE 11.17 Consequences of hurricane winds.

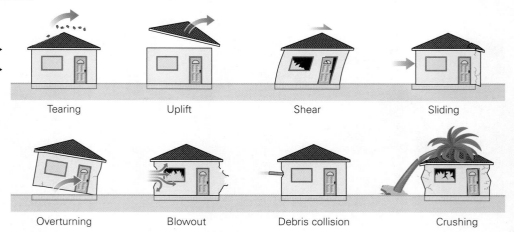

(a) Wind applies dynamic pressure, a push, to walls. But because wind follows curving paths around a building, air pressure isn't the same on all sides, so suction pulls on parts of the building.

(b) The action of wind causes many different types of damage to buildings.

When a hurricane makes landfall, disaster strikes. Here we see a cutaway view of a hurricane making landfall on a Caribbean island. The storm's forward motion is to the right, so the strongest winds are blowing onshore. Huge waves sink ships offshore, break apart reefs, and together with storm surge, overwhelm harbors and wash away coastal communities. Winds tear apart buildings and rip up trees. Inland, torrential rains drench hillsides, triggering flash floods, mudflows, and slumps. Large rivers cover their floodplains. In the wake of the storm, residents left amid the wreckage struggle to find food, clean water, and shelter.

FIGURE 11.18 The wind speed in a hurricane at any specific location (dots) is the sum of the hurricane's rotational wind speed (red arrows) and the overall forward motion of the hurricane (blue arrows). Therefore, as Hurricane Katrina moved north, the wind speed on the east side of the storm (green arrow) exceeded that on the west side (pink arrow).

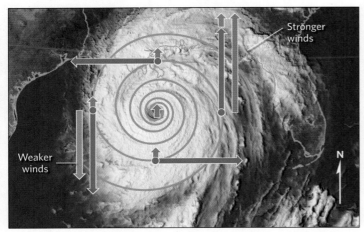

or air conditioning. Consequently, during a heat wave following the hurricane, many people endured illness and some died.

The wind speed within a hurricane at a given location depends both on the distance of the location from the eye and on the forward speed of the overall storm. The side of a hurricane rotating in the direction of to its forward motion causes more wind damage than the side rotating in the direction opposite to its forward motion. For example, if the rotating air beneath the eye wall on the right side of a counterclockwise-rotating hurricane (as viewed looking in the direction in which the hurricane is moving along its track) moves at 150 km/h, and the forward speed of the hurricane is 40 km/h, then the total wind speed will be 150 + 40 = 190 km/h. On the left side of this storm, the wind speed will be 150 − 40 = 110 km/h (FIG. 11.18). Notably, winds on the left side of a hurricane crossing the shore (as viewed looking toward the shore) are also slower than those on the right side, because the left-side winds have passed over land and have been slowed down by friction, while those on the right side are coming from the sea and have their maximum speed.

Storm Surge

A hurricane produces a dome-like mound of ocean water that can rise several meters above mean sea level. (*Mean sea level* is halfway between the levels at high tide and at low tide.) This **storm surge** develops mostly because the wind of the storm drives water in front of it, effectively building a pile of water (FIG. 11.19a). To a lesser extent, it develops because the low

FIGURE 11.19 Storm surge.

(a) Wind and low atmospheric pressure combine to produce a bulge of water.

(b) Storm surge arriving onshore, with waves on top.

(c) Due to storm surge, coastal areas can be submerged.

(d) Storm surge and waves together can devastate coastal communities.

atmospheric pressure beneath the storm reduces the weight of the atmosphere pushing down on the sea surface relative to that in locations away from the storm, causing ocean water to rise at the center of the storm, as water in a straw does when you place the end of the straw in water and suck. The waves built by a hurricane are on top of the storm surge, so the peak height of water where the storm comes ashore is the *amplitude* of the waves (half the vertical distance from crest to trough) plus the storm surge (**FIG. 11.19b**). Storm surge can cause flooding by ocean water well inland of the beach, and it may leave low-lying areas flooded with saltwater long after the storm has passed (**FIG. 11.19c**). The combination of storm surge and waves can wash away coastal communities completely (**FIG. 11.19d**).

FIGURE 11.20 Hurricane-driven waves.

The height of storm surge depends on the strength of the hurricane's winds and on position relative to the hurricane's eye. Storm surge is much greater where winds are blowing shoreward than where they are blowing seaward. The height of the surge decreases with distance outward from the eye wall because winds decrease, and air pressure increases, in that direction. Surge height also depends on the tide: if a hurricane arrives at high tide, when the sea surface is already higher than mean sea level, storm surge will be higher than if it arrives at low tide, a time when the sea surface is lower than mean sea level. Storm surge also becomes higher when water enters bays and estuaries, for the same reason that waves become higher as they move onshore (see Chapter 6). Storm surge produced by Hurricane Katrina along the Gulf Coast in 2005 reached as high as 8 m (26 ft), and surge produced by Hurricane Sandy in 2012 was as high as 4.2 m (14 ft) near New York City.

Wave Damage

The shear between a hurricane's winds and the sea surface can generate monster waves. Waves as high as 28 m (92 ft) have been recorded during some hurricanes. Such waves can damage offshore drilling rigs and swamp or damage ships (see Chapter 13). Indeed, *maritime disasters* have cost many lives and destroyed many ships over the centuries (**FIG. 11.20a**). Before modern weather prediction and communication technology allowed sailors to be aware of developing storms, a ship could be caught in a hurricane with little warning.

Hurricane-generated waves can travel well beyond the location of the hurricane. When they approach the shore and enter shallower water, they become higher, steeper, and more closely spaced (**FIG. 11.20b**). Such waves can destroy coral reefs and coastal wetlands, erode beaches

(a) Tall waves in the open ocean are a hazard to ships, and can cause maritime disasters.

(b) These waves, formed during Hurricane Florence (2018), damaged coastal structures.

(c) These pleasure craft were ripped from their moorings by waves and piled up against the shore.

and coastal dunes, wash over beaches and piers, and carry water into or over nearshore buildings. Nearer the eye wall of a landfalling hurricane, huge waves riding atop storm surge can destroy homes and other structures, mangle nearshore shipping and harbor facilities, or rip fishing fleets and other vessels from their moorings and bash and toss them onto shore (FIG. 11.20c).

Rain, Inland Flooding, and Landslides

Inland of the coast, tropical cyclones dump torrential rains, which can cause flash floods. In steep terrain, rainwater soaks into the ground and saturates the soil, causing unstable slopes to give way in devastating mudslides and debris flows (FIG. 11.21a). When storms stall over low-lying land, large areas can end up being submerged (see Fig. 11.1a).

The remnants of a hurricane can continue to move well inland of the coast and cause flooding in areas not normally viewed as vulnerable to hurricanes. For example, in 1972, the remnants of Hurricane Agnes drenched central Pennsylvania and New York State, where rising waters caused severe damage to riverside towns. The remnants of Tropical Storm Lee resulted in $2 billion in damage in the same region in 2011 (FIG. 11.21b). And in 2008, the remnants of Hurricane Ike followed a track across the midwestern United States (FIG. 11.21c), where it caused extensive flooding in Chicago. Hurricane Ida, in 2021, crossed the interior of the US along a path from the Gulf of Mexico to New England. It dumped record amounts of rain on New Jersey and New York, leading to the tragic deaths of people trapped in cars or basement apartments, and to the flooding and shutdown of highways, and of rail and subway lines.

> **Take-home message . . .**
> Violent winds, coastal flooding due to storm surge, inland flooding due to heavy rainfall, and landslides triggered by floods are the major causes of loss of life and property damage during hurricanes. The worst disasters are caused by flooding.
>
> **QUICK QUESTION** When a hurricane makes landfall, where along the shoreline is the storm surge greatest?

FIGURE 11.21 Inland hazards related to hurricanes.

(a) Landslides triggered by the rains of Hurricane Maria caused immense damage in Puerto Rico.

(b) Floodwaters due to rains from remnants of Tropical Storm Lee swamped towns in Pennsylvania and New York.

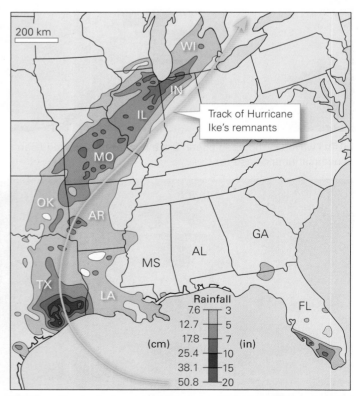

(c) A swath of rain from the remnants of Hurricane Ike caused flooding in the Midwest. This map shows rainfall amounts in the affected regions over September 8–15, 2008.

11.5 Hurricane Disasters: A World View

As we noted earlier, hurricanes form over ocean waters between latitudes of 5° and 30°, and have their strongest winds while over these tropical waters. Therefore, most devastation due to hurricanes (and weaker tropical cyclones) occurs in coastal regions through much of the tropics and along the east coasts of North America, Asia, Australia, and Africa. In this section, we take a world view of the disasters these storms leave in their wake.

Caribbean and Central American Hurricanes

The islands of the Caribbean all lie along a broad arc that separates the western Atlantic Ocean from the Caribbean Sea and the Gulf of Mexico (**FIG. 11.22a**). These islands are regularly struck by hurricanes that form over the Atlantic Ocean or the Caribbean Sea, but because of the gradual northward drift of hurricanes, islands on the northern side of the arc feel the force of hurricanes more frequently. The nature of hurricane devastation on Caribbean islands depends in part on the terrain and area of the island. Some islands have mountainous interiors (**FIG. 11.22b**), whereas others are built of sand and coral and rise only a few meters above sea level (**FIG. 11.22c**).

Hurricane winds coming off the sea can completely overwhelm small, low-lying islands that do not slow the wind at all. One such island, Barbuda, was made uninhabitable by storm surge and winds when the eye wall of Hurricane Irma, a Category 5 storm, made a direct hit on the island (**FIG. 11.23a**). In addition to obliterating structures, the wind stripped the trees and doused plants with salt spray, turning the lush green island brown (**FIG. 11.23b**). The storm inflicted damage on neighboring islands as well (**FIG. 11.23c**). Larger islands with mountainous interiors or volcanic edifices can slow winds, but they also redirect the air flow up mountainsides. This orographic lifting increases the intensity and the duration of rain, causing flash flooding and mudslides (**BOX 11.1**).

The impact of hurricanes striking Caribbean islands, in terms of human suffering, is worse on islands where poverty is widespread and construction quality is poor. Emergency services on these islands may be understaffed and under-equipped, with little access to resources for reconstruction. Haiti, on the western side of Hispaniola, is particularly vulnerable. When Hurricane Matthew struck in 2016 as a Category 4 storm, the country had not recovered from a 2010 earthquake that had left it in ruins (see Box 3.1). At least 200,000 homes were nearly or completely destroyed by the hurricane, and 1.4 million people required humanitarian aid. After the hurricane

FIGURE 11.22 Topography of Caribbean islands.

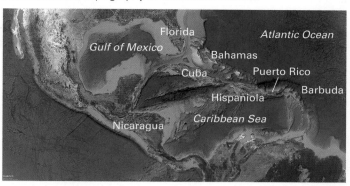

(a) The islands surrounding the Caribbean Sea are susceptible to relatively frequent hurricane strikes.

(b) Some Caribbean islands are mountainous. These mountains on Puerto Rico lie 7 km (4 mi) from the shore.

(c) Islands in the Bahamas are low lying and are fairly small. This one is less than 14 km (9 mi) across at its widest.

BOX 11.1 DISASTERS AND SOCIETY

Lasting consequences of Hurricane Maria

Hurricane Maria, which roared westward across the Atlantic and into the Caribbean in 2017, produced the greatest natural disaster ever recorded on the Caribbean islands of Dominica and Puerto Rico. The small island of Dominica has an economy largely dependent on tourism. On the night of September 18, Maria roared ashore as a Category 5 storm, wiping out the island's infrastructure. Violent winds blew down forests, tore off roofs, and trashed schools and homes. Floodwaters poured off hills and washed away roads, while winds snapped utility poles and cell-phone towers, leaving the island without power, clean water, or communications **(FIG. Bx11.1a)**. Before the sun rose the next morning, 31 people were dead, and the island had been reduced from a tropical paradise to complete poverty and ruin. Most tourist facilities had been destroyed, so the economy also lay in tatters.

Maria continued westward along its track and, 2 days later, it made landfall on the eastern tip of Puerto Rico as a Category 4 storm **(FIG. Bx11.1b)**. Nearly the entire island experienced the violent winds of the storm's eye wall, which caused the worst hurricane damage ever observed on the island. Not only were homes, hospitals, schools, and businesses ripped apart, but downed utility poles, wires, and trees were draped across roads, making them impassable. The island was left almost entirely without power or clean water. Many people died of injuries during or soon after the storm. Significantly, death rates remained far higher than average for over a month after the storm. The major causes were excessive heat at a time when there was no power for air conditioning or fans, and lack of access to medical treatment for injuries and disease. Estimates place the total storm-related death toll at nearly 3,000.

After the hurricane, Dominica managed a slow recovery. By the year's end, government services had resumed, some businesses had reopened, food had reappeared in markets, and most public schools were operating. Nevertheless, 80% of the population were still without electricity, and many remained unemployed and impoverished because of lost jobs in agriculture and tourism. Puerto Rico did not fare much better, despite being a United States territory **(FIG. Bx11.1c)**. Power had not been completely restored to the island, as long as a year after the disaster. In April 2018, Puerto Rico's Department of Education announced that it would close 283 of its 1,100 schools following a sharp drop in enrollment due to the economic slump and the departure of many families for the United States.

Unfortunately, as the aftermath of Hurricane Maria clearly demonstrated, many Caribbean islands do not have the resources to cope with hurricane disasters when they occur, in part due to their poverty and in part due to their inaccessibility. Arrival of the COVID-19 pandemic has made matters worse, and it will take many years for the islands to achieve full recovery.

FIGURE Bx11.1 The impact of Hurricane Maria in 2017.

(a) Hurricane destruction in Roseau, Dominica.

(b) Hurricane Maria over Puerto Rico. The colors indicate the temperature and height of clouds, which correlate with intensity of rain.

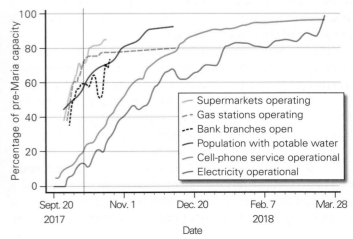

(c) Recovery of resources on Puerto Rico.

FIGURE 11.23 Hurricane Irma (2017) completely overwhelmed the island of Barbuda, and caused extensive damage on neighboring islands.

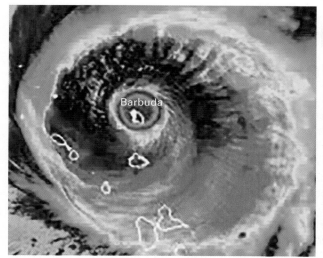

(a) In this radar reflectivity map view, the entire island of Barbuda can be seen within the eye of Hurricane Irma as it passes as a Category 5 hurricane. The highest point on the island is 38 m (125 ft).

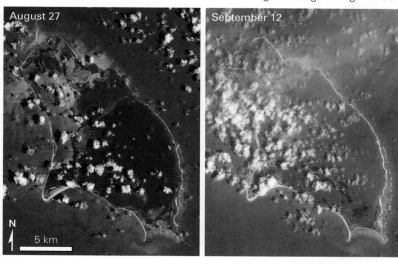

(b) The wind ripped apart, and salt spray killed, vegetation on Barbuda, causing the island's color, when viewed from space, to change from green to brown.

(c) Damage to nearby St. Maarten caused by Hurricane Irma.

subsided, flooding overwhelmed the sewage systems, leading to outbreaks of cholera, dysentery, and mosquito-borne illnesses.

Some hurricanes cross the Caribbean and strike the coast of Central America and Mexico. The torrential rains that such hurricanes drop on the slopes of inland mountainous areas can be catastrophic, especially if the regions have undergone deforestation. Hurricane Mitch (1998) became the deadliest Atlantic hurricane of the 20th century, not because it was particularly strong—it made landfall as a Category 2 storm—but because its slow movement led to extraordinary rainfall (up to 90 cm, or 35 in), which caused deadly mudslides and flash floods. A mudslide on the slope of a volcano in Nicaragua buried 10 communities (FIG. 11.24). The country was devastated: the overall death toll, including people who were never found, exceeded 19,000; 70% of crops were destroyed; and about 20% of the country's 1.5 million people became homeless.

Atlantic and Gulf Coast Hurricanes

Hurricanes that form over the western Atlantic or the Caribbean commonly enter the Gulf of Mexico, move over Florida, or drift northward along the east coast of North America. Those storms that move along the eastern seaboard of the United States can damage a long swath of the coastline before making landfall and heading inland. Landfall may happen anywhere between Florida and New England. Some hurricanes form directly over the Gulf of Mexico, eventually striking Gulf Coast states or eastern Mexico. Storm surge, waves, and wind have leveled towns and have caused severe coastal flooding (BOX 11.2), while rain from the remnants of hurricanes has inundated towns along streams and rivers.

During the past 150 years, about 300 hurricanes have reached the coast of the continental United States. Only five have struck the coast with Category 5 intensity, although several others had been classified as Category 5 earlier, while over the ocean. As will be discussed in Chapter 13, the hurricane that struck Galveston, Texas, in 1900 caused the greatest number of fatalities (up to 12,000) of any storm reaching the United States. Six other hurricanes have each caused over 1,000 fatalities. Because of the development of expensive communities along the coast, the financial cost of hurricanes has been substantial and rising: about 5% of the hurricanes have caused over $10 billion in damage, and Katrina (2005) and Harvey (2017) each caused an estimated $125 billion in damage. The extent of casualties and damage depends on

BOX 11.2 DISASTERS AND SOCIETY

Hurricane Katrina: Exposing the vulnerability of coastal cities to tropical cyclones

Meteorologists at the National Hurricane Center (NHC) knew that trouble lay ahead when, in late August 2005, Hurricane Katrina crossed southern Florida into the Gulf of Mexico and entered an area of very warm water south of Florida, called the Loop Current, a giant meander in the flow of surface water from the Caribbean to the Gulf Stream **(FIG. Bx11.2a)**. The rapid evaporation of this water provided enough energy to intensify Katrina into a Category 5 monster. As the storm headed toward New Orleans, the NHC issued the following sternly worded warning:

URGENT!

Hurricane Katrina . . . A most powerful hurricane with unprecedented strength . . . Rivaling the intensity of Hurricane Camille of 1969. Most of the area will be uninhabitable for weeks . . . Perhaps longer. At least one half of well constructed homes will have roof and wall failure. All gabled roofs will fail . . . leaving those homes severely damaged or destroyed. The majority of industrial buildings will become non-functional. Partial to complete wall and roof failure is expected. All wood framed low rising apartment buildings will be destroyed. Concrete block low rise apartments will sustain major damage . . . Including some wall and roof failure. High rise office and apartment buildings will sway dangerously . . . A few to the point of total collapse. All windows will blow out. Airborne debris will be widespread . . . And may include heavy items such as household appliances and even light vehicles. Sport utility vehicles and light trucks will be moved. The blown debris will create additional destruction. Persons . . . Pets . . . And livestock exposed to the winds will face certain death if struck. Power outages will last for weeks . . . As most power poles will be down and transformers destroyed. Water shortages will make human suffering incredible by modern standards. The vast majority of native trees will be snapped or uprooted. Only the heartiest will remain standing . . . But be totally defoliated. Few crops will remain. Livestock left exposed to the winds will be killed.

The eye of the hurricane crossed the shoreline just east of New Orleans, so while the storm flattened coastal towns in Alabama, the city of New Orleans, in the end, escaped the strongest winds. Sadly, the historic city didn't escape disaster. Storm surge driven by the hurricane's winds moved into Lake Pontchartrain, which isn't actually a lake, but rather a bay north of New Orleans that connects to the Gulf just east of the city. The influx of water caused Lake Pontchartrain's water level to rise, and the stage was set for catastrophe. Much of New Orleans lies as much as 2 m (7 ft) below mean sea level due to subsidence of the Mississippi River delta land on which the city was built (see Chapter 7). To keep the city from flooding, huge pumps work around the clock daily, draining water into a set of north-south-trending canals, lined by levees and dikes, that run from within the city to Lake Pontchartrain. The storm surge in 2005 caused water to rise in the canals nearly to the top of the levees **(FIG. Bx11.2b)**. Unfortunately, the foundations of some of the levees were anchored in a layer of weak sediment, and they couldn't handle the extra pressure. Some levees slipped on the weak layer and collapsed, and water poured into the city, quickly flooding whole neighborhoods. In the end, nearly 80% of the city was entirely or partially submerged as the bowl containing New Orleans filled to the level of Lake Pontchartrain **(FIG. Bx11.2c, d)**. Only the parts of the city built on the elevated land of the Mississippi's natural levees stayed out of the water. Houses in low areas filled to the ceiling with mud and debris, and even houses that did not fill with water became so humid that their walls became covered with mildew.

Some 1,836 fatalities were directly attributed to Katrina. Many of the victims had remained in the city because they did not have the means to evacuate, or wanted to protect their property. The disaster took on national significance as the trapped survivors sweltered without adequate food, drinking water, or shelter. With no communications or hospitals and few police, parts of the city descended into anarchy. It took days for help to arrive, and weeks for the floodwaters to be pumped out. Katrina displaced over a million people from New Orleans and surrounding regions of the central Gulf Coast. With their homes and property destroyed, many people moved away permanently, in the largest short-term migration of people ever recorded in the United States. Houston, Texas, one of the cities receiving Katrina evacuees, saw its population increase by 35,000. Years later, parts of New Orleans had still not recovered. The storm was also an environmental disaster, flattening trees, causing oil spills at facilities along Louisiana's coast, and destroying barrier islands (562 km^2 [217 mi^2] of land is now permanently under water). All told, the storm cost society $125 billion.

The consequences of Hurricane Katrina revealed deficiencies in the hurricane preparedness of a major American city. These vulnerabilities had been clear to engineers and emergency officials for years. For example, before Katrina, studies by the US Federal Emergency Management Agency (FEMA) and the Army Corps of Engineers demonstrated that a direct hit on New Orleans would lead to catastrophic floods. A 2001 article in *Scientific American*, "Drowning New Orleans," predicted much of the devastation that came 4 years later and the *Houston Chronicle* warned that a severe hurricane striking New Orleans "would strand 250,000 people or more, and probably kill one of 10 left behind as the city drowned under 20 feet of water. Thousands of refugees could land in Houston."

Could the disaster have been avoided? Perhaps. If the levee system had been properly constructed, if floodgates

FIGURE Bx11.2 Hurricane Katrina.

(a) The storm gained strength as it crossed warm water in the Gulf of Mexico.

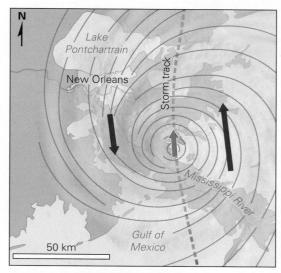

(b) The eye of the storm went west of New Orleans, but storm surge flowed into Lake Pontchartrain.

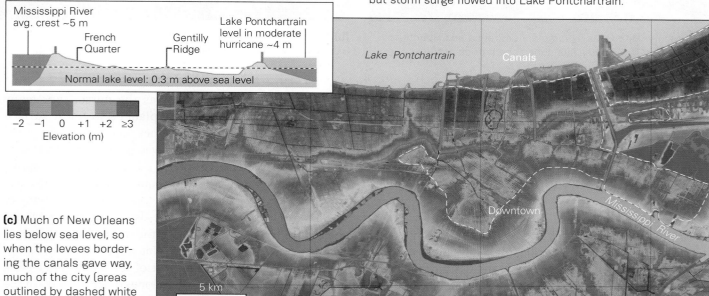

(c) Much of New Orleans lies below sea level, so when the levees bordering the canals gave way, much of the city (areas outlined by dashed white lines) was submerged.

(d) Water flowed over the levees and into neighborhoods.

had been installed to keep the canals from filling, if sufficient wetlands (which can absorb rain like a sponge) had been preserved, or if developers had not built in areas below known flooding levels, perhaps the city would have had only wind and rain damage. Unfortunately, despite our understanding of the threats associated with hurricanes, preparation for their inevitable strikes on urban areas seems to remain a low priority for society. The risk of random events is hard for people to fathom, and developers and emergency planners don't always agree on common goals. Hopefully, cities will invest in building seawalls and reinforcing levees, and widening wetlands and beaches, to help prevent destruction. Although Hurricane Ida in 2021 caused devastating power losses in New Orleans, minimal flooding in the city suggests that flood-prevention facilities that were updated following Katrina were successful.

FIGURE 11.24 Intense rain from Hurricane Mitch in 1998 caused a massive landslide in Nicaragua.

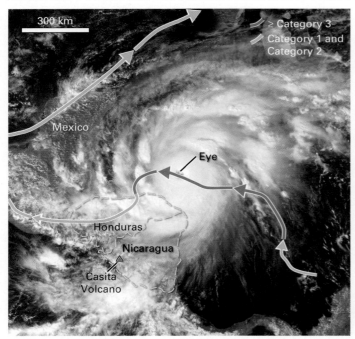

(a) Hurricane Mitch was as large as Nicaragua and Honduras put together. The curving line shows the hurricane track.

(b) The south flank of the Casita Volcano collapsed in a giant mudslide. The volcano is near the west coast of Nicaragua.

whether the storms come ashore over a city or over less developed coastlines. In the United States, because of better access to damaged areas and overall higher-quality emergency services, affected communities have been more resilient than those in other parts of the world, though the degree of resilience correlates strongly with community wealth.

Pacific Hurricanes and Typhoons

Hurricanes in the eastern Pacific, off the west coast of the Americas, develop in two ways. A few are Atlantic hurricanes that have drifted westward across Central America. The vast majority form in thunderstorm clusters that develop over the warm ocean just west of the highlands of Central America. In some cases, eastern Pacific hurricanes turn north, making landfall over northern Mexico, with flooding rains extending into California, Arizona, or Utah. Most drift westward across the Pacific, and a few strike the Hawaiian Islands. Hurricane Lane in 2018, for example, produced 133 cm (52 in) of rain on the slopes of the Hawaiian volcano Mauna Loa, causing massive floods.

Typhoons in the western Pacific normally form along the ITCZ in the northern hemisphere summer when the ITCZ lies north of 5° N latitude. Some typhoons track westward, crossing many islands, including the Philippines, and then proceed into China or Vietnam. Others turn north along the eastern margin of Asia, with many passing over Japan.

The western Pacific serves as the nursery for many typhoons because it hosts the planet's warmest surface seawater, especially in late summer and in La Niña years, when the Walker circulation drives warm water westward (FIG. 11.25a) (see also Chapter 8). Some build into **super typhoons**, the western Pacific's equivalent of a strong Category 4 or Category 5 hurricane (FIG. 11.25b). The persistence of warm water in the western Pacific also means that typhoons sometimes develop outside of the normal July–October typhoon season (see Fig. 11.6d). Not surprisingly, the Philippines, which lies just to the west of the warm-water pool, experiences the most tropical cyclones of any country in the world (FIG. 11.26a). The strongest tropical cyclone to strike the Philippines directly, Super Typhoon Haiyan, occurred late in the typhoon season, in November 2013 (FIG. 11.26b). Haiyan left total destruction in its wake and was responsible for the deaths of at least 6,300 people, over three times the number caused by Hurricane Katrina.

The impact of typhoons can be influenced by local topography. For example, a large mountain range runs along the axis of Taiwan (FIG. 11.27a), so heavy rainfall typically produces huge mudslides on the island, particularly on the eastern side, which, fortunately, is the less populated side. During Typhoon Morakot in 2009, more than 25,000 people evacuated before the heavy rainfall associated with the typhoon arrived, in order to avoid being caught by mudslides (FIG. 11.27b, c).

FIGURE 11.25 Western Pacific typhoons.

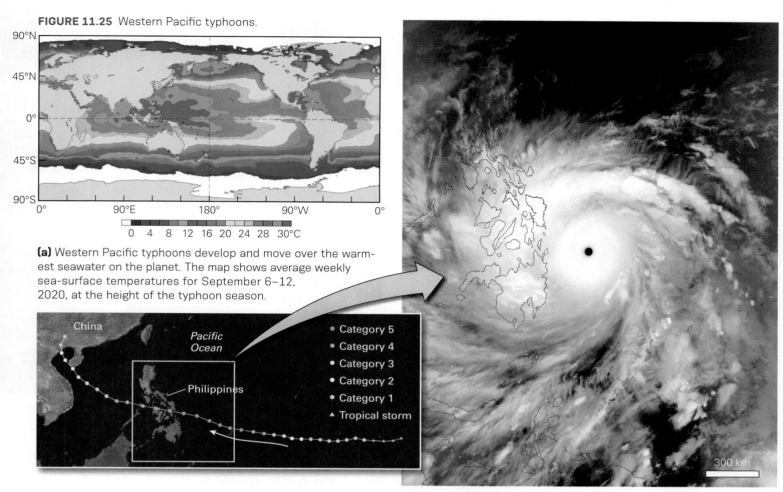

(a) Western Pacific typhoons develop and move over the warmest seawater on the planet. The map shows average weekly sea-surface temperatures for September 6–12, 2020, at the height of the typhoon season.

(b) Super Typhoon Haiyan (2013) packed winds of 290 km/h (180 mph) and devastated parts of the Philippines, killing at least 6,300 people.

FIGURE 11.26 Impacts of typhoons in the western Pacific.

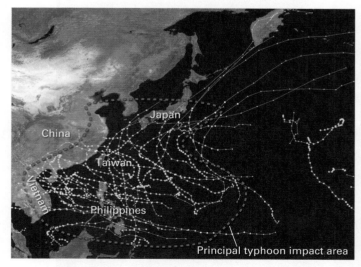

(a) The red dotted line outlines the principal area of typhoon impacts. Each year, typhoons track across the Philippines and Taiwan and into the coast of China and Vietnam. Some turn north over Japan.

(b) Destruction caused by Super Typhoon Haiyan (2013) in the Philippines.

11.5 Hurricane Disasters: A World View • **409**

FIGURE 11.27 Typhoon impacts on Taiwan.

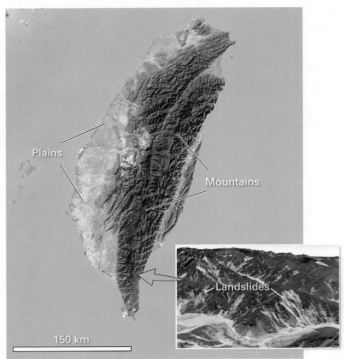

(a) Eastern Taiwan is mountainous. Typhoon rains can trigger landslides that cause destruction to the island's heavily populated plains.

(b) Typhoon Morakot (2009) just after it crossed Taiwan. Note that the principal rainband lies over the southern part of the island.

(c) Catastrophic effects of Typhoon Morakot on the eastern slopes of Taiwan.

Cyclones of the Indian Ocean and Southern Hemisphere

Cyclones are more frequent in the eastern Indian Ocean than in the west. Some of these storms head north into the Bay of Bengal. On November 12, 1970, an intense tropical cyclone moved northward from the Indian Ocean across the Bay of Bengal, and then over the Ganges River delta, at the head of the bay (FIG. 11.28a). The cyclone's winds drove storm surge that raised the ocean surface 5–6 m (16–20 ft) overall. Wind-driven waves up to 10 m (33 ft) high grew on top of the surge, allowing the sea to flood broad areas of the largely flat, and densely populated, Ganges River delta in what was then East Pakistan (now Bangladesh). At the time of this storm, now known as Cyclone Bhola, the countries affected had no capability to observe the storm, predict its path, or warn the population. As a result, when water inundated the delta, its population was trapped. More than 500,000 people died, and 100,000 additional people were never found. Cyclone Bhola holds the record as the deadliest tropical cyclone, and remains one of the deadliest natural disasters in all of human history. Twenty-one years later, in 1991, a similar storm took the lives of more than 140,000 people in the same region, and in 2008, yet another disastrous cyclone killed 138,000 people in Myanmar, along the east coast of the Bay of Bengal.

Why is the Ganges River delta so vulnerable to cyclone disasters? Many of the area's 100 million people inhabit its countless islands, each of which barely rises above sea level (FIG. 11.28b), so that when storm surge arrives, it submerges the landscape. Storm surge tends to rise particularly high in the Bay of Bengal because the bay narrows toward its north end, so when cyclones approach the delta from the south, storm surge funnels into a small area. Evacuation of the region represents a particular challenge. Given the distribution of people among islands, reaching vulnerable populations with information, persuading them to leave their property, and transporting them to safety presents enormous logistical challenges. Storms intensify rapidly in the bay, so by the time people board boats to seek higher ground, waterways have already become dangerous.

Cyclones in the southern hemisphere, which rotate clockwise (FIG. 11.29a), impact coastlines along the northern half of Australia and the eastern coast of Africa, including the island of Madagascar. Australia's most destructive storm, Cyclone Tracy, struck Darwin, a city on the north coast, on Christmas Day, 1974, with winds of about 250 km/h (155 mph). Though the storm had a small diameter, its winds and storm surge killed at least 71 people and injured thousands more. Of Darwin's population of 43,000, more than half were left homeless, and many moved elsewhere. Although other very strong Category 4 and 5 cyclones have made landfall on the Australian coastline, the low population densities of the areas affected have fortunately resulted in fewer casualties and less damage than has been caused by comparable storms in other regions of the world.

The portion of eastern Africa that lies in the southern hemisphere has felt the impact of Indian Ocean cyclones many times. Some of these storms form in the western Indian Ocean, and some in the Mozambique Channel, a strip of ocean 600 km (373 mi) wide separating the African continent from the island of Madagascar. Between 1948 and 2010, 94 cyclones formed over the channel. Once formed, the cyclones typically take one of two paths: some cross Madagascar and head back out over the Indian Ocean, whereas others drift over Mozambique, on the mainland of Africa (FIG. 11.29b). In 2019, back-to-back cyclones (Idai and Kenneth) caused

FIGURE 11.28 Cyclone Bhola (1970).

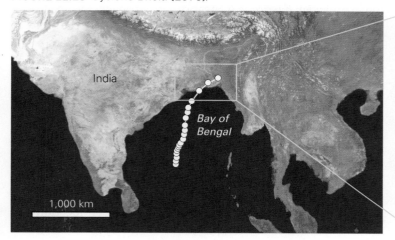

(a) The cyclone track led the storm directly to the heavily populated Ganges River delta.

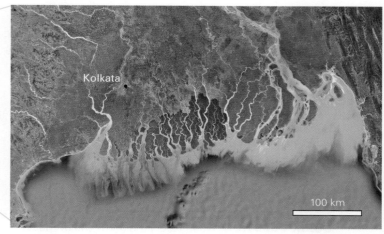

(b) Much of the land in the delta consists of islands that barely rise above sea level.

FIGURE 11.29 Southern-hemisphere cyclones in the western Indian Ocean.

(a) Southern-hemisphere cyclones rotate clockwise.

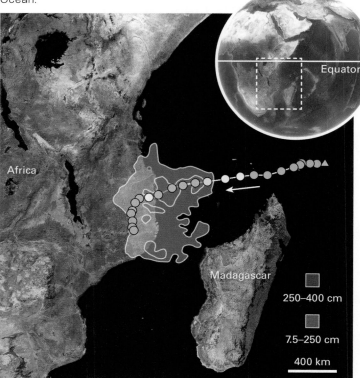

(b) Cyclone Kenneth tracked into Mozambique, on the coast of Africa, producing heavy rainfall.

(c) Flooding in Mozambique after Cyclones Idai and Kenneth in 2019.

catastrophic flooding in Mozambique (FIG. 11.29c). Due to the lack of adequate emergency resources and shelters, epidemics of cholera, dysentery, and malaria followed soon after.

> **Take-home message . . .**
>
> Hurricanes devastate coastlines across much of the world, particularly in the tropics. The effects of a hurricane at a given locality depend on the storm's intensity, size, and track, as well as on the topography of the landscape and the vulnerability of communities in the path of the storm.
>
> **QUICK QUESTION** What makes the north end of the Bay of Bengal so dangerous when a cyclone approaches?

11.6 Hurricane Impacts and Mitigation

Societal Disruptions

Where a *major hurricane* (Category 3 or stronger) has made landfall, the coast within 50–150 km (31–93 mi) of the eye becomes almost unrecognizable. Homes, businesses, factories, and public buildings lie in ruins; roads, railroads, harbors, and utility infrastructure are mangled, and water supplies are cut off. The vast damage caused by hurricanes disrupts the functioning of society. Destruction of manufacturing and transportation facilities may lead to unemployment and may interrupt supply chains. Repairs and reconstruction are likely to be costly for individuals, commercial organizations, and governments. Many individuals and organizations will face large debts. The degree of disruption depends on the vulnerability of communities, which, in turn, depends on location (affecting their access to supplies and emergency personnel), construction practices, and level of community organization (reflecting community wealth).

Predicting Hurricane Tracks

Because tropical cyclones are so dangerous, many countries have set up monitoring centers to detect and track the storms. In the United States, the National Hurricane Center (NHC), part of the NWS, identifies developing tropical depressions by collecting data on air pressure, wind, cloud cover, and sea-surface temperature, using satellites, radar, weather stations, and aircraft. When a tropical storm or hurricane develops, aircraft fly around the storm and release *dropsondes* (each consisting of a small package of instruments suspended from a parachute) to measure the background air flow in which the storm is embedded. The aircraft fly directly into the storm's eye to provide detailed measurements of its central pressure, wind speed, and location. All available data are input into computer models that predict hurricane tracks, dimensions, and future wind speeds. Such predictions help emergency planners provide appropriate warnings and evacuation recommendations.

Many variables can affect the strength, size, and track of a hurricane. Even with aircraft measurements, data from storms over the ocean remain sparse, and available data reflect natural uncertainty. In addition, researchers cannot completely model all physical processes operating in hurricanes. For all these reasons, no single computer model can perfectly predict the future track, size, and strength of a hurricane. Consequently, the NWS and meteorological organizations in other countries run *ensemble simulations*—that is, their meteorologists run a group, or ensemble, of computer models, each using slightly different values for variables (such as wind speed, sea-surface temperature, and air pressure) or slightly reformulated mathematical equations for calculations (**FIG. 11.30a**). The predictions made by all the models are assessed, and the NWS draws maps that show the most likely path of the hurricane (the average of the models' predictions) and a *cone of uncertainty* representing the range in the forecasts from the various models (**FIG. 11.30b**). Locations within the cone have a higher probability of being crossed by the storm than do locations outside the cone, but keep in mind that the cone indicates only probability, not certainty. Ensemble simulations generally provide a reasonably confident prediction for the next 1 to 3 days and a moderately well constrained prediction for an additional 2 days. The NHC runs its ensemble two to four times a day because as more data arrive, the predicted track typically shifts and becomes more certain. Similar ensemble simulations allow meteorologists to estimate the storm's wind speed and rainfall (**FIG. 11.31a, b**).

Because the cone of uncertainty can cover a large region of coast, public officials have a major challenge to determine where and when to issue evacuation orders. Using the track predicted by its ensemble simulations, the NHC identifies areas that could be subjected to hurricane-force and tropical storm–force winds. At least 48 hours before the storm's expected landfall, the NHC issues a **hurricane watch** or **tropical storm watch** for areas within the cone of uncertainty. Areas that are expected to experience hurricane-force or tropical storm–force winds within the next 36 hours receive **hurricane** or **tropical storm warnings** (**FIG. 11.31c**). *Storm-surge warnings* go out at the same time as hurricane warnings.

FIGURE 11.30 Prediction of the track for Hurricane Florence (2018), as provided by the NWS.

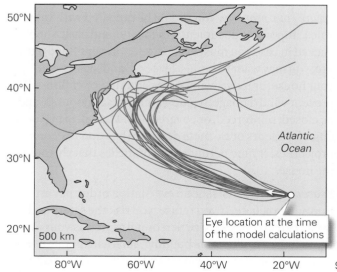

(a) Tracks predicted by an ensemble of computer models for Hurricane Florence's movement toward the northeastern US.

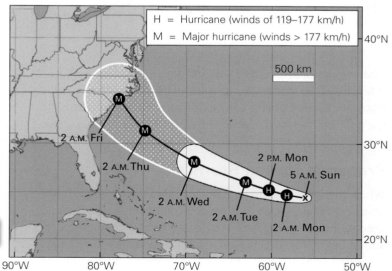

(b) The NWS combines results from the models to create prediction maps like this one, depicting the storm's possible track within a cone of uncertainty.

FIGURE 11.31 Prediction of wind speed and rainfall for hurricanes, as provided by the NWS.

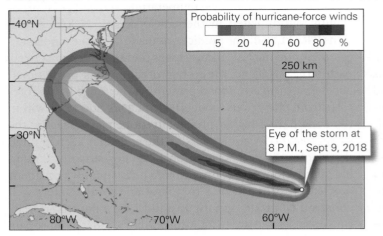

(a) A forecast of expected hurricane-force winds for the five-day period staring on September 9, 2018, for Hurricane Florence, as issued by the NWS.

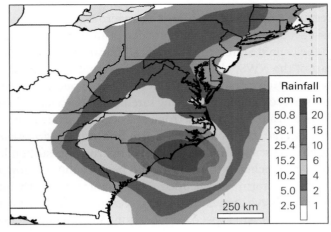

(b) A forecast of expected cumulative rainfall on the east coast of the United States for a seven-day period following the arrival of Hurricane Florence, beginning September 13, 2018. The heaviest rain was predicted for the Carolinas, where over 50 cm (20 in) fell.

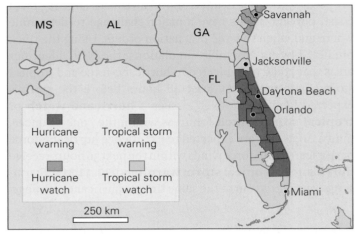

(c) A map issued to communities during Hurricane Isaias (2020) to show where hurricane alerts were in effect.

The urgency of a hurricane warning depends on the predicted intensity of the storm. The wording of hurricane warnings for Category 4 and 5 storms can be dramatic, in an effort to alert the public to the need to evacuate coastal and low-lying areas.

Preparing for the Storm's Arrival

When hurricane watches and warnings have been issued, public-safety officials in well-organized communities finalize and execute evacuation plans, businesses secure their property and prepare for evacuation, and utility companies stage resources (repair trucks; materials) in areas just outside the predicted damage zone. Similarly, when expecting a disaster declaration, the National Guard, FEMA, and various Non-Governmental Organizations (NGOs) prepare personnel and emergency supplies.

What should you do if hurricane winds threaten? If you live in an area with the potential to be struck by hurricanes, you should have an emergency plan in place.

- *Monitor the storm:* Be aware of what the storm is doing, and know its predicted track relative to your location and your planned evacuation route. TV, cell phones, the internet, and radio all provide this information. If other services are unavailable, radio may be your only resource, so be sure you have a battery-operated radio with plenty of batteries.

- *Secure your property:* You should be prepared with valuables, important documents, medicine, and basic supplies packed in case you must evacuate. To prevent flood damage, you may want to build a levee of sandbags around your house, or move your possessions to upper floors. To prevent wind damage, windows and garage doors should be closed, reinforced, or covered (FIG. 11.32a). Bring loose objects indoors or tie them down so they don't become projectiles. If time permits, trim branches that threaten the house.

- *Evacuate:* If officials issue an evacuation order, pay attention and evacuate. If you refuse, you risk the safety and lives of emergency service personnel as well as your own. If you live in a vulnerable building, such as a mobile home, you may be in extreme danger and must move to a safer location. Some people tend to ignore evacuation orders. They may assume that the storm won't be as dangerous as officials predict, or they may wish to protect property from the storm or from looting, or they may have

FIGURE 11.32 How can you prepare for hurricanes?

(a) Secure your property.

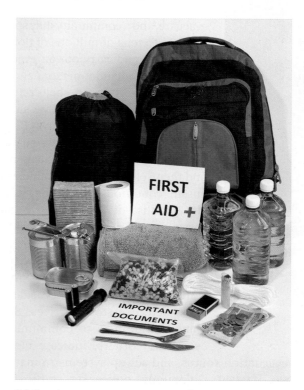

(b) Stock up on emergency supplies.

- *Shelter in place:* If circumstances require that you remain in your home during the storm, you can increase your chances of survival by taking important precautions. First, secure your property as described above. Then, assume that water supplies may be cut off, so fill tubs, pots, and large water containers with drinking water. Stock up on nonperishable food, for if electricity goes out, food in refrigerators or freezers will stay fresh for only a day or so (**FIG. 11.32b**). Be sure you have the medicines you need, as roads may become impassable and drugstores may be destroyed. Have an advance plan for how to communicate with and rejoin your family after the storm. Avoid overusing cell phones so you can maintain battery charge, in case electricity is cut off, and to prevent overloading networks. During the storm, assume that winds could reach tornado strength, and follow the same precautions that can save lives in a tornado: stay in an interior room that can protect you from flying debris.

Hurricane Evacuations

As we described earlier, when Cyclone Bhola struck the Ganges River delta in 1970, over 500,000 people lost their lives when storm surge flooded densely populated areas. But when Cyclone Amphan, an even stronger storm, struck the same region in 2020, the death toll was fortunately only 128. The difference was the result of a successful evacuation. When storm warnings went out 5 days before Amphan made landfall, ships nowhere to go, or no means of transportation. Unfortunately, in many cases, not evacuating leads to tragedy, as happened during Hurricane Katrina, when about 1,800 people who remained in New Orleans died.

FIGURE 11.33 Evacuating from a hurricane.

(a) Nearly all traffic was outbound from New Orleans in advance of Hurricane Katrina.

(b) A sign designating a hurricane evacuation route.

(c) A hurricane shelter in a school gym during an evacuation.

in the Bay of Bengal were warned to move out of its path, and thousands of shelters were set up. Two days before landfall, evacuations were ordered, and over 4 million people moved inland.

While timely evacuations can save lives, evacuation is expensive, time consuming, and potentially dangerous. Evacuations, on average, cost about $1 million per mile of coastline. An individual family may spend hundreds or thousands of dollars on transportation, lodging, and food when they evacuate, and a community may spend millions to tens of millions of dollars to organize and carry out an evacuation (costs include personnel and equipment, setting up shelters, and providing food and water). Evacuation of a large city can take days. Roads become clogged during an evacuation **(FIG. 11.33a)**, and late evacuees may find themselves trapped in vehicles in unsafe conditions when the storm hits. Because of the complexity and cost of evacuations, officials struggle with making a decision to evacuate.

In order for an evacuation to proceed smoothly, communities must work out comprehensive evacuation plans before a storm arrives. Such plans outline how to communicate evacuation orders, indicate clearly marked evacuation routes, and set up systems to manage traffic **(FIG. 11.33b)**. Commonly, police set traffic flow patterns along evacuation routes so that all lanes but one flow in the direction away from the storm. Furthermore, evacuation works only if people have safe places to go. Some people move in with relatives or friends outside of the storm track or stay in hotels at safe localities. Others use public shelters set up in schools or other public buildings **(FIG. 11.33c)**.

FIGURE 11.34 Hurricane-proof construction allowed this house to survive Hurricane Michael (2018).

Hurricane Damage Mitigation

Fortunately, approaches are available that can help mitigate the impact of a hurricane. Unfortunately, most are expensive, and thus are not always used, particularly in poverty-stricken communities.

- *Zoning:* Rules controlling minimum setbacks for buildings constructed along vulnerable shorelines and rivers can prevent buildings from being susceptible to damage by storm surge, high waves, and river flooding. Such zoning rules are often contentious because people like to live close to water.

- *Protective landscapes:* Building or preserving protective zones—such as beach dunes and coastal wetlands—along the shore can blunt the impact of waves before they reach vulnerable buildings. Preserved wetlands and prairie lands inland can absorb floodwaters, and preserving forests can help prevent mudslides (see Chapter 13).

- *Building codes:* Establishing building codes that require water- and wind-resistant construction can improve the likelihood that buildings in vulnerable areas will survive. For example, putting houses on reinforced concrete stilts can keep them out of the water during storm surge. Specialized construction methods can help buildings stand up to the onslaught of hurricane-force winds. During Hurricane Michael, a Category 5 hurricane that struck the Florida Panhandle in 2018, all buildings bordering Mexico Beach, save one, were completely destroyed (**FIG. 11.34**). The surviving building had been designed to exceed local building codes, and because of its reinforced concrete walls, strong foundation, and steel cables anchoring the roof, the building survived Michael's winds.

Unfortunately, despite a community's best efforts, damage will occur. The damage caused by a hurricane can be so expensive that normal home insurers can't afford to cover it. Many insurers refuse to offer coverage in coastal areas of hurricane-prone states. In the United States, homeowners' insurance policies cover only hurricane wind damage. Insurance for flooding must be obtained from the federal government through the National Flood Insurance program (see Chapter 12). Long court battles between homeowners and insurance companies over whether damage was caused by wind or water often follow a hurricane. Sadly, in impoverished communities, many structures are not insured, and survivors of a storm may find themselves homeless. In such cases, further suffering can be relieved only by humanitarian aid. The need for hurricane mitigation efforts increases every year, as the population density of vulnerable regions increases. This issue should not be ignored.

Take-home message . . .

To reduce loss of life and protect property, forecasters collect data and predict hurricane tracks, taking into account uncertainty due to varying data and models. This information, shared broadly in the form of watches and warnings, allows citizens and emergency managers to protect property and businesses and to save lives. If you are in a threatened area, follow the guidance of emergency officials regarding evacuation.

QUICK QUESTION What is a source of uncertainty in forecasting hurricane tracks?

Chapter 11 Review

Chapter Summary

- Tropical cyclones are the most destructive storms on the Earth.
- Weak tropical cyclones are called tropical depressions, and stronger tropical cyclones are called tropical storms. The strongest (with sustained winds over 119 km/h, or 74 mph) are called hurricanes over the Atlantic and eastern Pacific, typhoons over the western Pacific, and cyclones over the Indian Ocean and throughout the southern hemisphere.
- A hurricane's maximum sustained wind speed determines its classification on the Saffir-Simpson scale, which rates hurricanes from Category 1 (the least strong) to Category 5 (the strongest).
- Well-developed hurricanes have a nearly cloud-free eye surrounded by an eye wall, a ring of towering clouds extending from near the sea surface upward to the tropopause. Beyond the eye wall, spiral rainbands extend outward for hundreds of kilometers. Torrential rain and violent winds occur beneath the eye wall.
- Energy powering a hurricane comes from evaporation of water at the sea surface. Water vapor rises and condenses into clouds, at which time a vast amount of latent heat is released, warming the surrounding air, making air buoyant, and supporting thunderstorm development.
- A positive feedback for hurricane development occurs as winds build tall, frothy waves and produce sea spray, increasing evaporation rates by a factor 100 to 1,000, and providing more latent heat to the hurricane.
- A second positive feedback occurs as clouds moisten the troposphere in the storm's region of circulation, reducing evaporation and increasing rainfall.
- A cluster of thunderstorms must be located over tropical ocean waters to organize into a hurricane. These clusters most commonly form in easterly waves over the Atlantic, and within the ITCZ over the western Pacific.
- The Atlantic and eastern Pacific hurricane season extends from June through November, when tropical ocean water temperatures are at their warmest. The season is longer in the western Pacific, where ocean waters retain heat throughout the year.
- The energy of hurricanes comes from latent heat transferred to the atmosphere as water evaporates from warm seas and condenses to form clouds.
- In the tropics, hurricanes typically move westward with the trade winds while drifting poleward. As they cross into the mid-latitudes, where winds typically flow from west to east, they turn eastward.
- Air descends in the eye of a hurricane because the downward force of gravity overcomes the upward pressure-gradient force in the eye.
- The dimensions of a hurricane can be measured by the diameter of the eye wall, the extent of the storm's hurricane-force or tropical storm–force winds, or the diameter of its cirrus cloud shield as viewed from space.
- Winds are stronger, and destruction greater, on the right side of a hurricane, as viewed looking in the direction in which the hurricane is moving along its track.
- Storm surge is a mound of ocean water piled up by a hurricane, which can cause flooding and damage as it moves onshore.
- Winds, waves, storm surge, inland flooding, mudslides, and occasional tornadoes triggered by hurricanes cause the devastation associated with these storms.
- Tropical cyclones cause devastation along the coasts of many regions of the world. In the past, inadequate infrastructure for evacuation has led to tens of thousands of fatalities. Hurricane strikes may also be followed by health crises.
- A single computer model cannot perfectly predict the future track, size, and strength of a hurricane. Forecasters run ensemble simulations, using a group of computer models, to estimate the most likely hurricane path and a cone of uncertainty representing the range in the forecasts.
- When a hurricane approaches a coast, hurricane watches, followed by hurricane warnings, are issued for threatened areas.
- Surviving a hurricane requires adequate preparation of property, appropriate supplies, and often evacuation of coastal areas.

Review Questions

Blue letters correspond to the chapter's learning objectives.

1. What terms are used to describe tropical cyclones in different parts of the world? **(A)**
2. What is the difference in size between a tornado, a thunderstorm, and a hurricane? **(A)**
3. What is the Saffir-Simpson scale? What atmospheric property does it use to categorize a hurricane? **(B)**
4. Identify the key structural features of a hurricane. Where are its strongest winds found? **(C)**
5. Describe how the wind direction would change if you were located on the Gulf of Mexico coastline and a hurricane approached with the eye passing directly over you. What if the eye passed 40 km (25 mi) west of your location? **(C)**
6. Where does one find the heaviest rain in a hurricane? **(C)**
7. Label the eye and eye wall on the diagram. Where is the lowest pressure at sea level? Where is the largest pressure gradient? **(D)**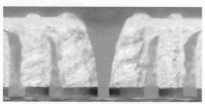
8. From where does a hurricane draw its energy? What is the primary way in which this energy is transferred from the ocean to the atmosphere? **(D)**
9. Describe at least one positive feedback that occurs as a hurricane intensifies. **(D)**
10. How does an eye develop in a hurricane? **(D)**
11. Would a hurricane be more likely to form over tropical oceans in a region where winds in the upper troposphere are weak or strong? Justify your answer. **(D)**
12. Convergence of air in the lower troposphere triggers thunderstorms over tropical oceans. In what atmospheric feature does convergence most commonly occur over the Atlantic Ocean? Over the western Pacific Ocean? **(D)**
13. Describe the life cycle of a typical Atlantic hurricane from the early formation of thunderstorms to its demise over land or cold waters. **(D)**
14. Where do tropical cyclones form, and what regions of the world are threatened by these storms? **(E)**
15. On the figure, draw the typical track of a hurricane that forms in the Atlantic and reaches the southeast United States. Why do hurricanes follow such tracks? **(E)**
16. Why do hurricane tracks typically curve to the right in the northern hemisphere? **(E)**
17. Does wind cause more damage on the right or left side of a hurricane, looking in the direction in which the hurricane is moving? Explain your answer. **(F)**
18. What is storm surge? What else causes destruction when a hurricane moves across a coastline? **(F)**
19. If a hurricane in the Gulf of Mexico moved toward the west and made landfall in Mexico, which side of the storm (north or south) would produce the strongest winds and the highest storm surge? **(F)**
20. Based on events of the past 20 years, which regions of the world do you think are most vulnerable to fatalities caused by hurricanes? How about property damage? Are these areas the same or different? Justify your answer. **(G)**
21. What is the purpose of using an ensemble of computer models to predict the path of a hurricane? Why does the cone of uncertainty get wider for times further in the future? **(H)**
22. Why might officials hesitate before ordering an evacuation? **(H)**

On Further Thought

23. What do you think the general trends in property losses and fatalities associated with hurricanes will be in the next 20 years? **(G)**
24. Of all the states on the east coast of the United States, Georgia receives the fewest hurricane strikes. Why might this be so? **(E)**
25. In Florida, you meet a hotel owner who claims that when the last hurricane, a Category 3, struck, her beach-front hotel sustained only minor wind damage. You ask the owner which way the winds were blowing during the hurricane, and she says, "From the north." Explain why her hotel was not severely damaged by the wind or flooded by storm surge. **(F)**

12

DANGEROUS DELUGES
Inland Flooding

By the end of the chapter you should be able to . . .

A. explain how streams develop and interact with sediment, and describe landscapes associated with streams.

B. distinguish among different types of streams and drainage networks, and describe how to characterize stream flow and depth.

C. interpret news mentioning the flood stage and crest of a stream.

D. distinguish among different types of floods.

E. describe the consequences of flooding, and discuss how flooding affects floodplains and deltas.

F. summarize the types of weather events responsible for floods.

G. interpret statements concerning the probability of flooding, maps of flooding hazards, and official statements about impending flooding.

H. outline advantages and disadvantages of flood-control measures.

I. explain how flood insurance differs from other types of insurance.

12.1 Introduction

In the 1880s, developers built a mud-and-gravel dam across the South Fork of Pennsylvania's Little Conemaugh River in order to trap a reservoir that could provide a pleasant setting for summer cottages. When torrential rain drenched the region on May 31, 1889, water in the reservoir rose until it started to flow over the dam. Suddenly, the soggy structure collapsed, releasing a ferocious flow of water that roared downstream, gathering a variety of debris that lay in its path. When the water—with its load of mud, logs, boulders, and even train cars—slammed into the city of Johnstown, it swept away homes, churches, and commercial buildings. Tragically, it killed 2,200 people, the worst loss of life from any American natural disaster until that time (FIG. 12.1). Johnstown had experienced a **flood**, an event in which water submerges normally dry land. The devastation shocked the nation and had lasting implications. Red Cross staff came to treat casualties, greatly enhancing the new organization's reputation, and arguments presented in flood-related lawsuits ultimately led to reinterpretations of liability laws related to flooding.

Through the ages, flooding has left its mark on society. In fact, memories of huge floods in prehistory evolved into legends that have been passed down in oral traditions for generations. For example, the Chinese myth of the Gun-Yu Great Flood describes the inundation of plains surrounding the Yangtze and Yellow (Hwang) Rivers. According to the myth, water remained until Prince Yu convinced a dragon to dig canals that let the water escape to the sea. Due to his success, Yu became emperor and founded the first of the traditional Chinese dynasties. Historians suggest that the myth commemorates a real event—namely, the first time that a leader organized labor crews to construct canals and alleviate flooding.

Floods can play a beneficial role in the Earth System by depositing nutrients and sediment and by washing away debris. But some floods are natural disasters that submerge population centers, farm fields, and industrial facilities (FIG. 12.2). Floods happen for many reasons. Elsewhere in this book, we describe *shore floods* associated with storm surges, storm waves, and tsunamis. In this chapter, we focus on *inland floods*, those due to an oversupply of freshwater from rainfall or melting snow in non-coastal regions. Examples include stream floods, when flowing water spills over the banks of a stream channel and spreads over surrounding land; areal floods, when low areas collect water and become submerged; urban floods, when city neighborhoods become inundated due to insufficient drainage; and floods due to the failure of dams or levees. In coastal areas, the consequences of shore floods may overlap with those of inland flooding. Floods can happen at different rates, so geologists and *hydrologists* (researchers who study water on and below the land) distinguish between *slow-onset floods*, which take days or weeks to develop, and *flash floods*, during which water rises in minutes to hours.

Because most inland floods result from stream flooding, we begin this chapter by describing how streams form and how they drain the land. Then we discuss various types of floods, factors that influence flooding severity, the consequences of flooding, and the weather systems that cause flooding. We conclude by examining how society can protect itself from floods.

⟨ After a heavy rain, a stream channel next to a town in Peru floods with turbulent water. The boulders that litter the stream's channel moved during even larger floods in the past. Such floods could potentially destroy streambank buildings.

FIGURE 12.1 In 1889, raging floodwaters destroyed Johnstown, Pennsylvania. Debris trapped by a bridge caught fire.

FIGURE 12.2 Floods are among the most deadly and costly natural disasters on the Earth. Here, neighborhoods are submerged by flooding in Townsville, Australia.

12.2 Draining the Land: Streams and Rivers

Runoff and the Formation of Streams

Water moves continually among reservoirs and realms of the Earth System in a *hydrologic cycle*, as we saw in Chapter 7. Water vapor that evaporates from the Earth's surface condenses or crystallizes in clouds to form rain or snow, which falls back to the surface. About 22% of this precipitation ends up on land. Some of this water eventually becomes **runoff**, water that flows out of an area of land in response to gravity. Runoff includes both *overland flow*, which flows across the land surface as a thin layer (*sheetwash*) or in trickles that seep around sediment grains or plant stems, and *stream flow*, which flows down a trough or *channel*. Researchers use the term **stream** generically for any flowing body of water in a channel. In everyday English, we often refer to medium-sized streams as creeks or brooks, and to large streams as **rivers**.

Why do streams form? Since the rate of erosion by water depends on the speed and volume of flowing water, the process begins at a location where overland flow happens to be faster, or involves more water than in adjacent areas. In such a location, water carves into the ground faster than it does in surrounding areas and produces a small trough (**FIG. 12.3a**). Eventually, the trough becomes deeper and wider and evolves into a stream channel. Over time, flowing streams deepen their channels and carve valleys or canyons by *downcutting* (carving into the substrate below). They also become longer by *headward erosion* (carving into the land at the stream's origin). Once a stream has been established, water enters it not only from overland flow, but also in the form of groundwater percolating out of soil or burbling up from springs (**FIG. 12.3b**).

Describing Streams

Streams are distinct landforms. We refer to the sides of a stream channel as its *banks* and to the floor of the channel as its *bed* (see Fig. 12.3b). The location where a stream begins to

FIGURE 12.3 The formation of streams, and the sources of water for streams.

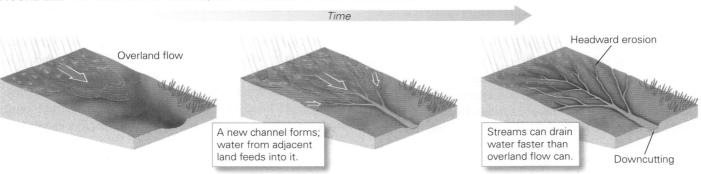

(a) Stream formation begins when overland flow focuses in a slight depression, so that the flow becomes faster and carves a channel.

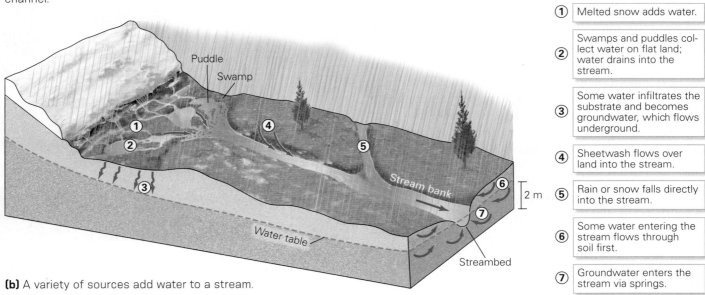

(b) A variety of sources add water to a stream.

flow is called its *headwaters*, the place where it empties into another body of water is its *mouth*, and a defined length along the stream is a *reach*. An upstream reach lies closer to the headwaters, and a downstream reach lies closer to the mouth. In order for a stream to flow at a location, the surface of the stream must have a slope, or **stream gradient**, in the downstream direction.

In 1804, Meriwether Lewis and William Clark, along with 40 other men, began their famed exploration of the Louisiana Territory by tracing the Missouri River from its mouth to its headwaters. They learned the hard way that a stream's gradient and channel characteristics change along the length of the stream. At its mouth, the Missouri is wide and has a very gentle gradient, so it's easily navigable. Upstream, its channel becomes narrower and rockier, and its gradient becomes steeper. Hydrologists represent a gradient change like the one that Lewis and Clark experienced by drawing a **longitudinal profile**, a line on a graph that plots elevation on the vertical axis and distance from the mouth on the horizontal axis **(FIG. 12.4a)**. A longitudinal profile of a typical stream displays a steep gradient in upstream reaches and a gentle gradient in downstream reaches. In detail, longitudinal profiles may include steps, representing interruptions by lakes, waterfalls, or hard-rock ledges.

FIGURE 12.4 A stream changes along its length.

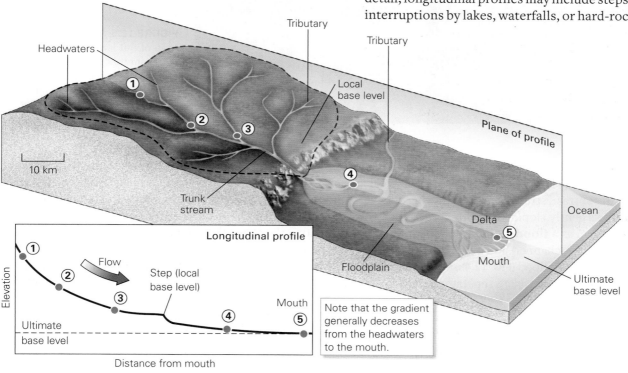

(a) A stream's gradient tends to decrease from its headwaters to its mouth, as indicated by a longitudinal profile. Points 1 to 5 refer to locations along the profile (inset).

(b) Streams tend to cut deep valleys near the upstream end of a longitudinal profile and to flow over broad plains at the downstream end.

12.2 Draining the Land: Streams and Rivers • **423**

FIGURE 12.5 A floodplain is the area of flat land on either side of a stream that will be partly to completely submerged during a flood. Like many floodplains, this one in Utah has fertile soil for farming.

Typically, the floodplain is underlain by a thick layer of alluvium.

Typically, stream channels flowing down a longitudinal profile's steep part lie within *V-shaped valleys* or vertical-walled canyons. In contrast, stream channels flowing down a longitudinal profile's gently sloping part are bordered by an area of flat land known as a **floodplain**. A floodplain is substantially wider than the channel, and it becomes partly or entirely submerged during a flood (**FIG. 12.4b; FIG. 12.5**). Commonly, small escarpments called *bluffs* separate the floodplain from higher land beyond.

Geologists refer to the elevation below which a stream's surface will not drop and allow water to continue flowing downstream as the **base level** of the stream (see Fig. 12.4a). A lake or resistant rock ledge can serve as a *local base level* for a stream, meaning a base level upstream of the stream's mouth. Where a **tributary** (a stream flowing into another stream) joins a larger stream, the surface of the larger stream acts as the base level for the tributary. Sea level defines the *ultimate base level* of a stream that flows into the ocean. Note that the water surface at the mouth of a stream can't be lower than that of the water body that the stream flows into. If it were, water in the stream would have to flow uphill to leave the stream's mouth, and that is impossible.

Erosion and Deposition by Streams

If you spray the ground with a hose and watch the water dig into the soil and carry it away, you are seeing erosion by flowing water. Erosion by a stream can take place because flowing water picks up loose sediment, breaks off and lifts up bedrock chunks, rasps rock surfaces, and dissolves minerals (**FIG. 12.6a**). Streams carry sediment as *dissolved load* (ions in solution), as *suspended load* (silt- or clay-sized grains that swirl along with moving water, giving the water a brownish tint), or as *bed load* (larger clasts that bounce or roll along the streambed) (**FIG. 12.6b, c**). All of these materials together constitute the stream's *sediment load*.

Hydrologists define the **competence** of a stream as the maximum clast size that the stream can carry. Competence depends primarily on a fluid's velocity, so a fast-flowing stream has greater competence than a slow-flowing one has. The density of a fluid also influences competence. For example, because a mixture of clay and water is denser than pure water, a boulder in muddy water is more buoyant than it would be in clear water, and the dynamic pressure exerted on it by moving muddy water is greater than that applied by clear water traveling at the same velocity.

The **capacity** of a stream refers to the total quantity of sediment the stream can carry. A stream with greater capacity can carry more sediment than can a stream with less capacity. Capacity depends on competence and on the volume of water in the stream. Both the competence and the capacity of a stream increase during a flood because flow velocity and flow volume both increase. Therefore, more erosion and sediment transport take place when streams are in flood than when they have normal flows.

When water in a stream slows down, the stream's competence decreases, so its sediment load settles out to form a layer of stream sediment, or **alluvium**. The sizes of the clasts that settle at a particular location depend on the flow velocity at that location. For example, coarse alluvium can settle from

FIGURE 12.6 Erosion, transport, and deposition by streams.

(a) During floods, turbulent water churns down this narrow canyon in Arizona. The sand and gravel carried by the water abrades the sandstone walls.

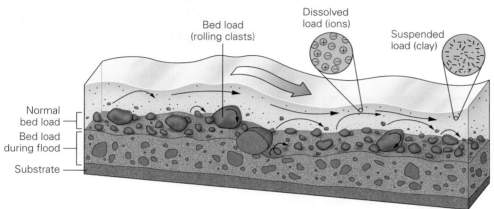

(b) Moving water carries sediment as dissolved load, suspended load, and bed load.

(c) When flow is low, this stream in Switzerland runs clear. When flow is high, the same stream contains muddy water.

(d) Coarse gravel is transported and deposited when flow velocity in this stream in Alaska (left) is high. Slower flow in this Canadian stream (right) deposits bars of sand and silt.

FIGURE 12.7 Deltas form where a stream enters standing water.

(a) A small delta formed where a stream drains into a lake.

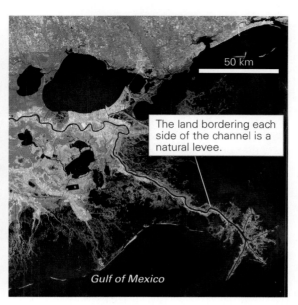

(b) The Mississippi delta has formed where the Mississippi River enters the Gulf of Mexico.

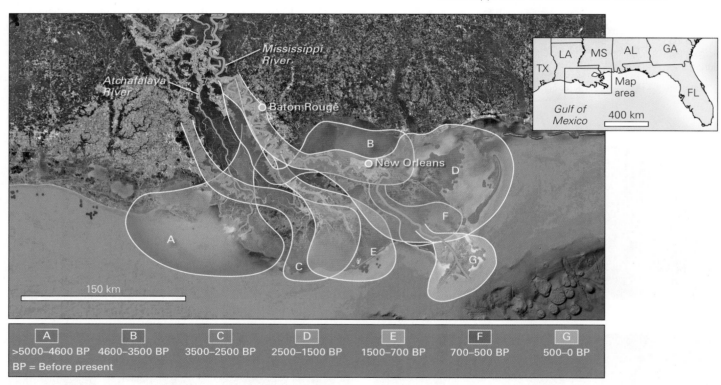

A	B	C	D	E	F	G
>5000–4600 BP	4600–3500 BP	3500–2500 BP	2500–1500 BP	1500–700 BP	700–500 BP	500–0 BP

BP = Before present

(c) The location of sediment deposition on the Mississippi delta has changed over time, building a broad delta plain. Each color represents a different stage of delta formation.

a fast-flowing stream, but fine alluvium can settle only out of slow or standing water (**FIG. 12.6d**). Consequently, gravel and sand tend to accumulate in elongate mounds called *bars* along or within the stream channel, or in *alluvial fans*, wedges of sediment deposited by floods, at the mouth of a canyon. Silt and clay, in contrast, tend to accumulate in *floodplain deposits* (layers of sediment on a floodplain) where water has slowed due to friction with the land surface. **Natural levees**, ridges of sediment built up on the banks of a stream channel, form from sediment deposited by the initial slowdown of water as it starts to spill out onto the floodplain.

Where a stream empties into a standing body of water, sediment can accumulate to form a wedge of sediment called a **delta** (**FIG. 12.7a, b**). As the delta grows outward from its mouth, the stream splits into smaller channels, or *distributaries*, that spread its water out over the delta. Over time, the surface of a large delta can become a broad **delta plain**, consisting of low land susceptible to flooding. The location of delta growth changes as different distributaries become dominant (FIG. 12.7c).

Types of Streams

Streams vary widely in character, reflecting differences in the amount of water a stream carries, the stream's gradient, and the nature of the substrate across which the stream flows. Let's look at some of this variation, for the nature of flooding depends on the character of the stream.

PERMANENT VERSUS EPHEMERAL STREAMS. When hiking in a region with a temperate or tropical climate, you're almost certain to find flowing water in the stream channels that you encounter. That's because at any given time, overland flow, tributary flow, and springs provide enough water to the stream to keep its bed submerged. Spring water is available because, as we saw in Chapter 7, the water table in temperate or tropical climates typically lies above the bed of a stream. In contrast, on a hike through a semiarid or arid region, you'll be able to cross stream channels on most days without getting your feet wet. The water table in such regions lies far below the streambed, so spring water does not enter the stream. Therefore, unless sufficient water is supplied from upstream, water in the channel infiltrates the streambed, and the stream dries up. To highlight this contrast, hydrologists distinguish between **permanent streams**, which flow all year, and **ephemeral streams**, which flow during only part of the year (FIG. 12.8). During the dry season in a semiarid climate, or between rare rains in a desert, flow in an ephemeral stream vanishes entirely, leaving an empty channel known as a **dry wash**, an *arroyo* (from Spanish), or a *wadi* (from Arabic).

MEANDERING VERSUS BRAIDED STREAMS. Some streams have fairly straight reaches, but some follow snake-like curves. A stream whose path contains many such curves is called a **meandering stream**, and each curve is a **meander** (FIG. 12.9a). Most meanders begin to form where streams have a very gentle gradient, and where the substrate of the stream banks is strong enough to resist collapsing into the stream. Typically, a broad floodplain made up of deep

FIGURE 12.8 The contrast between permanent and ephemeral streams.

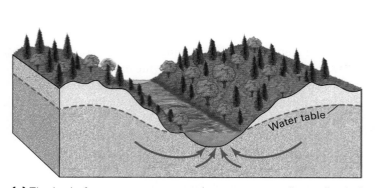

(a) The bed of a permanent stream in a temperate climate lies below the water table. Springs add water from below, so the stream contains water even between rains.

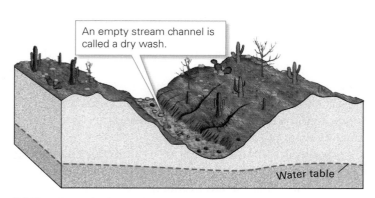

(b) The channel of an ephemeral stream lies above the water table, so the stream flows only when water enters the stream faster than it can infiltrate the ground.

FIGURE 12.9 The character and evolution of meandering streams.

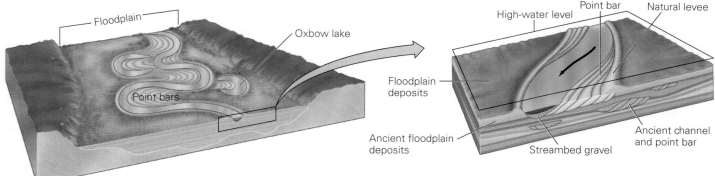

(a) Landforms along meandering streams include natural levees, point bars, and cut banks.

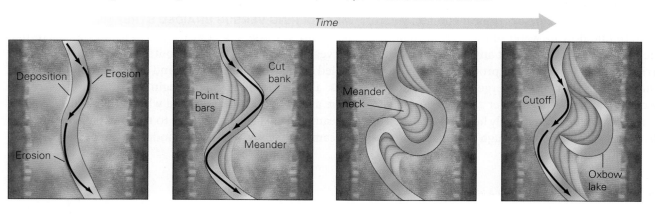

(b) Meanders evolve because erosion occurs on the outer edge of a curve, and deposition takes place on the inner edge. Eventually, a meander can be cut off and abandoned.

(c) An aerial view of a portion of the Mississippi River floodplain. The present channel (blue line) covers only a small part of the floodplain. Landscape features indicate where the channel was in the past.

alluvium deposits borders a meandering stream. But in places where the base level of a meandering stream has dropped relative to the land surface, the stream downcuts into bedrock, so the stream channel flows along the floor of a meandering canyon.

The course of a meandering stream flowing down a broad floodplain can change significantly over a time frame of just years to centuries. Why? Flow velocity is faster on the outer edge of a curving channel than on the inner edge. This difference causes the stream to erode along the outer curve to form a *cut bank*, while it deposits alluvium along the inner curve to form crescent-shaped *point bars* (FIG. 12.9b). Meanders evolve until eventually, only a narrow neck of land lies between two adjacent meanders. When the stream erodes through this *meander neck*, the channel takes a shortcut and abandons the former meander, which may become an *oxbow lake*. A map or aerial photo of point bars and abandoned meanders emphasizes that the course of a naturally meandering stream constantly changes (FIG. 12.9c).

Braided streams diverge into numerous small channels that flow around gravel bars and merge again downstream, resulting in an overall pattern that resembles braided hair (FIG. 12.10). They form where a stream carries a

428 • CHAPTER 12 • Dangerous Deluges: Inland Flooding

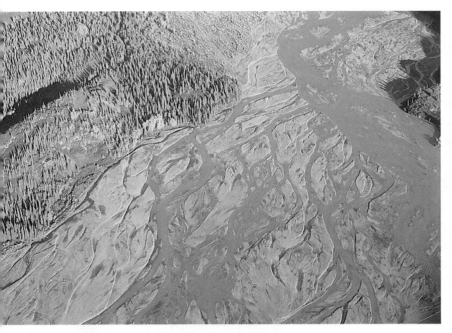

FIGURE 12.10 Braided streams carry large quantities of sediment when flow is high. When flow slows, it settles out to form elongate bars. The stream separates into small channels that flow around these bars.

large sediment load during times of high flow, and where the banks of the stream are so weak that they collapse frequently. When the flow slows and competence decreases, the sediment settles in bars. The remaining flow finds paths around bars of alluvium.

Drainage Networks and Watersheds

Look at a map of a land area and you'll see streams of many different sizes. Smaller tributaries drain into larger tributaries, which drain into still larger ones, which drain into a **trunk stream** that carries water out of the area. An array of streams that together provide water to the same trunk stream, and therefore drain water from a defined area, is called a **drainage network** (FIG. 12.11a). Geologists identify different geometries of drainage networks depending on how the tributaries are oriented with respect to one another and on the angle at which the streams intersect (FIG. 12.11b). The land area providing water to the drainage network of a specified trunk stream represents the stream's **watershed**. (A watershed can also be called a *catchment* or a *drainage basin*.) The elevated land that separates one drainage network from its neighbor is a **divide**; a *continental divide* separates drainage networks that flow into different oceans (FIG. 12.12).

We can define how a given stream fits within the hierarchy of streams in a drainage network by specifying its **stream order**. The most commonly used scheme for specifying stream order defines a *first-order* stream as a small stream without tributaries,

FIGURE 12.11 Drainage networks remove water from a watershed. Tributaries feed water into the trunk stream.

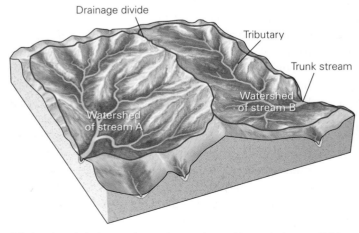

(a) Two local drainage networks separated by a drainage divide.

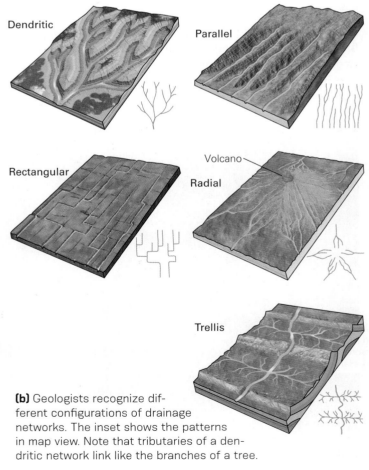

(b) Geologists recognize different configurations of drainage networks. The inset shows the patterns in map view. Note that tributaries of a dendritic network link like the branches of a tree.

FIGURE 12.12 Continental-scale drainage networks.

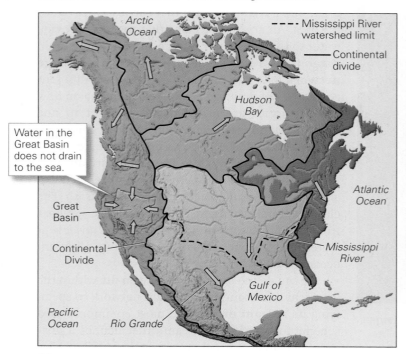

(a) Drainage networks of North America. Note the continental divides. The one labeled "Continental Divide" separates drainage heading to the Atlantic from drainage heading to the Pacific.

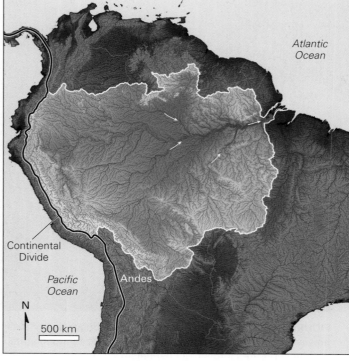

(b) The Amazon drainage network, the largest in the world, drains much of South America. The continental divide at its headwaters follows the crest of the Andes.

near the margin of a drainage network (**FIG. 12.13**). Two first-order streams merge to form a *second-order* stream, two second-order streams merge to form a *third-order* stream, and so on. A network's trunk stream is its highest-order stream. A similar ranking scheme can be applied to watersheds. Typically, a low-order watershed, meaning one whose trunk stream is a low-order stream, drains a small area, whereas a high-order watershed drains a large area.

FIGURE 12.13 The concept of stream order in a drainage network.

Stream Discharge and Stage

The amount of water moving down the Mississippi River, a high-order stream, as it flows past St. Louis, Missouri, is clearly greater than the amount moving down a low-order brook in a small mountain valley. Hydrologists describe the amount of water carried by a stream by stating its discharge. **Discharge** is defined as the volume of water that passes through a cross-sectional area of the stream (the area on a plane drawn perpendicular to the banks of the stream) in a given time, as summarized by the equation $D = A \times v$. In this equation, D is discharge, A is the cross-sectional area of the stream, and v is the average velocity at which water moves in the downstream direction through this cross section. Hydrologists report the discharge of a stream in terms of cubic meters per second or cubic feet per second.

To determine the discharge of a stream at a location where they have determined the cross-sectional area of the stream, hydrologists set up a

FIGURE 12.14 Measuring the discharge of a stream.

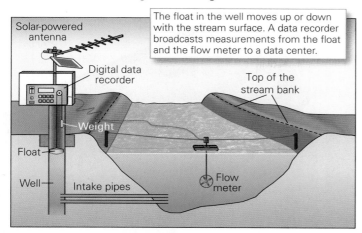

(a) At a gauging station, a well containing a float measures stage, and a flow meter measures flow velocity.

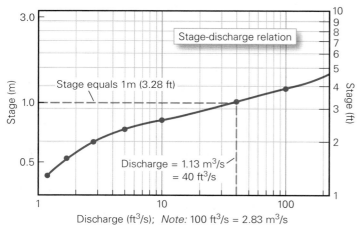

(c) A graph relating stage to discharge for a given gauging station.

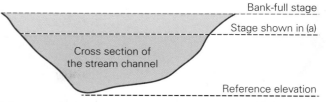

(b) A cross section of the channel shown in part (a). Bank-full stage occurs when the channel is completely full.

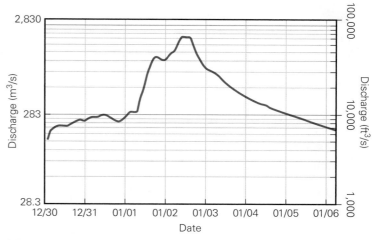

(d) USGS hydrograph for a location along the Cosumnes River, California, for the first week of 1997.

gauging station, which contains instruments that record both the stream's **stage** (the height of the water surface above a surveyed reference elevation, generally set just below the streambed) and the stream's average velocity **(FIG. 12.14a, b)**. Note that because of a stream's gradient, each reach will have a different reference elevation; sea level serves the purpose for reaches near the mouth of a stream that drains into the ocean. By knowing the shape of the stream channel, hydrologists can prepare a graph that converts a measurement of stage into a measurement of the cross-sectional area through which water is flowing **(FIG. 12.14c)**. Gauging stations record discharge at regular intervals, allowing hydrologists to plot a **hydrograph**, a chart that displays how discharge varies over time at a given location **(FIG. 12.14d)**.

The discharge of a stream varies along its length at any given time. In temperate and tropical climates, discharge tends to increase downstream as tributaries and springs add water to the stream (see Fig. 12.8a). In arid climates, discharge tends to decrease downstream as water seeps into the ground, evaporates, or is removed by humans, because tributaries and rainfall don't provide enough water to compensate for these losses (see Chapter 7). A trunk stream's average discharge near its mouth reflects the size of its watershed and the climate through which it flows. For example, the Amazon River, which drains a huge rainforest, has the largest average discharge in the world—about 210,000 m³/s (7,400,000 ft³/s) (see Fig. 12.12b). The Mississippi River's watershed is half the area of the Amazon's watershed, but its discharge is only about 8% of the Amazon's discharge because so much less rain falls in the temperate central United States than in the tropical Amazon rainforest.

Calculations of discharge are not precise. That's because, in practice, the average velocity of stream water may be difficult to calculate, for water in a stream doesn't all travel at the same velocity. Why? Because of friction, water near the banks or the *streambed* moves more slowly than water in the middle of the flow **(FIG. 12.15a)**, and because of *turbulence*—the twisting, swirling motion of a fluid—water in a stream doesn't follow a straight path down the channel **(FIG. 12.15b)**.

12.2 Draining the Land: Streams and Rivers • **431**

FIGURE 12.15 A variety of factors influence the velocity of flow in a stream.

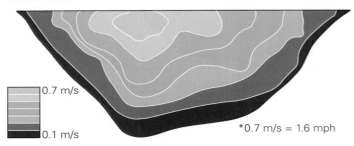

(a) Flow velocity in a stream channel slows near the streambed and the banks because of friction. This diagram shows flow velocities in a cross-sectional plane of the stream depicted in Figure 12.14a.

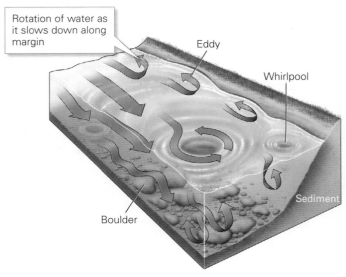

(b) Turbulence in flowing streams develops because of shearing between different volumes of water traveling at different speeds. Water swirls in eddies and whirlpools.

Take-home message . . .

Some of the water that falls on land as precipitation ends up in streams. The reaches of a stream near the headwaters have a steeper gradient than those near the mouth. Streams erode their substrate, transport sediment, and deposit alluvium. An array of tributaries that provides water to a trunk stream makes up a drainage network. A divide separates one drainage network from its neighbor. Stream discharge can be measured at a gauging station, and variation in discharge over time can be displayed on a hydrograph.

QUICK QUESTION Why does the velocity of water in a stream vary with location?

12.3 Natural Flooding in Streams and Rivers

Flood Stages and Crests

On the night of May 21, 2010, regular programing was interrupted in central Virginia for an announcement, "Alert! The New River is reaching flood stage, and will crest tomorrow morning." What does such an announcement mean? **Flood stage** occurs when a stream's discharge increases so much that water rises above the stream's banks in many places, and water submerges significant areas outside the stream channel. When a stream floods, both the stream's cross-sectional area and its average flow velocity increase. Notably, when a stream channel lies within a narrow valley or canyon, water cannot spread out, so a given increase in discharge results in a large increase in stage. However, when a stream flows across low-lying land, floodwaters spread out over a broad area, so a given increase in discharge causes relatively little increase in stage **(FIG. 12.16)**.

A hydrograph shows how a stream's discharge changes during a flood. As the flood develops, discharge increases, and as the flood recedes, discharge decreases, so the hydrograph, overall, resembles a bell-shaped curve **(FIG. 12.17)**. The highest point on the graph indicates the *peak discharge* of the flood. This is where the water surface reaches its highest stage, called the **flood crest**. The time difference between the event (such as a heavy rainfall) that triggers flooding and the flood crest defines the **lag time** of flooding.

Slow-Onset Floods

The Ganges and Brahmaputra Rivers, and their tributaries, carry runoff from rains and melting glaciers into the floodplains and delta plains of Bangladesh. Close to 80% of the country's nearly 170 million people live on these plains. When rains are heavy, vast areas of the country slowly become submerged, and millions of people must evacuate **(FIG. 12.18)**. A 1998 flood, for example, submerged up to 75% of Bangladesh for almost 3 months. When you read a news story about a flood where water rises for days or weeks, then takes weeks or months to subside, you're reading about a **slow-onset flood,** or a *downstream flood*. Such events generally affect high-order trunk streams in the downstream portion of a high-order drainage network, where streams have a gentle gradient. They are also known as *regional floods*, to emphasize that they affect broad areas. The hydrograph of a slow-onset flood displays a wide curve—floods rise slowly and subside slowly because different lower-order portions of the high-order drainage network provide water to the trunk stream at different rates and times.

FIGURE 12.16 The area affected by flooding for a given increase in stream discharge depends on the shape of the landscape being flooded, as illustrated by comparing the change in stage of a stream in a steep-sided valley with that of a stream in a broad, flat-floored valley, in response to the same change in discharge.

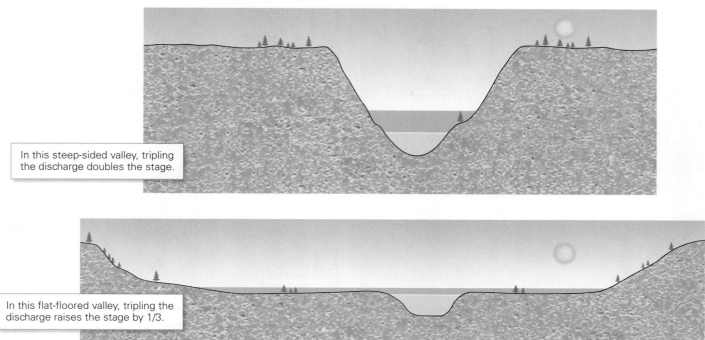

In this steep-sided valley, tripling the discharge doubles the stage.

In this flat-floored valley, tripling the discharge raises the stage by 1/3.

FIGURE 12.17 A hydrograph for a small stream. The lag time of flooding is the difference between the time of rainfall and the time at which discharge reaches its peak value (the flood crest). Flood stage is represented by a specified discharge.

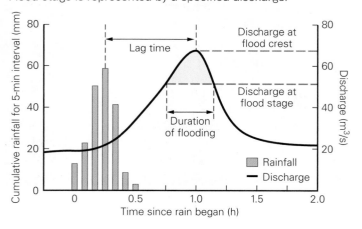

Major slow-onset floods involve huge volumes of water collected from across a broad, high-order drainage network. For example, near its mouth, the Mississippi River has an average discharge of about 16,800 m³/s (593,000 ft³/s), but during a major flood, its discharge can increase to almost 87,000 m³/s (3,000,000 ft³/s). As we'll see later in this chapter, slow-onset floods can be caused by persistent rains during a wet season, by a system of thunderstorms that inundate a broad region over a long time, or by rapid melting of a thick winter snowpack during heavy rains.

The consequences of major slow-onset floods can be devastating **(FIG. 12.19)**. For example, the 1931 flood of the Yangtze and Yellow Rivers in China, considered the worst flood in history, submerged vast areas of cities and croplands for months, ultimately triggering famine and disease. All told, the flooding and its aftermath led to the deaths of 3.7 million people and the displacement of 50 million more. This terrible event happened when water from the melting of heavy snowfall was supplemented not only by heavy seasonal (monsoon) rains, but also by rains from a succession of typhoons (hurricanes). At its peak, water in the Yangtze River was 16 m (52 ft) above flood stage. A similar flood affected the Yellow River in 1887, leading to the deaths of an estimated 2 million people.

The Mississippi and other large rivers in the midwestern United States have hosted many significant slow-onset floods over the years. For example, the *Great Mississippi Flood* of 1927 submerged about 65,000 km² (26,000 mi²) of land, covering parts of seven states. Water along a tributary to the Mississippi crested at about 17 m (56 ft) above flood stage, and the lower reaches of the Mississippi rose 9 m (30 ft) above flood stage. In places, the river's floodplain became a shallow sea nearly 130 km (81 mi) across, and the river channel itself was barely recognizable. It was visible only as a curve outlined by the tops of trees growing on natural levees. Over a million people—almost 1% of the United States population at the time—were displaced, and

FIGURE 12.18 Slow-onset floods are common in Bangladesh.

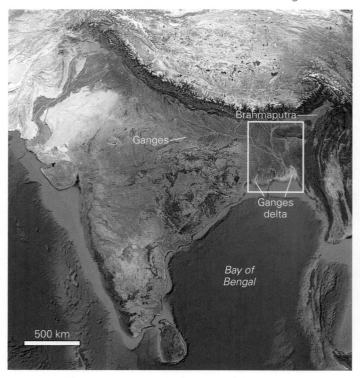

(a) The Ganges and Brahmaputra Rivers of India and Bangladesh flow into the Bay of Bengal.

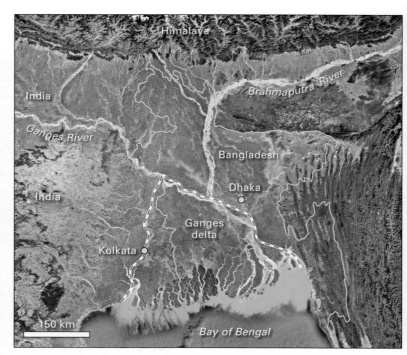

(b) Much of Bangladesh is a floodplain or delta plain. The yellow outline marks the country's borders, and the pink dashed line shows the Ganges River delta, also known as the Bengal delta.

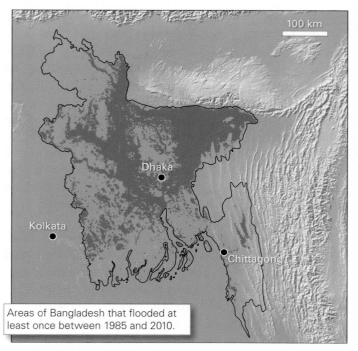

(c) Most of the country is susceptible to flooding.

Areas of Bangladesh that flooded at least once between 1985 and 2010.

(d) During floods, whole villages are swamped.

many never returned to the area (FIG. 12.20). The human tragedy of this legendary flood led to the assignment of national flood control responsibility to the US Army Corps of Engineers. Some Mississippi floods of recent decades have, unfortunately, been comparably devastating (BOX 12.1).

Flash Floods

On the afternoon of May 27, 2018, meteorologists at the National Weather Service (NWS) stared at the radar image of a thunderstorm developing over eastern Maryland. They realized that this wouldn't be just any storm: it was likely to

434 • CHAPTER 12 • Dangerous Deluges: Inland Flooding

FIGURE 12.19 Slow-onset flooding inundates broad areas of the landscape for days to months. In 1975, parts of Great Falls, Montana, were submerged by floodwaters of the Sun River.

FIGURE 12.20 A refugee camp for people displaced by the 1927 Mississippi River Flood, near Vicksburg, Mississippi.

produce torrential rain that could trigger deadly flooding in Ellicott City. So they broadcast an urgent message: "Move to higher ground now! Act quickly to protect your life." Their prediction was correct. Over the next 2 hours, 20 cm (8 in) of rain landed on the ground **(FIG. 12.21a)**. Within 15 minutes after the rain began, stream gauges detected that the nearby Patapsco River had started to rise. An hour and a half later, it had risen by 5 m (16 ft), to more than 2 m (7 ft) above flood stage, and muddy torrents flowing through Ellicott City's downtown streets were lifting cars and crumpling them against bridges **(FIG. 12.21b)**.

Hydrologists classify the flood that struck Ellicott City as a **flash flood**, defined as a flood during which a stream's discharge increases sufficiently to become hazardous in a time frame of less than 6 hours. Most flash floods affect a relatively small area along a low-order stream. Because the input of runoff happens nearby, stream discharge increases with a short lag time, and because the source of runoff has a short lifetime, the floodwaters recede quickly. Consequently, the hydrograph for a flash flood is typically a narrow bell-shaped curve. The water in a stream during a flash flood may reach flood stage or higher, as happened in Ellicott City, but in some cases, flash floods are dangerous simply because water rises so quickly that it overwhelms people or vehicles caught by surprise in the stream channel, even if the water does not overflow the stream's banks. Sometimes, the leading edge of the flood is a wall of water moving downstream at a rate faster than people can run, or even drive.

FIGURE 12.21 A flash flood struck Ellicott City, Maryland, on Sunday, May 27, 2018.

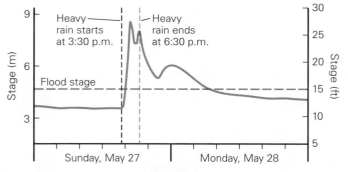

(a) A simplified hydrograph for the nearby Patapsco River. The lag time between the start of a downpour and the flood crest was about 1.5 hours.

(b) During the flood, Ellicott City's Main Street became a torrent.

12.3 Natural Flooding in Streams and Rivers • **435**

BOX 12.1 DISASTERS AND SOCIETY

Recent Mississippi floods

The Mississippi River watershed drains nearly half the continental United States (see Fig. 12.12a). Because of its central location and its navigability, the Mississippi River serves as a principal avenue of commerce in the central United States, so many cities and towns have grown up along the river and its tributaries. For 65 years after the 1927 flood, no comparable flood occurred. Then, for several weeks in 1993, extraordinary rains drenched the headwaters of the watershed. Waters rose until the upper reaches of the Mississippi River and its tributaries spilled onto their floodplains (FIG. Bx12.1a–c). The effects of the flood were far-reaching. All barge and rail transportation along the river stopped, halting the export of grain from midwestern fields for months. Over 75 towns were inundated. For example, in Davenport, Iowa, streets in the riverfront district lay beneath 4 m (13 ft) of muddy water, and in Des Moines, Iowa, 250,000 residents lost their supply of drinking water when floodwaters contaminated the municipal water supply with raw sewage and chemical fertilizer. Flooding in Valmeyer, Illinois, was so severe that the whole town relocated to a safer position on a nearby bluff after the flood. In the end, over 50,000 homes were destroyed or substantially damaged, and economic losses reached $20 billion.

In 2011, downpours fell over the central United States, causing late-season snowfall to melt rapidly. The combined runoff yielded the worst flood since 1927 in the lower reaches of the Mississippi (FIG. Bx12.1d). Hydrologists predicted that

FIGURE Bx12.1 Major recent floods along the Mississippi River and its tributaries.

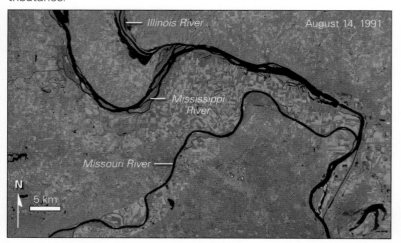

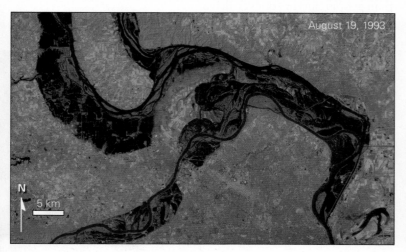

(a) The top satellite image shows the Mississippi and two tributaries with normal flow (in 1991). The bottom image shows the same rivers during the flood of 1993.

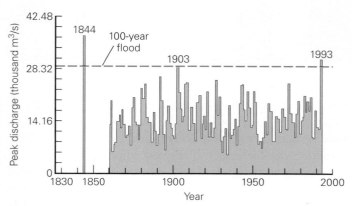

(b) Annual peak discharge for the Mississippi River at St. Louis, Missouri, between 1830 and 1993.

(c) Aerial photo of Missouri River floodwaters in 1993.

when the flood crested, water might overtop levees protecting large cities, including Memphis, Tennessee, and Baton Rouge and New Orleans, Louisiana. To protect these cities, engineers made the heartbreaking decision to divert some of the water into the floodplain, consequently inundating small communities and many farms. Severe flooding struck yet again in 2019, following the wettest 12-month period on record for the region. Several towns along the Mississippi and its tributaries saw the highest flood crests ever reported. The 2019 flooding was also remarkable because of its extended duration—in some locations, waters remained above flood stage for up to 226 days **(FIG. Bx12.1e)**. Due to the devastation, eleven states sought disaster-relief funds.

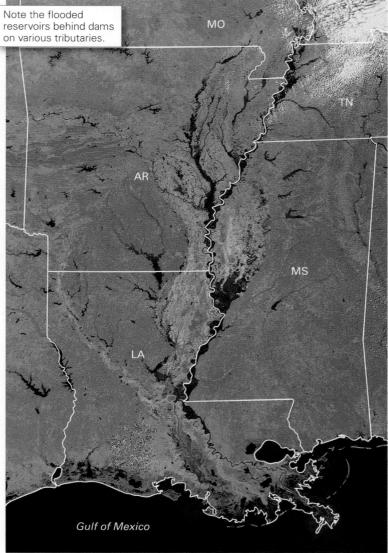

(d) The top satellite image shows the reaches of the rivers near St. Louis in 2006. The bottom image shows the same reaches during the flood of 2011.

(e) Normally, the Mississippi River channel is the width of the white line. In 2019, the river covered large areas of its floodplain.

The turbulent water of a flash flood can cause rapid erosion and can transport immense amounts of sediment. The dynamic pressure exerted by a flood's fast-moving, muddy water can knock down strong structures and lift heavy debris, and this debris can batter or bury areas farther downstream. Clearly, flash floods are a dangerous natural hazard (**FIG. 12.22**). Let's look at their causes.

FLASH FLOODS DUE TO EXTREME RAINFALL. Most flash floods happen when rain falls so fast that the ground can't absorb it all, and much of the water becomes overland flow that spills into stream channels. The flooding of Ellicott City serves as an example. Notably, rain-caused flash floods can move into downstream reaches of a stream that didn't receive a drop of rain. Such floods can tragically sweep away hikers walking along canyon floors in desert regions—the hikers have no idea that a flood is coming, until it suddenly hits them.

Flash floods can be a particular danger in mountainous areas, where valleys are narrow and rivers rise very quickly as discharge increases. Floods of the Big Thompson River in Colorado illustrate this hazard. The Big Thompson River has carved a deep canyon through the Front Range of the Rocky Mountains in Colorado. Despite the label "River," its discharge is generally only about 3 m^3/s (106 ft^3/s), and its depth is normally less than about 0.5 m (1.6 ft), so when it's not flooding, its shallow water froths over cobbles and around boulders on the streambed (**FIG. 12.23a**). A road follows the river on the canyon floor, and vacation cabins dot the river's banks. Sadly, on July 31, 1976, the canyon's peaceful landscape turned to one of terror. At 7:00 P.M., rain began to fall from towering thunderstorms in quantities that even longtime residents couldn't recall—19 cm (7.5 in) fell in the first hour, and an additional 17 cm (6.5 in) came during the next 3 hours.

The rain drenched the steep valley walls of the Big Thompson River's watershed. Because much of the watershed's ground surface exposes impermeable bedrock, and because the soil between outcrops was already wet from previous rains, most of the downpour became overland flow that had only a short distance to travel to reach the river. Within 2 hours, discharge at a gauging station along the river had increased to over 850 m^3/s (30,000 ft^3/s), over four times more than at any time in the previous century, and because the canyon was narrow, the river rose by several meters and almost completely submerged the canyon floor (**FIG. 12.23b, c**). The velocity and turbulence of the flow also increased dramatically, churning up so much sand and mud that the once-clear water became opaque. Rain-weakened rock and soil tumbled down the steep slopes bordering the river to add even more sediment to the flow. Boulders that had been landmarks for generations bounced along in the sediment-laden waters. Torrents undercut foundations and bridge pilings and washed away houses and bridges, and as the banks of the stream caved away, stretches of the road fell into the water. When the flood finally subsided, 144 people had lost their lives, and the canyon's landscape had changed forever (**FIG. 12.24**).

Recognition of flooding risk led residents and planners to relocate some communities in the canyon to higher ground and to reposition the road, where possible. But on a narrow canyon floor, there aren't many places to move to. So, unfortunately, tragedy visited Big Thompson Canyon again in September 2013, when 80% of a normal year's rainfall fell over a few days, increasing the river's discharge by a factor of 30. The 2013 flood spilled out of the mouth of the canyon into the plains beyond and destroyed houses, roads, pipelines, sewage treatment plants, and farmland. The tragedy claimed 8 lives and destroyed $2 billion in property.

FIGURE 12.22 A flash flood devastated Joso, Japan in 2015. Local residents had to be rescued from the surging waters by helicopter.

FLASH FLOODS DUE TO RAPID SNOW OR ICE MELT. When snow or ice melts quickly, it can contribute large quantities of water to overland flow, especially if the ground beneath the

FIGURE 12.23 The 1976 flood of Big Thompson Canyon, Colorado.

(a) Big Thompson Canyon is narrow and deep.

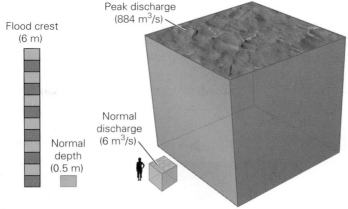

(b) Comparison of water depths and discharge for the Big Thompson River before and during the 1976 flood.

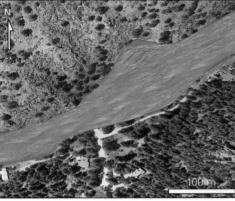

(c) Before the flood, the stream was narrow and flowed next to a road, close to buildings, and under a bridge. During the flood, the stream covered the entire canyon floor, and the road, bridge, and buildings vanished.

snow remains frozen, preventing water from infiltrating the subsurface. Such rapid melting can happen, for example, when heavy rain falls on snow or ice, because each raindrop causes some of the frozen water to melt. Volcanic eruptions can also cause rapid melting of ice and snow, both because the rise of molten rock in the volcano heats the volcanic edifice and because hot ash and lapilli can fall on snow or ice. Meltwater produced during eruptions can mix with ash to produce a *lahar*, a slurry of ash-laden water that flows down the volcano and into stream channels. Lahars can flood areas tens of kilometers from the volcano and can bury whole towns (see Chapters 4 and 5).

FLASH FLOODS DUE TO BREACHING OF NATURAL DAMS.
A *dam* is a wall across a stream channel that holds back water. Sudden breaching of a dam that retains a body of water can trigger a flash flood. Later in this chapter, we'll describe failures of dams that people have constructed. Here, we'll consider breaches of *natural dams*, those composed of landslide debris or ice.

Large landslides can be triggered by earthquakes or heavy rains (see Chapters 3 and 5). If debris from a landslide blocks a

FIGURE 12.24 Floodwaters from the Big Thompson Canyon flood in 1976 undercut this hillslope in Colorado.

12.3 Natural Flooding in Streams and Rivers • **439**

stream, water can quickly collect behind it. When this landslide dam breaks, flooding results. For example, when an earthquake-triggered landslide blocked an upstream reach of the Indus River in Pakistan, a lake 150 m (492 ft) deep and tens of kilometers across soon developed. In 1841, the dam gave way, and water rushed downstream, mangling towns along the river for a distance of several hundred kilometers.

An **outburst flood** happens when a dam made of ice or of ice-deposited sediment gives way. Outburst floods happen, for example, when a lobe of a glacier blocking the outflow of a valley suddenly fails. Such a flood happened in Alaska in 1986, when a lobe of the Hubbard Glacier broke up. A unique type of outburst flood occasionally takes place in Iceland when a volcano erupts under a glacier. Meltwater builds under the glacier until it breaks free at the toe of the glacier and rushes out as a flash flood that Icelanders call a *jökulhlaup* (FIG. 12.25a). The largest known outburst floods occurred at the end of the Ice Age, when dams formed by ice or moraines (piles of sediment deposited by glaciers) failed. Immense floods, called **glacial torrents**, ravaged large areas downstream. For example, on several occasions, an ice dam holding back Glacial Lake Missoula (a lake of glacial meltwater filling a now-dry valley near present-day Missoula, Montana) gave way. The resulting glacial torrents stripped eastern Washington State of most of its soil, leaving behind a landscape now known as the *Channeled Scablands* (FIG. 12.25b). At its peak, the discharge of this Missoula flood was over 80 times the average discharge of the Amazon River.

FIGURE 12.25 Floods can be caused by the failure of natural ice dams.

(a) This *jökulhlaup* in Iceland flowed out from the base of the glacier in the distance.

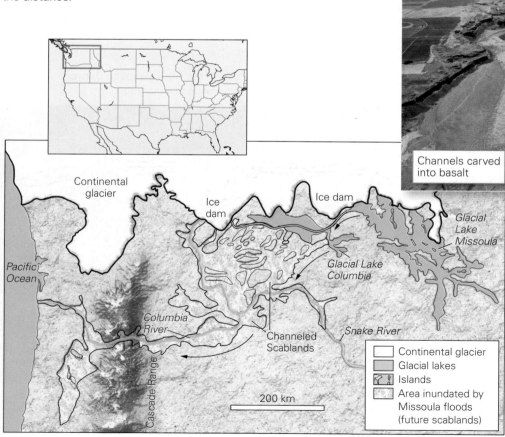

(b) The Great Missoula Floods. When ice dams broke, glacial torrents from Glacial Lake Missoula scoured portions of the Columbia River Plateau.

FIGURE 12.26 This ice jam forming on a river may soon lead to flooding.

A more common form of ice dam, better known as an **ice jam**, forms when winter ice covering a river starts to break up in the spring. The river current carries floating blocks of ice downstream until they reach a restriction or an obstacle in the channel that causes the ice to pile up (FIG. 12.26). This blockage can divert water from the river into the floodplain, rapidly submerging populated areas or fields. When the ice jam finally breaks, the backed-up water suddenly rushes downstream as a potentially damaging outburst flood.

Take-home message . . .

A river reaches flood stage when its water rises high enough to overtop its banks and submerge a significant area outside its channel. Slow-onset floods rise slowly and subside slowly. Such floods can inundate broad floodplains and delta plains. Flash floods are sudden and short-lived. Many develop due to intense rain concentrated in a relatively small drainage network. They can also be caused by rapid melting of snow and ice, as can happen during a heavy rain or a volcanic eruption, or by the failure of a landslide dam or an ice dam.

QUICK QUESTION Which typically produces more property damage overall—a slow-onset flood or a flash flood? Why?

12.4 Other Types of Floods

Areal Flooding

In a normal year, farmers in the corn belt of Illinois hope to plant their fields by early May, so that crops can mature in time for the fall harvest. Planting can't happen, however, until the soil is dry enough so that tractors don't get stuck in the mud and seeds can germinate before rotting. In 2019, only half of Illinois's fields had been planted by mid-June. Why? The fields remained too wet. In fact, in low-lying areas, even those that were distant from streams, standing water covered the soil, and large portions of fields looked like ponds (FIG. 12.27). Hydrologists refer to this situation, in which land becomes submerged due to heavy rain or rapid snowmelt without input from a nearby stream, as **areal flooding**. Such flooding can be a major problem not only in agricultural areas, but also in communities located on poorly drained, low-lying land.

Urban Flooding

A flood that inundates land in a *built environment* (a region where the landscape has been completely modified by human-built structures and surfaces) is an **urban flood** if it happens for either of two reasons: (1) the paved or building-covered land surface of the urban area is *impermeable* (cannot absorb water), and therefore, water that could infiltrate the ground in a natural environment remains at the land surface and can deepen enough to be a hazard; or (2) the constructed drainage system in the urban area (typically consisting of storm sewers and

FIGURE 12.27 Areal flooding of farm fields in Illinois, as viewed from an airplane. Sunlight reflects from the water.

concrete-lined culverts) is inadequate to remove water fast enough to prevent it from accumulating on the ground **(FIG. 12.28a)**. The area's drainage may be limited because debris or ice obstructs entrances to storm sewers or culverts, or due to insufficient capacity. Urban flooding can happen during the rise of streams or during heavy downpours, and in that regard, it can be a component of both slow-onset and flash floods. Devastating urban floods accompanied the passage of Hurricane Ida over New York City in 2021. Storm sewers were overwhelmed, and water in streets rose so high that it cascaded down stairways to the subway, where it submerged the tracks. But sometimes it is chronic and occurs during even normal rains. To minimize urban flooding, cities may build *flood-control channels* to expedite removal of stormwater from a city.

Unfortunately, water that does not enter storm sewers during an urban flood may bypass water treatment facilities. Consequently, it can carry trash and toxic chemicals (such as gasoline) washed from pavement surfaces directly into streams, causing environmental damage. Flooding from Hurricane Ida in 2021 led to contamination of rivers and coastal areas because overland flow carried off chemicals leaking from pipes and tanks at chemical plants and oil refineries. Furthermore, because of the lack of infiltration and the resulting increase in overland flow in an urban area, the lag time between rainfall and flood crest decreases, and the flood crest is higher, than happens downstream of a natural area **(FIG. 12.28b)**.

Failures of Constructed Dams and Levees

Heavy rains drenched central Michigan in May 2020, causing water in Wixom Lake, a reservoir held back by the Edenville Dam, to rise. The lake was already quite full, because a lawsuit had led to a requirement that the water level be kept high enough to preserve populations of freshwater mussels. The century-old packed-earth dam, located along the Tittabawassee River, was constructed for flood control and to provide a source of hydroelectric power. Unfortunately, it lacked an adequate *spillway*, a channel that would have allowed excess water to bypass the dam. As water in the lake rose, officials became concerned, and about 10,000 people were evacuated from low-lying towns downstream. These evacuations were challenging due to social-distancing rules to protect people from contracting COVID-19, so emergency housing at close quarters in shelters was not an option. In the late afternoon of May 19, the downstream face of the Edenville Dam slumped, opening a breach that released a torrent of brown water from the lake. The water rushed downstream into another reservoir, Sanford Lake, causing the Sanford Dam to fail. The water released by the Sanford Dam failure, in turn, flooded the downstream city of Midland to a depth of 3 m (10 ft). All told, flooding in Midland and in other communities along the river due to the cascade of dam failures caused $175 million in damage.

FIGURE 12.28 Examples of urban flooding.

(a) An urban flood submerges an intersection in Hong Kong.

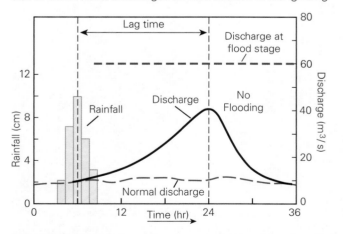
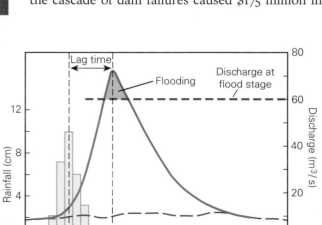

(b) Urbanization changes the characteristics of flooding. Before urbanization, rainwater infiltrates the ground, so discharge is less, and peak discharge occurs after a long lag time. After urbanization, rainwater flows directly into streams, so discharge is greater and lag time is less.

Fortunately, in contrast to the catastrophe of the Johnstown flood discussed at the start of this chapter, timely evacuations meant that no lives were lost.

As the Midland and Johnstown disasters demonstrate, the sudden failure of a constructed dam can lead to a flash flood. Such flooding can also occur due to damage to other types of human-built water-retention structures, including: **artificial levees**, long ridges of gravel, sand, and compacted clay (**FIG. 12.29**); and *floodwalls*, vertical concrete barriers—both of which are built along a stream to confine flow to its main channel. Three processes can cause these structures to fail:

- *Overtopping* happens when water rises high enough to flow over a dam or levee (**FIG. 12.29b**). When the overtopping water flows into a low or weak area of the structure, its velocity and volume, and thus its erosive power, increases, and it can carve a breach in the structure (**FIG. 12.29c**). A breach may also start if trees on an earthen dam or levee tip (causing their roots to disturb the soil surface), if floating debris collides with the structure and gouges out an opening, or if slumping takes place on the structure's walls.

- *Underseeping* or *undermining* occurs when water is higher on one side of the dam or levee than on the other (**FIG. 12.29d**). This difference produces pressure that can force water through the base of the structure, or through sediment below the structure. Concentration of underseeping in a relatively narrow pathway through the base

FIGURE 12.29 Three processes can cause a levee to fail.

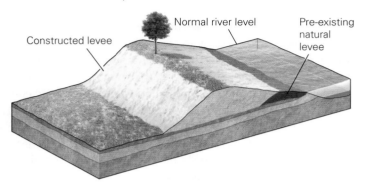

(a) Levees should be built so that they are at least 3 times as wide, at their base, as they are high. Some levees are built over an existing natural levee.

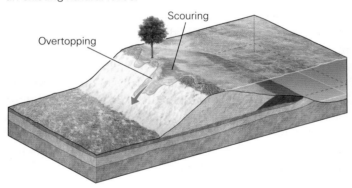

(b) Overtopping takes place when floodwaters reach the top of the levee, so river water spills down the levee. This flow can scour a channel in the levee, ultimately leading to the formation of a breach.

(c) Close-up of water rushing through a breached levee.

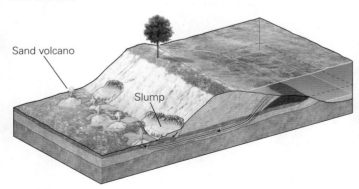

(d) Underseeping happens when the pressure of the deep water on the river side causes water to seep through or under the levee. Sediment carried by the water erupts in sand volcanoes on the land side of the levee. The process weakens the levee so it starts to collapse.

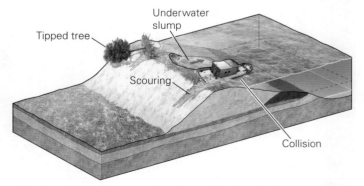

(e) Structural damage can happen for several reasons. When the material of the levee gets soggy, trees growing on top of the levee can tip over. Their roots pull off soil, so a channel can form. Collision of trees or ships with the levee can also cause a breach. Slumping can happen on either side of the levee.

of a dam or levee may cause water to start fountaining up out of the ground on the dry side. This fountain typically carries sand with it and produces a *sand volcano*. The presence of sand in the fountain indicates that underseeping is eroding material inside the dam or levee, and that a breach will eventually develop.

- *Structural failure* happens when the structure has a weak spot due to bad concrete, poorly compacted clay, or a weak foundation (FIG. 12.29e). Concrete barriers may crack and give way, or their foundations may be pushed out so that the structure's wall slumps or collapses, forming a breach through which water spills.

While dam failures typically focus flooding along the downstream channel, levee failures allow water to spill out onto the floodplain bordering the stream. Such failures are unfortunately frequent along major rivers during floods, as we'll see later in this chapter. Over the course of history, people have sometimes damaged water-retention structures for political or military gain. For example, in June 1938, when the Japanese army was advancing into central China during the Sino-Japanese War, Chinese Nationalist forces intentionally destroyed levees holding back the Yellow River, hoping to block the advancing Japanese soldiers. Unfortunately, this decision produced a disaster. Water from the released river covered about 55,000 km² (21,000 mi²) of eastern China, inundating thousands of villages, displacing about 4 million people, and ruining crops. Hundreds of thousands of people may have been killed by the resulting flooding, disease, and famine. Because World War II and extended political turmoil began shortly afterward, reconstruction of the levees was not completed until 1947.

Not all retention structures are for the purpose of holding back water alone. Power companies and mining companies, for example, build earthen dams and levees to contain *settling ponds*, in which slurries are held so that the solid can eventually drop out of the water. Settling ponds at coal-fired power plants contain slurries of *coal ash*, the residue left from burning coal, and settling ponds at mines contain slurries of *tailings*, the residue left from processing ore. Several significant failures of retention structures have happened in the past century, leading to the loss of more than 3,000 lives. Flash floods released by these failures swamp towns and fields with toxic chemicals. For example, 4,200,000 m³ (148,000,000 ft³) of coal-ash slurry released from a Tennessee settling pond in 2008, following the failure of an earthen levee, polluted the Emory and Clinch Rivers and covered 1.2 km² (0.5 mi²) of land (FIG. 12.30a). While the initial dam failure did not claim any victims, the cleanup tragically did: within 10 years of the disaster, 40 cleanup workers died due to illnesses that have been attributed to chemicals in the ash. In 2015, a dam holding a slurry of tailings at an iron mine near Mariana, Brazil, failed, releasing vast amounts of toxic materials that killed 19 people, destroyed many villages, and polluted hundreds of kilometers of streamside land (FIG. 12.30b). A similar tailings dam failed in Brazil in 2019, taking the lives of 270 people.

Concern about flooding due to the failure of water-retention structures is increasing as government officials and NGOs recognize that populations living downstream of such structures have grown dramatically in recent years. For example, by one estimate, 25,000 dams in the United States could be subject to failure that could cause devastation to downstream communities. Unfortunately, many of these dams have outlived their engineered lifetimes and have not been well maintained.

FIGURE 12.30 Failure of structures designed to contain settling ponds.

(a) Before and after satellite photos show what happened when failure of a levee retaining a slurry of coal ash from a Kingston, Tennessee, power plant flooded the downstream region with toxic sludge.

(b) Failure of a levee holding tailings from an iron mine near Mariana, Brazil, destroyed villages downstream and contaminated water supplies.

> **Take-home message...**
> Areal flooding takes place where low-lying areas become submerged during heavy rains, even if there is no stream nearby. Urban flooding happens where the land surface has been covered with impermeable material and where storm sewers and culverts are inadequate. Some floods are a consequence of dam or levee failure caused by overtopping, underseeping, or structural failure.
>
> **QUICK QUESTION** Why might the failure of a levee holding back a slurry of coal ash or mine tailings cause environmental destruction?

12.5 Consequences of Floods

Danger due to Moving Water

Trying to drive across a stretch of road that has been flooded by a fast-flowing creek can be a fatal mistake (**FIG. 12.31a**). Why? Recall from Chapter 6 that dynamic pressure, the pressure exerted by a moving fluid, increases with the square of the fluid's velocity. Therefore, fast-moving water exerts a lot of pressure! In fact, the dynamic pressure of knee-deep, rapidly moving water can easily sweep you off your feet, and the dynamic pressure of chest-high water can push a wall over. Water can also buoy up objects, whether the water is moving or not. So, for example, a car trying to cross a flooded highway may not only be pushed sideways by dynamic pressure, but may also be pushed upward by a *buoyancy force*. Impressively, moving floodwaters only 30 cm (1 ft) deep can carry a car off a road. It's no wonder that many of the casualties that occur during flash floods are people in cars. Deep floodwaters can lift and carry away mobile homes, and even houses anchored to foundations.

The moving water of a stream in flood can cause damage indirectly as well. Specifically, solid debris carried along with the water can batter dams, buildings, or bridge supports (**FIG. 12.31b**). And moving water can erode and undermine foundations and embankments (**FIG. 12.31c**). A single flood may cut into a riverbank or bordering escarpment to a level that may cause landslides that carry buildings down the bank and into the raging waters, where they break up and wash away. The alluvium carried by flooding streams can bury streets and buildings downstream (**FIG. 12.32**).

Damage due to Rising Water

Floodwaters can submerge fields, roads, railroads, forests, wetlands, homes, businesses, factories, sewage treatment plants, ranches, and wildlife habitats (**FIG. 12.33a**) (**VISUALIZING A DISASTER**, pp. 448–449.). And, sadly, if people and animals can't escape, they may drown in the high water (**FIG. 12.33b**). Rescue becomes challenging when roads are submerged, so boats become the main source of transportation. Water can also infiltrate homes, filling basements and sometimes even higher floors, soaking, staining, and ruining carpets, walls, and all types of personal possessions. And floodwaters can shut down power generation and transmission facilities, so that lights go out and pumps cease to operate. The loss of pumps can contaminate city water supplies and rural wells.

FIGURE 12.31 The force of moving water.

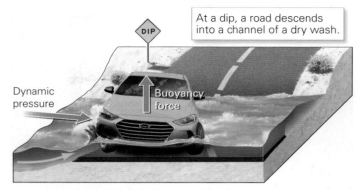

(a) Don't ignore a dip sign! Floodwaters crossing a dip apply both dynamic pressure and buoyancy force to vehicles. Even water that barely reaches above the wheel rims can carry a vehicle downstream.

(b) Flash floods washed away most of this bridge in China.

(c) Moving water can undermine streamside roads and foundations. This damage occurred in Big Thompson Canyon, Colorado, in 2013.

FIGURE 12.32 Sediment transport during floods.

(a) A resident digging out his mud-buried driveway after a 2010 flood in California.

(b) In April 2017, floods covered the streets of Moaca, Colombia, with boulders and debris.

FIGURE 12.33 Damage due to rising waters of a flood.

(a) Rising waters flooded this living room.

(b) Stranded livestock during a flood.

(c) After the water drains from a home, sediment remains.

After floodwaters recede, residents must deal with the mess left behind (**FIG. 12.33c**). While any remaining water can be pumped out or will eventually evaporate, the residue may contain garbage, vegetation, toxic chemicals, and sewage, which makes cleanup a Herculean task. In addition, when water rises and spreads out over a floodplain or delta plain, its competence decreases. Mud and silt settle out, and when the floodwaters recede, muck covers everything. As the muck dries, the organic material it contains starts to decay and smell.

Unfortunately, as the primary disaster of a flood wanes, secondary disasters may begin. Loss of access to drinkable water, and the spread of sewage and decay, can spread diseases such as cholera and dysentery, particularly in regions with inadequate resources and health care systems. The disruption of transportation and communications networks, as well as of the electrical grid, can make recovery from flooding very difficult. Soaked homes begin to grow mold on the exposed surfaces of walls and in the spaces between interior and exterior walls. Cleanup can take years.

> **Take-home message . . .**
>
> Flooding can cause damage in several ways. The dynamic pressure of moving water, and the buoyancy force produced when objects become submerged in water, can cause water to lift and transport objects, people, and animals. Moving water can batter structures with debris and erode stream banks. Submergence by floodwaters can saturate and ruin materials and can leave behind muck and debris. The impacts of flooding extend over time far beyond the flooding event itself.
>
> **QUICK QUESTION** What are the secondary disasters that can accompany flooding?

12.6 Weather Systems That Cause Flooding

Not every rain produces a flood. It takes specific types of weather systems to produce rains heavy enough to cause flooding. Systems that cause slow-onset flooding, affecting a broad region, are those that either transfer abundant moisture over the land or remain nearly stationary for an extended time. A local flash flood, in contrast, can be caused by a relatively short but intense local storm. Let's look more closely at some of the weather systems that can lead to flooding.

The Rainy Season in a Monsoonal Climate

Recall that a **monsoon** is a seasonally changing tropical wind circulation commonly manifested by an annual rainy season and an annual dry season (see Chapter 8). The most intense monsoonal rainy seasons, in southern Asia, begin when the intertropical convergence zone (ITCZ) drifts northward from its winter (December and January) location over the Indian Ocean, south of the equator, to its summer (June and July) location over the landmass of southern Asia, north of the equator. In this zone, the trade winds of the southern and northern hemispheres converge, causing updrafts that lead to towering thunderstorms. As the ITCZ passes over land, moist air blowing northward from the Indian Ocean crosses the land, warms, and rises, providing moisture to nourish the thunderstorms, so heavy rains fall during an extended period. Eventually, the storms reach the southern flank of the Himalayas. At the peak of the monsoonal rainy season, the storms extend across Pakistan, India, Bangladesh, and into China, dropping vast amounts of precipitation **(FIG. 12.34a)**. Discharge in rivers flowing out of the Himalayas and the Tibetan Plateau increases dramatically **(FIG. 12.34b)**.

Earlier in this chapter, we noted that Bangladesh often endures the brunt of monsoonal floods because it lies at the confluence of the Ganges and the Brahmaputra Rivers. Pakistan has endured similar devastation due to flooding of the Indus River **(BOX 12.2)**. China's summer monsoon-related floods occur to the east of the Tibetan Plateau and differ somewhat from those to the west. Over China, monsoon rains develop when a broad, cool, and dry polar air mass that develops over Siberia and Mongolia drifts southward and

FIGURE 12.34 Southern Asian monsoonal floods.

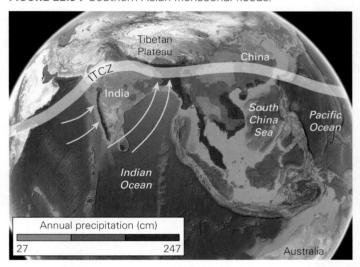

(a) The position of the heaviest rain, which occurs along the ITCZ, migrates during the season. When the ITCZ lies over the land, moist air blows in from the Indian Ocean and nourishes the storms. Regions of greatest total rainfall are highlighted.

(b) Intense monsoon rains can produce dramatic floods, turning city streets into raging rivers.

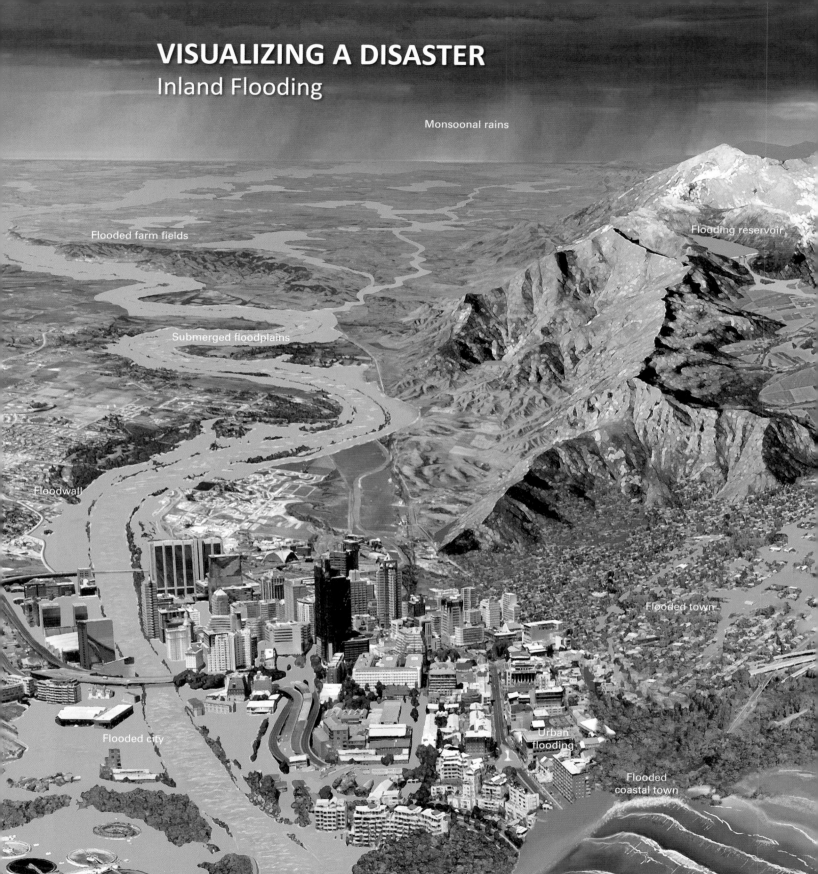

BOX 12.2 DISASTERS AND SOCIETY

The 2010 monsoonal floods in Pakistan

Almost a quarter of Pakistan's population lives on the floodplains of the Indus River and its tributaries. During 2010, particularly intense rains—up to 27 cm (11 in) of rain fell in a 24-hour period—caused these rivers to submerge about one-fifth of the country's land area **(FIG. Bx12.2)**. This flooding killed 2,000 people, damaged or destroyed the homes of up to 20 million people, decimated crops, and led to the deaths of nearly a half-million cattle. Millions of victims were left without safe water to drink or sufficient food to eat. Thousands of kilometers of roads and power lines were damaged, so transportation and communication were cut off. Inadequate access to food caused starvation. Outbreaks of malaria and cholera began, and these diseases remained endemic for years after the waters had subsided.

Nobel Peace Prize winner Malala Yousafzai, in her book *I Am Malala*, provided a dramatic account of the flood, which happened when she was 13 years old:

> We were in school when the floods started and were sent home. But there was so much water that the bridge across the dirty stream was submerged, so we had to find another way. The next bridge we came to was also submerged, but the water wasn't too deep so we splashed our way across. It smelled foul. We were wet and filthy by the time we got home. The next day we heard that the school had been flooded. It took days for the water to drain away and when we returned, we could see chest-high tide marks on the walls. There was mud, mud, mud everywhere! Our desks and chairs were covered with it. The classroom smelled disgusting. . . . Our own street was on a hill, so we were a bit better protected from the overflowing river, but we shivered at the sound of it, a growling heavy-breathing dragon devouring everything in its path.

FIGURE Bx12.2 Flooding in Pakistan during 2010 covered 62,000 km² (24,000 mi²) and affected 20 million people.

---

collides with a warm, moist tropical air mass moving north from the Indian Ocean and South China Sea. The boundary between these two air masses is a pronounced front, called the *Mei Yu front* (the Plum Rain front, in English). Along this front, northward-flowing warm, moist air rises over cooler, denser air, and clusters of monsoonal thunderstorms develop **(FIG. 12.35)**. In late June through July, the Mei Yu front tends to become nearly stationary over the Yangtze River watershed, so thunderstorms repeatedly dump heavy rain into the river's drainage network. Flooding here can be particularly devastating when typhoons drift in from the Pacific and add their rain to the monsoon rain, as in the case of the historic floods described earlier in this chapter.

In the western hemisphere, the southwestern United States and northern Mexico experience monsoonal floods when summer heating is at its maximum and moist air from the Pacific Ocean west of Mexico, as well as moisture from the Gulf of Mexico, flows northeastward, nourishing intense thunderstorms **(FIG. 12.36)**. Flooding due to these storms can cause normally dry washes in these desert areas to fill and overflow. Monsoonal floods also occur in the southern hemisphere, in Africa and along the northern coast of Australia, when the ITCZ reaches its southernmost extent. In Australia, the worst floods occur during the phase of the Indian Ocean Dipole (see Chapter 8) when warm surface waters appear along Australia's western coast at the same time as a La Niña in the Pacific causes warm waters to appear along the country's northeastern coast. Atmospheric moisture produced by evaporation from these warm waters leads to the growth of immense thunderstorms.

450 • CHAPTER 12 • Dangerous Deluges: Inland Flooding

FIGURE 12.35 Intense precipitation along the Mei Yu front in China triggers severe slow-onset floods.

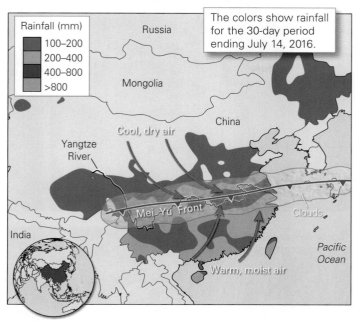

(a) The Mei Yu front forms where polar air collides with tropical air. Thunderstorms that develop along this front drop heavy precipitation over the Yangtze River watershed.

(b) This Yangtze River flood has submerged the houses and roads of this town in China

Tropical Cyclones

Not only do tropical cyclones devastate coastal areas with storm surge and towering waves, as we saw in Chapter 11, but they also produce some of the world's worst inland floods, especially when they arrive during the monsoon rainy season. These storms carry immense amounts of moisture into the lowlands of Bangladesh, India, and Myanmar as they move inland from the Bay of Bengal (FIG. 12.37a). Catastrophic floods in this region during the Bhola cyclone in 1970, combined with storm surge, caused nearly 500,000 deaths. A similar storm, Cyclone Amphan, struck India and Bangladesh in 2020, during the COVID-19 pandemic. Fortunately, improved weather prediction, along with better logistics, planning, and communication, allowed officials to evacuate millions of people, so only 121 lives were lost, but the concentration of evacuees in temporary camps may have increased the spread of COVID-19 infections.

Floods caused by tropical cyclones also regularly devastate regions in other parts of the world (FIG. 12.37b; FIG. 12.38a). For example, when Hurricane Mitch crossed Central America in 1998, rain caused severe flooding in the deforested highlands of Honduras and Nicaragua, killing nearly 18,000 people (see Chapter 11). Hurricane Harvey in 2017 stalled over Houston, Texas, and dropped over 127 cm (50 in) of rain on the metropolitan area. Not only did the water cause streams and swamps to overflow, but it led to widespread areal flooding in many low-lying areas of the city, which had sunk in recent decades due to land subsidence (FIG. 12.38b).

FIGURE 12.36 Moisture sweeping in from the Pacific Ocean and the Gulf of Mexico causes summer monsoon rains in the southwestern United States and in northern Mexico.

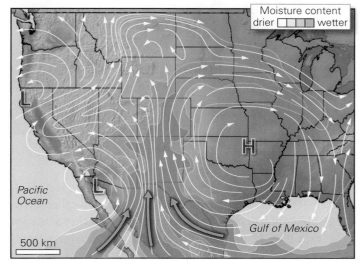

Mid-latitude Cyclones

Extreme rainfall can develop within a mid-latitude cyclone when a front in the system becomes stationary and winds in the middle troposphere flow nearly parallel to the front. With these conditions, thunderstorms that develop along the front move in succession over the same region, so that rain falls

12.6 Weather Systems That Cause Flooding • 451

FIGURE 12.37 Tropical cyclones can generate devastating inland floods, especially when they make landfall during the monsoon rains.

Cause: A cyclone

Cause: A typhoon

Consequences: Inland flooding

Consequences: Inland flooding

(a) Rain from a cyclone over the Bay of Bengal flooded rivers.

(b) Rain from a typhoon crossing Taiwan en route to China flooded cities.

FIGURE 12.38 Hurricane-related floods in North America.

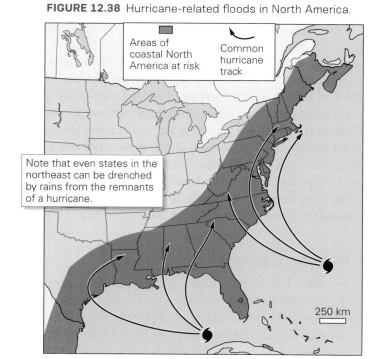

Note that even states in the northeast can be drenched by rains from the remnants of a hurricane.

(a) Areas affected by inland flooding from hurricanes extend well inland from the Atlantic Ocean and the Gulf of Mexico.

(b) A flooded highway in Houston during Hurricane Harvey.

over the region for an extended period (**FIG. 12.39a**). Meteorologists refer to this phenomenon as *training* because the storms are visually analogous to train cars passing over the same spot on a track (**FIG. 12.39b**). In August 2021, thunderstorm training along a stalled front dumped as much as 43 cm (17 in) of rain in less than 12 hours over the hilly watershed of rivers that flow through the middle of Tennessee. Rising water destroyed

FIGURE 12.39 Training of thunderstorms along a stationary front.

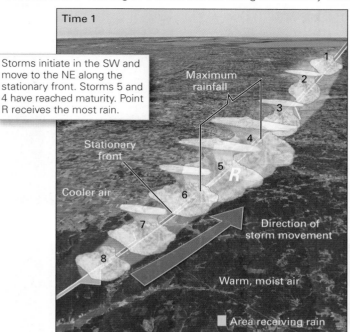
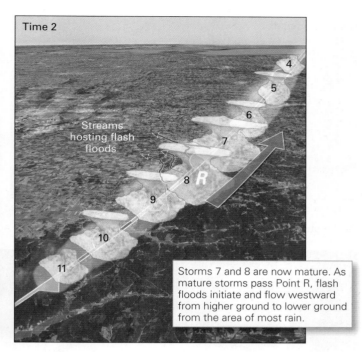

(a) Storms form on the south end of a stationary front, move northward, mature by Point R, where they dump the most rain, and dissipate further north.

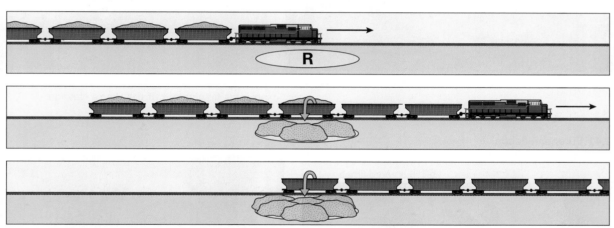

(b) The term "training" is an analogy to a freight train with full boxcars (storms) dumping a load (rain) at a location (Point R), while moving along a track (stationary front).

hundreds of houses and damaged hundreds more. The degree of devastation resulted, in part, from inadequate zoning rules to prevent construction in vulnerable areas. Catastrophic widespread flooding can also develop when lines of thunderstorms from several separate mid-latitude cyclones produce precipitation over a large watershed over a period of days or weeks.

Nearly all of the precipitation on the west coast of the United States falls during the winter, building a snowpack at high elevations in the Sierra Nevada and Cascade ranges and dropping rain at lower elevations. Sometimes winter precipitation becomes extreme and causes flooding. This happens when an **atmospheric river**, a relatively narrow band of moist air flowing along an extended cold front on the south side of a mid-latitude cyclone, flows into California, Oregon, or Washington from the Pacific (FIG. 12.40a). When the warm, moist air within an atmospheric river reaches the mountains, it rises and generates heavy rainfall, even at high elevations. A single atmospheric river can persist for days and dump over 75 cm (30 in) of rain, flooding mountain canyons and inundating the lower-lying areas to the west of the mountains (FIG. 12.40b, c).

Many drainage networks in Europe are susceptible to flooding caused by mid-latitude cyclones (FIG. 12.41a). Ireland, the United Kingdom, and western and central continental Europe endure floods when these storms,

12.6 Weather Systems That Cause Flooding • 453

FIGURE 12.40 Atmospheric rivers and their consequences.

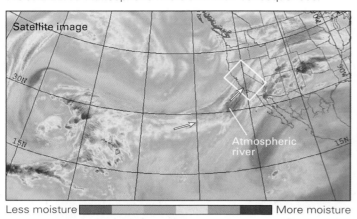
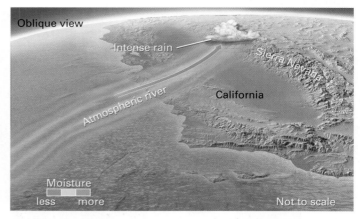

(a) An atmospheric river transporting moisture from the central Pacific northeast to California.

(b) Floodwaters rushing down a steep-gradient stream in the Sierra Nevada.

(c) The flooding of farmland in central California is fed by streams flowing out of the Sierra Nevada.

FIGURE 12.41 European floods.

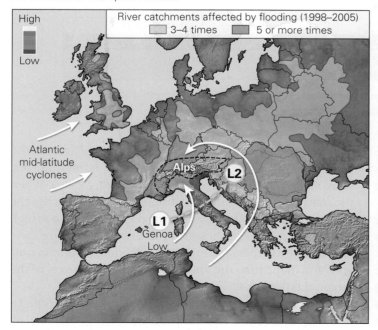

(a) Mid-latitude cyclones cause floods on the west side of Europe. Genoa Lows cause floods to the east. When the low-pressure center of the Genoa Low sits at L1, winds carry rain to the south side of the Alps, and when the center drifts to L2, winds sweep across eastern Europe and carry rain to the north side of the Alps.

(b) Floodwaters surround artworks in Florence, Italy, in 1966.

FIGURE 12.42 Mountain thunderstorms.

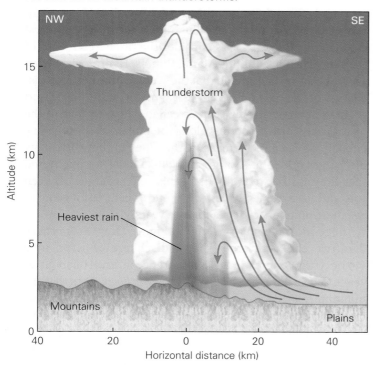

(a) A cross section of a thunderstorm at the boundary of a plain and a mountain range. Arrows indicate air flow.

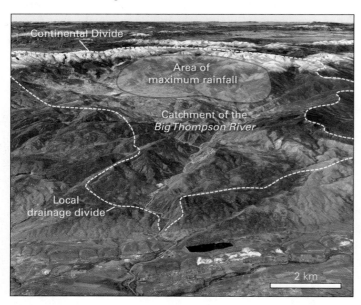

(b) The Big Thompson River watershed is outlined in white. The area of heaviest rain (more than 12 cm, or 5 in) during the 1976 Big Thompson Canyon flood is shaded in blue.

carrying vast amounts of moisture, blow in from the Atlantic Ocean. Southern Europe's floods, however, arise from *Genoa Lows*, mid-latitude cyclones that develop over the Mediterranean Sea. These storms drive moist air from the Mediterranean northward, where the air collides with European mountain ranges, rises, and cools to produce extreme rains. Floods caused by these storms have not only destroyed homes and businesses, but have also inundated museums and other public buildings, leading to the destruction of priceless artworks (FIG. 12.41b). Flooding made global headlines in July 2021, when heavy rains—heavier than any for at least 1,000 years—triggered immense floods with rivers rising by as much as 7 m (23 ft). Significantly, these rains were unrelated to Atlantic or Mediterranean cyclones. Rather, they were caused by a stalled front along which thunderstorm training took place. Such weather systems usually develop only in warmer regions, leading some researchers to suggest that the disaster was linked to global warming. Because such flooding had not happened for so long, hundreds of buildings built on vulnerable land washed away.

Mountain Thunderstorms

Most flash floods are a consequence of isolated but severe thunderstorms that develop along fronts within larger weather systems, as described in Chapter 8. In mountainous areas, however, flash floods can also develop where moist air has been forced upward after colliding with mountains. Such orographic lifting generates storms that drop intense rains, which fill river valleys and dry washes in the upstream portions of drainage networks. Particularly severe mountain thunderstorms develop in summer on the east side of the Rocky Mountains, when west-moving winds from the Great Plains bring moist air directly toward the mountains (FIG. 12.42a). Afternoon heating of the mountain slopes helps make the moist air unstable. It rises to produce towering thunderstorms, which dump torrential rain on the eastern flank of the Rockies, producing flash floods in the range's canyons. The rainfall may concentrate in a relatively small area, so only streams in a single low-order watershed may flood (FIG. 12.42b). Storms of this type caused the Big Thompson River floods, discussed earlier, as well as floods that swamped Boulder, Colorado, in 2013.

Take-home message . . .

Weather systems can produce floods if they can transfer abundant moisture over the land and/or can remain nearly stationary. Monsoon thunderstorms and tropical cyclones produce heavy rains in the tropics. In mid-latitudes, floods originate primarily from training thunderstorms, atmospheric rivers, and mountain thunderstorms as well as landfalling tropical cyclones.

QUICK QUESTION Why do mountain thunderstorms cause flash floods, but not necessarily regional floods?

12.7 Flooding Risk

When making decisions about investing in flood-control measures, issuing mortgages for homes, or offering flood insurance, planners need a basis for defining an area's exposure to flooding. Intuitively, they know that a plot of low-lying land close to a river will flood more frequently than one on higher ground farther from the river, and that a plot on a hill high above a river may never flood. How can these differences be discussed in a systematic way? To address this challenge, hydrologists estimate the likelihood that flooding will occur in a certain location within a given time range. Such calculations can help officials to prepare appropriate advisories to communities and develop maps that characterize flooding hazards.

Estimating Flooding Probability

To determine flooding probability for a given location along a stream, hydrologists examine discharge records for the stream over time. If the historical record from a gauging station goes back for a century or more, it will probably include many small floods, several medium-sized floods, and a few large floods. The record may or may not include any extreme floods, because such floods are rare.

With information on discharge, researchers can calculate the **recurrence interval** of flooding, meaning the average time in years between two floods of a given discharge or greater, at a particular location. (Note that the meaning of *recurrence interval* when researchers are discussing a flood differs subtly from their use of the term when discussing earthquakes; see Chapter 3.) To illustrate a recurrence interval estimate for flooding, let's consider the discharge record of the Embarras River at Ste. Marie, Illinois. Hydrologists from the USGS have determined that the average time between discharges of 880 m^3/s (31,000 ft^3/s) or greater at Ste. Marie is 10 years. A flood of that size can be called a *10-year flood*. If the historical record of flooding along the Embarras River is long enough, we can directly measure the time gaps between many discharges of 880 m^3/s or greater (FIG. 12.43). We might find that the gaps between floods of this size or larger between 1910 and 2010 are 11, 5, 7, 7, 4, 7, 28, 6, and 6 years, and if we average these numbers, we get about 9.2 years.

Let's look again at the data in Figure 12.43. From these data, we can determine that a discharge of 1,100 m^3/s (39,000 ft^3/s) or greater has a recurrence interval of about 30 years. Such a flood can be called a *30-year flood*. Note that the minimum discharge of a 30-year flood exceeds that of a 10-year flood. If we had enough data to see several 100-year floods, we would find that the discharge of a 100-year flood exceeds that of a 30-year flood.

Newspapers and other media commonly discuss flood size in terms of recurrence interval. Unfortunately, many people misinterpret the meaning of a recurrence interval. They may assume, incorrectly and dangerously, that floods are periodic. But if a location endured a 100-year flood, say, 12 years ago, it does not mean that the next one is 88 years in the future. Rather, a 100-year flood might happen again next

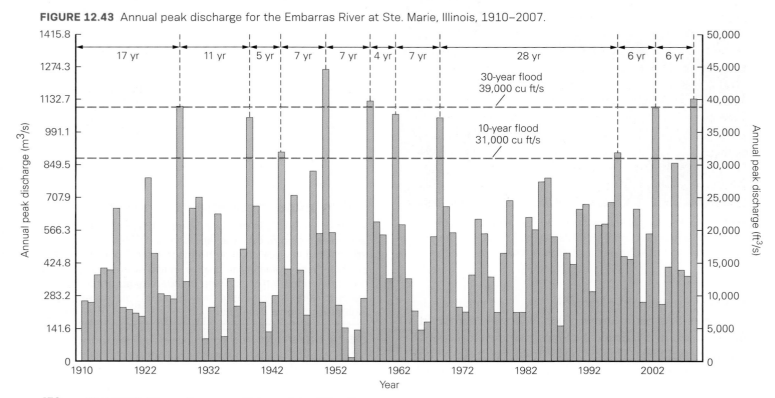

FIGURE 12.43 Annual peak discharge for the Embarras River at Ste. Marie, Illinois, 1910–2007.

month, next year, 32 years in the future, or 168 years in the future. To avoid such confusion, hydrologists prefer to represent the likelihood of flooding by stating the **annual exceedance probability** (**AEP**) of a flood. This number characterizes the likelihood that a flood of a given discharge or greater will happen in a given year. The AEP, given as a percentage, can be calculated from the recurrence interval (R) by using this equation: AEP = $(1 \div R) \times 100$. So, the likelihood that a 10-year flood will happen in a given year is 10%, because AEP = $\left(\frac{1}{10}\right) \times 100 = 10\%$. In words, an AEP of 10% means that there's a 1 out of 10 chance that such a flood will happen in a given year.

If hydrologists had discharge data for a river over many millennia, they would be able to obtain estimates of recurrence interval and, therefore, annual exceedance probability, for a 100-year flood, and maybe even for a 300-year flood, simply by measuring the time intervals between successive floods of the stated size or larger on a graph depicting annual peak discharge for the succession of years. Such data, however, do not exist. In fact, accurate data from gauging stations for a given stream typically cover only a few decades, or at most a century. During that time span, a flood with an AEP of 1% (a 100-year flood) might have happened only once or twice, or not at all, and a flood with an AEP of 0.33% (a 300-year flood) might never have occurred. To get around the problem of limited data, hydrologists use statistical methods that allow them to extrapolate from the data they have. **Box 12.3** illustrates how statistical estimates of flooding probability can be calculated.

Calculating the likelihood that a flood of a given discharge or greater will happen during a period longer than a year is more complicated. Intuitively, one would expect that this likelihood, known as the *cumulative probability*, will increase as the time window of interest extends further into the future. In other words, a 10-year flood is more likely to happen sometime during a 25-year time interval than in any 1-year time interval. Using complex equations, hydrologists have determined, for example, that the likelihood of a 10-year flood happening sometime during a 50-year period is 99.9%, and that the likelihood of a 100-year flood happening sometime during a 100-year period is about 63%. This result may sound strange at first, but keep in mind that the next 100-year flood (as defined by a given discharge) could happen next year, but might not happen for, say, 250 years.

Flood-Frequency Graphs and Flood-Hazard Maps

By calculating the recurrence interval of floods of any discharge at a point along a given stream (using the equations in Box 12.3), hydrologists can produce a graph in which the horizontal axis represents the recurrence interval in years and the

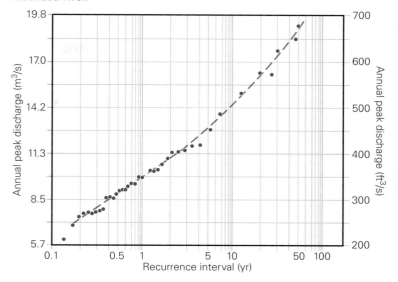

FIGURE 12.44 A flood-frequency graph showing the relationship between the recurrence interval and the peak discharge for an idealized river.

vertical axis represents the annual peak discharge (**FIG. 12.44**). The data don't scatter on such a graph, but rather cluster along a **flood-frequency curve**. For floods with relatively low discharge, data typically plot close to a straight line on this graph. For large floods, the line may change to a steeper slope, indicating that large floods are somewhat more likely than the initial straight-line relationship would suggest. Researchers can determine the recurrence interval for a flood of a given discharge simply by looking at a flood-frequency curve, and by extending the curve, they can estimate the recurrence intervals for floods larger than any that have occurred in the historical record.

Keep in mind that flood-frequency curves are based on limited data—namely, stream discharges measured and recorded in the past. As more data become available, the slope of the curve may change. In the future, as climate change progresses, a drainage network may become wetter or drier, and as land use changes, the amount of runoff in a given area may change. Such changes will affect the discharge of streams in the network, so the flood-frequency curve for the streams in the network may change. Note that if discharges increase because runoff rates are increasing, then floods currently designated as 10-year floods will start to happen more often, so they may be reclassified as, say, 8-year floods.

As we've seen, stream discharge correlates with the stage of a stream. Therefore, hydrologists can draw a line on a map tracing out areas whose elevations lie below a stream's stage for a flood of specified discharge, for example, areas that will be inundated by a flood with an AEP of 50% (a 2-year flood), an AEP of 10% (a 10-year flood), or an AEP of 1% (a 100-year flood) (**FIG. 12.45a**). This flooding risk can be portrayed on a **flood-hazard map** (**FIG. 12.45b**).

BOX 12.3 A CLOSER LOOK...

Calculating recurrence intervals for rare floods

Small floods happen much more frequently than large floods do. In fact, if we plot a graph displaying flood frequency on the vertical axis against flood discharge on the horizontal axis, we'll find that the frequency drops off at a roughly exponential rate **(FIG. Bx12.3)**. Using this relationship as a starting point, statisticians have determined that the recurrence interval of a flood of a given size or larger can be described by the following equation (named the *Weibull equation*, after a statistician): $R = (N + 1) \div M$. In this equation, R is the recurrence interval, N is the number of years that we have flood records for, and M is the "rank" of the flood, meaning its position in a list of floods for a given stream, ordered so that the largest flood lies at the top and the smallest flood lies at the bottom.

To see how to calculate recurrence interval, let's consider an example. Imagine that we have data on annual peak discharge at Smith Bridge, a gauging station on Clyborne Creek, for the past 10 years. First, list the floods in order of their peak discharge, with the highest discharge at the top and the lowest discharge at the bottom **(TABLE Bx12.3a)**. Now, assign a rank to each discharge: the highest discharge receives a rank of 1 and the lowest discharge receives a rank of 10. By applying the Weibull equation to these data, we find the recurrence interval of a flood with a discharge of, say, 7,000 m³/s or greater, which has a ranking of 8 in Table Bx12.3a, to be $R = (10 + 1) \div 8 = 1.4$ years. This result means that, given the data in Table Bx12.3a, a flood with a discharge of 7,000 m³/s or greater will happen, on average, once every 1.4 years. Using the same data, we'll find that the recurrence interval for a flood with a discharge

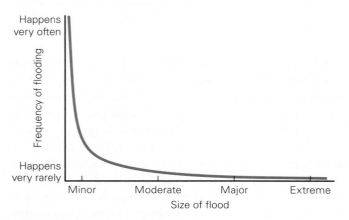

FIGURE Bx12.3 A graph representing the frequency of floods of different sizes.

TABLE Bx12.3a Ranking of Floods Recorded at Smith Bridge for 10 Years

Year	Discharge (m³/s)	Rank
2012	24,800	1
2015	21,450	2
2019	18,700	3
2014	18,500	4
2017	15,950	5
2020	14,350	6
2018	10,100	7
2011	7,000	8
2013	5,700	9
2016	4,200	10

TABLE Bx12.3b Ranking of Floods Recorded at Smith Bridge for 20 Years

Year	Discharge (m³/s)	Rank	Recurrence interval (yr)
2012	24,800	1	21.0
2001	24,000	2	10.5
2007	23,800	3	7.0
2015	21,450	4	5.3
2019	18,700	5	4.2
2014	18,500	6	3.5
2005	17,000	7	3.0
2006	16,750	8	2.6
2003	16,200	9	2.3
2017	15,950	10	2.1
2020	14,350	11	1.9
2010	12,950	12	1.8
2008	12,300	13	1.6
2018	10,100	14	1.5
2011	7,000	15	1.4
2002	6,800	16	1.3
2004	6,500	17	1.2
2013	5,700	18	1.2
2009	4,800	19	1.1
2016	4,200	20	1.1

of 24,800 m³/s or greater is about 11 years, because $R = (10 + 1) \div 1 = 11$ years. Note that the equation indicates that larger floods occur less often than smaller ones do, and also that we can use the equation to obtain an estimate of the recurrence interval for a flood of a size that occurred only once during the analyzed period.

Unfortunately, if the historical record is short, the results given by applying the Weibull equation can be inaccurate. To illustrate this problem, imagine that we find more data, maybe hidden in a dusty file cabinet, and can extend the record in Table Bx12.3a back for an additional 10 years **(TABLE Bx12.3b)**.

With the additional years in the table, a flood with a discharge of 24,800 m³/s or greater now appears to have a recurrence interval of $R = (20 + 1) \div 1 = 21$ years, and a flood of 7,000 m³/s or larger has a recurrence interval of $R = (15 + 1) \div 15 = 1.07$ years. If we had 100 years of records, our results would be even more accurate, and the largest flood might be larger than any shown in Table Bx12.3b. With a long enough record, the results of the equation would come close to the results obtained by averaging the time between floods.

We've already noted that the annual exceedance probability of a flood of a specified size can be determined from the equation $AEP = 1 \div R \times 100$. If we start with the Weibull equation as the basis for calculating AEP, then the equation becomes $AEP = [M \div (N + 1)] \times 100$. Thus, in a given year, the exceedance probability of a flood with 7,000 m³/s discharge at Smith Bridge (see Table Bx12.3b) would be $AEP = (15 \div 21) \times 100 = 71\%$.

FIGURE 12.45 Producing flood-hazard maps.

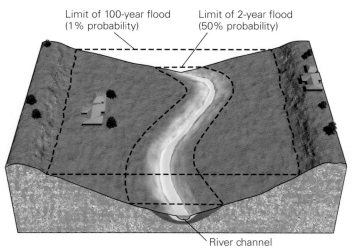

(a) Hydrologists draw lines on a map tracing out areas whose elevations lie below a stream's stage for floods.

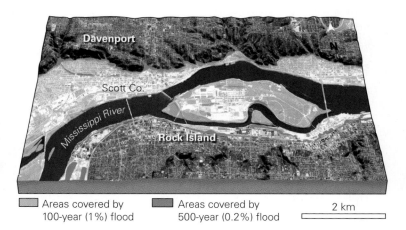

(b) Flooding risk can be portrayed on a flood-hazard map.

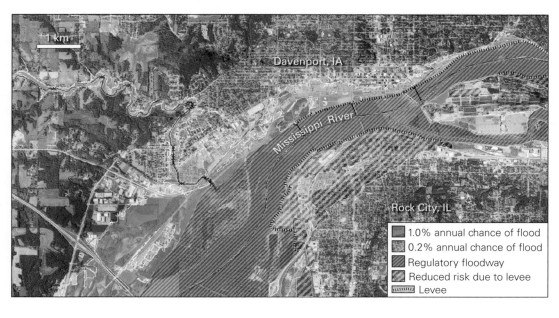

(c) FEMA produces detailed flood-hazard maps for the United States, which are accessible online.

12.7 Flooding Risk • **459**

FEMA provides most of the flood-hazard maps for the United States (FIG. 12.45c). Recently, other organizations have also produced such maps. Not all such maps agree—in fact, a set of maps released by a nonprofit foundation in 2020 suggests that for much of the country, FEMA's maps underestimate flooding hazard. These differences have major economic implications because flood-hazard maps help homeowners make decisions about building homes, insurers make decisions on insurance premiums, and governments make decisions about building levees.

How would you respond to a flood-hazard map? Would you build a house at a location with a 50% chance of flooding in any given year? What about a 1% chance? What about a 39% chance in the next 50 years? Answers to such questions will vary, for individuals have different degrees of risk aversion. If land where flooding has an AEP of 5% costs half as much as land on higher ground, one person may be willing to take the risk and build there, while another might not. But given that taxpayers contribute to the costs of flood control, flood relief, and flood insurance, communities have to ask the hard question of whether constructing homes on land with a high likelihood of flooding is fair to the public at large.

Flood Watches, Warnings, and Advisories

If you come to a closed or flooded road, turn around, don't drown! Driving into flooded roads is a good way to die! Better to be late than end up on the news.

The NWS will broadcast messages like this when weather radar shows a massive thunderstorm approaching. In the United States, officials of the NWS are responsible for keeping the public informed of potential or imminent flooding, so they constantly monitor rainfall rates and hydrographs. If rainfall is expected to be intense, or if stream gauges indicate that water is rising, the NWS issues watches or warnings on a county-by-county basis. A *flash flood watch* means that conditions are likely to lead to flooding within 12–36 hours, and a *flash flood warning* means that flooding has been detected. *River flood warnings* are issued when large rivers are expected to reach a level above flood stage during a slow-onset flood. Different levels of advisories and warnings are issued, depending on estimates of how deeply a land area will be submerged. Notably, the NWS distinguishes among minor, moderate, and major flooding risks in the watches and warnings that it issues (TABLE 12.1).

> **Take-home message . . .**
>
> Knowledge of an area's flooding risk helps people plan for flooding. Flooding hazard can be represented by the recurrence interval (the average time between floods of a given size or larger) or the annual exceedance probability (the likelihood of a flood of a given size or larger happening in a given year). Using a ranked table of annual peak discharge, hydrologists can apply statistical equations to calculate these numbers, and from such data, planners can produce flood-hazard maps.
>
> **QUICK QUESTION** On a flood-hazard map, your house lies in the 100-year flood zone. Will your house definitely flood during the next 50 years? Explain.

12.8 Flood Mitigation

Types of Flood Control

Mark Twain once wrote of the Mississippi River that we "cannot tame that lawless stream, cannot curb it or confine it, cannot say to it, 'go here or go there,' and make it obey." Was Twain right? Indeed, left to its own devices, the course of the Mississippi (or any meandering river) evolves over time as meanders get cut off and new ones form (see Fig. 12.9). Prior to the 19th century, most communities along the river were mobile, so when the river moved, communities moved. Over time, however, permanent towns and farms have been settled, and populations vulnerable to flooding by the river have dramatically increased. The Great Mississippi Flood of 1927 provided abundant proof that fixed communities and a wandering river are not compatible. Following this catastrophe, Congress passed the *1928 Flood Control Act*, authorizing the US Army Corps of Engineers to undertake numerous flood-control projects along the Mississippi and its tributaries. The projects were designed for two purposes: first, to intercept runoff before it could reach trunk streams, and thereby decrease the discharge of the trunk streams; and second, to stabilize the positions of trunk channels and prevent the natural evolution of meanders.

TABLE 12.1 NWS Designations of Predicted Flooding Risks

Minor	Low ground flooded, and low parts of roads underwater; no internal house flooding (maybe a few basements); water over roads won't stall vehicles.
Moderate	Numerous buildings flooded; some infrastructure affected; some people need evacuation; water makes roads impassable.
Major	Houses destroyed or floated off foundations; infrastructure destroyed; bridges washed away; severe erosion; extensive evacuations; National Guard generally called in.

FIGURE 12.46 Bull Shoals Dam, along the White River in Arkansas, has backed up a large reservoir. The jagged shape of the reservoir's shore reflects the shape of the dendritic drainage network that was submerged due to the dam's construction. The White River drains into the Mississippi.

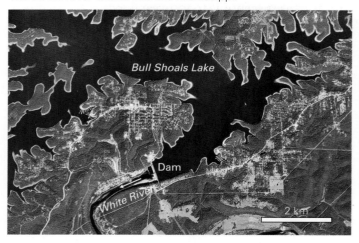

FIGURE 12.47 Examples of floodwalls, levees, and floodways.

(a) A floodwall along the Mississippi near Cape Girardeau, Missouri.

(b) Levees along a small tributary of the Mississippi at Galena, Illinois.

(c) The Morganza Spillway is a floodwall that interrupts a portion of a long levee along the Mississippi. The structure contains gates that can be opened during a flood to divert water from the river.

In succeeding years, the Corps built about 300 dams along the Mississippi and its tributaries to diminish discharge downstream (FIG. 12.46). The runoff that these dams trap in reservoirs can be released slowly later. Globally, tens of thousands of dams have been built to prevent flooding. Reservoirs trapped by these dams can also provide water for municipalities, irrigation, energy production, and recreation. Unfortunately, damming rivers can have significant negative consequences. For example, dams can cause towns, forests, and rapids upstream to be submerged; they can block fish migration and cause changes in water temperature and other physical properties downstream that damage ecosystems; they can prevent floodwaters from flushing out channel-blocking sediment, debris, or contaminants; they can trap nutrients; and they can prevent sediments from reaching downstream floodplains and deltas, causing those regions to undergo subsidence (see Chapter 7).

In addition to dams, the Corps has built 5,600 km (3,480 mi) of artificial levees, as well as concrete floodwalls along urban corridors (FIG. 12.47a). The levees, made primarily of compacted mud and gravel, range between 6.1–15.2 m (20–50 ft) high (FIG. 12.47b). Levees and floodwalls increase the channel's volume and isolate portions of the floodplain from the river.

Artificial levees work... until they don't! Due to expense, levees can't be built high enough to retain the highest floods, so they sometimes fail due to overtopping. Also, because they are made of earth, they are vulnerable to failure by underseeping. Due to the limitations of levees, the Corps has also constructed gigantic spillways in some places along the Mississippi levee system (FIG. 12.47c). These structures

12.8 Flood Mitigation • 461

FIGURE 12.48 People can increase the height of a levee by adding sandbags.

consist of a floodwall with many gates. During extreme floods, the Corps can open the gates, allowing water from the river to take a shortcut to another river channel, the sea, or a bay, or to spill into less populated land. Where levees aren't high enough, or don't exist, communities may respond to flooding by trying to reinforce existing levees or by building emergency earth levees using either earth-moving equipment or sandbags (FIG. 12.48).

Construction of an extensive floodwall and levee system increases the stage of a river for a given discharge because floodwaters can no longer spread out over the floodplain when discharge increases. Consequently, flooding risk can increase in regions downstream of levees.

Because of the challenges posed by levee development, officials have explored alternatives. Options include the construction of diversion canals, which shunt excess water to storage facilities or into other rivers, and the restoration of wetlands along rivers, since wetlands can absorb floodwater. In addition, planners may prohibit construction within designated land areas adjacent to a channel, creating a **floodway** that can be submerged without causing expensive damage (FIG. 12.49). Establishment of such natural sponge zones may have lessened the impact of 2021 flooding in rapidly urbanizing areas of China, where land area covered by impermeable surface has increased immensely in the past 20 years.

Zoning can play a broader role in flood mitigation: by preventing communities from building in areas that face severe flooding hazards, damage due to flooding can be minimized. Commonly, communities focus on restricting development within areas susceptible to 100-year floods.

Who Pays for Flood Damage?

In the aftermath of a flood, questions immediately arise concerning who should pay to rebuild, or whether rebuilding should occur at all. Most private insurance companies in the United States have stopped offering flood insurance because they don't have the financial resources to cover losses caused by flood disasters. Today, flood insurance can be obtained only through the *National Flood Insurance Program*, which is effectively subsidized by taxpayers or by growth in the federal deficit. But even with such subsidies, increasing flood-related disaster costs have caused premiums for such insurance to grow, so many people living in floodplains remain uninsured and must bear the brunt of financial losses themselves.

When flooding becomes sufficiently severe, state or federal disaster aid may be available to assist recovery. Taxpayers fund this aid, so it is in everyone's interest for flood costs to be controlled. Federal and local governments are beginning to address this problem with *property buyout programs*, in which the government buys flood-prone areas and converts them into parks or wetlands. These programs cost billions of dollars, but in the long term, they may save money by reducing the cost of disaster relief and restoration.

Flood Vulnerability

The vulnerability of communities to flooding depends on their proximity to flooding hazards and on the degree to which they are protected by flood-control structures. The extent of their vulnerability, unfortunately, generally reflects wealth. Because land is cheaper where flooding is more likely, less wealthy communities may be developed in flood-prone areas. Residents who are fortunate enough to have resources can take precautions to improve the resilience of their houses. A common approach is to build a house on stilts, so that if water rises, it won't submerge the living area. If that's not possible, houses can be designed to have a concrete-floored lower level with removable carpet. Ground-floor cabinetry can be installed so that it isn't anchored to the floor, and levels of the building can have separate electrical circuits. Unfortunately, urbanization also increases vulnerability to flooding hazards for everyone, for, as we have seen, paving the landscape with concrete and asphalt prevents water infiltration.

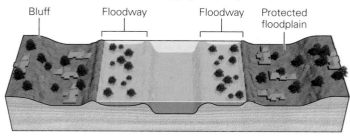

FIGURE 12.49 A floodway is an area left free of construction so it can be flooded without damaging property.

Labels: Bluff, Floodway, Floodway, Protected floodplain

Researchers have shown that global climate change tends to increase flooding in areas impacted by storms because atmospheric warming increases sea-surface temperatures, evaporation rates, and the atmosphere's capacity to store water vapor (see Chapter 8). Increased atmospheric moisture produces heavier rainfall in flood-producing weather systems. Studies suggest that heavy downpour events in some locations have increased by up to 70% since the 1950s. Flooding has been a major aspect of the Earth System since water first appeared on land. But dealing with the human and economic impacts of floods on society in coming decades will require significant resources and leadership.

Take-home message...

Hydrologists can use historical data to estimate the probability that a stream will flood. To prevent floods, engineers can build dams, levees, and reservoirs to store excess water. Generally, private insurers won't cover the cost of flooding, so many people need to buy federally subsidized flood insurance. Vulnerability to flooding tends to be increased by urbanization.

QUICK QUESTION How might climate change affect flooding hazards?

ANOTHER VIEW Engineers are constructing floodwalls along a stream in a mountainous area of western China.

Chapter 12 Review

Chapter Summary

- During an inland flood, areas that lie out of reach of rising seas can still be submerged by water falling as precipitation. Most inland floods happen when streams overflow.
- Streams are bodies of water that flow down a channel. The runoff that flows out of a watershed includes stream flow and overland flow.
- Streams slope from their headwaters to their mouths. The stream gradient tends to be steeper near the headwaters and shallower near the mouth. The surface of a stream cannot drop below the stream's base level.
- Streams erode land and transport sediment. Stream competence is the maximum clast size, and stream capacity is the total quantity of sediment that a stream can carry.
- When streams slow, sediment settles out to form alluvium. Alluvium can build gravel bars or form a layer of floodplain deposits. Where a stream empties into a standing body of water, a delta may accumulate.
- Meandering streams follow snake-like curves. Many such streams flow on floodplains. The course of a meandering stream changes as meanders get cut off and new ones form.
- Braided streams are streams that divide into many small channels separated by elongate bars.
- Permanent streams flow all year, whereas ephemeral streams flow for only part of the year. Dry washes contain no water at all most of the time.
- Water in a watershed flows in a drainage network. Within the network, tributaries flow into larger streams, eventually exiting the watershed in a trunk stream.
- The volume of water passing through a cross-sectional area of a stream defines the stream's discharge at that location. As discharge increases, the stage of the stream rises. A hydrograph shows change in a stream's discharge over time.
- When a stream reaches flood stage, water starts submerging areas outside the stream's channel. At the flood crest, the stream reaches its highest stage.
- Slow-onset (regional) floods take time to develop and generally involve high-order streams in the downstream region of a drainage network. Such floods may submerge floodplains or delta plains, causing casualties and damage.
- During a flash flood, water rises in less than 6 hours. These floods can be caused by extreme rainfall, rapid snow or ice melt, or the breaching of natural or constructed dams.
- During areal floods, water accumulates in low-lying areas. Urban floods happen in built environments where the land surface has been covered with impermeable material and drainage is inadequate.
- Failure of dams and levees can yield flash floods. Such failure can happen when water spills over the top or seeps underneath the structure. In some cases, failures release toxic slurries of coal ash or mine tailings.
- Flooding causes casualties and damage in several ways. The dynamic pressure of floodwaters can push structures over and, together with buoyancy force, move heavy objects. Flowing water can undermine riverbanks, roads, and foundations. Rising water can drown people and livestock, flood homes, fields, and infrastructure, and leave muck.
- Most slow-onset floods are associated with monsoons, tropical cyclones, and mid-latitude cyclones. Some flash floods are a consequence of thunderstorms.
- Flooding probability can be represented either by a recurrence interval or by the annual exceedance probability.
- Some people misinterpret the recurrence interval: it does not indicate how many years will pass between one flood of a given size and the next, because floods are not periodic.
- Larger floods happen less frequently than smaller ones. A flood-frequency curve shows the relationship between the recurrence interval and the annual peak discharge at a location on a stream. The area that a flood of a given discharge will cover can be represented on a flood-hazard map.
- Because of the hazard that floods represent, the NWS broadcasts flood watches and warnings.
- Flooding can be controlled, at least temporarily, by construction of dams, levees, and floodwalls. Levees effectively raise the banks of a stream and increase the channel's volume. But they are subject to failure.
- Construction of a dam can cause problems. For example, the dam may cause flooding of upstream communities and may restrict the supply of sediment and nutrients to reaches downstream.
- Flood damage is too expensive to be privately insured in most cases, so rebuilding must be funded by governments or individuals. Climate change may cause flooding to worsen in the future.

REVIEW QUESTIONS

Blue letters correspond to the chapter's learning objectives.

1. How can overland flow begin to produce a stream channel? Contrast downcutting with headward erosion. **(A)**
2. How does the gradient of a stream change along its longitudinal profile? What features can form local base levels and the ultimate base level along a stream? **(A)**
3. Distinguish between competence and capacity. What happens to the sediment of a stream when a stream's competence decreases? **(A)**
4. Describe the key characteristics of a meandering stream, and contrast them with those of a braided stream. How does a meandering stream change its course over time? **(B)**
5. What is a drainage network, and what landform separates one drainage network from another? **(B)**
6. Which stream has a lower order: a small mountain tributary or a broad trunk stream? **(B)**
7. Define the discharge of a stream, and explain how the discharge and the stage of a stream at a location are related. What information does a hydrograph provide? **(B)**
8. Which is higher: the flood stage or the flood crest? On the hydrograph, label the lag time and the discharge at the flood crest. **(C)**
9. Explain why slow-onset floods generally involve higher-order streams and tend to submerge broad areas. **(D)**
10. During what time frame does a flash flood develop? Compare the hydrograph of a flash flood with that of a slow-onset flood, and describe the various causes of flash floods. **(D)**
11. How does areal flooding differ from stream flooding? What factors cause urban flooding? Can urban flooding happen in response to a rainfall that would not cause flooding in a natural area? **(D)**
12. Describe the different processes by which a constructed dam or levee can fail and produce a flood. Which is depicted in this diagram? **(D)**

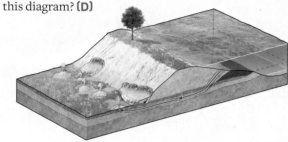

13. In what ways can floods cause casualties and damage? Can entering a flooded highway be hazardous even if the water flowing over the highway is only knee deep? **(E)**
14. Why does monsoonal flooding affect southern Asia mostly during June and July? How does the weather system responsible for monsoon rains in India differ from that responsible for monsoon rains in China? **(F)**
15. How can slow-onset regional floods develop in association with a mid-latitude cyclone? What is an atmospheric river, and how can it cause flooding in mountains near the west coast of North America? **(F)**
16. Why do flash floods develop in mountainous areas? **(F)**
17. Distinguish between the recurrence interval and the annual exceedance probability of a flood. How are these quantities related? If a 50-year flood happened last year, will the next one happen 49 years from now? **(G)**
18. What information does a flood-frequency curve provide? How is a flood-hazard map produced? **(G)**
19. Explain the difference between a flood watch and a flood warning. What does a river flood warning mean? **(G)**
20. Describe the various ways that people try to control flooding. Can they be 100% effective? Why? **(H)**
21. Why does the US federal government have to subsidize flood insurance? **(I)**

On Further Thought

22. How might climate change and urbanization affect the intensity of flooding, and the cost of flood insurance, in the future? What factors may lead to increased vulnerability of communities to this flooding? **(I)**

13

SAVAGE SEAS
Hazards of the Shore

By the end of the chapter you should be able to . . .

A. characterize the various ways in which ocean water circulates.

B. explain how wind-driven waves form and move, how big they can get, and how they change as they approach the shore.

C. interpret landforms that form along coasts.

D. discuss the hazards associated with nearshore currents, waves, and tides.

E. describe hazards associated with coastal erosion, and how they can be mitigated.

F. recognize the causes and consequences of oceanic algal blooms.

FIGURE 13.1 The 1900 hurricane in Galveston, Texas.

(a) Galveston is on a long, narrow island on the Gulf of Mexico.

(b) Waves and storm surge reduced the city to rubble.

13.1 Introduction

By the beginning of the 20th century, Galveston, Texas, had become a prosperous city of 38,000. It had been developed on a narrow island of sand that rises only 2 m (7 ft) above the adjacent Gulf of Mexico. Near midday on September 9, 1900, tourists enjoying the city's broad beach noticed that the surface of the Gulf seemed to be getting higher and had become quite rough. Over the course of the afternoon, clouds darkened the sky, the wind intensified, and the waves grew. At 8:00 P.M. that night, a hurricane made landfall. Towering waves and storm surge swept over Galveston, flooding its streets to a depth of 3 m (10 ft) and transforming its homes, stores, and warehouses into a chaotic pile of splintered wood and brick (FIG. 13.1). On that terrible day, up to 12,000 people died, and more than 10,000 were left homeless. Clearly, oceans and their borders are places of beauty, but as the residents of Galveston sadly learned, they can be places of terror as well.

Almost 71% of our planet's solid surface area lies hidden beneath a layer of *seawater* (consisting of 96.5% water and 3.5% salt) with an average depth of about 4 km (2.5 mi). Notably, most continents are bordered by a band of shallower water, the 50–300 km–wide *continental shelf*, above which water has a depth of generally less than 100 m (300 ft). Geographers divide this layer of water into distinct oceans, seas, and bays (FIG. 13.2). Over half of the world's human population lives in communities along oceanic **coasts**: land areas up to several kilometers wide in which the landscape and local climate are influenced by proximity to the ocean. *Oceanographers*, scientists who study the ocean, refer to the relatively narrow portion of a coast (less than a kilometer wide) that interacts directly with the water of the ocean as the ocean's **shore**. In everyday English, though, the terms *coast* and *shore* are often used interchangeably.

In this chapter, we explore the natural hazards of oceans and their shores. To provide context, we begin by characterizing the movement of ocean water in currents, tides, and waves, and by describing coastal landscapes and how they evolve. Next, we address the dangers that interactions between land and water at or near the shore can pose to individuals and communities—problems that are being exacerbated by sea-level rise. Note that while our discussion focuses on oceanic coasts, the same concepts apply to coasts of large lakes.

13.2 Circulation in the Ocean: Currents and Tides

Currents: The Rivers of the Sea

In 1992, a container holding plastic bath toys tumbled off a cargo ship in the middle of the Pacific Ocean. Over the next several years, yellow duckies started appearing on beaches all around the world. The amazing voyage of these toys emphasizes that seawater in the oceans constantly circulates, and that all oceans are interconnected. Oceanographers refer to a definable band of flowing water within a larger body of water as a **current**. Note that, unlike a river on land, a current does not occupy a physical channel. Therefore, water just inside the edge of a current—which may be moving at speeds of up to 10 km/h (6 mph)—shears against water outside the current. Currents flow at two levels in the ocean: *deep-ocean currents* affect water far below the surface and, in some cases, even at the ocean floor; whereas **surface currents** affect at most

◄ Storm waves batter a coastline in southern England. Such waves can alter coastal landscapes permanently, and can cause significant casualties and property damage.

FIGURE 13.2 Geographers have divided the ocean into named oceans and seas.

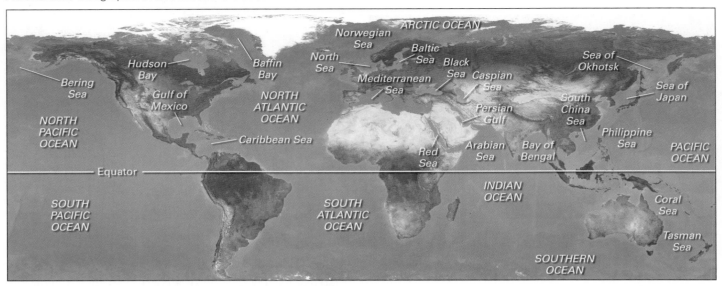

FIGURE 13.3 Currents circulate water around the world's oceans.

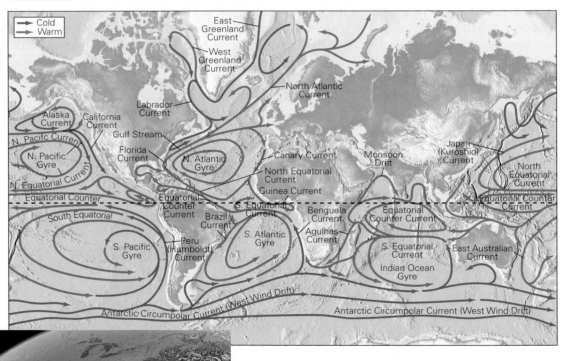

(a) This global-scale map of ocean surface currents distinguishes warm currents (red) from cold currents (blue).

(b) At a local scale, currents are more complicated because many eddies develop around them, as shown in this view of the Gulf Stream, which flows along the east coast of the United States.

only the upper 400 m (1,300 ft) of water (**FIG. 13.3a**). The Gulf Stream serves as an example of a surface current—it can be traced for thousands of kilometers, and has a width ranging between 100 and 200 km (60–120 mi). Because people interact with surface currents, not deep-ocean currents, they are our focus in this discussion.

Surface currents are initiated when wind applies a shear force against the surface of the ocean. Once surface water starts moving, both the Coriolis force and a pressure-gradient force (which develops because water movement produces broad but subtle mounds of elevated water) influence the direction of flow (see Chapter 8), resulting in the development of **gyres**, large surface currents that carry water entirely around the margin of an ocean (see Fig. 13.3a). Due to the complex shapes of shorelines, many smaller currents develop in addition to gyres, and some of these carry water close to shore. Locally, circular swirls, known as **eddies**, develop where a current shears against land or against a bordering volume of water (FIG. 13.3b).

The Tides Come In . . . The Tides Go Out

Fishermen hoping to sail from a shallow port must pay attention to the **tide**, the generally twice-daily rise and fall of the sea surface (FIG. 13.4a). At *low tide*, when the sea surface sinks to its lowest elevation, fishing boats may run aground, whereas at *high tide*, when the sea surface rises to its highest elevation, they can easily cruise to open water. Tides are caused by the *tide-generating force*, a combination of three other forces. These forces are the gravitational pull of the Moon (the largest component), the gravitational pull of the Sun, and centrifugal force caused by the rotation of the Earth-Moon system around its center of mass (FIG. 13.4b). The tide-generating force produces two *tidal bulges*, one on the side of the Earth nearer the Moon and one on the opposite side (FIG. 13.4c). The Earth's daily rotation brings a location on the planet beneath these two bulges in succession (FIG. 13.5a). When a location lies beneath a bulge, it experiences a high tide, and when it lies between bulges, it experiences a low tide. During times of the month when the Sun and the Moon lie on the same or opposite sides of the Earth (the times of a new Moon and full Moon, respectively), *spring tides* occur, during which high tides are higher than average and low tides are lower than average

FIGURE 13.4 Tides and their causes.

(a) High tide (left) and low tide (right) on the southern coast of England. The two photos were taken from the same viewpoint.

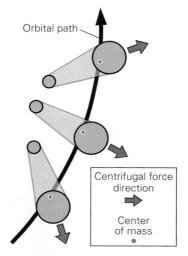
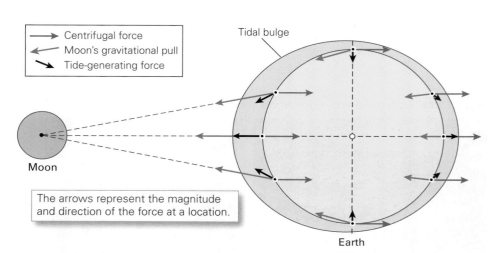

(b) The Earth and Moon are gravitationally linked. The Earth-Moon system rotates around a point called the center of gravity. The center of mass (red dot) moves along the black line as the system orbits the Sun. The system's rotation generates an outward centrifugal force everywhere on the Earth.

(c) The contributions of the gravitational pull of the Moon, the gravitational pull of the Sun (not shown), and centrifugal force produce the tide-generating force, which results in a tidal bulge on opposite sides of the Earth. The Moon contributes the most to this force, so the larger bulge is on the side facing the Moon.

FIGURE 13.5 Motion of the Earth relative to tidal bulges.

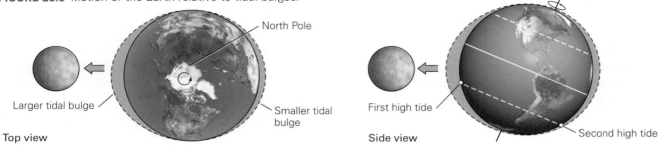

(a) The highest tide at a location in a 24-hour period occurs when the location passes under the larger bulge. Another, lower high tide occurs when the Earth passes under the smaller bulge.

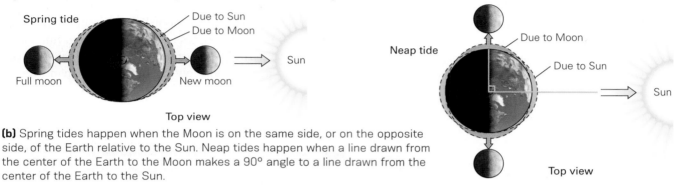

(b) Spring tides happen when the Moon is on the same side, or on the opposite side, of the Earth relative to the Sun. Neap tides happen when a line drawn from the center of the Earth to the Moon makes a 90° angle to a line drawn from the center of the Earth to the Sun.

(FIG. 13.5b). When a line drawn from the Earth to the Sun makes a right angle to a line drawn from the Earth to the Moon, *neap tides* occur. During these events, high tides are lower than average and low tides are higher than average.

The magnitude of the **tidal range**, the difference between sea-surface elevation at high tide and at low tide, varies not only with the time of the month, but also with location, because many factors—such as the position of a location relative to the tidal bulge, and the shape of the coastline—influence tides **(FIG. 13.6a, b)**. In some regions the tidal range is less than 50 cm (20 in), whereas in others it is so large that it can be measured in meters. The largest tidal range on this planet occurs in the Bay of Fundy, Nova Scotia, where water at high tide attains an elevation 16.3 m (53.5 ft) higher than the level of water at low tide. Note that the elevation designated as *mean sea level* for a particular location represents the midpoint between the average high tide and the average low tide there.

In the open ocean, you won't notice the tides, but they can affect coasts significantly. During a rising tide, or *flood tide*, the **shoreline**—the boundary between water and land—moves inland, and during a falling tide, or *ebb tide*, the shoreline moves

FIGURE 13.6 Tidal range and the intertidal zone.

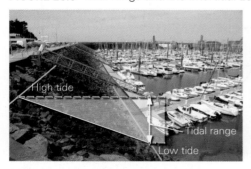

(a) Tidal range is the difference in elevation between low tide and high tide. This photo shows a French harbor at low tide.

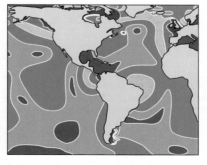

(b) A map showing how tidal range varies with location in the Atlantic and eastern Pacific.

(c) On this shore on Cape Cod, the high-tide elevation is marked by a line of seaweed that was carried in by the water.

470 • CHAPTER 13 • Savage Seas: Hazards of the Shore

seaward. The region between the two shoreline positions constitutes the **intertidal zone** (FIG. 13.6c). Its width depends on both the tidal range and the slope of the shore. Specifically, for a given tidal range, the width of the intertidal zone is much greater along a gently sloping shoreline.

> **Take-home message . . .**
> Ocean water circulates in currents. Surface currents form due to shear force applied by the wind. The tide is the generally twice-daily rise and fall of the sea surface. Tidal ranges vary with location, from less than 50 cm to over 16 m. The width of the intertidal zone depends on the tidal range and the slope of the shore.
>
> **QUICK QUESTION** What is the difference between a spring tide and a neap tide?

13.3 Wind-Driven Waves

Formation of Waves

Ocean waves—the up-and-down motions of the ocean surface—produce restless, ever-changing vistas (FIG. 13.7a). Waves don't occur in isolation, but are typically members of a **wave train**, a group of waves traveling together. Oceanographers use the same basic terminology of describing a wave train in water as are used for other types of waves (see Chapters 3 and 6): The highest part of a wave is its *crest*, the lowest part is its *trough*, the *wave height* is the vertical elevation difference between crest and trough, and the *amplitude* is half the wave height. The *wavelength* is the horizontal distance between two successive troughs or crests, the *wave period* is the time between the passage of two successive crests (or

FIGURE 13.7 The formation and character of ocean waves.

(a) Waves of many sizes wrinkle the surface of the ocean.

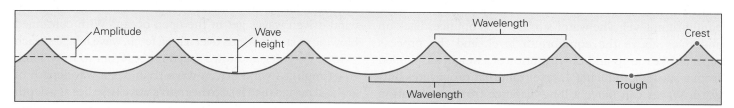

(b) A wave train consists of many similar waves moving together. Note that the crests of ocean waves are narrower than the troughs. We can characterize a wave train by specifying wave height and wavelength.

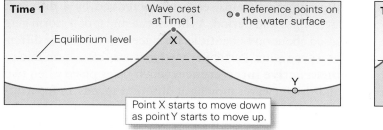 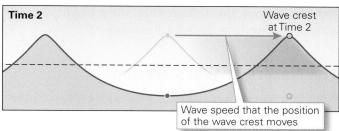

(c) At Time 1, as soon as the water in a wave is uplifted above the equilibrium level, gravity pulls it down below the equilibrium level to its position at Time 2. A restoring force then acts to push the surface of the water back up. Note that as the water surface moves downward, the position of the wave's crest moves horizontally at the wave speed (arrow).

troughs), and the *wave speed* is the horizontal velocity at which a crest or trough moves (**FIG. 13.7b**). Notably, in a water-wave train, the shape of crests differs from that of troughs: crests tend to be narrow and pointed, while troughs tend to be wide and smoothly curving.

Most waves that you see when you look out to sea from the shore, or from a ship's deck, are **wind-driven waves**. These waves come from the interaction between moving air and the surface of the ocean. How do such waves begin? To picture the process, imagine still water in a pond on a windless day. Recall from Chapter 6 that the horizontal surface of the water represents an *equilibrium level*—if you pushed the surface up or down with a paddle, it would return to that level (**FIG. 13.7c**). At the water surface, water molecules attract one another more strongly than they attract air molecules above. This attraction produces *surface tension*, which makes the water surface behave somewhat like an elastic sheet. When a breeze starts to blow, shear force causes this elastic surface to stretch. As soon as the surface stretches, elastic rebound causes it to twang back like a rubber band, and as a consequence, it wrinkles, producing small waves called *ripples*. Once ripples exist, wind can push against their sides, causing them to build still higher, lifting water farther above the equilibrium level. (Strong winds blow the tops of the waves over, causing water to mix with air, which produces *whitecaps*.) Once water has been uplifted in a wave, gravity pulls the water back down. Due to inertia, the sinking water surface descends below the equilibrium level. The water surface then bounces back up and rises above the equilibrium level, and the process repeats, like the up-and-down motion of a weight suspended from a spring, producing a wave train that propagates outward. Waves can move a long distance from the location where they were generated by the wind. Notably, as they propagate outward away from where the wind is blowing, each wave's energy spreads over a broader area, and friction opposes its movement. Therefore, wave height decreases with increasing distance from the source.

The Nature of Wave Motion

When you watch a wave travel, it looks as if a mass of water is moving with the wave. But if you place a cork in the water, you'll see that it only bobs up and down and back and forth as a wave passes. In fact, as an "ideal" ocean wave passes, the combination of back-and-forth and up-and-down motion

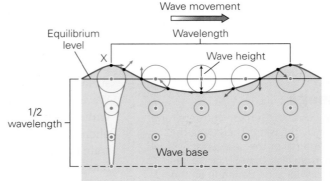
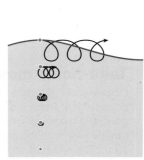

FIGURE 13.8 The movement of water in waves.

(a) Ideally, as each wave passes, water within the wave follows a circular path. The diameters of these paths decrease with depth to the wave base. The positions of crests and troughs move (here, to the right), but the water itself does not.

(b) In real waves, there is typically some lateral motion due to shear force applied by the wind. The water actually follows a spiral-like path.

means that water affected by the wave moves approximately in a circle. The diameter of such movement circles decreases downward until, at a depth called the **wave base**, no movement takes place at all (**FIG. 13.8a**). Consequently, submarines traveling below the wave base enjoy smooth water while ships endure rough seas above. The depth of the wave base is equal to about half the wavelength. In reality, waves aren't ideal, and some lateral motion of water accompanies the passage of a real wave, due to shear force applied by the wind. In fact, water typically moves forward by a distance of about one wavelength for every four waves that pass (**FIG. 13.8b**). It's because of this *wave drift* that surface currents develop.

The heights of waves generated by wind in the open ocean depend on the wind's speed, its **fetch** (the distance over which it blows), and its duration (**FIG. 13.9a**). How high can wind-driven waves get in the open ocean? With continued blowing of strong winds over a long fetch, wave heights reach 2–10 m (7–33 ft), and wave speed ranges from 10 to 50 km/h (6–31 mph). A wind-driven wave train has a wavelength of 40–200 m (130–656 ft), so the train's wave base lies at a depth of 20–100 m (66–328 ft). As wind-driven waves grow higher, their sides also become steeper (**FIG. 13.9b**).

When a storm blows over a region, it generates many different wave trains, each characterized by a different wavelength and period. A ship in the midst of a storm gets tossed about on a chaotic sea surface as it rides over different wave trains. Particularly large waves may form due to **constructive interference**, which takes place when two wind-driven waves moving at different speeds, or from different directions, come together in such a way that the wave crests overlap to form a single crest that is higher than that of either wave (**FIG. 13.10a**). Interference can produce giant waves at sea that can capsize boats (**FIG. 13.10b**).

FIGURE 13.9 Building large waves.

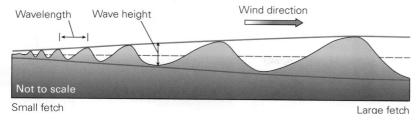

(a) Wave size depends not only on wind speed, but also on the wind's fetch. For a given wind speed, the longer the fetch, the larger the waves, up to a limit.

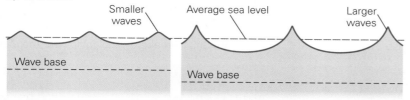

(b) As wind-driven waves in the open ocean grow higher, their wavelength increases, the waves become steeper, and the wave base becomes deeper.

As wave trains of different periods travel outward from a storm, they separate from one another because they travel at different speeds. Long-period waves, which maintain their energy longer, can travel for thousands of kilometers across the ocean at speeds of up to 50 km/h (31 mph). Such wave trains, which typically have fairly uniform heights and regular spacing, are called **swells** (FIG. 13.11).

FIGURE 13.10 Constructive interference.

(a) Constructive interference between two waves produces a larger wave.

(b) During the Fastnet yacht race of 1979, high waves produced by constructive interference capsized many sailboats, requiring dangerous rescues.

When Waves Approach the Shore

Out in the deep ocean, the wave base of a wind-driven wave is so far above the seafloor that even the largest such wave has no effect on the ocean floor. As a wind-driven wave enters shallower water near shore, however, the wave base touches the ocean floor, causing a slight back-and-forth motion of sediment. In very shallow water, wave motion at the seafloor becomes strong enough to churn up sediment. Shear between the seafloor and a wave slows the deeper portion of the wave, so instead of moving in a circle, water in the wave starts to trace out a tilted ellipse (FIG. 13.12a). Friction also causes waves in a train to slow down overall, though the period of the train remains the same, so wavelength decreases and wave height increases. As a wave moves into progressively shallower water, the top of the wave curves over its base and falls, so that the wave becomes a **breaker**, to the delight of surfers (FIG. 13.12b). The toppling water of a breaker mixes with air to form a turbulent white froth. After it breaks, a wave's height decreases, but the wave still moves shoreward. The band between the location where breakers form and the shoreline is called the *surf zone* (FIG. 13.12c). When the wave finally washes onto the shore, water surges up the shore's slope. This run-up, or **swash**, continues until friction and gravity bring water motion to a halt. Then, gravity draws the water back down the slope as **backwash** (FIG. 13.12d).

The shifting winds within a storm produce overlapping wave trains traveling in different directions, so if the storm is near the coast, chaotic breakers slam onto the shore. If, however, the waves were produced far offshore, they arrive as fairly evenly spaced swells of uniform height. When waves approaching the shore are oriented at an angle to the shore, they undergo bending, or **wave refraction**, as they enter shallow water. To understand why wave refraction happens, imagine a wave approaching the shore so that its crest initially makes an angle of 45° with the shoreline (FIG. 13.13a). The wave base under the end of the wave closer to the shore touches bottom first, so the overlying wave is slowed by friction. Meanwhile, the end farther offshore continues to move at its original velocity. This difference in speed swings the

FIGURE 13.11 Swells have long wavelengths and travel long distances, moving out of the area where they were produced.

FIGURE 13.12 Waves change as they wash onto the shore.

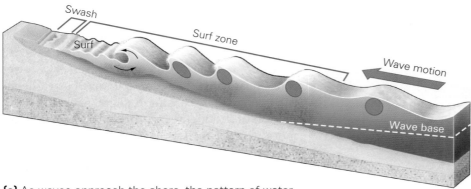

(a) As waves approach the shore, the pattern of water movement within a wave shifts from a circle to a tilted ellipse.

(b) Small breakers developing near the shore.

(c) In the surf zone, waves steepen and become breakers.

(d) Swash is the movement of water up onto the shore, and backwash is the movement of water down the beach.

FIGURE 13.13 Wave refraction occurs when waves approach the shore at an angle.

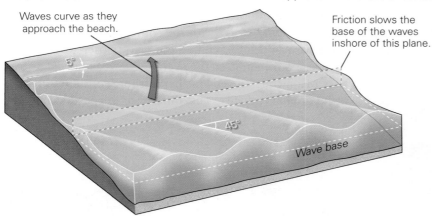

(a) The part of the wave that reaches the shore first slows down, so the part farther out swings around.

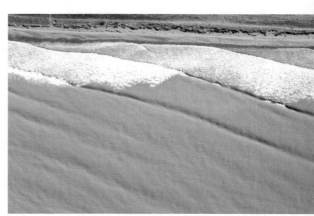

(b) An example of wave refraction as seen from the air.

whole wave around. Because of wave refraction, most waves end up making an angle of less than 5° relative to the shoreline at the landward edge of the surf zone (FIG. 13.13b).

> **Take-home message . . .**
>
> Wind, along with gravity, produces wave trains. Water in a deep-ocean wave follows a circular path. The amount of motion decreases with depth down to the wave base, below which wave motion doesn't occur. When waves approach the shore, they get taller and may eventually break, and they may undergo refraction.
>
> **QUICK QUESTION** What is the difference between swash and backwash?

13.4 Coastal Landforms

Welcome to the Coast

The shores of the Virgin Islands host a gently sloping fringe of sparkling sand, but along the coast of Brittany, France, cliffs of jagged rock rise straight out of the water (FIG. 13.14a, b). As these examples illustrate, shores, and coastal areas overall, vary dramatically in terms of their topography and associated landforms (FIG. 13.14c). As we will now see, these differences depend on three factors: the topography of coastal land near the shore; whether the coast is rising or sinking relative to sea level; and whether sediment is being added to or removed from the shore.

FIGURE 13.14 Contrasting types of shores.

(a) A sandy shore in the Virgin Islands.

(b) A rocky shore in Brittany, France.

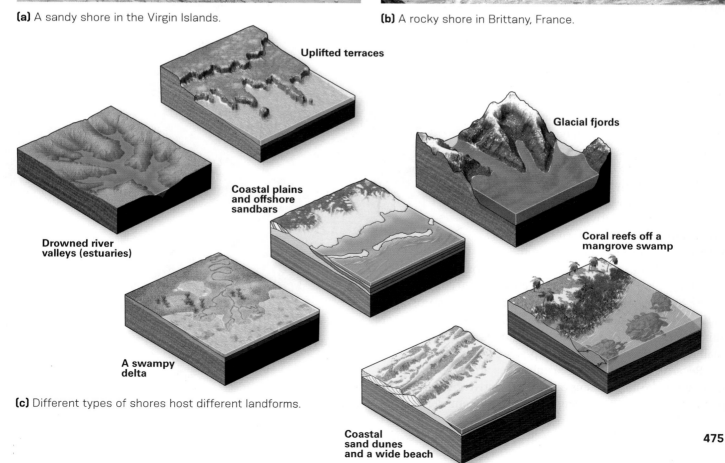

(c) Different types of shores host different landforms.

475

FIGURE 13.15 Factors that control the character of a coastal landscape.

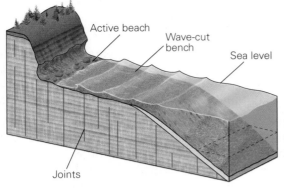
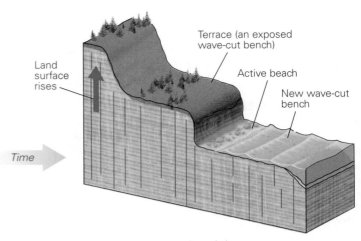

(a) Wave erosion produces a wave-cut bench. If the land surface rises relative to sea level, the bench becomes a terrace, and a new wave-cut bench forms.

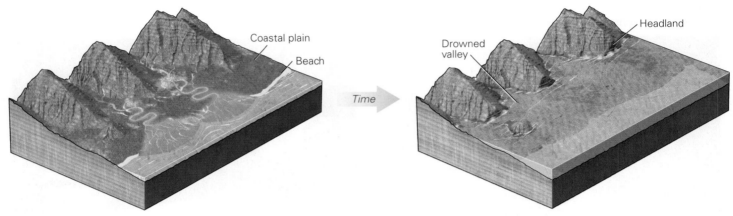

(b) If the land surface sinks relative to sea level, valleys are flooded and waves erode headlands.

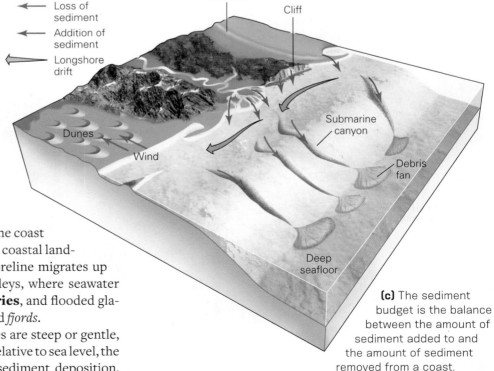

(c) The sediment budget is the balance between the amount of sediment added to and the amount of sediment removed from a coast.

The topography of a coast depends on its climate and its geologic character. For example, where mountains border the ocean, coasts have steep slopes, and where the ocean laps onto a coastal plain, the land rises in gentle slopes. Coastal landscapes can change over time due to changes in the elevation of land relative to sea level. In areas where land bordering the sea is rising relative to sea level, **wave-cut benches**, platforms carved by wave erosion, become uplifted and exposed (**FIG. 13.15a**). At places where the coast is sinking relative to sea level, however, coastal landforms become submerged and the shoreline migrates up valleys (**FIG. 13.15b**). Flooded river valleys, where seawater mixes with river water, become **estuaries**, and flooded glacially carved valleys become steep-sided *fjords*.

Regardless of whether coastal slopes are steep or gentle, or whether the land is rising or sinking relative to sea level, the character of a coast is influenced by sediment deposition,

476 • CHAPTER 13 • Savage Seas: Hazards of the Shore

mass wasting, **wave erosion** (the breaking up and carrying away of rock or sediment by the waves coming ashore), and in some cases, wind erosion. Notably, the competition among these processes determines the **sediment budget** of a coast, meaning the difference between the amount of sediment being added and the amount being removed **(FIG. 13.15c)**. Sediment may be brought in from deeper water by waves or carried from inland to the shore by streams or wind. Similarly, sediment can be removed by wind or waves, or by turbidity currents that carry it out into deeper water. *Sedimentary coasts* receive more sediment than they lose, so that a broad ribbon of sediment builds out from the coast. *Rocky coasts* lose more sediment than can be supplied, so as sediment gets stripped away, the shore becomes rocky. *Organic coasts* develop where living organisms cover near-shore land. Since different types of coasts host different hazards, let's look at the landscape characteristics of these coast types more closely.

Sedimentary Coasts

For millions of vacationers, holidays may include a trip to a **beach**, a gently sloping fringe of unconsolidated sand (commonly composed of quartz grains or shell fragments), pebbles, or cobbles that extends from a few meters above high tide to several meters below low tide along a sedimentary coast **(FIG. 13.16a)**. Beach sediments don't contain silt and clay because waves winnow out those finer clasts and carry them

FIGURE 13.16 Characteristics of beaches.

(a) The grain size of sediment on a beach depends on the source of the sediment, the intensity of wave action, and time.

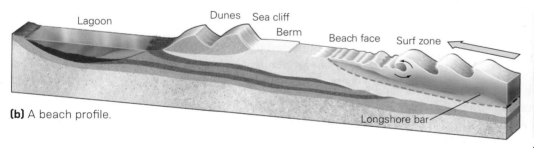

(b) A beach profile.

(c) A beach face and berm on a Cape Cod beach.

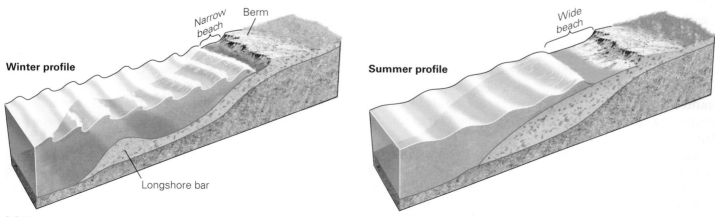

(d) The contrast between a winter beach and a summer beach in a temperate climate. In winter, when waves are stronger, sand moves offshore to build a longshore bar. In summer, the milder waves bring sand back to the beach.

offshore, where they settle in quieter water. The band of unvegetated and unconsolidated sediment that defines a beach exists because waves move sediment frequently enough that vegetation can't take root. Turbulent water in the surf zone, and in swash, picks up sediment and carries it onto and up a beach. As the swash slows, this sediment settles out. Then, as backwash starts streaming down the beach, it picks up sediment and carries it seaward, where it may be picked up by the next wave.

Erosional and depositional processes produce the distinct zones delineated on a **beach profile**, an idealized sketch illustrating how the surface of a beach changes from the shoreline inland (FIG. 13.16b). The portion closest to the surf zone is the *beach face*, the region in which swash and backwash operate. Upslope, the beach face ends at a small step defining the seaward edge of the *berm*, a region of sand that moves only during storms or very high tides (FIG. 13.16c). The berm terminates landward at the base of a belt of *sand dunes* or at a **sea cliff**, a steep wall of sediment or rock. Dunes may separate the beach from a **lagoon**, a pool of relatively quiet water separated from the ocean by a narrow strip of land or by a **reef**, a ridge of coral, rock, or sediment lying just below the water surface. In some locations, wave action carries enough sand offshore to build a submerged, elongate mound of sand, called a **longshore bar**, that is oriented parallel to the shore.

The location where erosional and depositional processes take place on a beach at a given time depends on the intensity of the waves. Large waves tend to shift sand offshore, while smaller waves tend to bring it up onto the beach. For that reason, beach profiles can change dramatically during a storm. Notably, in regions with temperate climates, beach profiles vary seasonally, for storms tend to be stronger in the winter (FIG. 13.16d).

Where waves roll onto a beach at an angle, grains of sand carried by water follow a blunted sawtooth pattern that results in a gradual net transport of the sediment parallel to the beach. Such movement, called **longshore drift**, happens because the swash of a wave moves perpendicular to the wave crest, so an oblique wave carries sediment diagonally up the beach (FIG. 13.17a). Backwash, however, flows straight down the slope of the beach due to gravity. Over time, longshore drift may move sand tens to hundreds of kilometers along a shore. Where the coastline indents landward, longshore drift can stretch a beach out into open water and produce a **sand spit** (FIG. 13.17b). A sand spit may eventually block the entire entrance to a bay or estuary, becoming a *baymouth bar* (FIG. 13.17c).

In some locations, long, narrow ridges of sand emerge above sea level by 1–2 m (3–6 ft) at a distance of a few hundred meters to as far as 30 km (19 mi) offshore (FIG. 13.17d, e). These **barrier islands** are popular with developers who want to build vacation homes or resorts. Some barrier islands form when a longshore bar accumulates enough sand to rise above sea level. Others form where a sand ridge that began as a beach or sand spit became isolated as relative sea level rose, and a lagoon or estuary formed and widened between the sand ridge and the shore.

Tidal flats—broad, nearly horizontal surfaces underlain by a mixture of mud, silt, sand, and organic debris—develop in lagoons and estuaries that receive a supply of fine clastic sediment and are protected from waves (FIG. 13.18a). At high tide, seawater submerges a tidal flat entirely, while at low tide, its mucky surface can extend for tens to thousands of meters out from the shoreline (FIG. 13.18b).

Rocky Coasts

More than one ship has met its end smashed and splintered in the spray and thunderous surf of a **rocky coast**, where bedrock sea cliffs border the water (FIG. 13.19a). Lacking the protection of a beach, rocky coasts feel the full impact of breakers and the wave erosion that they cause (FIG. 13.19b). The dynamic pressure generated by the impact of the moving water in a breaker can pick up boulders and smash them into bedrock, causing rock to shatter, and it can compress air in cracks with enough force to pry chunks free from bedrock. In addition, the turbulent water in waves washing onto a rocky shore carries suspended sand that can abrade rock.

Rocky coasts typically have an irregular shoreline, with **headlands** protruding into the sea and **embayments** set back from the sea (FIG. 13.19c). As a result of wave refraction, wave energy focuses on headlands. The curving waves gradually erode through the sides of a headland, producing a *sea arch* connected to the mainland by a narrow bridge. Eventually, the arch collapses, leaving an isolated **sea stack** offshore (FIG. 13.19d, e).

Organic Coasts

Along **organic coasts**, living organisms influence the landforms of shore and nearshore regions. For example, on the margins of tidal flats, estuaries, and lagoons, plants capable of surviving in saltwater during high tide can take root to produce coastal wetlands. In mid-latitude climates, coastal wetlands include *swamps* (dominated by trees) and *salt marshes* (dominated by grasses) (FIG. 13.20a). In tropical or subtropical climates, *mangrove swamps* grow. A mangrove tree has a broad network of roots that look like an octopus standing on its tentacles (FIG. 13.20b). Wetlands can protect coasts from wave erosion by absorbing the impact of waves, but unfortunately, they are vulnerable to pollution and development. Sadly, about a third of coastal wetlands worldwide have been destroyed in the last century (FIG. 13.20c). Because coastal wetlands also serve as spawning grounds for marine life, their loss has significant economic implications.

FIGURE 13.17 Longshore drift and its consequences.

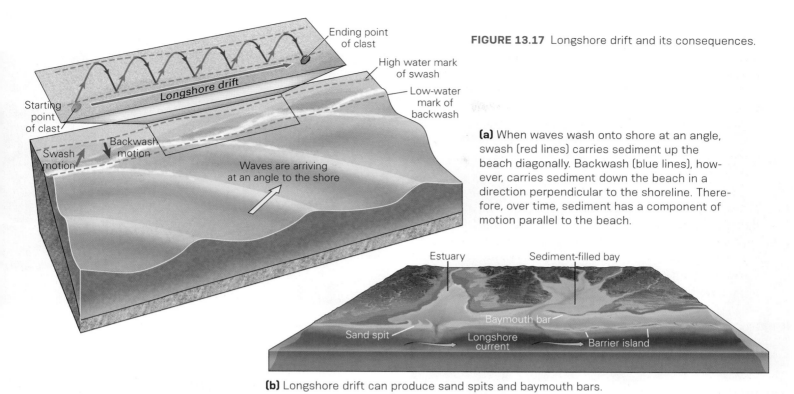

(a) When waves wash onto shore at an angle, swash (red lines) carries sediment up the beach diagonally. Backwash (blue lines), however, carries sediment down the beach in a direction perpendicular to the shoreline. Therefore, over time, sediment has a component of motion parallel to the beach.

(b) Longshore drift can produce sand spits and baymouth bars.

(c) Aerial view of a sand spit that is growing to become a baymouth bar.

(d) Barrier islands far off the coast of North Carolina, as viewed from space.

(e) Aerial view of a small barrier island.

FIGURE 13.18 Examples of tidal flats.

(a) A broad tidal flat is exposed at low tide around Mont-Saint-Michel, on the coast of France.

(b) At low tide, boats rest in the mud of a tidal flat on the coast of Brittany.

13.4 Coastal Landforms • **479**

FIGURE 13.19 Landscape features of a rocky coast.

(a) A rocky coast in southern England.

(b) A breaker smashes into a sea cliff 40 m (130 ft) high.

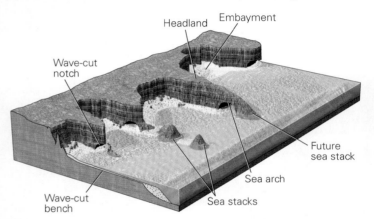

(c) The irregular shoreline of a typical rocky coast. Wave energy focused on a headland produces a sea arch. When the roof of the arch collapses, a sea stack remains.

(d) A sea arch on the southern coast of England.

(e) A sea stack off the coast of the Isle of Wight, England.

480 • CHAPTER 13 • Savage Seas: Hazards of the Shore

In warm and clear tropical waters, beautiful coral reefs develop (FIG. 13.21a, b). A **coral reef** is a mound of limestone formed by corals and other invertebrate animals (such as clams and sponges) that extract calcium and carbonate ions from seawater to grow hard mineral shells. At any given time, only the reef's surface is alive, while its interior consists of limestone made from the shells of previous generations of

FIGURE 13.20 Examples of coastal wetlands.

(a) A salt marsh along the coast of Cape Cod.

(b) Mangroves growing along the coast of southern Florida.

(c) Built areas are encroaching on this coastal wetland.

FIGURE 13.21 Examples of coral reefs.

(a) A coral reef along the shore of Puerto Rico. Note how the reef protects the shore from breakers.

(b) A living coral reef displays many colors and serves as a habitat for fish.

(c) Coral bleaching killed this coral reef.

13.4 Coastal Landforms • **481**

organisms. Some coral reefs fringe the shore; others lie offshore, separated from land by a lagoon. Living corals must remain submerged, and in many places the tops of the shallowest reefs lie just below the level of low tide. The shallow water causes waves to break on reefs and lose energy before they reach land.

Unfortunately, coral reefs are very sensitive to environmental stresses. Over half of coral reefs worldwide have been destroyed in the past three decades by pollution and the warming of seawater. Corals under severe stress expel the algae that live in their tissues, and consequently, they lose their color. The contrast between a reef that has undergone such *coral bleaching* and a healthy reef is dramatic (FIG. 13.21c). When such reefs die, they no longer protect the land from waves, resulting in increased rates of wave erosion on adjacent shores, and oceanic productivity also decreases, with major economic implications.

> **Take-home message . . .**
> The type of coast at a location depends on topography, whether the coast is rising or sinking relative to sea level, and on the sediment budget. Some coasts are rocky, and some are bordered by sediment. Wave erosion can significantly modify a coast. Organic coasts, where wetlands and reefs protect the shore from erosion, are under threat.
>
> **QUICK QUESTION** Why aren't beaches covered with grass?

13.5 Hazards of Currents, Tides, and Waves

The movement of water in the sea poses a danger both near shore and offshore, causing thousands of people to drown every year worldwide—vastly more than are killed by shark attacks (which cause about 10 deaths annually, on average). Using our background knowledge on how ocean waters move, we can examine the hazards posed by currents, tides, and waves.

Dangerous Currents

What makes a nearshore current dangerous? As a rule of thumb, if a current's water moves faster than you can easily swim, then the current is dangerous! Although competitive swimmers can go as fast as 8 km/h (5 mph), most people can't swim faster than about 3 km/h (2 mph)—less than half the speed of some currents—and can do that for only a short distance. As we look at these types of nearshore currents, let's consider ways that people can avoid being trapped by them.

- *Longshore currents:* Although wave refraction decreases the angle at which waves move close to shore, it does not eliminate that angle. In places where waves arrive at the shore obliquely, water in the nearshore region moves parallel to the shore. This motion, called a **longshore current**, causes swimmers just offshore to drift along, parallel to the shoreline (FIG. 13.22a). Such currents can become dangerous when they carry swimmers beyond the end of an accessible beach, or into a dangerous place like a rocky headland or a pier. If you find yourself moving with a longshore current, swim to shore.

- *Rip currents:* Nearshore waves incessantly push water toward the beach. An equal volume of water must flow back to the sea; otherwise, the water would accumulate on the beach. Where the return flow focuses over a low spot in a longshore bar, it can accelerate into a strong, narrow stream, called a **rip current**, that carries water out beyond the surf zone (FIG. 13.22b). The flow in rip currents, which affects water up to the sea surface, can drag unsuspecting swimmers out into deeper water. Unfortunately, when caught in a rip current, swimmers may instinctively try to swim directly to shore, against the current, and become exhausted before they make it back. It's safer to swim parallel to the beach to get out of the rip current, and then head to shore.

- *Nearshore regional or tidal currents:* In some locations, a *tidal current* (caused by incoming or outgoing tides) or a regional current causes water to flow rapidly near the shore (FIG. 13.22c). Such currents can carry swimmers or divers away from their destinations, and in the vicinity of coral reefs, they can rake swimmers against sharp coral.

- *Channel currents:* A *channel current* develops where a current flows through a relatively narrow channel of deeper water between a beach and an offshore seamount, island, or reef. Such focusing of a current causes flow to accelerate, just as water coming from a hose accelerates when you pinch the end of the hose (FIG. 13.22d). Swimmers heading to the opposite side of the channel risk getting caught in the current.

- *Structure-related currents:* The construction of piers or other structures that jut from the shore into the water can disrupt a longshore current, causing it to turn and flow parallel to the structure, toward deeper water (FIG. 13.22e). Swimmers jumping off a pier may be swept by this current into the pier's pilings (support columns) or into boats tied to the pier.

FIGURE 13.22 Examples of treacherous nearshore currents.

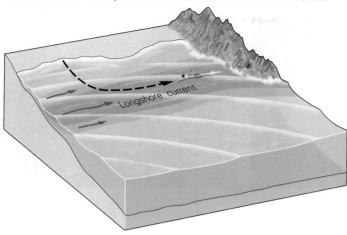

(a) A longshore current can carry a swimmer along a beach and into rocks.

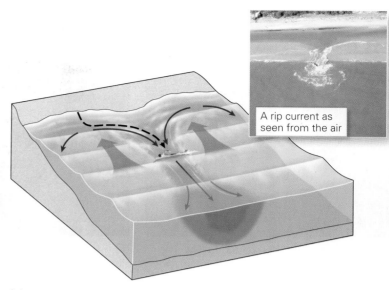

(b) A rip current can carry a swimmer out into deeper water.

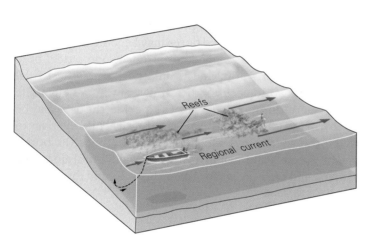

(c) A nearshore regional current, or a tidal current, can rake victims along reefs or carry them far from their boats.

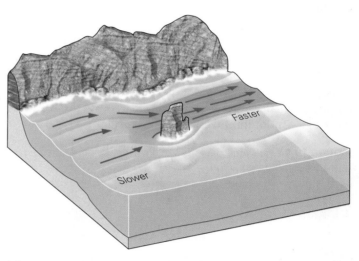

(d) A channel current forms where a broader current accelerates through a narrow passage.

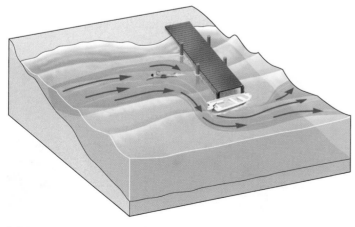

(e) A structure-related current moves along a pier and can carry a swimmer into a hazard.

Dangerous Tides

A rising tide can pose a hazard to those working or playing in the intertidal zone. People may find themselves immersed in cold seawater before they can make it back to shore, and thus become susceptible to drowning or hypothermia. Broad tidal flats are particularly dangerous because the shoreline at low tide may be more than a kilometer seaward of the shoreline at high tide. Nearly every year, shellfish hunters digging for clams in the mud of a tidal flat become victims of rising tides because they can't slog through the mud fast enough to reach dry land. Vehicles parked on tidal flats can become trapped (FIG. 13.23a).

Tidal range varies from day to day, and from month to month, at a given location because of the varying distances

FIGURE 13.23 Hazards associated with tides.

(a) A car stranded on a tidal flat. The tide came in while the driver was hiking in the dunes.

(b) A tidal bore moving up the Qiantang River estuary.

(c) A particularly high bore in 2016 washed over seawalls bordering the estuary and slammed into a crowd of spectators.

of the Sun and Moon from the Earth. On some days, a combination of factors will lead to an unusually high tide, called a **king tide**. Especially high king tides, known as **nuisance tides** (formally defined as tides during which water rises more than 50 cm, or 20 in, above the normal high-tide elevation), may inundate coastal property and streets with corrosive saltwater. The corrosion of concrete and steel by saltwater may have played a role in the 2021 collapse of a condominium complex in Florida.

FIGURE 13.24 Maritime disasters claim lives, ships, and sometimes valuable cargo.

(a) A painting depicting the sinking of SS *Central America*.

(b) If a ship sails onto a large wave, buoyancy force pushes up its middle while gravity pulls its ends down. These forces can cause the ship's hull to break.

In places where the tide moves against a river current in an estuary, or where it funnels up a progressively narrowing shallow bay, it can produce a **tidal bore**, a visible wall of water ranging from a few centimeters to a few meters high (**FIG. 13.23b**). In places, tidal bores may become large enough for surfers to ride—a skilled surfer once rode a tidal bore continuously for 1 hour and 10 minutes! The largest tidal bores in the world occur in the estuary of China's Qiantang River, where, during the arrival of a king tide, the bore may be up to 9 m (30 ft) high, and crowds of tourists line the river to watch the spectacle. Unfortunately, sometimes the bore's height exceeds the height of the seawalls built along the estuary, and the bore washes over the walls and into spectators (**FIG. 13.23c**).

Hazards of Waves

On September 3, 1857, the steamer SS *Central America* encountered hurricane winds off the coast of South Carolina. Towering waves crashed over the deck, leaked into the boiler room, and shut down the engine and the bilge pumps. Unable to navigate, the ship eventually broke apart and sank,

(c) MV *Prestige*, an oil tanker that snapped in the middle and eventually sank.

FIGURE 13.25 Rogue waves in the open ocean.

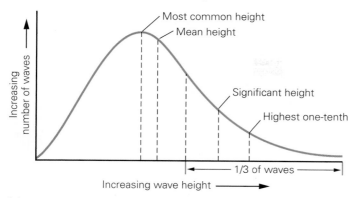

(a) This graph illustrates the statistical distribution of the heights of waves that occur at a location during a given time interval. Significant waves occur much less frequently than waves of the most common height. Rogue waves are even rarer.

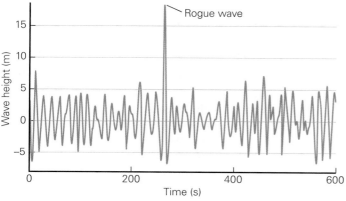

(b) A recording of waves in the North Sea over a 10-minute period. The significant wave height was about 5 m. The rogue wave was 18 m high.

(c) A rogue wave washing over the deck of a large ship.

causing the deaths of 425 people, the worst loss of life in a shipwreck until that time **(FIG. 13.24a)**. The sinking had global consequences as well, for the *Central America* was carrying over a half billion dollars' worth of gold (at today's prices) to replenish reserves held by banks in New York. The loss of this supply contributed to the start of a global economic downturn known as the Panic of 1857. This story of a **maritime disaster**—the loss of a ship at sea—emphasizes that towering storm waves can cause ships to fill with water, capsize, or suffer severe structural damage. In extreme cases, a ship may even snap in two **(FIG. 13.24b, c)**.

Waves passing a location over a given time interval vary in height. A graph plotting the number of waves on the vertical axis and the wave height on the horizontal axis yields a lopsided bell-shaped curve. The mean height of the highest third of the waves on this curve is called the **significant wave height** for the period of observation **(FIG. 13.25a, b)**. Waves that are more than twice the significant wave height are called **rogue waves**. Statistical studies show that, for a significant wave height of 10 m (33 ft), 1 wave in 1,000 will be a rogue as high as 19 m (62 ft). By one estimate, every major ship will encounter a rogue wave at least once during its 25-year life span. Rogue waves are probably caused by constructive interference, by local interaction between waves and strong currents, or by the interaction of waves with bathymetric features on the seafloor. While rogue waves occur throughout the world, they are notably more frequent in some localities. One such region is the ocean south of South Africa, where currents from the Atlantic collide with those from the Indian Ocean.

Encountering a rogue wave at sea can be terrifying, even for people on a large ship or a giant offshore drilling platform **(FIG. 13.25c)**. Between 2006 and 2010, at least 10 rogue waves encountered ocean liners. During one encounter, a wave 34 m (112 ft) high smashed through windows in a lounge high above the deck and killed two passengers. Over the centuries, many of the thousands of ships lost at sea without a trace may have been sent to the bottom by rogue waves.

Maritime disasters causing hundreds to thousands of casualties have happened when ships inadvertently sailed too close to shore or were blown to shore by strong winds **(FIG. 13.26a)**. In shallow water, not only are waves taller, but a ship can strike bottom, tearing a breach in the hull, or can be battered against a beach or rocky shoreline. Offshore bars or reefs can be particularly hazardous because their locations aren't well known and can change over time. Not surprisingly, the region of shifting sands off Cape Hatteras, North Carolina, where over 600 notable ships have foundered, has been called the Graveyard of the Atlantic.

For shore-based fishermen or swimmers, the surf zone represents the greatest hazard, because saltwater in a wave buoys a person up, while the moving wave face produces sufficient dynamic pressure to knock a person over. Even a medium-sized wave can, therefore, toss a person head over heels. If this tumbling takes place over soft sand, the person may get water up their nose, sand in their hair, and a few scrapes, but is likely to survive. Sometimes, though, a wave can slam victims into the sand hard enough to break bones. And someone who tumbles over a wave-cut bench or a jagged coral reef can be injured severely.

FIGURE 13.26 The hazard of waves near shore.

(a) The tanker MT *Phoenix* was pushed ashore and battered by waves off South Africa in 2011.

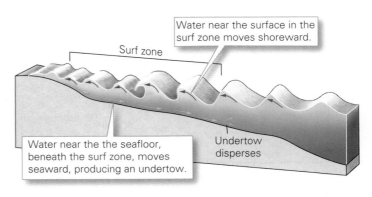

(b) In the surf zone, waves move water toward the shore. This water returns to the sea beneath the waves as an undertow, which can knock beachgoers off their feet. Undertow dies out at the seaward edge of the surf zone.

off piers, tourists off scenic rocky promontories, and sunbathers off beaches, causing dozens of fatalities over the years. Because wave height increases in shallow water, some shoreline rogue waves dwarf those encountered at sea. A wave that struck a lighthouse on the coast of Ireland in 1985, for example, was 47 m (154 ft) high. A recent study shows that rogue waves have tossed boulders weighing over 100 tons as far as 200 m (656 ft) inland along the west coast of Ireland.

Waves are a significant hazard to coastal communities built on low-lying land close to the shore. While normal waves stay within the confines of the beach and do not reach buildings inland of the beach, storm waves can carry water beyond the inland edge of a beach. This problem is worse during high tides, or when waves come in on top of storm surge. At such times, waves can completely wash over beaches, barrier islands, or sand spits. The dynamic pressure of waves and the debris they carry can smash buildings apart (FIG. 13.26c); the buoyancy of buildings in seawater can cause them to float off their foundations; and sediment carried by waves can bury property.

(c) Wave-battered houses along a beach in New Jersey after Hurricane Sandy.

As we've seen, water carried onto a beach by waves must return seaward. The return flow along the seafloor beneath the surf zone is known as **undertow** (FIG. 13.26b). It dies out at the seaward edge of the surf zone. Undertow plays a role in many nearshore drownings at beaches when the seaward-flowing water drags people off their feet and carries them outward. If the beach slope is steep, victims may suddenly find themselves in deeper water and stronger surf, be unable to regain their balance, and drown. Undertows and rip currents are related—a rip current forms in association with a particularly strong, focused undertow.

Rogue waves also pose a hazard for people standing or swimming near the shoreline—they have washed fishermen

Take-home message . . .

The movement of ocean waters in currents, waves, and tides can be dangerous to people and property along the shoreline. Nearshore swimmers face the hazards of longshore currents, rip currents, and other currents that move faster than a person can swim, as well as the hazard of undertow. Rogue waves can appear unexpectedly and cause casualties. Extreme tidal bores and nuisance tides can also be a hazard.

QUICK QUESTION How can large waves cause marine disasters?

13.6 Coastal Challenges

Numerous phenomena—both physical (such as nuisance tides, mass wasting, and wave erosion) and biological (such as algal blooms)—can cause casualties and property damage in coastal regions. Many of these problems are increasing as a result of sea-level change. Here, we examine these phenomena and describe possible responses to some of them.

Nuisance Tides

As we will discuss further in Chapter 15, global climate warming during the past two centuries has led to a rise in sea-surface temperature, and as seawater warms, it expands. This expansion, along with the addition of glacial meltwater, is causing global sea level to rise at a rate of about 32 cm (13 in) per century. Many of the world's large cities lie on coastal land less than a few meters above sea level, so sea-level rise is a stealth disaster that is beginning to disrupt society. In many locations, the first manifestation of problematic sea-level rise has come in the form of nuisance tides. High-tide elevations creep upward, year by year (FIG. 13.27a). In the United States, these changes are a bigger problem along the east and Gulf Coasts, which have gentle slopes, than they are along the steeper Pacific coast.

Until the past two decades, nuisance tides were relatively rare, happening only in association with storm surge. But along the east coast and Gulf Coast, nuisance tides now happen with increasing frequency, even on sunny days in the absence of storm surge (FIG. 13.27b). In fact, according to some estimates, the frequency of nuisance flooding is now 300%–500% greater than it was a half century ago. At places along the coast of Texas, nuisance flooding happens between 18 and 64 days each year. Nuisance tides, also known as *sunny-day floods*, have also become particularly problematic in Florida, South Carolina, Virginia, and Maryland (FIG. 13.27c). While not deadly in themselves, they have the potential to cause catastrophic property losses, because saltwater is corrosive to concrete and metal and deadly to plants. Notably, similar phenomena affect coastal areas of the Great Lakes, due not to tides but rather to lake levels rising from increased rainfall. Chicago has struggled to deal with runoff from rains and storm surge. Currently, it sends excess water into abandoned tunnels and quarries for temporary storage.

In coastal areas where bedrock consists of permeable limestone, building seawalls might not keep the saltwater out, for it can pass through the rock and bubble up onto the ground surface in springs. Those in affected communities must invest in pumps, sewers, and raised roadbeds, and at some point, they will need to raise their building foundations. Eventually, when sea level rises high enough, coastal areas will be submerged not just during nuisance tides, but permanently. By one estimate, 15% of the oceanic islands in the Pacific are already suffering from nuisance tides that keep the land submerged for a large part of the year (BOX 13.1). Chapter 15 discusses additional consequences of sea-level rise.

FIGURE 13.27 Effects of rising sea level on nuisance tides.

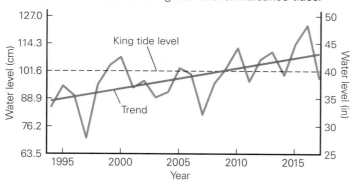

(a) The highest water elevation reached during the highest tide of the year has followed an upward trend over the last few decades. If a king tide becomes a nuisance tide at a particular elevation—say, 40 inches—then the frequency of nuisance tides is increasing.

(b) A nuisance flood swamps a nearshore road.

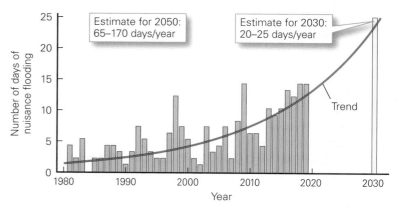

(c) The change in the number of nuisance tides recorded at Norfolk, Virginia, since 1980. Also shown is the projection for 2030. The number of flood days per year appears to be increasing exponentially, and in 2050, could be 65–170.

BOX 13.1 DISASTERS AND SOCIETY

Nuisance tides on Pacific islands

The province of Bohol in the Philippines includes several small islands, some only tiny strips of land rising no more than 3 m (10 ft) above sea level. These islands, surrounded by reefs, host small fishing communities whose residents live along the shoreline. In 2013, an M_W 7.2 earthquake violently shook the Bohol region. Fault movement caused some islands to undergo subsidence (see Chapters 3 and 7). Due to this subsidence, islands that had been completely above the high-tide elevation during spring tides were no longer so fortunate, and this change has disrupted the islands' populations forever. In some places, nuisance tides rise as much as 40 cm (16 in) above street level and submerge the floor levels of homes and schools for as long as 4 hours a day, on about 135 days over the course of a year **(FIG. Bx13.1a)**. When this happens, sewage seeps into the water and garbage floats in it, and homes, soaked for so much of the time, grow mildew. Beautiful stands of coconut palms that once lined the shores of these islands have died, for the trees can't survive long-term submergence in saltwater.

Sea-level rise will inevitably make nuisance flooding in Bohol Province even worse in the future. How have the affected communities responded to the flooding? In recent years, the national government offered to evacuate residents and move them to the high, dry slopes of larger islands. But most residents refused to move; subsistence fishing from their island homes was the only way of life they knew. Instead, remaining residents have adapted, some by raising their houses on stilts (which keep the houses out of the water, but make them more vulnerable to typhoon winds), and others by raising the foundations of their homes using limestone blocks that they break off the reef. Some homeowners who couldn't afford such changes simply extended the legs of their furniture so they could sit and sleep above the water, even when it covers their floors, and they built storage lofts in the rafters of their homes to keep clothing and food dry.

The Marshall Islands, east of the Philippines, are the eroded remnants of long-extinct hot-spot volcanoes. These islands are less than a few meters above sea level, so nuisance tides already submerge parts of the islands, making dry land scarce. In fact, on one island, the airport runway also serves as a community park, for it's the only land to be consistently above water. Wave erosion is eating away at the islands, despite sporadic sandbagging efforts, making the land areas a bit smaller every year. When high waves accompany nuisance tides, saltwater washes over parts of the islands. This salty flooding has ruined the soil for growing

FIGURE Bx13.1 Submergence of Pacific islands due to rising sea level.

(a) A nuisance tide floods a Philippine community.

(b) Seawater overwash swamps a community in the Marshall Islands.

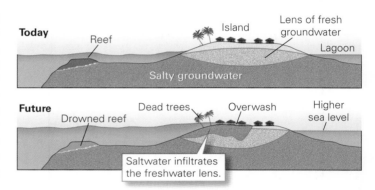

(c) Cross section illustrating how overwash damages the groundwater supply.

crops. It has also threatened and, in some cases, already destroyed groundwater supplies. Why? In the past, the islands were high enough that a lens of fresh groundwater, fed by infiltrating rainwater, accumulated beneath dry land. During nuisance tides, saltwater spreads over the islands **(FIG. Bx13.1b, c)** and infiltrates the ground. The saltwater sinks downward and displaces the freshwater, causing the lens of fresh groundwater to shrink. To replace groundwater supplies, residents now must trap rain in barrels.

These Pacific islands, in effect, provide a window into the future for mainland population centers on low-lying coastal land. First comes an occasional nuisance tide. Then, nuisance tides become a way of life. Eventually, the location becomes uninhabitable. Society then faces a stark choice: relocate or accommodate. On Pacific islands, some people make the first choice, which often involves painfully breaking family and cultural ties and learning new livelihoods—about 20% of the population has already emigrated. Others make the second choice, which means building defenses against the sea, identifying new sources of freshwater, and perhaps having a less comfortable existence. Time will tell whether those who live in large, low-lying urban areas on the mainland will choose to build expensive defenses (seawalls and pumps, for example) or prefer to relocate.

Sea-Cliff Retreat

Wave erosion can undercut sea cliffs, as we saw in Chapter 5. If the cliff consists of sediment, this erosion removes the toe of the slope. This loss of sediment decreases the support that had stabilized the upper part of the cliff and eventually causes slope failure and mass wasting **(FIG. 13.28a)**. If the cliff consists of rock, wave erosion may produce a **wave-cut notch (FIG. 13.28b)**, which widens until the overhanging rock becomes unstable, breaks away along a joint, and collapses into the surf zone, where it breaks up and gets carried away. Over time, successive slope failures cause the position of a sea cliff to migrate landward, a process called **sea-cliff retreat**, and as this happens, the surf zone shifts landward. In places where wave erosion cuts into bedrock, sea-cliff retreat leaves behind a wave-cut bench **(FIG. 13.28c)**.

Every year, property worth millions of dollars (buildings, land, and highways) that was built above sea cliffs tumbles down to the shore during the mass-wasting events associated with sea-cliff retreat **(FIG. 13.28d)**. If people in susceptible areas did not evacuate, these events would result in casualties. Sea-cliff retreat tends to be episodic, but averaged over many years, it may proceed at rates of up to 2 m (7 ft) per year, accelerated in some cases by sea-level rise. In response, homeowners along some coasts have lifted their houses and physically moved them landward to avoid ending up in the surf zone.

FIGURE 13.28 Hazards associated with sea-cliff retreat along coasts.

(a) A sea cliff made of weak sediment underwent slumping when its toe was undermined by wave erosion.

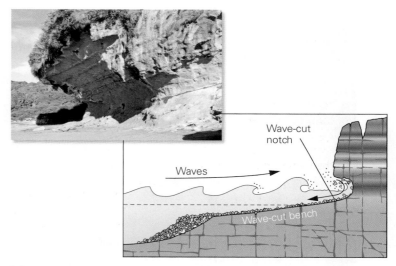

(b) Formation of a wave-cut notch. The overhanging rock can break along a joint and fall onto the beach below.

(c) Eventually, sea-cliff retreat produces a wave-cut bench.

(d) Houses collapsing into the Pacific due to sea-cliff retreat along the coast of California.

FIGURE 13.29 Examples of beach erosion.

(a) A wide beach on Cape Cod before erosion (left), and a close-up of the same beach after erosion (right). Now the shoreline lies right at the base of the sand cliff.

(b) Weakly cemented sand on the east coast of Florida is collapsing into the sea, making the beach narrower.

(c) A Florida shore in 2014 (top), and in 2016 (bottom) after Hurricane Matthew had removed most of its beach.

Beach Erosion

As we noted earlier, the width of a beach reflects the sediment budget of a coast: if the sediment supply decreases, or the sediment removal rate increases, **beach erosion**—a reduction in beach width and thickness—takes place, and the beach becomes a narrower and stonier skeleton of its former self (FIG. 13.29a, b). Sediment supply may decrease due to the loss of sediment input from a nearby river, or due to an interruption of longshore drift. Sediment removal may increase due to the loss of vegetation, such as beach grasses or shrubs, that had held sediment in place, or due to the loss of an offshore reef or mangrove swamp that absorbed wave energy. Storms greatly increase the rate of

FIGURE 13.30 The Cape Hatteras Lighthouse was moved inland in 1999, at great expense, to preserve it from beach erosion.

sediment removal, as the backwash of large waves sweeps vast quantities of sand seaward and into offshore longshore bars (FIG. 13.29c).

As sea level rises, beach erosion in susceptible locations can progress at an average rate of up to 2 m (7 ft) per year, so structures originally built well inland of a beach may eventually end up in the surf zone, unless they can be moved. For example, when built in 1870, the historic Cape Hatteras Lighthouse sat about 1.5 km (about a mile) inland from the shoreline of a barrier island in North Carolina. By 1970, it was only 40 m (130 ft) from the shoreline. Efforts to stabilize the beach prevented the lighthouse from being washed away for another two decades. Finally, however, in 1999, it was hoisted onto a trailer and moved 880 m (2,887 ft) inland to remove it from the surf zone (FIG. 13.30).

Protecting Coastal Property

Different types of coasts have different vulnerability to hazards. To represent these contrasts, the USGS developed a *coastal vulnerability index*, which takes into account tidal range, wave height range, the slope of the coastal landscape, the historical record of erosion or sedimentation, and local rates of sea-level change. Low-risk coasts may remain unchanged for millennia or more because they are composed of resistant rocks, have gentle slopes, or are protected from large waves. High-risk coasts may change drastically in a single day due to mass wasting or wave erosion. Because of the value of coastal land, people have developed many approaches to decrease the land's vulnerability, spending billions of dollars on such efforts.

HARD PROTECTION. Construction of **hard barriers**—structures composed of *riprap* (piles of boulders or blocks of concrete), reinforced solid concrete, steel, or stone—can protect a stretch of coast or a harbor from waves. But these barriers alter the natural sediment budget, and they can therefore drive changes in the shape of a beach, sometimes with undesirable results.

- *Groins*: **Groins** are concrete or stone walls that run from the shore outward at a high angle to the beach in order prevent longshore drift from removing sand (FIG. 13.31a). After construction, sand accumulates on the updrift side of the groin, forming a triangular wedge. But sand tends to erode from the beach on the downdrift side.

- *Jetties*: A **jetty** resembles a groin, but unlike groins, jetties are typically built to shelter a harbor, or along a river channel at the river's mouth (FIG. 13.31b). River-mouth jetties effectively extend the river channel into deeper water, and may lead to the deposition of an offshore sandbar at the entrance to the channel.

FIGURE 13.31 Hard protection of shores.

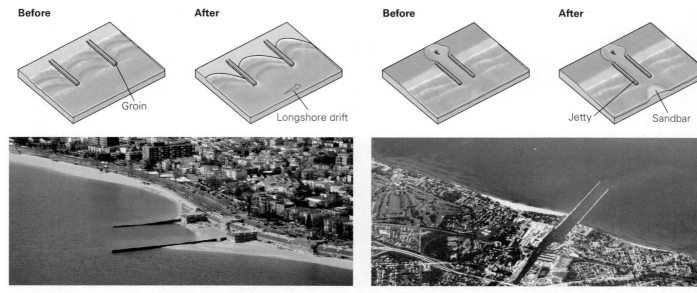

(a) Groins are intended to prevent sand from being carried away by longshore drift.

(b) Jetties extend a river channel, but they may produce a sandbar offshore.

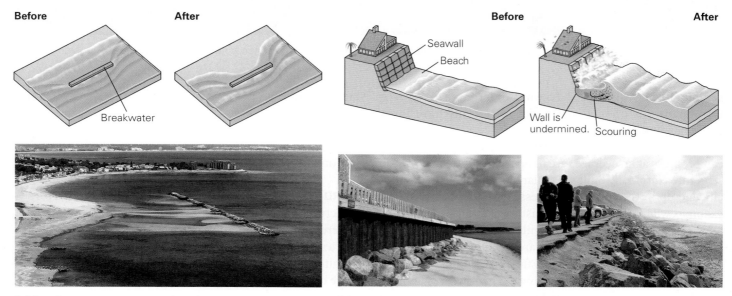

(c) Breakwaters can cause a beach to build outward toward the breakwater.

(d) Seawalls made of stone, steel, or concrete can protect the shore from waves, but in extreme storms, they can be undermined and destroyed.

- *Breakwaters*: An offshore wall oriented parallel to a beach is called a **breakwater**. Such structures prevent the full force of waves from reaching the beach (**FIG. 13.31c**). With time, sand builds up on the landward side of the breakwater, and the beach grows seaward.

- *Seawalls*: A **seawall**, unlike a breakwater, is built along the shoreline, or at the landward edge of a beach (**FIG. 13.31d**). While seawalls can slow erosion and sea-cliff retreat, they may reflect wave energy, causing constructive interference with incoming waves and, therefore, an increase in the height of the waves that impact them. As a result, they can be destroyed by extreme storms.

- *Storm-surge barriers*: A **storm-surge barrier** is a concrete or steel wall that remains open most of the time, but can be closed to keep out storm surge or king tides (**FIG. 13.32**). Such structures in the Netherlands, in the Thames River near London, and near New Orleans, Louisiana, have successfully prevented property damage, but they can cause undesired changes to the natural ecosystems of protected areas.

FIGURE 13.32 A storm-surge barrier at the mouth of an estuary in the Netherlands.

FIGURE 13.33 Soft methods of reversing or slowing beach erosion.

SOFT PROTECTION. Soft protection methods can reverse or slow beach erosion without the use of stone, concrete, or steel. Examples include beach nourishment and beach stabilization.

Beach nourishment is the process of replacing lost sand along a stretch of beach. The new sand may be brought in by trucks or barges from nearby dunes or deltas, or sucked up by dredges from longshore bars offshore and piped onto the beach **(FIG. 13.33a)**. In some places, contractors import sand from other countries. Once piled up on the beach, the sand is sculpted by bulldozers to reproduce a normal beach profile. Locations that rely on beach tourism for income have invested heavily in beach nourishment. For example, the city of Miami Beach, Florida, has used imported sand to widen the beach in front of its famous hotels by about 200 m **(FIG. 13.33b)**. Similar efforts—called *beach nourishment episodes*—have taken place

(a) Sand may be pumped onto a beach through a pipe from a dredge offshore.

(b) Before (left) and after (right) a beach nourishment episode in Miami Beach, Florida, in 2017.

(c) Laborers mine sand by hand on a beach in Sierra Leone. Such mining, like natural beach erosion, can destroy habitats and remove protection for areas inland.

13.6 Coastal Challenges

BOX 13.2 DISASTERS AND SOCIETY

Beach restoration following Hurricane Sandy

When Hurricane Sandy raked along the east coast of the United States during late October 2012, it devastated the shores of New Jersey and southeastern New York, damaging or destroying 650,000 homes, along with the region's infrastructure. The effects of the storm on coastal landscapes were dramatic. As the storm surge engulfed the shore, wave erosion not only flattened beachside structures, but also stripped away large areas of the beaches themselves, carrying some of the sand offshore and some of it inland. Indeed, some beaches lost as much as half their sand volume. Churning waters lifted sand from beach dunes and dumped it on the streets of beachside communities **(FIG. Bx13.2a)**, breached barrier islands, and carved new inlets **(FIG. Bx13.2b)**. In areas where dunes did not shield communities from the sea, storm-surge waters flooded homes. Floodwaters inundated wetlands to such depths that they couldn't drain during low tides, resulting in ecosystem destruction.

The catastrophe caused by Hurricane Sandy triggered immediate responses by emergency personnel to help with casualties, to clear streets and then homes of debris, and to re-establish infrastructure. Longer-term recovery and restoration, aided by financial resources made available by the US Congress through the Hurricane Sandy Relief Bill, involved collaborations among governments at the federal, state, and local levels, and others. The partners worked to develop a plan for the future that would provide the affected communities with greater resilience, so that they would be better able to respond to storms quickly and recover easily. Ideally, this plan would lead to more realistic risk assessment and management; numerous hazard mitigation efforts, such as estab-

FIGURE Bx13.2 Beach restoration efforts after Hurricane Sandy.

(a) Storm surge picked up dune sand and deposited it over streets.

(b) Storm surge and waves breached barrier islands. This breach is at Mantoloking Beach, New Jersey.

(c) A broad beach was built out where the breach had been.

ter during the storm than had those without such features, for storm waves weakened as they crossed the beaches, and storm surge did not overtop the dunes. This observation led them to conclude that beach nourishment and dune construction were, by and large, useful in decreasing the risk of inundation, at least in the short term. Consequently, the Army Corps of Engineers, along with private contractors, set to work to pump sand to the beaches from offshore bars.

Over the next few years, workers filled in breaches in barrier islands and baymouth bars, broadened and elevated the surfaces of beaches, and built new dunes **(FIG. Bx13.2c)**. As the efforts were underway, some beachside parks installed sand fencing, erosion-control mats, and seawalls made of sandbags or concrete. This work was impressively expensive: restoration of the beach at New York's Rockaway Park alone cost $140 million. Other erosion control efforts took place as well. For example, in appropriate locations, piles of shell debris were dumped in the water just offshore to stimulate the growth of oyster reefs, which can act as natural breakwaters. And to make wetlands more resilient, drainage channels were dredged so that floodwaters could escape more quickly.

Of course, the restoration efforts following Hurricane Sandy stimulated plenty of controversy, for beach nourishment, viewed on the scale of decades, is a temporary fix. People questioned whether spending billions of dollars in order to keep the shoreline in its present position, together with the expense of raising home foundations or placing houses on stilts, was the best public policy. Some communities instead considered *managed retreat*, meaning relocation of structures and streets to positions farther back from the shoreline, thereby allowing the beach to move landward naturally. Given the relentless rise of sea level, such solutions may eventually become inevitable.

lishing smarter electrical grids and better building codes; and more inclusive efforts to help the public understand storm risks and insurance limitations.

Dealing with the storm-ravaged beaches was no easy task. Ideally, such a process would feature thoughtful design of a stabilized shoreline for the future. But after Sandy, damage had to be repaired and beaches restored quickly, so planners had to move forward, realizing that some short-term decisions might later need re-evaluation. After the sand covering streets and filling basements had been removed, planners had to decide whether to undertake extensive beach nourishment efforts. To make that decision, they examined the outcomes of earlier beach nourishment projects. They found that communities bordered by broad, nourished beaches rimmed by constructed dunes had fared much bet-

at over 400 sites along the east coast of the United States. Extensive beach nourishment has also been used to restore beaches following severe storms **(BOX 13.2)**. Unfortunately, beach nourishment does not provide a permanent solution to beach erosion, for it doesn't stop the various causes of erosion. So, at many beach nourishment sites, new sand must be imported repeatedly.

On a global basis, beach nourishment efforts have become problematic. Mining sand, primarily for the production of concrete but also for beach nourishment, has become a lucrative industry, and sand is now being shipped worldwide, sometimes illegally **(FIG. 13.33c)**. Consequently, sand supplies from offshore and from riverbeds are being depleted in some regions at rates that lead to landscape destruction. In effect, there is a *sand crisis*, meaning that over the long term, the supply of sand won't keep up with demand. Sand mining operations can also have other negative consequences: dredging sand can destabilize nearshore slopes, and it can transfer toxins and clays that had settled in quieter water offshore back into nearshore water.

Beach stabilization, the process of installing temporary or living protective materials, has helped prevent the erosion of some beaches. Some efforts involve the installation of *erosion-control mats* made of biodegradable materials, *sandbag walls* (built of bags full of sand or soil), or *sand fencing* (wood-and-wire fences that trap blowing sand before it blows away) **(FIG. 13.34a, b)**. More commonly, beach stabilization involves the planting of native grasses and shrubs to hold sand in place, or marsh grasses and sea grasses to slow the escape of sediment to deeper water **(FIG. 13.34c)**.

FIGURE 13.34 Soft methods of beach stabilization.

(a) Plastic or fiber mats can hold sand in place.

(b) Planting of beach grasses can conserve sand.

(c) Sand fencing can prevent sand from blowing away.

Because of the hazards posed by beach erosion and sea-cliff erosion, many communities have started to establish zoning rules. Such rules may require construction to be set back from the shore, so that natural beach or wetland landscapes can serve as buffers between the shore and the community. They may require that nearshore houses be placed on stilts, so that when storm surge or high waves arrive, the houses remain above water **(FIG. 13.35)**.

FIGURE 13.35 A beach house on stilts may survive a king tide or storm surge.

Biological Hazards along Coasts

In addition to the many physical hazards of the shore, biological hazards sometimes pose threats (**VISUALIZING A DISASTER**, pp. 498–499). Single-celled algae, or *phytoplankton*, play an important role in marine ecosystems, for they serve as the foundation of the food chain. But sometimes conditions in coastal waters trigger **harmful algal blooms**, overwhelming growths of algae that can not only disrupt fishing industries and tourism, but in some cases, sicken people.

FIGURE 13.36 Consequences of harmful algal blooms.

(a) This satellite image shows surface concentrations of algae along the coast of Louisiana and Texas, near the mouth of the Mississippi River. The water underlying the red area will become a dead zone.

Oceanographers distinguish between two problems caused by harmful algal blooms:

- *Dead zones:* Nutrients from farm fertilizers that rivers have carried to the sea can stimulate harmful algal blooms. When the algae die, they sink to the seafloor, where bacteria feed on them. The bacteria deplete the water of its dissolved oxygen, causing many fish and shellfish to suffocate and die, so that the affected area becomes a **dead zone**. The dead zone near the mouth of the Mississippi River has ruined the livelihoods of many commercial fishermen operating in the Gulf of Mexico (**FIG. 13.36a**).

- *Red tides:* Algal blooms may include species of plankton that produce toxins. When concentrated in shellfish, these toxins can cause illness in their human consumers. Such blooms are called **red tides** because the phytoplankton that cause the problem turn the waters reddish brown (**FIG. 13.36b**).

Take-home message . . .
Society faces many challenges along coastlines. Sea-level rise is increasing the frequency of nuisance tides. Sea-cliff retreat and beach erosion along the shore can cause casualties and destroy property. To prevent such damage, people can build structures for hard protection, or they can try soft protection measures such as beach nourishment or beach stabilization. The economies and health of coastal communities can also be affected by harmful algal blooms in nearshore waters.

QUICK QUESTION How does a storm-surge barrier protect coastal property?

(b) A red tide washing onto the shore along the west coast of Mexico.

13.6 Coastal Challenges

Chapter 13 Review

Chapter Summary

- A global ocean of seawater submerges about 71% of our planet's surface in a layer with an average thickness of about 4 km (2.5 mi). Water over continental shelves is shallower.
- A coast is a belt of land bordering a body of water. A shore is the portion of a coast directly affected by interaction with water.
- Currents, definable bands of flowing water, occur at two levels in the oceans. Surface currents, driven by the wind, can be a hazard near the shore. The largest currents, called gyres, circulate water around an entire ocean basin.
- Tides, caused by the tide-generating force, result in the rise and fall of the sea surface, generally twice daily. The tidal range can vary significantly with location, as can the width of the intertidal zone.
- Wind-driven waves form due to the shear force of moving air against the water surface. They occur in wave trains, which can be characterized by wave height and wavelength. Within a wave, water follows a roughly circular path. Wave motion decreases downward, and no motion occurs below the wave base.
- Waves get higher, and their wavelength decreases, as they approach the shore until, in the surf zone, breakers form and collapse. Water runs up the slope of the shore as swash and flows back down as backwash. Waves refract when approaching the shore at an angle.
- A great variety of coastal landforms have developed on the Earth. The type of coast that develops depends on coastal topography, on whether the land is rising or sinking relative to sea level, and on the sediment budget.
- Along sedimentary coasts, broad beaches may develop. Several distinct landscape features may form on a beach, as displayed on a beach profile. Where waves arrive at an angle to the beach, longshore drift carries sand along the beach. Offshore, longshore bars and barrier islands may develop. In quiet water, broad tidal flats form.
- On rocky coasts, waves slam directly against bedrock. Wave erosion carves headlands and sea stacks.
- Along organic coasts, wetlands (including salt marshes and mangrove swamps) and coral reefs can blunt the force of incoming waves.
- Several types of nearshore currents, including longshore currents and rip currents, can be dangerous to people in the water.
- Tides can be a hazard to people trapped offshore while walking on tidal flats. Particularly high tides, called king tides, can become nuisance tides when they inundate coastal property with corrosive saltwater.
- High waves can cause maritime disasters both at sea and near the shore. Some of these disasters are due to particularly high rogue waves. Waves can also be a hazard to people and communities along the shore.
- Sea-level change is increasing the frequency of nuisance tides.
- Wave erosion can cause sea-cliff retreat, which may involve mass wasting along the shore, as well as beach erosion.
- To counter the consequences of wave erosion, coastal communities may use hard protection (groins, jetties, breakwaters, seawalls, and storm-surge barriers) and/or soft protection (beach nourishment and beach stabilization).
- Harmful algal blooms along coasts can produce dead zones and red tides, which affect the economies and health of coastal communities.

ANOTHER VIEW Waves pummel the coast of southern Portugal, causing cliff retreat and beach migration. Buildings constructed on the edges of cliffs, or along the beach, could be subject to damage by mass wasting or storm surge.

Review Questions

Blue letters correspond to the chapter's learning objectives.

1. What is a current? How do surface currents in the ocean form, and how fast do they typically flow? Distinguish between a gyre and an eddy. **(A)**
2. Why do tides form, and what is a tidal bulge? Are tidal ranges the same everywhere? Which can be higher, a spring tide or a neap tide? What factors determine the width of the intertidal zone? **(A)**
3. Describe how wind-driven waves form. What is the wavelength and the period of a wave train? Why don't wind-driven waves affect sediment on the floor of the deep ocean? **(B)**
4. How do waves change when they approach the shore? Why does this change happen? **(B)**
5. Distinguish among a coast, a shore, and a shoreline. **(C)**
6. Label the components of a beach profile as shown in the diagram. **(C)**

7. Why does longshore drift take place along some beaches? What landforms develop due to longshore drift? **(C)**
8. What's the difference between a longshore bar and a barrier island? What conditions allow a tidal flat to develop? **(C)**
9. What factors determine whether a particular coastline becomes a sedimentary coast or a rocky coast? How does wave erosion take place on a rocky coast? How do rocky coasts evolve over time? **(C)**
10. Describe various types of organic coasts. What role can an organic coast play in protecting the land from wave erosion? **(C)**
11. Contrast a longshore current with a rip current. Why do channel currents develop? What kind of current is a particular hazard to a swimmer jumping off a pier? Which type is depicted in the diagram? **(D)**

12. Why is it important to be aware of the timing of tides if you plan to explore the seaward edge of a broad tidal flat? What is a nuisance tide? **(D)**
13. Distinguish between a significant wave and a rogue wave. What phenomena may produce a rogue wave? How can rogue waves contribute to maritime disasters? **(D)**
14. Under what conditions do waves become a hazard to the structures in coastal communities, and how do such waves affect those communities? **(D)**
15. Describe some hazards that can become more problematic due to rising sea level. **(E)**
16. What is sea-cliff retreat, and why does it occur? In what way can it be a hazard to homeowners? **(E)**
17. What conditions can lead to beach erosion? What actions can people take to diminish the rate of beach erosion? Be sure to distinguish between hard protection and soft protection approaches. **(E)**
18. Explain how agricultural practices on land can contribute to triggering a harmful algal bloom offshore. Distinguish between a dead zone and a red tide. Explain how each can be a hazard to society. **(F)**

On Further Thought

19. Developers constructed two resort hotels on the north side of an east-west-trending beach. The longshore current flows toward the east. Before development, the beach was very wide and covered with beach grass. The developers pulled up the grass to expose sand. But a few years after construction, guests of the resorts started complaining that the beach had become too narrow. What had happened, and why? The owners of the resort at the east end of the beach decided to build a groin halfway between the east and west resorts. On the map below, draw what the shoreline looked like after the groin had been built. The western resort owners then sued the eastern resort owners. Why?

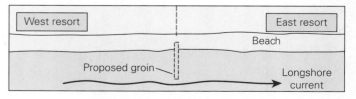

14

DRY, HOT, AND LETHAL
Drought, Heat Waves, and Wildfires

By the end of the chapter you should be able to . . .

A. explain the causes and impacts of drought around the world.

B. identify the signs of heat stress.

C. interpret the heat index reported by the media on excessively hot days.

D. identify the weather patterns associated with drought and heat waves.

E. discuss the nature of fire, how wildfires start and spread, and the role that wind plays in fire disasters.

F. assess why wildfires occur more commonly in some regions of the world than in others.

G. explain how firefighters contain wildfires, and what communities can do to limit fire risk.

14.1 Introduction

One morning in 2019, many residents of Australia woke to hear an emergency order to fill their bathtubs and other containers with drinking water, and to be prepared to evacuate. A **wildfire**—an uncontrolled blaze sweeping across the landscape—was roaring toward them. Across the country, wildfires (known as *bushfires* in Australia) were raging. The glow of flames lit the sky beyond the horizon, and a smoky haze filled the skies over thousands of kilometers (FIG. 14.1a). Many forests, scrublands, fields, and towns burned that year, with only Herculean efforts by firefighters, and a lucky turn of the winds, sparing some towns from disaster.

The Australian wildfires began in November 2019. By December, thousands of fires were burning, and it wasn't until the austral summer waned, in mid-February 2020, that the fires finally died out (FIG. 14.1b). At least 3,500 homes, as well as thousands of other buildings, were destroyed, and 34 people died. Studies estimated that the bushfires impacted a quarter of all businesses across Australia. The Australian tourism industry largely shut down, as it did again only a few months later when the COVID-19 pandemic started spreading. Insured losses due to the fires totaled approximately $1.3 billion, but the true loss of property, and the loss of income and productivity, was much greater. Environmental damage was catastrophic: 21% of the country's temperate forests were gone, including 80% of the Blue Mountains World Heritage Site (*World Heritage Sites* are areas designated by the United Nations as having special physical or cultural value). A large region in New South Wales, and half of the Gondwana Rainforests World Heritage Site in Queensland, burned. Scientists estimated that a billion animals—including a third of Australia's koala population—died in the fires.

How did these wildfires flare up to become a natural disaster in Australia? When November 2019 arrived, much of the country had been suffering through a dry period, or *drought*. In addition, the country was enduring a *heat wave*, a period during which air temperature is significantly higher than normal. In fact, 7 of the 10 hottest days on record in Australia occurred in a single week in January 2020, when nationally averaged temperatures exceeded 40°C (104°F) for 5 days in a row. Clearly, as the disaster that befell Australia in 2019-2020 demonstrates, wildfire, drought, and heat waves are closely related. The absence of rain during a drought and high air temperatures during a heat wave both dry out vegetation, priming it to burn.

◄ A firefighter trying to control one of the many wildfires to strike the western United States in 2020. Wildfires, droughts, and heat waves worsen as our planet faces increasingly hotter and drier conditions. All can cause disastrous fatalities and property damage.

Many areas of the world experience the triple threat of drought, heat, and fire. Frequently, these three conditions, individually or together, give rise to immense natural disasters. Their consequences are worst in countries with dense populations and poor infrastructure, and where ensuing famine and disease can lead to staggering numbers of fatalities. In this chapter, we examine disasters related to hot and dry conditions. Understanding the causes of these events requires an understanding of global atmospheric and oceanic circulations discussed in Chapter 8 and the hydrologic cycle discussed in Chapter 7. Unfortunately, society's vulnerability to these disasters continues to grow with the expansion of the world's population, which increases demand on water resources, and with climate change (see Chapter 15).

FIGURE 14.1 The 2019–2020 wildfires in Australia.

(a) The flames of a burning forest in southeastern Australia make the dense smoke from the fire glow.

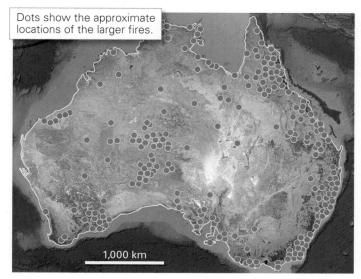

Dots show the approximate locations of the larger fires.

(b) 2019-2020 wildfires (bushfires) affected many parts of Australia, including the densely populated areas of the southeast.

FIGURE 14.2 The Dust Bowl drought of the 1930s in the west-central United States.

14.2 Drought: When the Land Dries Out

In *The Grapes of Wrath*, John Steinbeck's novel recounting the desperation of mid-American migrant farmers during the Great Depression of the 1930s, the fictional protagonist and his family abandon the parched plains of Oklahoma, where crops had failed, for the wetter and lusher agricultural areas of California. The suffering during the 1930s was at its worst in western Oklahoma and surrounding states in the southern Great Plains, a region that was experiencing a decade-long historic drought. During those years, huge dust storms blew across the plains, sweeping away topsoil in some places and burying crops in others **(FIG. 14.2)**. Dust infiltrated homes to such a depth that people had to use shovels to remove it, and

FIGURE 14.3 Types of drought.

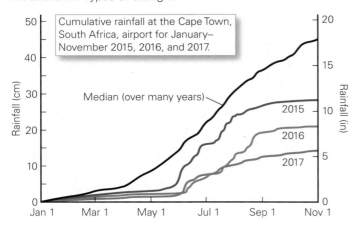

(a) Meteorological drought occurs when the cumulative rainfall over a long period is significantly less than the median value over the past several years or decades.

(b) Mudcracks may be a sign of an agricultural drought, during which soils dry out and plants wither.

(c) During a hydrological drought, water tables drop and reservoirs dry up.

(d) During a socioeconomic drought, livelihoods are destroyed and people barely subsist.

as the drought progressed, the region came to be called the *Dust Bowl* (see Box 7.3).

As bad as the 1930s Dust Bowl drought was, its consequences are dwarfed by the global impacts of other droughts. Famine and disease triggered by 19th-century droughts in southern Asia, for example, led to tens of millions of deaths. In 1928, 3 million people perished during a drought in China. Over 150,000 people died in northeastern Africa in 1983 due to a drought in Sudan. As recently as 2017–2018, drought in Somalia caused a food shortage affecting 6 million people; drought in Brazil led to a temporary cutoff of the water supply in major cities; and drought in southern Africa led several countries to declare a state of emergency. Drought is a stealth disaster that might not receive the headlines that follow major earthquakes or tsunamis. Yet it has caused more fatalities, resulted in more displaced people, and triggered more wars than any other type of natural disaster. When rains fail, food and clean water become rare, diseases spread, people migrate, and wars start over resources.

What Is a Drought?

Drought, in a general sense, refers to a deficiency of freshwater within a region lasting long enough to harm normal vegetation, crops, livestock, surface and underground water supplies, human health, and human activities. In this regard, *drought* is a relative term that implies a comparison of water availability at a given time and locality with normal water availability typical of the local climate. A deficiency of water sufficient to cause a drought in a region that normally features a wet climate will be different from a deficiency sufficient to cause a drought in a region normally hosting a dry climate.

Scientists distinguish three types of drought by the cause and manifestation of the water deficiency. A **meteorological drought** (also called a *climatological drought*) occurs when precipitation rates are less than normal for weeks to years. Meteorological droughts can be identified by comparing recorded precipitation over a specified time period with the average precipitation characteristic of the region during the same time period in other years (FIG. 14.3a). An **agricultural drought** occurs when soil moisture drops below a value that permits crop germination or growth (see Chapter 7) and remains below that value over an extended period. When this happens, the soil may turn hard and develop *mudcracks*, an array of small intersecting fissures (FIG. 14.3b). An agricultural drought may be a consequence of a meteorological drought, but it may also be caused by human activity. People may install too many drainage ditches and pipes, or they may plant crops that require more water than normal rainfall can provide. A **hydrological drought** develops when insufficient water flows into lakes, ponds, streams, or reservoirs. Surface-water levels drop, and as insufficient water sinks into the subsurface to sustain or recharge groundwater reserves, the water table drops (FIG. 14.3c). During a hydrological drought, which can be caused by a deficit in rainfall or a lack of spring snowmelt in a drainage network's headwaters, water supplies may become dangerously low as reservoirs and wells run dry. In some cases, hydrological droughts are triggered by human activities, such as diversion of rivers (see Chapter 7).

The three types of drought mentioned above can lead to a **socioeconomic drought** if water shortages reduce the supply of goods and services so that demand exceeds supply (FIG. 14.3d). For example, if a hydrological drought decreases a region's capacity to produce hydroelectric power, industrial production may be limited, and people may lose their livelihoods or food supply. A socioeconomic drought, such as the one that occurred in areas of Africa from 2011 through 2020, degrades living standards.

The duration of droughts varies significantly. Some last for less than a year, some for several years, and some for decades. Researchers refer to particularly long droughts, of two decades or more, as **megadroughts**. The drought afflicting the southwestern United States since 2000 has lasted for over 20 years, with few interruptions; it has become a megadrought. Records of past climate, as indicated by the widths of

FIGURE 14.4 The Earth's climates and agricultural zones.

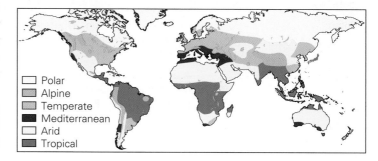

(a) The Earth hosts many different climate types, which scientists divide into climate groups. Some climate types are more susceptible to drought than others.

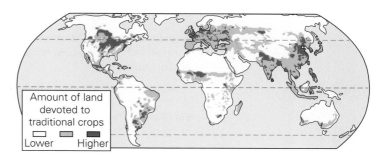

(b) The Earth's major agricultural areas occur in temperate and tropical climates, which include Mediterranean and monsoonal climate types, the most susceptible to drought.

annual tree rings (see Chapter 15), suggest that over the past two millennia, the southwest has experienced several megadroughts, some lasting as long as 50 years.

The Relation of Drought to Climate Zones

The Earth hosts several distinct *climate zones* or **climate types**. Each region is characterized by a particular range of temperatures, seasonality, and precipitation, as controlled by latitude, elevation, and proximity to the ocean, among other factors. Climate types can be identified by the vegetation that they host, because most plant species thrive only in the specific climates to which they have adapted (FIG. 14.4a). Researchers recognize several general *climate groups* (TABLE 14.1), each of which includes one or more types. The droughts with the greatest impact on society are those that develop in regions with a *monsoonal climate type* (within the tropical climate group), and a *Mediterranean climate type* (within the temperate climate group). These regions host major agricultural areas and, therefore, large populations (FIG. 14.4b). Let's examine conditions that can lead to droughts in these regions.

DROUGHT IN MONSOONAL CLIMATE ZONES. A *monsoonal climate* is a tropical climate zone in which the wet season generally features heavy rains and the dry season tends to have no rain at all. During a normal year, the rains of a monsoonal wet season saturate the soil, fill reservoirs, and flood streams (see Chapter 12). A drought happens when these rains are significantly less frequent and less intense than usual (FIG. 14.5a, b), so that during the dry season, high temperatures cause reservoirs, streams, and soil to dry out rapidly. When drought conditions develop, crops fail, water supplies become critically low, and livestock perish. The change from flood conditions to drought conditions can be very rapid in a monsoonal climate zone if weather patterns suddenly shift (FIG. 14.5c). Regions with monsoonal climates include densely populated areas of southern Asia and Africa. Droughts in these regions can lead to devastating numbers of fatalities, particularly when inadequate food-distribution networks trigger widespread famine (BOX 14.1).

Why do droughts develop in monsoonal climate zones? To address this question, recall from Chapters 8 and 12 that heavy monsoon rains fall from thunderstorms that develop along the *intertropical convergence zone* (ITCZ). This zone forms where trade winds of the northern hemisphere converge with those of the southern hemisphere, causing air to rise and form thunderstorms. The ITCZ forms along the line of maximum heating on the Earth and marks the boundary between the northern and southern Hadley cells. Over the course of a year, the position of the ITCZ migrates. It lies at latitudes south of the equator in January and at latitudes north of the equator in July (FIG. 14.6). The distribution of land and sea affects heating rates, so the ITCZ's specific latitude varies depending on the local positions of continents and oceans. A number of factors, and feedbacks among these factors, play a role in influencing the exact timing of ITCZ migration and the intensity of the rains that fall within the ITCZ. Two factors, in particular, trigger drought in monsoonal climates:

1. *Wind strength:* When trade winds are weaker than normal, the amount of moist air carried into the ITCZ, and thus available for the formation of storms, decreases. Winds may slow in years when high-altitude cloud

TABLE 14.1 Principal Climate Groups, and Selected Climate Types*

Group	Type	Characteristics
Polar		High-latitude regions where average temperature of the warmest month is ≤10°C (50°F), and snow and ice can survive all year.
	Alpine	A polar-like climate that occurs in high-elevation mountains in mid- to low-latitude regions. Average temperature of the warmest month is ≤10°C, and average temperature of the coldest month is ≤−3°C (27°F).
Arid		Receives very little rainfall over the course of the year. This group includes both hot and cold deserts.
Temperate		Average temperature of the warmest month is ≥10°C, and average temperature of the coldest month is between −3°C (27°F) and 18°C (64°F). Precipitation is distributed throughout the year.
	Mediterranean	A temperate climate characterized by wet winters, during which temperature remains above freezing, and dry summers.
Tropical		Average temperature of the coldest month is >18°C, and precipitation in the driest month is >6 cm (2.4 in). Precipitation is concentrated in a rainy season.
	Monsoonal	A tropical climate characterized by distinct wet and dry seasons.

*This table lists only climates mentioned in the text, and uses simplified terminology.

BOX 14.1 DISASTERS AND SOCIETY

Drought, famine, and infrastructure in Africa

Bradford Morse, former head of the United Nations Development Program and an advocate for aid to countries in need and to victims of famine in Africa, wrote in the foreword to the 1987 book *Drought and Hunger in Africa*, "Drought itself is not the fundamental problem in sub-Saharan Africa. After all, drought prevails in many parts of the world, and, in affluent societies, need be no more than a nuisance. The real problem in Africa is poverty—the lack of development—the seeds of which lie in Africa's colonial past and in unwise policy choices made in the early days of independence by national governments and external aid donors." Mr. Morse's statement remains true today.

In the Horn of Africa, along Africa's east-central coast, a drought began in 2011 **(FIG. Bx14.1a)**, and it continues at the time of this writing. Agricultural production deteriorated dramatically during the drought. In major crop-producing parts of the region, the driest years in decades decimated crops and livestock and contributed to the spread of famine and disease. At least 17 million people, especially children, faced *water stress* (a lack of sufficient clean water), and *food insecurity* (lack of regular access to food) as crops failed **(FIG. Bx14.1b, c)**. According to the United Nations, at least 5.5 million people became vulnerable to diseases caused by drinking contaminated water, the only kind available.

How can mass famine happen in an era when food in the developed world is so plentiful, and advanced transportation systems can move goods to any place in the world so rapidly? Morse, in his essay, laid the blame on lack of infrastructure and development in the poorest regions of Africa. He noted, for example, that when food aid arrives in port, not enough trucks are available to transport it to the backcountry regions where it is most needed. In some cases, criminal networks or corrupt officials divert supplies from people in need, or aid workers are attacked. To help solve these problems, Morse recommended efforts to promote rural development and local sustainable agriculture. Unfortunately, more than 30 years after the essay, his recommendations have been only minimally implemented, and people still endure the ravaging consequences of drought. Suffering due to drought contributes to sparking military conflicts in the region.

FIGURE Bx14.1 Drought in the Horn of Africa.

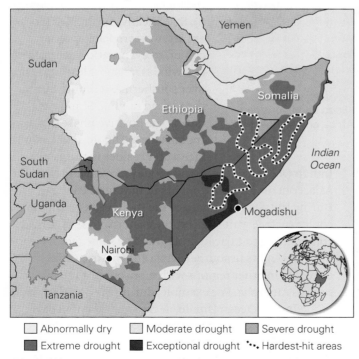

(a) Drought conditions have varied across the region. The least rain has fallen in the areas shown within the dotted lines. Human suffering has occurred within a broader area.

(b) An Ethiopian farmer inspects his fields of decimated crops.

(c) Victims of drought and food shortages include children.

FIGURE 14.5 In the monsoonal climate of India, overall average rainfall varies, but the variation is not uniform. In drought years, most of the country has a deficiency of rain, but some areas still have an excess.

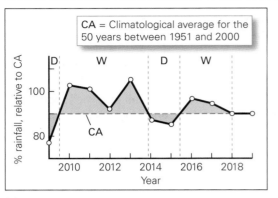

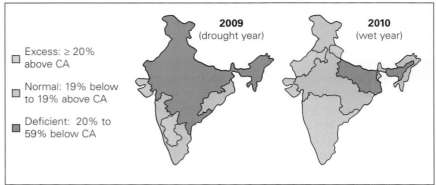

(a) From 2009 to 2019, India experienced droughts (D) and wet years (W) relative to its climatological average (CA).

(b) In the drought year of 2009, rainfall overall was 23% below the climatological average. In the wet year of 2010, rainfall was 2% above the climatological average.

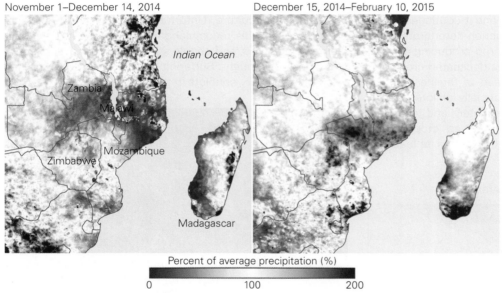

(c) In a monsoonal climate zone, the shift from drought conditions to flood conditions can happen rapidly, as shown by precipitation records from east-central Africa and Madagascar.

cover or volcanic aerosols block sunlight (see Chapter 4), so heating of the land beneath the ITCZ diminishes, or when winter snow covers more high-elevation areas (such as in the Himalayas), so that more of the solar energy striking the land reflects back into space. A decrease in surface heating weakens the low-pressure zone along the ITCZ and, therefore, decreases the steepness of the pressure gradient between the ITCZ and latitudes on either side of it. Since wind velocity depends on the pressure gradient, a less steep gradient produces weaker winds (see Chapters 8 and 11).

2. *Sea-surface temperature:* When sea-surface temperatures decrease in oceanic regions that provide moisture to the trade winds flowing toward the ITCZ, less moisture evaporates to fuel storms along the ITCZ. In cycles lasting several years, the location of cooler surface water seesaws between the east and west sides of the Pacific Ocean during El Niño/La Niña events, and between those of the Indian Ocean as a consequence of the *Indian Ocean Dipole* (see Chapter 8). Specifically, during El Niño events, western Pacific water is cooler, so evaporation decreases and winds transport less moisture over southern Asia, contributing to reduced rainfall. Similarly, during the positive phase of the Indian Ocean Dipole, cooler surface water resides in the eastern Indian Ocean, so less moisture may move over southern Asia, contributing to droughts in India and Pakistan. During the negative phase of the Indian Ocean Dipole, cooler surface water lies off the eastern coast of Africa, contributing to African droughts. The same factors leading to droughts in southern Asia and in Africa affect droughts in other regions with monsoonal climates, such as Australia. For example, during the positive phase of the Indian Ocean Dipole, decreases in rainfall occur along the western and northern coasts of Australia. During El Niño events, decreases in rainfall occur along Australia's northeastern and eastern coasts. When both occur simultaneously, Australia has its worst droughts, like the one described in the beginning of this chapter.

FIGURE 14.6 Over Africa and Asia, the ITCZ annually migrates from just south of the equator to as far as 26° N. The bands show the approximate ranges of its northern and southern limits. A cross section from north to south shows air circulation in the Hadley cells. (Not to scale.)

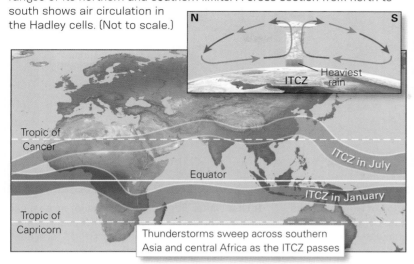

Thunderstorms sweep across southern Asia and central Africa as the ITCZ passes

DROUGHT IN MEDITERRANEAN CLIMATE ZONES. Several distinct climate zones occur within the temperate climate group. *Mediterranean climates* occur along the northern coast of the Mediterranean Sea, as well as along parts of the west coast of the United States and Mexico, in southern Africa, and in southern Australia. Regions with Mediterranean climates have wet winters and dry summers, but the wet season is not as intense as that in monsoonal climates, and rare rains can occur during the summer. Droughts in Mediterranean climates occur when the winters are significantly less wet than normal. For example, California's summer water supply depends on reservoirs fed by the melting snowpack of the Sierra Nevada and of the mountain ranges at the headwaters of the Colorado River. The state's worst drought in 1,200 years, which occurred between 2011 and 2017, happened in part because the melting snowpack provided insufficient water to streams (**FIG. 14.7**), which in

FIGURE 14.7 The California drought of 2011–2016.

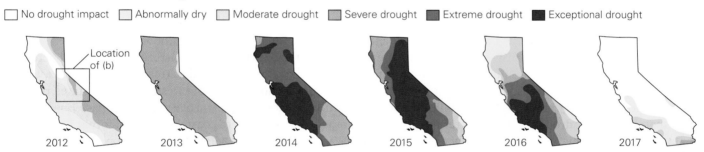

(a) The drought worsened between 2012 and 2016, and eventually affected most of California. Drought conditions decreased in 2017.

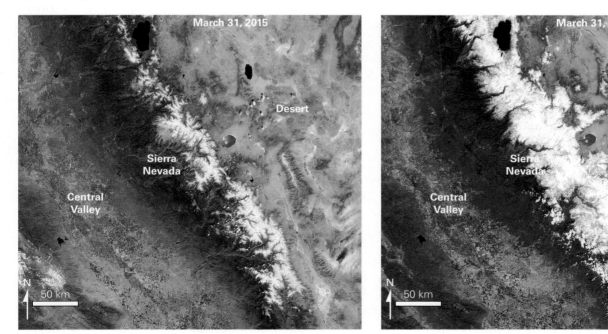

(b) Comparison of the snowpack in the Sierra Nevada in 2015, during the drought, and in 2017, after the drought was temporarily over. The location of these satellite photos is shown in (a).

FIGURE 14.8 A winter ridge in the upper troposphere can cause a drought, as happened in the western United States in 2021.

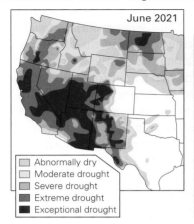

(a) Drought conditions returned to the southwest in 2021, after a dry winter.

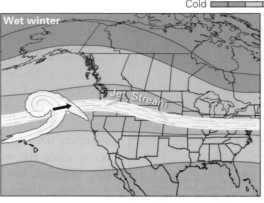

(b) If a ridge doesn't develop in winter, storms blow over the western states. Rain falls at low elevations and snow accumulates in the mountains.

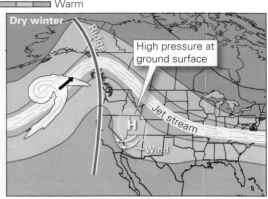

(c) If a winter ridge develops, the jet stream carries storms into Canada. A high-pressure zone of dry air hovers over the western US, and easterly winds blow into southern California.

turn led to excessive groundwater pumping and a drop in the water table (see Chapter 7). California survived the drought by passing laws to conserve water. Remarkably, urban Californians cut their water use by 22.5% between June 2015 and February 2017. California also borrowed funds to build new reservoirs, recycle water, and desalinate water. Nevertheless, the drought not only devastated crops and inconvenienced millions of people, but it also killed millions of trees. Unfortunately, extreme drought conditions returned in 2021 **(FIG. 14.8a)**. As California, Arizona, and other areas of the southwest baked, reservoirs dropped to all-time lows.

The rains and snows of Mediterranean climate zones come primarily when fronts associated with winter mid-latitude cyclones pass over those regions, so drought happens when these storms are less frequent. As we learned in Chapters 8 and 10, the formation and movement of mid-latitude cyclones are related to changes in the location and shape of the *jet stream*, the overall eastward-flowing circulation of air at high altitude. Recall that the positions of waves in the jet stream determine where divergence takes place in the upper troposphere, and that divergence triggers the formation of the low-pressure centers around which mid-latitude cyclones form. Mediterranean climates are located near the low-latitude end of the zone where mid-latitude cyclones influence weather. In years when the jet stream is displaced toward higher latitudes, the fronts associated with mid-latitude cyclones rarely reach Mediterranean climate zones, and they don't produce much rain or snow when they pass over those zones.

Much of the precipitation that falls on the US west coast and builds snowpack in its mountains comes from mid-latitude cyclones moving in from the Pacific during the winter. (Rain is rare during the summer, in most years.). In some years, fast northeastward-flowing streams of moist air form along the long comma tails of Pacific mid-latitude cyclones. Such streams are known as *atmospheric rivers* because they carry so much moisture ashore (see Chapter 12). During these wet years, the jet stream flows west to east, with little undulation, across the western United States, so mid-latitude cyclones that form over the Pacific pass, one after the other, over the western coastal states (California, Oregon, and Washington) **(FIG. 14.8b)**. Droughts develop when a persistent *ridge* develops over the western states. (A **ridge** forms when atmospheric circulation transports warm air northward in the lower and middle troposphere, causing isobaric surfaces in the upper troposphere to bow upward and produce high pressure in the upper troposphere in an elongate north-south zone; see Chapter 8.) The jet stream arches northward around the ridge **(FIG. 14.8c)**. When a large ridge exists over western North America, Pacific mid-latitude cyclones track northward on the west side of the ridge into Canada and Alaska and, as a result, they don't generate precipitation across western states.

Notably, during the late fall and into the winter, a high-pressure zone also develops in the lower troposphere over the relatively high-elevation deserts of the southwest (see Fig. 14.8c). This zone develops as a result of overnight radiative cooling over these deserts during the cool season, when days are short and the Sun does not rise high in the sky. (Note that the high-pressure zone in the upper troposphere, associated with a ridge, is a separate feature from the lower-troposphere high-pressure zone that forms over a relatively cool region of the Earth's surface.) Clockwise-circulating ground-level winds around the lower-troposphere high-pressure zone flow westward over

FIGURE 14.9 The difference between wet years and dry years can be represented by the standardized precipitation-evapotranspiration index (SPEI), which compares the amount of water lost from the ground and from plants with the amount of water supplied by rain or snow. A positive SPEI means a water surplus; a negative SPEI means a water deficit. The graph shows California's SPEI figures over the last 120 years.

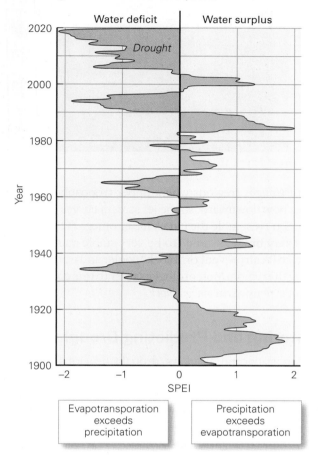

California's eastern mountains and down into the low-elevation agricultural lands of California. Air within these *downslope winds* dries and warms as it descends down the mountain slopes and undergoes compression, thereby worsening drought conditions in California's valleys. Over time, the western states experience alternating periods of wet years and dry years, depending on the jet stream's orientation. A succession of dry years leads to a drought **(FIG. 14.9)**.

Other regions with Mediterranean climates experience droughts for similar reasons. For example, lands bordering the Mediterranean Sea experience floods or droughts depending on the amount of winter precipitation they receive, either from mid-latitude cyclones that blow in from the Atlantic, or from Genoa Lows (see Chapter 12) that form over the Mediterranean Sea.

DROUGHTS IN OTHER TEMPERATE CLIMATE ZONES

Across the Earth's mid-latitudes, many regions with temperate climates are not identified as occupying Mediterranean climate zones because they receive precipitation year-round. These zones, which occur in central North America east of the Rockies as well as in many parts of Europe, Asia, and South America, include many major agricultural districts as well as large cities (see Fig. 14.4). Unlike in Mediterranean zones, where droughts reflect primarily the lack of winter rain, droughts in other temperate-climate zones occur when normally frequent summer rains do not happen.

The reasons why summer rains fail in these temperate regions vary across the world. Let's focus here on summer rainfall deficits in the central and eastern United States. During a normal, wet summer, a subtropical high-pressure center called the *Bermuda High* forms at low altitudes over the North Atlantic **(FIG. 14.10a)**. (At higher altitudes, air

FIGURE 14.10 The position of the Bermuda High, a semi-permanent high-pressure system over the North Atlantic Ocean, influences drought conditions in the central and eastern United States.

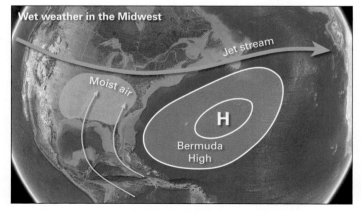

(a) When the Bermuda High lies close to the east coast, winds flowing clockwise around the high bring in moist air from the ocean.

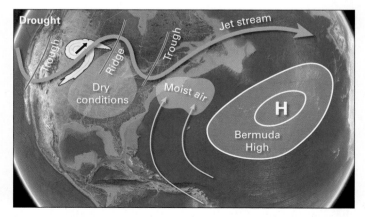

(b) If changes in sea-surface temperatures cause the Bermuda High to shift eastward, the flow of moist air misses the central and eastern United States, so the region becomes dry.

14.2 Drought: When the Land Dries Out • **511**

FIGURE 14.11 Dry, unvegetated soil radiates more heat into the air than does wet, vegetated soil.

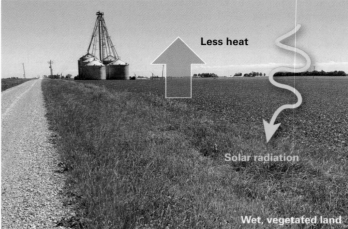

flow overall is from west to east and is most rapid in the jet stream.) Air flows clockwise around the Bermuda High and carries moisture northward from the ocean into the eastern and central United States, fueling thunderstorms. In summer, weak disturbances in the jet stream trigger these thunderstorms, which then produce widespread rains. Dry summers happen when the Bermuda High shifts eastward due to changes in the distribution of sea-surface temperatures, so that the northward flow of air on the western side of the high shifts to the east of the Atlantic coast (**FIG. 14.10b**), and less moisture is carried over eastern and central North America.

POSITIVE FEEDBACKS THAT AMPLIFY DROUGHTS Hot, dry weather can trigger a positive feedback that makes a developing drought even worse. Why? Heat increases the rate of evaporation, causing the ground to dry out faster. Once dry, the ground provides less moisture (from evaporation) to the air, so humidity decreases. Lack of moisture also causes plants to wither and die. Thus they not only return less moisture to the air (by transpiration), but also cast less shade, so more solar energy reaches the ground (**FIG. 14.11**). As a result, the ground warms up even more, and it becomes drier, radiating more infrared energy back into the base of the atmosphere, making the air even warmer.

The development of a warm area on land causes the overlying atmosphere to expand. In the upper troposphere, this expansion bows up isobaric surfaces and produces a ridge, as we noted earlier. Consequently, the path of the jet stream shifts northward (see Fig. 14.10b). Since mid-latitude cyclones, and the thunderstorms they harbor, tend to form west of the ridge and move northward, hardly any rain falls in the central United States under the ridge, making drought conditions there worse. As shown in **Figure 14.10b**, the shape of the jet stream as it curves around a ridge, and the adjacent *troughs* (regions where upper-troposphere isobaric surfaces bow downward, so the jet stream curves southward; see Fig. 8.31) resemble the Greek letter omega (Ω). When the ridge is very broad, it tends to be very stable and nearly stationary, so it can prevent or "block" cooler and wetter air from entering the region. Meteorologists refer to this configuration of the jet stream as an *omega block*.

Monitoring and Predicting Drought

Unlike most natural disasters, drought develops slowly and can persist for years. Furthermore, the consequences of drought are cumulative—conditions worsen each day as weeks and months pass without adequate rainfall. To monitor drought, the US National Weather Service (NWS) developed the **Palmer Drought Severity Index** (**PDSI**). The PDSI measures moisture deficiency, relative to average local moisture conditions, by comparing the supply of water available from precipitation and stored reserves with the depletion of water by evaporation, infiltration, and runoff. By using the PDSI, scientists can compare the severity of drought in one region with that in another. The PDSI is cumulative, so that each period's index value reflects incremental changes from prior values (**FIG. 14.12a**). The *US Drought Monitor*, a regularly updated map based on measurements of climatologic, hydrologic, and soil conditions as well as on reported drought impacts and observations, provides a visual summary of drought information (**FIG. 14.12b**). Similar drought-monitoring maps are produced for Europe by the European Drought Observatory (**FIG. 14.12c**), and for the world by the US National Integrated Drought Information System. Such indices and maps show that drought is a global phenomenon (**FIG. 14.13**).

Can droughts be predicted? If the answer were yes, vulnerable regions could take steps to prepare, such as identifying alternative sources of water, stockpiling food, and

FIGURE 14.12 Various kinds of drought indices and maps.

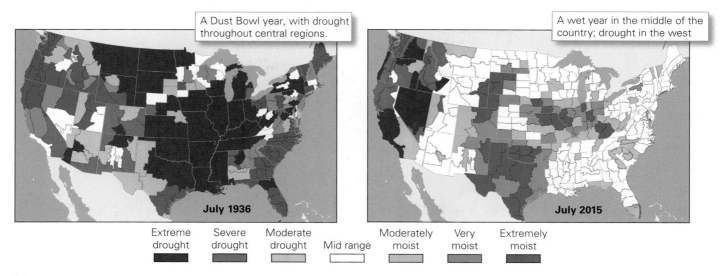

(a) In the United States, the NWS assigns a Palmer Drought Severity Index value to each county. These maps show PDSIs for two different years.

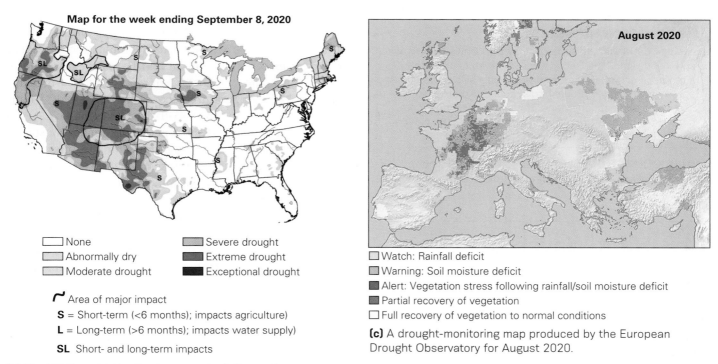

(b) The US Drought Monitor is a map depicting regions experiencing drought in the United States.

(c) A drought-monitoring map produced by the European Drought Observatory for August 2020.

planting drought-resistant crops. Unfortunately, the answer at present is no. While measurements of sea-surface temperatures hint at whether drought-triggering conditions will develop in the next month or two, predictions made from these measurements are not consistently correct, and predictions beyond a month are not sufficiently accurate that actions can be taken. For example, some researchers have tried to tie the timing of droughts to the distribution of cool seawater caused by La Niña, and to comparable phenomena in other oceans, but such correlations have not proven robust. The challenge of predicting droughts, as with many phenomena in the Earth System, is that they are consequences of interactions among many overlapping factors and various feedbacks. Computer modeling of trends in rainfall over periods from 1 to 10 years is currently a very active area of research.

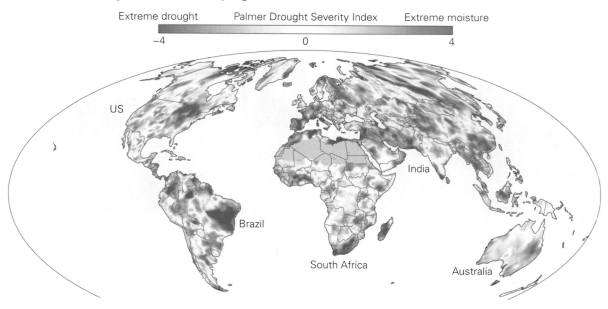

FIGURE 14.13 Drought occurred in many regions of the world in 2017.

Take-home message...

Drought develops when a deficiency of freshwater over a sufficiently long time leads to adverse effects on water supplies, crops and other vegetation, animal life, and human activities. Prolonged droughts can cause catastrophic casualties and economic loss. The causes of drought are different in different climate zones. While droughts can be monitored, they can't be predicted accurately.

QUICK QUESTION What are the differences among meteorological, agricultural, and hydrological droughts?

14.3 Heat Waves

The Danger of High Temperatures

You may be surprised to learn that one of the leading causes of death from weather-related natural disasters in the United States is heat. Heat-related fatalities typically occur during **heat waves**, periods ranging from 2 days to several weeks of abnormally hot conditions that make most people uncomfortable and some very sick. The worst heat-related disaster in recent US history, the 1995 Chicago heat wave, resulted in at least 739 heat-related deaths in only 5 days (FIG. 14.14). Heat-related deaths in the United States were more frequent in the early 20th century, before air conditioning was generally available—for example, over 9,500 deaths were attributed to heat in 1901. But fatalities in the United States today pale in comparison to those caused by heat waves in other countries. Europe's 2003 heat wave caused more than 70,000 deaths (FIG. 14.15a). The combination of a heat wave with pollution or smoke frequently amplifies health problems. For example, the deaths of 56,000 people in western Russia in 2010 were attributed to a combination of wildfire smoke and extreme heat (FIG. 14.15b). Heat waves made headlines in the summer of 2021 when record hot air—breaking previous high-temperature records by 6°C (10°F)—hung over Oregon, Washington, and British Columbia.

Why is heat a problem for people? Humans constantly produce *metabolic heat*, a by-product of the biochemical reactions that keep us alive. Like all mammals, our bodies regulate the production and removal of metabolic heat to maintain a nearly constant core (internal) temperature. Normal body temperature varies slightly among individuals, but generally ranges between 36.1°C and 37.3°C (97.3°F and 99.1°F), regardless of whether a person is resting or exercising. To maintain this temperature, the body shivers to stay warm or sweats to stay cool. Sweating cools a human body

FIGURE 14.14 During the 1995 heat wave, the number of deaths in Chicago increased dramatically soon after the temperature topped 38°C (100°F).

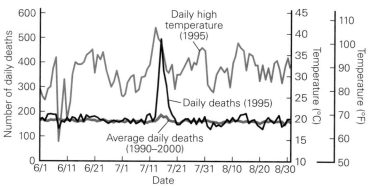

FIGURE 14.15 Killer heat waves in Eurasia have claimed the lives of thousands.

Land-surface temperature anomaly (°C)
−12　　0　　12

(a) Europe endured temperatures much higher than normal during a 2003 heat wave.

(b) Many places in Russia had temperatures as much as 12°C (22°F) above normal during a 2010 heat wave.

because evaporation of water in the sweat transfers body heat into the air, leaving the body cooler. When the body can't cool itself sufficiently and its core temperature rises beyond the normal range, *heat stress* begins, leading to *heat illnesses*. If the core temperature of the human body reaches 39°C (102.2°F), *heat exhaustion*, manifested by extreme sweating, weakness, cramps, a fast pulse, and breathing difficulty, sets in. If a person's core temperature rises to 40°C (104°F), *heat stroke* may occur, characterized by fainting, a very fast pulse, headache, vomiting, and confusion. Victims of heat stroke need to be hospitalized (FIG. 14.16a). When core temperature reaches 42°C (107.6°F), a person may become delirious or comatose, and may suffer convulsions. At a temperature of 43°C (109.4°F), victims suffer brain damage leading to death.

Many factors determine whether you will develop a heat illness or not. Some depend on your personal characteristics and some on your surroundings. These include:

1. *Physical condition and activity*: Factors such as body size and shape, which influence both the ratio of surface area to volume in your body and the amount of energy needed to move, determine your metabolic heat. Heat production also depends on whether you are stationary or exercising, because you burn more energy when you move. Your body's ability to lose heat depends on the rate at which you sweat, which will be influenced by your hydration state and by your clothing.

2. *The intensity of solar radiation*: The amount of radiant heat that reaches your skin depends on the Sun's position in the sky (which in turn depends on latitude, season, and time of day), on whether you are standing in shade, in the direct Sun, or under a cloudy sky, and on the extent to which clothing covers your skin. Radiant heat can increase your skin's temperature, relative to your body's core, by as much as 8°C (15°F). Hot skin limits heat

FIGURE 14.16 Human bodies must maintain a core temperature within the normal range. Otherwise, heat illnesses can strike.

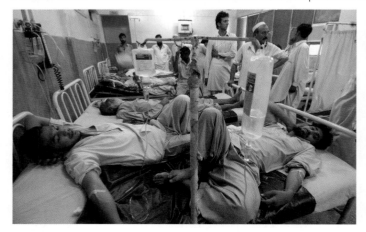

(a) Hospital wards can fill with people suffering from heat stroke during a heat wave.

(b) During a heat wave, cities may use mist-generating fans or fountains to help people cool down.

transfer from inside a human body into the air, because heat always flows from hotter to colder material.

3. *Current weather conditions*: Air temperature, relative humidity, and wind speed all affect the rate at which you can exchange heat with your surroundings. When air temperature rises, the rate of heat flow from your body to the air slows—in fact, if air temperature exceeds body temperature, the air transfers heat into the body. When relative humidity rises, sweat evaporates more slowly, so your body can't cool itself as efficiently. When the wind speed increases on a moderately hot, dry day, the air removes a thin, sweat-saturated air layer that develops next to your skin, so that you can sweat more and therefore feel cooler. (If the air is humid, wind has little or no effect.) On an extremely hot day, though, wind can feel like a blast from a furnace. It won't cool you, and you will become dehydrated faster. Clearly, the **apparent temperature**, the temperature that a person feels, isn't necessarily the same as the air temperature that a thermometer shows. Apparent temperature depends not only on the air temperature, but also on relative humidity, wind speed, and the amount of shade.

On any given hot day, some people will suffer heat stress, usually because they exercise more than is safe for the situation, without drinking enough water or without replacing electrolytes (the salts that also seep out with sweat). During a heat wave, many more people than normal will endure this fate. What can communities do to prevent heat stress during a heat wave? The first action should be to broadcast warnings that encourage people to stay out of the heat. For those suffering from heat, dousing yourself with cool water, standing in front of a fan or a stream of mist, or sitting in an air-conditioned room can make a heat wave survivable **(FIG. 14.16b)**, but without these amenities, the heat can cause illnesses. Availability of cooling centers is particularly important overnight. If people remain at temperatures over 33°C (91°F) all night, their core temperature cannot return to a safe range during sleep. Unfortunately, dangerously hot nights continue to increase in formerly temperate regions.

Poorer communities suffer most from heat waves because they often lack air conditioning. Perhaps surprisingly, populations living in cooler, temperate climates experience higher heat-related mortality rates than do populations living in climates where hot weather is common. People who live in hotter regions have developed cultural adaptations, such as taking a siesta, to avoid overheating.

When Is Hot Weather Considered a Heat Wave?

Whether or not a spell of hot weather in a region can be called a *heat wave* depends, in part, on how high the air temperature gets relative to the normal range, or *climatological average*, for the region, as well as on relative humidity.

Consider two US cities: Phoenix, Arizona, and Seattle, Washington. During June, Phoenix, in the hot desert of southern Arizona, experiences an average high temperature of 40°C (104°F). Everyone would agree that this is hot weather. In Phoenix, however, residents expect such conditions, air conditioning within the city is nearly universal, and heat-related health problems are relatively rare at these temperatures. Meteorologists would not characterize a series of 38°C (100°F) days during the summer in Phoenix as a heat wave. Yet, if these same conditions occur along the coast of Oregon, Washington, and British Columbia, where cooling breezes from the Pacific normally cap June high temperatures at 21°C (69°F), and many homes aren't air conditioned, multiple 38°C days would be considered a heat wave, and could cause many heat-related health problems. When temperatures in this region peaked at 47°C (117°F) during a record-breaking heat wave in June 2021, hundreds died.

Can cities located in hot deserts, such as Phoenix, experience a heat wave? Yes, when conditions exceed that area's climatological average. During the summers of 2020 and 2021, for example, temperatures in Phoenix exceeded 43°C (110°F) for weeks, exceeded 46°C (115°F) for many consecutive days, and at times, rose to 48°C (118°F), even in the early part of the summer. Low-paid workers laboring in construction to satisfy booming demand for new luxury homes, and in the fields to provide fruit and vegetables, could not avoid the heat. Consequently, hospitals reported increased admissions for heat illnesses and a rise in heat-related fatalities. During a 2017 heat wave, airlines canceled flights out of the Phoenix airport because the air was so hot, and the air density, therefore, so low, that planes could not get enough lift to take off before reaching the end of the runway.

Given the complexity of communicating the dangers of heat, meteorologists have developed a **heat index** to convey the threat of heat exposure to the public in a straightforward way. Because shade and wind speed can vary significantly depending on where a person is standing outdoors, this index assumes a standard wind speed of 8 km/h (5 mph), and assumes that the person is in the shade, so that the apparent temperature that a person feels depends only on air temperature and relative humidity. Using a *heat index table* **(FIG. 14.17a)**, we can see that a person experiencing an air temperature of 32°C (90°F) and a relative humidity of 80% will feel as hot as a person experiencing an air temperature of 45°C (113°F) and a low relative humidity. If the second person moves into direct sunlight, the same combination of air temperature and relative humidity could result in an apparent temperature as much as 8°C (15°F) higher, for a total of 53°C (127°F). Humans cannot tolerate these conditions very long before developing a heat illness.

FIGURE 14.17 Defining a heat wave.

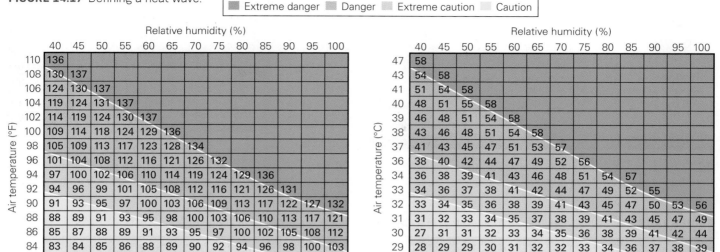

(a) The heat index depends on air temperature and relative humidity, assuming 8 km/h (5 mph) wind and shade. "Danger" means that heat stroke is possible.

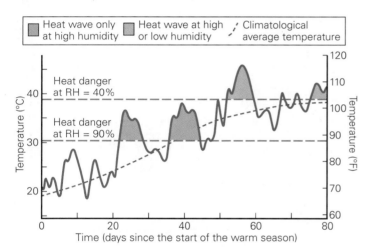

(b) Heat waves are recognized by comparison with the climatological average for a region. The dashed red lines on the graph represent the limit of heat waves at two different relative humidities (RH).

Given the roles that both air temperature and relative humidity play in determining the apparent temperature, a graph showing when heat waves occur and how long they last must account for both factors (FIG. 14.17b). As the graph shows, an air temperature that would be dangerous (capable of causing heat illness) under high humidity might not be dangerous under low humidity.

In the United States, the National Weather Service has not set a specific temperature at which it warns the public about heat because human responses to heat vary with location and acclimatization. Rather, the NWS issues a **heat advisory** when it predicts that a regionally defined threshold value of the heat index will be reached or will remain for 2 or more consecutive days. It issues an **excessive heat warning** when prolonged periods (4 or more days) of high heat index values are expected or when extremely high heat index values are expected for a single day. Some countries have implemented *heat-health warning systems* tailored to individual large urban areas. These systems require media to broadcast warning announcements along with information on how to avoid heat-related illnesses, encourage people to set up a "buddy system" so that friends, relatives, or neighbors take responsibility to check on elderly and vulnerable persons, and activate a "heatline" telephone number that people experiencing heat illnesses can call to receive assistance. In large cities, air-conditioned shelters may also be set up.

Urban Heat Islands

On a given day during warm weather, temperatures in urban areas are hotter than those in surrounding suburban or rural areas. Meteorologists refer to this phenomenon as the **urban heat island** effect (FIG. 14.18a). During a heat wave, the temperature difference can be as high as 3°C–5°C (5°F–10°F), or even greater (FIG. 14.18b). Significantly, low-income urban communities commonly suffer more from the urban heat island effect than do wealthier ones because they tend to have fewer trees or grassy areas, higher population densities, and less air conditioning.

Why are urban areas hotter than rural areas? Several factors contribute to this effect. Roads, walkways, and buildings in cities are made of concrete, asphalt, bricks, and other materials that readily absorb solar radiation during the day and re-emit this energy as infrared radiation at night, warming the air. (Sometimes, asphalt becomes so hot that roads

FIGURE 14.18 The urban heat island effect.

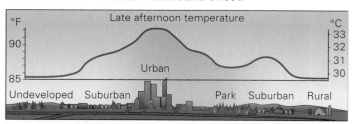

(a) Air temperatures vary across a city, with the warmest temperatures in the downtown area.

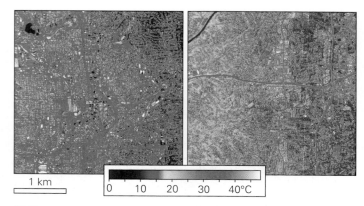

(b) Temperatures in urban Atlanta (left) are much warmer than in suburban Atlanta at the same time (right), as seen in satellite imagery.

FIGURE 14.19 Increasing the amount of vegetation in a city can decrease the urban heat island effect.

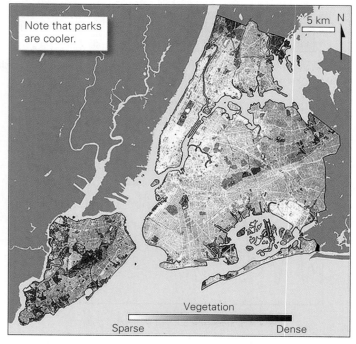

(a) Temperatures (left) compared with vegetation distribution (right) in New York City.

(b) Green roofs, such as these from Portland, Oregon, can keep a building (and thus the city) cooler.

buckle.) In addition, urban surfaces do not absorb rainwater, so the ground does not contain moisture that can absorb heat and evaporate. Cities also house machinery that vents heat into the air, such as car and truck engines and air-conditioning units. In contrast, in suburban and rural areas plants on the ground absorb solar radiation as they undergo photosynthesis and evapotranspiration, and thus cool the air. Rain sinks into the soil in these settings, and some of the incoming solar energy works to evaporate soil moisture rather than to raise the soil temperature. Evaporation continues after sunset, so air cooling continues. Because vegetation is so important in keeping temperatures low, many cities plant trees and encourage the development of green roofs (rooftop vegetation cover) or painted white roofs (to reflect radiation) on large buildings **(FIG. 14.19)**.

Weather Patterns Associated with Heat Waves

A variety of meteorological patterns can initiate or amplify heat waves. Different patterns affect different climate zones, as we now see.

TEMPERATE-CLIMATE HEAT WAVES. In temperate climates of the Earth's mid-latitudes, several interrelated meteorological conditions can trigger heat waves:

1. *A broad ridge in the upper troposphere:* We noted earlier that a ridge develops in the upper troposphere where isobaric surfaces at high elevation bow upward (see Fig. 8.31). On a map, the trace of the jet stream, as it curves around a ridge and adjacent *troughs* (places where isobaric surfaces in the upper troposphere bow downward), forms a wave-like shape (**FIG. 14.20a**). Also, as we noted earlier, when a broad ridge develops in central North America, it can be stable and stationary for an extended period, producing an omega block that prevents cooler air from flowing into the region beneath the ridge. Therefore, heat can build in the lower troposphere under the ridge, potentially causing a heat wave. News stories often refer to a long-lived upper-troposphere ridge that develops during the summer as a *heat dome*, to emphasize that it can trigger a heat wave.

2. *Increased radiation reaching the ground:* When skies are clear, intense solar radiation reaches the ground, where it is absorbed by materials at the Earth's surface. The materials return this energy upward, in the form of infrared radiation, which heats air in the lower troposphere. If clear skies last for many days, such heating can cause, or amplify, a heat wave. Notably, the development of long-lasting clear skies during summer in the central United States is associated with the formation of a broad

FIGURE 14.20 Weather conditions required for a heat wave at mid-latitudes.

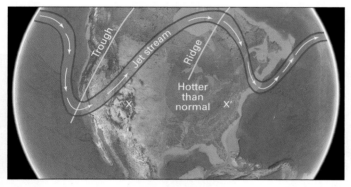

(a) When a broad ridge develops over central North America, an omega block forms, preventing cooler air from reaching the area under the ridge, so the region heats up.

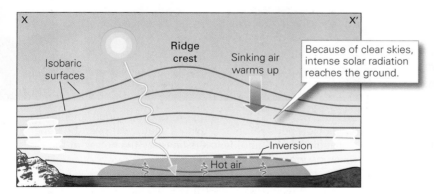

(b) Air sinks on the east side of the ridge, depicted by cross section XX′ (see map in (a) for location). This air compresses, warms, and dries, so clouds can't form, and intense solar energy reaches the ground. The ground releases infrared radiation that heats the lower troposphere.

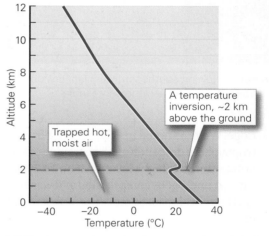

(c) Air above an inversion is warmer than air below, as represented by the blue line. The trapped air beneath the inversion absorbs infrared radiation rising from the ground, and heats up.

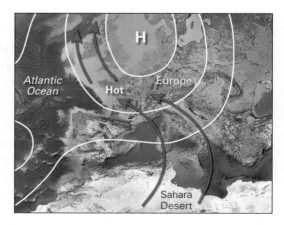

(d) When a ground-level high-pressure center sits over northern Europe, winds carry hot air from the Sahara Desert into central Europe.

FIGURE 14.21 Heat waves in tropical climates.

(a) Hot air flowing southeast from high, hot desert areas in Pakistan brings high temperatures into India.

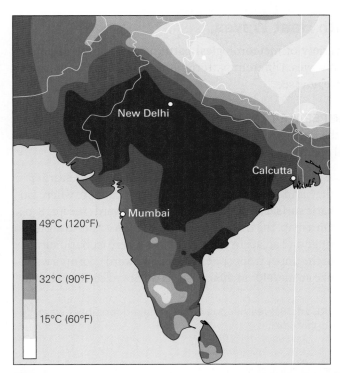

(b) During this heat wave, temperatures were highest in northern and central India.

ridge. Why? As noted in Chapter 8, air from the upper troposphere descends on the east side of a ridge, due to flow patterns at high elevation **(FIG. 14.20b)**. As this air sinks, it undergoes compression and warms. The sinking air, therefore, absorbs moisture, causes relative humidity to decrease, and prevents clouds from forming.

3. *Development of an inversion:* Because sinking air on the east side of the ridge warms as it sinks (see Fig. 14.20b), an **inversion** may develop at an elevation of about 2 km **(FIG. 14.20c)**. This means that the air just above this elevation is somewhat warmer and less dense than the air just below. Consequently, air from below the inversion cannot rise buoyantly into the upper troposphere. Without such vertical flow, hot, humid air gets trapped at low elevations. (Note that this inversion lies just above the near-surface hot air depicted in Fig. 14.20b.)

4. *An influx of tropical air:* Humid air exists all year in the tropics. When winds blow this air northward from the Gulf of Mexico and southwestern Atlantic into the central United States, humidity rises, so the apparent temperature rises, and heat illness becomes a risk. The problem becomes worse when this air is trapped beneath an inversion (see Fig. 14.20c). Notably, not all the humidity in this air comes from the sea—transpiration from plants in forests and farm fields adds as much as half of the water vapor.

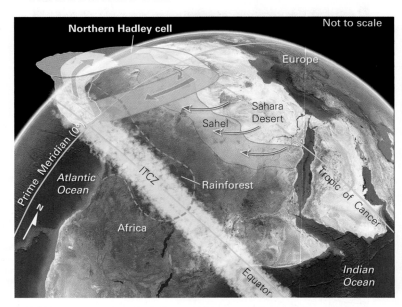

(c) Heat waves in sub-Saharan Africa can happen when trade winds blow hot air from the Sahara southwest over the Sahel. Commonly, a low-altitude inversion traps the hot air at the Earth's surface.

5. *An influx of desert air:* Disastrous heat waves in Europe can happen when winds carry hot air of the Sahara Desert northward, and across the Mediterranean Sea. Such winds develop when a surface high-pressure zone center becomes established over northern Europe, for air

circulates clockwise around this zone (FIG. 14.20d). A similar phenomenon occasionally happens along the coast of the western United States, when air from desert lands flows westward and pushes ocean-cooled air offshore.

6. *Downslope winds:* When winds carry air from higher elevations down to lower elevations, the air undergoes compression and warms (see Chapter 8). The increase in temperature can be substantial. For example, the air carried by winds from a high desert, at an elevation of 1.5 km, down to the coast at sea level, will warm by about 15°C (27°F). So air that may be a comfortable 27°C (80°F) in the desert will be a stifling 42°C (107°F) when it reaches the coast.

7. *Drought conditions:* In regions where drought conditions have developed, vegetation and soil dry out. Consequently, an influx of solar radiation heats the ground more than it would during non-drought conditions. The additional infrared radiation released by this warmed ground raises the temperature in overlying air. Some of the extreme heat waves of the desert southwest in the United States are due to this amplification.

In many cases, heat waves result from a combination of the aforementioned factors. For example, the heat wave that struck coastal Oregon, Washington, and British Columbia in June 2021—causing record-breaking temperatures of up to 47°C (117°F)—was due in part to the development of a ridge, as warm air from the south moved northward, in part to the warming of west-flowing air as winds carried it from high deserts down to sea level, and in part to the pattern of air circulation over the adjacent Pacific Ocean, which caused cool winds from over the ocean to flow parallel to shore, rather than onto land.

TROPICAL-CLIMATE HEAT WAVES. Heat waves that affect the world's tropical climates result from different meteorological conditions than those that affect temperate climates. Heat waves in monsoonal climate zones can develop when hot winds descend from nearby regions. For example, severe heat waves in India occur when air flows southeastward off the hot desert areas of northern Pakistan and out across India's northern plains (FIG. 14.21a, b). In tropical Africa, heat waves can develop when surface winds blow southward from the Sahara (FIG. 14.21c). These winds develop beneath the northern part of the Hadley cell, where air descends from high elevation, and consequently has undergone compression, which in turn causes air to heat up. As air heats, its relative humidity decreases. Consequently, clouds can't develop in this air, so skies remain clear for weeks on end and intense radiation heats the ground.

> **Take-home message . . .**
>
> Heat waves occur when uncommonly hot temperatures last for a sufficiently long period that human health is threatened. Prolonged heat can lead to thousands of deaths, especially in areas of the world where hot temperatures are not common and air conditioning is rare. Heat waves are caused by weather conditions that allow more solar energy to reach the ground, or that cause hot air to blow in from warmer climates. Conditions considered to be a heat wave in a warm climate differ from those in a cool climate.
>
> **QUICK QUESTION** What factors are considered in calculating the heat index?

14.4 Wildfires: Flames of Destruction

In recent years, the western United States endured some of the most destructive fires ever. Each year, it seemed that the fires couldn't get worse—and then they did. In states already reeling from the COVID-19 pandemic and its economic fallout, thousands of fires have ignited, destroying large regions—in some cases, more than 20,000 km² (7,700 mi²)—and causing economic disaster before the rains of winter could douse the blazes (FIG. 14.22a, b). Along with scenic forests, brush, and grasslands, many homes and parts of some towns went up in flames. The fires caused numerous injuries and fatalities, and required the evacuation of tens of thousands of people. Smoke blanketed much of California, Oregon, and Colorado. In cities such as San Francisco—already suffering from a record heat wave—people developed health problems ranging from eye irritation to breathing difficulty (FIG. 14.22c). The only bright spot in the headlines came from stories of dramatic rescues. In 2020, over 200 people, trapped along the shores of Mammoth Pool reservoir with flames all around, were rescued when National Guard helicopters heroically plunged through the smoke, crammed everyone into their cargo holds, and airlifted them to safety. Fires of equal ferocity burned worldwide. Some places that rarely have fires, such as the Pantanal, a normally moist swamp and grassland in Brazil, endured massive fires due to a regional drought. Similarly, major fires devastated areas of Siberia's tundra.

This section discusses wildfires—self-sustaining fires that start and spread out of control and often move into or through populated areas and agricultural areas. Wildfire disasters occur when these fires destroy homes, farms, infrastructure, parks, and valuable resources. Wildfires impact lives, livelihoods, and property, and they can turn

FIGURE 14.22 Devastating wildfires struck the US west coast in 2020.

(a) A fire burns to the edge of a highway in California.

(b) Devastation in Talent, Oregon, after a wildfire.

(c) Smoke in San Francisco made 10:00 A.M. seem like dusk.

once-beautiful scenery into a sea of ash. We begin by discussing the nature of fire and why wildfires happen. Then we consider different types, causes, and consequences of wildfires.

What Is Fire?

Strike a match and you'll see **fire**, a familiar phenomenon, during which gases chemically react with oxygen to produce heat and light **(FIG. 14.23a)**. In technical terms, fire results from the very rapid oxidation of flammable molecules. This process, known as *burning*, transforms an original set of chemicals (reactants) into a new set of chemicals (products) and releases energy in the form of heat and light. **Flames**, the visible part of fire, consist of glowing superheated gases—including carbon dioxide, carbon monoxide, nitrogen, oxygen, water vapor, and various organic chemicals—as well as glowing microscopic particles, including carbon. In the case of burning wood, grass, or brush, the gases come from the breakdown of *cellulose*, an organic molecule that makes up the walls of plant cells, along with various other organic chemicals, such as fats, oils, and sugars, within plant cells. The color of flames depends on their temperature (blue is hottest), which in turn depends on which gases are burning and on whether the combustion is complete or incomplete **(FIG. 14.23b)**. The **smoke** that rises from a fire contains these gases, after they have cooled and stopped glowing, along with **soot** (tiny specks of carbon) and **ash** (flaky particles of carbon and other material, the remnants of incomplete burning). In places where fires consume houses, smoke may also contain chemicals released by the burning of materials such as plastic and vinyl.

Burning, in effect, represents the reverse of photosynthesis, the process by which plants effectively trap incoming energy from the Sun and use it to construct organic chemicals. This energy is stored as potential energy in the chemical bonds that hold atoms to one another within these chemicals. When burning takes place, the chemicals break apart to form new molecules. The bonds in the new molecules store less energy, so the extra energy is released as heat and light—chemists call such reactions *exothermic*. Simplistically, using "sugar" to represent organic chemicals in plants:

Photosynthesis: Carbon dioxide + water + solar radiation → sugar + oxygen

Burning: Sugar + oxygen → carbon dioxide + water + heat and light

Fire can't burn without sufficient oxygen, and it can't spread without access to a supply of **fuel**, flammable material such as trees, shrubs, grasses, decaying plant matter, peat, or houses. The availability of oxygen determines whether combustion is complete or incomplete, so it also affects the temperature in a flame.

Igniting a Wildfire

How does a fire begin? The first step, called the *preheating stage*, happens when a heat source raises a fuel's temperature (see Fig. 14.23b). When the fuel reaches a high enough temperature, the *pyrolysis stage* begins, in which molecules in the fuel decompose to produce gases that rise into the air. (*Pyrolysis* comes from the Greek *pyro*, meaning fire, and *lysis*, meaning separate.) Finally, during the *combustion stage*, the gases ignite, meaning that they suddenly start

FIGURE 14.23 The nature of fire.

(a) Logs in a campfire produce flames as they burn. The flames are glowing gases and glowing particles of carbon.

(b) Production of fire involves multiple steps. Heating of the wood evaporates water, pyrolysis yields flammable gases, and the gases then react with oxygen.

reacting rapidly with oxygen. Put simply, in order for a wildfire to begin, the environment must host three components—fuel, heat, and oxygen—which together constitute the **fire triangle** (FIG. 14.24).

Raising the temperature of living or recently dead plants sufficiently to cause combustion takes a lot of heat because plants contain a lot of water—about 50% of a living tree, for example, consists of water. Consequently, heat applied to wood warms not only flammable cellulose, but also the water in the wood. Water has a high *heat capacity*, meaning it can absorb a lot of heat while producing only a small change in temperature. In addition, when water reaches 100°C (212°F), it boils. It remains at 100°C, less than the ignition temperature of wood, until it has completely evaporated. As a result, wet wood doesn't burn easily, because the heat applied to the wood goes first toward raising the temperature of, and then toward boiling, the water within the wood, before raising the temperature of the wood itself. Wood can become hot enough—120°C (248°F)—for pyrolysis to take place rapidly only after most of its water has evaporated. If kept at its pyrolysis temperature, wood slowly turns to charcoal, which is solid carbon, and doesn't burn. Only when temperatures rise to 260°C–450°C (500°F–842°F) will the gases produced by pyrolysis ignite. If enough oxygen is present at the ignition temperature, *flaming combustion* takes place, and the temperature can rise to about 1,100°C (2,012°F). With an abundance of dry fuel and plenty of oxygen, the temperature may even reach 1,500°C (2,732°F), hot enough to melt metal and glass—a wildfire can turn aluminum automobile engines into pools of liquid metal. If the oxygen supply is limited, as happens when burning wood gets buried by a layer of ash, *smoldering combustion* takes place, and the burning wood glows but doesn't flame. Once a fire has started, heat from the fire moves into the interior of the fuel by *conduction*, a process of heat transfer during which heat in one part of a material flows into the adjacent part of that material, so that pyrolysis breaks down progressively more fuel, providing more flammable gases to maintain the fire.

Because fires won't ignite unless the fuel is relatively dry, a forest's or grassland's susceptibility to fire depends on how long vegetation and trees have been drying, as well as on air temperature, since plant material can absorb water vapor from the air, and on relative humidity. Thus, the chance of fire varies by season, and also depends on whether recent temperatures have been higher or lower than average. During a drought, a heat wave, or both, vegetation becomes drier and more flammable.

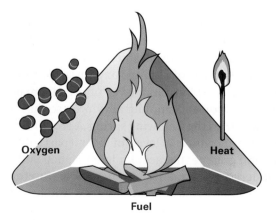

FIGURE 14.24 The fire triangle emphasizes the three components needed for a fire to ignite.

14.4 Wildfires: Flames of Destruction

What triggers a wildfire? The heat source that starts a wildfire can be natural or caused by human activity. Lightning ignites more fires than any other natural source. When lightning strikes an object, it provides enough energy to evaporate water in wood, cause pyrolysis, and trigger ignition almost instantly. During the 2020 fire season in the western United States, tens of thousands of lightning strikes occurred across forests of the region **(FIG. 14.25a)**. News media referred to the lightning as *dry lightning* because it accompanied rain that fell through dry air, most of which evaporated before reaching the ground. Fires triggered by dry lightning burn beyond the time the thunderstorm passes, and they continue to spread. Rarely, *spontaneous combustion* can result when a mass of rotting organic material generates enough heat on its own to reach its combustion temperature. Even less commonly, lava or volcanic ash ignites fires. During the Earth's history, some meteorite impacts have ignited fires (see Chapter 16).

As many as 90% of forest fires in the United States are caused by human activity, mostly unintentionally. Accidental wildfires have been triggered by sparks from fallen power lines, malfunctioning electrical transformers, the steel wheels of trains, burning debris, hot automobile exhaust pipes, chains that drag behind a trailer, untended campfires, discarded cigarettes, and celebratory fireworks **(FIG. 14.25b)**. Fires that have been set to clear land may cause wildfires if they go out of control. Unfortunately, some wildfires are set by arsonists.

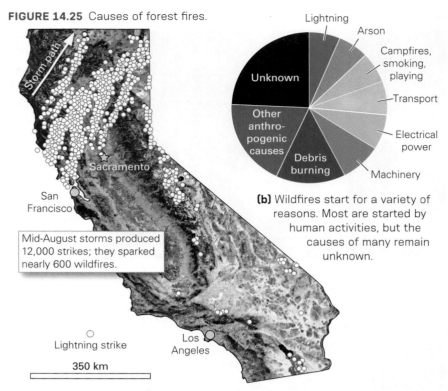

FIGURE 14.25 Causes of forest fires.

Mid-August storms produced 12,000 strikes; they sparked nearly 600 wildfires.

(a) On August 17, 2020, 7,393 lightning strikes occurred in California. The map shows where they clustered.

(b) Wildfires start for a variety of reasons. Most are started by human activities, but the causes of many remain unknown.

Types of Wildfires

Wildfires can spread across the landscape in different ways, depending on the nature and location of the fuel that feeds them **(FIG. 14.26a)**. **Ground fires** occur within roots and buried organic material, and they typically smolder for long periods because of a limited oxygen supply **(FIG. 14.26b)**. Some ground fires burn in a layer of organic matter in or just below the soil and smolder all winter, so that they trigger new wildfires the following spring. A *peat fire* is a type of ground fire in which a layer of peat (partially decayed mosses and grasses that accumulate in a bog and then dry out) undergoes smoldering combustion. **Surface fires** burn low-lying vegetation such as grass and brush. In forests, surface fires consume *undergrowth* (consisting of short plants, shrubs, and young trees), as well *forest litter* (accumulated dead leaves and branches) on the ground **(FIG. 14.27a)**. **Ladder fires** burn undergrowth and forest litter as well as vines and medium-sized trees that grow between ground vegetation and the forest's *canopy* (the crowns of trees) **(FIG. 14.27b)**. **Crown fires** (or *canopy fires*) burn through the canopy and can send flames high above the tops of trees. Occasionally such fires jump from crown to crown without involving undergrowth.

Fire severity represents the percentage of the biomass that burns during the fire. *Fire intensity* for a given area—a measure of energy released during burning—depends on the amount, type, and moisture content of fuel as well as on the availability of oxygen. Grassland fires, for example, are typically less intense than forest fires because less biomass is available to burn. Fire intensity affects the *flame length*, the distance from the base to the end of a flame **(FIG. 14.28)**. In a very intense forest fire, the flame length can be two or three times the height of tall burning trees, so in a mature forest, it can reach 150 m (492 ft). If there is no wind, flames rise straight up, so flame length equals *flame height* (the distance from the ground to the top of the flames). When the wind blows and flames are angled, the flame length exceeds the flame height.

The Spread of Wildfires

As we've seen, an intense heat source (such as lightning) can ignite a wildfire instantly. More often, however, wildfires start

FIGURE 14.26 Wildfire classification.

(a) Wildfires vary in character depending on the type of fuel available and the fire's intensity.

(b) Examples of different types of wildfires. Ground fires and surface fires can burn in grasslands or forests. Ladder fires bring fire to the canopy. Crown fires are the most intense.

FIGURE 14.27 Examples of fuel in a forest.

(a) When plants die and dry out, becoming part of forest litter, they become more flammable.

(b) Forest litter and undergrowth provide fuel for a wildfire.

when a small heat source (a cigarette butt or a campfire spark) comes in contact with flammable *tinder* (dry grass, small sticks, or thin bark). Tinder has a large surface area relative to volume, so it undergoes pyrolysis quickly and ignites. Once tinder starts burning, conduction preheats fuel in direct contact with the already burning fuel. Radiation preheats and dehydrates wood on all sides of the fire, and convection and radiation preheat branches above the fire (**FIG. 14.29**). When this fuel becomes hot enough, and pyrolysis has produced enough flammable gas, the fuel ignites.

Once it starts, the wildfire spreads from its *ignition point*, moving outward over a progressively larger area as heat from the existing fire ignites previously unburned fuel. The heat transforms to new fuel by convection,

14.4 Wildfires: Flames of Destruction • 525

FIGURE 14.28 Flame length compared with flame height in the presence of wind.

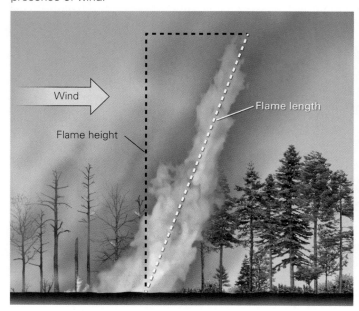

by radiation, and by blowing sparks and burning embers (FIG. 14.30a). Let's consider each of these fire-spreading mechanisms in turn.

1. *Convection:* In the context of wildfires, **convection** refers to the vertical circulation of heat in air. Convection takes place because fire produces hot air and flaming gases. These are less dense than surrounding cooler air, so they rise buoyantly. The rising air transports heat upward from the ground toward the canopy in forests, or up a slope in forests or grasslands. In the absence of a *background wind* (wind due to weather conditions at the time of the fire), heated air and flaming gases rise vertically over a fire. As they rise, cooler fresh air rushes inward around the perimeter of the fire to replace them. This inward air movement, or *convergent wind*, ventilates the fire and maintains the oxygen supply.

2. *Radiation:* Radiation, electromagnetic energy similar to the Sun's light, heats the ground on a cloudless day, and warms you when you sit by a campfire. Wind doesn't affect the radiation emitted by hot objects and glowing gases in a fire, so radiative heating can spread a fire regardless of wind direction. Fortunately, the intensity of radiation produced by a fire decreases rapidly with increasing distance from the fire. Only materials within about 10–35 m (33–115 ft) of a wildfire will receive enough radiation to ignite.

3. *Firebrands:* The rising plume of hot air and flaming gases above a fire produces a strong enough updraft to loft *sparks* (small fiery particles) and *embers* (glowing fragments of wood, leaves, or ash). These materials, called **firebrands**, can land on fresh fuel far beyond the perimeter of the fire. If the material where they land is already dry, new **spot fires** can ignite, which can then spread outward.

The rate at which a wildfire spreads depends on fuel availability, weather (especially wind), and terrain (topography). Fuel availability, in turn, depends on the volume of flammable material available, its vertical distribution, and its moisture content. Sparse or wet fuel, for example, produces a less intense fire, and only if fuel is present at several levels in a forest can a ladder fire develop. The amount of fuel in a location depends on the climate, as different climates favor different types of vegetation and support different volumes of biomass. Other factors include the time since the last fire (a longer time interval allows more fuel to accumulate) and the season (plants have higher moisture content in spring than in fall). Notably, fuel availability increases when disease, insect infestation, or drought kills trees. Huge numbers of trees in the Rocky Mountains have been killed by pine-bark beetles, and these dry, dead trees are much more flammable than they were when alive. The extreme flammability of forests in California stems in part from the drought of 2011–2017, which killed over 160 million trees.

FIGURE 14.29 A large wildfire can start from a small heat source, such as a campfire spark or cigarette butt, if the heat source happens to come into contact with tinder. Once the fire starts, it acts as a heat source, preheating and drying nearby wood, then causing pyrolysis, and finally ignition.

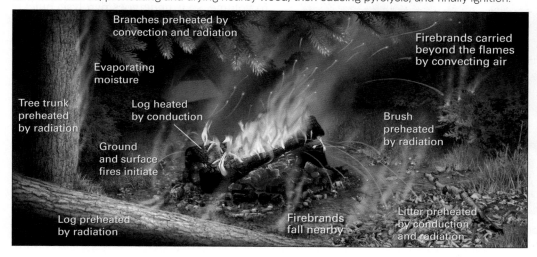

FIGURE 14.30 A wildfire's speed and direction depend on the availability of fuel, the wind, and topography.

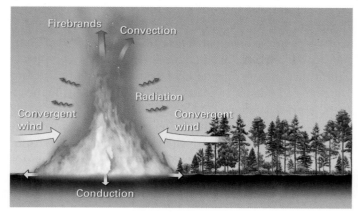

(a) If there is no background wind, heat rises vertically by convection, so the fire spreads slowly. Radiation can ignite surrounding fuel. Conduction can ignite fuel adjacent to the burning fuel.

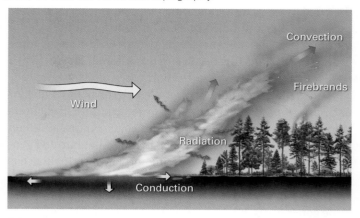

(b) With background wind, the flames tilt, so convection and radiation ignite fuel downwind, and firebrands can ignite spot fires.

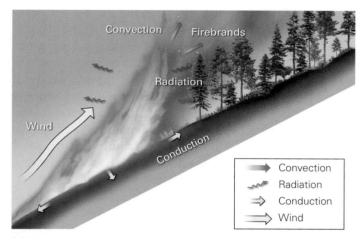

(c) On a slope, convection and radiation heat the fuel upslope, so the fire spreads mostly upslope. Burning debris may ignite trees downslope.

Weather affects the spread of fire by influencing the moisture content of fuel and the availability of oxygen. In very hot and dry air, vegetation dries out and can ignite more easily. Strong winds bring oxygen-rich air to replace the oxygen consumed during burning, so the fire doesn't suffocate and burning continues. Winds also tilt flames so that radiation heats unburned areas, and winds transport firebrands over unburned areas, where fresh fuel can be set afire (FIG. 14.30b). During particularly heavy winds, firebrands can travel downwind for several kilometers, igniting spot fires far from the initial fire.

Terrain, particularly the slope of the land, affects the spread of fire because flaming gases and hot air convect upward. Fire generally migrates faster up slopes than down slopes (FIG. 14.30c), unless downslope winds are strong. In some cases, burning debris can roll downhill and start new fires.

As a wildfire grows and hot air rises, convergent wind becomes stronger, fans the flames, and further intensifies the fire. This process can lead to a **firestorm**, an intense conflagration driven by convergent wind. In some cases, rising air interacts with converging air to produce a *fire whirl*, a vortex of fire (FIG. 14.31). In particularly hot fires, a *fire tornado*, rising to a height of 1 km and hosting 200 km/h (125 mph) winds, can develop.

When a wildfire starts, officials who dispatch emergency personnel assign a name to the fire, to avoid confusion. Names are usually based on geographic locations or features in the area where the blaze began. Because of the importance of wind, firefighters and emergency planners describe the parts of a fire relative to the wind direction (FIG. 14.32a). The *fire perimeter* delineates the boundary around the actively flaming area. A *heading fire* forms along the part of the perimeter moving in the direction in which the wind is blowing—this tends to be the most intense and fastest-advancing part of the fire. A

FIGURE 14.31 Because convection causes hot air to rise, convergent wind rushes in to replace the rising air. This wind fans the flames and can generate a firestorm. In some cases, a fire whirl rises high above other flames.

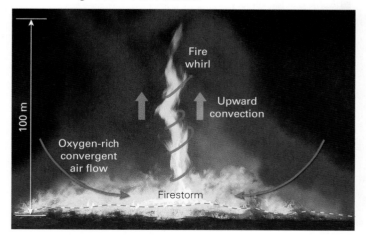

backing fire forms along the perimeter on the side from which the wind blows. This tends to be the least intense and slowest-advancing part of the fire. *Flank fires* occur on the sides of the burnt area and move at rates between those of the heading fire and the backing fire. Because of the different rates at which the parts of the fire advance, the burnt area tends to be elliptical. A fire can spread for weeks unless it runs out of fuel, or until cool, wet weather causes it to die out, or until firefighters stop it—some fires spread for more than 2 months. When looking at a map of a wildfire, keep in mind that burning generally happens only along the perimeter. The interior area has already burned, though it may still be smoldering.

In some cases, two or more fires may be ignited in the same general area by the same general cause (such as a thunderstorm), or embers carried from one fire may be wafted downwind to start new spot fires outside the perimeter of the initial fire. An array of such fires produces a **wildfire complex**. Eventually, separate fires in a wildfire complex may merge into a huge fire. The 2018 Mendocino Fire Complex broke records to become the largest wildfire complex in California's history, ultimately burning a total of 1,858 km² (717 mi²) and causing over $250 million in damage. But just 2 years later, the August Fire Complex, which started as 38 separate fires ignited by lightning on August 17, 2020, was nowhere near being put out when its area surpassed that of the Mendocino Fire Complex (FIG. 14.32b). The fire continued to burn well into November. The Dixie Complex of 2021 also grew to become larger than the Mendocino complex.

The Special Challenge of Downslope Winds

Downslope winds, which flow from higher to lower elevations, are a threat because they can spread fires toward population centers near the base of mountain ranges. Such winds are a particular problem along west-facing slopes near the west coast of the United States and along the eastern slope of the Rocky Mountains. These winds are known as *Santa Ana winds* (named for the Santa Ana Canyon) in southern California, *Diablo winds* (for the Diablo Mountain Range, from the Spanish word for devil) in northern California, and *chinook winds* on slopes along the east flank of the Rocky Mountains, from Colorado north into Alberta, Canada.

Both Santa Ana and Diablo winds develop when a ridge of high pressure lies over a region of high desert in Utah, Nevada, and southern California, particularly as these regions start to cool in the fall (FIG. 14.33a). At that time, low pressure often lies over the lower elevations to the west, and off the coast. Winds flow outward from the high-pressure zone and westward toward the low-pressure zone. These winds become particularly strong if the pressure gradient is large. They pour through gaps in mountain ranges and down canyons toward the coast, accelerating as they flow downslope. The descending air undergoes compression and becomes warmer and drier as it moves to lower altitudes. The winds cross forests, and grasslands at lower elevations, drying and heating vegetation and forest litter, making them more flammable, and driving flames forward when fires develop (FIG. 14.33b). Although chinook winds form for somewhat different reasons—they blow when high pressure lies just west of the Rocky Mountains and low pressure develops on the plains to the east—they also fan the flames of forest fires. Unfortunately, the flanks of scenic mountain ranges are prime real estate because of the beauty of the countryside and the cool air (relative to lower elevations). In the past few decades, many homes have been built in the paths of fires that have been intensified by downslope winds (BOX 14.2).

FIGURE 14.32 Fire perimeters and spreading directions.

(a) The boundary around an active fire is the fire perimeter. When the wind blows, a fire migrates faster downwind from the ignition point, so the fire perimeter is roughly elliptical.

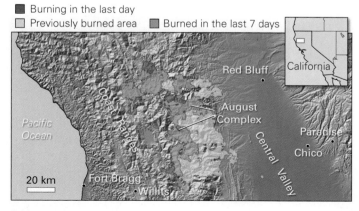

(b) The August Fire Complex of 2020 in northwestern California, as it was advancing to the west.

FIGURE 14.33 Downslope winds fan the flames of wildfires in west-coast states.

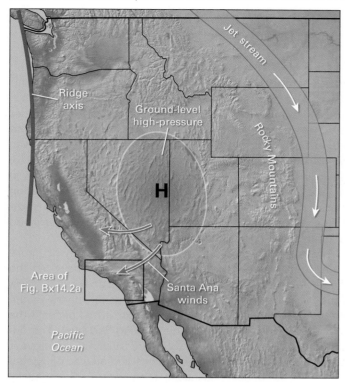

(a) When a high-pressure air mass lies over the high deserts of Nevada, Utah, and California, winds spiral outward. Some of these winds blow west and flow downslope to the coast.

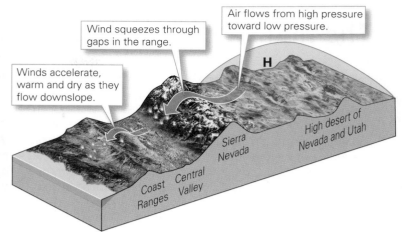

(b) When the winds reach west-coast mountains, they flow down canyons and accelerate. As they descend, they become warmer and drier. Diablo winds cross the Sierra Nevada, fanning fires in the foothills, then cross the Central Valley and the Coast Ranges, fanning fires along the Pacific coast.

Climate Controls on Wildfires

As we discuss in Chapter 15, climate change has produced hotter and drier conditions in some highly vegetated areas, leading to larger and more frequent fires, some of which grow into megafires like the fire complexes of California. Wildfires occur around the world, but not everywhere. They are particularly common in monsoonal and Mediterranean climate zones. Both zones are characterized by distinct wet seasons, during which abundant rains trigger plant growth and build a fuel supply, and long dry seasons, during which relentless sunshine and heat dry out the plant material, making it susceptible to ignition.

Notably, most wildfires in monsoonal climates begin when people set fires intentionally. The purpose of such fires is to control undergrowth, clear forests to make land available for farming or ranching, reduce the biomass of crops such as sugarcane before harvest, prevent grasslands from becoming forests, or stave off unwelcome wildlife. Intentional burning occurs to a dangerous extreme in Brazil, where every year nearly 100,000 fires destroy rainforest and grasslands **(FIG. 14.34)**.

FIGURE 14.34 Many wildfires in monsoonal climates result from fires set intentionally.

(a) A wildfire in the Amazon rainforest.

(b) The aftermath of an Amazon rainforest fire.

14.4 Wildfires: Flames of Destruction • **529**

BOX 14.2 DISASTERS AND SOCIETY

The human cost of downslope winds in California

California's canyons experience strong downslope windstorms every year. Air flows westward from the high desert of Nevada and eastern California, down the western slopes of the Sierra Nevada, San Bernardino, and San Gabriel mountain ranges, and into the Central Valley and the Los Angeles basin. These warm, dry winds are notorious because they frequently spread wildfires into population centers. As many as 20 downslope windstorms, each lasting about 24–36 hours, happen in some years. On occasion, the winds persist much longer—in September 1970, Santa Ana winds fanned wildfires for 10 days.

During the west-coast wet season (December through March), grasses, shrubs, and trees normally flourish. During the rest of the year, almost no rain falls, and vegetation dries out, producing an immense volume of fuel. Conditions favorable for downslope winds typically occur between September and March, but they can begin earlier, as they did in 2020. The most dangerous times are the dry months of September through November, when the vegetation becomes desiccated before the seasonal rains arrive.

Because towns and suburbs have expanded to the edge of and into wildlands, wildfires blown by downslope winds have incinerated many homes and, in some cases, have destroyed whole towns. For example, in 1991, a blaze known as the Oakland Firestorm was swept down the Berkeley Hills and into the city of Oakland by 100 km/h (62 mph) winds. Embers carried by the winds lit spot fires ahead of the initial fire, so even eight-lane highways did not prevent the fire's advance. Soon, convergent wind became so strong that the conflagration evolved into a firestorm. Firefighters from all over California and adjacent states came to fight the fire. Unfortunately, their efforts were stymied in many ways: power failures cut off the water supply, hoses from fire departments outside Oakland didn't fit the city's fire hydrants, and car-filled streets blocked firefighters' access to the fire. At one point, houses were igniting at a rate of about 25 every 5 minutes. Despite the efforts of 1,500 firefighters, the blaze lasted for 3 days and destroyed 3,000 homes. In the aftermath, firefighters worked to establish standard practices for fighting such fires, and they developed better water-supply distribution systems so that future fires could be extinguished more quickly.

Disasters related to downslope wind–driven fires have happened repeatedly in California over the past few decades. The worst ones took place between 2017 and 2020. Plant growth during the wet winter of 2017 provided large amounts of fuel. That fall, over 9,100 wildfires rampaged through California, destroying more than 10,800 structures, burning over 5,000 km^2 (1,930 mi^2), including famous wine-producing areas, and killing at least 46 people. The October fires in northern California resulted in more than $9.4 billion in insured property losses. In December, Santa Ana winds blew fires westward across southern California, causing the destruction of thousands of structures and over 1,000 km^2 (386 mi^2) of forest, and the evacuation of hundreds of thousands of people. The Thomas Fire, which started in December 2017, was the largest fire in California history until that time **(FIG. Bx14.2a–d)**. It started from a power-line spark and was fanned by Santa Ana winds. It burned until March 2018, destroying almost 1,100 buildings and causing $2 billion in damage. It cost $230 million to control, led to the evacuation of over 100,000 people, and burned an area almost the size of Los Angeles. The fires throughout the region strained national firefighting resources to their limits.

Fire returned to plague California in 2018, with more record-breaking conflagrations. One of these, known as the Camp Fire, sparked by damaged power lines on the west flank of the Sierra Nevada, reduced several towns to cinders. By the time it finally was contained 17 days later, it had become the most destructive fire in state history **(FIG. Bx14.2e)**. The fire started upslope, but 56 km/h (35 mph) Diablo winds drove it downslope. Fueled by dry vegetation, it spread into populated areas, killing 85 people and leading to the destruction of nearly 19,000 buildings. Nearly 95% of two towns, Paradise and Concow, vanished in firestorms. When the tens of thousands of evacuees were finally able to return, many found only the ash-covered foundations of their homes and empty metal hulks of what had been their cars. Many lost everything they owned, as well as their livelihoods. The fire resulted in about $16.5 billion in damage, of which about 25% was uninsured. The utility company blamed for the faulty power line admitted guilt for 84 counts of manslaughter. It paid out over $13 billion in settlements, after declaring bankruptcy.

When wind and fire returned in 2019 and 2020, utility companies took the unprecedented step of cutting off power across wide areas, sometimes for days, to keep power lines from triggering new fires while downslope winds blew. But the precautions did not prevent disaster from returning. All of the 2018 season's fire records were broken again in 2020. Fires sprouted up in California and Oregon, and by September 10, long before the end of the fire season, close to 8,000 fires had struck. At times, nearly 400 fires were active at the same time. Over 5 million acres burned, more than the area of the state of Connecticut. Satellite images showed the plume of smoke moving far out over the Pacific and spreading eastward, eventually over much of the United States **(FIG. Bx 14.2f)**. In cities near the fire, such as San Francisco, smoke settled like a choking fog, sickening many people.

Recovery from disastrous wildfires can take years. Residents may pick up and move elsewhere, or make slow progress toward rebuilding as they seek funds from aid agencies or insurance companies. Each fire disaster inspires communities to plan better, but the cost of fire avoidance projects can't always be borne by local tax bases.

Why does California face such an overwhelming fire challenge? Large areas of forest remain choked with forest litter and underbrush due to years of extreme fire suppression and lush plant growth during wet winters. The warming climate, with the associated reduction in rainfall and earlier melting of snowpack, has made vegetation drier during the dry season, so fires ignite and spread faster. Compounding these problems, dead trees are widespread due to the multi-year drought in the region. And finally, population centers have continued to expand and intrude into wildlands, so the number of vulnerable communities has increased.

FIGURE Bx14.2 Downslope winds and the 2017–2020 California fires

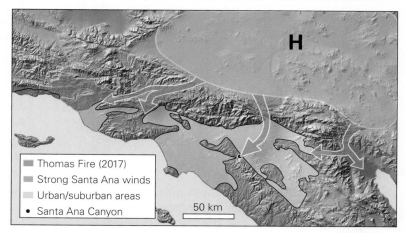

(a) Santa Ana winds blow from the high Mojave Desert region through canyons in the mountains bordering the northeastern side of the densely populated Los Angeles Basin.

(b) Smoke from several fires, including the massive Thomas Fire of December 2017, blows offshore, carried by Santa Ana winds.

(c) The Thomas Fire burned across the suburban-wildland divide.

(d) The Thomas Fire completely incinerated many homes.

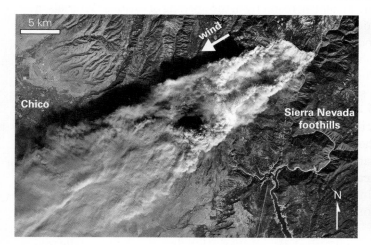

(e) The 2018 Camp Fire, one of the most destructive in California history.

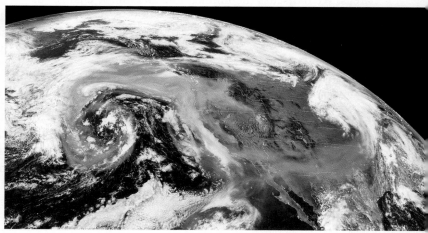

(f) Smoke from the 2020 fires on the west coast spread out over the Pacific and eastward across the United States.

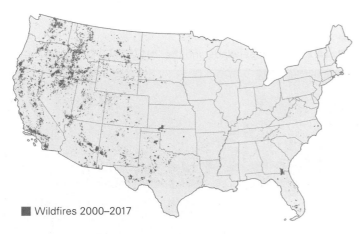

FIGURE 14.35 Wildfire distribution in the United States varies from year to year. This compilation of wildfire locations for 2000–2017 shows which states have the most wildfires.

Wildfires are not unique to monsoonal and Mediterranean climate zones, however. They can occur anywhere that forests and grasslands endure droughts or heat waves. In North America, for example, wildfires have raged across western forests as far north as Alaska, over grasslands of the Great Plains, and through coniferous and deciduous forests of the western, southern, and eastern United States (**FIG. 14.35**). Fires burned through 17,000 acres of the Everglades, the wetlands of southern Florida, during a drought in 2019.

Because of the relationship between wildfire susceptibility and climate, climate change can affect the frequency and intensity of fires. For example, California has seen an overall increase in average temperature by about 2.7°C (3°F) in recent years. While this change may seem small, it has a significant effect on the rate at which vegetation dries out. As noted in Box 14.2, because of this increased desiccation, fires ignite more easily, burn more intensely, and spread more easily. Climate change may also influence global weather patterns in ways that lead to an increase in fire-threatened areas. We'll discuss the implications of climate change further in Chapter 15.

> **Take-home message . . .**
>
> Wildfires are self-sustaining fires that spread out of control and often move into or through populated areas and agricultural areas. For a wildfire to begin, the environment must provide fuel, heat, and oxygen—the three components of the fire triangle. Once burning, wildfires spread as the result of convection, radiation, the lofting of firebrands, and the action of strong winds. Wildfires are most common in monsoonal and Mediterranean climate zones, but can occur in other temperate climate regions.
>
> **QUICK QUESTION** What are the differences among ground, ladder, and crown fires?

14.5 Consequences and Mitigation of Wildfires

Impacts of Wildfires

Tragic consequences result when wildfires spread into populated areas, injuring or killing people and destroying homes and businesses. Although the overall death toll from wildfires tends to be lower than that from other severe natural disasters, it can be horrific, as we've seen. The worst loss of life due to a single fire complex in the United States occurred in 1871. Over 1,500 people were killed by the Peshtigo Fire in Wisconsin when, driven by strong winds coming off Lake Michigan, it burned 30,000 km² (11,583 mi²) of forest, including 17 towns. A similar fire struck the border area between Wisconsin and Minnesota in 1918, killing over 1,000 people.

The cost of a wildfire includes the value of the forest, grass, or crops it consumes as well as the buildings it destroys. The 2018 fire season in California, which affected many towns and burned 35,000 km² (13,514 mi²), is estimated to have cost over $24 billion. The cost of California's wildfires during 2017 and 2018 combined was a staggering $40 billion. Notably, a substantial proportion of a fire's cost is the expense of containment and suppression—the costs of fighting a single fire may be over $200 million. In 2018, the costs of suppression borne by the federal government exceeded $3.1 billion. Adding the expenditures by state and local governments raised that total substantially.

Major wildfires can also trigger secondary disasters. The first type happens due to the smoke produced while the fire itself is active. *Smoke inhalation* can cause health problems, since smoke contains many chemicals (both unburned chemicals released by pyrolysis and chemicals produced by combustion), some of which are toxic (**FIG. 14.36a**). Toxic chemicals can result from the burning of various materials in houses and cars, especially plastics. Very fine particles of soot and ash in smoke can cause serious problems if they are inhaled deep into the lungs, where they lodge permanently. The short-term effects of exposure to smoke include eye and respiratory tract irritation, breathing problems, and bronchitis, while long-term exposure to smoke can lead to asthma, cancer, and premature death. Most people survive the short-term effects of smoke, but they can be particularly dangerous for older people and those with compromised health. Long-term exposure is problematic for everyone. Unfortunately, smoke from large fires can spread across a continent, so the effects of smoke may be widespread.

The second type of secondary disaster develops because fire not only destroys vegetation, but it can change the texture and composition of soil. Without overlying leaves,

FIGURE 14.36 Secondary disasters associated with wildfires.

(a) Choking smoke rises above a wildfire near Yosemite National Park.

(b) Slope erosion after a wildfire.

more water reaches the ground when it rains, so the volume of overland flow increases. (If soil permeability decreases, as sometimes happens during intense fires, less water infiltrates into the ground, so overland flow for a given rain event increases even further.) In addition, the direct impact of raindrops lifts ash and debris so it can become suspended in overland flow. Overall, therefore, burned land may become subject to accelerated *slope erosion*, as soil erodes and new gullies form **(FIG. 14.36b)**. Consequently, significantly more sediment will enter the drainage network. This sediment, which may contain toxic substances, can clog or contaminate the water supply for communities downstream. If the fire has destroyed plant roots that hold soil together, rainwater that infiltrates the ground can make regolith on slopes unstable. Potential slumps and debris flows threaten downslope areas during the next rainy season. Mass wasting from fire-ravaged hills can block stream channels and cover roads and homes. The debris can also be carried in flash floods that can devastate communities many kilometers downstream of where the fire happened.

Fighting Wildfires

When a wildfire has been detected, officials get to work to assess the fire, determine the directions in which it likely to spread, and plan how to address the safety of communities in its possible path (**VISUALIZING A DISASTER**, pp. 534–535). They also name the fire, to avoid confusion in communication. The highest priority of fire management officials is to protect human life, so early decisions determine safe zones and evacuation plans. While addressing the safety of threatened populations, officials develop plans for deploying firefighting crews, and send out calls for personnel and equipment.

Fighting wildfires focuses on removing fuels ahead of an advancing fire, cooling fuels below their combustion temperature, and cutting off the oxygen supply to fuels. Small fires can be smothered with water, dirt, or foam. Larger fires can't be extinguished, so instead firefighters develop a strategy for **containment**—that is, to prevent the fire from spreading beyond a certain limit. Reports of firefighting progress describe the success of containment as a percentage. When news reports state that a fire has been "50% contained," that means that a **firebreak**, a natural or built barrier that prevents the fire from spreading, has been completed around half of the fire's perimeter. When a fire has been 100% contained, firebreaks completely surround the perimeter, so that the fire no longer threatens unburned areas. Once it is contained, the fire consumes all available fuel, and by doing so, it burns itself out. In the case of huge wildfire complexes, months may pass before they are completely contained. Success may come only when the weather cooperates by providing rain and cooler temperatures. Notably, even an increase in humidity can help slow a wildfire's spread, as unburned plant material absorbs water vapor and becomes less likely to burn.

Firefighters attempt to slow the advance of a large fire by spraying chemical **fire retardants** on fuel near the fire's

FIGURE 14.37 Fighting wildfires involves hard labor and specialized equipment.

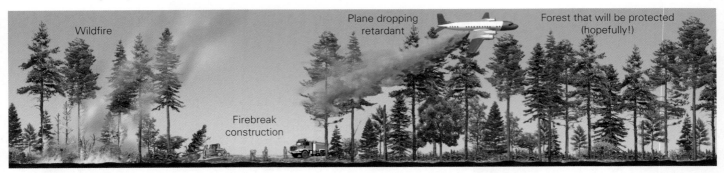

(a) Firefighters produce a firebreak on the ground in the path of a wildfire. Aircraft drop fire retardant on the opposite side of the firebreak from the fire.

(b) Frontline firefighters must wear protective clothing when getting close to fire to clear away brush.

(c) An aircraft drops fire retardant ahead of a wildfire.

perimeter (**FIG. 14.37**). Retardants work in different ways, depending on their chemical composition: they may cool the fuel, form a coating that prevents the fuel from igniting, release water or carbon dioxide when burned, or inhibit oxidation reactions. While some firefighters work to slow or extinguish fires, others produce firebreaks (**FIG. 14.38a**). Firebreaks may follow previously cleared areas, such as roads, or natural features, such as rivers or ridgelines. Where no such firebreak exists, firefighters clear brush and trees along a line in advance of the fire perimeter, using shovels, axes, bulldozers, and even explosives. Unfortunately, wind can carry firebrands over a firebreak, and radiation and convective heating from a fire can be so intense that the fire "jumps" the firebreak, igniting fuel on the far side. To prevent a fire from jumping the firebreak, firefighters spray water or retardants on the side of the break farther from the fire's perimeter. On the side closer to the fire, firefighters may ignite controlled fires called **backfires**, using **drip torches**, canisters from which a burning liquid spills on the ground (**FIG. 14.38b**). Ideally, convergent wind generated by convection over the primary fire will draw the backfire toward the primary fire. The backfire consumes fuel in the primary fire's path so that when the two fires meet, the primary fire stops advancing.

The success of wildland firefighting depends on the expertise, fitness, and bravery of firefighters. Elite firefighters, known as *hotshots*, tackle the worst wildfires. Hotshots receive specialized training in safety practices, risk management, fire behavior, communications, and fire-suppression techniques. While firefighters always plan possible escape routes, rapidly changing weather conditions can sometimes trap them. If trapped, a firefighter can deploy a portable personal fire shelter designed to contain breathable air while reflecting radiant heat and protecting against convective heat. Needless to say, fighting wildfires is extremely dangerous.

Firefighting in the wild requires equipment ranging from shovels and chain saws, to familiar fire trucks, to bulldozers and tractors, to specialized aircraft. Helicopters can rescue people trapped by a fire, carry crews and equipment, provide reconnaissance, and drop water or retardant. Fixed-wing aircraft carry teams of *smoke jumpers*, firefighters who parachute into remote areas to reach wildfires early and extinguish them before they spread. *Air tankers* carry water or fire retardant; some jumbo jets serve as giant air tankers

FIGURE 14.38 Producing a firebreak by controlled burning.

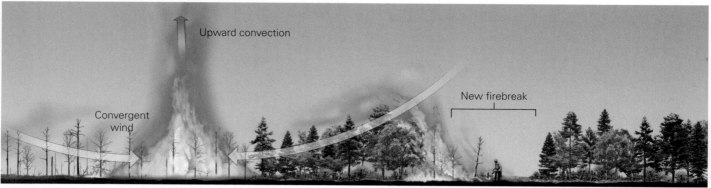

(a) The positioning of a backfire takes advantage of convergent wind to consume the fuel between the wildfire and unburned forest. A backfire, in effect, produces a broad firebreak.

(b) Setting a backfire with a drip torch.

and can drop as much as 75,000 L (20,000 gal) of retardant (see Fig. 14.37c).

In the United States, numerous federal and state agencies share responsibility for fighting wildfires. These agencies cooperate under the joint supervision of the National Wildfire Coordinating Group (NWCG) and the National Interagency Fire Center (NIFC). The NWCG publishes standards, guidelines, qualifications, training courses, and other information to enable operations. The NIFC serves as the dispatch center for air tankers, helicopters, and other firefighting aircraft, as well as for fire crews, management and command teams, military resources, communications, and weather forecasters.

Mitigating Wildfire Hazards

Wildfires are terrifying and destructive. In the late 19th and early 20th centuries, the focus of government agencies was to prevent wildfires from starting or spreading. Supporters of **wildfire suppression** believed that wildfires did much more harm than good, and in the United States, national advertising campaigns featured Smokey Bear, who said, "Remember, only YOU can prevent forest fires!" **(FIG. 14.39)**. A major change of thought began to take hold in the 1970s, as research demonstrated that fires prepare new seedbeds, encourage germination, and help tree seedlings survive. Fires recycle nutrients, remove excess undergrowth, nurture the natural succession patterns of plants, create new food sources, lessen the impacts of insects and disease, and reduce a region's fuel load. Put simply, fire is a natural process that helps sustain the health of forest ecosystems. Studies following the enormous wildfires that raged through Yellowstone National Park in 1988, burning 36% of the park's area, confirmed this conclusion **(FIG. 14.40)**. Agencies responsible for managing wildfire hazards now

FIGURE 14.39 A traditional Smokey Bear sign urging fire suppression.

FIGURE 14.40 The 1988 fires in Yellowstone National Park and their aftermath.

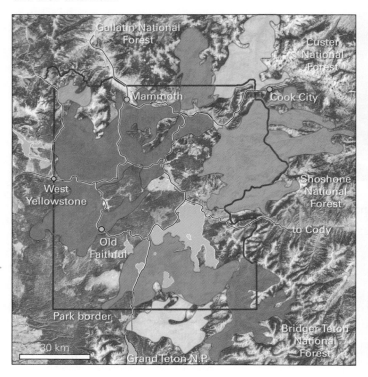

(a) Over a third of the park's area burned. The colors denote different fires during the 1988 season.

(b) The fires cleared growing vegetation from patches of forest.

(c) Several years after the fires, the forest began to recover.

FIGURE 14.41 A firebreak may be constructed prior to a fire. This example in the Sierra Nevada foothills is 40 m (131 ft) across and follows the path of a power line.

allow fires in forests and rangelands to take their natural course if they do not threaten human life or developed areas. Since 2001, Smokey Bear has used the term *wildfire*, instead of *forest fire*, to emphasize that the fires we should prevent are unintentional fires that threaten human life and property.

Intentionally set fires, called **prescribed burns** or *controlled burns*, are designed to reduce fuel loads and prevent future catastrophic wildfires. Prescribed burns prevent the accumulation of excess forest litter and underbrush, so that any fires that do start will be less intense and more easily contained. In a few cases, officials have brought in herds of goats to eat underbrush. Preventing wildfires from spreading can also be accomplished by building and maintaining firebreaks to contain future fires and provide firefighters with access to those fires **(FIG. 14.41)**. Public education helps decrease the number of fires set accidentally by individuals, and proper maintenance of power lines can prevent electrical sparking from igniting fires.

Officials can undertake many steps to lessen wildfire hazards. States can prepare *wildfire hazard maps* so residents can see their risks, and they can put systems in place to coordinate statewide response to wildfires **(FIG. 14.42)**. Emergency responders can receive comprehensive training in wildfire

FIGURE 14.42 Wildfire hazard maps.

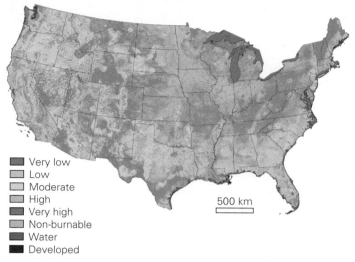

- Very low
- Low
- Moderate
- High
- Very high
- Non-burnable
- Water
- Developed

(a) A wildfire hazard map for the United States. Wildfire hazard is greatest in the western and southern states.

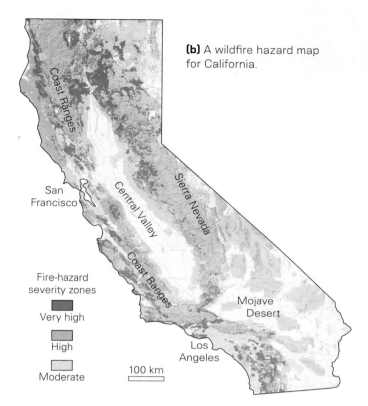

(b) A wildfire hazard map for California.

Fire-hazard severity zones
- Very high
- High
- Moderate

management and containment. Individuals can help make their homes less vulnerable by ensuring that an area of cleared land surrounds the house to act as a firebreak, that the roof is covered with fireproof materials, and that windows are glazed with dual-pane tempered windows, which resist fire four times as effectively as single-pane windows **(FIG. 14.43)**. Homes at the tops of slopes need to be set back at least 30 m (100 ft) from the slope edge, so that convecting hot air coming up the slope doesn't ignite the house. Individuals should always be prepared to evacuate. As more and more people live within areas subject to wildfire hazards, and as climate change continues to cause potential wildfire fuel to become drier, efforts at prevention become ever more important.

Take-home message...

The costs and societal consequences of wildfires are increasing as more people move into vulnerable areas. Fighting a wildfire requires removing fuel, cooling fuel, or cutting off the oxygen supply. Mitigation of wildfire hazards begins with education, followed by actions such as building firebreaks and modifying houses to make them more resistant to ignition.

QUICK QUESTION What is a hotshot?

FIGURE 14.43 Safeguarding a house from wildfire in a vulnerable location. Taking steps during construction and landscaping can help prevent ignition.

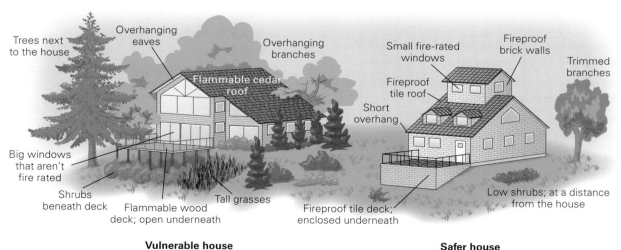

Vulnerable house — Trees next to the house; Overhanging eaves; Overhanging branches; Flammable cedar roof; Big windows that aren't fire rated; Shrubs beneath deck; Flammable wood deck; open underneath; Tall grasses

Safer house — Small fire-rated windows; Fireproof brick walls; Trimmed branches; Fireproof tile roof; Short overhang; Fireproof tile deck; enclosed underneath; Low shrubs; at a distance from the house

14.5 Consequences and Mitigation of Wildfires

Chapter 14 Review

CHAPTER SUMMARY

- Drought in a region is a deficiency of freshwater that lasts long enough to harm normal vegetation, crops, livestock, water supplies, human health, and human activities.
- Historically, drought has led to the worst natural disasters experienced by humanity.
- A meteorological drought occurs when precipitation over an extended period is less than the average for a region. An agricultural drought develops when there is insufficient moisture in the soil layers to support crop growth. A hydrological drought is a prolonged deficiency of groundwater or stream flow. Socioeconomic drought occurs when water shortages affect the well-being of people and reduce economic output.
- Drought's worst impacts occur in regions with distinct wet and dry seasons. Failure of rains in a wet season can lead to crop failures, starvation, and disease.
- Rainfall in monsoonal climate zones is associated with the passage of the ITCZ across the region. Drought develops when the rains fail to produce sufficient water.
- The impacts of drought increase when poverty, poor infrastructure, and disease make relief efforts difficult.
- Mediterranean climate zones experience rainfall during the winter and dry conditions in the summer. Rain tends to occur during the passage of mid-latitude cyclones in winter. When these storms are infrequent, drought develops.
- Drought in the western United States occurs when a persistent ridge in the jet stream lies over the west coast, so that mid-latitude cyclones move north of coastal states.
- Causes of drought in the central and eastern United States relate to complex interactions among the Bermuda High, sea-surface temperatures in the Atlantic, the position of the jet stream, soil moisture, and plant cover.
- In the United States, drought is monitored by the NWS using the Palmer Drought Severity Index. The US Drought Monitor is a regularly updated map of drought conditions.
- Heat waves are periods of abnormally hot, and often humid, conditions that cause heat stress in humans.
- Humans experience heat illnesses when the core temperature rises about 3°C (5°F) above normal, and death can result from a core temperature rise of about 5°C (10°F).
- Heat stress in humans depends on the apparent temperature, which is determined by air temperature, relative humidity, wind speed, and exposure to the Sun's radiation.
- The heat index is a common measure reported in the media to communicate the combined effects of air temperature and relative humidity on human comfort.
- Heat stress is a particular risk in low-income urban areas, where temperatures are higher than elsewhere and relief may not be available from air conditioning.
- Heat waves occur in mid-latitude regions due to a ridge in the jet stream, an influx of tropical air, dry ground with little growing vegetation, and inversion.
- In monsoonal climate zones, heat waves result when winds bring in hot, dry air.
- Fire results from the rapid oxidation of a fuel. The visible part of a fire consists of glowing superheated gases.
- Heat, oxygen, and fuel are all required to maintain wildfires. Fuel includes trees, grasses, shrubs, decaying organic material, and other flammable material.
- Most natural wildfires are caused by lightning. Human activities cause about 90% of all US wildfires.
- Depending on the nature of fuel, wildfires may spread as ground fires, surface fires, ladder fires, or crown fires.
- Fire behavior and rate of spread are determined by weather, topography, and fuel availability and type.
- Wildfires spread through the processes of convection, radiation, and firebrand transport to unburned areas.
- Wildfires in California are driven by a unique combination of climate and weather conditions that support a large supply of fuel and strong downslope windstorms that fan flames.
- Wildfires in monsoonal climate zones often start as controlled burns to clear land for agriculture. In Mediterranean climate zones, winter rains support rapid growth of vegetation, while the summer dry season dries the plants, priming them to act as fuel.
- Various techniques are used to control wildfires, such as dropping water or chemical retardant from aircraft and establishing firebreaks and backfires on the ground.
- Fire is now understood to be a natural part of a healthy forest ecosystem. When wildfires do not threaten human life or infrastructure, they are sometimes allowed to burn.

REVIEW QUESTIONS

Blue letters correspond to the chapter's learning objectives.

1. Explain why agricultural drought normally precedes hydrological or socioeconomic drought. **(A)**
2. Why are droughts more likely to trigger famine in countries with monsoonal climate zones than in those with Mediterranean climate zones? **(A)**
3. What external factors, aside from weather and climate, contribute to disaster and famine when a drought occurs? **(A)**
4. What meteorological feature of the atmosphere is responsible for the wet-season rainfall in monsoonal climate zones? **(D)**
5. When the Bermuda High is in the position shown, can a drought develop in the central United States? Explain your answer. **(D)**

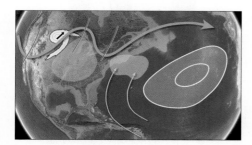

6. The average temperature of the human body's core is 37°C (98.6°F). How much warmer does the core have to get for serious health issues to appear? **(B)**
7. Heat waves typically lead to more fatalities in Chicago, Illinois, than in Phoenix, Arizona, even though the July average high temperature in Phoenix is far higher than that in Chicago. Why is this the case? **(C)**
8. What four factors influence the rate at which the human body heats when outdoors in hot weather? Which of these factors are included in the heat index? **(C)**
9. How do solar radiation and wind affect the human body? What roles do these factors play in the body's responses to heat when the outdoor temperature is 40°C (104°F)? **(C)**
10. Why are heat waves more likely when the ground is dry from lack of rain? **(D)**
11. What is an urban heat island? **(D)**
12. Describe the basic chemical reaction that takes place during burning, and explain how it differs from photosynthesis. **(E)**
13. Describe the changes that take place as a piece of wood undergoes heating and, eventually, ignition. **(E)**
14. What is the fire triangle? Describe the sources of each component shown in the figure. What are the common causes of wildfires? **(E)**

15. By what processes do wildfires spread? Which of these is likely to trigger spot fires ahead of the main fire? **(E)**
16. How does wind affect wildfires? What is a convergent wind, and how does it develop? What conditions cause the winds that can push fires downslope? **(E)**
17. Why are wildfires most common in monsoonal and Mediterranean climate zones? **(F)**
18. Are Mediterranean climate zones more likely to experience wildfires in spring or fall? **(F)**
19. What does it mean if a headline states, "The Bear Ridge Fire has been 50% contained"? **(G)**
20. How does fire retardant work? Why are firebreaks and prescribed burns important in controlling wildfires? **(G)**
21. What precautions can you take to minimize the chance that your house will be destroyed by a wildfire? **(G)**

On Further Thought

22. Throughout history, severe droughts have caused human populations to undertake mass migrations to avoid starvation and seek better living conditions. Identify one such event and its impact on the migrating population. **(A)**
23. Although wildfires have always been part of California's landscape, their destructive potential and financial costs have increased over the decades. Why might this be so? **(G)**

15

THE STEALTH MEGADISASTER
Climate Change

By the end of the chapter you should be able to . . .

A. interpret the factors that control a region's prevailing climate.

B. explain causes of long-term and short-term climate change.

C. evaluate evidence concerning modern-day global warming and its causes.

D. assess the impacts of future global climate change on society.

E. discuss the various options for mitigating or adapting to climate change.

15.1 Introduction

It seems that almost every day, news media carry a story about melting glaciers, rising sea level, unbearably hot temperatures, frequent wildfires, devastating floods, or worsening algal blooms (FIG. 15.1). Do these phenomena signal that our planet's overall climate is becoming different? Citizens, politicians, special-interest groups, and opinion writers continue to argue about whether *climate change* is happening, whether human activities cause it, and what the consequences of change will be. It's no wonder that many people are confused about how society must act to reduce the rate of change.

Thousands of articles in scientific journals conclude that indeed, climate change is, and has been, happening, that its impacts are potentially catastrophic, and that human society plays a major role in causing it. Concerned citizens and most world governments worry that if climate change continues unabated, many of the natural disasters discussed in this book will become more frequent or more severe, producing challenges to society that far exceed those caused by the COVID-19 pandemic. But despite the evidence documenting climate change and its causes, the topic remains contentious in some political circles. The controversy exists because climate change is a stealth disaster. The consequences of this disaster are developing so slowly that they can be hard to detect and easy to dismiss. Slowing climate change will require society to alter the ways it produces and uses energy, land, commodities, and resources.

In this chapter, to characterize climate change and its influence on natural disasters, we first examine the Earth's current climate and the processes that control it. Next, we describe how climate has changed over our planet's history, how changes in the past two centuries differ from earlier changes, and how human activities contribute to recent changes. We then explore linkages between climate change and natural disasters. We conclude on a more optimistic note by considering what people can do to slow climate change and mitigate its consequences.

15.2 Controls on the Earth's Climate

◀ A coal-burning power plant near the Grand Canyon, Arizona, emits CO_2 into the atmosphere. Globally, CO_2 emissions have changed the composition of the Earth's atmosphere, which may cause or amplify natural disasters.

Why can the Earth support life, while the Moon and neighboring planets cannot? Life as we know it depends on liquid water, so for the planet to support life, its temperature range must allow liquid water to exist. This temperature range occurs for two reasons. First, the Earth lies within the Sun's *habitable zone*, defined as the distance from the Sun where radiation intensity is neither too strong nor too weak. And second, the composition of the Earth's atmosphere traps just the right amount of heat for water, and therefore, for life, to exist. To understand these statements, let's look at how the Sun's energy interacts with the Earth's surface and atmosphere. We can then explore how climates naturally vary around the world today, and why. This knowledge provides a basis for understanding climate change.

Heating the Atmosphere: Solar Radiation and the Greenhouse Effect

All objects—the Sun, the sand on a beach, the chair you're sitting on—emit *electromagnetic radiation*, energy that passes through space as electromagnetic waves traveling at the speed of light. Electromagnetic waves come in a broad spectrum of wavelengths (FIG. 15.2). The wavelengths that an object emits depend on the temperature of the object: very hot objects emit mostly shortwave radiation (ultraviolet light, visible light, and some infrared light), whereas cooler objects emit longwave radiation (infrared light). *Solar radiation*, the energy that arrives at the Earth from the very hot Sun, consists almost entirely of shortwave radiation. If the Earth were to absorb all that incoming solar radiation without sending any energy back into space, it would quickly become hot enough to vaporize. But that doesn't happen because, over time, the amount of shortwave energy arriving at the Earth from the Sun exactly balances the amount of longwave energy returned to space from the relatively cool Earth.

Air is essentially transparent to visible light—which is why we can see through air—so most solar radiation that arrives at the top of the Earth's atmosphere reaches either its surface or its clouds. About 70% of that energy is absorbed by soil, rock, vegetation, surface water, or water

FIGURE 15.1 Warming water, along with an influx of nutrients, has stimulated toxic algal blooms in this pond, which was crystal clear 20 years ago.

FIGURE 15.2 The electromagnetic spectrum. Electromagnetic wavelengths vary from the width of an atomic nucleus to the length of a football field. Solar radiation consists mostly of ultraviolet and visible light. The Earth's surface emits infrared radiation.

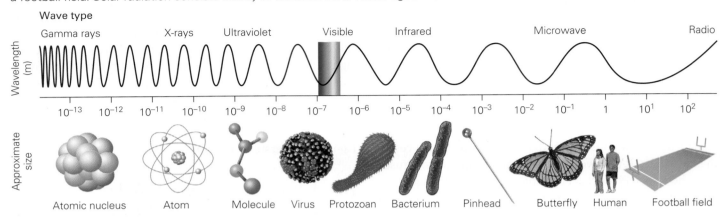

within clouds. The remaining 30% reflects back into space because the Earth's surface and clouds have a relatively high reflectivity, or **albedo** (measured as the percentage of light reflected by a surface).

The solar energy absorbed by the Earth's surface is reradiated into the atmosphere as infrared radiation. Recall that air consists mainly of oxygen and nitrogen gas (O_2 and N_2). Neither gas absorbs this infrared radiation. But trace gases, such as carbon dioxide (CO_2) and methane (CH_4), as well as water vapor, absorb infrared radiation. The absorbed energy causes molecules of these gases to vibrate. This vibration increases their thermal energy, which is then reradiated—some of it upward toward space, but most of it right back to the Earth's surface (**FIG. 15.3**). The ability of these gases to trap energy in this way keeps the air, and therefore the Earth's surface, warm. Without this trapping, our planet's average global surface temperature would be well below freezing, despite being in the Sun's habitable zone, and life might never have evolved.

The way our atmosphere traps heat is known as the *atmospheric greenhouse effect*, or just the **greenhouse effect**, and the gases responsible for trapping heat are called **greenhouse gases**. The analogy to a greenhouse isn't perfect, however, because a real greenhouse traps warm air, not

FIGURE 15.3 Energy exchange between the Earth and its atmosphere. Arrow thicknesses represent an amount of energy. The numbers are arbitrary units of energy. Note that if 100 units of radiation come from the Sun, the amount absorbed by the atmosphere (19 + 4 + 23) and the Earth (47 + 7) equals 100. Similarly, to maintain energy balance, the overall amount of outgoing radiation (23 + 7 + 49 + 9 + 12) also equals 100, and the energy input to the surface (47 + 98) equals the energy release from the surface (24 + 5 + 116).

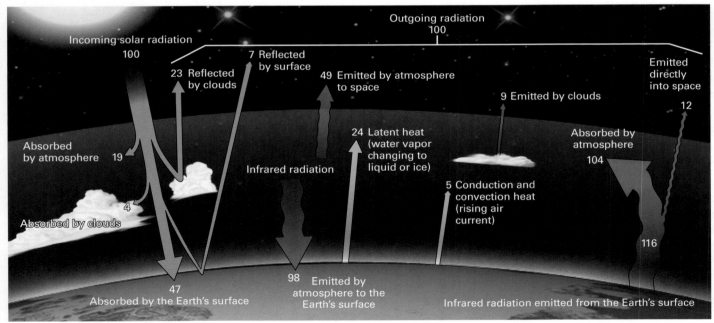

544 • CHAPTER 15 • The Stealth Megadisaster: Climate Change

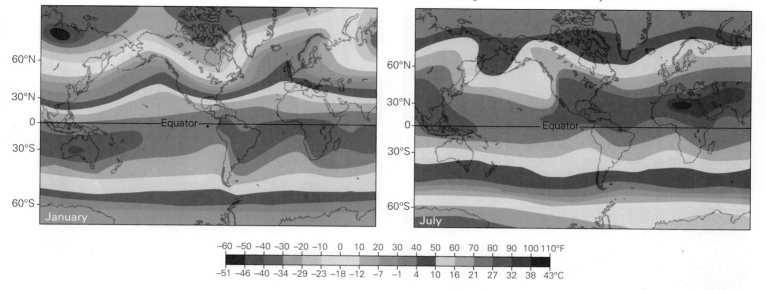

FIGURE 15.4 Average monthly surface temperatures vary with latitude and change over the course of a year.

radiation. Of the three most important greenhouse gases, water vapor is the most abundant (humid air can contain up to 4% water vapor molecules). As a result, water vapor contributes the most to the greenhouse effect (61%–87%), but it is less efficient at trapping energy than other gases. Carbon dioxide (0.04% of air) is much less abundant than water vapor, but it is 20 times more efficient in trapping energy. Methane (0.00018% of air) is much less abundant than CO_2, but it is 70 times as efficient as water vapor in trapping energy. Even though CO_2 and CH_4 occur in very low concentrations, they contribute significantly to the greenhouse effect: 9%–30% and 4%–9%, respectively. Other trace gases, such as ozone (O_3) and nitrous oxide (N_2O), also contribute to the greenhouse effect.

Although most of the energy transfer between the Earth's surface and its atmosphere occurs as a result of radiation, a small amount also transfers from the Earth's surface to the atmosphere by *conduction*, the direct transfer of heat within a material or between materials in contact; by *convection*, the vertical movement of fluid caused by differences in fluid density; and by *latent heat transfer*, the addition of heat to the atmosphere when water vapor transforms into liquid or ice in clouds (see Fig. 15.3).

Classifying Climates

Should you bring a coat, umbrella, or sunscreen when you go outside? You'll know if you follow the weather report. **Weather** refers to the condition of the atmosphere (temperature, air pressure, relative humidity, wind speed, cloud cover, and precipitation) at a given location at a given time. The weather report for a town in Alaska on a particular day will be different from that for a town in the Amazon rainforest because these locations have different climates. **Climate** describes the average weather conditions, and the range of those conditions, at a given location over the course of years to decades. A region's climate, therefore, depends on average temperature and temperature range, amount and distribution of precipitation, seasonal variation in temperature and precipitation, and characteristic weather extremes and storms.

To understand why climate varies with location, let's first look at average monthly temperature maps for January and July (**FIG. 15.4**). In both maps, you'll see that warmer temperatures dominate at lower latitudes and cooler ones at higher latitudes, but you'll also notice that the temperature at most latitudes varies over the course of a year. Because our planet's axis of rotation tilts by 23.5° with respect to its orbital plane, **insolation**, the amount of solar energy that arrives at a location, varies over the course of a year as the planet revolves around the Sun. The North Pole, for example, receives no insolation in January, but receives sunlight all day in June. This variation causes seasons (**FIG. 15.5a**). In addition, because our planet is a sphere, the Sun's rays strike low latitudes at a steep angle and high latitudes at a shallow angle. Consequently, a square meter of the Earth's surface at low latitude receives more energy than an area of the same size at high latitude (**FIG. 15.5b, c**). Put another way, insolation is greater at low latitudes than at high latitudes.

Latitude isn't the only factor controlling temperature. At a given latitude, warmer temperatures occur at lower elevations and cooler temperatures at higher ones because air temperature decreases with altitude (see Chapter 8).

FIGURE 15.5 The amount of insolation depends on the angle at which sunlight strikes the Earth.

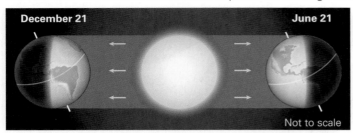

(a) The Earth has seasons because of the tilt of its axis. In June, when the northern hemisphere tilts toward the Sun, that part of the planet experiences summer. In December, when it tilts away from the Sun, it experiences winter.

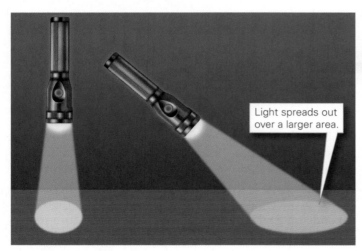

(c) The difference between the intensity of a vertical flashlight beam and a tilted one is an analogy for why insolation varies with latitude.

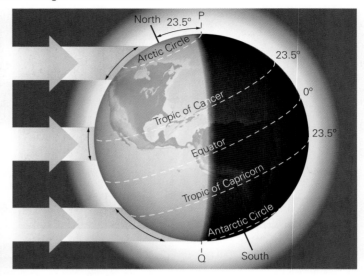

(b) High latitudes receive less insolation per square meter than do low latitudes.

Temperature also depends on proximity to oceans. Ocean water has high *heat capacity*, meaning that a given amount of water can absorb a large amount of heat without much change in its temperature. In addition, because light penetrates transparent water, a relatively thick water layer can absorb solar energy. In contrast, solar radiation striking opaque rock or soil heats only material close to the surface, and soil and rock have lower heat capacities than water. Therefore, oceans undergo relatively small changes in temperature seasonally. As a consequence, the annual temperature ranges of coastal locations are smaller than those of inland locations, and regions close to warm ocean currents have a milder climate than those near cold currents.

The global distribution of precipitation also varies significantly with latitude and season (FIG. 15.6). For example, large quantities of rain fall along the *intertropical convergence zone* (*ITCZ*) at the lower-latitude edges of the Hadley cells as the ITCZ shifts north and south over the course of a year, whereas very dry air persists in deserts beneath the higher-latitude edges of the Hadley cells (see Chapter 8).

To describe the Earth's climates, scientists classify them into *climate groups*, each with characteristic temperature ranges, precipitation, storms, and seasonality. The most common classification distinguishes among tropical, arid, temperate, cold, and polar climate groups (see Fig. 14.4a and Table 14.1).

Take-home message . . .

The Sun provides energy to the Earth and its atmosphere as electromagnetic radiation. Visible light and some ultraviolet light pass through the atmosphere and are absorbed by the Earth's surface, which in turn reradiates this energy as infrared radiation. Greenhouse gases trap some of the reradiated energy, so they keep the atmosphere warm enough for liquid water—and life—to exist on the planet. The Earth hosts many different climates. A region's climate primarily reflects temperature and precipitation, and their variation over the course of a year.

QUICK QUESTION If there were no atmosphere, how would the temperature of the Earth's surface be different from what we experience?

15.3 Climate Change and Its Causes

What Is Climate Change?

When your town endures higher-than-average temperatures for a few days in summer, is that evidence of a warming

FIGURE 15.6 The distribution of average monthly global precipitation in January and in July.

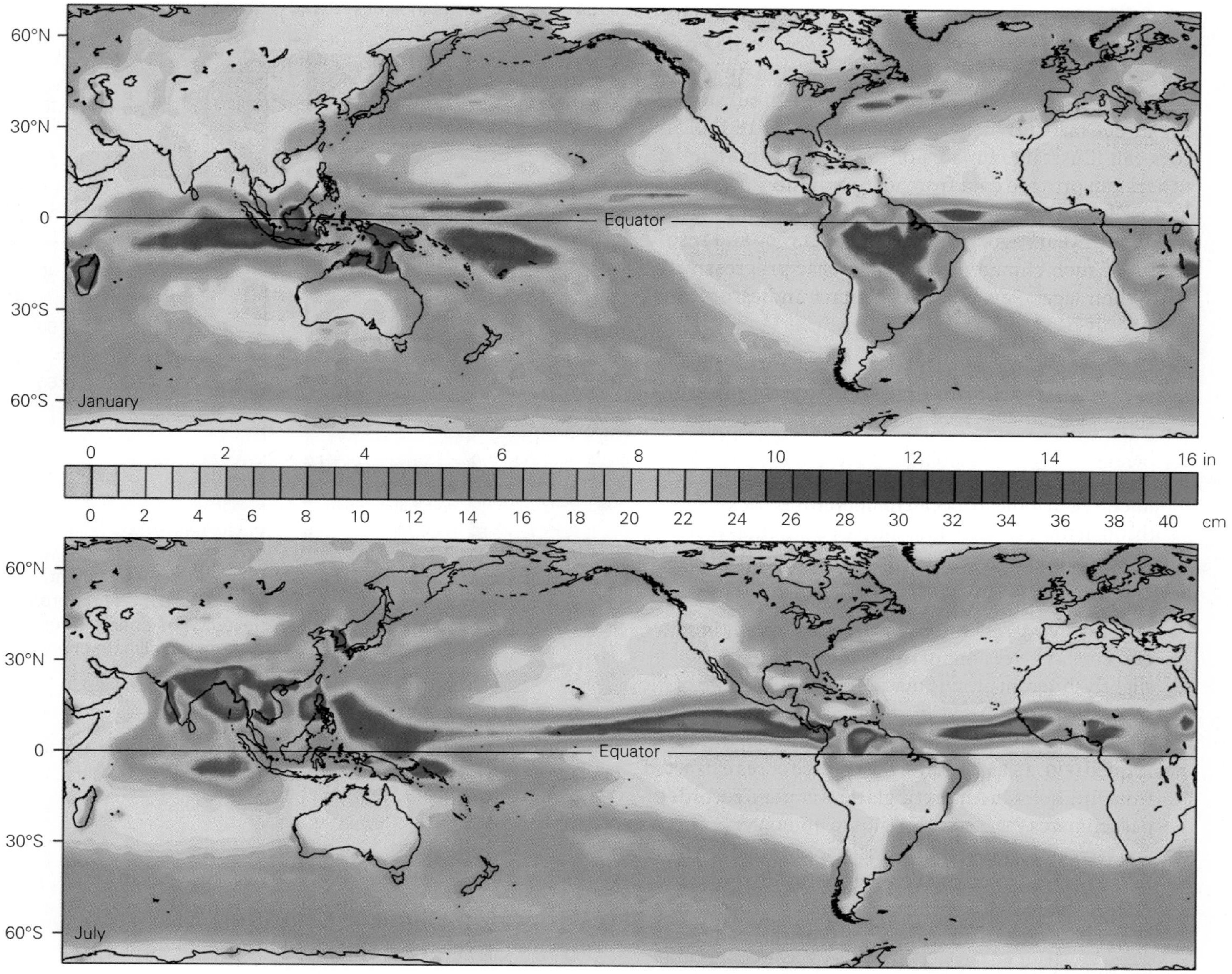

climate? If the town endures lower-than-average temperatures for a few days in winter, is that evidence of a cooling climate? The answer in both cases is no. **Climate change** refers to a shift in one or more climate characteristics over years, decades, centuries, or even millennia. Differences from average temperatures in a region may simply represent weather variability, not climate change. But when the average number of days per year that have very high temperatures changes over the course of decades, that is evidence that climate change has happened. Commonly, consideration of climate change focuses on changes over time in average annual global temperature within the lower atmosphere: an increase represents **global warming**, whereas a decrease represents **global cooling**.

Paleoclimate Indicators

For the past two centuries, researchers have used *instrumental measurements* made with thermometers, rain gauges, and other instruments at weather stations to characterize weather and climate. Satellites have provided additional data since the 1960s. Because instrumental measurements document climate only since the mid-19th century, researchers must turn to other tools to develop a longer record. Historical records of notable events (such as droughts, floods, or heat waves) can characterize aspects of climate in the past, as far back as centuries or a few millennia, depending on location. To consider even earlier climates—called **paleoclimates**—from times before historical records were available, researchers can study sediments, sedimentary rocks, and preserved organic materials. Certain characteristics of these materials, called **paleoclimate indicators** or *paleoclimate proxies*, can provide information about the climate at the time they formed, and this information can substitute for direct measurements. Some paleoclimate indicators can illustrate climates of the past few millennia; others can provide data from up to a million years ago; and still others can reach as far back as hundreds of millions of years ago. However, the accuracy and resolution of such climate records decrease progressively with their age. Several paleoclimate indicators are commonly used today:

- *Fossils and microfossils:* Different species of organisms live in different climates. The fossil record can tell us about past climates back through the Phanerozoic Eon—that is, the past 539 million years. *Microfossils*, remains of plankton species preserved in marine sediments, can indicate ocean temperatures back through the Cenozoic Era, the past 66 million years. Fossil pollen from young sediments on land can provide similar information **(FIG. 15.7)**.

- *Oxygen-isotope ratios:* Oxygen occurs naturally in several *isotopes*—versions of the oxygen atom with slightly different atomic masses. The ratio of two of these isotopes, ^{18}O and ^{16}O, in glacial ice can provide information about the temperature at which the ice formed **(FIG. 15.8a, b)**. Layers within ice cores extracted from drillholes in Antarctic glaciers contain records of past climates that go back almost a million years. Oxygen-isotope ratios within the calcite of plankton shells found in cores of marine sediment provide climate records extending even farther into the past **(FIG. 15.8c)**. Oxygen isotopes preserved in the minerals of limestone provide paleoclimate data extending back through the Phanerozoic.

- *Air bubbles in ice:* Bubbles trapped in glacial ice provide samples of ancient air whose composition can be measured.

- *Growth rings:* Some organisms and biological materials—such as trees, clam shells, and coral reefs—develop a distinct ring during each year of growth because the rate of growth varies with the seasons. The relative widths of such **growth rings** can depend in part on temperature and precipitation, so measuring the rings allows researchers to characterize past climates on a yearly basis. Notably, the pattern of tree rings (thicker for good years, thinner for bad years) serves like a fingerprint that allows researchers to correlate the patterns of the oldest rings in still living trees with the patterns of rings

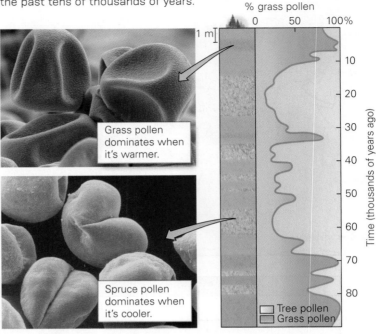

FIGURE 15.7 Fossil pollen can be used to track paleoclimate over the past tens of thousands of years.

(a) Different species of plants produce very different pollen.

(b) In a core of sediment, the proportions of different pollen types change over time as the climate changes.

preserved in trees that died long ago, to produce a composite record that can go back for up to 13,000 years **(FIG. 15.9)**.

How Has Climate Changed Naturally over Earth History?

Paleoclimate studies show that climate has changed over both long-term (millions to tens of millions of years) and short-term (decades to hundreds of thousands of years) time scales. Over the long expanse of geologic time, the Earth has alternated between warmer times, known as *hothouse intervals*, and colder times, known as *icehouse intervals* **(FIG. 15.10a)**. For example, at 70 Ma, during the Age of Dinosaurs, Antarctica and Greenland had no glaciers. But at 700 Ma, glaciers covered vast expanses of the Earth, even near the equator, so that interval has come to be known as *snowball Earth*. A higher-resolution record of climate change exists for the present era, the Cenozoic, starting at 66 Ma **(FIG. 15.10b)**. Note that this graph depicts **temperature anomalies**, meaning differences between the temperature at a particular time and the average temperature determined for a given time period.

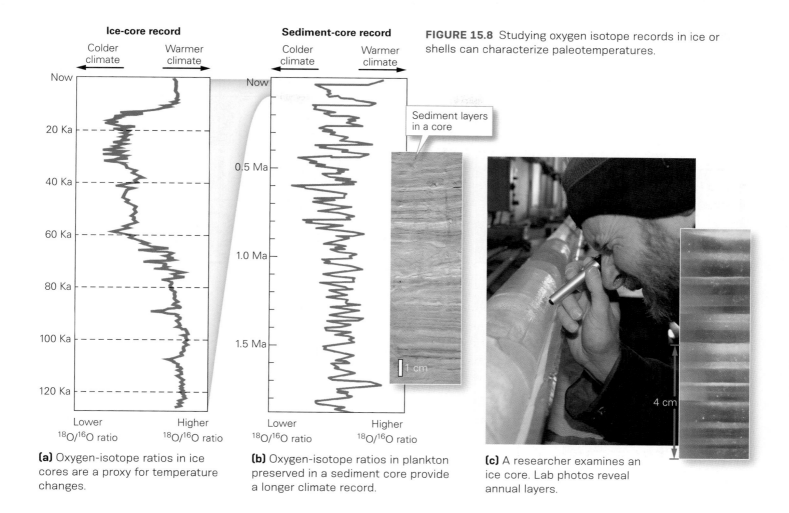

FIGURE 15.8 Studying oxygen isotope records in ice or shells can characterize paleotemperatures.

(a) Oxygen-isotope ratios in ice cores are a proxy for temperature changes.

(b) Oxygen-isotope ratios in plankton preserved in a sediment core provide a longer climate record.

(c) A researcher examines an ice core. Lab photos reveal annual layers.

FIGURE 15.9 Tree rings provide a record of past climate. More growth, leading to wider rings, happens in warm, wet years than in cold, dry years.

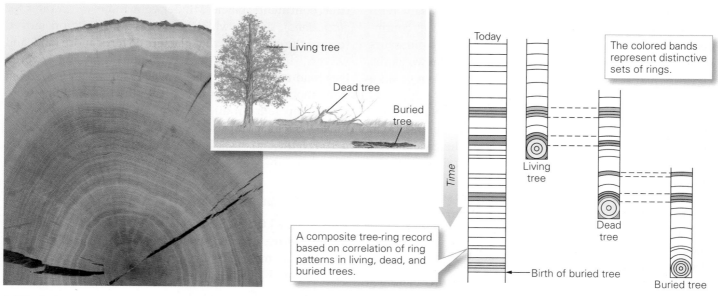

(a) Each ring represents growth in one year. Ring width reflects average precipitation and length of the growing season.

(b) By correlating distinctive ring patterns, researchers can obtain a composite tree-ring record going back thousands of years.

15.3 Climate Change and Its Causes • **549**

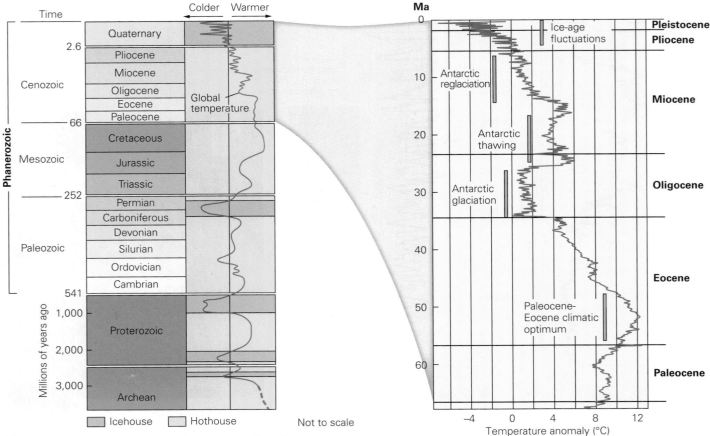

FIGURE 15.10 Long-term climate change.

(a) Icehouse and hothouse intervals, of varying lengths, have occurred over geologic time. The red line denotes relative temperature.

(b) Researchers can provide a higher-resolution record of climate change during the most recent era, the Cenozoic.

Global temperatures were fairly high during the Cretaceous, the latest period of the Age of Dinosaurs. They continued to warm during the early Cenozoic, after the dinosaurs went extinct. The peak of this warm interval, known as the *Paleocene-Eocene climatic optimum*, occurred at about 55 Ma, a time when the average global temperature may have been 5°C–8°C (9°F–14°F) higher than it is today. Temperatures then began to fall, and by about 34 Ma, the Earth became cold enough for the Antarctic ice sheet to form. There were a few ups and downs between 34 Ma and 15 Ma, but since then, the temperature has mostly declined.

The record of the past few million years shows that we are in an interval of geologic time that features many profound short-term changes in climate. For example, just 18,000 years ago, the ground beneath what is now the city of Chicago was covered by a glacier as tall as a 60-story building. The Earth at that time was in the grips of an **ice age**—a time when vast glaciers, or ice sheets, cover large areas of continents. This interval, called the *Pleistocene Ice Age*, began around 2.6 million years ago **(FIG. 15.11a)**. In that time, continental glaciers underwent about 30 *advances* (or *glaciations*), during which the limit of glaciation moved toward the equator so that glaciers covered more land, and *retreats*, during which glaciers melted back and covered less land **(FIG. 15.11b)**. The most recent glaciation ended only about 11,000 years ago, a time when humans were already present on all the continents except Antarctica. Since then, the planet has been in an *interglacial*, an interval between the end of the last retreat and the start of the next advance.

What Causes Climate Change?

CAUSES OF LONG-TERM CLIMATE CHANGE. Several phenomena drive climate change at long time scales. Some changes in climate pertain to events that have happened only once during Earth history, others to events that have happened many times.

- *Greenhouse gas concentrations:* If the concentration of greenhouse gases in the atmosphere increases, the

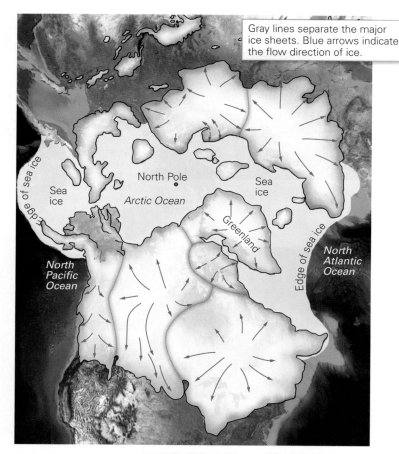

(a) At times, ice sheets (continental glaciers) covered large areas of land.

Gray lines separate the major ice sheets. Blue arrows indicate the flow direction of ice.

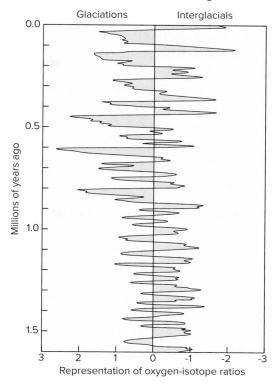

FIGURE 15.11 The Pleistocene Ice Age.

(b) The oxygen-isotope record in marine sediment indicates that glaciers advanced and retreated many times.

atmosphere traps more infrared radiation, leading to a warmer climate. Similarly, if the concentration of greenhouse gases decreases, the atmosphere traps less infrared radiation, and the Earth cools.

- *Ocean formation:* When the oceans filled with water early in the Earth's history, vast amounts of H_2O moved into the oceans from the atmosphere. Once the oceans existed, most atmospheric CO_2 dissolved in ocean water. Some of this CO_2 eventually became incorporated in shells of organisms. When the organisms died, the shells in turn became incorporated in sediment, and later lithified to become limestone. This limestone, effectively, sequestered (stored) CO_2 that had once been in the atmosphere. If water vapor and CO_2, two potent greenhouse gases, had remained within the Earth's atmosphere, it would be as dense as that of Venus, and like Venus, the Earth would be much too hot for life to have appeared.

- *Immense volcanic eruptions:* Volcanic eruptions emit CO_2. As we'll see, the amount added by volcanoes today is relatively small. But when *large igneous provinces* erupted (see Chapter 4), the quantities of CO_2 and aerosols that entered the atmosphere influenced climate.

- *Uplift of land:* When land surfaces are exposed to the air, land undergoes chemical weathering (see Chapter 2). Certain weathering reactions absorb significant amounts of CO_2. During uplift and mountain building, atmospheric CO_2 declines.

- *Evolution of life:* Photosynthetic organisms remove CO_2 from the atmosphere. The presence of these organisms as life evolved, therefore, reduced the concentration of CO_2 in the atmosphere. When shell-producing marine organisms evolved, they sequestered CO_2 from the oceans to build their calcite shells. As a result, more CO_2 gas could move from the atmosphere into the oceans. Photosynthesis also produces oxygen. In the stratosphere, oxygen interacts with ultraviolet light from the Sun to produce ozone, another greenhouse gas.

- *Changes in the intensity of solar radiation:* The amount of radiation produced by the Sun has changed over time as the Sun has slowly evolved. Notably, early in Solar System history, the Sun produced much less radiation than it does today. During the time of the *faint young Sun*, the Earth remained warm enough for liquid water to exist because enough CO_2 remained in the atmosphere to cause a strong greenhouse effect.

- *Distribution of continents:* Due to plate motion, the distribution of continents on the Earth's surface varies over time, causing climates on land to change. For example, when more land lies at high latitudes, more land can be colder, and an ice age can ensue.

- *Configuration of ocean currents:* Ocean currents distribute heat around the globe, so if their arrangement changes, heat distribution changes, affecting the locations of climate groups.

CAUSES OF SHORT-TERM CLIMATE CHANGE. What natural phenomena can cause a glacial advance or retreat on a time scale of centuries to a few thousand years? An answer to this question came from a Serbian astronomer, Milutin Milanković. In 1920, Milanković discovered that three characteristics of the Earth's orbit and tilt change cyclically. These characteristics (**FIG. 15.12a–c**) are now known as **Milankovitch cycles**, using the English spelling of the astronomer's name.

1. *Eccentricity (orbital shape):* The Earth's orbit gradually changes from a more circular shape to a more elliptical shape, and back again, over a period of about 100,000 years.

2. *Tilt of the Earth's axis:* As we've seen, seasons exist because the Earth's axis is tilted. The tilt angle varies between 22.5° and 24.5° over a period of 41,000 years.

3. *Precession:* You've probably noticed that the axis of a spinning toy top slowly rotates, a movement called *precession*. The Earth's axis precesses over a period of about 23,000 years.

Milanković examined how these three cycles might affect total annual insolation, and focused on the seasonal distribution of insolation at mid- to high latitudes. He found that the three cycles together can cause insolation to change by as much as 25% at higher latitudes in some seasons (**FIG. 15.12d**), and he suggested that glaciers advance during times when high latitudes receive the smallest amount of insolation during the summer, preventing ice from the previous winter from melting completely. Geologists have found that Pleistocene glacial advances correlate with times of minimum insolation, and that retreats correlate with times of maximum insolation, closely matching the timing predicted by Milanković. Because of this correlation, researchers conclude that Milankovitch cycles indeed play a significant role in short-term climate change.

While the Milankovitch cycles appear to influence the advances and retreats of glaciers during an ice age, they might

FIGURE 15.12 Milankovitch cycles cause the amount of insolation received at high latitudes to vary over time.

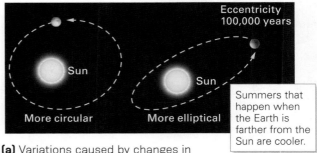

(a) Variations caused by changes in orbital shape. The shapes of the orbits are exaggerated.

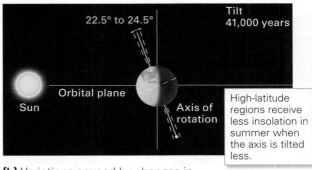

(b) Variations caused by changes in the tilt of the Earth's axis.

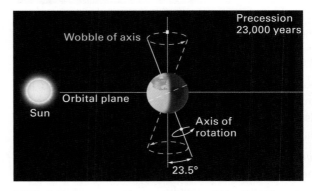

(c) Variations caused by precession of the Earth's axis.

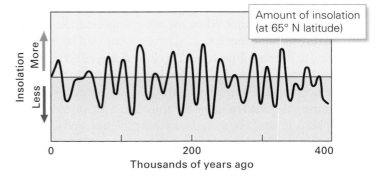

(d) Combining the effects of eccentricity, tilt, and precession produces varying amounts of insolation during the summer at mid- to high latitudes.

not be strong enough to cause ice ages on their own. Other factors, which act on time scales of years to millennia, may contribute to the triggering of ice ages:

- *Changes in the Earth's albedo:* The albedo, or reflectivity, of the land surface can increase due to increases in snow cover and vegetation cover, or when light-colored volcanic ash falls over a large area. The albedo of the atmosphere also increases if global cloud cover, or the concentration of volcanic aerosols in the atmosphere, increases (see Chapter 4).

- *Changes in oceanic thermohaline circulation:* The configuration of ocean currents can change if freshwater enters the ocean rapidly due to the release of an ice-dammed glacial lake (see Chapter 12) or increased rainfall or glacial melting. These changes can affect climate if they slow or stop **thermohaline circulation**, the oceanic circulation that transports heat held in water from the tropics to high latitudes (FIG. 15.13). Thermohaline circulation involves both surface and deep currents. It occurs because cooler, saltier water is denser than surrounding water—cold, salty seawater sinks to the bottom of the ocean at high latitudes, then flows toward lower latitudes along the seafloor. An influx of freshwater at the sea surface dilutes surface seawater. That water, now less dense, no longer sinks, causing the thermohaline circulation to cease.

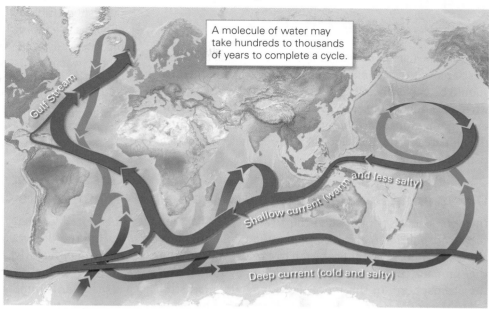

FIGURE 15.13 Thermohaline circulation involves both surface and deep currents. This circulation transfers heat from low latitudes to high latitudes.

concentration of water vapor in the atmosphere increases, amplifying the warming trend. If the temperature cools, water vapor condenses and rains out of the atmosphere, so the concentration of water vapor decreases, amplifying the cooling trend. Water vapor molecules have a short residence time in the atmosphere (*residence time* is the average time spent by any material within a component of the Earth System), so this feedback serves as a quick response to changes in global temperature. In contrast, CO_2 and CH_4 have longer residence times, because once added, these gases mix thoroughly with the whole atmosphere. A long-term increase in water vapor in the atmosphere due to a temperature increase represents a positive feedback. The addition or removal of CO_2 and CH_4 directly impacts global temperature, so such activity is a primary driver of climate change.

Feedbacks Affecting Climate Change

Regardless of the time frame over which they operate, any of the mechanisms we've just discussed may cause a **climate feedback**, a phenomenon triggered by a change in climate that further modifies that change. A *positive feedback* amplifies the phenomenon that caused it, and a *negative feedback* reduces the phenomenon that caused it.

Do changes in water vapor concentration in the atmosphere control climate change, or do they represent a feedback? It would seem at first glance that, since water vapor contributes so much to the greenhouse effect, its concentration would be a primary driver of climate change, not a feedback. In fact, that's not the case, because the water vapor concentration in the atmosphere depends on temperature. If the Earth's temperature warms, the

Take-home message . . .

The Earth's climate changes naturally. Long-term natural climate change (on the scale of millions to tens of millions of years) reflects long-term changes in the composition of the atmosphere linked to ocean formation, volcanism, and the evolution of life, as well as to the distribution and elevation of land and the pattern of ocean currents. Short-term natural climate change (on the scale of decades to hundreds of thousands of years) is related to the Milankovitch cycles, to changes in the Earth's albedo, and to changes in thermohaline circulation.

QUICK QUESTION Describe the phenomena that constitute the Milankovitch cycles.

15.4 Evidence of Recent Climate Change

How has the Earth's climate changed over the past few centuries? Analysis of Milankovitch cycles suggests that the Earth should be experiencing a cooling trend. That, however, isn't what researchers observe. Rather, the Earth appears to be undergoing global warming at a pace that is unprecedented in the climate record. Let's look at some of the evidence scientists have found.

Discovering Recent Climate Change

In 1896, scientists published the first articles describing the atmospheric greenhouse effect and explaining how it works. In time, researchers began to explore the possibility that humans could cause global warming by adding CO_2 to the atmosphere. In the 1960s and 1970s, detailed studies led to the conclusion that **recent climate change**, beginning with the industrial revolution of the 18th and 19th centuries, might already be taking place. Researchers focused on documenting evidence of this change and began to explore its implications.

It became apparent by the 1980s that recent climate change was a stealth megadisaster with profound consequences, and the topic entered the public arena. Scientific journals began to publish so many climate-change studies that individual researchers struggled to keep track of them all. Consequently, in 1988, the World Meteorological Organization, in collaboration with the United Nations Environment Program, founded the **Intergovernmental Panel on Climate Change** (**IPCC**). The IPCC evaluates published climate-related studies and summarizes its conclusions in a report intended for a broad audience, publishing a new edition every 5 years. The IPCC's conclusion that global warming is underway, and that human activities contribute to the problem, has increased in certainty with each edition of the report. Below, we examine evidence for recent global warming, as summarized by the IPCC and others.

Observed Recent Temperature Change

We noted earlier that the average global temperature during the Pleistocene (2.6 Ma to 15 Ka) bounced up and down significantly, causing advances and retreats of continental glaciers. Zooming in on the past thousand years, researchers found that the average global temperature went up and down within about 1°C (1.7°F) until 1880, when it began to rise relatively rapidly (**FIG. 15.14a**). The original graph of this change became known informally as the **hockey-stick diagram** because of its shape. Extending the record back for another thousand years reveals that around 1000 C.E., climates were relatively mild—during this *Medieval Warm Period*, people were able to settle in Greenland (**FIG. 15.14b**). Then, beginning around 1400, global cooling began in the northern hemisphere, and during the next four centuries, now known as the *Little Ice Age*, the average global temperature was 0.2°C–0.4°C (0.4°F–0.7°F) colder than in the previous centuries. During

FIGURE 15.14 Change in average global temperature over the past 2,000 years.

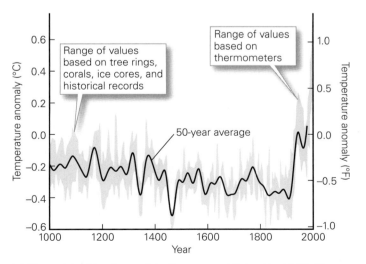

(a) The original hockey-stick diagram, published in 1998. The data are for the northern hemisphere. The temperature anomaly is indicated relative to the average 1961–1990 temperature.

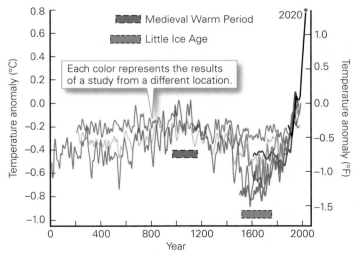

(b) Expanding the graph back to 2,000 years ago reveals the Medieval Warm Period and the Little Ice Age.

FIGURE 15.15 Changes in average annual global temperature since 1880.

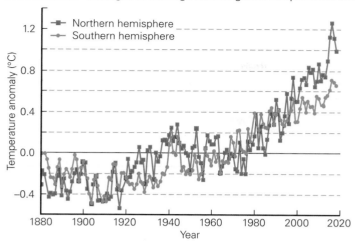

(a) Average annual temperature has increased by a maximum of about 1.6°C (2.9°F). Note that the change is a little less in the southern hemisphere.

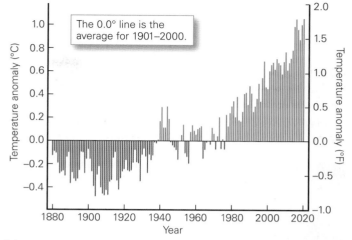

(b) This plot of the annual global surface temperature change for land and oceans emphasizes that since 1980, temperatures have been hotter than the average for the 20th century.

this time, mountain glaciers in Europe and Asia grew and canals in the Netherlands froze over.

If we zoom in closer to the present, focusing on 1880–2020, we see that the Earth has experienced an average temperature increase of about 0.07°C (0.13°F) per decade, and that this rate has doubled since 1980. In fact, the global temperature has climbed a total of about 1°C over the last 40 years, relative to the long-term average between 1901 and 2000 (FIG. 15.15a). Significantly, the warming hasn't been continuous; occasional multi-year intervals show no significant warming. Concern about global warming is based on examining trends lasting 20 years or more. Notably, in every year since 1980, temperatures have been higher than the long-term average (FIG. 15.15b). The 10 warmest years since 1880

occurred after 1998, and the top 7 since 2015. In fact, 2020 was tied with 2016 as the warmest year on record, and recent statistics show that what had been considered to be "normal" temperature has increased significantly to a "new normal."

Maps of temperature change emphasize that that amount of warming varies with location. The northern hemisphere has warmed more than the southern hemisphere, and air over continents has warmed more than air over oceans, because large bodies of water moderate temperature change (FIG. 15.16). The greatest warming during the past half century has taken place in northern North America, northern Asia, and the Arctic. In fact, in 2021 rain fell at the highest point on the Greenland ice sheet for the first time in recorded history.

FIGURE 15.16 Changes in temperature (temperature anomalies) vary with location.

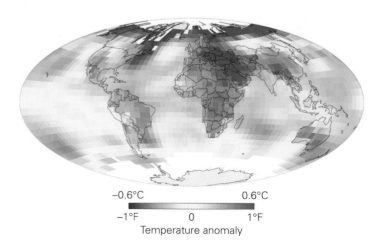

(a) Global temperature change between 1990 and 2019. Note that continents and Arctic regions are warming the most.

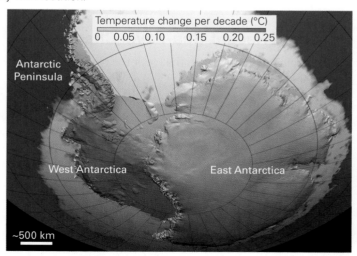

(b) Warming of Antarctica between 1957 and 2006. West Antarctica is at a lower elevation, overall, than East Antarctica.

15.4 Evidence of Recent Climate Change

FIGURE 15.17 A model of changes in relative sea level over geologic time.

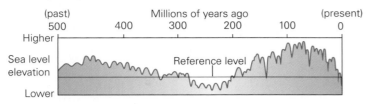

(a) Sea level has changed over the past 500 million years. The reference level is the current value of sea level.

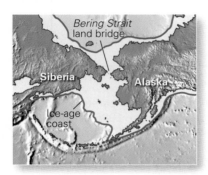

(b) At the peak of the most recent glaciation, sea level was so low that people could walk from Siberia to Alaska.

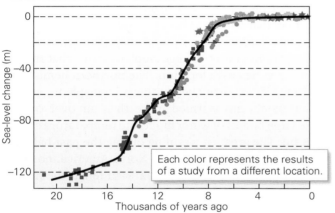

(c) Sea level rose rapidly during the most recent glacial retreat as continental glaciers melted.

Observed Recent Sea-Level Change

The geologic record indicates that over the course of geologic time, global sea level has gone up and down significantly relative to the land surface. At times, shallow seas covered large areas of continents (see Chapter 2) (**FIG. 15.17a**). Significant changes in sea level have happened during the past few million years. As an example, during Pleistocene glacial advances, a huge volume of water evaporated from the ocean and became trapped in continental glaciers. Sea level went down so far that large areas of continental shelves were exposed (**FIG. 15.17b**). During the most recent glacial retreat, which started around 15,000 years ago, the return of water to the ocean caused sea level to rise by about 120 m (394 ft) (**FIG. 15.17c**). This rise tapered off about 8,000 years ago, when the last North American and Asian ice sheets had completely melted.

FIGURE 15.18 Changes in sea level in the past 140 years.

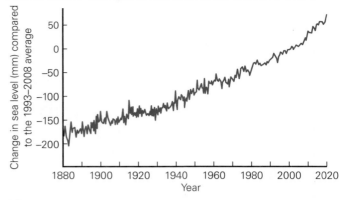

(a) Records from tidal gauges show that sea level has risen fairly steadily during the past 140 years. The rate has increased since 1970. This graph shows sea level relative to the average for 1993–2008.

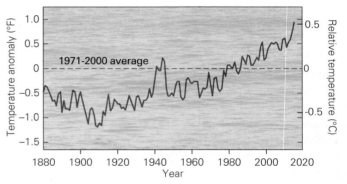

(b) Average global sea-surface temperatures have increased since 1880. The temperature anomaly is indicated relative to the average 1971–2000 temperature.

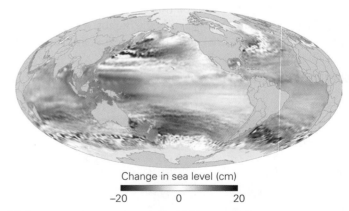

(c) Sea-level changes between 1993 and 2019. Regional variation occurs because winds and currents affect locations of sea level rise.

In the past 140 years, global sea level has started to rise again. Sea level rose at a rate of about 1.4 mm (0.06 in) per year between 1880 and 2006; since then, the rate of rise has increased to 3.6 mm (0.14 in) per year. Consequently, sea level today measures about 23 cm (9 in) higher than it did in 1880 (**FIG. 15.18a**). Researchers link this rise to global warming because rising temperature affects sea level in two ways. First, liquid water expands as it warms, and measurements of sea-surface temperature show that the oceans have warmed (FIG. 15.18b). Given that oceans are confined on their bottoms and sides by the seafloor, this *thermal expansion* forces the sea surface upward. The warming hasn't been uniform, however, and currents and winds transport water around the ocean, so sea-level change varies across ocean basins (FIG. 15.18c). Second, when temperatures rise, glaciers melt, so water that had been locked within ice on land returns to the oceans.

Melting Glacial and Sea Ice

Most mountain glaciers—such as those of the Alps, Andes, Himalayas, and Rockies—have retreated and thinned dramatically since the Little Ice Age (FIG. 15.19). In many locations, glaciers that once filled valleys have disappeared entirely. For example, in 1850, Glacier National Park, Montana, had 150 glaciers, but now it has only 25, and at current rates of retreat, the remaining glaciers will be gone in a few decades. The relatively few mountain glaciers that have advanced owe their movement to local precipitation increases, or to local increases in ice-flow rates as ice warms.

Are the vast ice sheets covering Greenland and Antarctica also shrinking? To answer this question, researchers use data from the GRACE and GRACE-FO satellites, whose instruments can measure the gravitational pull of the Earth very precisely. The resulting measurements indicate that the pull exerted by the masses of ice in the Antarctic and Greenland ice sheets has decreased (FIG. 15.20a). By converting these changes in gravitational pull into changes in ice mass, and then ice volume, researchers estimated that the Greenland ice sheet lost about 90 km^3 (22 mi^3) of ice in 2003 and 535 km^3 (128 mi^3) in 2019, and that it's now thinning by about 1 m (3 ft) per year (FIG. 15.20b). This result confirms conclusions from studies of satellite images, which show that the summer melt zone of Greenland's ice sheet has widened and now lasts longer. Meltwater first pools in lakes on the ice surface, then

FIGURE 15.19 Photos taken at different times provide clear evidence of glacial retreat.

(a) The Muir Glacier in Alaska retreated 12 km (7 mi) between 1941 and 2004.

(b) The Bear Glacier in Alaska retreated by about 2 km (1 mi) in just 7 years, between 2012 and 2019.

FIGURE 15.20 Changes in the Antarctic and Greenland ice sheets.

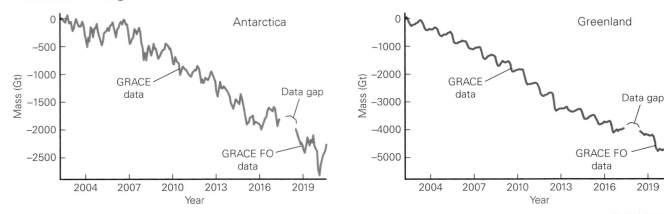

(a) Ice-mass loss in Antarctica and Greenland, measured in gigatons (billions of tons, abbreviated Gt), as measured by the GRACE and GRACE FO satellites for 2002–2020.

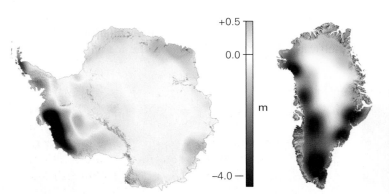

(b) Ice thickness has decreased significantly along ice-sheet margins in Antarctica and Greenland over the course of a decade. The scale shows the change in ice thickness.

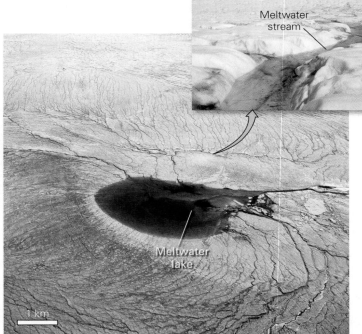

(c) Meltwater lakes and streams on the Greenland ice sheet often disappear suddenly as they drain through cracks to the base of the ice sheet.

drains through cracks to the base of the ice sheet, where it flows through tunnels to the sea **(FIG. 15.20c)**. Antarctica's ice sheets have also been diminishing—the continent now loses about 270 km³ (65 mi³) of ice per year.

In the 19th century, much of the Arctic Ocean was covered by *sea ice* (a layer of ice formed by freezing of the ocean surface) all year, and explorers' goal of finding a "Northwest Passage" that would allow a ship to sail north of North America from the Atlantic to the Pacific was unattainable. So much ice now melts away in summer that cruise ships can now make the journey **(FIG. 15.21a)**, and long-term trends in midsummer sea-ice cover show a definite decrease **(FIG. 15.21b)**. If the overall downward trend continues, it may be possible to sail to the North Pole in summer within the next few decades. Sea ice has also been disappearing in Antarctica. Most notably, *ice shelves* (regions where ice sheets have flowed over the sea adjacent to the coast) have been breaking up. In 2002, the Larsen B Ice Shelf, with an area of 3,250 km² (1,255 mi²), broke up **(FIG. 15.22)**, and in 2017, a 5,800 km² (2,239 mi²) section of the Larsen C Ice Shelf—85% larger than Rhode Island—separated from the ice sheet and drifted out to sea.

Combining data from observations of glacial and sea-ice loss shows that the Earth's total ice volume has been diminishing in the past half century **(FIG. 15.23)**. Since it takes time for glaciers to grow or melt, changes in the overall volume of glacial ice on land and sea ice in the ocean effectively represent temperature trends averaged over several decades. This clearly indicates warming.

Biological Indicators of Climate Change

As we noted earlier, some organisms that are sensitive to temperature and the length of seasons serve as *biological indicators* of climate change. Biological indicators provide

FIGURE 15.21 Sea-ice cover in the Arctic Ocean has decreased.

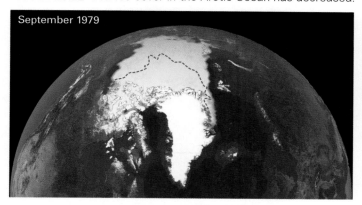

(a) A comparison of satellite images of Arctic Ocean sea ice in September 1979 and in September 2020. The blue dashed line on the 1979 image is the approximate trace of the 2020 ice extent.

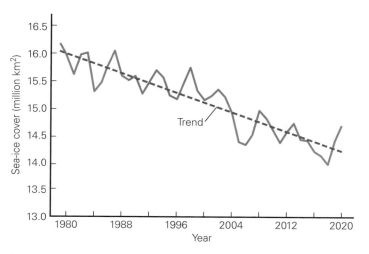

(b) A graph of sea-ice cover relative to the average for 1979–2020 shows fluctuations, but also shows an overall downward trend (blue line).

evidence that global warming is happening. For example, *plant hardiness zones*, defined by the US Department of Agriculture to help farmers and gardeners determine which plants will successfully grow in their region, have been migrating northward in the United States (FIG. 15.24). Other biological indicators of global warming include the date when maple sap starts to flow in spring, the date when leaves change color in fall, the date at which cherry blossoms bloom, and the latitudinal ranges of birds, fish, and insects, including some species of mosquitoes. In some locations water temperature has increased enough to kill fish and shellfish, ruining local industries.

Changes in Global Weather Patterns

Researchers have also detected changes in weather patterns due to climate change. For example, during the past century, overall precipitation over land areas has increased by about

FIGURE 15.22 A large portion of the Larsen B Ice Shelf on the coast of the Antarctic Peninsula disintegrated in 2002.

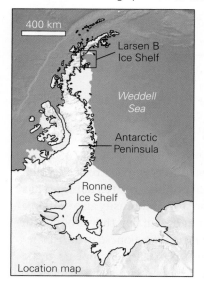
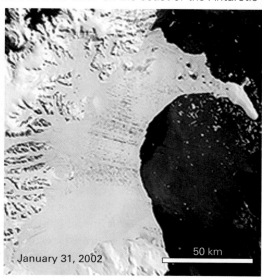

15.4 Evidence of Recent Climate Change • 559

FIGURE 15.23 Global ice loss (from mountain glaciers and continental glaciers combined), given in water-equivalent depth. This measure is obtained by converting ice mass to water volume, then dividing by the total area of ice.

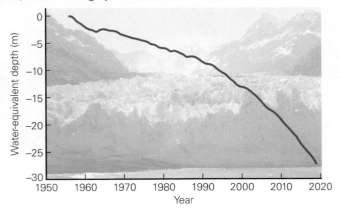

2%, and the distribution and character of precipitation have changed—some parts of the world have become drier, while others have become wetter—note the dramatic change in the US **(FIG. 15.25a, b)**. In addition, the ratio of rainy days to snowy days in winter in the northern hemisphere has increased.

Recent studies now show that the severity of storms has also increased. For example, while the total number of tropical cyclones per year (see Chapter 11) has not changed much,

FIGURE 15.24 The boundaries between plant hardiness zones, as defined by the US Department of Agriculture, shifted northward between 1990 and 2012. Blues denote shorter growing seasons, and reds indicate longer growing seasons.

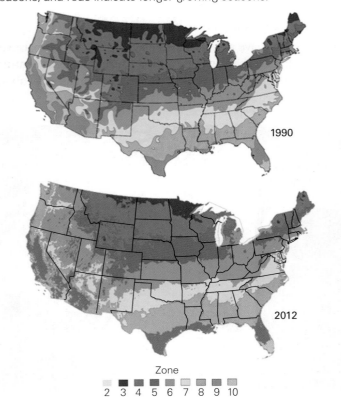

FIGURE 15.25 Changes in precipitation patterns.

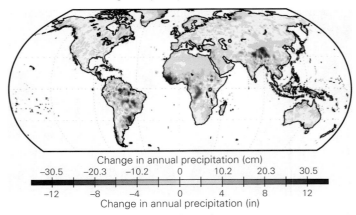

(a) Changes in average annual precipitation 1986–2015. Brown areas are getting drier, and green areas are getting wetter.

(b) Precipitation changes between 1991 and 2020, as compared to the average temperature of 1981–2010. Brown areas are getting drier, and green areas are getting wetter.

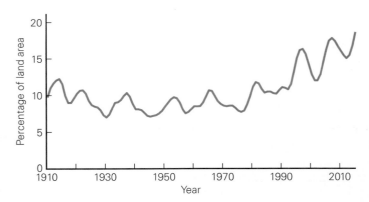

(c) The percentage of land area in the continental United States experiencing extreme one-day precipitation rates each year. An increase is evident beginning in the 1970s.

the number of major cyclones (Category 3 or stronger, which together cause 85% of tropical cyclone–related damage) has grown. The rate at which these storms intensify also seems to be increasing, as has the amount of rain that they produce. In fact, the land area experiencing extreme one-day hurricane-related rainfall events in the United States appears to be increasing **(FIG. 15.25c)**. Researchers relate such changes to global warming, because warmer sea-surface temperatures enhance evaporation, which puts more water vapor (the fuel of storms) into the atmosphere.

> **Take-home message . . .**
> Evidence for warming appears directly in the global temperature record, and is consistent with the observed rise in sea level, the melting of glaciers and sea ice, and other measures.
>
> **QUICK QUESTION** What two factors cause sea level to rise?

15.5 Causes of Recent Climate Change

We've seen the evidence that global warming and other aspects of climate change are occurring today. Now, we turn our attention to the causes. Global warming does not correspond to an increase in solar radiation, which has not changed significantly during past millennia. The 11-year periodicity of the *sunspot cycle* (changes in radiation intensity related to the appearance of dark-colored patches on the Sun, discussed in Chapter 16) has only a minor effect on temperatures on the Earth, and no long-term trend exists. There is, however, a clear correlation between observed global warming and an observed increase in the concentrations of greenhouse gases (particularly CO_2 and CH_4) in the atmosphere. Here, we first consider the evidence for this increase, then look at the sources of these gases.

Recent Changes in Atmospheric CO_2 Concentrations

INSIGHT FROM THE KEELING CURVE. In 1955, an American scientist, Charles Keeling, developed an instrument to measure the concentration of CO_2 in the atmosphere very accurately. To better understand variations in global CO_2 concentrations, Keeling set up a measuring station on Mauna Loa, a Hawaiian volcanic peak that rises 3.4 km (2.1 mi) above the central Pacific Ocean. He chose this locality—far from cities, factories, wildfires, and other sources of CO_2—so that his measurements would detect only gas that had mixed thoroughly within the atmosphere. When Keeling looked over the data from the first year of his study (1958), he found that the CO_2 concentration ranged between 312 and 318 ppm (parts per million, by volume), with an average value of 315 ppm (= 0.0315%). It dropped to lower values in summer, when plants of the northern hemisphere absorb the gas during photosynthesis, and rose in winter, when those plants shed their leaves, which then decay and release CO_2. In effect, Keeling had observed the Earth System inhaling and exhaling CO_2!

Keeling collected measurements on Mauna Loa several times each year until 2005, and since then, others have continued to collect the data. As a result, we now have a direct record of atmospheric CO_2 concentrations extending over more than 60 years. The graph of these concentrations, known as the **Keeling curve**, reveals that the peak annual CO_2 concentration has been increasing without interruption **(FIG. 15.26a)**. In May 2021, the atmosphere contained 419 ppm of CO_2, an increase of over 30% since Keeling's first measurements, and an increase of almost 50% since the mid-19th century. The Keeling curve, together with a graph of CO_2 concentrations over a longer time scale **(FIG. 15.26b)**, correlates with the graph

FIGURE 15.26 Changes in atmospheric carbon dioxide concentrations over time.

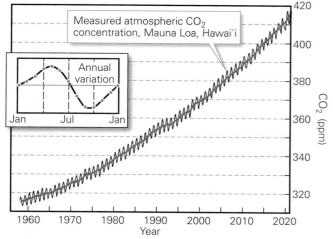

(a) The Keeling curve, showing changes since 1958. The red line shows change in the average annual concentration of CO_2.

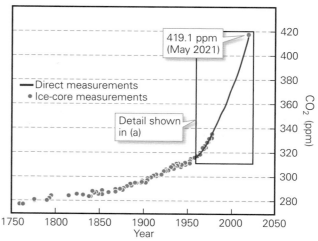

(b) Since the industrial revolution, the atmospheric CO_2 concentration has steadily increased.

FIGURE 15.27 Studies of air bubbles from glacial ice cores show that the atmospheric CO_2 concentration varied between 180 and 300 ppm over the past 800,000 years—until about 150 years ago. In 2021, it was 419.1 ppm.

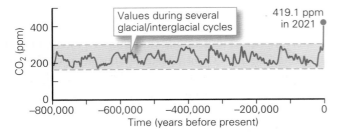

of global temperature increase (see Fig. 15.15). Such correlations imply that an increase in atmospheric greenhouse gas concentrations is the main cause of global warming since the start of the industrial age.

How do the CO_2 concentrations observed by Keeling compare with those from farther back in time? By measuring CO_2 concentrations in air bubbles within the layers of ice in cores from Antarctic glaciers, researchers have obtained a record of atmospheric CO_2 concentrations extending back 800,000 years **(FIG. 15.27)**. This record indicates that until the past two centuries, CO_2 concentrations cycled between about 180 and 300 ppm, dropping during glaciations and rising during interglacials. Researchers suggest that changes in CO_2 concentrations between glacial and interglacial periods are related to changes in the amount of oceanic phytoplankton, which cycle carbon between the atmosphere and ocean and are sensitive to global temperature. The Earth is currently in an interglacial, during which natural CO_2 concentrations were close to 300 ppm before the industrial revolution. The current value, about 40% above the range of natural variation during the past 800,000 years, indicates that CO_2 added to the atmosphere during the industrial era contributes to global warming.

INSIGHT FROM RISING CH_4 CONCENTRATIONS. Precise instrumental measurements of CH_4 concentrations in the atmosphere became available only in the early 1980s. The data reveal that the concentration of this gas has climbed from about 1,630 ppb (parts per billion, = 0.000163%) in 1985 to a 2020 value of about 1,892 ppb, an increase of 15% **(FIG. 15.28a)**. Looking back over 2,000 years, measurements of paleoclimate proxies indicate that the CH_4 concentration remained fairly constant (at about 600 ppb) until about 1750, when it began to increase almost exponentially **(FIG. 15.28b, c)**. The increase since the industrial era began exceeds 200%, an increase that also correlates with recent global warming.

INSIGHT FROM GLOBAL CLIMATE MODELS. For the past two decades, researchers have used sophisticated programs run on high-speed computers to simulate the behavior of climate over time scales of decades to a century or more. These programs, called **global climate models (GCMs)**, use equations that describe physical and chemical interactions among the atmosphere, oceans, and land. They are similar to models used for weather forecasting, but are designed to cover much longer time scales.

Using GCMs, researchers can simulate the way the climate would have behaved if greenhouse gas concentrations had remained at preindustrial levels while other natural processes, such as the Milankovitch cycles, sunspot cycles, and volcanic eruptions, took place as observed. They can then compare that behavior with the way the climate has actually behaved as greenhouse gas concentrations have increased and other climate-altering changes in land use, stratospheric ozone (see Chapter 16), and air pollution have occurred. Such simulations show that if greenhouse gas concentrations had stayed within natural ranges, the Earth's temperature would have experienced no clear trend over the last century. Models that include observed greenhouse gas concentrations predict the warming trend that

FIGURE 15.28 Changes in atmospheric methane concentrations over time.

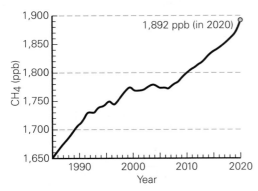

(a) Changes in atmospheric CH_4 concentrations since 1985.

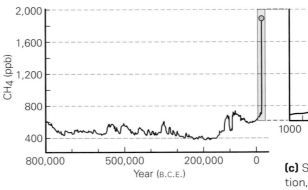

(b) Changes in atmospheric CH_4 concentrations over the past 800,000 years.

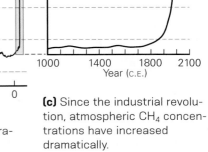

(c) Since the industrial revolution, atmospheric CH_4 concentrations have increased dramatically.

FIGURE 15.29 GCMs can be set to calculate global temperature changes since 1880, both for a situation in which greenhouse gas concentrations did not change, and for a situation in which they have changed by the amounts recorded. Observed temperature changes closely match those of the latter situation.

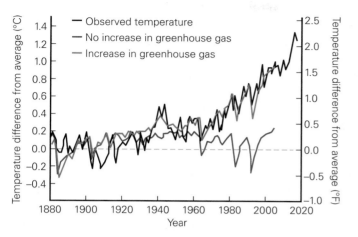

we have actually experienced (**FIG. 15.29**). The results of these simulations support the proposal that greenhouse gas increases are driving global warming.

Are Rising CO_2 and CH_4 Concentrations Natural or Anthropogenic?

Atmospheric concentrations of CO_2 and CH_4 have risen to levels much higher than any that have existed during the previous 800,000 years. Where do the excess greenhouse gases come from?

INSIGHT FROM STUDIES OF THE CARBON BUDGET. In Chapter 2, we introduced the **carbon cycle**, a biogeochemical cycle during which various forms of carbon move among various reservoirs (the atmosphere, biosphere, hydrosphere, and geosphere) in the Earth System (**FIG. 15.30a**). To understand changing concentrations of

FIGURE 15.30 The carbon budget of the atmosphere and human contributions to it.

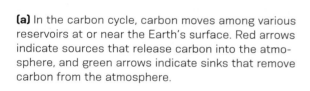

(a) In the carbon cycle, carbon moves among various reservoirs at or near the Earth's surface. Red arrows indicate sources that release carbon into the atmosphere, and green arrows indicate sinks that remove carbon from the atmosphere.

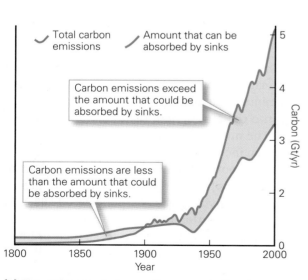

(b) The difference between the amount of carbon released into the atmosphere from all sources and the amount that can be absorbed by natural sinks.

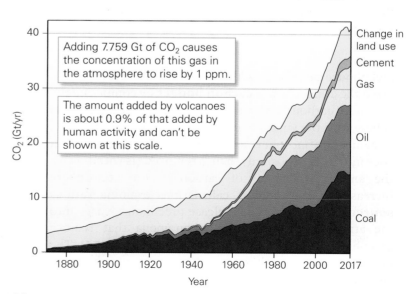

(c) Increases in CO_2 input into the atmosphere from anthropogenic sources during the 20th century.

TABLE 15.1 Sources and Sinks Involved in the Carbon Budget of the Atmosphere

SOURCES

Fossil fuel combustion	Coal, oil, and natural gas sequester immense amounts of carbon in geologic reservoirs. Burning these fossil fuels combines this carbon with oxygen to release CO_2.
Volcanic gas	About 10%–40% of the gas vented by volcanoes consists of CO_2.
Respiration	Organisms break down organic chemicals from food, a process that releases CO_2.
Digestion	Digestion of food by cattle and other organisms releases CH_4.
Decomposition	Some bacteria release CO_2 as they consume dead organisms; others release CH_4.
Dissolution of limestone	When limestone dissolves in acidic water, the reaction releases CO_2.
Venting natural gas	Fractures in rock allow natural gas to escape from geologic reservoirs. Extracting and refining oil and natural gas can release CH_4.
Coal-bed fires	When coal in mines catches fire, it smolders underground and releases CO_2 and CH_4.
Cement production	The cement in concrete consists mostly of lime (CaO), made by heating limestone, a rock made of calcite ($CaCO_3$). Heating calcite releases CO_2.
Industrial processes	Production of certain materials in factories involves chemical reactions that release CO_2.
Food production	Organic decay in rice paddies and production of cattle feed release CH_4.
Deforestation	Burning of biomass (forests and grasslands) releases CO_2.
Gas hydrate melting	CH_4 produced by organic decay that was trapped in ice-like gas hydrate is released during melting.
Wetland decay	Decay of organic material in wetlands (including melted permafrost) releases CH_4.

SINKS

Dissolution	Atmospheric CO_2 can dissolve in ocean water.
Growth of biomass	The growth of trees, phytoplankton, and other photosynthetic organisms absorbs CO_2.
Burial of organic matter	The burial of organic matter, in swamps or on the seafloor, sequesters CO_2.
Incorporation in soil	Soil formation incorporates organic matter, so it sequesters CO_2 and CH_4.
Shell growth	Many invertebrates and plankton make shells by extracting carbonate (formed from dissolved CO_2) from water. Burial of their shells, or growth of reefs, sequesters the carbon.
Weathering of rock	Chemical weathering of some silicate rocks absorbs CO_2 to produce new carbonate minerals.
Chemical reactions	Greenhouse gases react with other gases in the atmosphere to produce non-greenhouse gases.

greenhouse gases in the atmosphere, researchers have analyzed the atmosphere's **carbon budget** by comparing the volume of carbon (as CO_2 and CH_4) released into the atmosphere with the volume removed during the carbon cycle. Imagine the atmosphere as a bank account: if more carbon goes into it (is deposited) than is removed (is withdrawn), the amount (balance) of carbon in the atmosphere increases. Carbon budget calculations examine various **sources**, phenomena that release greenhouse gases into the atmosphere, and **sinks**, phenomena that remove greenhouse gases from the atmosphere (**TABLE 15.1**). The process of removing these gases from the atmosphere and storing them in the hydrosphere or geosphere, through a sink, is called **sequestration**.

By carefully quantifying the sources and sinks listed in Table 15.1, researchers have found that since the beginning of the 20th century, significantly more greenhouse gases are flowing into the atmosphere than are being removed (**FIG. 15.30b**). The only change in the carbon budget since 1900 has been an increase in carbon input from anthropogenic sources, supporting the interpretation that human activities are responsible for the observed rise in greenhouse gases that is driving global warming (**FIG. 15.30c**). In effect, people are extracting carbon that has been locked underground in geologic reservoirs (as fossil fuels or limestone) over millions of years, and are transferring it into the atmosphere much faster than it can be removed by natural sinks.

FIGURE 15.31 Rainforest destruction releases CO_2 into the atmosphere.

(a) An immense amount of biomass, which grew by extracting CO_2 from the atmosphere, is stored in the Amazon rainforest.

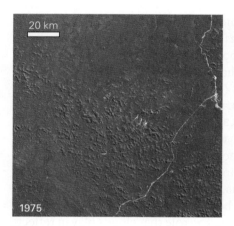

(b) A comparison of satellite photos taken 39 years apart shows the consequences of rainforest destruction in this area of Amazon rainforest in Brazil.

To make matters worse, human activities are decreasing the capacity of sinks. For example, conversion of natural biomass (living and recently dead organisms) into crop land or range land by *rainforest destruction* (slashing and burning of trees) **(FIG. 15.31)**, as well as by plowing up long-established prairies, releases CO_2 directly into the air. Also, it decreases the amounts of CO_2 that are both absorbed by photosynthesis to produce leaves or branches, and stored in soil and roots. Deforestation alone accounts for 8%–10% of anthropogenic inputs.

ISOTOPIC STUDIES TO IDENTIFY CARBON SOURCES.

Carbon has three isotopes, ^{14}C, ^{13}C, and ^{12}C. Of these, the isotope ^{12}C is the most common, and makes up most of the carbon incorporated by plants during photosynthesis. The isotope ^{13}C is heavier and is commonly found in volcanic gases. The radioactive isotope ^{14}C, which makes up a tiny proportion of all carbon, is found in once-living matter up to about 50,000 years old, but not in volcanic gases or fossil fuels. It is produced in the atmosphere continually when nitrogen converts to carbon due to cosmic ray bombardment from space.

Different ratios of the three carbon isotopes are found in living plants, in volcanic gases, and in fossil fuels. These ratios allow scientists to determine the source of the carbon in the atmosphere. Fossil fuels, which formed millions of years ago, contain no ^{14}C, and their $^{13}C/^{12}C$ ratio is less than that in the atmosphere. The $^{13}C/^{12}C$ ratio of volcanic gases is greater than that in the atmosphere because volcanic gases have a relatively high amount of ^{13}C. Thus, the products of fossil fuel combustion have a distinct CO_2 fingerprint with no ^{14}C and a low $^{13}C/^{12}C$ ratio, whereas volcanic CO_2 has a high $^{13}C/^{12}C$ ratio. Burning plant material such as wood produces CO_2 with a specific $^{14}C/^{12}C$ ratio. The atmospheric $^{13}C/^{12}C$ ratio has been declining over time, pointing to fossil fuels as the source of the additional CO_2.

Take-home message...

The concentration of carbon dioxide in the atmosphere has been increasing throughout the industrial age. Global climate models show that the increase in atmospheric CO_2 is strongly tied to global warming. Strong evidence shows that the source of the additional CO_2 is anthropogenic.

QUICK QUESTION What is the Keeling curve, and what does it tell us about trends in atmospheric composition over the past 60 years?

15.6 Impacts and Hazards of Climate Change

Unlike most other disasters described in this book, which happen rapidly and often with little warning, climate change takes place very slowly, and it will affect society's future well beyond our individual lifetimes. Here, we examine how researchers determine possible future changes in the Earth's climate, then consider what their predictions tell us.

FIGURE 15.32 Using GCMs to make predictions of future global warming.

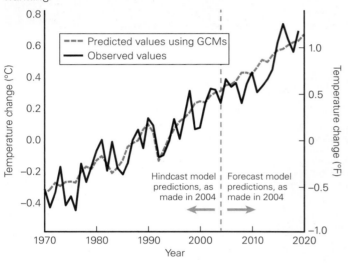

(a) In 2004, researchers used GCMs to predict temperature changes from 1970 to 2020. Comparison of the predictions with the observed values shows a good match. Temperature changes are indicated relative to the average 1980–1999 temperature.

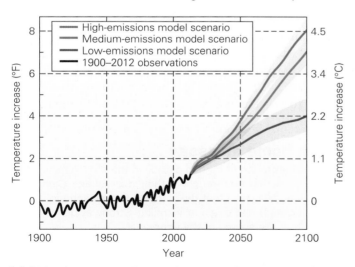

(b) GCM projections of temperature change, made in 2012 assuming different rates of anthropogenic CO_2 emission. All three simulations suggest a significant global temperature increase by 2100.

Using GCMs to Predict the Future

How do climate researchers envision the future? One approach involves using GCMs—since these models successfully reproduce observed past changes in the Earth's climate, they can also help us picture future changes (**FIG. 15.32a**). By adjusting their assumptions about the rate at which greenhouse gas concentrations will increase or decrease in the future, researchers can simulate a range of possible future scenarios. These include a *high-emissions scenario*, in which future anthropogenic greenhouse gas emissions increase at a rate consistent with the current increase in emissions (sometimes called a *business-as-usual* scenario); a *medium-emissions scenario*, in which the rate of increase in greenhouse gas emissions is somewhat reduced from the current rate; and a *low-emissions scenario*, in which the rate of greenhouse gas emissions decreases significantly below current rates (**FIG. 15.32b**). In the first scenario, GCMs show that atmospheric CO_2 concentrations could rise to 1,000 ppm by 2100; in the second, to about 620 ppm; and in the third, to about 475 ppm. The medium-emissions scenario predicts that atmospheric temperatures will increase by 2°C–4°C (3.6°F–7.2°F) over the next century, with the greatest warming taking place at high latitudes during winter (**FIG. 15.32c**).

The accuracy of these predictions depends on our understanding of atmospheric processes around the globe, but that understanding is not complete. The climates of some remote regions, such as those over the Southern Ocean, have not yet been characterized sufficiently to be modeled with precision. The role that clouds play in reflecting and absorbing solar radiation, which varies with the type of cloud and location, has not yet been fully constrained.

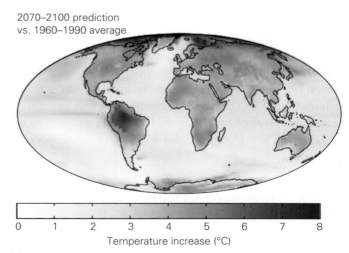

(c) Global warming varies with location. This map shows a GCM prediction of possible regional variation in temperature change by 2070–2100, relative to the average global temperature for 1960–1990.

FIGURE 15.33 Predictions of future sea-level change.

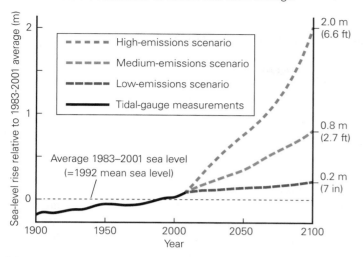

(a) GCMs suggest that sea level will continue to rise, increasing as much as 2.0 m by 2100.

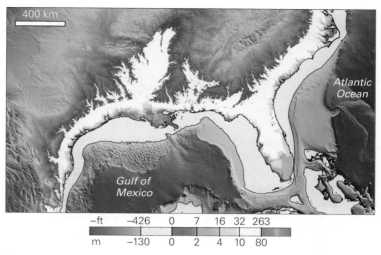

(b) Land elevations are low along the US Gulf and Atlantic coasts. The lowest of these areas will be flooded by rising seas.

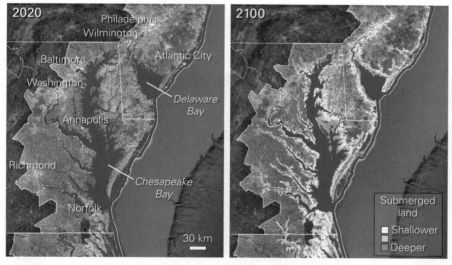

(c) Sea-level rise predicted for a high-emissions scenario will affect the densely populated area of Chesapeake and Delaware Bays by 2100. Dark blue shows present seas; light blue shows the area that will be covered by higher seas.

And many climate feedbacks remain poorly understood. A worldwide effort to improve understanding of these issues has been underway for decades. As more data become available, researchers refine GCMs. So far, refined models continue to indicate that climate will warm.

Future Sea-Level Rise

Sea-level rise has already impacted low-lying coastal communities—nuisance tides are happening more frequently, storm surge is leading to submergence of broader areas, and shorelines are migrating inland, as we saw in Chapter 13. GCMs predict an additional rise in sea level of 0.2–2.0 m (0.6–7 ft) by 2100, depending on the emissions scenario used (**FIG. 15.33a**). Given that storm surge can be as high as 8 m (26 ft), and that large waves occur on top of the surge, seawater may flood areas as much as 16 m (52 ft) above the present sea level in areas threatened by tropical cyclones under a high-emissions scenario. Because 10% of the world's population lives near a coast on land less than 10 m (33 ft) above sea level, large populated areas could be threatened along coastlines where tropical cyclones come ashore. Even a sea-level rise of 1.5 m (5 ft) would cause storm surge to flood major coastal cities and destroy beaches and shoreline properties (**FIG. 15.33b, c**) (**BOX 15.1**). Projecting further into the future, sea-level rise could inundate much of the world's low-lying coastal area (**VISUALIZING A DISASTER**, pp. 570–571). In addition to flooding populated areas, sea-level rise will damage coastal wetlands, which protect shorelines from flooding and provide habitats for many types of plants and animals, and will cause saltwater to replace fresh groundwater in some underground aquifers. Overall, the economic impacts of sea-level rise will be staggering.

Redistribution of Water Resources

People need clean water in order to live. Our major sources of freshwater are rivers, lakes, underground aquifers, and the snow and ice of mountains (see Chapter 7). Each of these sources has already shrunk, and GCMs indicate that they will shrink even more in the future.

BOX 15.1 DISASTERS AND SOCIETY

New York City, urban flooding, and sea-level rise

In 2012, Hurricane Sandy ravaged the northeastern United States, taking 233 lives and causing $75 billion in damage. In New York City, Sandy caused major destruction. Areas of the city flooded as storm surge 4.25 m (14 ft) high inundated homes, buildings, roadways, boardwalks, and transit terminals. Water cascaded into the 9/11 Memorial construction site at Ground Zero and into many subway tunnels **(FIG. Bx15.1a)**. Many areas of the city lost basic services, some for more than a week **(FIG. Bx15.1b)**. A major firestorm in one neighborhood burned out of control, for no hydrant water was available.

How can New York respond to such threats in the future, as sea level rises? The City Planning Commission has been charged with developing plans to minimize the impact of floods. Following the Hurricane Sandy disaster, the group began a study to better understand the threat that global sea-level rise might pose for the city. The results of this study were frightening **(FIG. Bx15.1c)**. In one scenario, hundreds of city blocks of skyscraper-studded Manhattan that were not affected by Sandy could be inundated by flooding in a storm whose surge was the same height, if such a storm were to happen in 2100. Impacts on New York's economy would be catastrophic. Fortunately, 2021's Hurricane Ida had lower surge than Sandy; Ida's flood damage was due to rain.

Are such risks unique to New York? The answer, unfortunately, is no, and some governments are already taking action. A study commissioned by Singapore's Building and Construction Authority, for example, developed a national plan for coastal protection measures. At present, however, Singapore is an exception among coastal cities. The costs of infrastructure changes are prohibitive, and most municipalities are reluctant to spend resources on protection against what they perceive as distant future events. The flooding of Hurricane Sandy was a grim reminder of what the future may bring.

FIGURE Bx15.1 Challenges like those New York City faced during Hurricane Sandy may become worse due to sea-level rise.

(a) A New York subway tunnel during Hurricane Sandy.

(b) A major power outage during Sandy darkened lower Manhattan.

(c) The left map shows areas that would be flooded in 2020 by storm surge equivalent to that during Hurricane Sandy. The right map shows areas that would be flooded under one scenario if storm surge of the same magnitude arrived in 2100.

FIGURE 15.34 Mountain snowpacks store water, so a reduction in snowpacks diminishes water supplies.

(a) If the elevation of the snow line rises by a relatively small amount, the volume of the snowpack can decrease substantially.

(b) Diminished snowpack on a mountain in Switzerland.

SHRINKING MOUNTAIN SNOWPACKS. As global temperatures warm, the average *snow line* (the elevation above which it's cold enough for snow to cover the ground) on mountains will move farther upslope **(FIG. 15.34)**. Over the course of winter, snow that accumulates above the snow line builds a **snowpack**, so a rise in the snow line will cause the volume of the snowpack to shrink. Snowpack volume will also shrink because the cold season during which snow accumulates will be shorter. Warming temperatures will also mean that snowpacks will melt earlier, so more meltwater will enter streams in winter and early spring, and less will be available in summer and fall. Such changes could lead to more frequent spring floods and worsening summer and fall droughts.

Loss of mountain snowpack is especially serious for Mediterranean climate zones, such as California, where virtually all precipitation arrives in the winter months. In such climates, mountain snowpack serves as a major water source for the rest of the year, so snowpack reduction has led to water shortages, as demonstrated by the 2021 water crisis in California's agricultural districts.

CHANGES IN THE DISTRIBUTION OF PRECIPITATION. GCMs predict that precipitation in arid subtropical regions and Mediterranean climate zones will decrease by about 20%–40%, while precipitation in temperate and polar climates and along the ITCZ may increase by 10%–60% **(FIG. 15.35)**. Models also predict that precipitation will decrease in the southwestern and southeastern United States, and will increase in the Midwest. Clearly, the water supplies of many large cities, such as Los Angeles, Las Vegas, and Phoenix, will be stressed even more than they are already, and the global water crisis that we described in Chapter 7 will become worse.

Precipitation changes will impact the water supplies of different climate groups differently. In arid climates, surface water will become scarce, so demand for groundwater will increase, leading to acceleration of the rate at which the water table drops and increasing all the problems that result from such drops (such as land subsidence; see Chapter 7). In regions where precipitation increases, more water, sediment, nutrients, and pollutants will enter streams during heavy rainfalls. Not only will this increase cause floods, which can disrupt water-supply and water-treatment facilities, but it will decrease surface-water quality.

Future Changes in the Arctic

GCMs indicate that the planet's most dramatic warming will happen in the Arctic and subarctic regions during winter **(FIG. 15.36a)**. This phenomenon, known as **polar amplification**, occurs because warming air temperatures are melting sea ice and snow, and therefore causing a significant drop in albedo, and an associated increase in absorption of solar energy. As spring comes to polar regions earlier in the year, and the transition from fall to winter happens later, there will be a longer period of **permafrost thawing**, melting of the layer of peat, soil, and sediment that remains frozen most of the year. When permafrost thaws, organic decay can happen, which releases CH_4—a positive feedback—increasing the

FIGURE 15.35 GCM prediction of regional variation in precipitation change by 2081–2100, relative to average global precipitation for 1981–2000, under a high-emissions scenario.

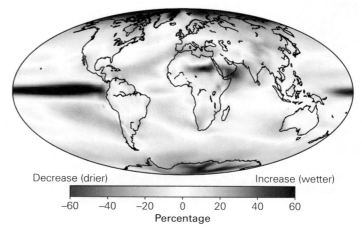

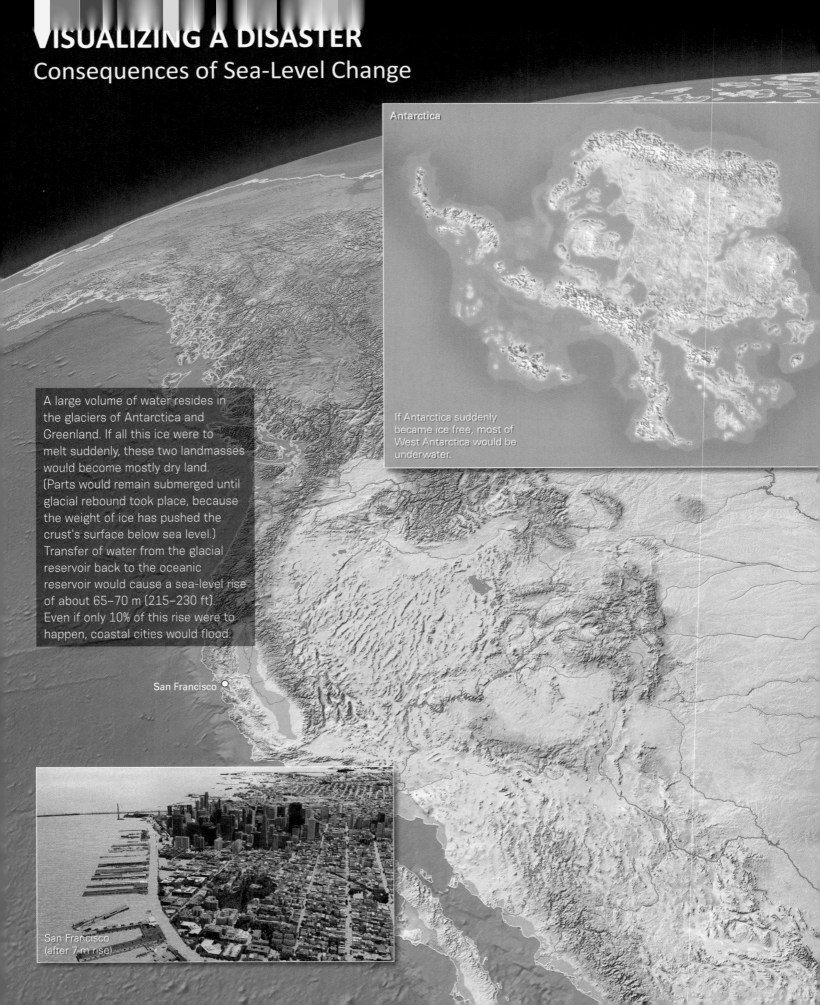

VISUALIZING A DISASTER
Consequences of Sea-Level Change

A large volume of water resides in the glaciers of Antarctica and Greenland. If all this ice were to melt suddenly, these two landmasses would become mostly dry land. (Parts would remain submerged until glacial rebound took place, because the weight of ice has pushed the crust's surface below sea level.) Transfer of water from the glacial reservoir back to the oceanic reservoir would cause a sea-level rise of about 65–70 m (215–230 ft). Even if only 10% of this rise were to happen, coastal cities would flood.

If Antarctica suddenly became ice free, most of West Antarctica would be underwater.

San Francisco (after 7-m rise)

Measured sea-level rise varies with location. The maximum current observed rate is about 1 m per century. At this rate, the 7-m rise depicted in the images will not happen for several centuries, but even a 1-m rise will lead to a costly increase in nuisance flooding and storm-surge damage. A sea-level rise of 70-m would flood large areas of coastal land worldwide (areas shown in purple). In fact, most of Florida would be underwater, and New York City buildings would be submerged up to about the 20th floor. Even a 7-m (20-ft) rise will turn streets in many major cities into coastal canals, as shown in the images.

A suddenly ice-free Greenland would host a central lake.

New Orleans (after 7-m rise)

New York (after 7-m rise)

FIGURE 15.36 Warming of the Arctic, and its consequences.

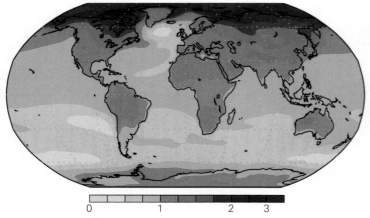

Predicted temperature change by 2100, relative to the average of 1981–2000 (°C).

(a) GCMs predict that the Arctic will warm more than lower latitudes, particularly under a high-emissions scenario, as shown here.

(b) Permafrost melting makes the ground uneven and has caused the foundation of this house to sink.

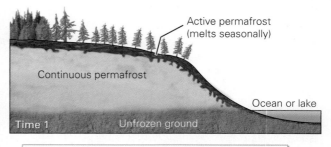

Before warming, permafrost forms a continuous layer down to a depth where the Earth's internal heat keeps ground temperatures above freezing.

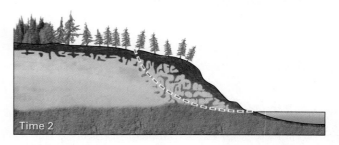

When the climate warms, the permafrost layer thins, and the upper part starts to break up. Slopes underlain by permafrost become unstable.

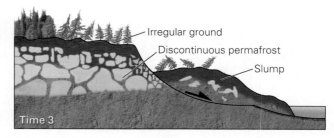

Eventually, permafrost breaks up entirely and becomes discontinuous. The land surface becomes irregular, and slumps carry melted permafrost downslope.

(c) When permafrost melts, the land sinks irregularly, so trees tilt; slumps develop because melting weakens sediment.

susceptibility of tundra and bordering forests to pests and wildfires. In places where people live on permafrost, thawing damages foundations for homes, roadbeds, and infrastructure, because it results in uneven sinking of ground surfaces and can cause landslides **(FIG. 15.36b, c)**. In Alaska alone, permafrost thawing is expected to add between $3 and $6 billion to the costs of maintaining public infrastructure over the next 20 years.

As the extents of sea ice and permafrost decreases, Arctic regions will become more accessible to trans-Arctic shipping, oil and gas exploration, and tourism. Therefore, risks to people and ecosystems from oil spills and maritime accidents will increase. Politically, concerns about security and sovereignty over the Arctic may lead to international disputes and conflicts over access to Arctic resources.

Future Changes in the Biosphere

The northward migration of plant hardiness zones in the United States, discussed earlier in this chapter, demonstrates that climate zones are slowly migrating poleward. GCMs suggest that climate zone migration will continue into the future, so that forests currently growing at temperate latitudes will grow in areas that are now tundra **(FIG. 15.37a)**. In fact, by 2100, Illinois, in the north-central United States, may have a climate resembling that of Louisiana, a south-central state on the Gulf Coast. The implications of such *climate zone migration* for agriculture are profound. For example, crops such as corn and soybeans that now grow in the US Midwest will be better suited for the temperatures of southern Canada. But southern Canada has hosted cool climates for thousands of years, and thus it has not developed thick soils like those of the midwestern prairies. Therefore, the future productivity of areas with appropriate climates for economically viable crops might be substantially less than the productivity of the areas that now produce those crops.

Wildlife will also feel the threat of climate change. Populations of climate-sensitive species have already been shrinking, and in some cases going extinct, as their habitable range diminishes. For example, declining sea ice in the Arctic has already resulted in a reduced survival rate for polar bears, because these predators need sea ice to hunt seals, their primary food source (**FIG. 15.37b**). While land-based herds of caribou may benefit from global warming as a longer growing season provides more food, they will also face a longer season of exposure to dangerous parasitic insects (**BOX 15.2**).

Future Changes in the Oceans

Unfortunately, the world's oceans, and the creatures that live within them, have already been substantially harmed by human activities, including pollution and overfishing. Climate change will cause other changes in the future.

FIGURE 15.37 Changing climates will affect the biosphere.

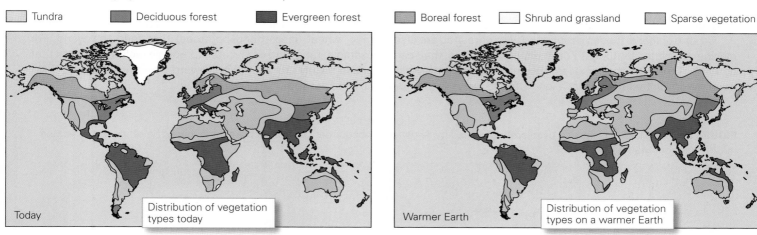

(a) Global warming will cause the latitude of climate zones to move poleward, and arid areas to expand.

(b) Polar bears rely on sea ice to gain access to food sources. Sea-ice cover has diminished greatly in the past few decades, so polar bear populations have decreased.

BOX 15.2 DISASTERS AND SOCIETY

The global poleward spread of diseases and pests

The most dangerous animal in the world is neither a tiger nor a crocodile—it's a mosquito! Two species, *Aedes aegypti* and *Anopheles*, have been responsible for human misery, illness, and death across a wide range of the Earth's tropics. *Anopheles* transmits the parasite that causes malaria, while *Aedes aegypti* transmits yellow fever, dengue fever, Zika, and other diseases. These mosquitoes survive only in warm temperatures, which is why they have been confined to the tropics—until now. With the increase in average global temperature associated with climate change, their ranges have been extending into the subtropics **(FIG. Bx15.2)**. One study predicts that by 2061–2080, a half-billion additional people or more could be at risk from mosquito-borne diseases.

Other tropical diseases associated with parasites are also moving to higher latitudes. For example, sandflies that can carry the parasite that causes leishmaniasis, a potentially deadly and disfiguring disease once found only in tropical rainforests, have been found in Texas and Oklahoma. And in Europe, disease carriers once restricted to areas that traditionally hosted Mediterranean climates are moving northward. For example, ticks that transmit Lyme disease were once found only in southern Europe, but now they are common in northern Europe. In some cases, it's not the range of the pest that expands, but the season during which the pest remains active. For example, tick season in Russia has increased over the past two decades, leading to a 23-fold increase in the rate of tick-borne encephalitis.

Agricultural pests, such as insects, and crop pathogens, such as bacteria, viruses, and fungi, have also been moving poleward with climate change, damaging the major crops (rice, wheat, maize, and potatoes) that feed the world. For example, fungi and oomycetes (parasitic molds and rusts that cause blight) are moving northward by an average of 2.7 km (1.7 mi) per year.

FIGURE Bx15.2 Malaria is spread by microbes carried by *Anopheles* mosquitoes. As global warming occurs, the range of these mosquitoes will increase.

OCEAN ACIDIFICATION. Chemists express acidity using a measure called *pH*. A material with a pH of less than 7 is an *acid*, and one with a pH of greater than 7 is a *base*. Pure water has a pH of 7, so it's *neutral*. Seawater, prior to the industrial revolution, had an average pH of about 8.25, meaning that it was slightly basic. In the past few decades, however, the pH of seawater has dropped to about 8.1, a phenomenon known as **ocean acidification**. (This term emphasizes that ocean water has become more acidic, but does not mean that it has become an acid.)

Why has ocean acidification occurred? An estimated 30%–40% of the CO_2 added to the atmosphere by human activities dissolves in the oceans, which act as a sink for CO_2, as we have noted. When CO_2 dissolves in water, it becomes carbonic acid, the same acid found in fizzy drinks. While this process, fortunately, decreases the concentration of greenhouse gas in the atmosphere, it unfortunately harms certain marine organisms by depressing their metabolic rates, their shell production, and their immune responses. The oceans still have the capacity to dissolve more CO_2, though not quickly enough to prevent CO_2 increases in the atmosphere. But, as a consequence, the pH of the oceans may have dropped to 7.7–7.8 by 2100 (**FIG. 15.38**).

FIGURE 15.38 Ocean acidification is taking place worldwide.

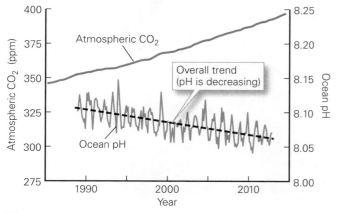

(a) As the CO_2 concentration in the atmosphere rises, more CO_2 dissolves in the ocean, where it produces carbonic acid.

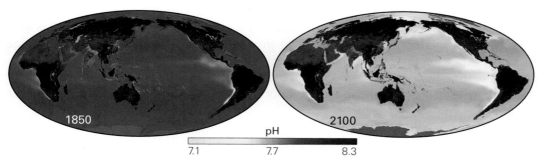

(b) Changes in ocean pH will vary with location. But by 2100, ocean water is predicted to be below pH 7.8 worldwide.

FIGURE 15.39 Coral bleaching.

(a) A living reef with various colored corals.

(b) A dead reef with bleached corals.

CORAL BLEACHING. As we saw in Chapter 13, many coral reefs are now undergoing **coral bleaching**, a process that occurs when corals under environmental stress expel the algae that live in their tissues, and turn white. Although corals can survive initial bleaching, they will die if the conditions that cause the bleaching persist. Rising water temperatures are probably the main cause of coral bleaching, but changes in pH and other changes in water chemistry may also play a role. The rates of bleaching and reef death are astounding. For example, in 2005, half the coral reefs in the Caribbean died in a single year due to a massive bleaching event. A 2019 survey of Australia's Great Barrier Reef shows that close to half the reef died in just a few years, and that there has been a nearly 90% drop in the number of new corals growing. A huge percentage of the loss happened in one year, 2016, during a heat wave that lasted many weeks (**FIG. 15.39**). The future of coral reef systems, and of the marine organisms that depend on them, will be further threatened as ocean temperatures rise.

15.6 Impacts and Hazards of Climate Change • **575**

FIGURE 15.40 The predicted redistribution of fish populations due to the warming of seawater as indicated by change in maximum catch potential predicted for 2051–2060, under a moderate- to high-emissions scenario, as a percentage of the average maximum catch potential for 2001–2010. These predictions do not account for impacts of overfishing or ocean acidification, both of which will cause decreases in nearly all species.

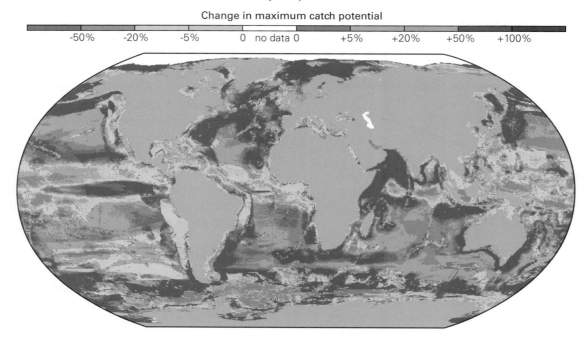

DEGRADATION OF FISHERIES. Zooplankton form the base of the marine food chain, and their numbers and locations are changing with the climate. Ocean acidification makes it more difficult for these tiny organisms to form their calcite shells, because the solubility of calcite in seawater increases as pH decreases. In addition, as oceans warm, cool-water zooplankton species migrate to higher latitudes. Fish and other marine animals that feed on zooplankton must move with the organisms. These changes have led to widespread changes in the productivity and distribution of fish populations. Continued warming will amplify these changes, so fishing grounds will have to shift to new locations. The shift can be visualized by considering the change in *maximum catch potential* (the maximum exploitable catch of a species) **(FIG. 15.40)**. Many areas, like the Southern Ocean, show steep declines, while others, like the North Atlantic, show increases due to migration of warm-water fish to higher latitudes. Unfortunately, overfishing has already decimated populations of many species, so the maximum catch potential is unlikely to be achieved and the economic impacts of these changes and overfishing worldwide will be significant.

MELTING OF METHANE HYDRATE. *Methane hydrate* is an ice-like compound consisting of CH_4 molecules surrounded by water molecules. It forms when anaerobic bacteria in seafloor sediments produce CH_4 that bubbles into cold seawater. Under the temperatures and pressures found at depths of 90–900 m (295–2,950 ft) below sea level, CH_4 reacts with seawater to produce crystals of methane hydrate. Huge quantities of CH_4—perhaps more than exists in all the world's oil and gas supplies—are trapped in methane hydrate on the seafloor **(FIG. 15.41)**. If the temperature of seawater at

FIGURE 15.41 Methane hydrate is an ice-like substance, composed of CH_4 molecules surrounded by H_2O molecules, that lies buried in sediment on the seafloor. If it melts, it releases CH_4.

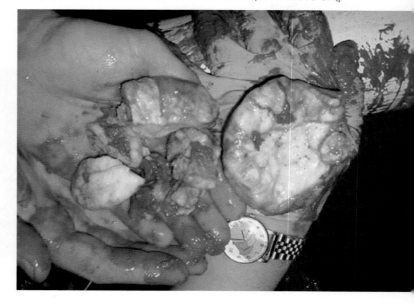

depths where methane hydrate forms were to rise by just a few degrees, large quantities of methane hydrate might melt and release CH_4, thereby increasing the methane concentration in the atmosphere. This process would be a positive feedback, accelerating global warming.

INTERRUPTION OF THERMOHALINE CIRCULATION. As we've seen, the thermohaline circulation transfers heat across latitudes (see Fig. 15.13). According to some models, if global warming melts enough polar ice, or if precipitation increases significantly, freshwater will dilute surface seawater at high latitudes so much that the water will not be dense enough sink, and the thermohaline circulation will be shut off. By some estimates, the thermohaline circulation in the Atlantic has already slowed by about 15% since the mid-20th century, perhaps due to melting of the Greenland ice sheet. If the thermohaline circulation did stop, there might be a substantial cooling of high-latitude regions.

Future Changes in Hazardous Weather

GCMs suggest that in the future, the number of storms probably won't increase, but their intensity and duration will. As we've noted, warming of the sea surface will increase the input of water vapor into the atmosphere, where it will serve as fuel for storms. Some calculations suggest that each 0.55°C (1°F) increase in average global ocean temperature may cause a 7% increase in the potential maximum wind speed of tropical cyclones. Such a change will increase the destructive capability of these storms.

As global warming progresses, and the Arctic and subarctic regions warm more than the tropics, the average temperature gradient across the polar front (see Chapter 8) will decrease during winter, and the polar front will shift northward. Researchers expect that, as a result, mid-latitude cyclones tracking across continents of the northern hemisphere will decrease in number. By some estimates, 5%–10% fewer winter mid-latitude cyclones will occur per year during the late 21st century, compared with the 20th century. The storms may also decrease in intensity over land. However, coastal mid-latitude cyclones, such as the nor'easters discussed in Chapter 10, may tap energy from warmer seas and increase in strength.

Continued global warming will probably increase the intensity of heavy rain events worldwide, since warmer air can hold more water vapor. Consequently, rainfall in many flood-producing weather systems will be more intense. Although researchers cannot attribute any single weather event to climate change, society has endured several record rainfalls in recent years, an expected consequence of a warming climate.

Future Increases in Heat Waves and Droughts

During a heat wave, the temperature remains significantly above normal for days or weeks (see Chapter 14). Heat waves become news when the temperature rises high enough for many people to suffer from heat illnesses. The number of heat waves has increased in recent years, which isn't surprising, given that recent years have seen records for average global temperature (**FIG. 15.42a**). More global warming will further increase the number of heat waves. To see why, consider a graph showing the range of temperatures for a given location, recorded over many years, with each temperature plotted against the number of days during which that temperature occurred (**FIG. 15.42b**). The result is a bell curve, with the right-hand limb representing extreme heat. If global warming takes place, the whole curve shifts to the right, so the number of days with extreme heat increases. GCMs predict that under a high-emissions scenario, large areas of the United States will suffer from frequent heat waves by 2100 (**FIG. 15.42c**).

Researchers who have examined the probability of heat waves conclude that global warming is the reason why the heat wave that struck Europe in 2003 (and caused 70,000 deaths) was 10 times more likely in 2003 than it would have been in 1900. Similarly, they predict that the number of days with heat index values over 41°C (105°F) will increase threefold by 2050 in the north-central United States, and that places that now have only a few such days per year may have more than 40 such days per year by the end of the century. Even places that are normally hot, such as Phoenix, Arizona, are setting temperature records, as we saw in Chapter 14. Such temperature increases will have many effects on society in addition to increased rates of heat illness, including decreased worker productivity, increased wildfire hazard, depleted water supplies, and more frequent crop failures.

As heat waves become more frequent or extreme, some regions of the Earth will become difficult to live in. Currently, about 1% of land areas on our planet are **hot zones**, defined as having average annual temperatures exceeding 29°C (84°F). These areas are currently confined to the southern Sahara and have few human inhabitants; the small communities that live in such places have developed cultures adapted to high temperatures. GCMs predict that under a high-emissions scenario, these hot zones will grow to encompass 20% of global land areas by 2070. As their sizes increase, hot zones will include regions, such as parts of India, with large populations. The difficulty of adapting to such temperatures may drive people to migrate to more comfortable climates.

A continued rise in global temperature will also worsen droughts because more soil moisture will evaporate, which will cause crops to wither. Loss of plant cover will act as a

FIGURE 15.42 Heat waves will be more common in the future due to global warming.

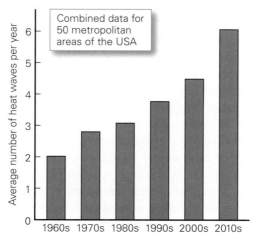

(a) The average number of heat waves per year in the United States has nearly tripled since the 1960s.

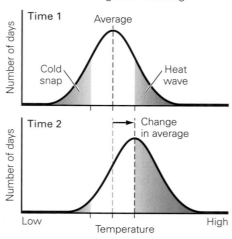

(b) If average global temperature increases slightly, the number of heat waves increases a lot.

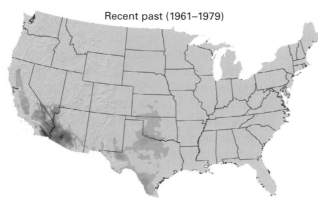

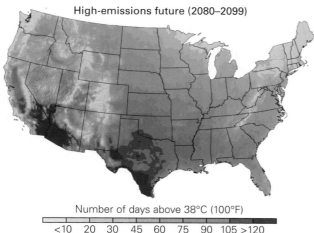

(c) GCMs indicate that the number of 100°F days will increase significantly by 2100.

positive feedback, as it will allow more radiation to reach and heat the ground, causing still more soil moisture to evaporate. Furthermore, global warming will increase the land areas affected by drought, for two reasons. First, the north-south width of the Hadley cells (see Chapter 8) will increase because, as oceans warm, more air will rise along the ITCZ, so air will spread outward to higher latitudes at the tropopause. This *Hadley cell expansion* will mean that hot, dry air will descend at latitudes that are now semiarid, causing those regions, including many places that are already poverty-stricken, to become increasingly arid. Second, the northward shift of the polar front in the northern hemisphere will cause western Pacific storm tracks to shift farther north. As a result, the atmospheric rivers that supply rain and snow to the western United States in winter will also shift northward, reducing water supply to the region, especially to California and Oregon (see Chapter 14).

> **Take-home message . . .**
> The impacts of climate change are far-reaching. The potential economic and social costs of sea-level rise, the spread of disease, degradation of the Arctic and the global ocean, and changes in weather systems pose real threats to future generations.
>
> **QUICK QUESTION** How is the range of the world's most dangerous animal related to climate change?

15.7 Can Climate Change Impacts Be Mitigated?

Protecting society from future disasters related to climate change will require significant efforts by individuals and by governments. Society has three options to deal with climate change: (1) reduce anthropogenic emissions of greenhouse gases; (2) remove greenhouse gases from the atmosphere; and (3) adapt to a new, warmer world. A fourth option, decreasing the amount of solar radiation absorbed by the Earth's surface (by changing the Earth's albedo, either by covering the ground with reflective material or by injecting large quantities of reflective aerosols, such as SO_2, into the stratosphere) remains highly speculative and is probably too risky because of unanticipated feedbacks.

REDUCING EMISSIONS. If society could reduce its **carbon footprint**, meaning the amount of CO_2 and CH_4 that it releases into the atmosphere, we could reduce the rate of global warming.

The most significant reduction of emissions would come if we substantially decrease the amounts of fossil fuels we burn. Reducing fossil fuel use is challenging, however, for several reasons. World demand for energy has increased dramatically over the past few decades, and it will continue to increase as more countries industrialize and as standards of living rise. At present, fossil fuels are the most widely used source of energy for transportation, producing electric power, and other applications (FIG. 15.43). Fossil fuels remain popular because they have a high *energy density* (the amount of energy contained in a kilogram of a fuel), because they're easily transportable, and because they're inexpensive. Two approaches could help to decrease the amount of fossil fuel burned—reducing demand and switching to alternative energy sources.

Demand for energy can be decreased in a number of ways. Such *energy conservation* measures include improving vehicle gas mileage, increasing mass transportation, developing more energy-efficient machines and lighting, turning off equipment or lights when not in use, and improving building insulation. Switching to *alternative energy sources*—those with a smaller carbon footprint than fossil fuels—is another way of decreasing fossil fuel use. **Carbon-neutral** energy sources—those with no carbon footprint—include renewable sources such as solar, tidal, geothermal, and wind power, hydropower, and biofuels, as well as nonrenewable sources such as nuclear power. Intense efforts have gone into developing carbon-neutral sources of energy. *Biofuels* made from perennial plants, such as grasses, are carbon-neutral because the carbon released by burning them in one year is absorbed when they regrow the next year. Notably, while alternative energy sources may be carbon-neutral, production of the equipment needed to produce the energy might not be—for example, a lot of cement goes into the construction of a hydroelectric dam, and cement production emits CO_2. Furthermore, dams flood large tracts of land and may cause environmental problems.

A switch to alternative energy sources has already started. Electric and hybrid cars are becoming increasingly popular; generation of electricity using carbon-neutral energy sources has increased substantially; and use of coal, the fossil fuel with the largest carbon footprint, has decreased (FIG. 15.44a). But why don't we see more progress, given the urgency of slowing global warming? Many practical issues stand in the way of switching to alternative energy sources. Energy infrastructure based on fossil fuel use has already been built and will be expensive to replace. In addition, the supply of energy produced by some alternative sources is variable. The efficiency of wind power, for example, depends on wind strength, and that of solar power varies with cloud cover, time of day, and season. Harvesting solar or wind energy requires large installations, but the land needed for these installations may not be readily available, and it may not be near locations with the greatest energy demand (FIG. 15.44b, c). Technologies don't yet exist to store significant amounts of electricity inexpensively, or to regulate its distribution through the electrical grid in ways that accommodate rapid variations in supply. Many people remain suspicious of nuclear power because it produces radioactive waste that can be difficult to store safely, and because plants in a few cases have leaked radioactivity. New nuclear-power technologies that are safer and produce less waste have been proposed, but they have not yet been implemented. Despite the obstacles, there is good reason to be optimistic that carbon-neutral sources will become major energy resources for the world in the future.

Efforts have already gone into a possible engineering approach to decreasing emissions from fossil fuel use and other CO_2-emitting activities. This process, called **carbon capture and sequestration**, involves trapping CO_2 at its emission source (such as a smokestack), condensing it into liquid form, and then pumping it deep underground at high pressure into porous sedimentary rocks, where it can permeate the pores of the rocks and remain trapped underground. This technology faces two challenges:

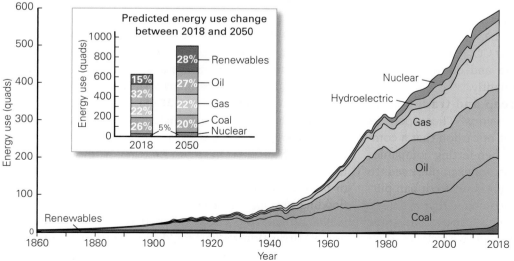

FIGURE 15.43 Historical and projected global energy demand in quads (quadrillions of British thermal units). Energy use has increased dramatically since 1800, and most of that energy has come from fossil fuels. The inset shows the predicted increase in energy demand by 2050.

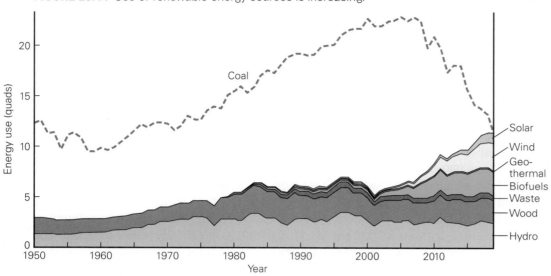

FIGURE 15.44 Use of renewable energy sources is increasing.

(a) Between 2000 and 2019, renewable energy use increased significantly in the United States, while use of coal dropped. Most coal has been replaced by natural gas.

(b) Giant wind turbines in England.

(c) A large solar farm in Nevada.

first, it takes significant energy to extract, condense, and pump CO_2, and second, sedimentary rocks contain natural fractures that may provide pathways for the gas to seep back to the ground surface. Another approach currently being explored involves injecting CO_2 into basalt, a common type of igneous rock. Tests have shown that CO_2 reacts with basalt to produce crystals of solid calcite and dolomite, which can remain locked in the rock permanently.

Some countries have instituted **cap-and-trade** policies as a way of reducing carbon emissions. These policies involve setting a cap on the amount of carbon that a company or other entity can emit, and then letting companies buy and sell *carbon credits*. If a factory, for example, needs to produce more carbon than its cap allows it to, it can purchase carbon credits from another company that is able to produce less carbon than its allowance. The cap-and-trade policy provides companies with an incentive to lower their carbon emissions.

REMOVING GREENHOUSE GASES FROM THE ATMOSPHERE. Atmospheric CO_2 occurs at such low concentrations that extracting the gas directly from the air would take a lot of energy. Nature, fortunately, has already designed a process that removes CO_2 from the air efficiently: photosynthesis. By increasing the amount of biomass carrying out photosynthesis, the concentration of CO_2 could be decreased, for it would be transferred from the atmospheric reservoir into the biomass. By one estimate, increasing the area of forest on land by 25% could remove about 25% of the CO_2 in the atmosphere. Discussions have taken place regarding spreading nutrients over large areas of the oceans to stimulate phytoplankton growth. At present, the broader consequences of such action aren't known, so the idea is too risky to test. There is also current interest in mixing CO_2-absorbing minerals, such as those in basalt, into soils so that the soils will absorb more CO_2.

ADAPTING TO A NEW, HOTTER WORLD. If our carbon footprint can be decreased quickly enough to keep the average global temperature from rising more than 2°C (3.6°F) above preindustrial values, many of the worst effects of global warming can be mitigated (**FIG. 15.45**). But if current trends in carbon emissions continue, society will have to adapt to a changing world. This challenge could be met with careful planning of, and investment in, strategies to mitigate some of the impacts of climate change. For example, building seawalls or broadening wetlands could reduce the effects of coastal flooding; developing more efficient air-conditioning systems could mitigate the effects of increasing heat waves; less wasteful irrigation systems could minimize the consequences of water shortages; and development of new varieties of crops

FIGURE 15.45 Mitigation measures to reduce greenhouse gas emissions and to remove CO_2 from the air could significantly slow global warming.

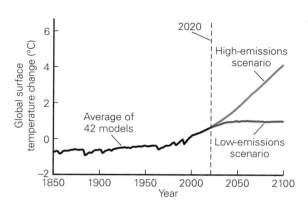

(a) With immediate mitigation, average global temperature could be kept from rising more than 2°C above preindustrial values, which would prevent the worst consequences of global warming.

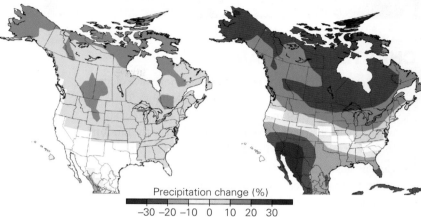

(b) Precipitation changes predicted for North America under a low-emissions scenario (left) and a high-emissions scenario (right). With immediate mitigation, precipitation changes in the United States by 2100 could stay under 20%. Without it, they would be much more.

that are more resistant to heat and drought could mitigate food shortages (**FIG. 15.46**). The alternative would be simply to react to catastrophes when they happen, such as making emergency repairs after a flood. Such a strategy puts people at risk and could end up being much more expensive than advance planning. As a sign of the times, Greece has appointed a "heat officer," charged with the task of keeping Athens habitable as its average temperature rises.

INTERNATIONAL ACCORDS. Climate change is a global problem that requires international cooperation. The *Kyoto Protocol* of 1997 committed 192 countries to decreasing their emissions of greenhouse gases. It was followed in 2015 by the *Paris Agreement*, which committed 196 countries to keeping the average global temperature increase to less than 2°C. While the spirit of these agreements has led to meaningful efforts to address global warming, the agreements have been controversial and have not been fully enforced.

Unfortunately, because climate change is a stealth disaster, if society lacks the political will to accept that a problem exists and to prepare to solve it, amplification of natural disasters due to climate change remains a real possibility. We may already be seeing the consequences of such amplification in large, low-latitude cities such as Karachi, Pakistan (population 20 million): the city has recently endured devastating floods, relentless heat waves, and losses of food supplies due to drought. Similar hardships have forced people to migrate to find more hospitable homes, and the World Bank estimates that by 2050, more than 150 million people will have moved to avoid the consequences of climate change, suggesting that the "great climate migration era" has already begun. Such large migrations potentially lead to unrest, terrorism, and war. Each individual's effort, and the efforts of their governments, are required to create the political will to address climate change so as to diminish these future risks.

Take-home message . . .

Mitigating the impacts of climate change will require reducing emissions through energy conservation, use of alternative energy sources, and carbon capture and sequestration; removing greenhouse gases from the atmosphere; or adapting to a warmer world.

QUICK QUESTION What challenges exist in generating and delivering electricity using solar and wind power?

FIGURE 15.46 This giant seawall, 32 km (20 mi) long, is being built around parts of Jakarta, Indonesia, to protect it from the sea as the city subsides.

15.7 Can Climate Change Impacts Be Mitigated?

Chapter 15 Review

CHAPTER SUMMARY

- The Sun provides energy to the Earth as shortwave electromagnetic radiation (ultraviolet and visible light). The Earth emits longwave (infrared) radiation. The amount of shortwave radiation arriving from the Sun balances the amount of longwave radiation leaving the Earth.
- Certain gases in the atmosphere, specifically water vapor, carbon dioxide, methane, ozone, and nitrous oxide, absorb infrared radiation and radiate some of it back to the Earth's surface, keeping the surface warmer than it would be without the gases. These gases are called greenhouse gases, and the warming they cause is called the atmospheric greenhouse effect.
- Weather refers to the condition of the atmosphere (temperature, air pressure, relative humidity, wind speed, cloud cover, and precipitation) at a given location at a given time. Climate describes the average weather conditions, and their range, for a region over the course of years to decades.
- Climate change refers to a shift in one or more climate characteristics over periods of years, decades, centuries, or even millennia.
- A region's climate depends on the region's average temperature, temperature range, annual amount of precipitation, seasonal variation in temperature and precipitation, characteristic weather extremes, and characteristic storms.
- A location's average temperature, and its range of temperatures over a year, are impacted by its latitude, elevation, and proximity to an ocean.
- The Earth's climates can be classified into five groups: tropical, arid, temperate, cold, and polar.
- Various paleoclimate indicators allow researchers to characterize the climates of the past. Paleoclimate studies indicate that climate change has occurred throughout the course of Earth history.
- Long-term climate change, on time scales of millions to tens of millions of years, is related to changes in atmospheric composition, volcanic activity, distribution of continents and ocean currents, uplift of land, and the evolution of life.
- Climate change on time scales of thousands of years results from changes in the eccentricity of the Earth's orbit, and in the tilt and precession of its axis. These phenomena, as well as the periodic changes in the insolation that a location on the Earth receives due to them, are called Milankovitch cycles.
- Other factors affecting short-term climate change, on the scale of decades to hundreds to thousands of years, include changes in the Earth's albedo and changes in the thermohaline circulation.
- A climate feedback is a phenomenon triggered by a change in climate that further modifies the change. A positive feedback amplifies change, and a negative feedback reduces it.
- Changes in average global atmospheric and ocean temperatures are now occurring on time scales of years to decades. These changes have affected sea-level rise, glacial retreat, Arctic sea ice cover, global weather patterns, and distributions of various organisms.
- Strong evidence shows that recent rapid increases in temperature during the past two centuries are due to increases in atmospheric concentrations of greenhouse gases emitted during fossil-fuel combustion and other human activities. This evidence comes from decades of measurements of greenhouse gases, studies of atmospheric gases trapped in glacial ice, and studies of carbon isotope ratios.
- If climate change continues at its current rate, society will face rising sea levels and flooding of coasts; diminishing water resources; melting of permafrost and loss of summer sea ice; reduction or extinction of some species; ocean acidification and degradation; losses of fisheries and bleaching of coral reefs; poleward spread of tropical diseases and pests; and increases in extreme weather events such as hurricanes, floods, droughts, and heat waves.
- Society has three options to deal with climate change: reduce emissions of greenhouse gases to levels that can better sustain the current climate; develop engineering solutions to remove anthropogenic carbon from the atmosphere and oceans; and adapt to the conditions that climate change will bring.
- Because climate change is a stealth disaster, the lack of political will to prepare for it, or even to accept that it exists, will continue to hamper progress toward solutions.

REVIEW QUESTIONS

Blue letters correspond to the chapter's learning objectives.

1. What is the difference between shortwave and longwave radiation, and which does the Earth's surface emit? **(A)**
2. Explain the cause of the atmospheric greenhouse effect and how it affects the Earth's energy balance. **(A)**
3. Distinguish between *weather* and *climate*. What factors control a region's climate? **(A)**
4. What factors control a region's temperature and temperature variation? **(A)**
5. What factors control a region's precipitation? **(A)**
6. How does the proximity of an ocean affect climate in a particular region? **(A)**
7. Explain the difference between natural long-term and natural short-term climate change. **(B)**
8. What tools or methods allow researchers to characterize the Earth's past climates? What is the difference between icehouse conditions and hothouse conditions? Was the temperature in the Eocene greater than or lesser than that of today? **(B)**
9. Describe the key factors that play a role in causing natural long-term climate change and those involved in natural short-term climate change. **(B)**
10. What are the Milankovitch cycles, and which is represented by the figure? How do they impact climate? **(B)**

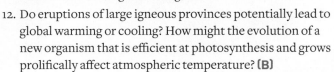

11. Based on the Milankovitch cycles, should the current climate be warming or cooling? **(B)**
12. Do eruptions of large igneous provinces potentially lead to global warming or cooling? How might the evolution of a new organism that is efficient at photosynthesis and grows prolifically affect atmospheric temperature? **(B)**
13. What evidence leads researchers to conclude that the Earth's climate is warming at an unprecedented pace? **(C)**
14. When was the Little Ice Age, and how has the volume of glaciers changed since then? What do changes in glaciers and ice sheets of the past century tell us about global temperatures? **(C)**
15. The Keeling curve in the figure shows that the carbon dioxide concentration increases and decreases every year, but trends upward. What causes the changes within a year? What causes the upward trend? **(C)**

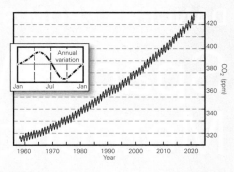

16. What is a global climate model? **(C)**
17. What is a climate feedback, and why is the concentration of water vapor in the atmosphere considered to be an example of one? **(C)**
18. Use the figure to explain where on the Earth global temperature increases are expected to be the greatest. Give reasons why the impacts are large in this region.

19. How can global warming cause sea level to rise? What impacts do you think a rise in sea level of 1–2 m (3–7 ft) might have on society? **(D)**
20. How might global warming increase the frequency of both floods and drought? **(D)**
21. What are the potential effects of global warming on life in the oceans? **(D)**
22. What efforts may help to diminish human society's carbon footprint? **(E)**
23. What carbon-neutral energy resources does society have as alternatives to fossil fuels? **(E)**

On Further Thought

24. If the CO_2 concentration in the atmosphere doubles during the next century, how might this affect water resources? **(D)**
25. Mosquitoes carry diseases (malaria, dengue fever, Zika) that are dangerous to humanity. The species of mosquitoes that carry such diseases prefer warmer climates. If current observed trends in global temperature change continue, will human populations susceptible to these diseases increase or decrease during the next century? Explain your answer. **(D)**

16

DANGERS FROM SPACE
Space Weather and Meteorite Impacts

By the end of the chapter you should be able to . . .

A. distinguish among different types of emissions from the Sun, explain their origin, and explain their relationship to space weather on the Earth.

B. discuss how the Earth's stratospheric ozone layer and magnetic field allow life to exist on land.

C. describe why solar storms have become a danger to society, and how their consequences can be mitigated.

D. explain the differences among an asteroid, a comet, and a meteoroid.

E. explain why some meteoroids glow and may explode when passing through the Earth's atmosphere, and why some reach the Earth's surface.

F. discuss the manifestations of past impacts with the Earth, and why some impacts are considered to be extinction-level events.

G. describe actions being taken to protect the Earth from a future asteroid strike.

16.1 Introduction

March 13, 1989, was an unlucky day for people in the Canadian province of Quebec. Three days earlier, an intense eruption of energy and matter burst from the Sun's surface. This *solar storm* sent billions of tons of very energetic charged particles (isolated protons and electrons, as well as helium nuclei) racing across space at 1.6 million km/h (1 million mph). When the charged particles reached the Earth, they disturbed the planet's magnetic field, which in turn caused electric currents to surge through the transmission lines of Quebec's *electrical grid* (the interconnected network of power stations, transformers, and transmission lines that provides the province with electricity). The surge shut down the grid, leaving 6 million people in the dark for the next 9 hours (**FIG. 16.1a**).

Only a week later, on March 22, 1989, our planet had another close call, but of a different type. A block of rock 800 m (0.5 mi) in diameter—an *asteroid*—swept past the Earth at a speed of about 74,000 km/h (46,000 mph). Fortunately, it never got closer than 1.7 times the distance between the Earth and the Moon, but given the vastness of space, astronomers considered the flyby to be a very near miss (**FIG. 16.1b**). The path this asteroid followed was particularly frightening: it passed through the exact position where the Earth had been *only 6 hours earlier*! If the asteroid had arrived 6 hours earlier, it would have struck the Earth and produced an explosion 400 times larger than the atomic bomb that leveled the city of Hiroshima in 1945.

So far, this book has focused on natural hazards and disasters arising from phenomena within the Earth System. But the failure of electrical grids due to solar storms, and the near misses by asteroids, are reminders that society also needs look skyward and be prepared for dangers coming from space. In this chapter, we first consider hazards posed by the Sun. Without the visible light that comes from the fiery star at the center of our Solar System, life couldn't exist on the Earth. But the Sun also emits dangerous ultraviolet (UV) radiation, and during solar storms, it spews bursts of charged particles into space at high velocity. The flow of these charged particles near

◀ A bolide lights up the sky over a city in India.

FIGURE 16.1 Two hazards from space: space weather and asteroid impacts.

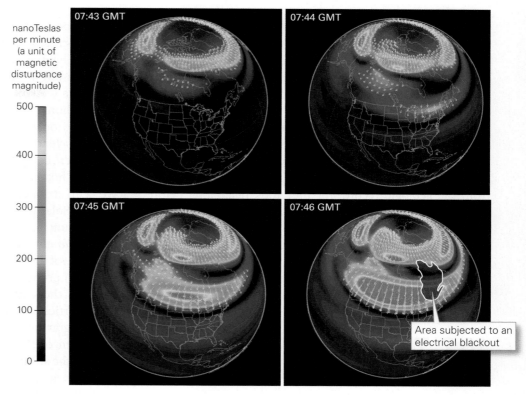

(a) Intense space weather in March 1989 disturbed the Earth's magnetic field and shut down Quebec's electrical grid. The images show the rapidly developing superstorm over a four-minute period on March 13, 1989. The colors indicate intensity, with red being the highest. Images and data provided by the Metatech Corporation, Applied Power Solutions Division (March 13, 1989 - 4 Minutes of a Superstorm).

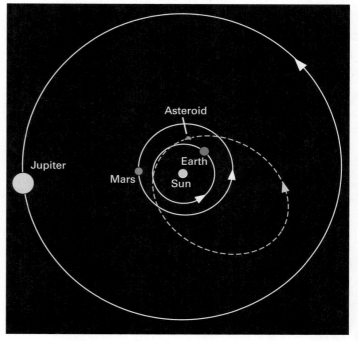

(b) If an asteroid crosses the Earth's orbit, it has the potential to strike the Earth.

the Earth constitutes **space weather**. Space weather events are also known as *geomagnetic storms*. Until the last 150 years, people didn't need to worry about space weather, but our modern society depends on electricity, electronic equipment, and satellites. Therefore, intense space weather can wreak havoc, shutting down electrical grids, damaging computers, interrupting radio and navigation signals, and disrupting satellites.

Second, we consider the consequences of collision with space objects whose orbits cross the Earth's orbit. Although people have witnessed dramatic encounters with such objects over historic time, truly catastrophic impacts are known only from the geologic record, for they took place long before humans inhabited the planet. This record shows that collision with an object just 15 km (9 mi) across could lead to the extinction of most life on this planet. In our discussion of dangers from space, we'll explain how these hazards originate, what their consequences are, and what society can do to prevent them from causing future disasters.

16.2 The Solar Inferno and Its Products

The Sun's Energy Source and Structure

Under the extreme temperature and pressure conditions deep inside the Sun, hydrogen nuclei (consisting of one proton) collide so forcefully that they bond together, a process called **nuclear fusion**. After a succession of steps, some of which convert protons into neutrons by releasing very tiny particles, a new helium atom forms—its nucleus contains two neutrons and two protons (FIG. 16.2a). During each step of this process, a tiny amount of matter converts into a huge amount of energy. About 99% of the Sun's energy comes from nuclear fusion in the *solar core*, the innermost zone of the Sun, where temperatures reach an inconceivable 15 million °C (27 million °F). That energy moves upward as electromagnetic radiation, through the overlying *radiative zone* and into the outer layer of the Sun, the *convective zone* (FIG. 16.2b). This outer layer consists of rapidly convecting **plasma**, an extremely hot gas-like substance composed of particles that have been stripped of their electrons due to the high temperature. Solar plasma consists of hydrogen ions (free protons), helium nuclei, and free electrons.

The *solar atmosphere* surrounds the Sun's convective zone. At the base of the solar atmosphere, glowing gases at 5,700°C (10,300°F) produce the light that radiates from the Sun. At the top of the solar atmosphere, in a region known as the *corona*, particles of plasma travel fast enough to escape the immense gravitational pull of the Sun and fly off into space. The stream of these particles flowing away from the Sun makes up the **solar wind** (FIG. 16.3). Solar wind particles typically travel at speeds between 400 and 750 km/s (900,000–1,700,000 mph), so the slowest of these particles take about 4.3 days to reach the Earth, while the fastest traverse the distance in about 2.3 days. Like wind in air on the Earth, solar wind consists of moving particles, but unlike wind in air, the particles are charged, and the solar wind has such low density (only 5 particles per cubic centimeter) that it can't be felt.

FIGURE 16.2 Energy production in the Sun, and the Sun's structure.

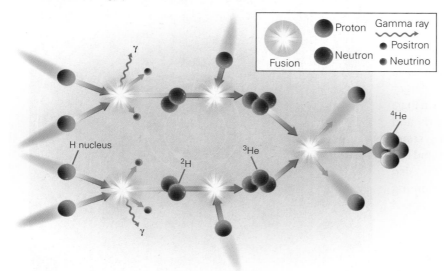

(a) The energy of the Sun comes from a sequence of fusion reactions that ultimately produce helium. Each step releases electromagnetic radiation and subatomic particles (called positrons and neutrinos).

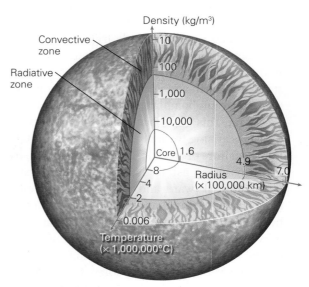

(b) The Sun has a core, a radiative zone, and a convective zone. The density and temperature of the Sun increase from surface to center.

The Sun's Magnetic Field and Sunspots

Moving electrically charged particles produce a *magnetic field*. Not surprisingly, therefore, the rapidly convecting plasma in the Sun produces a very strong magnetic field. The magnetic field lines of the Sun trend at a small angle to the solar equator, unlike those of the Earth, which run from pole to pole, as we'll see shortly. The Sun's magnetic field lines point in one direction in the northern hemisphere and in the opposite direction in the southern hemisphere. This unusual orientation exists because the Sun is a fluid, and its rotation rate—the time its surface takes to spin on its axis—decreases from equator to poles. (On the solid Earth, in contrast, the rotation rate is the same everywhere.) Shear caused by this variation in rotational speed wraps magnetic field lines around the Sun (**FIG. 16.4a**).

FIGURE 16.3 The solar wind consists of charged particles escaping from the Sun and streaming into space. In this image, brightness represents the solar wind's density. A filter blocks the bright inner part of the Sun's atmosphere so that solar wind is visible.

FIGURE 16.4 The Sun's magnetic field and the formation of sunspots.

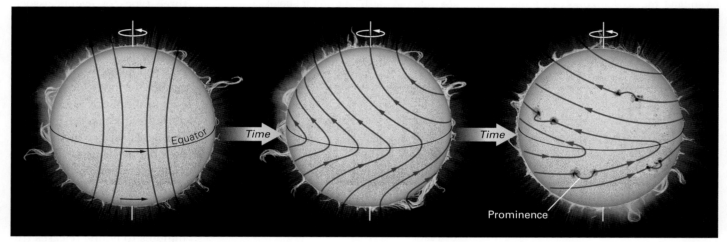

(a) The rotation rate of the Sun is faster at the equator than at the poles, so over time, its magnetic field lines bend and become almost parallel to the equator.

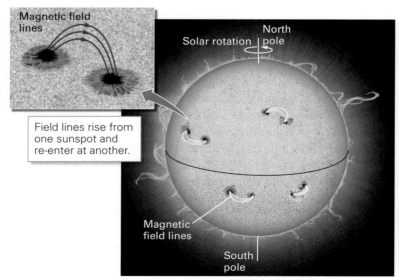

(b) The magnetic field can become so intense that its field lines arc out from the solar surface.

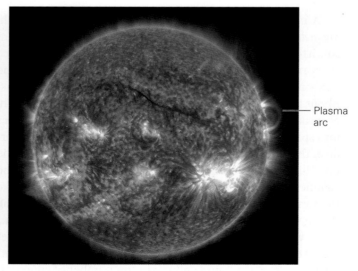

(c) A photo of the Sun showing arcs of plasma associated with sunspots.

16.2 The Solar Inferno and Its Products

FIGURE 16.5 Sunspots and the solar cycle.

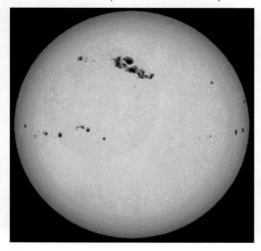

(a) Sunspots appear as dark patches. At any given time, they cover 0.0%–0.4% of the Sun's surface.

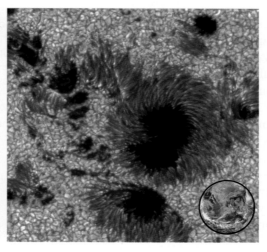

(b) A close-up of sunspots, which can be bigger than the Earth (shown for scale).

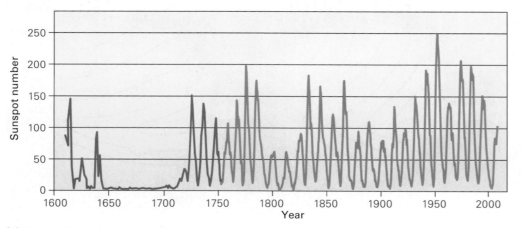

(c) The number of sunspots changes over time. The solar cycle is about 11 years long, but the number of sunspots varies among cycles.

A magnetic field has two poles, called north and south. *Magnetic polarity* is represented by an arrow that points from south to north. The Sun's magnetic field reverses polarity once every 11 years. Just after a reversal happens, the strength of the Sun's magnetic field decreases until it becomes very weak. About halfway through the cycle, the field strengthens again, eventually becoming so intense that in some locations, it arcs into space. Where these arcs leave and re-enter the Sun's surface, the magnetic field grows strong enough to inhibit the upwelling of hot plasma (**FIG. 16.4b, c**). These locations become significantly cooler than the rest of the Sun's surface and take the form of dark patches known as **sunspots** (**FIG. 16.5a, b**). Sunspots always occur in pairs—one member of the pair forms where the magnetic field lines arc upward and the other where they arc downward. The 11-year-long pattern, alternating between a maximum and minimum number of sunspots, is known as the **solar cycle** or *sunspot cycle* (**FIG. 16.5c**).

Solar Storms

Occasionally, local intensification or disruption of the Sun's magnetic field triggers the ejection of particularly large amounts of high-energy particles from the convective zone into space. These explosion-like events, called **solar storms**, eject plasma that is significantly denser and hotter than that in the normal solar wind. They generally take place during sunspot maxima within the solar cycle.

Astronomers distinguish three types of events associated with solar storms. During a **solar prominence**, outbursts of glowing gas and plasma emerge from the Sun's surface at one sunspot in a pair and follow arcing magnetic field lines, returning to the Sun's surface at the other member of the pair (**FIG. 16.6a**). Solar prominences can last for hours, days, or weeks. Occasionally, a **solar flare**, a sudden flash of light that can last minutes to hours, releases tremendous amounts of electromagnetic energy. Traveling at the speed of light, that energy takes 8 minutes to reach the Earth (**FIG. 16.6b**). The energy from a solar flare can disrupt the part of the Earth's atmosphere through which radio waves travel, degrading or, at worst, temporarily blacking out navigation and communications signals. A third type of event, called a **coronal mass ejection** (**CME**), hurls solar plasma directly into space (**FIG. 16.6c**). The plasma of a CME moves much more slowly than the speed of light, but it travels significantly faster than the typical solar wind—some reaches speeds of over 3,000 km/s (6.7 million mph). To picture the difference between a solar flare and a CME, imagine the firing of a cannon—the flash of light at the cannon when it fires is comparable to a solar flare, while the cannonball flying away from the cannon resembles the CME.

CMEs send large amounts of plasma streaming into space in specific directions (**FIG. 16.6d**). They affect the Earth only when our planet is in the path of the ejected particles (**FIG. 16.6e**). When the particles from CMEs reach the Earth,

FIGURE 16.6 Examples of events caused by solar storms.

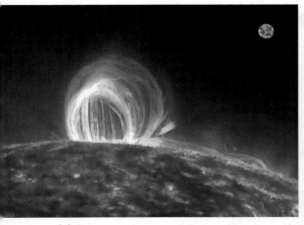

(a) Solar prominences follow arcing magnetic field lines and therefore form a loop. The Earth is shown for scale.

(b) Solar flares are giant flashes of light.

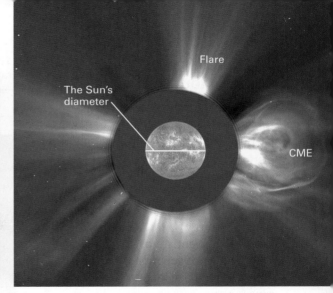

(c) Coronal mass ejections (CMEs) release a huge number of charged particles into space.

(d) CMEs eject plasma in specific directions.

they produce strong disturbances of the Earth's magnetic field. In some events, CMEs can send so much high-energy plasma into space that they pose a significant hazard to human activities, as we discuss later in this chapter.

(e) If the burst of charged particles from a CME reaches the Earth, severe space weather develops.

Take-home message . . .

The Sun produces huge amounts of energy by nuclear fusion. Plasma from the Sun (made up of charged particles) heads into space as solar wind. Convection of plasma in the Sun generates an intense magnetic field. Sunspots form where the magnetic field arcs into space. Solar prominences are arcs of plasma that flow between sunspots. Particularly intense disruptions of the Sun's magnetic field yield solar flares and coronal mass ejections.

QUICK QUESTION Why aren't all CMEs dangerous to the Earth?

16.3 The Earth's Protective Shields

If the Earth's surface were not protected, at least partially, from dangerous UV radiation and charged particles, its land surface would be barren of life. Fortunately, the Earth, like the Sun, hosts both a magnetic field and an atmosphere, which together protect its surface from the Sun's full radiative power. Let's explore each of these protections.

The Earth's Atmosphere: A Shield against UV Radiation

In Chapter 8, we learned that the Earth's atmosphere includes four distinct layers: the troposphere, stratosphere, mesosphere, and thermosphere (see Fig. 8.12). In the stratosphere, shorter-wavelength (higher-energy) UV radiation, known by scientists as *UVC*, gets absorbed by oxygen molecules (O_2) (**FIG. 16.7a**). This input of energy causes some O_2 molecules to break apart, forming individual oxygen atoms (O), which then immediately combine with remaining O_2 molecules to form ozone (O_3). Ozone molecules, too, absorb UVC, as well as longer-wavelength (lower-energy) UV radiation, known as *UVB* and *UVA*. When they absorb energy, the molecules split apart to form O_2 and individual O atoms. The reactions involving O_2, O_3, and O are constantly happening, as O_2 and, particularly, O_3 molecules continuously absorb UV radiation. Nevertheless, some UVB energy, and even more UVA energy, reaches the Earth's surface. UVA rays cause long-term skin damage, such as wrinkles, and some skin cancers. UVB rays have slightly more energy than UVA rays, and they can cause direct damage to the DNA in skin cells. These two types of rays cause sunburns.

The reactions of oxygen with UV radiation lead to the formation of an *ozone layer* within the stratosphere, at altitudes between 17 and 35 km (11–22 mi) (**FIG. 16.7b**). Without the ozone layer, so much UV radiation would reach the Earth's surface that organisms couldn't survive on land or in the air, and the biosphere would be restricted to the oceans. Unfortunately, human activities emit pollutants that diminish the concentration of O_3 in the stratosphere, producing an *ozone hole* over the Antarctic region, and posing a hazard of ozone reduction across the stratosphere (**BOX 16.1**).

Incoming UV radiation also knocks electrons off molecules high in the atmosphere, producing positively charged molecular ions and freely circulating electrons. Scientists refer to the resulting layer of charged particles in the upper atmosphere as the **ionosphere**. It exists above altitudes of about 60 km (37 mi) during the day and 100 km (62 mi) at night, and extends upward to the outer edge of the atmosphere.

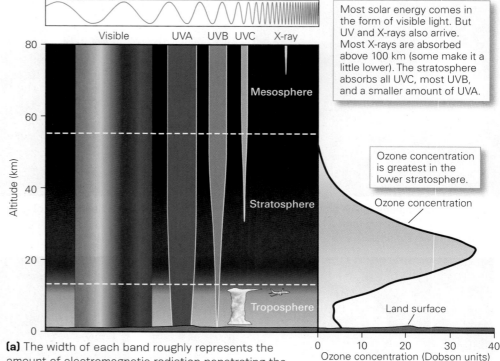

FIGURE 16.7 The Earth's atmosphere absorbs dangerous UV radiation from the Sun. Ozone in the stratosphere plays an important role in absorbing this energy.

(a) The width of each band roughly represents the amount of electromagnetic radiation penetrating the atmosphere. Most incoming UV radiation is UVA, some is UVB, and a minor amount is UVC. UVC is most dangerous to life.

(b) Variation in ozone concentration with altitude, measured in Dobson units (higher values = more ozone).

The Earth's Magnetosphere: A Shield against Space Weather

As we noted in Chapter 2, the Earth's outer core consists of iron alloy that is hot enough to flow like a liquid. This flow generates a magnetic field, which can be represented by magnetic field lines curving through space around the Earth (**FIG. 16.8a**). The Earth's magnetic field isn't symmetrical; its overall shape resembles a teardrop with its narrow end, the *magnetotail*, pointing away from the Sun. This shape exists because the solar wind distorts the field, flattening it on the side closer to the Sun and extending it on the side away from the Sun. Most solar wind particles do not travel fast enough to penetrate the magnetic field, so the outer edge of the field deflects those particles, which flow past the Earth (**FIG. 16.8b**). In this way, the magnetic field acts as a shield against solar-wind particles. Scientists refer to the region of the magnetic field inside this shield as the **magnetosphere**.

The flow of charged particles in the solar wind produces an electric current along the boundary of the magnetosphere. Electric currents create magnetic fields, and changing magnetic fields, in turn, can induce other electric currents. Thus, the current generated by the solar wind perturbs the Earth's

BOX 16.1 A CLOSER LOOK...

The ozone hole and the Montreal Protocol

When emitted into the atmosphere, specific anthropogenic chemicals, such as *chlorofluorocarbons* (CFCs), react with and break up O_3 molecules. These ozone-destroying chemical reactions happen most rapidly on the surfaces of tiny ice crystals in polar stratospheric clouds. So, when the Sun rises over the poles in spring, providing the necessary UV radiation, chlorine atoms derived from CFCs react with and destroy ozone. The resulting **ozone hole** is a region of diminished stratospheric ozone concentration, particularly over Antarctica **(FIG. Bx16.1a)**. If the area of ozone depletion remained confined to the Antarctic, it would pose little problem for humanity, since few people live there. But sometimes the hole becomes wide enough to extend over populated regions, especially later in the season. Global reductions in ozone can occur when air from the ozone hole mixes with the rest of the stratosphere, reducing the ozone concentration worldwide.

CFCs began to be used in refrigerants, air-conditioning units, and spray cans only in the 20th century, so this loss of O_3 is quite recent in human history. Ozone depletion allows higher levels of UVB and UVA radiation to reach the surface of the Earth, potentially increasing the likelihood of skin cancer.

Because of the importance of stratospheric ozone to human health, a 1987 international summit in Montreal, Quebec, led to an agreement to reduce CFC emissions globally by phasing out production of CFCs and replacing them with new substances that do not react with O_3. This agreement, formally known as the *Montreal Protocol on Substances That Deplete the Ozone Layer*, has succeeded in slowing the release of ozone-destroying chemicals **(FIG. Bx16.1b)**. However, the global recovery to previous ozone concentrations in the stratosphere will take time. Model calculations suggest that the ozone layer will return to 1980 levels sometime between 2050 and 2070 **(FIG. Bx16.1c)**. The Montreal Protocol, which has been assessed every four years since its adoption, was hailed by Kofi Annan, the former secretary general of the United Nations, as "perhaps the single most successful international agreement to date."

FIGURE Bx16.1 The ozone hole and its mitigation.

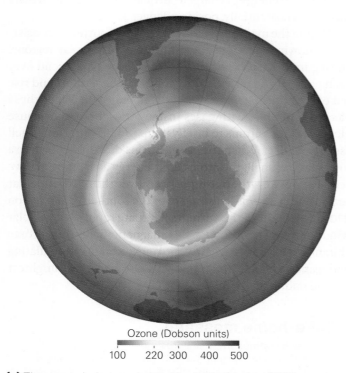

(a) The ozone hole over Antarctica in September 2020.

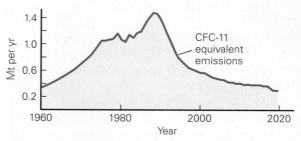

(b) Due to the Montreal Protocol, the quantity of ozone-destroying chemicals released into the atmosphere has decreased since the 1990s. (Mt = Megatonnes)

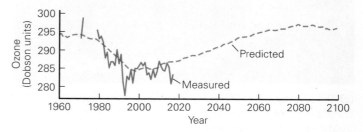

(c) Predicted (blue dashed line) and measured (red line) global concentrations of stratospheric ozone. Ozone destruction continued to the end of the 20th century, but the ozone concentration has now stabilized and is expected to increase during the 21st century.

FIGURE 16.8 The Earth's magnetic field shields the planet from charged solar particles.

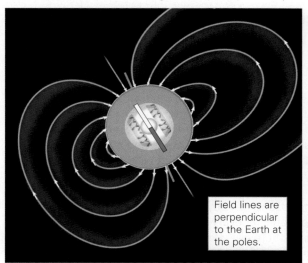

(a) A magnetic field, represented here by magnetic field lines, surrounds the Earth.

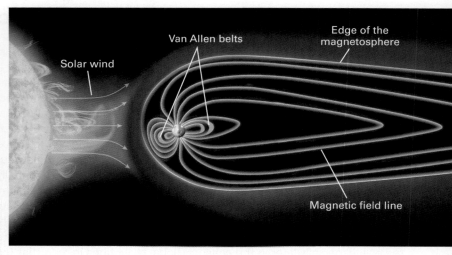

(b) Most solar wind particles are deflected by, and flow along the surface of, the magnetosphere. Some energetic particles are trapped in the Van Allen belts.

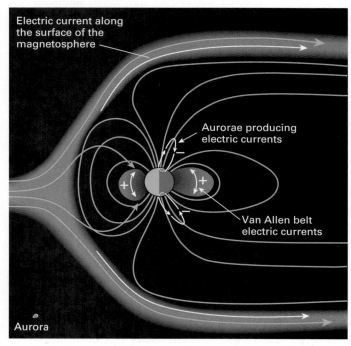

(c) The passage of the solar wind causes perturbations in the Earth's magnetic field, which cause positively charged particles to oscillate between the polar regions in the Van Allen belts, and cause electrons to form currents that dip into the upper atmosphere in the polar regions, triggering aurorae.

magnetic field and generates new currents within the field itself. These currents exist within two zones. The first zone, known as *Van Allen radiation belts*, lies between 1,000 and 12,000 km (620–7,456 mi) from the Earth. Induced currents within these belts are associated with trapped positively charged particles that oscillate rapidly back and forth between the polar regions (FIG. 16.8c). The particles within the Van Allen belts normally do not descend into the Earth's atmosphere. During periods of intense space weather, however, the quantity and energy of charged particles in the Van Allen belts increases substantially, posing a danger to astronauts.

Within the magnetotail, a second pair of currents, consisting of flowing electrons, develops above the polar regions. During intense space weather, these electrons descend over each polar region on the side facing the magnetotail, and rise out of the atmosphere on the side facing the Sun. When these electrons interact with gases in the ionosphere, they cause electrons in gas atoms to briefly jump to a higher orbital shell (see Appendix). When the gas electrons drop back down to a lower orbital shell, the atoms emit energy as light, which we see as **aurorae**—the *aurora borealis* over the northern polar region and the *aurora australis* over the southern polar region (FIG. 16.9). Because the currents fluctuate rapidly as the solar wind interacts with the Earth's magnetic field, the aurorae shimmer and look like gauzy moving curtains of undulating green and red color in the sky. The aurorae appear brightest around midnight.

Take-home message...
The Earth's atmosphere absorbs most dangerous electromagnetic radiation from the Sun, but some wavelengths of UV light do get through. The Earth's magnetic field protects the Earth's surface from charged particles in the solar wind, but energetic particles descend into the ionosphere and cause aurorae.

QUICK QUESTION Are all wavelengths of UV light absorbed equally in the stratosphere?

FIGURE 16.9 Aurorae appear as curtains of shimmering light.

(a) Aurora borealis over Alaska.

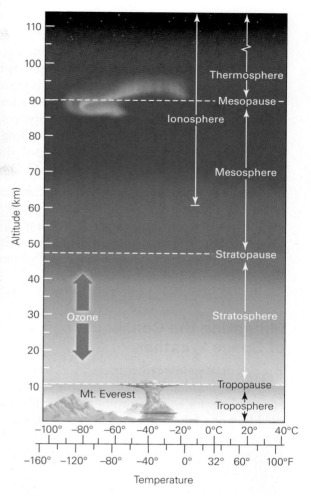

(b) Aurorae occur in the ionosphere.

16.4 Space Weather

Consequences of Space Weather

The concentration and speed of charged particles (mostly protons emitted from the Sun) in the region surrounding the Earth determines space weather at a given time. Although the Earth's surface is shielded from these particles, they can pose a hazard to astronauts and satellites. In fact, during intense space weather in 1989, astronauts in the space shuttle reported a burning sensation in their eyes, so Mission Control ordered them to retreat to the most shielded part of the spacecraft. While the particles from the solar wind and solar storms themselves are not a direct health risk to people on the Earth, they have notable consequences for human activities:

- *Low-latitude aurorae:* During intense space weather, incoming solar particles have enough energy to produce aurorae much farther from the poles than usual. For example, during the 1989 event, aurorae were visible in Florida and Texas.

- *Heating of the thermosphere:* The particles and radiation of a solar storm, when they reach the Earth, add energy to the thermosphere, the layer of the atmosphere that lies above about 90 km (56 mi). This energy causes the thermosphere to expand outward, since gases expand when heated. Consequently, the concentration of air molecules increases at the altitudes where as many as 1,000 satellites, and more than 10,000 fragments of *space junk* (defunct satellites and loose pieces of equipment), currently orbit. As satellites and space junk interact with air molecules, they slow down and lose altitude. When this happens, their orbits become less predictable, and collisions between objects may be more frequent. In fact, concern is growing that some satellites or pieces of space junk may slow so much that they will fall toward the Earth, where they will become hazards to surface inhabitants.

- *Disruption of radio transmissions:* Solar flares generate radio waves that interfere with radio transmissions from the Earth. Radio noise generated by solar flares can disrupt the communications of satellites, cell phones, and the global positioning system (GPS). GPS not only serves as the basis for navigation, but also provides essential time signals to numerous computer applications.

- *Electrical discharges on satellites:* The influx of charged particles causes a buildup of static electricity in and on satellites, eventually leading to sparks that may damage the satellites.

- *Disruption of power grids:* Electric currents produced within the Earth's magnetic field by moving solar particles can induce rogue electric currents in electrical transmission lines, producing a power surge that can trigger a shutdown of an electrical grid. Because the electrical grids of different regions are interconnected, one failure can lead to a cascade of failures affecting a very broad region. In fact, the 1989 failure of the grid in Quebec almost knocked out grids across North America. Loss of power shuts off lights, refrigerators, computers, and anything else that runs on electricity. The resulting blackout can trigger chaos at airports, crises in financial markets, and looting in cities.

- *Damage to electronic equipment:* Electric currents generated by intense space weather can cause power surges, putting electronic equipment out of commission. The consequences of such widespread failure are almost inconceivable, for nearly every aspect of our modern society relies on computers and the internet.

Mitigating the Consequences of Space Weather

The risk that space weather poses to society increases every year, as we become more and more dependent on electric power, electronic equipment, and the internet. Imagine what would happen if there were a sudden and simultaneous blackout of electric power, GPS, nuclear power plant cooling systems, radar, computers, and the internet. The lives of millions of people would be at risk, and by some estimates the cost of such a disaster could be over $2 trillion. Society has not experienced intense space weather since the advent of worldwide computing and the internet. Fortunately, the energy of the huge CMEs that have taken place in the last half century have not been directed at the Earth. But potentially catastrophic space weather has happened in the past (BOX 16.2), and it will happen again in the future.

What can be done to prevent a space weather catastrophe? Two precautionary physical measures may help. First, satellite-based electronic equipment can be *radiation hardened*, meaning that it can be protected from an intense influx of charged particles and the currents they generate. Methods to achieve this include mounting computer chips on insulating surfaces, covering chips with a resistant coating, or encasing equipment in resistant shells. On the Earth's surface, electrical grids can be designed to be smarter, so that surges can be stopped or excess electricity shunted into capacitors (electronic devices that store energy). Power companies can prepare replacement transformers (devices that transfer electrical energy from one electrical circuit to another) so that they are in place and operable, ready to be activated the instant an active transformer gets knocked out.

The easiest and cheapest way of preventing catastrophe is simply to shut down electronic equipment in advance of the arrival of intense space weather. Of course, such preparation

FIGURE 16.10 At the Space Weather Prediction Center, observers track the intensity of the solar wind and watch for solar flares and CMEs.

requires enough lead time. To meet this challenge, the US National Weather Service (working with the military, NASA, and other agencies) has set up a *Space Weather Prediction Center* to watch for solar storms and send out warnings to utilities and government agencies so that they can take action to prevent damage to electrical grids, electronic equipment, and satellites (FIG. 16.10). Researchers estimate that there is a 50% chance that intense space weather will happen by the end of this century. Hopefully, at that time, society will be ready.

> **Take-home message...**
>
> Space weather is determined by the concentration and speed of charged solar particles near the Earth. These particles can disrupt our planet's magnetic field, producing electric currents that disrupt electronic equipment on the Earth's surface. Intense space weather produces aurorae at low latitudes and disrupts radio transmissions, computers, and electrical grids. Mitigation of these effects involves hardening protection on electronics and detecting solar storms in advance so that sensitive equipment can be shut down.
>
> **QUICK QUESTION** How might space weather affect the navigation capabilities of airplanes?

16.5 Dangerous Objects Lurking in the Solar System

Our Solar System consists of the Sun, eight planets, at least 200 moons, several dwarf planets, and countless smaller objects, most of which are too tiny to see from the Earth (TABLE 16.1). Because of their orbits, the planets, their moons, and the known dwarf planets will never cross the Earth's orbit, and therefore they can never collide with the Earth. But

BOX 16.2 DISASTERS AND SOCIETY

Intense space weather of the past

Since the 19th century, several episodes of intense space weather caused by strong solar storms have disrupted human activities. Space weather in 1903 caused a power surge in telegraph lines and blew fuses; a 1921 event shut down telephone lines; a 1967 event caused radar stations in the Arctic to go offline, leading the US Air Force to initiate preparations for a nuclear attack; a 1972 event set off underwater magnetic mines; and as we've seen, an event in 1989 knocked out the electrical grid of Quebec. Space weather in 2003 knocked out communications with several satellites.

When intense space weather happens, the flow of charged solar particles produces electric currents that cause major disruptions in the Earth's magnetic field, which can be measured by magnetometers on the Earth (see Fig. 16.1). Because the concentration of particles reaching the Earth varies rapidly, the magnetic field strength detected by magnetometers varies rapidly as well, as demonstrated by measurements of field-strength variations during the 1989 geomagnetic storm **(FIG. Bx16.2a)**. Space weather also causes aurorae to occur at lower latitudes than usual, and their occurrence can be detected by viewers on the Earth and astronauts in space **(FIG. Bx16.2b)**.

On September 1–2, 1859, particles from a huge coronal mass ejection collided directly with the Earth's magnetic field, producing the most intense space weather ever recorded. This event is now known as the **Carrington event**, named for an English amateur astronomer who reported the CME that triggered it. The event was recorded by magnetometers of the time **(FIG. Bx16.2c)**. Because of the intensity of the CME, solar particles were traveling particularly fast—up to 8.5 million km/h, or 5.3 million mph—and took only 17.2 hours to reach the Earth. When the charged particles arrived, they triggered aurorae so bright that people could read by them in the northeastern United States. Aurorae were observed as far south as sub-Saharan Africa, and there were even reports of aurora sightings in Colombia, a country at equatorial latitudes. In 1859, electrical grids, telephones, and computers didn't exist, so the impact on society was minimal. But telegraph lines were affected—the surge through these lines knocked out many telegraph stations, shocked operators, and started fires.

FIGURE Bx16.2 Space weather impacts.

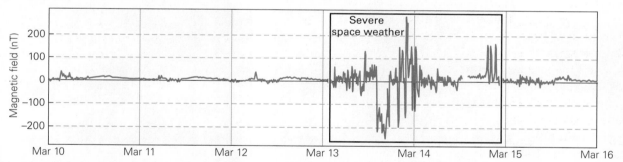

(a) Severe space weather between March 13 and 15, 1989, caused large, rapid changes in the strength of the Earth's magnetic field, which in turn induced power surges that shut down the electrical grid in Quebec. Magnetic field strength is measured in nano-Teslas (nT).

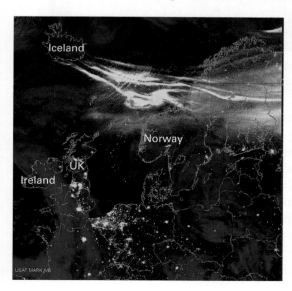

(b) The 1989 event also caused lower-latitude aurorae. This satellite image shows the aurorae as they spread across northern Europe.

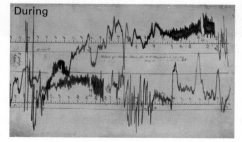
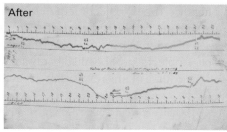

(c) Magnetic field disturbances caused by the Carrington event were recorded by magnetometers. During the event, magnetic field strength fluctuated wildly. After the event, it was steady.

TABLE 16.1 Objects in the Solar System

Planet	A spherical object, in orbit around the Sun in the Sun's equatorial plane; planets have cleared their orbit of other matter; the four inner planets (Mercury, Venus, Earth, and Mars) are relatively small and consist of rock and metal; the four outer ones (Jupiter, Saturn, Uranus, and Neptune) are huge and consist mostly of gas and ice.
Moon	An object that orbits a planet; very small moons are irregularly shaped, whereas larger ones are spherical; the largest moon is bigger than Mercury, the innermost planet.
Dwarf planet	A very small planet that has not cleared its orbit of other matter or does not orbit in the Sun's equatorial plane; Pluto is now considered to be a dwarf planet.
Kuiper Belt	A zone containing millions of icy objects, including several dwarf planets, that orbit the Solar System outside of the orbit of Neptune **(FIG. 16.11a)**.
Oort Cloud	A diffuse cloud of ice objects outside of the Kuiper Belt; its outer edge may be 1.5 light-years away from the Sun **(FIG. 16.11b)**, far beyond the edge of the heliosphere.
Asteroid	A small rocky or metallic object ranging from about 100 m to dwarf planet size; most follow an orbit between Mars and Jupiter; some have orbits that cross that of the Earth.
Comet	A spherical or irregularly shaped mass of ice, rock and dust that evaporates and sends out a tail as it approaches the Sun; most comets follow highly elliptical orbits not within the equatorial plane of the Sun; most comets come from either the Kuiper Belt or Oort Cloud.
Meteoroid	A relatively small object, composed of rock, metal, dust, or ice, traveling through space; meteoroids can be small individual remnants of Solar System formation, or fragments broken off comets, asteroids, planets, or moons.

FIGURE 16.11 The Kuiper Belt and Oort Cloud.

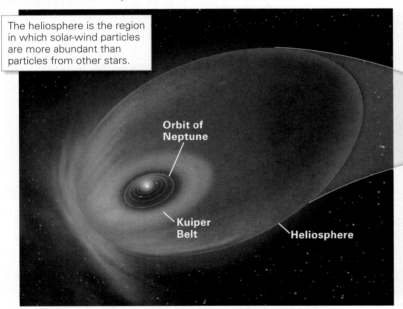

(a) The ring-shaped Kuiper Belt lies beyond Neptune, but is entirely within the heliosphere.

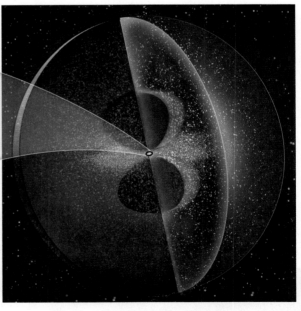

(b) The Oort Cloud extends far outside the heliosphere, but remains gravitationally bound to the Sun.

thousands of smaller space objects do collide with the Earth every year. Most vaporize as they pass through the atmosphere, but occasionally, one survives and lands on the planet's surface. During historic times, objects large enough to cause damage have arrived only rarely, and none of the resulting impacts has caused a disaster. During geologic time, however, some arriving space objects have been so big that their impacts have changed the course of life's evolution. If such an impact were to happen today, it would cause a catastrophe far greater than any we have yet discussed in this book. Specifically, it could be an **extinction-level event**, one that threatens the existence of life on the Earth. In this section, we introduce the categories of objects that could potentially threaten our planet. Then, in the remainder of the chapter, we describe the consequences of impacts with those objects and the risks that they represent to society.

FIGURE 16.12 Asteroids and their distribution.

(a) The largest asteroid, Ceres, which is spherical, has a diameter of 945 km (587 mi).

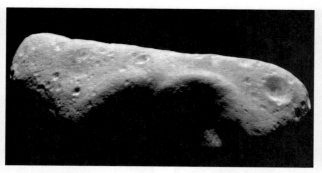

(b) Some asteroids are oblong. Eros is 21 km (13 mi) long.

(c) Lutetia, a nonspherical asteroid, has a diameter of 121 km (75 mi).

Asteroids

By convention, astronomers consider a rocky or metallic object larger than about 100 m (328 ft) across to be an **asteroid**. Some asteroids are fragments of solid blocks that never attached to others to become a larger body during the formation of the Solar System. Others are remnants of bodies that became large enough to warm up inside, so that they differentiated into a metallic core and a rocky mantle. Most of these differentiated bodies later collided with others and broke apart into fragments. Rocky asteroids came from the mantle of a differentiated object that was once larger, and metallic ones came from the core. Only a few larger asteroids are now spherical. These softened enough inside for gravity to reshape them, but did not break apart subsequently. The largest of these, Ceres, has been classified as a dwarf planet (FIG. 16.12a). All the others are irregularly shaped and generally cratered (FIG. 16.12b, c). There are many millions of small asteroids—about 1.5 million have diameters greater than 1 km (0.6 mi), about 200 have diameters greater than 100 km (62 mi), and only 4 have diameters greater than 400 km (249 mi).

While the orbits of some asteroids pass within the orbit of Mars, most occupy the **asteroid belt** between the orbits of Jupiter and Mars. Four small groups of asteroids come relatively close to the Earth, and two of these, known as the *Aten asteroids* and *Apollo asteroids*, have been nudged by Jupiter's gravity into paths that cross the Earth's orbit (FIG. 16.12d). Astronomers have so far identified approximately 1,500 Atens and 10,500 Apollos, of which about 2,000 may represent a threat to our planet. An individual Aten or Apollo asteroid survives for only a few million years before it collides with a planet or is gravitationally thrown out of its orbit, but the tug of Jupiter's gravity on other objects constantly replenishes the supply.

Comets

Comets mystified ancient observers because unlike other celestial objects, they produce a glowing tail as they travel

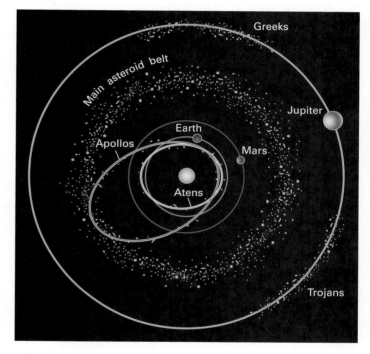

(d) The Apollo and Atens asteroid groups follow orbits that cross that of the Earth.

across the sky, over the course of weeks to months, relative to the backdrop of stars (FIG. 16.13a). A **comet** differs from an asteroid because it follows a very elliptical orbit around the Sun (FIG. 16.13b), consists of a relatively loosely bound aggregate of rock, dust, and various ices (water ice, frozen methane, dry ice, and frozen ammonia), and emits gas and dust that form tails (FIG. 16.13c–e). Tails form when a comet gets closer to the Sun than about four times the radius of the Earth's orbit. At that distance, solar radiation causes the comet's interior to start undergoing **ablation**, the transformation of a solid directly into a gas. The gas seeps out of cracks in the comet (a process called *outgassing*) and carries dust out with it. At first, the gas and dust form a haze of ionized gas called a *coma*, which surrounds the comet's solid

FIGURE 16.13 Comets consist of ice, rock, and dust.

(a) The comet Hale-Bopp, a particularly dramatic comet, was visible for 18 months beginning in May 1996.

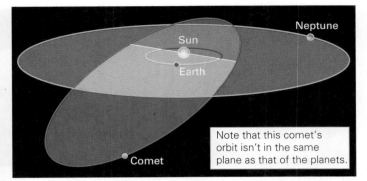

(b) Comets follow elliptical orbits that bring them from far beyond Neptune into the inner Solar System.

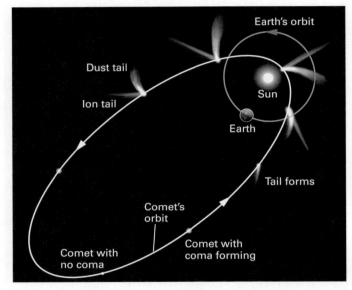

(c) When a comet approaches the Sun, it develops two tails. The ion tail points straight away from the Sun, whereas the dust tail curves.

(d) The irregular solid nucleus of a comet.

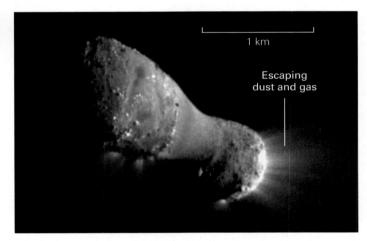

(e) Outgassing releases gas from the comet, which carries dust along with it.

nucleus. Gas from the coma streams outward to form an *ion tail* oriented parallel to the direction of the solar wind, and dust forms a curving *dust tail* oriented partway between the comet's orbit and the solar wind.

Astronomers classify comets based on the time they take to orbit the Sun. *Short-period comets*, which complete an orbit in less than 200 years, come from the Kuiper Belt, while *long-period comets*, which complete an orbit in more than 200 years (in some cases, in thousands of years), come from the Oort Cloud (see Fig. 16.11). Objects from the Kuiper Belt or Oort Cloud can become comets when gravitational tugs or collisions with other bodies send them careering toward the inner Solar System. At present, 106 comets are known to have orbits that bring them relatively close to the Earth's orbit.

Meteoroids and Meteorites

Astronomers refer to small space objects (less than 100 m across) as **meteoroids**. Meteoroids whose paths cross the Earth's orbit may enter the atmosphere and collide with the

FIGURE 16.14 Examples of meteorites.

(a) A chondrite, a type of stony meteorite, contains chondrules.

(b) Metal crystals inside an iron meteorite.

(c) An iron meteorite with an ablated surface.

(d) A stony-iron meteorite contains dark metal and lighter-colored silicate minerals.

Earth. If a meteoroid makes it through the atmosphere without vaporizing, it strikes the Earth. The remnant of a meteoroid on the Earth is called a **meteorite**. Researchers recognize three classes of meteorites (**FIG. 16.14**): 93% are *stony meteorites*, which consist of silicate rock; 6% are *iron meteorites*, which consist of metallic iron-nickel alloy; and 1% are *stony-iron meteorites*, which contain both metal and rock. Most stony meteorites (87%) are *chondrites*, so named because they contain colorful, rounded fragments, known as *chondrules*, that represent the original material from which the Solar System formed. The remaining stony meteorites are formed from the mantles of differentiated asteroids, and a small minority are fragments of the Moon or Mars. Iron meteorites and stony-iron meteorites are also fragments, probably from the cores of differentiated bodies.

Over the past few centuries, researchers have collected about 50,000 meteorites. Those located beneath a place where an object was observed as it sped through the atmosphere are called *meteorite falls*, and those obtained by examining ancient impact sites, or by searching for rocks that differ from terrestrial rocks, are called *meteorite finds*. About 1,180 meteorite falls have been documented—the largest of these was a 23,000 kg (25 ton) iron meteorite that landed in Russia in 1947. The largest meteorite find, dug out of a Namibian farm field in 1920, weighed 60,000 kg (66 tons). Some meteorites have a complex shape and a smooth surface. Meteoroids heat up as they pass through the atmosphere, and the intense heat causes rapid ablation of the surface as pieces of the meteoroid spall, or peel off. Meteorites are easiest to discover in places with little vegetation, such as deserts and farm fields. The

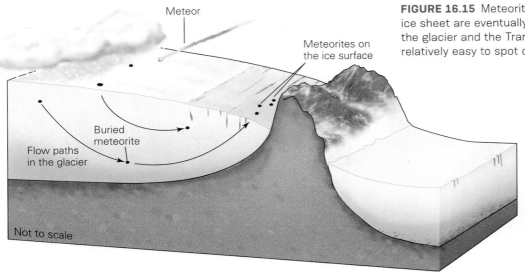

FIGURE 16.15 Meteorites that fall on the East Antarctic ice sheet are eventually carried to the boundary between the glacier and the Transantarctic Mountains. They are relatively easy to spot on the ice surface.

edge of the East Antarctic ice sheet that borders the Transantarctic Mountains has turned out to be one of the best places to find meteorites. Meteorites that land over a vast area of the ice sheet are eventually carried to the mountain front, where they accumulate **(FIG. 16.15)**.

Take-home message . . .

In addition to planets and moons, millions of smaller objects orbit the Sun. Some of these cross the Earth's orbit and therefore might collide with our planet. Potentially dangerous objects include certain groups of asteroids, numerous comets, and countless meteoroids.

QUICK QUESTION What is the difference between an asteroid and a comet?

16.6 Meteors, Fireballs, and Bolides

Most meteoroids, comets, and asteroids miss the Earth entirely. Some arrive at a very gentle angle, graze the atmosphere like a stone skipping on a pond, and head back into space. Those that arrive at a steeper angle penetrate the atmosphere. All told, about 16 million kg (17,640 tons) of meteoroids enter the atmosphere every day. Of these, the vast majority are so small that they "burn up," meaning that they vaporize almost completely in the atmosphere. In a typical year, only about 90,000 kg (100 tons) of meteorite mass actually survives the journey through the atmosphere and reaches the surface of the Earth.

When a meteoroid first enters the Earth's atmosphere, it's traveling at **cosmic speed** (the speed at which it moves through space relative to the Earth), ranging from 11 km/s (25,000 mph) to 72 km/s (160,000 mph). The cosmic speed of a given meteoroid depends on the object's orbital velocity, and this, in turn, depends on its orbit and on the direction from which it approaches the Earth. If an object is chasing the Earth, the Earth's orbital speed subtracts from the object's orbital speed, whereas if the object collides with the Earth head on, the two speeds add together. As the object approaches the Earth, acceleration due to the Earth's gravity also affects its speed. Note that even slow meteoroids travel 50 times faster than a commercial jet plane.

At an altitude of 80–120 km (50–75 mi), air is dense enough that a meteoroid can start intensely compressing the air in its path. This **ram compression** of air heats it to more than 1,700°C (3,100°F). As the meteoroid plunges through the air, this immense heat causes the solid surface of the object to undergo rapid ablation. In addition, each impact with an air molecule causes individual atoms on the surface of the object to break free, a process called *sputtering*. A skin of melt and superhot rock forms on the surface of larger meteoroids and then spalls off.

Glowing gases from the meteoroid vapor, along with glowing superheated air molecules, produce a blazing light streak behind the object **(FIG. 16.16a, b)**. This light streak then spalls (flakes) off. Notably, friction due to the shear of a meteoroid against air molecules does not generate sufficient heat to produce a meteor—most heat comes from ram compression.

Glowing vapor and particles leaving a meteoroid, together with glowing superheated air molecules, produce a blazing light streak as a meteoroid traverses the atmosphere. This atmospheric phenomenon is called a **meteor**. (In everyday English, such light streaks are sometimes called *shooting stars*, but this name is misleading, as meteors have nothing to do with stars.) Very small meteoroids, the size of a few grains of sand, can be seen for only about a second before the meteoroid vaporizes entirely or slows down so much, due to air resistance, that it no longer glows. In that brief moment, they may descend to an elevation of about 50–95 km (31–59 mi). Pebble-sized meteoroids descend farther, and produce longer-lasting, brighter meteors.

FIGURE 16.16 Meteors in the night sky.

(a) This meteor moved from the top to the bottom of the image. The light streak is kilometers to tens of kilometers long, but only about a meter wide.

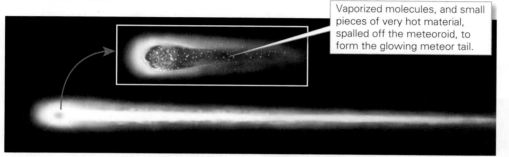

Vaporized molecules, and small pieces of very hot material, spalled off the meteoroid, to form the glowing meteor tail.

(b) At the head of a meteor is a solid meteoroid undergoing ablation and sputtering. The temperature is so high that the resulting vapor and dust glow.

(c) Several meteors of the Perseids meteor shower. Note that the meteors look as if they are all originating near the same point.

During a **meteor shower**, an observer can see from 10 to over 100 meteors per hour, all of which emerge from a common area in the sky **(FIG. 16.16c)**. Most meteor showers, which last for 1–3 days, happen when the Earth passes through the remnant dust tail of a comet. The Earth crosses certain cometary orbits at the same time every year, producing named showers such as the Perseids in August and the Leonids in November. If the Earth passes through a particularly dense patch of a comet's tail, a *meteor storm* can happen, during which over 1,000 meteors streak across the sky every hour. In one intense meteor storm during the Leonids in 1833, an estimated 100,000 meteors were visible each hour. Not all meteoroids arrive in showers or storms, however. Astronomers refer to those that arrive independently as *lone meteoroids*.

A meteoroid that produces a meteor typically ranges from sand-sized to golf-ball-sized—because of air resistance, dust-sized grains slow down too much to glow, and they simply float down to the Earth's surface. When a meteoroid exceeds a few centimeters in diameter, it produces a **fireball**, a light brighter than Venus, that can leave a visible smoke-like trail across the sky. A meteoroid tens of centimeters to meters across produces an exceptionally bright fireball, and may explode while still several kilometers above the Earth. Astronomers refer to an exploding fireball as a **bolide (FIG. 16.17)**. As it passes through the atmosphere, a high-speed meteoroid, asteroid, or comet produces **shock waves** of extreme pressure that head to the Earth's surface. For this reason, a meteor may produce an audible sonic boom. A strong shock wave can produce pressures greater than 100 times normal atmospheric pressure. A very large meteoroid pushes air in front of it, producing an intense wind. Researchers suggest that if the object is permeable, high-pressure air squeezes into it, causing it to blow apart while still in the atmosphere. Such an explosion of a very large meteoroid produces an **air blast** (or *air burst*), a sudden pulse of extreme wind, like the blast produced by a bomb.

Because of its interaction with the atmosphere, the **impact velocity**, the speed at which an object hits the Earth's surface, and its cosmic velocity are not necessarily the same. The difference between these two velocities depends on the size of a meteoroid. A meteoroid that weighs less than about 7,000 kg (7.7 tons), or whose diameter is less than 1 m (3 ft), slows down so much as it passes through the air that for the final part of its journey, it falls due to the pull of gravity alone, reaching a terminal velocity of up to about 500 km/h (311 mph). Larger meteoroids retain part of their cosmic speed. In fact, a meteoroid heavier than 1 million kg (1,100 tons), or with a diameter of about 4 m (13 ft), will reach the ground at about 70% of cosmic speed. Meteoroids heavier than about 100 million kg (110,000 tons), or larger than about 15 m (49 ft) across, slam into the Earth at full cosmic speed.

FIGURE 16.17 Fireballs are particularly bright meteors. Bolides are fireballs that explode.

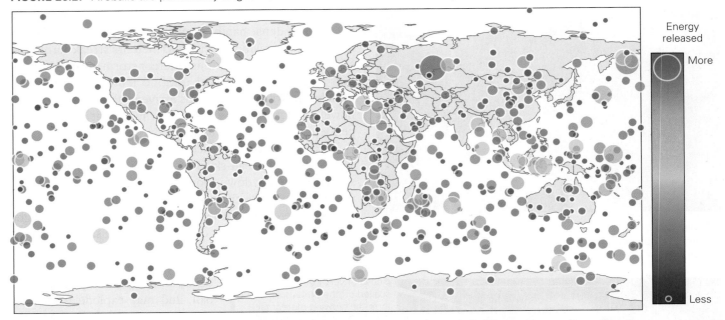

(a) Fireballs and bolides observed by US government ultrasound sensors in 1988–2020. These sensors detect low-frequency sound, which travels a long distance.

(b) A 2018 bolide streaks across Oregon and Washington.

The likelihood of a person being struck by a meteorite is extremely small—but it can happen. For example, in 1954, a woman living on a farm in Alabama was napping on a couch when a surviving fragment of a fireball bashed through the roof of her house, bounced, and struck her, causing a very large bruise (FIG. 16.18a). In 1992, a 12.4 kg (27.3 lb) meteorite, also a survivor of a fireball, crushed the trunk of a car in Peekskill, New York—the car is now a museum display (FIG. 16.18b). In 2003, debris from a fireball crashed into a village in India, setting several houses on fire and injuring residents.

Notably, not all observed impacts have taken place on the Earth. Many impacts on the Moon's surface have been observed with telescopes, and one impact, in 2013, produced an explosion so bright that people saw it from the Earth without telescopes. The most spectacular extraterrestrial impact happened on Jupiter in July 1994. A string of 21 comet fragments, ranging from a few hundred meters to 2 km (1.2 mi) across, plunged into the atmosphere of the giant planet over the course of 6 days (FIG. 16.19a). These objects were remnants of a 5 km (3 mi) comet called Shoemaker-Levy 9, which had been captured and then broken apart by the gravitational pull of Jupiter. The largest impact produced an explosion that released 6 million megatons of energy and sent a **vapor plume** (an upward-convecting cloud of gas) 3,000 km (1,864 mi) above the planet (FIG. 16.19b). Material from this plume settled onto the top of Jupiter's atmosphere as a visible ring 12,000 km (7,456 mi) across (FIG. 16.19c). Notably, the impact blast was 120,000 times larger than that produced by the Soviet Union's detonation of the *Tsar Bomba*, which, at 58 megatons—3,800 times as powerful as the bomb that destroyed Hiroshima—was the most powerful nuclear weapon ever exploded on the Earth.

No historic meteorite impact caused widespread destruction—until the beginning of the 20th century. Since then, two destructive events have been observed that could have resulted in disasters had they landed on a large city. Both, coincidentally, took place in Siberia (BOX 16.3).

FIGURE 16.18 Small meteorites have come close to striking people.

(a) In 1954, a meteorite crashed through the roof of an Alabama home, bounced, and bruised a resident. A similar event happened in 2021, when a 1.3 kg (2.8 lb) meteorite came to rest next to the pillow of a sleeping woman in western Canada.

(b) In 1992, a meteorite struck the trunk of this car in Peekskill, New York. Many people observed the fireball that preceded it. In this photo, the car is being delivered to a museum.

FIGURE 16.19 The observed impact of comet Shoemaker-Levy 9 on Jupiter, July 1994.

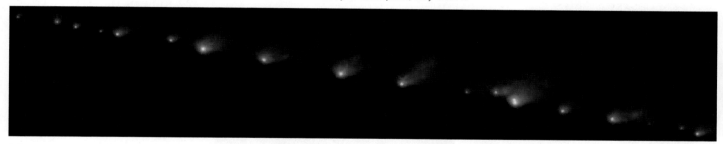

(a) The comet fragmented into a train of 21 pieces, the largest of which was 2 km (1.2 mi) across. The train was 1.1 million km (684,000 mi) long. This view comes from the Hubble Space Telescope.

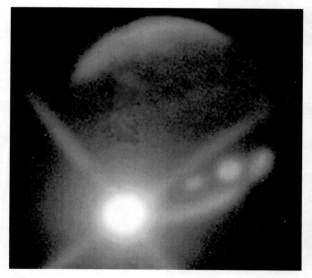

(b) The impact of one of the fragments appeared as a bright light. The scars of other impacts are also visible.

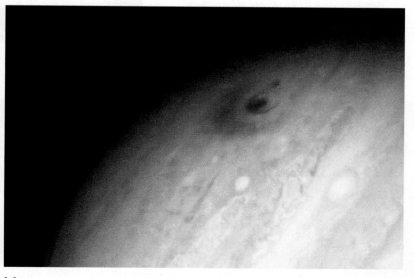

(c) Material ejected by the impact settled on the surface of Jupiter's clouds, forming a huge ring.

BOX 16.3 DISASTERS AND SOCIETY

The Tunguska and Chelyabinsk events

On the morning of June 30, 1908, an immense explosion took place over the Stony Tunguska River, in Siberia **(FIG. Bx16.3a)**. This explosion, known now as the **Tunguska event**, occurred when an asteroid 50–200 m (164–656 ft) in diameter either passed through, or exploded as a bolide in, the atmosphere at an altitude of 5–10 km (3–6 mi). In either case, the event yielded about 10–15 megatons of energy—1,000 times more than the Hiroshima atomic bomb—and produced shock waves that registered as an M_W 5 earthquake. The explosion flattened 80 million trees covering an area of about 2,150 km² (830 mi²)—12 times the area of Washington, D.C.—in a remote forest **(FIG. Bx16.3b, c)**. Because of the isolated location, there were no known human casualties. If it had exploded over a major city, that city could have been destroyed.

Over a century later, on February 15, 2013, a 10 million kg (11,000 ton), 17 m (56 ft) meteoroid crashed into the Earth's atmosphere over Chelyabinsk, east of the Ural Mountains, at an estimated speed of 60,000–69,000 km/h (37,000–43,000 mph). It left a blazing trail of light and smoke before exploding as a bolide—momentarily brighter than the Sun—releasing about 30 times as much energy as the Hiroshima atomic bomb **(FIG. Bx16.3d, e)**. The explosion, which was recorded on video, vaporized most of the object and shattered the remainder into tiny fragments and dust. It occurred about 19–24 km (12–15 mi) above the Earth's surface, and the air blast it produced, when it arrived in the city of Chelyabinsk 2 minutes later, blew out windows and shook buildings in an area 100 km (62 mi) wide **(FIG. Bx16.3f)**. Shattered glass and falling objects injured nearly 1,500 people. If the bolide had exploded closer to the ground, casualties would have been much worse.

FIGURE Bx16.3 The Tunguska and Chelyabinsk events.

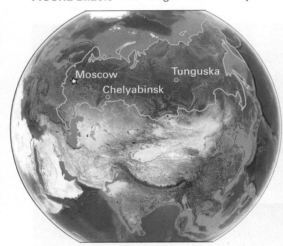

(a) Locations of the Tunguska and Chelyabinsk events.

(b) A forest blown down by the Tunguska shock waves.

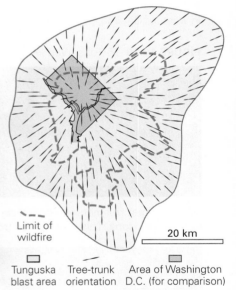

(c) Comparison of the Tunguska blast area to the area of Washington, D.C.

(d) The Chelyabinsk bolide explosion was as bright as the Sun.

(e) The smoke trail behind the Chelyabinsk meteoroid. The inset shows a tiny meteorite beneath the explosion site, sitting on the snow.

(f) Damage caused by the air blast from the Chelyabinsk explosion.

> **Take-home message...**
> When a very small meteoroid enters the Earth's atmosphere, it produces a meteor. Larger ones produce fireballs, some of which explode as bolides. Only objects greater than a certain size land on the Earth's surface.
>
> **QUICK QUESTION** Why might the meteors in a meteor shower all emerge from the same region of the sky?

16.7 What Does an Impact Do to the Crust?

Meteorite Craters

When a sufficiently large meteoroid smashes into a planetary surface at high speed, it produces a **crater** (from the Greek word for bowl), a depression surrounded by a raised rim. The size and character of a crater depend on the size, density, and coherence of the object as well as on the impact velocity, the angle of impact, and the composition of the planetary surface at the point of impact. Very small meteorites leave no craters. Small and relatively slow meteorites produce craters less than 8 times the diameter of the meteorite, whereas large and fast meteorites can produce craters at least 16 times the diameter of the meteorite.

Few meteorite craters have been preserved on the Earth. Therefore, to understand the processes of crater formation, researchers study craters on the Moon and other planets, and they simulate crater formation using laboratory and computer models. Results of this work distinguish between two general types of craters (FIG. 16.20):

SIMPLE CRATERS. A **simple crater** develops in response to a relatively small impact. On the Earth, simple craters are less than about 2 km (1.25 mi) in diameter. On the Moon, because of its weaker gravity, simple craters can be up to 20 km (12 mi) in diameter. Crater formation begins when the impacting object compresses the surface of the *target* (the larger object that is being struck) and sends shock waves into the target (FIG. 16.21a). An instant later, during the *excavation stage*, the impact's energy explosively fragments the target's rocks, shooting **ejecta**, composed mostly of target-rock fragments, outward. It also extensively fractures rock below the point of impact. Some of the impact energy is converted into intense heat, which can be sufficient to produce *impact melt* (completely liquefied rock) and vapor from both the meteoroid and the target rock. By the end of this excavation stage, the force of impact has also pushed the rock below the impact site downward, forming an indentation called a *transient crater*.

Moments after the excavation stage, the *modification stage* begins, during which the transient crater rebounds upward; thus, the final crater dimensions are controlled primarily by the amount of ejecta that has been excavated.

FIGURE 16.20 Contrasts between a simple crater and a complex crater.

Simple crater

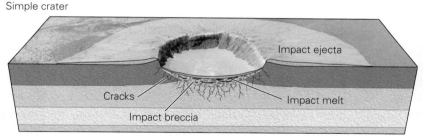

(a) A simple crater is bowl-shaped, with no central uplift.

Complex crater

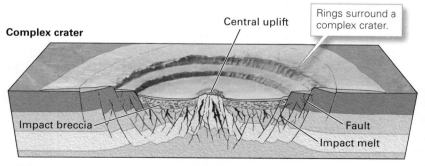
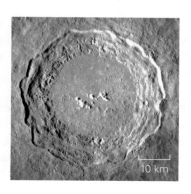

(b) A complex crater has a central uplift, is underlain by many faults, and is rimmed by terraces due to slumping on ring-shaped normal faults.

FIGURE 16.21 Stages in crater formation.

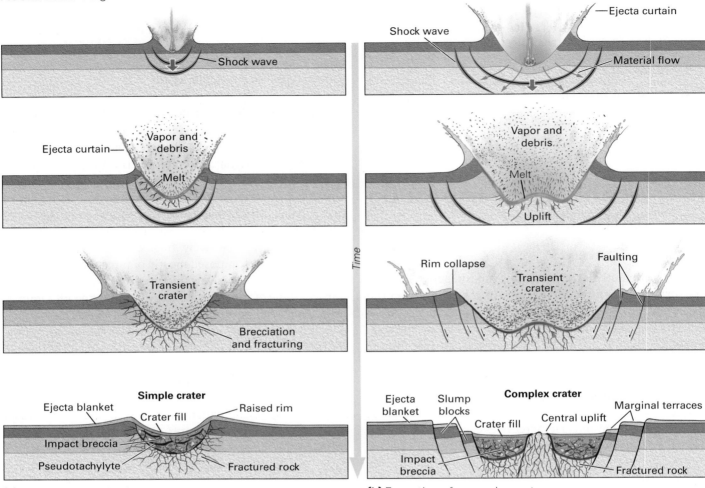

(a) Formation of a simple crater.

(b) Formation of a complex crater.

This rebound—much like the twanging back of a spring—also causes the crust of the target adjacent to the crater to move upward, forming a *raised rim*. Some ejecta and impact melt fall onto the floor of the crater, where they form a thick lens of *impact breccia* mixed with solidified impact melt, and some form an *ejecta blanket* that covers the raised rim and spreads over the surrounding land. Finer-grained ejecta, which takes longer to settle, eventually covers the crater itself. Impact melt solidifies so quickly that it produces a glassy rock called *pseudotachylyte*, so called because it resembles a type of volcanic glass called tachylyte. In some cases, rock layers dragged upward along the edge of the crater collapse downward to form an upside-down layer along the raised rim.

COMPLEX CRATERS. The formation of craters larger than 2 km on the Earth, and 20 km on the Moon, involves additional phenomena, so these craters are called **complex craters**. The energy produced by these larger impacts is so great that it not only excavates a crater, but also fractures and heats the rock beneath the crater so intensely that this rock can flow almost like a fluid (**FIG. 16.21b**). Therefore, during the modification stage, as the transient crater rebounds, the interior of the crater rises to form a *central uplift*. (To visualize the process, throw a pebble into a puddle and observe how the water surface bulges up an instant after being indented by the pebble.) Numerous faults develop in association with the formation of the central uplift, in which layers of rock that were once deep beneath the crater rise to the Earth's surface and may be tilted steeply. Because so much impact melt and impact breccia fill the impact site, the floor of the crater becomes flat. In addition, the margins of the crater collapse downward along normal faults that circle the crater, producing terraces ringing the crater.

Other Evidence of Large Impacts

Very large impacts damage the underlying crust to sufficient depth that even after significant erosion, evidence of the impact still remains. Ancient, eroded impact structures that

FIGURE 16.22 Manifestations of crustal deformation in an astrobleme.

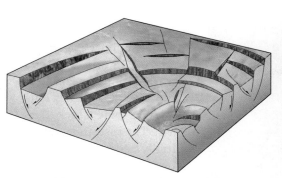

(a) Faulting caused by a large impact.

(b) Faults (black lines) exposed in a quarry in the Kentland impact structure, Indiana.

(c) Tilted strata in the Kentland impact structure.

(d) Fractured bedrock in the Sudbury Basin. The cracks are filled with pseudotachylyte.

(e) Impact breccia at the Sudbury impact site. Large fragments are surrounded by pseudotachylyte.

(f) Shatter cones in limestone from the Kentland site.

Quartz (density = 2.62 g/cm³)   **Coesite** (density = 2.91 g/cm³)   **Stishovite** (density = 4.29 g/cm³)

(g) Shock metamorphism transforms quartz into denser minerals (coesite and stishovite). The areas of the squares represent density.

can be recognized by the damage that occurred beneath the crater are known as **astroblemes**. Astroblemes can be identified by the presence of several features:

- Numerous faults that accommodate displacement of the crust; some of these faults are curved and are parallel to the rim of the crater, while others extend from the center of the crater outward (**FIG. 16.22a, b**).
- A central uplift, in which layered rocks that were once horizontal tilt steeply (**FIG. 16.22c**).
- Countless fractures and broad zones of brecciation, producing rocks of many sizes (**FIG. 16.22d**).
- Large quantities of impact melt (**FIG. 16.22e**), which has been injected into fractures to form dikes of **pseudotachylyte**, has filled pores within impact breccia, or has sprayed outward with ejecta and frozen into glass beads called *microtektites*.
- Broad zones in which bedrock contains *shatter cones*, distinct cone-shaped fractures nested one inside the other, and oriented so that the cones' peaks point toward the direction from which the impacting object came (**FIG. 16.22f**).
- Minerals produced by *shock metamorphism*, a sudden change into a denser crystal structure caused by the pulse of extremely high pressure as a shock wave passes; shock metamorphism, for example, changes quartz (SiO_2) into two unusual minerals called coesite and stishovite (**FIG. 16.22g**).

The energy released during the impact of a large meteoroid vaporizes most of the impacting object itself along with some of the rock beneath the target, and the vapor rises in a plume to high altitudes, where winds spread it around the world. This vapor includes elements, such as *iridium*, that no longer occur in the Earth's crust—the Earth's iridium sank to become part of the core early in Earth history. Elements in the impact vapor eventually settle back to the Earth's surface and become incorporated, along with microtektites and other fine ejecta, in sedimentary layers being deposited at the time of impact.

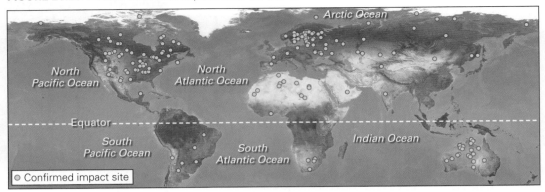

FIGURE 16.23 Known meteorite impact sites on the Earth.

Examples of Impact Sites on the Earth

There are about 190 known impact sites on the Earth (FIG. 16.23). Only some of these are marked by craters at the Earth's surface—most are astroblemes identified by deeper impact damage exposed by erosion. By looking at examples of past impacts, we can gain appreciation for the danger that a future impact represents.

VISIBLE CRATERS. The most spectacular impact crater on the Earth, Meteor Crater (also known as Barringer Crater) in Arizona, is 1.2 km (3,900 ft) across and 174 m (571 ft) deep. It was formed about 50,000 years ago by the impact of an iron meteoroid roughly 50 m (164 ft) across (FIG. 16.24). At the moment of impact, the object was traveling at least 12.8 km/s (29,000 mph), and possibly faster. The impact explosion vaporized more than half of the meteoroid, but surviving fragments—called the Canyon Diablo meteorites—have been found nearby. The impact on horizontally bedded Mesozoic and Paleozoic sedimentary strata at the site produced a simple crater surrounded by a raised rim. Meteor Crater remains clearly visible because the impact site is now a desert, so erosion occurs slowly, and vegetation is sparse.

Craters that formed in polar climates also tend to remain distinct for a long time, because unless they are overridden by a glacier, they erode very slowly. For example, Pingualuit crater, in the tundra of Arctic Quebec, stands out clearly in the landscape (FIG. 16.25a). In contrast, Kaali Crater, which formed about 2,500 years ago in Estonia, has become forested and is gradually filling with mud (FIG. 16.25b).

Only 10 confirmed craters worldwide are younger than Meteor Crater, and all but one are much smaller, less than 300 m (984 ft) across. The youngest, in the desert of Saudi Arabia, may have formed just over 200 years ago. All the others are more than 2,000 years old. The smallest known crater is only 15 m (49 ft) across.

EXPOSED LARGE ASTROBLEMES. Fourteen confirmed large astroblemes, with diameters ranging from 50 to 300 km (31–186 mi), have been found on continents. Let's consider a few examples.

The Sudbury Basin in western Ontario formed about 1.85 billion years ago when an asteroid 10–15 km (6–9 mi) wide stuck the Earth. The resulting crater initially had a diameter of 150–250 km (93–155 mi), but the preserved remnant is 60 km (37 mi) long by 30 km (19 mi) wide—the crater is asymmetrical because it was compressed horizontally by mountain-building processes after it formed (FIG. 16.26). The impact caused decompression melting (see Chapters 2 and 4) of the underlying mantle, producing magmas that rose and intruded the crust. These magmas carried valuable metals with them, so the Sudbury area now hosts one of the world's most economically valuable *ore deposits*—hundreds of billions of dollars worth of nickel, copper, gold, and silver have been extracted from rocks in the astrobleme. Outcrops provide abundant examples of fractures, shatter cones, and pseudotachylyte (see Fig. 16.22d–e).

About 1,050 km (652 mi) northeast of Sudbury, in the wilderness of Quebec, an even larger astrobleme, known as Manicouagan Crater, was formed by a collision about 215 million years ago (FIG. 16.27a). The impact produced a complex crater over 100 km (62 mi) in diameter that filled with hundreds of cubic kilometers of impact melt. The central uplift of this crater brought up rock that had been as deep as 10 km (6 mi) below the surface. The circular structure visible today isn't the actual crater, which was stripped away by erosion. Rather, it's a landscape controlled by crustal structures produced by the impact—the circular low area surrounding the central uplift is composed of weaker rock that preferentially eroded away. Construction of a hydroelectric dam impounded water in this low area to form the circular Manicouagan Reservoir.

The Vredefort Dome of South Africa was formed about 2 billion years ago by the impact of a 10–15 km (6–9 mi) asteroid (FIG. 16.27b, c). It was 300 km (186 mi) across when it formed, which makes it the largest confirmed astrobleme on the Earth. Today, now that 6 km (4 mi) of crust has eroded away, the remnants of the impact have a diameter of about 70 km (43 mi). This astrobleme was a complex crater with a central uplift of granite. The overlying sedimentary strata tilt away from the uplift, so the impact site is a structural dome (a fold with the shape of an overturned bowl). The instant after impact, the transient crater was 40 km (25 mi) deep.

FIGURE 16.24 Meteor Crater in Arizona.

(a) Aerial view of Meteor Crater, which is 1.2 km (0.75 mi) across.

(b) View from the edge of Meteor Crater.

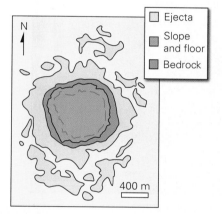

(c) A simplified geologic map of the crater. The light yellow area is land that isn't currently covered by ejecta.

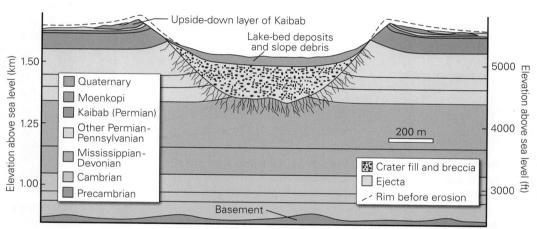

(d) A simplified cross section of the crater. Erosion has lowered its rim, and sediment has covered the ejecta blanket on the floor of the crater.

FIGURE 16.25 Other examples of visible craters on the Earth.

(a) Pingualuit Crater has filled with water to become a circular lake. It is about 3.4 km (2.1 mi) wide, and has a raised rim 160 m (525 ft) high.

(b) Kaali Crater, in Estonia, is only 110 m (361 ft) wide. It is one of eight small craters formed about 2,500 years ago by a bolide whose explosion became incorporated into myth.

16.7 What Does an Impact Do to the Crust? • **609**

FIGURE 16.26 The Sudbury Basin, Ontario, is a large Precambrian astrobleme.

FIGURE 16.27 Examples of very large astroblemes.

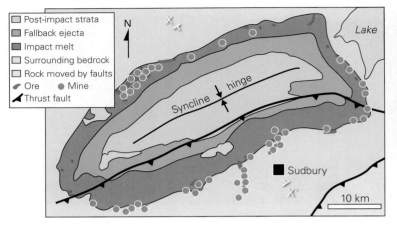

(a) Long after the impact, the central region of the impact site become a sediment-filled basin. Later, mountain building squashed the basin, making it elliptical in map view. Major ore deposits occur along its edge.

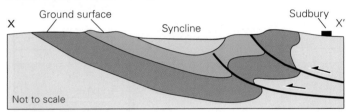

(b) A cross section along the line X–X' emphasizes that the basin is a trough-shaped fold (curve), known as a syncline. Mountain building also caused thrust faults to cut rocks of the syncline.

BURIED ASTROBLEMES. Some ancient impact sites are not visible at the Earth's surface because they have been buried beneath younger strata. For example, the Kentland impact structure, in western Indiana, underlies farm fields. Because the central uplift brought limestone close to the surface, the impact structure is now a working quarry (see Fig. 16.22b, c, f). Unlike the horizontal layers of sedimentary bedrock in the surrounding region, the rock of the Kentland quarry is full of shatter cones, and in places, its bedding tilts steeply. A larger buried impact structure underlies the southern end of Chesapeake Bay. The Chesapeake Bay crater, first discovered by drilling in 1983, was formed about 35 million years ago by an object that plunged into shallow water and the underlying sediments of the continental shelf **(FIG. 16.28)**. The impact not only disrupted the entire sedimentary column, but it also fractured underlying granite down to a depth of about 8 km (5 mi), and left a crater about 85 km (53 mi) wide. The crater filled with impact breccia and was then buried by shallow marine strata. Because of the porosity of the impact breccia, it has undergone more compaction than the adjacent rock, so the region has been subsiding (sinking) faster than surrounding areas (see Chapter 7). This subsidence produced Chesapeake Bay, a large estuary.

(a) Manicouagan Crater, in Quebec, is an astrobleme. A reservoir fills the circle of weaker rock surrounding the central uplift that has eroded away.

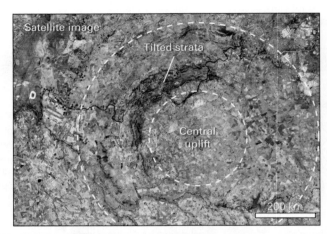

 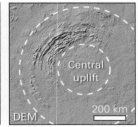

(b) The Vredefort Dome is an astrobleme. Tilted strata surrounding the central uplift form distinct curving ridges. The region of the central uplift, the site of impact, is now covered by farm fields. The digital elevation model (DEM) of the present-day ground surface emphasizes the curving ridges.

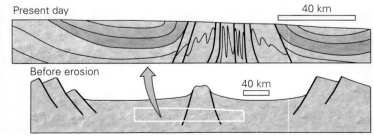

(c) The present-day exposure of the Vredefort Dome is much smaller than the original crater, whose floor was kilometers above the present-day land surface, as shown by these cross sections.

FIGURE 16.28 The southern end of Chesapeake Bay overlies an impact crater 35 million years old.

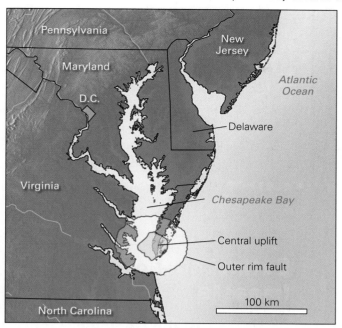

(a) The crater is vastly larger than Washington, D.C.

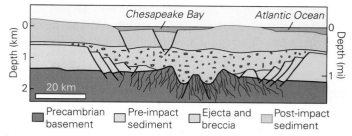

(b) This cross section shows the underground form of the crater. The bay formed because the impact breccia can undergo more compaction than surrounding strata.

The **Chicxulub crater** lies completely hidden beneath thick, flat-lying limestone layers on the Yucatán Peninsula in Mexico. Oil exploration surveys first suggested its existence in the 1970s, but it wasn't confirmed as a crater until the 1990s, when a modern gravity survey outlined its shape clearly (**FIG. 16.29a**). The crater, 150 km (93 mi) in diameter and 16 km (10 mi) deep, was formed by the impact of a chondritic asteroid that was about 15 km (9 mi) in diameter

FIGURE 16.29 The Chicxulub crater lies buried beneath flat-lying strata on the Yucatán Peninsula, Mexico.

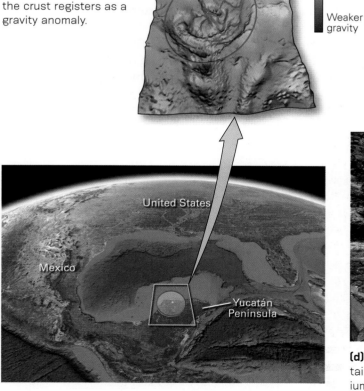

(a) A gravity survey of the crater. Damage to the crust registers as a gravity anomaly.

(b) The location of the crater. On land, a ring of sinkholes delineates its shape.

(c) A layer of microtektites deposited with ejecta that landed in Haiti.

(d) A distinct layer of shale, containing a high concentration of iridium, marks the time of the impact. This layer has been found at many locations worldwide.

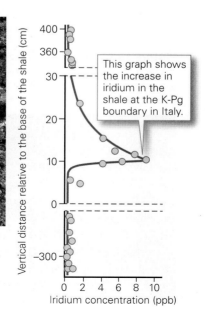

16.7 What Does an Impact Do to the Crust? • **611**

(FIG. 16.29b). The impact distributed iridium and microtektites, along with other ejecta, worldwide (FIG. 16.29c, d). The impact is estimated to have had 100 million times the explosive force of the Tsar Bomba, which made it about 330 billion times the size of the Hiroshima explosion. As we'll see, geologists consider this impact to be an extinction-level event responsible for ending the Age of Dinosaurs.

Why Aren't There More Impact Sites on the Earth?

The 190 confirmed impact sites on the Earth represent a vastly smaller number than are visible on other bodies in the Solar System. The Moon, for example, has about 5,200 craters with diameters greater than 29 km (12 mi), and hundreds of millions of smaller ones (FIG. 16.30). Why does the Earth have so few craters? A consideration of the various processes and cycles of the Earth System provides an answer. Most impacts in the Solar System took place over 3 billion years ago. On the Moon, which has no plate tectonics, no atmosphere, and no hydrosphere, craters from these very old impacts have remained largely unchanged since their formation. On the Earth, however, due to plate tectonics, all craters of this age on the seafloor (70% of the planet's surface) have long since been subducted. Of the craters that formed on continental crust, some have been destroyed by mountain building, some have been buried by younger sedimentary strata, and some have been removed by erosion. Therefore, the remnants of all impacts, except those that are relatively recent and occurred where erosion is slow, or those that were large enough to affect the crust many kilometers deep, are gone. In fact, it should be no surprise that of the approximately 60 asteroids or comets with a diameter greater than 5 km (3 mi) that probably have struck the Earth in the last 600 million years, craters from only three have been found.

FIGURE 16.30 Craters of all sizes completely cover the Moon's surface. If the Earth didn't have plate tectonics, running water, glaciers, and an atmosphere, its surface would also be pockmarked with craters.

> **Take-home message...**
>
> When a large space object hits the Earth's surface at cosmic speed, it produces a crater. Smaller objects produce simple craters, and larger ones produce complex craters with a central uplift. The force of impact also severely damages the crust beneath the crater, producing, for example, faults, shatter cones, and impact melt. An impact site where damaged crust is exposed due to erosion of the overlying crater is an astrobleme. About 190 craters and astroblemes have been identified on the Earth, some of which are buried.
>
> **QUICK QUESTION** Why does the Moon host so many more craters than the Earth?

16.8 Extinction-Level Impacts

Mass Extinctions of the Geologic Past

Extinction—the death of the last organism of a given species—is forever, in that once a species goes extinct, it never reappears. Extensive studies of the fossil record indicate that during the course of geologic time, the Earth's *biodiversity*, represented by the number of species and genera (groups of related species), has decreased catastrophically at distinct times. Such **mass extinctions** were recognized even before modern radioisotopic dating techniques could pin numerical ages on these events, for they define the major boundaries between subdivisions of the geologic time scale (FIG. 16.31). For example, a mass extinction at the end of the

FIGURE 16.31 Mass extinctions, as represented by the percentage of genera lost, have occurred at distinct times in Earth history.

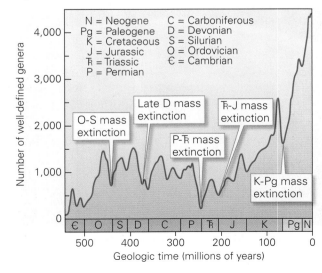

Permian Period defined the end of the Paleozoic Era (252 Ma), and a mass extinction at the end of the Cretaceous Period (66 Ma) defined the end of the Mesozoic Era. At least five mass extinctions have occurred over Earth history, during which 70%–96% of species disappeared.

Most researchers have concluded that the mass extinction at the end of the Mesozoic, during which all dinosaur species vanished, along with 70%–95% of other species, was caused by the impact that produced the Chicxulub crater. Because people tend to be fascinated with dinosaurs, this event, now known as the *K-Pg mass extinction* (K stands for Cretaceous, and Pg stands for Paleogene, the first period of the Cenozoic Era), has captured the public's imagination.

Convincing evidence that the K-Pg mass extinction resulted from an asteroid impact was published in 1980 by Walter Alvarez, an American geologist, his father, Luis, a physicist, and colleagues. Their work showed that a shale layer deposited exactly at the K-Pg stratigraphic boundary contained unusually high quantities of iridium, an element not normally found in the Earth's crust, but often found in meteorites (see Fig. 16.29d). Further study showed that the shale layer also contained microtektites, as well as grains of shocked quartz (coesite and stishovite), materials also produced by large impacts, and also that the layer occurs worldwide. Ten years later, geologists proved that the Chicxulub crater was an astrobleme that formed at exactly 66 Ma—the K-Pg boundary. This discovery confirmed that the unusual materials in the shale layer resulted from the impact that produced the crater, and led the researchers to associate the demise of the dinosaurs with the formation of Chicxulub.

Studies continue to evaluate whether meteoroid impacts caused any other mass-extinction events. Evidence suggests that they possibly caused some, probably not all of these events. Causes of mass extinctions in the geologic past have long been a subject of debate. In addition to impacts, these events have also been attributed to climate change, the appearance of new predators in the food chain, and the eruption of large igneous provinces (LIPs). Evidence doesn't always lead to universally accepted explanations. For example, some researchers suggest that the Permian mass extinction was caused by an impact, but others attribute it to the eruption of an LIP in Siberia. We may be living in the midst of a **sixth extinction** today, for human activities are causing species to disappear at rates that are hundreds to thousands of times higher than the rates that have occurred during the rest of the Cenozoic. By some estimates, about 25% of species will disappear by the end of the 21st century. These species extinctions are due to lost habitat as natural areas are converted into farmland, ranches, or human communities; to parasites and diseases that have become more prevalent; and to anthropogenic climate change.

How Might an Impact Cause Mass Extinction? The Chicxulub Example

We've noted that the impact of a Tunguska-sized object could wipe out a major city, if it exploded over the city. But when the Tunguska event happened, it had no major effects beyond the blast area. The 15 km (9 mi) asteroid that formed the Chicxulub crater, however, appears to have caused a mass extinction by affecting the entire planet (**VISUALIZING A DISASTER**, pp. 614–615). Despite its large size, the Chicxulub asteroid was tiny compared with the whole Earth. How could the impact of a 15 km asteroid produce enough energy to wipe out most life on a planet 12,743 km (7,918 mi) in diameter? The answer comes from recognizing the impact speed of the asteroid (20 km/s, or 45,000 mph). Moving objects have *kinetic energy*, which is proportional to an object's mass and the square of the object's velocity. Consequently, if the object's velocity is high, its kinetic energy becomes immense. On impact, that energy instantly transforms into heat, shock waves, and air blast.

How would the sudden input of energy from an asteroid affect the Earth? Geologic and modeling studies of the Chicxulub impact suggest the following scenario. First, shock waves propagated through the air, across the land surface, and deep into the Earth (**FIG. 16.32a**). Ground shaking was comparable to an M_W 10–M_W 13 earthquake—put another way, the impact caused shaking vastly greater than the largest megathrust earthquake possible (see Chapter 3). Intense heat associated with a high-velocity air blast immediately followed, flattening everything on the Earth's surface within a distance of about 1,500 km (932 mi), and hurricane-force winds continued out to a distance of perhaps 3,000 km (1,864 mi) (**FIG. 16.32b**). The impact also produced megatsunamis that spread outward from the impact site. These waves rose to a height of 1.5 km (1 mi) near the impact, and they may still have been 300 m high (almost the height of the Empire State Building) when they crossed into what is now Texas and Louisiana. Because the Gulf of Mexico coastal plain has a low elevation, tsunamis may have traveled inland as far as what is now southernmost Illinois.

The impact vaporized the asteroid, as well as some crust at the point of impact, producing a vapor plume that rose hundreds of kilometers into the atmosphere. Solid particles precipitating from this plume—including particles containing iridium—were distributed worldwide. In addition to vapor, solid ejecta—from dust-sized to large building-sized—spewed high into the atmosphere and then fell back to the Earth. Ejecta piled up to a depth of a few kilometers near the impact side, and formed an ejecta blanket up to 0.3 km (0.2 mi) thick for a distance as far as 600 km (373 mi) away. Significant quantities of ejecta reached a distance of at least 4,000 km (2,485 mi), and fine-grained ejecta spread worldwide (as indicated by the presence of coesite,

VISUALIZING A DISASTER

Not all the natural hazards that society faces originate on the Earth—some come to us from space. Solar flares and coronal mass ejections send bursts of highly energetic particles into space. When they reach the Earth, these particles not only cause intense aurorae, but they can severely disrupt electrical and communication grids. Solid objects—meteoroids, asteroids, or comets—collide with the Earth quite frequently. When very small meteoroids enter the atmosphere, they produce meteors. Slightly larger ones may become fireballs. Fortunately, only rarely are meteoroids large enough to cause significant damage on the Earth's surface. Some damage the planet by exploding as bolides in the atmosphere, thereby producing strong shock waves. Some strike the planet, thereby producing a crater, damaging the rock below, and sending debris and vapor skyward. Particularly large impacts, such as those that have left astroblemes, could result in a planet-wide catastrophe, and even in mass extinction.

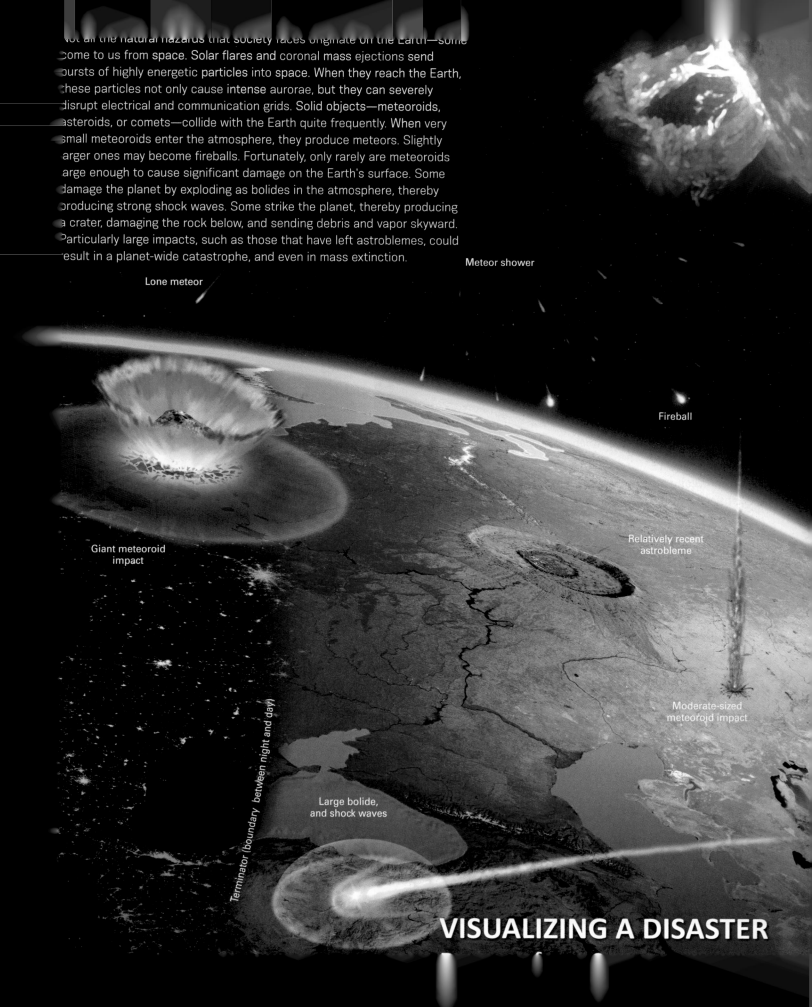

FIGURE 16.32 The Chicxulub impact as a model for what happens when a small asteroid collides with the Earth.

(a) An artist's image of the moment of impact at Chicxulub.

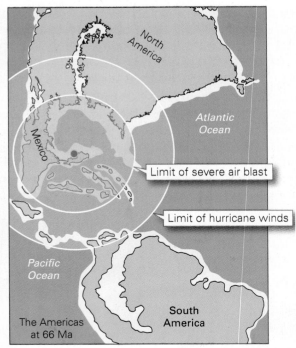

(b) The area affected by the air blast, extreme winds, and megatsunamis.

stishovite, and microtektites throughout the sediment layer deposited at the K-Pg boundary). All told, the impact produced more than 10,000 km³ (2,400 mi³) of ejecta. As this debris fell through the air, it was as if millions of meteoroids were falling at the same time. The fall of each piece heated and dried the air, so the net result of the ejecta fall was significant heating of the atmosphere for a period of at least several minutes. This heat, along with the landing of incandescent particles, led to widespread wildfires (FIG. 16.33), which sent billions of tons of soot into the air. Remnants of this soot have been found in shale formed from sediment deposited at the time of the impact.

Because the Chicxulub impact site was a shallow sea underlain with sediments that included thick evaporites (deposits of salt and gypsum), the vapor plume included large volumes of chlorine and bromine, which destroyed the Earth's stratospheric ozone layer and allowed intense UV light to bathe the planet's surface. This UV light probably decreased the reproductive capacities of many species. Vaporization of evaporites also produced *sulfate aerosols*, which dispersed worldwide. These aerosols, together with soot from wildfires and dust from ejecta, blocked sunlight, plunging the Earth into winter-like cold for years. (A similar effect, called *nuclear winter*, would take place after a nuclear war.) The absence of sunlight effectively shut down photosynthesis worldwide, thereby interrupting the food chain and causing countless species of plants to perish, and animals to die of starvation. When sulfate aerosols eventually washed out of the atmosphere in rain, severe acid rain resulted worldwide for several years.

The Chicxulub impact not only filled the air with dust and aerosols, but it also roasted the thick layers of limestone beneath the impact site, sending vast quantities of CO_2, a greenhouse gas, into the atmosphere. Because CO_2 has a long

FIGURE 16.33 An artist's image of dinosaurs hit by the blast from the Chicxulub impact.

residence time in the atmosphere, its presence caused significant global warming after the initial period of global cooling due to blocked sunlight came to a close. By some estimates, global temperatures may have increased by 5.0°C–7.5°C (9°F–13.5°F). Such greenhouse conditions could have persisted from a few thousand to a hundred thousand years.

Clearly, an impact on the scale of the Chicxulub event would cause immense destruction at the impact site and for a significant distance around it, so life within an area of up to 1,500 km (932 mi) from the site would be destroyed directly. But worldwide extinction did not result from that direct destruction. Rather, it was the interruption of the food chain caused by fires, then global winter, then longer-term climate change that drove the worldwide extinction event.

> **Take-home message . . .**
> Mass extinctions have happened at five times during Earth history, and we may be living through a sixth. The mass extinction at the Cretaceous-Paleogene boundary, which included the extinction of the dinosaurs, was almost certainly caused by the Chicxulub impact. This impact would have caused immense destruction within a distance of about 1,500 km, but would also have interrupted the global food chain by causing climate changes.
>
> **QUICK QUESTION** How can an asteroid 15 km wide cause damage over a much broader area on impact?

16.9 Mitigating Impact Risk

Identifying Near-Earth Objects

Clearly, the potential for a catastrophe caused by objects arriving from space is real. Could any **near-Earth objects** (**NEOs**)—defined as asteroids, comets, and large meteoroids that come within 50 million km (31 million mi) of the Earth's orbit—collide with the Earth within the next century? To answer this question, in 1992, the US Congress commissioned a study, the *Spaceguard Survey Report*, to clarify the risks of impact. It then asked NASA to locate 90% of NEOs with diameters greater than 1 km (0.6 mi) by 2005. When the fragments of comet Shoemaker-Levy 9 collided with Jupiter in 1994, public awareness of the potential for a catastrophic impact increased, and the International Astronomical Union encouraged Spaceguard-like efforts worldwide. In 2005, Congress expanded the Spaceguard Survey goals and asked NASA to identify 90% of potentially hazardous objects with a diameter greater than 140 m (459 ft). Efforts continue under the auspices of the *Center for Near-Earth Object Studies*, overseen by NASA's *Planetary Defense Coordination Office*, which is also responsible for issuing warnings about possible impacts. In the years since the systematic search of NEOs began,

FIGURE 16.34 The search for near-Earth objects.

(a) The Pan-STARRS telescope in Hawai'i has discovered more NEOs than any other telescope.

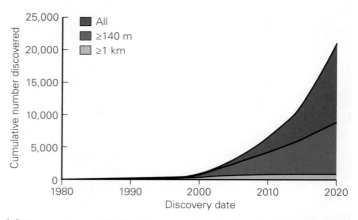

(b) The number of NEOs discovered has increased dramatically since the search began.

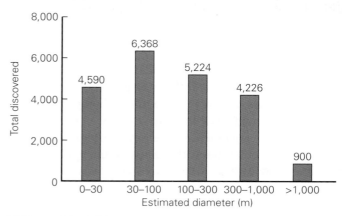

(c) Most of the NEOs discovered range from 30 m to 100 m (98–328 ft) in diameter. There are probably more small ones, but they are hard to detect.

astronomers have documented over 97% of objects larger than 1 km, and almost 90% of those larger than 140 m. Over 20,000 NEOs have been identified, and new ones are discovered every year **(FIG. 16.34)**. About 10% of NEOs might

FIGURE 16.35 Finding the orbits of NEOs.

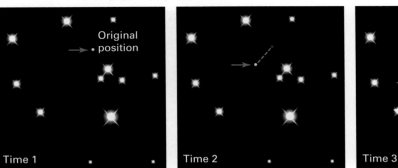

(a) By comparing the positions of an object at different times, astronomers can calculate its orbit.

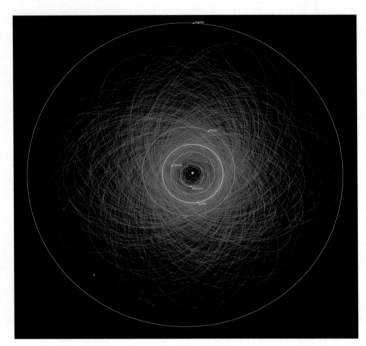

(b) The orbits of thousands of NEOs have been calculated.

represent a threat to our planet, and of these, approximately 5,000 have diameters greater than 300 m.

To track NEOs, astronomers compare photographs of the night sky taken at different times to see if any of the points of light have moved, documenting the views systematically with computers and sometimes with the help of amateur telescope enthusiasts (**FIG. 16.35a**). Since stars are so far away, they do not appear to move with respect to one another, so any object that does move, and is not a known planet, moon, or asteroid, becomes a focus of attention. By comparing the positions of the object at different times, astronomers can calculate its orbit, and can decide whether the object qualifies as an NEO. The accuracy of orbital calculations depends on how far away the object is from the Earth and on the time between observations, for these factors determine the change in position of the object between sequential images. In most cases, initial calculations are not accurate, as the observation window for seeing small objects from the Earth is minimal, but as new data come in from subsequent near approaches of the object, orbital calculations become more accurate (**FIG. 16.35b**). For example, in 1998, when astronomers discovered an asteroid 1 km in diameter, initial calculations of its orbit suggested that it might collide with the Earth in 2028. Needless to say, a media circus followed, raising public awareness of destructive impacts. However, refined calculations showed that the object, fortunately, will not collide with the Earth.

The Consequences and Frequency of Impacts

How often do impacts happen on the Earth? To answer this question, researchers have studied the record of impacts that we can see on the Moon, for which we have much more complete information. By looking at cross-cutting relations (applying the principle that if one crater cuts across another, the crater that has been cut must be older), and the degree of *space weathering* (blunting of features due to influxes of micrometeorites, and alteration of minerals by solar wind and cosmic rays), astronomers can estimate the relative ages of meteorite craters. They conclude that most of the impact craters that we can see on the Moon were formed over 3 billion years ago, and that since that time, the rate of impact has been fairly constant. Researchers have also studied the paths of NEOs, and have concluded that most will survive for only a few million years before they strike another object, are gravitationally flung toward the Sun by another object, or are accelerated by grazing another object and leave the Solar System. Therefore, in order for impact rates to remain fairly constant, new asteroids and meteoroids must be nudged regularly into near-Earth orbits—and they are, primarily by the gravitational force of Jupiter. In other words, the supply of NEOs is constantly being replenished.

How much damage will an impact cause? Researchers suggest that a meteoroid smaller than a car (about 10 m, or 33 ft, in diameter) generally burns up or blows up in the atmosphere, so that only small pieces of the object land on the Earth's surface. Shock waves from explosions of such objects, however, could shatter windows. An object on the order of 15–20 m (49–66 ft) across (the size of a small house) could produce an explosion comparable to that of the Hiroshima atomic bomb, and could flatten an area 1.5 km (1 mi) in diameter. The impact of the 50 m (165 ft) meteoroid that formed

Meteor Crater could produce a ground-level fireball extending out to a distance of 10 km (6 mi), a deadly shock wave (100 times atmospheric pressure) extending out more than 24 km (15 mi), an air blast traveling 2,000–2,500 km/h (1,243–1,553 mph) out to 2.5–3.0 km (1.6–1.9 mi), and hurricane-force winds out to 40 km (25 mi). The impact of an object 100–200 m (330–660 ft) in diameter could wipe out a large city, and the impact of any object greater than about 0.8 km (0.5 mi) across could affect global climate. Recent evidence has emerged, for example, suggesting that fragments of a large (>4 km), disintegrating asteroid or comet struck several locations worldwide between 12,800 and 11,700 years ago, triggering the Younger Dryas cooling period, which interrupted the warming trend following the most recent glaciation. The impact and explosion of a 1 km (0.6 mi) asteroid or comet landing in the ocean would vaporize nearly 500 km^3 (120 mi^3) of water and would release 10 times the total energy of all the nuclear weapons on the Earth during the Cold War. An object the size of a mountain (7–15 km, or 4–9 mi, across) would produce a crater 160 km (100 mi) wide and cause mass extinction.

How often do impacts occur? NASA's observations of cratering on other planets and of the frequency of fireballs and bolides on the Earth, along with studies of the geologic record of prehistoric impacts on the planet, make it possible to determine rough recurrence intervals for impacts of various sizes **(TABLE 16.2) (FIG. 16.36)**.

Overall, NASA estimates that the likelihood of a strike by a potentially destructive object anywhere on the Earth in any given year is about 0.1% (which means that it's actually vastly more likely than any one individual winning the lottery). Given that 70% of the Earth's surface is water, and that only about half of its land is hospitable to dense habitation, the likelihood that an object will strike a heavily populated area is much smaller. But with the availability of data from Spaceguard surveys, researchers try to represent the risk of collision with a specific asteroid or comet over the next century on the *Torino scale* **(TABLE 16.3)**. Only one object has ever gained a rating of 4 on the scale. So far, as more data have become available about the orbital path of any given object, the threat it represents has decreased—the object that once held a rating of 4 now rates 0. Since 2015, NASA has been developing the Asteroid Terrestrial-impact Last Alert System (ATLAS), which employs two robotic telescopes that survey the sky and can trace smaller NEOs a few days to a few weeks before impact. On two occasions, meteoroids have been followed from space and into the atmosphere, where they became fireballs or bolides.

Deflecting Near-Earth Objects

In 2017, 56 objects passed within 1 lunar distance of the Earth (the average distance between the centers of the Earth and the Moon). Of these, 4—ranging from 2 to 14 m (7–46 ft) in diameter—passed within 36,000 km (22,369 mi) of the Earth—closer than the orbits of weather satellites. In 2019, an asteroid 130 m (427 ft) long—longer than a football field—passed within 65,000 km (40,389 mi) of the Earth. Near-misses have also happened in 2004, 2011, and 2020. On November 13, 2020, a meteoroid 5–11 m (16–36 ft) in size passed 383 km (238 mi) above the Earth's surface, closer than the International Space Station, traveling at a speed of 48,303 km/h (30,014 mph).

How can society respond to an impact threat? NASA's defense strategies include developing techniques for deflecting or redirecting threatening objects that are

TABLE 16.2 Recurrence Intervals and Consequences of Impacts

Diameter of impacting object	Estimated recurrence interval (years)	Consequences
4 m (13 ft)	1–2	Object burns up as a fireball; small pieces may land at relatively slow speed.
7 m (23 ft)	5–10	Object explodes in upper atmosphere with force similar to the Hiroshima atomic bomb; pieces may land at relatively slow speed.
20 m (66 ft)	50–100	A house-sized block, 20% bigger than the Chelyabinsk object; could flatten an area around 1.5 km in diameter, if it lands.
60 m (200 ft)	1,500	Comparable to the object that produced Meteor Crater or flattened the forest of Tunguska; explodes on impact with the force of a 10-megaton nuclear bomb.
100 m (328 ft)	10,000	Could wipe out a large city.
1 km (0.6 mi)	500,000	Could have global climate implications.
5 km (3 mi)	18 million	A major disaster, approaching extinction level.
15 km (9 mi)	50–100 million	The size of a large mountain; produces a crater 160 km in diameter; an extinction-level event.

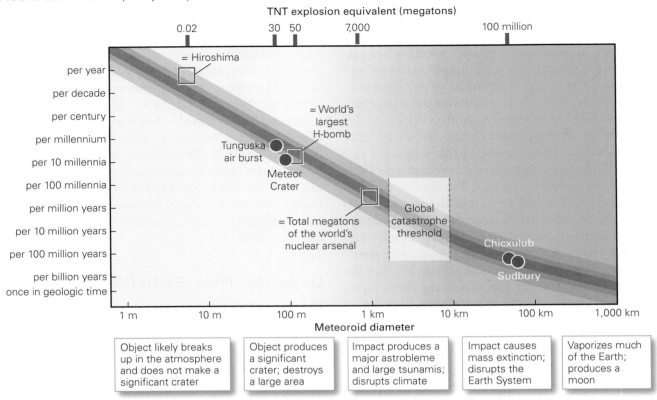

FIGURE 16.36 The frequency of impacts of different sizes.

determined to be on a collision course with the Earth. NASA defines two categories of deflection techniques. *Impulsive techniques* provide a lot of force nearly instantaneously, while *slow-push techniques* apply less force over a long time. Impulsive techniques include exploding conventional or nuclear devices on or near the surface of the object, or striking the object with a high-velocity spacecraft, called a *kinetic impactor*. NASA is conducting experiments to determine the maximum size that an impact can be without shattering the object. And a mission to send a kinetic impactor into a nonthreatening 160 m (525 ft) asteroid to see if it can be nudged is already in the planning stage. New studies are also exploring the possibility of launching a spacecraft that begins by mining tons of rock from an asteroid, then crashes into the asteroid with its load of rock. Slow-push techniques include using a mirror to focus solar energy onto the surface of the risky object to vaporize part of it, or using a laser to vaporize part of an object. Another technique involves attaching a device that mines material from an object and ejects it at high velocity, or ejects vaporized fuel from the spacecraft. Other ideas include flying near an object to nudge it with gravity, and coating part of an object to change its albedo (making one side emit more photons than the other). Studies suggest that the most effective method of deflecting an object would be to set off a nuclear device, but researchers worry that the explosion might simply fragment the object. Slow-push techniques will probably work only for small objects (<200 m), and in cases where push can be applied over decades. The greater the distance between an object and the Earth, the less deflection would be needed, so clearly, the farther from the Earth that deflection takes place, the better.

Unfortunately, technological solutions to an impending impact have not yet been designed and tested thoroughly, so whether they can be successful remains a topic of speculation. NASA officials estimate that 4–5 years could pass while a rocket armed with a device capable of deflecting an incoming object was set up, and as those years passed, the object would get closer and would be harder to deflect. Furthermore, the officials don't think that existing rockets are sufficiently powerful to gain enough speed to reach a distant NEO, so a spacecraft would have to use the gravitational assist of another planet or moon.

NASA has tried to estimate how well the consequences of an impact could be predicted. In one scenario, researchers could define a belt on the Earth's surface about 100 km (62 mi) wide and 5,000 km (3,107 mi) long in which the impact might occur. With this belt defined, they could then estimate how many people might be affected by an air blast or tsunami. If deflection or redirection is not possible, society would have to develop emergency response plans unlike any developed in the past (**FIG. 16.37**). We can only hope that no threatening objects will be detected until deflection technologies really become feasible.

TABLE 16.3 The Torino Scale of Potential Earth Impacts (Simplified)

0	White	No hazard	Likelihood of a collision is zero, or the object is so small that it will burn up before impact.
1	Green	Normal	An observed object will pass near the Earth, but poses no danger.
2	Yellow	Merits attention	A close encounter that merits attention from astronomers, but no cause for concern.
3	Yellow	Merits attention	There is a 1% or greater chance that a collision could cause local destruction; officials should be notified if an encounter is less than 10 years away.
4	Yellow	Merits attention	There is a 1% or greater chance that a collision could cause regional destruction; officials should be notified if an encounter is less than 10 years away.
5	Orange	Threatening	A close encounter has a significant probability of causing regional devastation; astronomers should pay attention, and planning may be necessary.
6	Orange	Threatening	A close encounter could pose a serious threat of catastrophe; astronomers should pay attention, and planning may be necessary.
7	Orange	Threatening	A close encounter with a large object could cause a global catastrophe; international planning should take place, and astronomers should study the object carefully.
8	Red	Certain collision	A collision is certain and will cause local destruction or a tsunami.
9	Red	Certain collision	A collision is certain and will cause regional devastation or major tsunamis.
10	Red	Certain collision	A collision is certain and will cause global climate catastrophe that will threaten the future of civilization.

FIGURE 16.37 Consequences of an asteroid strike on a major city.

(a) Itokawa, an Apollo asteroid that crosses the Earth's orbit, is about 540 m (1,772 ft) long, which makes it taller than the 443 m Empire State Building.

(b) If Itokawa, or a comparable asteroid, were to land on the Empire State Building, it would produce a crater 10 km (6 mi) wide. At a distance of 100 km (62 mi), wind would be blowing at 400 km/h (250 mph), and the heat would be so intense that everything would instantly ignite.

Take-home message...

Geologic and more recent physical evidence shows that asteroid impacts happen often on the Earth, although globally catastrophic impacts are exceeding rare. To guard against future calamities, astronomers are cataloging and determining the orbits of all near-Earth objects with Earth-crossing orbits.

QUICK QUESTION What might happen if an asteroid the size of the Tunguska object crashed directly into the ocean?

Chapter 16 Review

Chapter Summary

- The Sun's energy arises from nuclear fusion in its core, during which hydrogen nuclei bond to become helium nuclei.
- The Sun's outer convective zone consists of plasma, made up of electrically charged protons, helium nuclei, and free electrons. The plasma generates a strong magnetic field as it moves within the convective zone.
- Charged particles continually flow outward from the Sun at high speeds. This flow is called the solar wind.
- The Sun's magnetic field lines sometimes arc into space from the Sun's surface. Where they do, sunspots and solar prominences occur.
- Large bursts of energy from the Sun produce solar flares and coronal mass ejections, which send energetic particles into space.
- The Earth's magnetic field protects our planet against charged particles arriving from the Sun, while its atmosphere provides a natural protective shield against solar ultraviolet radiation (UVA, UVB, and UVC).
- Oxygen molecules in the stratosphere (and mesosphere) protect the Earth from UVC radiation, while stratospheric ozone absorbs most UVB and some UVA radiation, shielding the Earth. UVC radiation that does reach the Earth's surface can cause sunburn and skin cancer.
- Specific chemicals such as chlorofluorocarbons (CFCs) react with and destroy ozone molecules in the stratosphere. CFCs were phased out under the Montreal Protocol to protect the stratospheric ozone layer.
- The solar wind warps the Earth's magnetic field into the shape of a huge teardrop surrounding the Earth.
- Currents of electrons, caused by solar wind, interact with gas atoms in the ionosphere above the poles to produce aurorae.
- Space weather refers to the concentration and speed of charged solar particles in the region of space surrounding the Earth. Intense space weather can hobble satellites, communications networks, electrical grids, computer systems, cell phones, and GPS navigation.
- Because of the threats posed by space weather, the NWS operates a Space Weather Prediction Center so that action can be taken to prevent damage to vulnerable equipment following solar storms.
- The Solar System contains countless objects in addition to planets and moons. Objects larger than 100 m in diameter are classified as asteroids or comets. Smaller objects are called meteoroids.
- The orbits of some objects have the potential to cross the Earth's orbit and collide with our planet. Collision with the largest of these objects would be catastrophic.
- Aten and Apollo asteroids are two groups whose paths cross the Earth's orbit. Some of these represent a threat to our planet.
- Most comets come from the Kuiper Belt or the Oort Cloud. Some take as long as thousands of years to orbit the Sun.
- The bright streak of light a meteoroid produces when it vaporizes in the Earth's atmosphere is called a meteor, and remnants of the object on the ground are called meteorites.
- When the Earth passes through a remnant comet tail, meteor showers are visible in the night sky.
- Meteoroids entering the Earth's atmosphere travel at cosmic speeds ranging from 11 km/s (25,000 mph) to 72 km/s (160,000 mph).
- Larger meteoroids can produce a very bright fireball that leaves a visible smoke-like trail as they traverse the sky. If it explodes, a fireball becomes a bolide.
- When a large object strikes the Earth, the Moon, or another body, it produces a crater. Craters can be simple or complex, depending on the size and speed of the object.
- Astroblemes are ancient impact structures on the Earth. They can be identified by rock structures uniquely formed by high impacts.
- There are 190 known impact sites on the Earth. Some are craters, but most are astroblemes. Erosion and geologic processes have eliminated most other impact sites.
- Exceptionally large impacts, like the impact that killed the dinosaurs, have the potential to produce, tsunamis, fires, global climate change, and mass extinctions.
- NASA monitors near-Earth objects, tracking those that have the potential to cross the Earth's orbit and collide with the planet in the foreseeable future. Efforts are underway to develop techniques to deflect objects whose orbital path may threaten the planet in the future.

Review Questions

Blue letters correspond to the chapter's learning objectives.

1. What is the difference among a solar prominence, a solar flare, and a coronal mass ejection? Which is most likely to interfere with radio communications, and which is most likely to cause a power outage? **(A)**
2. When a coronal mass ejection occurs, which arrives at the Earth faster, electromagnetic radiation or particulate radiation? **(A)**
3. How does the Sun's magnetic field differ from that of the Earth? Why are the Sun's near-surface magnetic field lines nearly parallel to the Sun's equator? Why is the Earth's magnetic field shaped like a teardrop? On the figure below, point to the magnetosphere, and to the Van Allen belts. How does the magnetic field protect the Earth from solar particles? **(A)**

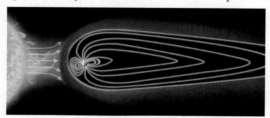

4. What roles do the atmosphere and the magnetosphere play in allowing living organisms to survive on land on the Earth? **(B)**
5. What is the difference among UVA, UVB, and UVC radiation, and which is completely removed by the atmosphere before reaching the Earth's surface? **(B)**
6. What type of radiation is responsible for sunburns? **(B)**
7. What is the ozone hole? Where in the Earth's atmosphere does this "hole" occur? **(B)**
8. If the concentration of ozone in the stratosphere were suddenly reduced by half, what might happen to humans and other organisms? **(B)**
9. What is space weather, and how is it manifested? **(C)**
10. Contrast the consequences of a coronal mass ejection for the people during the Roman Empire with the consequences of a similar event today for people in New York City. **(C)**
11. What is the difference among a meteoroid, a comet, and an asteroid? Which does the photo show? What is the difference between an Apollo or Aten asteroid and other types of asteroids? **(D)**

12. Very few meteors result in meteorites. Why? Describe the different kinds of meteorites, and indicate their source. **(D)**
13. Why aren't all space objects traveling at the same speed when they strike the Earth? Why does impact speed play a major role in determining damage? **(D)**
14. How does a meteor form, and why do most occur in meteor showers? How does a meteor differ from a fireball or a bolide? Which type of object caused the Tunguska event? **(E)**
15. Contrast a simple crater with a complex crater. Which does the image below show? What is an astrobleme? Why are so few craters and astroblemes found on the Earth? **(F)**

16. What is the relationship between the size of an impacting object and the recurrence interval at which such objects strike the Earth? About what size of object could cause an extinction-level impact? **(F)**
17. What is the evidence that the K-Pg boundary was caused by an asteroid impact? Describe the phenomena that might have happened when the object that formed the Chicxulub crater struck the Earth, and how they could have contributed to the extinction of the dinosaurs and other species. **(F)**
18. Explain how the impact of an asteroid 15 km (9 mi) wide obliterated the dinosaurs and led to the extinction of many other species on the Earth. **(F)**
19. What two strategies are being developed to divert an asteroid on a collision course with the Earth? **(G)**

On Further Thought

20. Why was life restricted to the ocean until significant quantities of oxygen began to accumulate in the atmosphere? **(B)**
21. Relate the threats from space discussed in this chapter to local threats astronauts face when in orbit. **(C, F)**
22. Contrast the potential catastrophic outcomes of an asteroid impact in the middle of the Pacific Ocean with those of a similar impact in the center of Asia. **(E)**

EPILOGUE

Our journey to understand the physical hazards that we face—on land, on the sea, in the air, and from space—has come to a close. We hope that this book has helped you to appreciate that the internal and external natural forces affecting the Earth yield not only the beautiful features of our planet, but also hazardous ones. While we can't eliminate all hazards, understanding them can decrease the risk that they will become disasters, affecting people. Hopefully, this book will provide you with such an understanding, as well as with an increased awareness of how to decrease risk for yourself and for your community—knowledge that is especially important in an era of climate change. All people should be able to focus on appreciating the beauty of our home planet and helping to keep it safe for future generations.

A view of the Grand Canyon reveals the consequences of several natural hazards, but also emphasizes the power of the Earth to uplift vast volumes of rock, and to carve them over time.
And . . . it's beautiful to look at!

APPENDIX: MATTER AND ENERGY

A.1 Introduction

Many readers of this book will have learned the basic terminology for describing matter and energy in a high-school science course. But for those who haven't, and for those who would benefit from a quick refresher, this appendix reviews the relevant concepts from physics and chemistry that are used throughout this book.

A.2 The Nature of Matter

Matter, the material substance of the Universe, consists of extremely tiny particles called **atoms**. Internally, an atom consists of even smaller *subatomic particles*: *protons*, with a positive charge; *electrons*, with a negative charge; and *neutrons*, with no charge (i.e., neutrons are neutral). In this context, *charge* refers to the electrical behavior of the particles. Simplistically, you can think of a positive charge as the plus end of a battery, and a negative charge as the minus end of a battery.

In detail, an atom consists of a nucleus containing tightly packed protons and neutrons, surrounded by an open cloud of much tinier electrons (**FIG. A.1**). Physicists refer to the "glue" holding protons and neutrons together in the nucleus as a **nuclear bond**. Each type of atom can be characterized by its **atomic number**, the number of protons in its nucleus, and by its *atomic mass*, roughly the sum of the number of protons and neutrons in the atom.

A material composed of only one type of atom is called an **element**—examples include hydrogen, carbon, oxygen, iron, sulfur, and uranium. Physicists recognize 92 naturally occurring elements. Note that an atom represents the smallest piece of an element that still has the properties of the element. Each element can be represented by a symbol—for example, we represent hydrogen by H, oxygen by O, carbon by C, and iron by Fe.

If the number of electrons in an atom matches the number of protons, then the atom has a neutral charge. If, however, an atom loses or gains electrons, it becomes an electrically charged **ion**. For example, an iron atom that has lost two electrons has an excess positive charge of 2, so it becomes an Fe^{+2} ion.

Two different versions of an element that differ not in terms of the number of electrons but rather in terms of the number of neutrons are called **isotopes** of the element. Different isotopes of an element have the same number of protons—they must, to be considered atoms of the same element—and therefore they have the same atomic number. But because they have different numbers of neutrons, they have different atomic masses.

Atoms can attach to one another by another type of "glue" called a **chemical bond**. (Keep in mind that a chemical bond holds whole atoms to one another, whereas a nuclear bond holds subatomic particles together in the nucleus of one atom.) Chemical bonds can involve the sharing or transferring of electrons in the atoms' electron clouds. We refer to a particle consisting of two or more atoms bonded together as a **molecule**. Some molecules contain atoms of only one element—for example, two hydrogen atoms can bond together to form a molecule of hydrogen, which we can represent in shorthand form (a *chemical formula*) as H_2. Similarly, two atoms of oxygen can bond to form a molecule of oxygen (O_2). Many types of molecules, however, contain atoms of two or more different elements. For example, water molecules contain

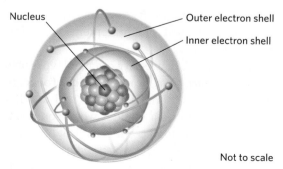

FIGURE A.1 Schematic image of an atom.

two hydrogen atoms and one oxygen atom (H_2O). Similarly, molecules of common salt have the formula NaCl, for they contain a sodium atom bonded to a chlorine atom. A material, such as water or salt, whose molecules contain more than one element is called a **compound**. Note that a molecule represents the smallest piece of a compound that still has the characteristics of the compound.

A.3 States of Matter

Scientists distinguish four **states of matter**, which differ from one another in the way they behave (**FIG. A.2**). In a **solid**, atoms or molecules remain fixed in position, so a solid retains its shape regardless of changes in the size of its container. In a **liquid**, atoms or molecules can move relative to one another, but remain in contact, so a liquid can flow and conform to the shape of its container. In a **gas**, atoms and molecules move around freely, so a gas can not only flow, but can also expand or contract when the size of its container changes. In other words, a gas will spread out and completely fill whatever container it occupies. A **plasma** resembles a gas in that atoms can move freely, but in a plasma, atoms have been stripped of electrons, so all the particles are ions. Plasmas form only at very high temperatures.

When we say that one object contains more *mass* than another, we mean that it contains more matter. For example, by saying that the Earth's mass is 81 times that of the Moon, we imply that there's lots more matter in the Earth than in the Moon. From your everyday experience, you know that some materials feel heavier than others—for example, it takes more strength to lift a block of iron than it does to lift a block of wood of the same size. That's because the two materials have different densities. **Density** is the amount of mass within a given volume. Put another way, density equals mass (a measure of the amount of matter) per unit volume (i.e., density = mass ÷ volume). A *vacuum* is a volume that contains hardly any mass at all, and therefore has a very low density.

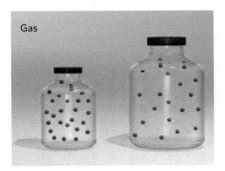

FIGURE A.2 Three of the four states of matter.

A.4 Force and Energy

Isaac Newton defined a **force** as simply a push or pull that causes the velocity (speed or direction) of an object to change. In your everyday experience, you constantly see or feel the effects of forces. Forces can speed objects up or slow them down or make them change direction. Forces can also tear, stretch, squash, spin, and twist objects and can make them float or sink. Any **system** (a defined volume of space and everything in it) contains **energy**, meaning that it has an inherent ability or capacity to do work. In this context, **work** refers to the product of a force and the distance over which the force is acting (i.e., $W = F \times D$). You do work when you lift this book 10 cm, and you do more work when you lift it 20 cm. When you do work, energy transfers from one part of a system to another, but the total amount of energy in the system remains the same, because energy cannot be created or destroyed. This statement is called the *law of conservation of energy*.

Scientists distinguish between two general types of forces. The first type, a *contact force*, or **mechanical force**, results when one mass moves and comes in contact with another. You apply a mechanical force to a boulder when you push on it (**FIG. A.3a**), and the wind applies a mechanical force to a sail when it blows. The second type, a *noncontact force*, or **field force**, applies across a distance; gravity and magnetism serve as examples.

FIGURE A.3 Examples of the forces of nature in everyday life.

(a) A person applies a mechanical force to push a boulder.

(b) Gravity, a field force, pulls a person down a zip line.

(c) A magnet produces a field force sufficient to hold onto these clips.

Gravity is the force of attraction between two objects—it is what holds you to the surface of the Earth and pulls objects from higher elevations to lower ones (**FIG. A.3b**). The strength of gravity depends on the masses of the two objects and on the distance between them. For example, you feel a much stronger gravitational pull to the huge Earth than you do to a small baseball. *Weight* is the force that an object exerts due to gravitational pull, so the stronger the gravitational pull, the greater the weight of an object. For example, on the Moon, you weigh much less than on the Earth because the Moon contains less mass, so it exerts less gravitational pull. The region in which on object feels a significant gravitational attraction is called a **gravitational field**. Note that in a gravitational field, a less dense material experiences an upward **buoyancy force** when placed in a denser material. For example, a cork has a density of 0.2 g/cm³, so if you place it in water, which has a density of 1.0 g/cm³, a buoyancy force causes the cork to rise.

Magnetism, simplistically, is the field force generated by electricity flowing in a wire, or by special materials called magnets. Unlike gravity, a magnetic force can be attractive (pulling objects together) or repulsive (pushing them apart). Over short distances, the magnetic force of even a small magnet can be larger than the gravitational force produced by the Earth, so you can overcome gravity by lifting objects with a magnet (**FIG. A.3c**). The region around a magnet in which the magnetic force has a significant effect is called a **magnetic field**. Magnetic fields are *dipolar*, meaning that they have a north end and a south end. A magnetic field can be represented by magnetic field lines that curve through space, from one pole of the magnet to the other. Like poles repel, unlike poles attract.

Scientists also distinguish between two general types of energy. **Kinetic energy** is the energy that an object has when it's moving. For example, a boulder bounding down a hill has kinetic energy. **Potential energy**, in contrast, is the energy held in an object. For example, potential energy exists when a field force acts on the object. A boulder sitting at the top of a hill stores potential energy because it is being pulled on by the Earth's gravity, but isn't moving. Energy of one type can be transformed into energy of another. When the boulder starts rolling, its potential energy transforms into kinetic energy.

Physicists distinguish among **thermal energy**, which is the energy resulting from the vibration or movement of atoms or molecules in a material; **electromagnetic energy**, such as light and X-rays, that moves from one location to another in the form of invisible waves; **chemical energy**, which is the energy stored in chemical bonds in a material; and **nuclear energy**, which is the energy stored in nuclear bonds. These forms of energy can be converted into other forms of energy. For example, electromagnetic energy can be converted into thermal energy when it is absorbed by a material and makes the material hot. And it can be converted into kinetic energy when, for example, electricity from a battery moves a car. Chemical energy can be converted into heat and light, as happens when you strike a match and it bursts into flame.

Potential energy can be stored in matter, in the form of the bonds that hold matter together, and this energy can be released when such bonds break or form. Chemical bonds, as we've seen, hold the atoms of a molecule together. Chemical bonds form and break during a **chemical reaction**, in which some molecules break apart and others form. Some chemical reactions release heat and light.

Potential energy can be stored in the form of the bonds, which hold together subatomic particles in the nucleus of

FIGURE A.4 Nuclear reactions.

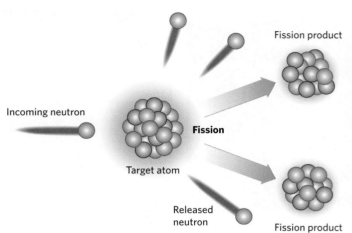

(a) Fission takes place when an atom is split by an incoming particle, producing several smaller fragments of the original atom.

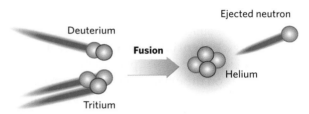

(b) Fusion happens when colliding particles combine to form a new, larger nucleus. (Deuterium and tritium are isotopes of hydrogen nuclei.)

an atom. During a **nuclear reaction**, nuclear bonds break or form, and a small amount of matter converts into energy according to Einstein's famous equation, $E = mc^2$. (In this equation, E stands for energy, m stands for mass, and c stands for the speed of light.) Nuclear reactions during which nuclei break apart are called **fission reactions**, whereas those during which the nuclei of two atoms fuse together are called **fusion reactions** (FIG. A.4). Fission reactions generate energy in nuclear power plants and in the explosion of an atomic bomb, whereas fusion reactions power the Sun and take place during the explosion of a hydrogen bomb.

A.5 Heat, Temperature, and Heat Transfer

In everyday English, the words *heat* and *temperature* may seem interchangeable, but in scientific discussion, they have different, distinct definitions. Scientists refer to the total kinetic energy contained in matter due to the movement of its particles as *thermal energy*, or **heat**. A jar of gas that has been warmed in an oven contains more heat than a jar of the same size that has been sitting in a refrigerator. **Temperature**, in contrast, is a measure of the average velocity of particle movements. These movements can include the vibration of atoms or molecules in place or the physical movement of atoms or molecules from one location to another. The faster the particles are moving, the higher the temperature. Therefore, if you take a jar containing helium gas and add heat to the gas by putting the jar in the oven, the temperature of the gas rises, because the gas particles start to move faster. We use a variety of scales—the *Celsius scale* (also called centigrade scale) in the metric system and the *Fahrenheit scale* in the English system—to represent temperature. Water freezes at 0°C, or 32°F, and boils at 100°C, or 212°F. Note that a single degree on the centigrade scale represents a larger temperature change than does a single degree on the Fahrenheit scale.

The coldest a material can be is the temperature at which its atoms or molecules stand still. We call this temperature *absolute zero*, or 0 K, where K stands for Kelvins (after Lord Kelvin, 1824–1907, a British physicist), another unit of temperature. Degrees in the Kelvin scale represent the same increment of temperature change as degrees in the Celsius scale. You simply can't get colder than absolute zero, meaning that you can't extract any thermal energy from a material at 0 K (−273.15°C).

Heat can be transferred from one object to another. It can be measured in **calories**, defined such that 1,000 calories can heat 1 kg of water by 1°C. Heat can be transferred from one place or material to another. When heat is added to a material, the material warms, in that its molecules start to vibrate or move more rapidly. When a material cools, the motion of its molecules slows. There are four ways in which **heat transfer** (movement of heat from one material or location to another) takes place in the Earth System: radiation, conduction, convection, and advection.

Radiation is the process by which electromagnetic waves transmit heat into a body or out of a body (FIG. A.5a). For example, when sunlight heats the ground during the day, radiative heating takes place. Similarly, when heat is radiated from the ground at night, radiative heating also occurs, but this time in the base of the atmosphere.

Conduction takes place when you hold the end of an iron bar in a fire (FIG. A.5b). The iron atoms at the fire-licked end of the bar start to vibrate more energetically; they gradually incite atoms farther up the bar to start jiggling, and these atoms in turn set atoms even farther along in motion. In this way, heat slowly flows along the bar until you feel it with your hand. Conduction does not involve actual movement of atoms from one place to another.

Convection happens when you set a pot of water on a stove (FIG. A.5c). The heat from the stove warms the water at

FIGURE A.5 The four processes of heat transfer.

(a) Radiation from sunlight warms the Earth.

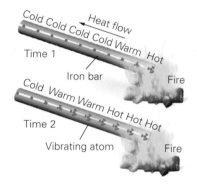

(b) Conduction occurs when you heat the end of an iron bar in a flame. Heat flows from the hot region toward the cold region as vibrating atoms cause their neighbors to vibrate.

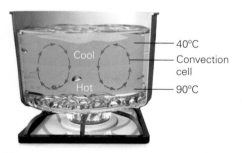

(c) Convection takes place when moving fluid carries heat with it. Hot fluid rises while cool fluid sinks, setting up a convective cell.

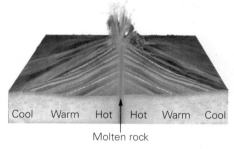

(d) During advection, a hot fluid (such as molten rock) rises into cooler material, and heat then conducts from the hot liquid into the cooler material.

the base of the pot by making the molecules of water vibrate faster and move around more. As a consequence, the density of the water at the base of the pot decreases, for as you heat a liquid, the atoms move away from one another and the liquid expands. For a time, cold water remains at the top of the pot, but eventually the warm, less dense water becomes buoyant relative to the cold, denser water. In a gravitational field, a buoyant material rises if the material above it is weak enough to flow out of the way. Since liquid water can flow easily, hot water rises. When this happens, cold water sinks to take its place. The new volume of cold water at the base of the pot then heats up and rises. Thus, during convection, the actual flow of the material itself carries heat. The trajectory of flow defines **convective cells**.

Advection is a less familiar process. In geology, the term has been used in reference to situations where heat carried by a moving fluid warms the material that surrounds it. For example, when molten rock rises into cooler crust, it heats up the surrounding crust (**FIG. A.5d**). Similarly, when hot groundwater passes into cooler rock or sediment, it warms the rock or sediment.

GLOSSARY

A

'a'ā A basalt lava flow with a rubbly surface.

ablation The transformation of a solid directly into a gas.

abyssal plain A broad, relatively flat region of the seafloor that lies 4–5 km (2.5–3 mi) below sea level.

acceptable risk A risk that an individual, business, or government is willing to accept without seeking protection or insurance, or making alternate plans.

accretionary prism A wedge-shaped mass of sediment and rock scraped off the top of the downgoing plate and accreted onto the overriding plate at a convergent boundary.

active fault A fault that has slipped in relatively recent geologic time and will probably slip again in the future.

active margin A continental margin that is also a plate boundary and, therefore, hosts seismicity.

active volcano A volcano that has erupted within the past few centuries and will likely erupt again.

adiabatic expansion An expansion of a volume of air during which no mass or energy is exchanged with the environment surrounding the volume.

advection The transfer of heat by a fluid flowing through another material; in rocks and sediments, the fluid (hot groundwater or magma) flows through cracks and pores.

aerosols Microscopic solid particles or liquid droplets small enough to remain suspended in the atmosphere.

aftershock One of a series of smaller earthquakes that follows a major earthquake.

agricultural drought A drought in which soil moisture drops below a value that permits crop germination or growth and remains below that value over an extended period.

air The mixture of gases that make up the Earth's atmosphere.

air blast A sudden pulse of extreme wind, resembling the blast produced by a bomb; also called an *air burst*.

air mass A large volume of air within which temperature and humidity are relatively uniform.

air pollution Gases and aerosols in the air that endanger human health or the environment.

air temperature A measure of air warmth, made by a thermometer, that represents the average speed at which air molecules move.

albedo The reflectivity of a surface.

alluvium Sorted sediment deposited by a stream.

amplification The process by which a phenomenon becomes more intense. If the phenomenon involves transfer of energy by waves, amplification causes wavelength to decrease and wave amplitude to increase.

amplitude Half of the vertical distance between the trough and crest of a wave (i.e., half the wave height).

angle of repose The angle of the steepest slope that a pile of uncemented granular material can attain without collapsing from the pull of gravity.

annual exceedance probability (AEP) The likelihood that an event of a given size or larger will take place in a given year.

annual probability (AP) The likelihood that an event, such as a flood or an earthquake, of a given size will happen in a given year, expressed by a percentage equal to 1 divided by the recurrence interval.

Anthropocene A name that has been proposed for the last few centuries or millennia, during which human activity has had a major influence on the environment, landscapes, and climate.

anthropogenic hazard A hazard produced by the activities of people.

anvil cloud The upper part of a large cumulonimbus cloud that spreads laterally at the tropopause to form a broad, flat top.

apparent temperature The temperature that a person feels, which depends on four factors: the actual air temperature, the relative humidity, the wind speed, and the amount of shade.

aquifer Sediment or rock, with high porosity and permeability, that can hold a significant amount of groundwater and can transmit water easily.

aquitard Sediment or rock that does not transmit groundwater easily, and therefore retards the motion of groundwater.

areal flooding An event during which the land surface becomes submerged due to heavy rain or rapid snowmelt without input from a nearby stream; it occurs on poorly drained, low-lying land.

arrival time The time at which a specific seismic wave, such as a P-wave or an S-wave, arrives at a given seismograph.

arsenic contamination The presence of trace amounts of arsenic, a toxic chemical; groundwater may contain dangerous amounts of arsenic due to the natural presence in sediment of minerals that contain this element.

artificial levee A long ridge of gravel, sand, compacted clay, or concrete built along a stream to confine flow to its main channel.

ash (1) Very fine particles of glass or pulverized rock erupted by a volcano. (2) Flaky particles of carbon and other materials produced by incomplete burning.

asteroid A rocky or metallic object, larger than about 100 m (330 ft) across that orbits the Sun; it may be a planetesimal or a fragment of a planetesimal or protoplanet.

asteroid belt The zone between Mars and Jupiter where most asteroids reside.

asthenosphere The layer of the mantle that lies directly below the lithosphere; it is relatively soft and can flow very slowly.

astrobleme An ancient, eroded impact structure that can be recognized by evidence of damage to the Earth's crust that extended so deep that it has not eroded away.

atmosphere A layer of gases that surrounds a planet.

atmospheric instability Conditions in the atmosphere that allow lifted air to become less dense than the air that surrounds it, so that it can continue to rise buoyantly, once it undergoes initial lifting.

atmospheric pressure The force applied by air on a surface of a specified area; equivalent to the weight of a column of air above that area; the overall push that still air applies to its surroundings.

atmospheric river A relatively narrow band of moist air flowing along an extended cold front on the south side of a mid-latitude cyclone; because of the moisture it carries, areas beneath an atmospheric river can receive heavy rainfall.

atom An extremely tiny particle consisting of even smaller subatomic particles (protons, neutrons, and electrons); the smallest piece of an element that has the characteristics of the element.

atomic number The number of protons in the nucleus of an atom; each element has a unique atomic number.

aurora An undulating curtain-like band of varicolored light that appears across the night sky when electrons interact with ions in the ionosphere.

avalanche chute A path commonly followed by avalanches due to the shape of the land surface; it may be identified by the absence of trees.

B

backfire A controlled fire ignited by firefighters to consume fuel in a wildfire's path and stop its progress.

backwash The gravity-driven flow of water back down a beach slope, after swash has finished carrying the water up the slope.

barrier island An offshore sand bar that rises above the mean high-water level, forming an island.

base isolation A component of earthquake-resistant building design; it decouples the motion of the ground from that of the superstructure (the portion of the building above the basement), so that the building sways less during an earthquake.

base level The lowest elevation that a stream's surface can reach at a given locality.

bathymetry Variation in depth, comparable to topography on land.

beach A low-relief band of sediment, along the shore, that undergoes sorting and shifting by the swash and backwash of waves.

beach erosion A reduction in beach width and thickness, caused by a decrease in the sediment supply or an increase in the sediment removal rate.

beach nourishment The process of adding sediment, generally pumped or dredged from offshore, to replace beach sediment that has been eroded away.

beach profile The variation in elevation of a beach as measured along a line perpendicular to the shoreline.

beach stabilization The installation of temporary, permanent, or living protective materials to prevent beach erosion.

bed A layer of sediment or sedimentary rock.

bedrock Rock still attached to the Earth's crust.

biogeochemical cycle The exchange of chemicals between living and nonliving reservoirs in the Earth System.

biosphere The region of the Earth System inhabited by life; it stretches from a few kilometers below the Earth's surface to a few kilometers above it.

black smoker A hydrothermal vent along a mid-ocean ridge that spews hot, mineral-rich water, whose dissolved sulfide components instantly precipitate when the water mixes with seawater and cools to produce a dark cloud of tiny mineral particles.

blizzard A snowstorm in which winds exceed 56 km/h (35 mph) and visibility remains at less than a quarter mile (400 m) for at least 3 hours.

block A large, angular chunk of pyroclastic debris.

body wave A seismic wave that passes through the interior of the Earth.

bolide An exploding meteoroid in the atmosphere.

boundary layer The layer of the atmosphere adjacent to the Earth's surface, in which friction significantly affects air movement.

braided stream A stream in which the water slows and deposits coarse sediment, which forms numerous elongate gravel and sand bars, choking the channel and forcing the stream to divide into strands that weave back and forth among the bars.

breaker A water wave in which water at the top of the wave curves over its base.

breakwater An offshore wall, oriented parallel to a beach, that prevents the full force of waves from reaching the beach.

brittle deformation The cracking and fracturing of a material subjected to stress.

buoyancy force A force that causes a less dense material within a denser material to rise.

C

calcite A mineral composed of calcium carbonate ($CaCO_3$); it is the major mineral in limestone and marble and is a carbon reservoir in the Earth System.

caldera A large circular or elliptical depression with steep walls and a fairly flat floor, formed after a volcanic eruption as the summit of the volcano collapses into the drained magma chamber below and/or gets blasted away during an explosive eruption; large calderas form after large volcanic explosions.

calorie The energy required to raise the temperature of 1 g of water by 1°C.

capacity In the context of streams, the total quantity of sediment a stream can carry; the capacity of a stream depends on both its competence and its discharge.

cap-and-trade A government policy that involves setting a cap on the amount of carbon that a company or other entity can emit, and then letting companies buy and sell carbon credits, as a way to motivate the reduction of carbon emissions.

carbon budget The balance between additions and subtractions of carbon within various reservoirs of the Earth System.

carbon capture and sequestration The process of trapping CO_2 produced by large sources, such as power plants, liquefying it, and pumping it into reservoir rocks deep underground to keep it from entering the atmosphere.

carbon cycle The biogeochemical cycle during which various forms of carbon move among the Earth System's carbon reservoirs.

carbon footprint The total amount of greenhouse gases produced by a given process or activity.

carbon-neutral Having no net carbon footprint.

Carrington event The most intense space weather ever recorded; it was due to a huge coronal mass ejection and caused a severe geomagnetic storm during September 1–2, 1859.

Cascadia The Pacific Northwest region of the United States and adjacent regions of Canada.

casualty A death or injury caused by an event.

catastrophe A disaster that affects a broad region, causing large losses of life and property, major economic losses, and long-term disruption of society.

cave network A network consisting of many caverns connected by narrow passages.

cavern A sizable underground open space that never receives direct sunlight.

cement In the context of sedimentary rock, mineral material that precipitates from water in the spaces between clasts, holding the clasts together.

charge separation In clouds, the movement of positively and negatively charged particles into different regions of a cloud by air currents; its development may lead to the occurrence of lightning.

chemical bond A force holding atoms together in a molecule; it generally involves sharing or transferring electrons.

chemical energy Potential energy stored in the chemical bonds of a material.

chemical reaction The process of forming or breaking chemical bonds when two or more materials come in contact; it leads to the breakup of some molecules and the formation of new ones, and may release or absorb energy.

Chicxulub crater A circular depression buried beneath younger sediment on the Yucatán Peninsula; geologists suggest that it formed due to impact of a meteorite there at 66 Ma.

chronic health disaster A type of disaster caused by widespread, continuous use of materials, such as contaminated water, that are dangerous to human health.

cinder cone A small volcanic edifice consisting of a cone-shaped pile of tephra.

cinders Lapilli formed from solidified clots ejected by a lava fountain.

clastic rock Rock consisting of grains or fragments, derived from the weathering of pre-existing rock, that have been packed together and cemented to one another.

clay A weak material composed mostly of extremely tiny platy crystals formed when minerals react with water and air.

climate The average, range, daily variation, and seasonality of weather conditions in a given region over the course of decades to millennia.

climate change A long-term shift in the average of one or more climate conditions.

climate feedback A phenomenon, triggered by a change in climate, that further modifies climate change.

climate type A region with a particular range of temperatures, weather seasonality, and precipitation amounts; a climate zone hosts characteristic natural vegetation.

climate zone See *climate zone*.

cloud A visible mass of condensed water vapor (consisting of tiny water droplets, ice crystals, or both) floating in the atmosphere, typically above the ground.

cloud cover The proportion of the sky that is covered by clouds.

cloud-to-cloud lightning Lightning extending between clouds, or from one part of a cloud to another, that does not reach the ground.

cloud-to-ground lightning A lightning stroke that extends from a cloud to the ground.

coast The land area, up to several kilometers wide, in which the landscape and local climate are influenced by proximity to the ocean.

cohesion A component of a material's shear strength, or of a potential sliding surface's shear strength, caused by the attraction of grains to each other and by other factors; it helps prevent materials from falling apart, or from detaching and sliding.

cold front A boundary between a cold air mass and a warm air mass where the cold air advances forward, lifting the warmer air mass.

cold wave A relatively rapid, prolonged fall in temperature that requires people to take precautions against cold.

collapse sinkhole A sinkhole that forms when weak rock in the roof of a cavern breaks off and falls into the cavern; eventually, the roof becomes too thin and weak to support itself, and it falls downward suddenly.

collision In a geologic context, the process by which two relatively buoyant pieces of crust come in contact and push toward each other after intervening oceanic lithosphere has been completely subducted.

comet An object, composed of ice and dust, that orbits the Sun in a highly elliptical orbit; it produces tails of gas and dust as it nears the Sun.

competence In reference to a stream, the maximum particle size of sediment that a stream can carry.

complex crater A large impact crater characterized by a central uplift, a flat floor, and terraces ringing the crater.

complex landslide A mass-wasting event that involves more than one type of movement.

compound A material whose molecules contain more than one element.

conduction The transfer of heat in a material that occurs when vibrating atoms in a warmer part of the material incite neighboring atoms in the cooler part of the material to vibrate; this causes heat to flow slowly from warmer to cooler material without the atoms themselves moving from one place to another.

constructive interference The overlapping of two waves moving at different speeds, or from different directions, in such a way that the wave crests overlap and form a single crest higher than that of either wave.

containment In the context of wildfire, the prevention of a wildfire's spread beyond a certain defined limit.

continent A large block of land 2,000–11,000 km (1,240–6,800 mi) across.

continental arc A volcanic arc formed where an oceanic plate subducts beneath a continent.

continental drift The idea that continents slowly move relative to each other across the Earth's surface.

continental shelf A broad region of shallow sea bordering a continental margin; the widest continental shelves occur over passive-margin basins.

convection The transfer of heat that occurs when a material is warmed, expands, becomes less dense than the cooler material around it, and therefore rises, while the cooler material sinks to take its place.

convective cell A circulation pattern caused by convection.

convergence In the context of air flow, a net inflow of air molecules into a region of the atmosphere; convergence in the jet stream leads to an increase in surface air pressure; convergence at the surface, as in locations where wind speeds slow or winds from different directions collide, causes air to rise.

convergent boundary A plate boundary, delineated by a deep-sea trench, where one plate (the downgoing plate) sinks beneath another (the overriding plate).

coral bleaching Loss of color in a coral reef, resulting from the expulsion of the corals' symbiotic algae.

coral reef A realm of shallow water underlain by a mound of coral and coral debris and associated organisms.

core The innermost of the Earth's concentric layers.

Coriolis effect The deflection of a moving object or material relative to the surface of a spinning object beneath it; in the northern hemisphere of the Earth, moving objects deflect to the right, whereas in the southern hemisphere, they deflect to the left.

Coriolis force The apparent force that causes the Coriolis effect.

coronal mass ejection (CME) A type of intense solar storm during which the Sun hurls plasma from its corona into space.

cosmic speed The speed at which an object moves through space relative to the Earth.

crater (1) A bowl-shaped depression at the peak of a volcano. (2) A bowl-shaped depression, surrounded by a raised rim, formed on a planetary surface by the impact of a meteorite.

creep In the context of mass wasting, the slow, gradual downslope movement of regolith; see also *fault creep*.

critical zone The surface and near-surface realm of the Earth System that contains the resources and maintains the conditions that sustain life.

crown fire A wildfire that burns within the forest canopy; it can send flames high above the tops of trees; also known as a canopy fire.

crust The outermost layer of the Earth, extending down to the Moho.

cryosphere The frozen component of the hydrosphere in the Earth System; it includes glaciers and permafrost.

crystal A single, continuous piece of a mineral inside which atoms, molecules, or ions are fixed in an orderly arrangement.

crystalline rock A rock with a texture characterized by interlocking crystals that grew together.

crystalline texture A rock texture characterized by the presence of interlocking crystals; the coherence of the rocks comes from the jigsaw-puzzle-like intergrowth of crystals with their neighbors.

current A band of flowing water that moves distinctly faster than adjacent water.

cyclone (1) A large system of winds rotating around an area of low atmospheric pressure; it circulates counterclockwise in the northern hemisphere, and clockwise in the southern hemisphere. (2) A strong tropical cyclone over the Indian Ocean or Australia, in which sustained wind speeds reach 119 km/h (74 mph) or higher; see also *tropical cyclone*, *mid-latitude cyclone*.

D

DART buoy (for Deep-Ocean Assessment and Reporting of Tsunamis) A floating beacon that receives an acoustic signal from a pressure sensor on the seafloor when the sensor detects the pressure pulse associated with a tsunami, and in turn transmits a radio alert to a receiver on land.

dead zone A region of the sea where nutrients brought in by a river cause algal blooms; the subsequent death and decomposition of the algae deplete the water of oxygen so that it cannot sustain oxygen-breathing marine organisms.

debris flow A downslope movement of mud mixed with larger rock fragments.

decompression melting Melting that occurs when pressure decreases but temperature remains unchanged; it occurs in the Earth where hot mantle rock rises to shallow depths.

deep-sea trench An elongate trough, marking a convergent boundary, where the ocean floor may reach depths of 7–11 km (4–7 mi).

deforestation The removal of trees over a broad area.

deformation A change in the shape, position, or orientation of a material by bending, breaking, or flowing.

delta An apron of sediment formed at the mouth of a stream where the stream enters standing water and its current slows, so that sediment settles out.

delta plain The low-lying, low-relief land surface of a large delta.

density The amount of mass within a given volume; density = mass ÷ volume.

deposition The process by which sediment settles out of a transporting medium.

derecho A severe straight-line windstorm that can extend over a large geographic region; a derecho can be produced in front of a squall line.

disaster prediction An estimate of the character, location, and timing of a disaster.

discharge In the context of streams, the volume of water that passes through a cross section of a stream in a given time (usually 1 second).

displaced person A person forced from their home by a disaster, whether they have left home but remain within the borders of their country (an *internally displaced person* or *evacuee*), or have traveled to another country (a *refugee*).

displacement In the context of geologic structures, the amount of movement or slip across a fault.

divergence In the context of air flow, a net outflow of air molecules from a region of the atmosphere, caused by increasing wind speed or changes in wind direction; divergence in the jet stream leads to a decrease in atmospheric pressure at the Earth's surface.

divergent boundary A plate boundary where two oceanic plates move apart by the process of seafloor spreading; also called a *spreading boundary*.

divide The higher land that separates a drainage network from a neighboring drainage network.

Doppler radar A radar system that detects precipitation and wind by measuring shifts in the frequency of the returned energy.

dormant volcano A volcano that has not erupted for hundreds to thousands of years, but may have the potential to erupt again in the future.

downdraft A downward-moving air flow.

downslope wind A strong flow of wind down the leeward side of a mountain range; North American examples include the chinook and Santa Ana winds.

drainage network An array of interconnecting tributary streams that all flow into the same trunk stream and carry water out of a watershed.

drawback In the context of waves, a drop in the sea surface due to the arrival of a wave trough, so that the water moves down the beach; drawback due to the arrival of a large tsunami may cause the water to drop far below the normal shoreline, exposing offshore areas that are normally submerged.

drawdown Sinking of the water table that occurs when the rate at which water is removed from the ground exceeds the rate at which groundwater is replenished.

drip torch A canister from which a burning liquid can be spilled onto the ground, used by firefighters to set backfires.

drought An extended period of little or no rain that causes a deficiency of freshwater.

dry adiabatic lapse rate The rate (~10°C/km) at which an unsaturated parcel of air will change temperature if it is displaced vertically in the atmosphere without exchanging mass or heat with its environment.

dry wash The channel of an ephemeral stream when empty of water; also called an *arroyo* or a *wadi*.

Dust Bowl A region of the southern Great Plains of the United States where a drought in the 1930s resulted in dust storms and extreme soil erosion.

dust storm An event during which strong winds strip fine-grained sediment from unvegetated soil and send it skyward to form rolling dark clouds that block out the Sun.

dynamic pressure The push caused by a moving fluid; it increases with the fluid's viscosity (or density) and the square of its velocity.

E

earthquake An episode of ground shaking caused by the sudden breaking of rock, or by frictional sliding along a pre-existing fault in the Earth.

earthquake engineering The development and implementation of designs, and the use of special materials, to construct buildings and facilities that can withstand ground shaking.

earthquake hazard map A map that portrays spatial variations in earthquake risk.

earthquake-resistant design A plan for buildings or infrastructure, intended to minimize damage in the event of an earthquake.

earthquake zoning Restrictions on the location and construction of buildings, based on a region's susceptibility to earthquake-related damage.

Earth System The geosphere, hydrosphere, atmosphere, and biosphere, along with the intricate ways in which these realms interact with one another over time.

easterly wave A north-arching map-view curve in the general east-west flow path of the trade winds near the Earth's surface; its presence causes convergence lifting and can produce a tropical depression.

economic impact In the context of disasters, the estimated cost of replacing or repairing buildings, property, and infrastructure damaged by a disaster, together with the value of lost economic activity caused by the disaster.

eddy An isolated swirl or vortex in a water current or in the wind's flow direction.

effusive eruption A volcanic eruption that produces mainly lava; it is more common when eruptions produce low-viscosity mafic lava.

ejecta In the context of volcanism, the debris blasted skyward during an eruption; in the context of meteorite impacts, the fragments, composed mostly of target rock, explosively shot outward during the excavation stage of crater formation.

elastic deformation A change in the shape of a solid that takes place, without brittle or plastic deformation, in response to an applied stress; it is accommodated by stretching and bending chemical bonds, and can be reversed by removal of the stress that caused the change.

elastic-rebound theory The idea that elastic deformation builds up in rock adjacent to a fault prior to an earthquake, and that the rock elastically rebounds when the fault slips, generating seismic waves.

electromagnetic energy Energy that moves in the form of waves through a vacuum or a substance from one location to another; examples include light, X-rays, and radio waves.

element A material composed of only one type of atom.

El Niño A flow of warm water eastward in the equatorial Pacific Ocean that reverses the upwelling of cold water along the western coast of South America and causes significant global changes in weather patterns.

embayment In the context of discussing coasts, a location where the shoreline curves inland.

energy In physical systems, the ability to do work, meaning it can produce a force that can move or change matter.

Enhanced Fujita (EF) scale A scale for classifying the intensity of a tornado based on the damage it causes, using correlations of observed damage with wind velocities.

ENSO (for El Niño/Southern Oscillation) The overall seesaw pattern of shifts in sea level and atmospheric pressure across the Pacific Ocean in association with the Walker circulation, together with its consequences.

environmental lapse rate The rate at which the environmental temperature changes with height in the atmosphere.

ephemeral stream A stream that flows during only part of a year.

epicenter The point on the Earth's surface located directly above an earthquake's focus.

erosion The grinding or breaking away and then removal of materials at the Earth's surface, due to moving water, air, or ice.

eruptive jet Material, composed of hot pyroclastic debris and some lava, driven skyward up to several hundred meters during an explosive volcanic eruption.

eruptive style The specific manner in which material comes out of a volcano in a particular volcanic eruption; geologists name eruptive styles (e.g., Hawaiian, Strombolian, Plinian).

estuary An area at a river's mouth where seawater and river water mix; large examples form where coastal valleys flood, due to either rising sea level or land subsidence.

eutrophication The transformation of a well-oxygenated body of water into a poorly oxygenated one by an influx of nutrients that causes an algal bloom.

excessive heat warning A warning issued by the National Weather Service when high heat index values are expected for 4 or more days, or when extremely high heat index values are expected for a single day.

expansive soil Soil that contains significant amounts of a clay mineral called smectite, which absorbs water; such soil expands when it gets wet and shrinks when it dries.

explosive eruption A violent volcanic eruption that produces clouds and avalanches of pyroclastic debris.

exposure In the context of disasters, the potential casualties, economic losses, and social disruption that a community faces due to its proximity to a natural hazard.

external energy Energy that travels to the Earth from the Sun, where it enters the Earth System.

extinction-level event A disaster that threatens the existence of a majority of living species on the Earth.

extinct volcano A volcano that was active in the distant past, but that will not erupt in the future because the geologic conditions that led to eruption no longer exist.

extrusive igneous rock Rock that forms when lava freezes after it extrudes onto the Earth's surface and comes into contact with the atmosphere or ocean.

eye The vertical, downward-tapering vertical cylinder of relatively calm, clear air in the center of a hurricane.

eye wall The violently rotating, funnel-shaped cylinder of clouds that surrounds the eye of a hurricane; it hosts the storm's strongest winds.

F

failure The breaking, flowing, or collapsing of a material; the term can apply to solid rock or to regolith.

failure surface The plane of weakness on which a mass of material moves downslope during mass wasting.

far-field tsunami A tsunami that crosses an ocean and reaches a shore far from its source; also called a *distant tsunami* or *teletsunami* or *orphan tsunami*.

fault A fracture on which one body of rock slides, or slips, past another.

fault creep Slip on a fault that takes place slowly, perhaps by plastic deformation, rather than in seismic pulses.

Federal Emergency Management Agency (FEMA) A US government agency that deploys staff and supplies to assist victims during and after a disaster.

fetch The distance across a body of water along which a wind blows.

field force A type of force that applies across a distance; also called a *noncontact force*.

fire The very rapid oxidation of flammable molecules, producing heat and light.

fireball A meteor produced by a meteoroid larger than a few centimeters in diameter; it is brighter than Venus and can leave a visible smoke-like trail across the sky.

firebrand A spark or ember, lofted by the updraft of a fire, which can land on and ignite fresh fuel beyond the fire's perimeter.

firebreak A natural or built barrier that prevents a wildfire from spreading.

fire retardant A chemical sprayed by firefighters near the perimeter of a wildfire to slow the fire's advance.

firestorm A huge fire whose updraft generates strong winds, which fan the flames and make the fire even larger.

fire triangle The three components—fuel, heat, and oxygen—that must be present for a fire to begin.

fission reaction A nuclear reaction during which nuclei break apart.

fissure A vertical, downward-tapering, open crack in the ground.

fissure eruption The ejection or flow of lava from an elongate crack, rather than from a circular vent.

flame The visible part of fire, consisting of glowing superheated gases and microscopic carbon particles.

flank collapse In the context of mass wasting, a sudden, catastrophic slump that removes the side of an oceanic island or a volcano.

flash flood A flood that occurs within 6 hours of unusually intense rainfall, a dam failure, or other cause; during such floods, water can rise so quickly that escape from the path of the water may be impossible.

flood An event during which water covers significant areas of land that are normally dry; a stream floods when its discharge becomes so great that water overtops the stream's banks and submerges areas outside its channel.

flood basalt A huge volume of low-viscosity mafic lava that erupts over a relatively short time and spreads in a layer over a large area.

flood crest The highest stage that water reaches during a flood.

flood-frequency curve A line on a graph that can be used to predict the recurrence interval (return period) of a flood of a given discharge for a specified stream; the horizontal axis of the graph indicates a flood's recurrence interval in years, and the vertical axis indicates discharge.

flood-hazard map A map depicting areas that will become submerged during a flood of a specified discharge.

floodplain The broad, flat land, bordering a stream, that becomes partially or entirely covered with water during a flood.

flood stage The stage at which a stream rises above its banks in many places and submerges significant areas outside the stream channel.

floodway A mapped region likely to be flooded, in which people avoid constructing buildings.

flux melting The transformation from hot solid to liquid that occurs when volatile material is added to the solid; it occurs where volatiles released from the downgoing slab enter the overlying asthenosphere.

focus The location where a fault slips during an earthquake; also called a *hypocenter*.

fog A cloud at ground level.

fold A geologic structure manifested as a curve in rock layers; formed by bending or by plastic deformation.

foliation Parallel surfaces or layers that develop in a rock as a result of flattening, shear, or pressure solution during deformation, generally under metamorphic conditions; schistosity is an example.

force A push, pull, or shear that causes the velocity (speed or direction) of an object to change.

forecast In the context of meteorology, a prediction of the future character of the weather at a locality; it may include a storm prediction.

foreshock One of a cluster of smaller earthquakes that precede a major earthquake.

fracture zone A narrow band of vertical fractures in the ocean floor, lying roughly at right angles to a mid-ocean ridge; the actively slipping part of a fracture zone is the transform fault that links two ridge segments.

fragmental texture The texture of igneous rock formed either when pyroclastic debris welds together while still extremely hot, or when cool pyroclastic debris accumulates and later undergoes compaction and cementation.

freezing rain Precipitation that reaches the ground in liquid form but freezes when it contacts a cold solid surface.

freshwater Generally, water that contains less than 0.05% salt.

freshwater depletion A reduction in available surface freshwater that develops when the amount of water leaving (by evaporation, infiltration, or diversion into pipes or canals) exceeds the amount being supplied (by rain over the watershed or by groundwater seepage).

friction Where solids are in contact, the resistance to sliding caused by bumps and irregularities on a solid surface that dig into an adjacent solid surface; where fluids flow over a solid, the resistance caused at the contact between the fluid and the solid surface; where two fluids are in contact, the resistance caused by interaction of water molecules.

front A boundary between air masses that differ in temperature and humidity.

frontal squall line A line of thunderstorms forming along a frontal boundary; such squall lines commonly develop along the cold front of a mid-latitude cyclone.

frostbite Actual freezing of body tissues subjected to freezing temperatures.

fuel A substance that burns or undergoes a reaction that produces heat.

fumarole A vent at an active volcano that emits volcanic gases and steam.

fusion reaction A nuclear reaction during which the nuclei of two atoms fuse together.

G

gas A state of matter in which atoms and molecules move around freely. A gas can expand or contract as the size of its container changes.

geologic structure A feature that has formed as a consequence of deformation in the lithosphere; examples include faults, folds, joints, and foliation.

geologic time The span of time since the formation of the Earth.

geologic time scale A scale that delineates intervals of geologic time and indicates the numerical ages of boundaries between intervals.

geologist A scientist who studies the Earth and processes that affect the Earth.

geosphere The solid Earth, from its surface to its center.

geostrophic balance The balance that exists in a flowing fluid when the pressure gradient force and the Coriolis force are equal and opposite.

geostrophic wind A wind that flows parallel to isobars when air in the atmosphere is in geostrophic balance (i.e., the Coriolis force balances the force caused by pressure gradients).

glacial torrent A catastrophic flood caused by the sudden release of water when an ice dam, or a moraine dam, fails.

glacier A sheet or stream of ice that flows slowly across the land surface and lasts year-round.

glass An inorganic solid in which atoms are not arranged in an orderly pattern.

glassy texture The texture of igneous rock formed from a melt that solidifies quickly, before the ions or molecules can organize into crystals.

global change Transformations or modifications of the Earth System over time; climate change is one aspect of global change.

global climate model (GCM) A numerical model consisting of mathematical equations used to simulate changes in global climate conditions; GCMs are generally run on supercomputers.

global cooling A decrease in average global atmospheric temperature.

global positioning system (GPS) A satellite system that allows people to locate their position on the Earth's surface; it can also be used to measure rates of movement of portions of the Earth's crust relative to one another.

global warming A rise in the average global atmospheric temperature.

global water crisis The stealth disaster developing in many regions of the world as a result of the scarcity of clean freshwater.

grain Generally, a mineral crystal, or a fragment of a mineral crystal or rock; grains can be within an intact rock or can be loose; coarse grains are larger and fine grains are smaller.

grassland destruction The conversion of grasslands into farm fields or suburbs.

gravitational field A region in which an object feels a gravitational attraction.

gravity A field force that one object exerts on another; its magnitude depends on the masses of the objects and the distance between them.

greenhouse effect The trapping of heat, reradiated from the Earth's surface in the form of infrared radiation, in the Earth's atmosphere by carbon dioxide, water vapor, methane, and other greenhouse gases.

greenhouse gas A gas that traps heat in the Earth's atmosphere by absorbing infrared radiation.

groin A concrete or stone wall that runs outward from a beach at a high angle in order to prevent erosion of beach sand.

ground blizzard Blizzard conditions that develop after a snowfall has stopped, because a strong surface wind lofts freshly fallen snow into the air.

ground fire A wildfire that burns roots and buried organic material and typically smolders for long periods.

groundwater Water that resides under the surface of the Earth, mostly in pores or cracks in rock or sediment.

groundwater depletion The permanent removal of groundwater by human activities, leading to a drop in the water table.

growth rings Concentric layers that develop in trees and other organisms whose growth rates vary with the seasons.

gust front The leading edge of the outflow, formed where a thunderstorm's rain-cooled downdraft reaches the ground and turns to move horizontally, whose passage is often marked by a sudden increase of wind speed.

gyre A large ocean surface current with a roughly circular or elliptical path in map view. Five major gyres occur worldwide—in the North Atlantic, South Atlantic, North Pacific, South Pacific, and Indian Oceans. Each gyre is bounded on three sides by the margins of the ocean basin in which it forms, and on the remaining side by the equator.

H

Hadley cell A low-latitude convection cell in the troposphere that extends from the equator to a latitude of about 25° in each hemisphere.

hail Precipitation in the form of spherical or irregularly shaped lumps of solid ice that fall from a thunderstorm.

hailstone A frozen precipitation particle whose diameter ranges from 3 mm to 20 cm, resulting from the growth of ice particles as they collect supercooled liquid droplets in the strong updrafts of a thunderstorm.

hailstorm A period during a thunderstorm when hail reaches the ground.

hard barrier A structure composed of piles of boulders, blocks of concrete, reinforced solid concrete, steel, or stone; it can protect a stretch of coast or a harbor from waves, but may also alter the natural sediment budget.

harmful algal bloom Overwhelming growth of algae that can create dead zones and red tides, which can disrupt fishing industries and tourism, and, in some cases, can sicken people.

hazard mitigation The development of approaches that can prevent or decrease the consequences of a hazardous event.

hazard potential map A map that shows where disasters have a high probability of taking place, based on historical data on hazardous events and assessments of preconditions that make a locality susceptible to a given hazard.

hazard preparedness The development of protocols and procedures to prepare a community for hazardous events; also called *disaster-risk reduction*.

haze Microscopic droplets in the atmosphere, formed under high-humidity conditions, whose presence causes an overall decrease in visibility; the droplets form when aerosols capture water molecules and then may dissolve in the water.

headland Coastal land that protrudes into the sea.

head scarp A cliff or step in the land surface formed by the exposed upslope edge of the failure surface of a slump.

heat Thermal energy that can be transferred from warm materials to cooler materials.

heat advisory A notification issued by the National Weather Service when it predicts that a regionally defined threshold value of the heat index will be reached and will remain for 2 or more consecutive days.

heat flow The movement of heat from one location to another, such as from the Earth's interior to its surface.

heat index The relationship among human comfort, air temperature, and relative humidity, used to convey the threat of heat illness.

heat transfer Movement of heat from one material or location to another.

heat-transfer melting Melting that results from the transfer of heat from a hotter magma to a cooler rock with a lower melting temperature.

heat wave A prolonged period of abnormally high air temperatures.

high-pressure system The circulation surrounding a center of high pressure, generally associated with sinking air and clear skies.

hockey-stick diagram An informal name for the graph of average global temperature over the past thousand years, which demonstrates that temperatures rose sharply at about 1880; also called *hockey-stick graph*.

hot spot A location in a plate interior where volcanoes form, probably where a mantle plume causes melting just below the base of the lithosphere.

hot-spot track A chain of extinct volcanoes transported off a hot spot by the movement of a lithosphere plate.

hot spring A source where very warm to boiling groundwater spills out of the ground.

hot zone A region of the Earth where average annual temperatures exceed 29°C (84°F), making the area potentially too hot to sustain large human populations.

hummocky An adjective used to describe a bumpy, irregular land surface with many small depressions and hills.

hurricane A strong tropical cyclone over the North Atlantic or eastern Pacific Ocean, in which sustained wind speeds reach or exceed 119 km/h (74 mph).

hurricane season The time of year when ocean water becomes warm enough to provide energy to fuel tropical storms whose winds can accelerate to hurricane speed.

hurricane track The path a hurricane follows, as portrayed on a map.

hurricane warning A warning issued by the National Hurricane Center when it expects an area to experience hurricane-force winds within the next 36 hours.

hurricane watch An alert issued by the National Hurricane Center at least 48 hours before it expects a hurricane to make landfall.

hydrograph A chart that displays how a stream's discharge varies over time at a given location.

hydrological drought A drought during which the amounts of water flowing into lakes, ponds, streams, or reservoirs are insufficient to maintain normal surface-water levels and, consequently, the normal water table.

hydrologic cycle The constant flow of water from reservoir to reservoir, over time, in the Earth System.

hydrosphere The Earth's solid and liquid water realm, including surface water (lakes, rivers, and oceans), groundwater, glaciers, and permafrost.

hypocenter See *focus*.

hypothermia A drop in core temperature of an animal's body below its safe range, giving rise to symptoms from shivering and minor motor impairment to death.

I

ice age An interval of geologic time during which the climate was colder than it is now, when glaciers advanced to cover large areas of the continents, and mountain glaciers grew; an ice age can include many glaciations and interglacials.

ice jam An ice dam that forms when blocks of winter ice break up in the spring, are carried downstream in a river, and pile up against an obstacle.

ice storm A winter storm that produces an accumulation of ice at the ground surface as a result of prolonged freezing rain.

igneous activity The overall process during which rock deep underground melts, producing magma that rises up into the crust and, in some cases, extrudes from volcanoes at the Earth's surface.

igneous rock Rock that forms when hot magma or lava cools and freezes solid, or when pyroclastic debris accumulates and becomes lithified.

ignimbrite Igneous rock formed from a pyroclastic flow that comes to rest, compacts, and undergoes lithification; also called an *ash-flow tuff*.

impact velocity In the context of space hazards, the speed at which an object from space strikes the Earth's surface.

Indian Ocean Dipole The seesaw change in surface water temperatures that occurs between the western and eastern sides of the Indian Ocean over a period of a few years.

induced seismicity Seismic activity caused by human actions.

inertia The tendency of an object to remain in motion with the same velocity unless acted on by an outside force.

infrasound Low-frequency sound below 20 Hz.

InSAR (for *Interferometric Synthetic Aperture Radar*) A tool that allows geologists to measure very subtle vertical movements of the ground.

insolation The amount of solar energy that arrives at a location on the Earth's surface.

insurance In the context of property, a contract between a company and a policyholder in which the policyholder agrees to make payments to the company and, in return, the company agrees to cover the cost of the policyholder's lost or damaged property if such loss should occur.

intensity In the context of earthquakes, an indication of the degree of ground shaking at a locality, based on people's perception of the shaking and on the observed damage caused by the shaking; it is calibrated by the Modified Mercalli Scale; in the context of storms, an indication of the storm's strength as manifested by wind speed and precipitation.

Intergovernmental Panel on Climate Change (IPCC) An international assembly of scientists charged with determining how the Earth's climate is changing, and how humans are influencing those changes; it evaluates the quality of independent research on the subject and summarizes the findings.

internal energy Thermal energy that rises from the interior of the Earth.

intertidal zone The area of land along the shore that is exposed at low tide and submerged at high tide.

intertropical convergence zone (ITCZ) The zone where the trade winds (the surface flow of the Hadley cells) of the northern and southern hemispheres converge in the vicinity of the equator, forcing some air upward.

intraplate earthquake An earthquake that occurs away from plate boundaries.

intrusive igneous rock Rock formed by the freezing of magma underground.

inundation distance In the context of tsunamis, the horizontal distance between the normal shoreline and the point at which the advancing water of a tsunami stops.

inversion In the context of the troposphere, the presence of warmer, less dense air above slightly cooler, denser air below; this condition prevents the lower air layer from convecting upwards, so it remains trapped at low elevation.

ion An atom that has an electrical charge because its number of electrons differs from its number of protons.

ionosphere The region of the atmosphere, at elevations of 60–100 km (37–62 mi), that contains a high concentration of ions.

island arc A chain of volcanic islands that builds up on the seafloor where one oceanic plate subducts beneath another.

isobar A contour line on an air-pressure map at a specified altitude in the atmosphere; all points on an isobar have the same atmospheric pressure.

isotope One of two or more versions of an element that differ in their number of neutrons.

J

jet stream A fast-moving current of air that flows at high elevations.

jetty A concrete or stone wall that runs outward from the shore at a high angle to shelter a harbor or protect a beach; it may also serve to extend the length of a river channel at the river's mouth.

joint In the context of geologic structures, a naturally formed crack in rock.

K

karst terrain A region underlain by caves formed in limestone bedrock; the collapse of caves may create a landscape of sinkholes separated by higher topography or of limestone spires separated by low areas.

Keeling curve A graph, first developed by Charles Keeling, showing the ongoing change in the atmospheric concentration of carbon dioxide since the late 1950s.

kinetic energy The energy of a moving object, which is proportional to the object's mass and the square of its velocity.

king tide An unusually high tide.

L

ladder fire A wildfire that burns undergrowth and forest litter, as well as vines and medium-sized trees that grow between ground vegetation and the forest canopy.

lagoon A body of shallow seawater separated from the open ocean by a barrier island, sand bar, or reef.

lag time In the context of flooding, the difference between the time of a heavy rainfall and the time that water rises above flood stage.

lahar A slurry composed of volcanic ash and debris mixed with water, flowing either down river channels or down the flank of a volcano.

lake-effect snowstorm A snowstorm over and immediately downwind of a large lake, triggered by the flow of very cold air over relatively warm lake water; the warm water supplies moisture and destabilizing heat to the atmosphere.

land An area of the Earth's surface not covered by water.

landform A distinctive natural feature or shape of the land surface.

landslide An informal term for sudden movement of rock and debris down a slope.

landslide potential map A map that ranks regions according to the likelihood that mass wasting will occur, taking into account variations in slope steepness, ground composition, and the occurrence of phenomena that can destabilize slopes.

La Niña A strengthening of the Walker circulation and the trade winds that increases the upwelling of cold water along the western coast of South America; the phase of the Southern Oscillation opposite to an El Niño event.

lapilli Pyroclastic particles 2–64 mm (0.08–2.5 in) in diameter (i.e., marble- to golf ball–sized), consisting of frozen lava clots, fragments of pre-existing rock, or ash clumps.

large igneous province (LIP) A region of extremely voluminous volcanism that can produce abundant flood basalts or immense quantities of ash.

latent heat The energy required for, or released by, a phase change.

lateral spreading Mass wasting that occurs when an area of land overlying a horizontal plane of weakness is unsupported on one side, so that the land fails and moves horizontally in a slump-like manner along the plane of weakness.

lava Molten rock that has flowed out onto the Earth's surface.

lava dome A dome-like mass of rhyolitic lava that accumulates above a volcanic vent.

lava flow A sheet or mound of lava that flows onto the ground surface or seafloor in molten form and then solidifies; some lava flows resemble rivers in that the flow is much longer than it is wide; the term is used for actively moving lava and for lava that has solidified.

lava fountain An eruption of lava that is under pressure, during which the lava spurts as high as a few hundred meters.

lava lake A pool of lava tens to hundreds of meters deep that has collected around a volcanic vent.

lava tube The empty space left when a lava tunnel drains; this happens when the surface of a lava flow solidifies while the inner part of the flow continues to stream downslope.

laze (for *lava haze*) An atmospheric hazard caused when lava entering seawater produces a cloud of steam containing hydrochloric acid as well as fine particles of ash.

lifting mechanism A process that causes air parcels to rise in the atmosphere. Examples include the ascent of air at a frontal boundary, air flow up mountain flanks, and convergent winds near the Earth's surface.

lightning Electrostatic discharges in the atmosphere that occur when negatively and positively charged particles become separated.

lightning stroke A giant spark and pulse of current produced by a thunderstorm; also called a *lightning bolt*.

liquid A state of matter in which atoms or molecules can move relative to one another, but remain close; a liquid can flow and can conform to the shape of its container.

lithification The transformation of loose sediment into solid rock through compaction and cementation.

lithosphere The relatively rigid outer layer of the Earth, 100–150 km (62–93 mi) thick, consisting of the crust and the top part of the mantle.

lithosphere plate One of the distinct pieces of the Earth's lithosphere that move relative to one another.

longitudinal profile A cross-sectional image showing how the elevation of a stream changes from its headwaters to its mouth; it can be represented as a line on a graph whose vertical axis is the elevation of a stream and whose horizontal axis is the distance from the stream's mouth.

longshore bar An elevated underwater ridge of sand, oriented parallel to the shoreline, composed of sand carried offshore by the surf.

longshore current A near-surface flow of water parallel to the shore just offshore, caused when waves approaching the shore are not parallel to the shore.

longshore drift The net transport of sediment laterally along a beach that occurs when waves wash up a beach at an angle.

low-pressure center A location at which the atmospheric pressure is less than the pressure at all surrounding points.

low-pressure system The circulation surrounding a low-pressure center; in a low-pressure system, air rises, generally leading to the production of clouds and precipitation.

M

magma Molten rock beneath the Earth's surface.

magma chamber A space below ground, filled mostly with magma.

magnetic field The region affected by the force emanating from a magnet.

magnetism A field force generated by electric currents, or by special materials called magnets.

magnetosphere The region of the Earth's magnetic field inside which the Earth is mostly shielded from solar wind particles.

magnitude A number representing the amount of energy released by an earthquake.

mainshock The main or largest earthquake of a sequence; it may be preceded by foreshocks and is always followed by aftershocks.

major disaster declaration A declaration of a major disaster by the US president, at the request of the governor of a state; it authorizes mobilization of federal funds and resources to assist victims.

mantle The thick layer of dense rock below the Earth's crust and above its core; most of it is warm enough to undergo slow convection.

mantle plume A column of very hot rock that is very slowly rising upward through the mantle.

maritime disaster A disaster that occurs at sea, or is caused by the sea.

mass extinction An interval of geologic time when vast numbers of species abruptly vanish.

mass wasting The gravitationally caused downslope transport of rock, regolith, snow, or ice.

matter The material substance of the Universe; it is composed of atoms.

meander A snake-like curve in the map-view path of a stream channel.

meandering stream A stream that flows around many meanders.

mechanical force A type of force that results when one mass moves and comes in contact with another; also called a *contact force*.

megadisaster A subjective term for a very rare, particularly immense catastrophe that has affected, or could affect, society worldwide.

megadrought A particularly long drought, of two decades or more.

megathrust earthquake An earthquake of M_W 8.6 or greater, caused by slip along a relatively shallow thrust fault that delineates the boundary between the base of the overriding plate and the top of the downgoing plate at a convergent boundary.

megatsunami A tsunami with a wave height of over 100 m (328 ft).

mesocyclone The strong and long-lasting, rotating updraft within a supercell thunderstorm; its diameter is typically 5–10 km (3–6 mi); tornadoes may form within a mesocyclone.

mesosphere The layer of atmosphere overlying the stratosphere; temperature decreases with increasing elevation in the mesosphere.

metamorphic rock Rock that forms when pre-existing rock undergoes significant change (such as growth of a new mineral assemblage or development of foliation) without first becoming a melt or a sediment.

meteor The incandescent trail produced by a small piece of interplanetary debris as it travels through the atmosphere at very high speeds.

meteorite A piece of rock or metal alloy that has fallen from space and landed on the Earth.

meteorite impact The point of collision of an object from space with the Earth.

meteoroid A small fragment of planetary debris (less than 1 m, or 3 ft, across) that enters the Earth's atmosphere.

meteorological drought A drought in which precipitation is below average for weeks to years; also called a *climatological drought*.

meteorologist A person who studies and forecasts weather and its consequences.

meteor shower The arrival of many more meteors than is normal during a given time; meteor showers can occur when the Earth intersects the dust tail of a comet.

mid-latitude cyclone A large, comma-shaped system of clouds and storms that develops around a low-pressure center; in the northern hemisphere, the clouds and storms rotate in a counterclockwise direction around the low-pressure center; generally, a long cold front defines the comma's tail, and a warm or occluded front underlies much of the comma's head; mid-latitude cyclones form along the polar-front jet stream at latitudes between 30° and 60°, and they generate many types of weather, some of which can be hazardous; also known as an *extratropical cyclone* or a *wave cyclone*.

mid-latitudes The regions between latitudes 30° and 60,° which host temperate climates.

mid-ocean ridge A submarine mountain belt that forms along a divergent boundary.

Milankovitch cycle Any of three cycles of change in the Earth's orbital shape, axial tilt, and wobble, that occur over tens to hundreds of thousands of years, and influence climates on the Earth.

mineral A naturally occurring, homogeneous, crystalline solid with a definable chemical composition that, in most cases, is inorganic.

Modified Mercalli Intensity (MMI) scale A scale for assessing an earthquake's intensity by the damage it causes and people's perception of the shaking; intensity on the scale is specified by Roman numerals; regional variations in intensity for a given earthquake can be represented by contours on a map, with each contour representing a value from the MMI scale.

Moho The seismic-velocity discontinuity that defines the boundary between the Earth's crust and mantle.

moist adiabatic lapse rate The rate (°C per km) at which a saturated air parcel will cool, as a result of expansion, during its ascent, if there is no exchange of heat or mass with the surrounding environment.

molecule A particle consisting of two or more atoms bonded together; a molecule can consist of atoms of the same element, or of atoms of more than one element; the term also refers to the smallest piece of a compound that still has the characteristics of that compound.

moment magnitude scale A logarithmic scale for assessing the magnitude of an earthquake using measurements of the amplitudes of seismic waves and the dimensions of the slipped area and the displacement on the fault; it has replaced the Richter scale in modern descriptions of earthquake size.

monsoon   A seasonally changing air circulation in the world's tropical and subtropical regions in which summer winds blow from the ocean toward the land, bringing heavy rain, and winter winds blow from the land toward the ocean, causing drier weather.

mountain belt   An elongate region of significantly uplifted crust; typically, mountain belts are sculpted by erosion and have rough topography.

mountain building   The set of events that produces a mountain belt.

mountain snowstorm   A snowstorm that results from orographic lifting on the windward side of a mountain range.

mudflow   A downslope movement of mud at slow to moderate speed.

N

National Weather Service (NWS)   The US government agency that monitors storms and other weather phenomena.

natural disaster   A natural hazardous event that causes many human casualties, extensive property destruction, and economic loss so significant that victims need outside help and resources to survive and recover.

natural hazard   A phenomenon or process that exists independently of human activity, but that nevertheless has potential to harm people or property, and therefore represents a threat to society.

natural hazardous event   A disturbance caused by a natural hazard that damages and destroys features on a portion of the Earth's surface, disrupts the environment, and may cause human casualties and destruction of property; if it harms society, an event can be considered a natural disaster.

natural levees   A pair of low ridges that develop on either side of a stream, formed from sediment that accumulates naturally during flooding.

near-Earth object (NEO)   An asteroid, comet, or large meteoroid whose orbit intersects the Earth's orbit.

near-field tsunami   A tsunami that reaches shore close to its source; also called a *local tsunami*.

nor'easter   A mid-latitude cyclone that affects the east coast of North America and intensifies rapidly as it draws in moisture from the Atlantic Ocean, resulting in strong northeast-blowing winds in a region north of its low-pressure center.

normal fault   A dip-slip fault on which the hanging-wall block moves down the slope of the fault.

normal stress   The push or pull that is perpendicular to a surface.

nuclear bond   A bond that holds together protons and neutrons in the nucleus of an atom, or holds together subatomic particles within protons and neutrons.

nuclear energy   Potential energy stored in nuclear bonds.

nuclear fusion   See *fusion reaction*.

nuclear reaction   The formation or breaking of nuclear bonds, during which a small amount of matter is converted into energy.

nuisance tide   A tide during which water rises far enough above the normal high-tide elevation to produce coastal flooding.

numerical age   The age of an Earth material or geologic feature, specified in years.

O

occluded front   A boundary between air masses that forms when a cold front overtakes a warm front and lifts the cold air behind it, so that the warm front no longer intersects the ground surface and the two cold air masses come into contact.

ocean   A large body of saltwater between continents.

ocean acidification   The drop in the average pH of seawater since the industrial revolution; this happens because CO_2 is dissolving in the ocean faster than it can be removed, and when CO_2 reacts with water, carbonic acid results.

oceanic cyclone   An extremely strong wintertime mid-latitude cyclone that develops over the open ocean.

ocean wave   A wave moving within ocean water.

ordinary thunderstorm   A thunderstorm that does not rotate; it forms in an environment where the winds do not change substantially with altitude; also called a *single cell thunderstorm* or an *air mass thunderstorm*.

organic coast   A coast along which living organisms control landforms along the shore.

orographic lifting   The forced ascent of air on the windward side of a mountain range.

outburst flood   A flood that occurs when a dam made of ice or of ice-deposited sediment suddenly gives way.

outcrop   An exposure of bedrock at the Earth's surface.

overshooting top   An upward bulge from the top of a thunderstorm's anvil cloud; it forms above the updraft within the cloud when the updraft is strong enough to push moist air across the tropopause and into the base of the stratosphere.

ozone hole   An area of diminished ozone concentration in the stratosphere over the Antarctic polar region.

P

pāhoehoe   A lava flow with a surface texture of smooth, glassy, rope-like ridges.

paleoclimate   The past climate of the Earth.

paleoclimate indicator   A characteristic of sediments, sedimentary rocks, or preserved organic materials that can provide information about the climate at the time those materials formed; also called a *paleoclimate proxy*.

paleoseismicity   Earthquakes that happened in the past; various features, such as buried sand volcanoes, may serve as indicators of paleoseismicity.

Palmer Drought Severity Index (PDSI)   An index developed by the US National Weather Service that measures moisture deficiency, relative to average local moisture conditions, by comparing the supply of water available from precipitation and stored reserves with the depletion of water by evaporation, infiltration, and runoff.

pandemic   The spread of a disease over many countries, so that it affects a large part of the world and vast numbers of people.

Pangaea   A supercontinent that assembled at the end of the Paleozoic Era and broke apart during the Mesozoic.

passive margin   A continental margin that is not a plate boundary.

peak ground acceleration (PGA)   The maximum acceleration of the ground during an earthquake.

peridotite   A dense, coarse-grained ultramafic rock; most of the Earth's mantle consists of peridotite.

permafrost   A layer of regolith in which water remains frozen all year; most permafrost occurs at high latitudes.

permafrost thawing   Melting of regolith that was previously frozen all year, due to global warming.

permanent stream   A stream that flows year-round.

permeability   The degree to which a material allows fluids to pass through it via an interconnected network of pores, cracks, and conduits.

phase change A change in the state of a substance, such as freezing or evaporation; in the context of mineralogy, a transformation of one mineral into another with the same chemical composition but different crystal structure.

phreatic eruption An explosive volcanic eruption resulting from the outward pressure that develops when heat from magma rapidly transforms liquid water into water vapor that expands and blasts out pyroclastic debris, but not lava.

phreatomagmatic eruption An explosive volcanic eruption that results from the outward pressure that develops when magma heats water to form water vapor that expands and blasts out lava and pyroclastic debris.

planet A large spherical body that directly orbits a star and has incorporated all the matter that lies in or near its orbit.

plasma A state of matter consisting of ionized atoms and free electrons; it forms only at very high temperatures.

plastic deformation A process by which materials subjected to stress permanently change shape without cracking or breaking.

plastic flow The very slow movement of a solid material as plastic deformation takes place.

plate boundary The border between two adjacent lithosphere plates.

plate interior A region away from plate boundaries.

plate tectonics See *theory of plate tectonics*.

polar amplification An increase in temperature that is greater in polar regions, relative to the planetary average, occurring as global warming takes place.

polar climate The climate of regions between latitudes of 60° and 90°; in such a climate, temperatures stay near or below freezing for most or all of the year.

polar front The boundary between the cold air mass located over the polar and subpolar regions and the warm air mass located over the tropics and subtropics; it encircles the globe in each hemisphere.

polar-front jet stream A band of very fast winds that occurs at the top of the troposphere above the polar front, and which encircles each hemisphere at middle or high latitudes; it is most prominent during the winter months, but exists in all seasons; it tends to trace out a wavy path in map view.

polar vortex The circumpolar flow of cold air at latitudes higher than the polar-front jet stream.

population explosion A sudden, large increase in the population of a species.

pore A small, open space within sediment or rock.

pore collapse The packing of grains in sediment or weakly cemented sedimentary rock that occurs when air replaces water in pores and the weight of overburden compresses those pores, decreasing porosity.

porosity The total volume of empty space (pore space) in a material, usually expressed as a percentage.

positive feedback A process that leads to a change that, in turn, increases the rate at which the original process occurs.

potential energy The energy stored in an object being acted on by a field force.

precipitation (1) The process by which atoms dissolved in a solution come together and form a solid; (2) atmospheric water that condenses to a liquid or solid state and falls from the sky.

precursor In the context of natural disasters, a phenomenon that indicates that a hazardous event is starting or will soon start.

preparation stage The time, before a predicted a natural disaster has begun, during which evacuations take place, emergency personnel are brought in, and protections are added to buildings and infrastructure.

prescribed burn An intentionally set fire designed to reduce fuel loads and prevent future catastrophic wildfires; also called a *controlled burn*.

pressure Force per unit area, or the push acting on a material, in cases where the push is the same in all directions.

pressure gradient The rate of pressure change over a given horizontal distance.

primary disaster The casualties and destruction that result directly from a natural hazardous event.

pseudotachylyte Natural glass formed when fault slip or meteorite impact suddenly melts rock, and the melt freezes quickly, before mineral crystals have time to form; such glass tends to occur in cracks adjacent to faults, in rock at and below an impact site, or in impact ejecta.

pumice A glassy igneous rock, formed from frothy felsic lava, that contains abundant small vesicles.

P-wave A compressional seismic body wave that causes back-and-forth motion within rock parallel to the direction of wave motion; also called a *primary wave*.

pyroclastic debris Ash, lapilli, blocks, bombs, and other fragments blasted explosively from a volcano.

pyroclastic flow A fast-moving, intensely hot avalanche that occurs when scalding volcanic ash, gas, and debris mix with air and flow down the side of an erupting volcano.

Q

quartz A common mineral that consists of SiO_2.

R

radiation The emission of energy as electromagnetic waves or as moving subatomic particles.

radioisotopic dating A tool used by geologists that can provide the age of a rock in years; also called *radiometric dating*.

rain Precipitation consisting of liquid water in drops.

raindrop A sphere or blob of liquid water that is heavy enough to fall through the atmosphere.

ram compression The pressure in a fluid caused by a solid body moving through the fluid; the ram compression caused by a meteoroid passing through the atmosphere produces a meteor.

rapid-onset disaster A natural disaster that begins so quickly that people do not have time to evacuate or take precautions.

realm One of several components of the Earth System that exchange materials and energy with other such components; examples include the geosphere and the hydrosphere.

recent climate change Global change in the Earth's climate since the industrial revolution, caused by the addition of greenhouse gases to the atmosphere by human activities.

recovery stage The stage of dealing with a natural disaster that follows the response stage; it involves cleaning up, finding or building longer-term shelter, re-establishing health care systems, and providing essential services such as access to drinking water, food, sanitation, electricity, and communication.

recurrence interval (RI) The average time between successive hazardous events of a particular size or magnitude; in some cases, it refers to the average time between events of a particular size or greater; also called a *return period*.

red tide   A harmful marine algal bloom that includes species that produce toxins; the name is used because the organisms can turn the water reddish brown.

reef   An underwater mound or ridge overlain by an area or belt of very shallow water; see also *coral reef*.

reflection   The return of light or other types of waves that have bounced off a boundary between materials with different properties.

refraction   The bending of light or other types of waves when the waves pass through a boundary between materials with different properties.

regolith   A general term for unconsolidated material at the Earth's surface, including soil, uncemented sediment, and weathered rock.

reinsurance   Insurance that a primary insurer buys from other companies to help cover catastrophic losses; it can spread the financial risk of insurance among multiple companies.

relative humidity (RH)   The ratio of the amount of water vapor in the atmosphere to the atmosphere's capacity for holding water vapor at a given temperature (expressed as the vapor pressure divided by the saturation vapor pressure).

relief   In the context of topography, the difference in elevation between adjacent high and low regions on the land surface.

resilience   The ability of a community to respond, recover, and restore itself in a timely manner after a disaster.

resonance   An increase in the amplitude of a periodic motion caused by the addition of energy at the same frequency.

resonance disaster   Greater than expected damage to structures during an earthquake, due to amplification of their movements as a consequence of resonance.

response stage   The stage of dealing with a natural disaster during which officials and volunteers rescue survivors, treat the injured, provide emergency shelter and security, and prevent the effects of the primary disaster from triggering a secondary disaster; this stage may begin even before the event causing the disaster is over.

restoration stage   The stage of dealing with a natural disaster during which officials turn their attention to rebuilding damaged buildings and infrastructure and re-establishing economic stability; this stage may take months to years.

reverse fault   A dip-slip fault on which the hanging-wall block moves up the slope of the fault.

Richter scale   A logarithmic scale that calculates earthquake magnitude based on the amplitude of the largest ground motion recorded on a seismogram.

ridge   (1) In the context of meteorology, an elongate area of high atmospheric pressure; in the upper troposphere, a ridge coincides with the location where the polar-front jet stream bows toward the poles; (2) in a topographic context, a narrow, elongate area of higher terrain; (3) a shortened form of the term *mid-ocean ridge*.

rift   A distinct linear belt where rifting has occurred.

rifting   The process of stretching and breaking a continent apart.

Ring of Fire   An informal term for the belt of convergent boundaries along the margin of the Pacific Ocean; so-named because of the many subaerial volcanoes that lie within it.

rip current   Return flow from a beach to the ocean, focused over a low spot in a longshore bar, that accelerates into a strong, narrow stream extending beyond the surf zone.

risk   In the context of natural disasters, the probability that an individual or community will be subject to injury or loss due to a hazardous event.

river   An everyday term referring to a large stream.

rock   A coherent, naturally occurring solid, consisting of an aggregate of mineral grains or crystals or, much less commonly, a mass of glass.

rock cycle   A succession of processes that cause the material in a rock to transform into part of another rock of the same or a different group over geologic time; for example, as a result of weathering, erosion, and deposition, the atoms that were in an igneous rock can become incorporated in sedimentary rocks.

rockfall   Mass wasting that involves the sudden drop of a mass of rock from a vertical cliff or overhang, so that for part of its downward journey, the rock free-falls through air.

rockslide   Mass wasting that occurs when a mass of rock detaches from its substrate on a failure surface and moves down a nonvertical slope.

rocky coast   An area of coast where bedrock rises directly from the sea and beaches are absent.

rogue wave   In the context of maritime hazards, an unexpected wave that is more than twice the significant wave height.

runoff   Water that flows downslope over the land surface as overland flow or in stream channels.

run-up elevation   The vertical distance between the normal shoreline and the elevation of the point at which the advancing water of a wave stops.

S

safety factor   A number that represents the stability of a slope, defined as resistance stress (or shear strength) divided by downslope shear stress; if this number is less than 1, the slope is unstable.

Saffir-Simpson scale   A scale that assigns intensity categories to hurricanes on the basis of their maximum sustained wind speed.

salinification   A rise in the salinity of freshwater to a level that makes the water unusable; it can be caused by the addition of salts, or by a decrease in water inflow so that the rate of evaporation becomes faster than the rate of freshwater replenishment.

saltwater   Water that, on average, contains between 3.0% and 5.0% dissolved salt; such water occurs in the ocean and in saline lakes.

sand spit   A ridge of sand, bordered on both sides by water, that is connected to land at one end; it may grow over time due to longshore drift.

saturation vapor pressure   The vapor pressure at which air is saturated (contains as much water vapor as possible) at a given temperature.

sea cliff   A steep wall of sediment or rock along the shoreline.

sea-cliff retreat   The landward migration of the position of a sea cliff over time, due to erosion and slope failure.

seafloor   The surface of the geosphere that lies hidden beneath the oceans.

seafloor spreading   The gradual widening of an ocean basin as new oceanic lithosphere forms at a mid-ocean ridge axis and then moves away from the axis.

sea level   The elevation that the sea surface would have if there were no waves. Sea level on a map usually refers to mean sea level, which is the average level over time, lying roughly halfway between high tide and low tide.

sea-level cycle   A cycle of global change during Earth history, manifested by a change in the elevation of the sea surface relative to the land surface.

seamount   A volcanic edifice, built on the seafloor, whose peak lies below sea level.

sea stack   A chimney-shaped column of rock produced by wave erosion along a rocky coast.

seawall   A wall built along the shoreline, or at the landward edge of a beach, to slow erosion.

secondary disaster   A hazardous event triggered by, and following, a primary disaster.

sediment   A deposited accumulation of loose grains that are not cemented together; examples include sand and gravel.

sedimentary rock   Rock that forms at or near the Earth's surface by the compaction and cementation of rock fragments or shells and shell fragments, by the accumulation and alteration of organic matter, or by the precipitation of mineral crystals from water solutions.

sediment budget   The difference between inputs of sediment into a region and the removal of sediment from a region.

sediment liquefaction   A phenomenon that happens when seismic shaking or the rapid input of water causes wet sediment that has been consolidated, but not cemented, to disaggregate and form a slurry of sediment and water that cannot support weight.

seismic belt   One of the bands or areas of the Earth where most earthquakes occur.

seismic gap   A section of a seismic zone that has not hosted a major earthquake for a long time, relative to adjacent portions that have.

seismicity   Earthquake activity.

seismic retrofitting   The process of strengthening existing buildings or infrastructure that were not originally designed to withstand an earthquake.

seismic waves   Vibrations generated by an earthquake that pass through the Earth or along its surface.

seismogram   The record of an earthquake produced by a seismograph.

seismograph   A device that can measure and record ground motion; it consists of a seismometer (a device that detects motion) and a rotating paper roll or a digital recording apparatus.

seismology   The scientific study of earthquakes.

sequestration   In general, the removal and storage of a material; carbon, for example, can be sequestered in the hydrosphere, biosphere, or geosphere; see also *carbon capture and sequestration*.

severe thunderstorm   A thunderstorm with the potential to produce hail with diameters of 2.5 cm (1 in) or larger, winds that exceed 50 knots (92 km/h, or 56 mph), or a tornado.

shear stress   A stress that moves one part of a material sideways past another part.

shield volcano   A subaerial volcanic edifice with the shape of a broad, gentle dome; it forms from many successive layers of low-viscosity mafic lava, and contains relatively little pyroclastic debris.

shock wave   An intense, abrupt change of pressure in a narrow region, resulting from sudden mechanical force such as that caused by an explosion or by a meteorite impact; it travels through a medium, such as air or rock, at a speed faster than sound.

shore   The relatively narrow portion of a coast (less than 1 km wide) that interacts directly with the water of an ocean or lake.

shoreline   The boundary between water and land.

significant wave height   The mean height of the highest third of the waves passing a given location over a given observation period.

simple crater   An impact crater caused by a relatively small meteorite; it has a bowl-like shape and no central uplift.

sink   In the context of climate change, a reservoir for storing carbon in the geosphere, biosphere, or hydrosphere so that the carbon doesn't remain in the atmosphere.

sinkhole   A circular depression in the land surface that forms when the surface collapses into an underground space; the space can be a cavern, or a zone of liquefied sediment; in karst terrains, the term also applies to depressions formed by dissolution.

sixth extinction   An informal term for the vast number of species extinctions currently occurring, due to human civilization and its environmental consequences; use of this term emphasizes that current rates of extinction resemble those of the five other mass extinctions recorded in geologic history.

sleet   Precipitation consisting of frozen raindrops; also called *ice pellets*.

slope   The tilted surface between locations at different elevations.

slope failure   The downslope movement of material on an unstable slope.

slow earthquake   Slip on a fault that releases energy over a time frame of hours to months (i.e., faster than fault creep, but slower than a seismic event).

slow-onset disaster   A natural disaster that takes days or weeks to develop, providing people with time to prepare.

slow-onset flood   A flood that develops over several days and takes days to weeks to subside, as can occur when snowmelt and substantial rainfall enter a large watershed over a period of time; it usually affects the trunk stream or larger tributaries of a drainage network. Also called a *downstream flood*, *regional flood*, or *seasonal flood*.

slump   A type of mass wasting that involves the displacement of rock or regolith above a spoon-shaped failure surface; it takes place at slow to moderate rates.

smoke   Material rising from a fire that contains gases that have cooled and stopped glowing, along with soot, ash, aerosols, and other remnants of incomplete burning.

snow avalanche   The downslope movement of a mass of snow.

snowdrift   A mound or ridge built from windblown snow.

snowflake   A composite of hexagonal ice crystals that forms in the atmosphere.

snowpack   A layer of snow, formed from the accumulation of several snowfalls; the term is generally applied to relatively thick layers that accumulate at high elevation and survive for some time after snowfalls have ceased; melting of snowpack contributes to spring runoff.

socioeconomic drought   A drought during which water shortages reduce the supply of goods and services, thereby negatively affecting the quality of life.

soil   Sediment that has undergone changes at the surface of the Earth due to reactions with air, water, and life, and has mixed with organic material.

soil conservation   A combination of practices intended to diminish soil erosion and soil depletion.

soil contamination   A component of soil degradation that involves modification of the composition of soil by the addition of toxic components.

soil degradation   The loss of soil due to soil erosion, soil contamination, and loss of nutrients and carbon in soil.

soil depletion   A component of soil degradation that involves modification of the composition of soil by the removal of carbon and nutrients, so that the soil is less fertile.

soil erosion   A component of soil degradation that involves the removal of soil by water or wind.

soil moisture   Water that clings to minerals in soil or gets absorbed by organic matter in soil.

solar cycle   The 11-year cycle of sunspot frequency that occurs as a consequence of reversals in the Sun's magnetic field.

solar flare A sudden release of electromagnetic energy from the Sun's surface, manifested by a detectable brightening of the Sun that can last minutes to hours.

solar prominence An event associated with a solar storm in which a large, bright arc of plasma extends outward from the Sun's surface between a pair of sunspots.

solar storm An explosion-like event in which the Sun ejects particularly large amounts of high-energy particles into space; this can result in a disturbance of the Earth's magnetic field called a geomagnetic storm; see also *coronal mass ejection*.

solar wind A stream of charged particles emerging from the Sun with enough energy to escape from the Sun's gravity and flow outward into space.

solid A state of matter in which atoms or molecules remain fixed in position; a solid can maintain its shape indefinitely.

solifluction A type of creep characteristic of tundra regions; it occurs during summer when the uppermost layer of permafrost melts and the resulting soggy, weak layer of ground flows slowly downslope in overlapping sheets above frozen permafrost below.

soot Very fine particles of carbon that are products of incomplete burning.

source In the context of climate change, a reservoir or process that releases carbon to the atmosphere.

Southern Oscillation The shift in sea-level atmospheric pressure back and forth across the Pacific Ocean in association with the Walker circulation; it is associated with El Niño.

space weather Conditions in the region of space surrounding the Earth that result from interactions with charged particles coming from the Sun; intense space weather is also called a *geomagnetic storm*.

spiral rainband One of the nearly continuous arcing (in map view) cloud banks, consisting of tall, aligned thunderstorms that curve outward from the eye of a hurricane; these cloud banks define the spiral shape of the storm, and are so named because they are the source of torrential downpours.

spot fire A new wildfire, ignited by a firebrand that has been carried by the wind ahead of the original fire and lands on dry fuel.

spring A natural outlet from which groundwater flows to the ground surface.

squall-line thunderstorm One of a distinct line of thunderstorms that marks the edge of a cold front or gust front; several of these thunderstorms together can be called a *multicell storm*.

stable slope A slope on which materials tend to stay in place because the shear strength of the materials of the failure surface (i.e., the resistance stress) exceeds the downslope shear stress.

stage The height of a stream's surface relative to a surveyed reference elevation (generally set just below the streambed).

state of emergency declaration A declaration by the governor of a US state that allows authorities to mobilize state police, National Guard, and various state-employed emergency personnel in response to a hazardous event or a disaster.

state of matter One of four conditions of matter: solid, liquid, gas, or plasma.

stationary front A boundary between air masses where the colder air mass is neither advancing nor retreating at the Earth's surface.

stealth disaster A disaster that takes so long to develop that its onset may not be noticed for months to decades; also called a *creeping disaster*.

stick-slip behavior Stop-start movement along a fault over time; it occurs because friction prevents movement until stress builds up sufficiently to overcome it; the slip event generates seismicity due to elastic rebound.

storm An episode of strong wind and heavy precipitation, sometimes accompanied by lightning.

storm surge A local rise of sea level, developing primarily because the winds of a strong storm (especially a hurricane) push water into a mound, and secondarily because of the drop in atmospheric pressure at the sea surface near the storm's center; when a storm surge reaches the shore, it can cause inundation of land that is normally above sea level.

storm-surge barrier A concrete or steel wall along the shoreline that remains open most of the time, but can be closed to keep out storm surge or king tides.

straight-line wind A strong wind produced by a downdraft that blows in a single direction at the ground surface.

strata A succession of sedimentary beds.

stratosphere The layer of the Earth's atmosphere directly above the troposphere, in which the temperature increases with elevation.

stratovolcano A large, cone-shaped, subaerial volcanic edifice consisting of alternating layers of lava and pyroclastic debris; also called a *composite volcano*.

stream A body of water flowing down a channel.

stream gradient The slope of a stream surface in the downstream direction.

stream order The category of a stream within the hierarchy of streams in a drainage network; in one common classification, a tributary into which no other tributary flows has an order of 1, and the trunk stream of the network has the highest order.

strength The ability of a material to avoid failure (breaking, flowing, or collapsing).

stress The push, pull, or shear that a specified area of material experiences when subjected to a force; formally, the force applied per unit area.

strike-slip fault A fault along which one block of crust slides horizontally past another (i.e., in a direction parallel to the strike line of the fault surface), so that no relative vertical motion takes place.

subaerial volcano A volcano that erupts into the air.

subduction The process by which one oceanic plate bends down and sinks into the asthenosphere beneath the edge of another plate.

subduction zone The region along a convergent boundary where one plate sinks beneath another.

submarine mass wasting Mass wasting that occurs underwater; also called a *submarine slide*.

subsidence The slow vertical sinking of the Earth's surface in a region.

subsoil The layer of a soil that lies beneath the topsoil and represents the zone of accumulation; it corresponds with the B-horizon.

subtropics Areas at latitudes between 23° and 35°, which are warm, but drier than the tropics.

sunspot A relatively cool, and therefore darker, region on the solar surface, produced where loops of the Sun's magnetic field rise up from, or reenter, the surface of the Sun and locally inhibit convection; a sunspot can last for weeks to months.

supercell thunderstorm A thunderstorm containing a mesocyclone; supercells often become severe and can produce hail, strong straight-line winds, and tornadoes.

supercontinent cycle A cycle of global change during which blocks of continental crust collect into supercontinents and later break up to form smaller continents.

super typhoon A tropical storm in the western Pacific equivalent in wind velocity to a strong Category 4 or Category 5 hurricane in the Atlantic.

supervolcano A volcano that ejects more than 1,000 km³ (240 mi³) of volcanic material during an explosive eruption; none have erupted during recorded human history.

surface current An ocean current in the upper 100–400 m (330–1,300 ft) of water.

surface fire A wildfire that burns low-lying vegetation such as grass and brush, forest undergrowth, and forest litter.

surface water The portion of the hydrosphere that occurs in oceans, lakes, streams, rivers, swamps, snow, and glaciers.

surface wave In the context of earthquakes, a seismic wave that travels along the Earth's surface.

swash The surge of water up the slope of a beach.

S-wave A seismic body wave that causes up-and-down motion in rock perpendicular to the direction of wave motion; S-waves are shear waves (also known as transverse waves); S stands for secondary, arriving after P-waves.

swells Long-wavelength, periodic ocean waves that have traveled far from their source and are not driven by the local wind.

system In the context of Earth science, a group of interacting or interrelated physical elements.

T

talus pile An accumulation of fallen rock fragments along the base of a cliff, typically in a steep apron whose surface is at the angle of repose.

tectonic activity A geologic process that produces mountain belts, crustal deformation, volcanoes, or earthquakes; most tectonic activity is related to the motion of lithosphere plates, but some occurs in association with hot spots.

temperate climate The climate of mid-latitude regions, characterized by weather that changes overall as seasons progress, and by weather that can change during the course of a single day when a front passes.

temperature A measure of the average velocity of atomic and molecular movements in a material; these movements can include the vibration of atoms or molecules in place or the physical movement of atoms or molecules from one location to another; in common experience, the temperature indicates how hot or cold a material is.

temperature anomaly The difference between the temperature at a particular time and the average temperature for a given time period.

tephra Unconsolidated accumulations of pyroclastic material.

tertiary disaster A disaster that follows a secondary disaster. For example, an earthquake (a primary disaster) may make water supplies toxic (a secondary disaster), which may in turn trigger a disease outbreak (a tertiary disaster).

texture In the context of geology, the way in which the grains of a rock are held together, along with factors such as the size and relative orientation of the grains.

theory of plate tectonics The comprehensive, well-supported idea stating that the outer layer of the Earth (the lithosphere) consists of separate plates that move with respect to one another and interact along plate boundaries; these interactions are responsible for most volcanism, seismicity, and mountain building; also known simply as *plate tectonics*.

thermal energy Energy resulting from the vibration or movement of atoms or molecules in a material; see also *heat*.

thermohaline circulation A global oceanic circulation that involves both surface and deep-sea currents, formed by the rising and sinking of ocean water and driven by contrasts in water density, which are due in turn to variations in temperature and salinity.

thermometer An instrument used to measure temperature, often constructed by partially filling a sealed glass tube with a liquid such as alcohol.

thermosphere The hot, outermost layer of the atmosphere containing very few gas molecules.

thunder The booming sound created by the rapid expansion and then collapse of air along the path of a lightning bolt.

thunderstorm A cumulonimbus cloud that produces lightning and thunder.

tidal bore A visible wall of water produced when a flood tide heads upstream in an estuary so the river current opposes the tidal current.

tidal flat A broad, nearly horizontal plain of mud and silt that is exposed or nearly exposed at low tide but totally submerged at high tide.

tidal range The difference in sea-surface elevation between high tide and low tide at a given location.

tide The rising or falling of sea level that generally occurs twice daily and is a consequence of the tide-generating force.

topography The shape of the land surface caused by variations in elevation.

topsoil The top layer of soil (the O- and A-horizons or, in some cases, the E-horizon), which lies within the zone of leaching and can be dark and nutrient-rich.

tornado A nearly vertical, funnel-shaped cloud in which air rotates violently; these range from ~100 m to 1 km (328–3,280 ft) across, and occur in association with the updrafts of strong thunderstorms; as they move along the ground, they leave a track of destruction.

tornado outbreak An occurrence of many tornadoes within a relatively short period of time, in association with a particularly strong front in a mid-latitude cyclone.

tornado track The path a tornado takes across the ground.

total people affected The human toll of a disaster, including casualties, plus displaced people, plus people who have not moved but have suffered significant economic loss or distress.

trade winds Surface winds formed at the base of the Hadley cells, found throughout the tropics and flowing toward the equator; they curve to blow nearly from east to west at low latitudes, due to the Coriolis force.

transform boundary A plate boundary where one plate slips sideways relative to the other along a strike-slip fault, so that no new lithosphere forms and no old lithosphere subducts.

transform fault The strike-slip fault system marking a transform boundary; most transform faults are the actively slipping segment of a fracture zone between two segments of a mid-ocean ridge, but some cut through continental crust and can be traced for hundreds of kilometers.

tributary A smaller stream that flows into a larger stream.

tropical cyclone A large, spiral-shaped storm that originates over warm tropical ocean waters and has an organized rotation (counterclockwise in the northern hemisphere and clockwise in the southern hemisphere) around a low-pressure center; intense tropical cyclones are called hurricanes, typhoons, or cyclones, depending on location.

tropical depression A large storm system that evolves in the tropics from a tropical disturbance when winds begin to circulate around a low-pressure center and accelerate to reach speeds between 37 and 62 km/h (23–38 mph).

tropical disturbance A cluster of thunderstorms that grows in the tropics and lasts for more than a day.

tropical storm A tropical cyclone with sustained winds of between 63 and 118 km/h (39–73 mph).

tropical storm warning A warning issued by the National Hurricane Center when it expects an area to experience tropical storm-force winds within the next 36 hours.

tropical storm watch An alert issued by the National Hurricane Center at least 48 hours before it expects a tropical storm to make landfall.

tropics Latitudes between 23° N and 23° S, where cold weather rarely occurs.

troposphere The layer of the Earth's atmosphere that extends from the Earth's surface to the tropopause and contains all of the Earth's weather.

trough In the context of meteorology, an elongate area of low atmospheric pressure. In the upper troposphere, a trough coincides with the location where the polar-front jet stream bows toward the equator.

trunk stream The single larger stream into which an array of tributaries flow.

tsunami A wave produced by the sudden movement of a mass against water; the wavelength of tsunamis greatly exceeds that of storm waves, and they travel much faster; a large tsunami is very broad and can involve an immense amount of water.

tsunami elevation The greatest vertical distance between the crest of a tsunami and sea level as the tsunami reaches the shore.

tsunami warning A warning issued by NOAA's tsunami warning centers if a DART buoy detects a tsunami; it states that tsunamis are approaching the notified areas.

tsunami watch An alert issued by NOAA's tsunami warning centers, stating that an event happened that could trigger tsunami formation.

tsunami wave train A group of tsunamis generated by a push of a mass against a body of water.

tuff A fine-grained pyroclastic rock composed mainly of volcanic ash.

Tunguska event A bolide explosion that took place over the Stony Tunguska River, east of the Ural Mountains, on June 30, 1908, and destroyed vast areas of forest.

typhoon A strong tropical cyclone over the northwestern Pacific Ocean in which sustained wind speeds reach 119 km/h (74 mph) or higher; it is equivalent to a hurricane, but occurs in the northwestern Pacific.

U

undercutting Removal of supporting material at the base of a cliff that triggers a slide or slump.

undertow Return flow of water from a beach to the ocean beneath the surf zone and near the seafloor.

uniformitarianism The geologic principle stating that the same physical processes that we observe today happened in the past, and that they occurred at roughly the same rates; simply put, the present is the key to the past.

unstable slope A slope whose materials tend to move downslope because the downslope shear stress exceeds the resistance stress.

updraft An upward-moving air flow.

uplift The upward vertical movement of the Earth's surface in a region.

urban flood A flood that inundates land in a built environment because the ground surface in the area is impermeable, or because the constructed drainage system in the area is inadequate to remove water fast enough to prevent it from accumulating on the ground.

urban heat island A built area that experiences temperatures hotter than those in surrounding suburban or rural areas, partly due to the presence of large amounts of concrete that absorbs and reradiates heat, and partly due to the lack of vegetation to shade the ground.

urbanization The migration of people from the countryside to cities, and the consequent expansion of cities.

V

vapor plume In the context of an asteroid collision with a gas-giant planet, an upward-convecting cloud of gas.

vapor pressure That part of the total atmospheric pressure exerted by water vapor molecules.

vector A quantity that has both magnitude and direction; it can be represented by an arrow.

velocity The speed and direction of an object's motion.

viscosity The resistance of material to flow.

visibility In the context of meteorology, the maximum distance from which a person with normal vision can distinguish objects when looking through the atmosphere.

vog (for *v*olcanic sm*og*) Air pollution formed when sulfur-bearing volcanic gases react with atmospheric water to produce acidic aerosols, which are then trapped near the ground by an inversion.

volatile An element or compound, such as H_2O or CO_2, that can melt at relatively low temperatures and can exist in gaseous forms at the Earth's surface.

volcanic arc A chain of active volcanoes formed adjacent to a convergent boundary.

volcanic bomb A large piece of pyroclastic debris that is still soft when erupted and attains a streamlined shape during eruption.

volcanic edifice A hill or mountain composed of solidified lava and/or pyroclastic debris that builds up around a volcanic vent.

volcanic eruption An event during which lava and/or pyroclastic debris is expelled from a volcanic vent.

volcanic explosivity index (**VEI**) A logarithmic scale used to classify explosive volcanic eruptions by size, based on the volume of erupted tephra as measured in cubic kilometers.

volcanic hazard assessment map A map that delineates areas near a volcano that lie in the path of potential lava flows, lahars, or pyroclastic flows.

volcanic unrest Changes in the state of a volcano that differ from its normal state and that may indicate that it will soon erupt. Evidence of unrest may include increased seismicity, heat flow, gas emission, and shape change.

volcano (1) A vent out of which lava or pyroclastic debris spews onto the planet's surface or into the air; (2) a volcanic edifice.

volcano observatory One of five stations overseen by USGS at which staff monitor volcanoes for seismicity, shape changes, gas emissions, and other signs of unrest, and from which rapid-response teams can be deployed to land on a volcano and monitor an eruption in real time.

vulnerability   The characteristics of a community that affect its ability to cope with a hazardous event.

W

Wadati-Benioff zone   A sloping band of seismicity, within or along the downgoing plate, that extends to a depth of 660 km (410 mi) at convergent boundaries.

Walker circulation   An atmospheric circulation in the equatorial regions of the Pacific Ocean, characterized by rising air over the western Pacific, eastward air flow near the tropopause, sinking air over the eastern Pacific, and westward air flow near the Earth's surface.

wall cloud   A region of rotating clouds that extends below the main clouds of a supercell thunderstorm and surrounds the updraft of the storm. The formation of a wall cloud often precedes tornado formation.

warm front   A boundary between air masses where the colder air mass is retreating and the warmer air mass is advancing at the Earth's surface.

warning   In the context of meteorology, an alert issued by the US National Weather Service when a hazardous weather event develops, to areas that will be affected within a time frame of minutes to hours.

watch   In the context of meteorology, an alert issued by the US National Weather Service when predictions indicate that conditions are right for a hazardous weather event.

water pollution   The contamination of water with undesirable materials (chemicals, nutrients, garbage, bacteria) in quantities that exceed those that would normally occur, and which natural systems cannot quickly remove.

watershed   The land area providing water to the drainage network of a specified trunk stream; also called a *catchment* or *drainage basin*.

water stress   A lack of access to safe and affordable water for communities due to physical or economic water scarcity.

water table   The underground boundary, approximately parallel to the Earth's surface, that separates substrate in which groundwater completely fills the pores (the saturated zone) from substrate in which air partially fills the pores (the unsaturated zone); the saturated zone lies below the water table and the unsaturated zone lies above.

water vapor   Water in its gaseous form.

wave base   The depth in a body of water, approximately equal in distance to half a wavelength, below which there is no wave movement.

wave-cut bench   A shelf of rock, cut by wave erosion, at the low-tide line, left behind by sea-cliff retreat.

wave-cut notch   A notch cut by wave erosion in a rocky sea cliff, which widens until the overhanging rock becomes unstable and breaks away.

wave erosion   The grinding or breaking away of the coastline by the action of water waves.

wavelength   The horizontal distance between two adjacent wave troughs or two adjacent wave crests.

wave refraction   The bending of waves; in the context of water waves, waves that approach a beach refract as they enter shallower water so that their crests make no more than a 5° angle with the shoreline; in the context of seismic waves, refraction takes place as waves pass into materials of different material properties.

wave train   A series of waves that travel together.

weather   Local-scale atmospheric conditions as defined by temperature, atmospheric pressure, relative humidity, wind speed, and precipitation.

weathering   The processes that gradually modify and weaken rock exposed to air and water, eventually transforming it into sediment; mechanical or physical weathering involves fracturing and fragmentation, whereas chemical weathering involves chemical reactions.

weather station   A facility or site at which an array of instruments gathers data such as air temperature and pressure, wind speed and direction, and relative humidity, in order to make weather predictions; they are typically set up by a government agency and are calibrated so that the data they collect can be compared to data from elsewhere.

well   A hole in the ground dug or drilled in order to obtain underground fluids such as water or oil.

wildfire   An uncontrolled, self-sustaining blaze sweeping across the landscape that may consume forests, grasslands, and neighboring communities.

wildfire complex   An array of two or more wildfires ignited in the same general area, or when embers from one fire start spot fires at a nearby location; the separate fires may merge into a single fire.

wildfire suppression   An effort to prevent wildfires from starting or spreading, under the assumption that wildfires do more harm than good.

wind   The horizontal movement of air.

wind chill temperature   The temperature that human skin feels due to heat loss caused by the combined effects of both cold and wind.

wind direction   The direction from which the wind blows.

wind-driven wave   A wave formed by the interaction of moving air with the surface of a water body.

wind speed   The speed at which air moves horizontally across the Earth's surface, as measured by an anemometer.

work   The energy transferred to or from an object or material via the application of force along a displacement; lifting an object, for example, constitutes work.

CREDITS

TEXT

CHAPTER 1
Fig 1.14a: Figure 3.1 From *Global Estimates 2015: People Displaced by Disasters*. Copyright © 2013 - 2019 Internal Displacement Monitoring Centre (IDMC). https://www.internal-displacement.org/sites/default/files/inline-files/20150713-global-estimates-2015-en-v1.pdf

Fig 1.15b: Adapted from Kevin A Borden and Susan L Cutter: "Spatial Patterns of Natural Hazards Mortality in the United States," International Journal of Health Geographics 2008, 7:6. © 2008 Borden and Cutter; licensee BioMed Central Ltd.

CHAPTER 6
Fig 6.9b: Republished with permission of The Royal Society of London; Figure 12a from "Can small islands protect nearby coasts from tsunamis? An active experimental design approach," T. S. Stefanakis, E. Contal, N. Vayatis, F. Dias and C.E. Synolakis; *Proceedings of the Royal Society A: Mathematical, Physical, and Engineering Sciences*, 470:2172 © 2014; permission conveyed through Copyright Clearance Center, Inc.

CHAPTER 7
Fig 7.38: From "Global Soil Erosion: An Assessment of the global impact of 21st century land use change on soil erosion" by the Joint ResearchCentre /European Soil Data Centre (ESDAC), https://esdac.jrc.ec.europa.eu/themes/global-soil-erosion

CHAPTER 10
Fig 10.6, BX10.1a, Fig10.14c: From *Severe and Hazardous Weather 5e*, by Robert Raube, John Walsh, Donna Charlevoix. Used by permission of Kendall Hunt Publishing Company

CHAPTER 11
Fig 11.3c: Figure "RHI plot from the Miami radar during hurricane Andrew;" Republished with permission of The American Meteorological Society, from "Small scale spiral bands observed in hurricanes Andrew, Hugo and Erin," Gall, R., Tuttle, J., and Hildebrand, P.; *Monthly Weather Review* 126:7 ©1998; permission conveyed through Copyright Clearance Center, Inc.

Fig 11.4a: Figure "Schematic illustration of radar reflectivity in a Northern Hemisphere tropical cyclone with a double eyewall;" "Republished with permission of The American Meteorological Society, from "Clouds in Tropical Cyclones," Robert A. Houze Jr. *Monthly Weather Review* 138:2 ©2010; permission conveyed through Copyright Clearance Center, Inc.

CHAPTER 12
Fig 12.18c: Figure 5 from Ward et al., 2013: "A framework for global flood risk assessment;" *Hydrology and Earth System Science*, 17:5.

CHAPTER 13
Bx13.1c: (C. Storlazzi, USGS)

CHAPTER 14
Fig 14.7a: The U.S. Drought Monitor is jointly produced by the National Drought Mitigation Center at the University of Nebraska-Lincoln, the United States Department of Agriculture, and the National Oceanic and Atmospheric Administration. Map courtesy of NDMC.

Fig 14.12b: The U.S. Drought Monitor is jointly produced by the National Drought Mitigation Center at the University of Nebraska-Lincoln, the United States Department of Agriculture, and the National Oceanic and Atmospheric Administration. Map courtesy of NDMC.

Fig 14.28: Figure 3 Republished with permission of CSIRO Publishing, from "Modelling the effects of surface and crown fire behaviour on serotinous cone opening in jack pine and lodgepole pine forests," Miguel G Cruz, *International Journal of Wildland Fire* 21(6) © 2012; permission conveyed through Copyright Clearance Center, Inc.

Fig. 14.42b: Map: Fire Hazard Severity Zones (Adopted in 2017) created by the Conservation Biology Institute from data provided by California Department of Forestry and Fire Protection, Fire and Resource Assessment Program (CALFIRE - FRAP). https://databasin.org/datasets/fbb8a20def844e168aeb7beb1a7e74bc/

PHOTOS

FRONT MATTER
Page ii: Stephen Marshak; p. ix: Alfredo Martinez / Getty Images; p. x (top): Stephen Marshak; (bottom): Nik Wheeler / Getty Images; p. xii: Westend61 GmbH / Alamy Stock Photo; p. xiii: Xinhua / Alamy Stock Photo; p. xiv (top): AP Photo / The Yomiuri Shimbun, Atsushi Taketazu; (bottom): Stephen Marshak; p. xv: Stephen Marshak; p. xvi (top): Cultura Creative Ltd / Alamy Stock Photo; (bottom): Design Pics Inc / Alamy Stock Photo; p. xvii: FotoKina / Shutterstock; p. xviii: Stephen Marshak; p. xix: Bob Sharples / Alamy Stock Photo; p. xx (top): JOSH EDELSON / AFP via Getty Images; (bottom): Stephen Marshak; p. xxi: Prasenjeet Yadav.

CHAPTER 1
Page 2: Alfredo Martinez / Getty Images; p. 3 (a): JIJI Press / AFP / Getty Images; (b): Horizon Images / Motion / Alamy Stock Photos; p. 4 (top left): Stephen Marshak; (top right): Henriette Olsbo Foye; (bottom a): U.S. Geological Survey; (bottom b): U.S. Forest Service Pacific Northwest Region; (bottom c): John Barr / Liaison / Getty Images; p. 5 (top a): Stephen Cheatley / Alamy Stock Photo; (top b): Cultura Creative Ltd / Alamy Stock Photo; (top c): US Army Photo / Alamy Stock Photo; (bottom a): NOAA; (bottom b): AP Photo / David Goldman; p. 6: (a): Jane Tyska / Bay Area News Group / MediaNews Group via Getty Images; (b): Topical Press Agency / Hulton Archive / Getty Images; (bottom): MARK RALSTON / AFP via Getty Images; p. 8 (a): Michael Darden / ZUMAPRESS.com / Newscom; (b): Mike Newman / Alamy Stock Photo; (c): BackyardBest / Alamy Stock Photo; (d): Mike Goldwater / Alamy Stock Photo; (e): TopFoto / Alamy Stock Photo; (f): Avalon / Photoshot License / Alamy Stock Photo; p. 9: AP Photo / Dan Joling, File; p. 10 (a): KIKO HUESCA / EPA-EFE / Shutterstock;

(b): REUTERS / Jeenah Moon / Newscom; p. 11: BuildPix / Construction Photography / Avalon / Getty Images; p. 12 (left a): Narendra Shrestha / EPA / Shutterstock; (right a): AP Photo / La Prensa Grafica; (b): Sankei Archive / Getty Images; (c): AGF Srl / Alamy Stock Photo; p. 13 (a): NOAA; (inset): NASA / Jeff Schmaltz, MODIS Land Rapid Response Team; p. 15: ABDIRAZAK HUSSEIN FARAH / AFP via Getty Images; p. 18 (a): Bryan Denton / The New York Times / Redux; (b): Copyright © Michael DeBock; p. 19: Stephen Marshak; p. 20: ANDERLEI ALMEIDA / AFP via Getty Images; p. 21 (a): Citizen of the Planet / Alamy Stock Photo; (b): AP Photo / Stephen Wandera; p. 23 (a): Planetpix / Alamy Live News; (b): michelmond / Alamy Stock Photo; (c): Carmen K. Sisson / Cloudybright / Alamy Stock Photo; (d): Stephen Marshak; (e): MARK RALSTON / AFP via Getty Images; p. 24: NWS Fort Worth; p. 25 (a): Judi Saunders / Alamy Stock Photo; (b): Stephen Marshak; p. 26: Professor John W. van de Lindt The NEESWood Project Colorado State University; p. 27: Stephen Marshak; p. 28 (all): Emma Marshak; p. 29: Cultura Creative Ltd / Alamy Stock Photo.

CHAPTER 2
Page 30: Stephen Marshak; p. 31: NASA; p. 32: R. Stockli, A. Nelson, F. Hasler, NASA / GSFC / NOAA / USGS; p. 33: NASA; p. 34 (top): ESA / NASA; (bottom): Google Earth; p. 35 (top): Google Earth; (bottom): NASA; p. 36 (a): Tom Pfeiffer / Alamy Stock Photo; (b): Stephen Marshak; p. 38: Stephen Marshak; p. 39 (left a): Stephen Marshak; (left b): sciencephotos / Alamy; (bottom all): Stephen Marshak; p. 40 (top both): Stephen Marshak; (granite): Stephen Marshak; (basalt): Joyce Photographics / Science Source; (gabbro): Dreamstime; p. 41 (all): Stephen Marshak; p. 42 (all): Stephen Marshak; p. 43 (shale): Science Stock Photography / Science Source; (schist): Breck P. Kent / Shutterstock; (right both): Stephen Marshak; p. 45: Wikimedia, public domain; p. 47: Deep Time Maps; p. 48: Deep Time Maps; p. 49: Bruce C. Heezen, Marie Tharp, Maurice Ewing and the Lamont Geological Observatory, Columbia University (now the Lamont-Doherty Earth Observatory); p. 55 (top): Elliot Lim, Cooperative Institute for Research in Environmental Sciences, NOAA National Geophysical Data Center (NGDC) Marine Geology and Geophysics Division; (bottom): Google Earth; p. 57 (top): Google Earth; (bottom right): Kevin Schafer / Alamy; p. 58 (b): Google Earth; (c): Nature Picture Library / Alamy Stock Photo; p. 59: Google Earth; p. 60: Google Earth; p. 62: Stephen Marshak; p. 64 (both): Stephen Marshak; p. 65 (all): Stephen Marshak; p. 66 (both): Stephen Marshak; p. 69 (top): Deep Time Maps; (bottom): Stephen Marshak.

CHAPTER 3
Page 72: Nik Wheeler / Getty Images; p. 73: North Wind Picture Archives / Alamy Stock Photo; p. 75 (a): Stephen Marshak; (b): USGS; (c left): Stephen Marshak; (c right): Stephen Marshak; p. 76 (a): Stephen Marshak; (b): NOAA / NGDC, USGS; (c): Photo courtesy of Paul "Kip" Otis-Diehl USMC, 29 Palms CA; p. 87 (b): USGS; (c): Robert Williams and Eric Jones, USGS; p. 92 (b): National Archives; (c): AP Photo / Paul Sakuma, file; p. 94 (a): Guillaume Payen / ZUMA Wire / ZUMAPRESS.com / Alamy Live News; (b): Tim Dirven / Panos Pictures; p. 96 (a): Environmental Canterbury Regional Council, New Zealand; (b): Dr. Ross W. Boulanger; p. 97: The Asahi Shimbun via Getty Images; p. 98 (a): Joseph Sohm / Shutterstock; (b): MANAN VATSYAYANA / AFP via Getty Images; p. 99 (a): NOAA / National Geophysical Data Center (NGDC); (b): Nicola Litchfield, GNS Science / Earthquake Commission; (c): AP Photo / New Zealand Herald, Geoff Sloan; (bottom): Ulet Ifansasti / Getty Images; p. 101 (c): Francois Gohier / Science Source; (bottom): National Geophysical Data Center (NGDC); p. 102 (c): Brendan Hoffman / Alamy Stock Photo; (d): Stocktrek Images, Inc. / Alamy Stock Photo; p. 103 (a): Dimas Ardian / Getty Images; (b): AP Photo; p. 107: USGS; p. 110: Dr. Yuri Fialko; p. 111 (a): USGS; (b): USGS; (bottom): Giardini, D., Grünthal, G., Shedlock, K. M. and Zhang, P.: The GSHAP Global Seismic Hazard Map. In: Lee, W., Kanamori, H., Jennings, P. and Kisslinger, C. (eds.): International Handbook of Earthquake & Engineering Seismology, International Geophysics Series 81 B, Academic Press, Amsterdam, 1233–1239, 2003; p. 116 (top b): JAVIER CASELLA / AFP / Getty Images; (bottom): M. Celebi, U.S. Geographical Survey; (bottom b): Stockbyte / Getty Images; (bottom c): AP Photo / Str; p. 119: USGS.

CHAPTER 4
Page 122: Westend61 GmbH / Alamy Stock Photo; p. 123 (a): Yale Center for British Art, Paul Mellon Collection / Bridgeman Images; (b-c): Jack Repcheck;

p. 126: USGS; p. 128: Stephen Marshak; p. 129 (rhyolite): geoz / Alamy; (granite): Stephen Marshak; (andesite): Siim Sepp / Alamy; (diorite): Stephen Marshak; (basalt): Joyce Photographics / Science Source; (gabbro): Ekaterina Kriminskaya / Dreamstime; (bottom all but obsidian): Stephen Marshak; (obsidian): vvoennyy / Fotosearch / Agefotostock; p. 131 (b): Chris German, ©Woods Hole Oceanographic Institute; p. 132 (top all): Google Earth; (bottom b): Photo by Cynthia Ebinger; p. 133 (b): Google Earth; (c): Stephen Marshak; p. 134: Stephen Marshak; p. 136 (a): Stocktrek Images, Inc. / Alamy Stock Photo; (b): Stephen Marshak; p. 137 (a): Stephen Marshak; (b): Stephen Marshak; (c): Vittoriano Rastelli / Corbis via Getty Images; (d): Stephen Marshak; (bottom left): Stephen Marshak; (bottom right): INTERFOTO / Alamy Stock Photo; p. 138 (a): Alvaro Vidal / AFP / Getty Images; (inset a): Photo by Suzanne MacLachlan, British Ocean Sediment Core Research Facility, National Oceanography Centre, Southampton; (b, d, inset d): Stephen Marshak; (c): USGS; p. 139 (inset): Travel Pix / Alamy Stock Photo; p. 140 (a): fboudrias / Shutterstock; (b): N. Banks / USGS; (bottom a-c): Stephen Marshak; (bottom d left): GeoEye / Science Source; (bottom d right): USGS; p. 141: (a): Tom Pfeiffer / www.VolcanoDiscovery.com; (b): Tom Bean / Alamy Stock Photo; p. 142 (a): ARCTIC IMAGES / Alamy Stock Photo; (b): USGS; p. 143 (a): LOTHAR SLABON / AFP / Getty Images; (b): Xinhua / Alamy Live News; p. 144 (a): Google Earth; (d): Chronicle / Alamy Stock Photo; p. 145 (left): Tom Pfeiffer / www.volcanodiscovery.com; (right): Magnus T. Gudmundson, University of Iceland; p. 146 (a): USGS; (b): Tarko SUDIARNO / AFP via Getty Images; p. 147 (a): Dr. John Crossley, www.americansouthwest.net; (b): Google Earth; (c): Earth Sciences and Image Analysis Laboratory, NASA; p. 148 (top): Lyn Topinka / USGS; (bottom): Stephen Marshak; p. 149: David R. Frazier / Science Source; p. 151 (a): Google Earth; (bottom): NOAA; p. 152 (b): Jeremiah Osuna; (c): Science History Images / Alamy Stock Photo; (d): Michael Darden via ZUMA Wire / ZUMAPRESS.com / Newscom; (e): USGS via AP Photo; p. 154 (a): Wikimedia, public domain; (b): Library of Congress; p. 155 (top a): Cesar Santana / Alamy Stock Photo; (top b): DAVID CORTES SEREY / AFP via Getty Images; (bottom a): Bruce Omori / Paradise Helicopters / EPA-EFE / Shutterstock; (bottom b): Stephen Marshak; (bottom c): Mike Goldwater / Alamy Stock Photo; p. 156: Rudianto / ZUMA Press / Newscom; p. 157 (top both): reproduced with permission of ITO World, Radar Virtuel, Flightradar24; (bottom): Gueffier Franck / Alamy Stock Photo; p. 158 (a): USGS; (b): Matthew Oldfield Editorial Photography / Alamy Stock Photo; p. 159 (b): Thierry Orban / Sygma via Getty Images; (c): Peter Turnley / Corbis / VCG via Getty Images; (bottom a): USGS; (bottom b): Stephen Marshak; (bottom c): Stephen Marshak; p. 162: Stephen Marshak; p. 164: NASA Image Collection / Alamy Stock Photo; p. 165 (a): Fiona D'Arcy; (b-c): USGS; p. 168: AB Forces News Collection / Alamy Stock Photo; p. 169 (a): Vittoriano Rastelli / Corbis / Getty Images; (b): FLPA / S Jonasson / Agefotostock; (c): Steve Marshak; p. 170 (a-b): Google Earth; (c): Wikimedia, public domain; (d): Stephen Marshak; Data from Lirer, L., Petrosino, P., and Alberico, I, 2010, Hazard and risk assessment in a complex multi-source volcanic area: the example of the Campania Region, Italy: Bulletin of Volcanology, v. 72, p. 411–429; p. 171: Westend61 GmbH / Alamy Stock Photo; p. 173: Tarko SUDIARNO / AFP via Getty Images.

CHAPTER 5
Page 174: Xinhua / Alamy Stock Photo; p. 175: Science History Images / Alamy Stock Photo; p. 176 (both): Stephen Marshak; p. 177 (all): Stephen Marshak; p. 178 (all): Stephen Marshak; p. 180 (all): Stephen Marshak; p. 181 (both): Stephen Marshak; p. 182: Stephen Marshak; p. 183 (left): National Museum of Forest Service History; (inset): Google Earth; p. 185 (all): Stephen Marshak; p. 186 (b): Marli Miller, University of Oregon; (bottom left): Mick Jack / Alamy Stock Photo; (bottom inset): PA Images / Alamy Stock Photo; p. 187 (c-d): Stephen Marshak; (e):Photo by Marta Jurchescu, first published in the volume, "Recent Landform Evolution in the Romanian Carpathians and Pericarpathian Regions" (Springer, 2012); (bottom a): Nasir Waqif / EPA / Shutterstock; (bottom b): Stephen Marshak; p. 188: Stephen Marshak; p. 189: Google Earth; p. 190 (a): Stephen Marshak; (b): BrazilPhotos.com / Alamy Stock Photo; (c): Cascades Volcano Observatory / USGS; p. 191 (a): Google Earth; (b): AP Photo / Kevork Djansezian; (both c): USGS; p. 192 (a): Stephen Marshak; (b): Hemis / Alamy Stock Photo; p. 193: Mike Crozier; p. 194: Stephen Marshak; p. 195 (top both): Stephen Marshak; (bottom a): AP Photo / Keystone, Bruno Petroni; (bottom c-e): Stephen Marshak; (bottom f): AP Photo; p. 196 (a): Wikimedia, public

domain; (b): Stephen Marshak; 197 (a): Jacques Lange / Getty Images; (b): swissdrone / Shutterstock; (c): Alaska Stock / Alamy; (d inset): Gaume, J., Gast, T., Teran, J., van Herwijnen, A., & Jiang, C. (2018). Dynamic anticrack propagation in snow. *Nature communications*, 9(1), 1–10; (e): Stephen Marshak; p. 198: Jerome Neufeld and Stephen Morris, Nonlinear Physics, University of Toronto; p. 199 (a): USGS / Barry W. Eakins; (b left): USGS, Geologic Investigations Series I-2809 by Barry W. Eakins, Joel E. Robinson, Toshiya Kanamatsu, Jiro Naka, John R. Smith, Eiichi Takahashi, and David A. Clague; (b right): Google Earth; (c): Google Earth; p. 200: David Wald, USGS; p. 201: AP Photo / Scanpix, Norway; p. 202 (all): Google Earth; p. 204: AP Photo / The Deseret News, Ravell Call, File; p. 205 (a): Stephen Marshak; (b): Ventura County Sheriff's Office via AP; p. 208: Google Earth; p. 209 (bottom a-b): NASA; p. 210 (a): AP Photo / Ted S. Warren; (b): Ralph A. Haugerud / USGS; (c both): Google Earth; p. 213: Stephen Marshak; p. 215: Stephen Marshak.

CHAPTER 6

Page 216: AP Photo / The Yomiuri Shimbun, Atsushi Taketazu; p. 217 (a): FLHC 4 / Alamy Stock Photo; (b): Orville T. Magoon, NOAA; p. 218 (a): AFP / Getty Images; (b): David Rydevik, 2004. Wikimedia, public domain; p. 220: Eric L. Geist, Vasily V. Titov, Diego Arcas, Fred F. Pollitz, and Susan L. Bilek, Implications of the 26 December 2004 Sumatra-Andaman Earthquake on Tsunami Forecast and Assessment Models for Great Subduction-Zone Earthquakes, Bulletin of the Seismologica; p. 221 (a): NASA Earth Observatory images created by Jesse Allen and Robert Simmon using data provided by Tony Song (NASA / JPL); (left c): NASA; (right c): NOAA / PMEL / Center for Tsunami Research; p. 227 (c): Kyodo News via AP, File; (bottom): STR / AFP via Getty Images; p. 229 (a): Google Earth; (b): NOAA / NWS; p. 230 (a): John R. Foster / Science Source; (b): Atwater, B.F., Musumi-Rokkaku, S., Satake, K., Tsuji, Y., Ueda, K., and Yamaguchi, D.K., 2015, The orphan tsunami of 1700 — Japanese clues to a parent earthquake in North America, 2nd ed.: Seattle, University of Washington Press, U.S. Geological Survey Professional Paper 1707, 135; p. 232 (b): Jessica Wilson / NASA / Science Source; (c): CHOO YOUN-KONG / AFP via Getty Images; p. 233 (a): JIJI PRESS / Getty Images; (b): NHK / SIPA / Newscom; (c): Horizon Images / Motion / Alamy Stock Photo; (bottom a): The Tokyo Electric Power Company / dpa / picture-alliance / Newscom; (bottom b): ZUMA Press, Inc. / Alamy Stock Photo; p. 234 (a): Google Earth; (b): FLHC 46 / Alamy Stock Photo; p. 236 (top): Google Earth; (a): Harvey Greenberg, University of Washington, from USGS data; (c): Google Earth; p. 237 (a): Google Earth; (b): PALANI MOHAN / Fairfax Media via Getty Images; (d): Eric Geist / USGS; p. 238 (a): Stephen Marshak; (b): Ricardo Ramalho; p. 239 (b): Steven Ward; p. 241 (a): REUTERS / Newscom; (b): Azizul Moqsud, Ph.D.; (c): Dr. Hermann M. Fritz; p. 244 (a): US Navy via Wikimedia; (b): AP Photo / Kyodo News; p. 245: NOAA; p. 246 (a): Ko Sasaki / The New York Times / Redux; (b-c): Potential Tsunami Inundation, Bandon Oregon, by Daniel E. Coe, courtesy of Oregon Department of Geology and Mineral Industries (DOGAMI); p. 247 (top both): Stephen Marshak; (bottom): Carl Court / Getty Images; p. 249: Google Earth.

CHAPTER 7

Page 250: Stephen Marshak; p. 251 (a): REUTERS / Mike Hutchings / Newscom; (b): AP Photo; p. 252: (a): NASA; (c): Emma Marshak; (d): Stephen Marshak; p. 253: Stephen Marshak; p. 255: (b): Stephen Marshak; (c): Google Earth; (d): Stephen Marshak; (e): Photo courtesy of the Bureau of Reclamation; p. 256: (left): Stocktrek Images, Inc. / Alamy; (center and right): USGS; (bottom): BDR / Alamy; p. 257 (a): Les Gibbon / Alamy Stock Photo; (b): Stephen Marshak; p. 259: Stephen Marshak; p. 260: Vince Streano / Getty Images; p. 261 (a): Stephen Marshak; p. 264 (a): Bryan Denton / The New York Times / Redux; (b): ARUN SANKAR / AFP via Getty Images; p. 265 (top a): World Resources Institute / Wikimedia; (top b): Map copyright IWMI, reprinted with permission; (center a): Prashanth Vishwanathan / Bloomberg via Getty Images; (bottom d): Lawrence Migdale / Science Source; p. 266 (a): USGS; (b): Google Earth; p. 267 (b): Stephen Marshak; (c): Terry Allen / Alamy Stock Photo; p. 268 (inset): Francesco Tomasinelli / Science Source; p. 269 (top b): David Wall / Alamy Stock Photo; (top c): Red Huber / Orlando Sentinel / MCT via Getty Images; (center left): Google Earth; (center right): courtesy Philip S. Prince, data from the USGS; (bottom b): GEO View; (bottom c): Clint Kromhout, Florida Geologic Survey; p. 270 (a): San Antonio Fire Department / via AP, File; (b): AP Photo / Moises Castillo; (c): Steven Harmon; p. 272 (c): Google Earth; (d): Arizona Geological Survey - Brian F. Gootee; p. 273 (a): Justin Brandt, USGS; (b): ESA / NASA-JPL / Caltech / Google Earth; (c): Gabrysch, R.K., and Neighbors, Ronald, 2005, Measuring a century of subsidence in the Houston-Galveston region, Texas, USA, in Proceedings of the Seventh International Symposium on Land Subsidence, Shanghai, PR China, October 23–28: p. 379–387. Thompson; (d): Josh Haner / The New York Times / Redux; (e): NASA / Trent Schindler and Matt Rodell; p. 275 (a): USGS; (d): Ken James / California Department of Water Resources; p. 276 (both a): Google Earth; (b): Stephen Marshak; (c): Reuters / Manuel Silvestri / Newscom; p. 277: DON EMMERT / AFP via Getty Images; p. 278 (a-b): Stephen Marshak; p. 280 (top): Philippe Rekacewicz, UNEP / GRID-Arendal; https://www.grida.no/resources/7424; (bottom): Kevin Foy / Alamy Stock Photo; p. 281 (top): European Soil Data Centre; (a, c): Stephen Marshak; (d): Patrick Pleul / picture / alliance / dpa / AP Images; p. 282 (a): Image courtesy Jacques Descloitres MODIS Rapid Response Team / NASA; (b): AP Photo / Ross D. Franklin; p. 283 (c): RGB Ventures / SuperStock / Alamy Stock Photo; (e): Library of Congress; (f): USDA; p. 284 Stephen Marshak; p. 285: Stephen Marshak.

CHAPTER 8

Page 290: Stephen Marshak; p. 291 (a): ZUMA Press / Newscom; (b): Smiley N. Pool-Pool / Getty Images; (c): Zuzana Dolezalova / Alamy Stock Photo; (d): Richard Rowe / Reuters / Newscom; p. 293 (a-b): Stephen Marshak; (b insert): Nigel Cattlin / Alamy Stock Photo; (c): Jeffrey Frame; (insert c): Ted Kinsman / Science Source; (bottom a): James Anderson, Arizona State University; (bottom b): Luis Sinco / Los Angeles Times via Getty Images; (bottom c): Travelart / Alamy Stock Photo; p. 294 (a): REUTERS / Newscom; (b): blickwinkel / Alamy Stock Photo; p. 298 (warm): Stephen Marshak; (foggy): Dan Lloyd / Alamy Stock Photo; (cool): Stephen Marshak; (hot): Stephen Marshak; p. 299: Cameron Beccario; p. 300 (a): © 2016 UCAR, photo by Scott Ellis; (b): Robert Rauber; p. 301 (a-b): NOAA / CIRA, courtesy Professor Chris Kummerow; p. 303: NASA; p. 307 (a): JeffG / Alamy Stock Photo; (b): Greg Vaughn / Alamy Stock Photo; p. 312 (top): Frans Lemmens / Alamy Stock Photo; (center): FLPA / Alamy Stock Photo; (bottom): Hemis / Alamy Stock Photo; p. 313: Image Courtesy GOES Project Science Office / NASA; p. 314 (b, d): NOAA; p. 317 (b): STR / AFP / Getty Images; (c): SAM PANTHAKY / AFP / Getty Images; p. 321: NASA; p. 323 (a): Cameron Beccario; (c): © Dennis MacDonald / age footstock; (d): Mira / Alamy Stock Photo; (e): Terry Livingstone / Alamy Stock Photo.

CHAPTER 9

Page 326: Cultura Creative Ltd / Alamy Stock Photo; p. 327 (a): NOAA; (b): NOAA; p. 328: Jeffrey Frame, University of Illinois at Urbana-Champaign; p. 329: NASA image by Marit Jentoft-Nilsen, based on data provided by the Global Hydrology and Climate Center Lightning Team; p. 330: Iowa Environmental Mesonet, Iowa State University; p. 331: Dr. Steven A. Rutledge, Colorado State University; p. 332 (top): Wild Horizons / UIG via Getty Images; (bottom): Stephen Marshak; p. 334 Andrew Fox / Alamy Stock Photo; (inset): Judith Collins / Alamy Stock Photo; p. 335: Ryan McGinnis / Alamy Stock Photo; (inset) Domenic Di Girolamo; p. 337 (top): Iowa Environmental Mesonet, Iowa State University; (bottom a): Modified from base image by Gregory Carbin, NOAA Storm Prediction Center; (bottom b): Wade Sisler; p. 339 (a): NOAA; (b): NASA; (c): Howard Bluestein / Science Source; p. 340 (a): Robert Rauber; (b): Mike Hollingshead / Science Source; p. 343 (a): Mark Bassett / Alamy Stock Photo; (b): Stock Connection Blue / Alamy Stock Photo; (c): © Dr. Mary Ann Cooper; (bottom): sciencephotos / Alamy Stock Photo; p. 344: NOAA Legacy Photo ERL / WPL; p. 345 (a): NOAA; (b): Tribune Content Agency LLC / Alamy Stock Photo; (c): NOAA; p. 346 (a-b): courtesy Steve Nesbitt, University of Illinois; p. 347 (a): Roger Hill / Science Source; (b): Dr. Jeffrey Frame; p. 348: Terra satellite, part of NASA's Earth Observing satellite system (NASA); p. 351: NOAA / Science Source; pp. 352–353: (background spread): Google Earth; p. 352 (EF0): Brendan Fitterer / Zuma Press / Newscom; (EF1): Gary Coronado / ZUMA Press / Newscom; (EF2): Benjamin Simeneta / Dreamstime; p. 353 (EF3): NOAA; (EF4): US Department of Labor / ZUMA Press / Newscom; (EF5): National Weather Service, Birmingham, AL; p. 354 (b): NOAA; (c): UPI / Alamy Stock Photo; (d): AP Photo / Mike Gullett, File; p. 355 (top): From "Understanding basic tornado radar signatures" by Kathryn Prociv, 2/14/2013; (bottom): AP Photo / Gerald Herbert; p. 356: Jeffrey Frame; p. 357: NOAA.

CHAPTER 10

Page 358: Design Pics Inc / Alamy Stock Photo; p. 360 (a): Daryl Herzmann; (b): AP Photo / David Tulis; (c): Google Earth; p. 361: ullstein bild / ullstein bild via Getty Images; p. 364 (a): philipus / Alamy Stock Photo; (b): Thomas Bethge / Alamy Stock Photo; (c): MANDEL NGAN / AFP / Getty Images; (d): North Dakota Department of Transportation; p. 366: NASA; p. 367 (b): AP Photo / Joel Page; (c): AP Photo / Jacques Boissinot; p. 368 (a): Enigma / Alamy Stock Photo; (b): Emma Marshak; (c): Bloomberg / Getty Images; p. 369 (all but bottom): NOAA; (bottom): AP Photo / Elaine Thompson; p. 370: PA Images / Alamy Stock Photo; p. 371: (a): NASA / NOAA; (d): NOAA; (e): John G. Walter / Alamy Stock Photo; p. 372: Darroch Donald / Alamy Stock Photo; p. 373: Imaginechina via AP Images; p. 374: Truba7113 / Shutterstock; p. 375 (a): AP Photo / The Bismarck Tribune, Tom Stromme; (b): AP Photo / John Raoux; p. 380: Stephen Marshak; p. 381: NASA.

CHAPTER 11

Page 382: FotoKina / Shutterstock; p. 383 (a): U.S. Air National Guard photo by Staff Sgt. Daniel J. Martinez; (b): Erika P. Rodriguez / The New York Times / Redux; (c): AP Photo / Carlos Giusti; p. 385 (both): NOAA; p. 386: NOAA; p. 387 (inset): NASA; (b): Cameron Beccario; (right): NASA / Science Source; p. 390 (c): Cameron Beccario, earth.nullschool.net, using data from GFS / NOAA; (d): Cooperative Institute for Research in the Atmosphere (CIRA) at Colorado State University, and the National Oceanic and Atmospheric Administration (NOAA); p. 391 (c): NASA Earth Observatory image by Joshua Stevens, using VIIRS data from LANCE / EOSDIS Rapid Response; p. 393 (b): NASA Goddard MODIS Rapid Response Team; (c): NASA Goddard MODIS Rapid Response Team; p. 396: NASA; p. 400 (top): NASA / Goddard Space Flight Center Scientific Visualization Studio; (b): Guy Reynolds / Dallas Morning News / MCT / Newscom; (c): U.S. Air Force photo / Staff Sgt. James L. Harper Jr; (d): Smiley N. Pool / Rapport Press / Newscom; p. 401 (a): NOAA; (b): Travis Long / The News & Observer via AP, File; (c): NOAA; p. 402 (a): Erin Bessette-Kirton, USGS; (b): AP Photo / Hans Pennink; p. 403 (all): Google Earth; p. 404 (a): Roosevelt Skerrit, released into the public domain; (b): Weather.gov; p. 405 (a): NASA / NOAA / UWM-CIMSS, William Straka III; (both b): NASA; (c): Gerben Van Es / Dutch Defense Ministry via AP; p. 407 (a): NASA; (d): POOL / AFP via Getty Images; p. 408 (a): NASA / NOAA; (b): USGS; p. 409 (a): NOAA; (b and inset): NASA; (bottom b): REUTERS / Newscom; p. 410 (a): Newscom; (b): NASA image courtesy the MODIS Rapid Response Team at NASA GSFC; (c): REUTERS / Su Sheng-bin / Pool / Alamy Stock Photo; p. 411 (both): Google Earth; p. 412 (a): NASA; (c): ADRIEN BARBIER / AFP via Getty Images; p. 413 (a): Levi Cowan www.tropicaltidbits.com; (b): NOAA; p. 414 (a-b): NOAA; p. 415: Roger Brown / Alamy Stock Photo; p. 416 (a): Rick Wilking / Reuters / Newscom; (b): Stephen Marshak; (c): Yin Bogu / Xinhua / Alamy Live News; p. 417: JOHNNY MILANO / The New York Times / Redux.

CHAPTER 12

Page 420: Stephen Marshak; p. 421 (top): Wikimedia, public domain; (bottom): DAVE ACREE / EPA / EFE / Shutterstock; p. 423 (b left): Stephen Marshak; (b right): Stuart Sly / EyeEm / Getty Images; p. 424: Stephen Marshak; p. 425 (all): Stephen Marshak; p. 426 (a): Stephen Marshak; (b): LWM / NASA / LANDSAT / Alamy Stock Photo; (c): NASA; p. 427 (a-b): Stephen Marshak; p. 428: Stephen Marshak; p. 429: Stephen Marshak; p. 434 (a,c): Google Earth; (d): Abdul Momin / Solent News / Shutterstock; p. 435 (top left): Stephen Marshak; (top right): Science History Images / Alamy Stock Photo; (bottom): Libby Solomon / The Baltimore Sun via AP; p. 436 (both a): NASA; (c): The Missouri Department of Transportation; p. 437 (all): NASA; p. 438: The Asahi Shimbun via Getty Images; p. 439 (top all): Google Earth; (bottom): NOAA; p. 440 (a): Oddur Sigurðsson, Iceland Meteorological Office. Public domain; (b): National Geographic Image Collection / Alamy Stock Photo; p. 441 (top): ako / Alamy Stock Photo; (bottom): Stephen Marshak; p. 442 (a): SCMP / Newscom; p. 443: AP Photo / The News-Star, Margaret Croft; p. 444 (left a): TVA.gov; (center a); (b): REUTERS / Alamy Stock Photo; p. 445 (b): Stephen Marshak; (c): DOD Photo / Alamy Stock Photo; p. 446 (top a): REUTERS / Alex Gallardo / Alamy Stock Photo; (top b): LUIS ROBAYO / AFP / Getty Images; (bottom a): Karen Warren / Houston Chronicle via AP; (bottom b): Stephen Marshak; (bottom c): Diarmuid / Alamy Stock Photo; p. 447 (a): NASA; (b): AP Photo / Shakil Adil; p. 450 (inset): Caro / Stefan Trappe / Newscom; p. 451: Imaginechina via AP Images; p. 452 (top a): NASA; (bottom a): Staff Sergeant Val Gempis (USAF) - Image from the Defense Visual Information Center; (top b): NASA; (bottom b): STR / AFP via Getty Images; (bottom right): Marcus Yam / Los Angeles Times via Getty Images; p. 454 (b): Kelly M. Grow / California Department of Water Resources; (c): Randy Pench / The Sacramento Bee via AP; (bottom): David Lees / The LIFE Picture Collection / Shutterstock; p. 455: Google Earth; p. 459: FEMA; p. 461 (left): Google Earth; (a-b): Stephen Marshak; (c): U.S. Army Corps of Engineers; p. 462: PJF Military Collection / Alamy Stock Photo; p. 463: Stephen Marshak.

CHAPTER 13

Page 466: Bob Sharples / Alamy Stock Photo; p. 467 (a): Google Earth; (b): Library of Congress; p. 468: NASA / SVS; p. 469 (both): www.michaelmarten.com; p. 470 (both): Stephen Marshak; p. 471 (a): Stephen Marshak; p. 473 (left): IWM / Royal Navy Archive, courtesy of the UK Ministry of Defence; (right): Stephen Marshak; p. 474 (all): Stephen Marshak; p. 475 (both): Stephen Marshak; p. 477 (all): Stephen Marshak; p. 479 (c): David Wall / Alamy Stock Photo; (d): Google Earth; (e): Stephen Marshak; (bottom a): Wikimedia, public domain; (bottom b): Stephen Marshak; p. 480 (a,d,e): Stephen Marshak; (b): Tone Coughlin / Shutterstock; p. 481 (left a-c): Stephen Marshak; (right a, c): Stephen Marshak; (right b):Reinhard Dirscherl / ullstein bild via Getty Images; p. 483 (inset): NOAA; p. 484 (top a): Keith Morris / Alamy Live News; (top b): Reuters / Newscom; (top c): AP Photo; (center a): FALKENSTEINFOTO / Alamy Stock Photo; (bottom c): AP Photo / Douanes Francaise / Avion Polmar II; p. 485: NOAA; p. 486 (a): Chris Coates; (b): Archive Image / Alamy Stock Photo; p. 487: Avpics / Alamy Stock Photo; p. 488 (a): MICHAEL LIGALIG / AFP via Getty Images; (b): HILARY HOSIA / AFP via Getty Images; p. 489 (a): AP Photo / CP,Tara Brautigam; (b): Universal Images Group / Getty Images; (c): Stephen Marshak; (d): Aerial Archives / Alamy Stock Photo; p. 490 (both a): Stephen Marshak; (b): Stephen Marshak; (both c): USGS; p. 491: Nature and Science / Alamy Stock Photo; p. 492 (all): Stephen Marshak; p. 493 (top): frans lemmens / Alamy Stock Photo; (a): imageBROKER / Alamy Stock Photo; (both b): Google Earth; (c): Tommy Trenchard / Panos Pictures; p. 494 (both a): USGS; (both b): NASA Image Collection / Alamy Stock Photo; p. 495: U.S. Army photo by Mary Markos; p. 496 (a): Woods Hole Group; (b): Stephen Marshak; (c): Hillebrand Steve, U.S. Fish and Wildlife Service; (bottom): Mira / Alamy Stock Photo; p. 497 (a): Goddard SVS / NASA; (b): Cathyrose Melloan / Alamy Stock Photo; p. 500: Stephen Marshak.

CHAPTER 14

Page 502: JOSH EDELSON / AFP via Getty Images; p. 503: Glen Morey via AP, File; p. 504 (top): NOAA George E. Marsh Album; (b): M. Brodie / Alamy Stock Photo; (c): Justin Sullivan / Getty Images; (d): Xinhua / Alamy Live News; p. 507 (b): AP Photo / Mulugeta Ayene; (c): AP Photo / Rebecca Blackwell; p. 508 (both): NOAA Climate.gov; p. 509 (both b): NASA; p. 512 (both): Stephen Marshak; p. 513 (both a): NOAA / NCEI; (b): Richard Tinker, CPC / NOAA / NWS / NCEP; (c): Copyright European Drought Observatory (EC-JRC) 2021; p. 514: NOAA; p. 515 (top both): NASA; (bottom a): ARIQ MAHMOOD / AFP via Getty Images; (bottom b): Khalil Dawood / Xinhua / Alamy Live News; p. 518 (top b both): NASA Goddard's Scientific Visualization Studio; (center both): NASA; (bottom): Nick Higham / Alamy Stock Photo; p. 522 (a): AP Photo / Noah Berger; (b): Nathan Howard / Getty Images; (c): Benjamin Rothstein; p. 523: Stephen Marshak; p. 525 (Surface): NPS / Mike Kessler; (Ground): Chris Lowie / USFWS; (Ladder): Bob Hilscher / Alamy Stock Photo; (Crown): A. T. Willett / Alamy Stock Photo; (a): Stephen Marshak; (b): Stephen Marshak; p. 527: U.S. Fish and Wildlife; p. 529 (a): Loren McIntyre / Stock Connection Blue / Alamy Stock Photo; (b): Paralaxis / Alamy Stock Photo; p. 531 (b): NASA; (c): Daniel Dreifuss / Alamy Stock Photo; (d): Joseph Sohm / Visions of America, LLC / Alamy Stock Photo; (e): NASA; (f): CSU / CIRA and NOAA / RAMMB; p. 532: map courtesy of Headwaters Economics; p. 533 (a): U.S. Department of Agriculture; (b): Shay Levy / PhotoStock-Israel / Alamy Stock Photo; p. 536 (b): Patti McConville / Alamy Stock Photo; (c): U.S. Air Force photo by Staff Sgt. Daryl McKamey; p. 537 (b): USDA / Forest Service photo by Cecilio Ricardo; (bottom): Stephen Marshak; p. 538 (b): NPS; (c): Michele and Tom Grimm / Alamy Stock Photo; (bottom): Google Earth; p. 539 (a): Dillon, Gregory K; Gilbertson-Day, Julie W. 2020. Wildfire Hazard Potential for the United States (270-m), version 2020. 3rd Edition. Fort Collins, CO: Forest Service Research Data Archive. https://doi.org/10.2737/RDS-2015-0047-3.

CHAPTER 15

Page 542: Stephen Marshak; p. 543: Stephen Marshak; p. 547 (both): Cooperative Institute for Climate, Ocean, and Ecosystem Studies (CICOES); p. 548 (top a): Bob Sacha / Getty Images (bottom a): Scimat / Science Source; (b inset): NOAA; p. 549 (b inset): Professor Jessica Conroy (c): Karim Agabi / Science Source; (c inset): NOAA; (bottom left): Stephen Marshak; p. 555 (a): NOAA / NCEI; (b): NASA / GSFC Scientific Visualization Studio; p. 556 (b): Stephen Marshak; (c): NOAA; p. 557 (left a): USGS; (right a): USGS photograph by Bruce Molnia; (bottom both): NPS / Deborah Kurtz; p. 558 (both b): NASA; (c): Nature Picture Library / Alamy Stock Photo; (c inset): Stephen Marshak; p. 559 (top left): NASA; (top right): NASA's Scientific Visualization Studio; (bottom both): NASA / Goddard Space Flight Center Scientific Visualization Studio; p. 560 (top left): Stephen Marshak; (top right a): NOAA NCEI and CICS-NC; (center b): NOAA and CiSESS NC;[FL4] p. 565: (top): Stephen Marshak; (both): Google Earth; p. 567 (top): Geological Society of America, map is based on the NOAA Etopo2.1 dataset and printed by Emanuel Soeding, Kiel University for https://www.eurekalert.org/pubreleases/2012-11/gsoa-wsa110112.php; (c both): NOAA; p. 568 (a): REUTERS / Mike Segar / Newscom; (b): REUTERS / Alamy Stock Photo; (both c): Powered by Esri and Earthstar Geographics; p. 569 (top all): Stephen Marshak; (bottom): figure created by Zeke Hausfather / Carbon Brief using CMIP5 climate model data accessed from KNMI climate explorer; p. 570–571: Google Earth; p. 572 (a): Courtesy of Vladimir Ka1tsov. Figure 1 of 'The Polar Amplification Model / lntercomparison ProJect (PAMIP) contribution to CMIP6 Investigating the causes and consequences of polar amplification" from GeoSci Model Dev, 12, 1139–1164, 2019 https://gmd.copernicus.org/articles/12/1139/2019; (b): Professor Vladimir Romanovsky, University of Alaska Fairbanks; p. 573: McPHOTO / picture alliance / blickwinkel / M / Newscom; p. 574: GRID-Arendal; p. 575 (center): IGBP, IOC-UNESCO and SCOR; (bottom a): Martin Strmiska / Alamy Stock Photo; (bottom b): Doug Perrine / Alamy Stock Photo; p. 576 (top): Figure 2.6 (panel a) from IPCC, 2014: Topic 2 - Future Climate Changes, Risk and Impacts. In: Climate Change 2014 - Synthesis Report. Contribution of Working Groups I, II and III to the Fifth Assessment Report of the Intergovernmental Panel on Climate Change, 2021; (bottom): William J. Winters / USGS; p. 578: NOAA; p. 580 (b-c): Stephen Marshak; p. 581 (top): Walsh, J., D. Wuebbles, K. Hayhoe, J. Kossin, K. Kunkel, G. Stephens, P. Thorne, R. Vose, M. Wehner, J. Willis, D. Anderson, S. Doney, R. Feely, P. Hennon, V. Kharin, T. Knutson, F. Landerer, T. Lenton, J. Kennedy, and R. Somerville, 2014: Ch. 2: Our Changing Climate. Climate Change Impacts in the United States: The Third National Climate Assessment, J. M. Melillo, Terese (T.C.) Richmond, and G. W. Yohe, Eds., U.S. Global Change Research Program, 19–67. doi:10.7930/J0KW5CXT; (bottom): Anton Raharjo / Pacific Press / LightRocket via Getty Images.

CHAPTER 16

Page 584: Prasenjeet Yadav; p. 585: Metatech Corporation in Goleta, California; p. 587 (both): NASA / SDO; p. 588 (a): Courtesy of the SOHO-MDI consortium. SOHO is a project of national cooperation between ESA and NASA; (b): Observed with the Swedish 1-m Solar Telescope (SST) by the Institute of Solar Physics; (b inset): NASA; p. 589 (a): NASA / SDO; (b): NASA / SDO / STEREO / Duberstein; (c): B.A.E. Inc. / Alamy Stock Photo; (d): NASA / SDO / STEREO / Duberstein; p. 591: NASA Earth Observatory image by Joshua Stevens, using data courtesy of NASA Ozone Watch and GEOS-5 data from the Global Modeling and Assimilation Office at NASA GSF; p. 593: Suranga Weeratuna / Alamy Stock Photo; p. 594: NOAA National Centers for Environmental Information, Katie Palubicki; p. 595 (b): US Air Force; (c both): Permit Number CP21/021 British Geological Survey © UKRI 2020. All rights reserved; p. 597 (a): NASA / JPL-Caltech / UCLA / MPS / DLR / IDA; (b): NASA / JHUAPL; (c): Science History Images / Alamy Stock Photo; p. 598 (a): Gerard Lodriguss / Science Source; (d): Science History Images / Alamy Stock Photo; (e): NASA / JPL-Caltech / UMD; p. 599 (a): Mirko Graul / Shutterstock; (b-d): Stephen Marshak; p. 601 (a): Michiko Smith / Alamy Live News; (c): Alexandros Maragos / Getty Images; p. 602: Howard Edin, Okie-Tex Star Party; p. 603 (top a): The University of Alabama Museums, Tuscaloosa, Alabama; (top b): dpa picture alliance archive / Alamy Stock Photo; (center): NASA Image Collection / Alamy Stock Photo; (bottom b): Mount Stromlo and Siding Spring Observatories, ANU / Science Source; (bottom c): H. Hammel, MIT and NASA; p. 604 (b): ITAR-TASS News Agency / Alamy Stock Photo; (c): AP Photo / AP Video; (d): Alex Alishevskikh; (d inset): Björn Wylezich / Alamy Stock Photo; (e): Yulia Airikh / Itar-Tass / ABACA / Newscom; p. 605 (a inset): NASA / JPL / UArizona; (b inset): NASA; p. 607 (all): Stephen Marshak; p. 609: (top a): USGS / D. Roddy; (top b): Stephen Marshak; (bottom a): Science History Images / Alamy Stock Photo; (bottom b): Alex Polo / Alamy Stock Photo; p. 610 (all): Google Earth; p. 611 (a-b): Google Earth; p. 611 (c): Norm Lehrman; (d): Kirk Johnson; p. 612: © Japan Aerospace Exploration Agency / JAXA / NHK; p. 616 (top): DETLEV VAN RAVENSWAAY / Science Source; (bottom): Mark Garlick / Science Photo Library / Alamy Stock Photo; p. 617: University of Hawaii Institute for Astronomy / Rob Ratkowski; p. 618: NASA / JPL-Caltech; p. 621: © Japan Aerospace Exploration Agency / JAXA; p. 623: Science History Images / Alamy Stock Photo; p. 624: Stephen Marshak.

APPENDIX

A-3 (all): Stephen Marshak; A-5: Stephen Marshak.

INDEX

Note: Page numbers in *italics* refer to illustrations, tables, and figures. Those in **boldface** indicate terms boldfaced in the text.

A

ʻaʻā, **135**, *136*, 152
Aberfan disaster, 189
ablation, **597**, 599, 601
absolute zero, 295–296, A4
abyssal plains, **49**, *49*
acceleration, 97
acceptable risk, **19**
accretion prism, **55**, *56*
accumulated depth of snow, 300
accumulation, zone of, *279*, 279
acid, 575
acidification of ocean, **575**, *575*
acid rain, 294
active faults, 64, **74**
active margins, *49*, **52**
active volcano, **161**, *162*
actuaries, 27
adiabatic expansion, **331**, *332*
adiabatic lapse rate
  dry, **332**
  moist, **332**–333
advances, 550, *551*
advection, **A5**
Aedes aegypti, 574
AEP (annual exceedance probability), 14, **456**–**457**
aerosols, 158, 160, **292**, *292*, 293
Africa
  and cyclones, 411–412, *412*
  droughts in, 506–508, *507*–*509*
aftershocks, **78**, *79*, 84
agglomerate, volcanic, 130
agricultural activity, soil degradation due to, *280*, 280–281, *281*
agricultural damage, in hazardous winter weather, 375–378
agricultural drought, 504, **505**
agricultural pests, 574
Agulhas Current, 468
air
  composition of, 292, *292*, 293
  defined, **33**, **291**
  dry, 292
  forces acting on, 306–309, *307*, *308*
  lifting of, 303, *303*
  moisture of, **297**, *297*–298, *298*, 313
  saturated *vs.* unsaturated, 332
  stable *vs.* unstable, 331, 333, *333*, 389

air blast, **601**, *604*, *616*
air bubbles in ice, 548
air burst, 601, *604*, *616*
aircraft in hazardous winter weather, *364*, *374*, *374–375*
air-fall tuff, 145
air flow in hurricane, *394*, 394–395, *395*
air masses, **317**–318, *318*
air-mass thunderstorms, 333–335, *334*
air parcel, 331–333, *332*
air pollution, 294, **294**–295
air pressure, 33, 295, *296*, **296**–297, *297*
  effect of heating and cooling on, *309*, 309–310
Air Quality Index, 295
air tankers, 535, *536–537*
air temperature, **295**–296, *296*
  and heat illnesses, 516
  and heat wave, 516, *517*
  and saturation vapor pressure, *297*, 297–298
air travel
  blizzards and, 8
  effect of volcano on, 156–157, *157*, *167*
Alaska Current, 468
Alaska tsunami, *229*, 229
albedo, **544**
  and climate change, 553
Aleutian Arc, *132*, *162*
Aleutian Trench, *112*
algal blooms, harmful, **497**, *497*, 545, *545*
alluvial fans, 426
alluvium, **424**–426, *425*, 428
alpine climate, *505*, *506*
Alpine fault, 90–91
alternative energy sources, *579*, *580*
altitude
  and atmospheric pressure, 296–297, *297*
  and temperature, *301*, 301–302
altocumulus clouds, *304*
altostratus clouds, *304*
Alvarez, Luis, 613
Alvarez, Walter, 613
Amazon River
  discharge and watershed of, 431
  drainage network of, *430*
ampere (amp), 340
amplification of seismic waves, **82**, *82*, 96
amplitude
  of seismic waves, **80**, *82*, 88
  of waves, 401, *471*
Anak Krakatoa, *144*, 149
Andean Arc, *132*
Anders, Bill, 31, *31*
andesite, *129*, 132

anemometers, 300
angle of repose, **182**, *182*
angular momentum, 394
Annan, Kofi, 591
annual exceedance probability (AEP), 14, **456**–**457**
annual probability (AP), **14**
Anopheles mosquitoes, 574, *574*
Antarctica
  ice sheet covering, 557–558, *558*
  ozone hole over, 590, *591*, 591
  sea ice in, 558, *559*
Antarctic Circumpolar Current, 468
Anthropocene, global change during, **69**, *69*
anthropogenic hazard, **7**, 294
anvil cloud, **334**, *334*
AP (annual probability), 14
Apollo 8 spacecraft, 31, *31*
Apollo asteroids, 597, *597*, 621
Appalachian Mountains, 95
  and cold air damming, *370*, 370
apparent temperature, **516**
aqua alta, 276, *276*
aquifer, **258**, *260*
aquitard, **258**, *260*
Ar (argon), in air, 33, 292, *292*
Arabian Sea, 468
Aral Sea, 256, *256*
Arctic, future changes in, 569–572, *572*
arctic air masses, *318*, 318
Arctic Ocean, 468
areal flooding, **441**, *441*
argon (Ar), in air, 33, 292, *292*
arid climate, *505*, *506*
Armenian earthquake, 116, 118–119
arrival time, 84, **85**, 96
arroyo, *427*, 427
arsenic in groundwater, 11, **262**
artesian wells, 260
artificial levees
  alternatives to, 462
  building of, *461*, 461–462
  failures of, **443**, 443–444, *444*
  reinforcement of, *461*, 462
asbestos, 11, *11*
ash
  coal, 444, *444*
  from fire, **522**
  volcanic, **138**, *139*, 160, 293
ash clouds, 35, *35*, 128, *138*
  hazards due to, *147*, 156, 156–157
ash falls, 24, 128, *153*, 166
  hazards due to, *147*, 156, 156–157
ash-fall tuff, 145

I-1

ash flow, *128*
ash-flow tuff, *145*
ash umbrella, *145*
Asia
    hazardous winter weather in, *372, 373, 373*
    monsoons in, *316, 316–317, 317, 447, 447–450, 449, 450*
assimilation of wall rock, *126*
asteroid(s), *597, 614*
    Apollo, *597, 597, 621*
    Aten, *597*
    classification of, *597, 597*
    defined, *596,* **597**
    impact of, *585, 585, 613–617, 616*
    near-Earth, *614*
    number of, *597*
    orbit of, *585, 585, 597, 597*
    size and shape of, *597, 597*
    source of, *597*
asteroid belt, **597,** *597*
Asteroid Terrestrial-impact Last Alert System (ATLAS), *619*
asthenosphere, **46,** *46*
    in plate tectonics, *51, 51, 52*
astroblemes, **607,** *607, 615*
    buried, *610–612, 611*
    exposed large, *608, 610*
Aten asteroids, *597, 597*
Atlanta, Georgia, winter storm in, *359, 360*
Atlantic hurricanes, *405–408*
Atlantic Ocean, *468*
ATLAS (Asteroid Terrestrial-impact Last Alert System), *619*
atmosphere, *290–325*
    clouds in, *302–304, 303, 304*
    composition of, *292, 292, 293*
    defined, *31,* **33–34,** *34,* **291**
    energy exchange between Earth and, *544, 544–545*
    layers of, *301, 301–302*
    of mid- and high latitudes, *317–320, 318–320*
    mid-latitude cyclones in, *320–322, 321–323*
    overview of, *291, 291–292*
    polluted, *294, 294–295*
    precipitation in, *304–306, 305*
    properties of, *295–301, 296–301*
    as shield against UV radiation, *590, 590, 591, 591*
    solar, *586, 587*
    of tropics and subtropics, *311–317, 312–317*
    vertical structure and heat sources in, *301, 301–302*
    winds in, *306–311, 307–311*
atmosphere (atm), *296*
atmospheric conditions, measurement of, *300–301, 301*
atmospheric density, *296–297, 297*
atmospheric greenhouse effect, *544–545*
atmospheric hazards, *292*
atmospheric instability, development of, *329,* **331–333,** *332, 333*
atmospheric pressure, *295, 296,* **296–297,** *297*
atmospheric rivers, **453,** *454, 510*
atom(s), *A1,* **A1**
    radioactive, *36*
atomic mass, *A1*
atomic number, **A1**
attenuation of seismic waves, *80–82, 82, 96*
August Fire Complex, *528, 528*

aurora(e), *32,* **592,** *592, 593*
    australis, *592*
    borealis, *592, 593*
    low-latitude, *593, 595, 595, 614*
Australia
    and cyclones, *411*
    wildfires in, *503, 503*
avalanche(s), *196–197, 206*
    debris, *197*
    dry, *196, 197*
    mass wasting due to, *175*
    mountain snowstorms and, *372, 372*
    slab, *196, 197*
    snow, **196,** *197, 372, 372*
    wet, *196, 197*
avalanche chutes, **196,** *197, 206*

B

backfires, **533, 534, 536,** *537*
background wind, *526*
backing fire, *528, 528*
backwash, **473,** *474*
Baffin Bay, *468*
ballistic blocks, **138,** *138, 144*
Baltic Sea, *468*
Bangladesh
    chronic health disaster in, *11*
    cyclone in, *11, 16–17*
    flooding in, *434, 447*
bank(s)
    cut, *428, 428*
    of stream, *422, 422*
bank-full stage, *431*
bar(s)
    of atmospheric pressure, *296*
    point, *428, 428*
    sand, *223*
    in stream, *426, 429*
Barbuda, and Hurricane Irma, *403, 405*
barometers, *300*
barrier islands, **478,** *479*
Barringer Crater, *608, 609*
basalt(s), *40, 40, 129*
    flood, **134, 134–135**
basalt sills, *161*
base, *575*
base isolation, in earthquake engineering, **117,** *117*
base level of stream, *423,* **424**
Basin and Range Province, *94, 112*
Basin-and-Range Rift, *162*
bathymetric maps, *48–49, 49*
    of submarine slides, *198, 199*
bathymetry, **34,** *48–49, 49*
bay(s), names of, *467, 468*
baymouth bar, *478, 479*
Bay of Bengal, *468*
beach, *475, 477–478*
    active, *476*
    characteristics of, *477, 477–478, 479*
    defined, **477**
    pebble vs. sand, *477*
beach erosion, *490,* **490–491,** *491, 499*
beach face, *477, 478*
beach grasses, *495, 496*
beach houses on stilts, *496, 496*
beach nourishment, *493–495,* **493–495,** *499*
beach nourishment episodes, *493–495*

beach profile, *477,* **478**
beach restoration, *494, 494–495, 495*
beach stabilization, **495–496,** *496*
Bear Glacier, *557*
bed(s)
    sedimentary, **41**
    of stream, *422, 422, 431, 432*
bedding, sedimentary, *41, 42*
bed load, *424, 425*
bedrock, **39,** *39*
benches
    excavation of, *211, 213*
    wave-cut, **476,** *476, 480, 489, 489, 498*
Benguela Current, *468*
Bering Sea, *468*
berm, *477*
Bermuda High, *511, 511–512*
Bhola, Bangladesh cyclone, *11*
Big Thompson River, Colo., flash flooding of, *438–439, 439, 445, 455, 455*
Bingham Canyon copper mine, rock avalanche in, *203*
biodiversity, *612*
biofuels, *579*
biogeochemical cycles, **68**
biological disaster, *7, 8, 9*
biological hazards along coasts, *497, 497*
biological indicators of climate change, *558–559, 560*
biosphere, *31,* **33,** *33*
    future changes in, *572–573, 573, 574, 574*
Black Death, *10*
black lung disease, *11*
Black Sea, *468*
black smokers, *32,* **131,** *131*
blizzard, *363–364*
    and air travel, *8*
    deaths and economic impact of, *16*
    defined, **363,** *364*
    ground, **364**
    and mid-latitude cyclone, *323*
    whiteout conditions due to, *291, 363*
blizzard warning, *359*
BLM (Bureau of Land Management), *22*
blocks, **138,** *138, 144*
bluffs, *191, 424*
body temperature, normal, *514–515*
body wave(s), **80**
body-wave magnitude (M_B) scale, *88*
bolide, **601,** *602, 615*
bolts, in earthquake engineering, **117,** *117*
bora, *372*
Borman, Frank, *31*
boundary(ies), lifting along less distinct, *330*
boundary layer, *307*
bracing, in earthquake engineering, **117,** *117*
braided streams, **428–429,** *429*
Brazil, shantytowns in, *20, 20*
Brazil Current, *468*
breach, *494, 494, 495, 495, 498, 499*
breaker, **473,** *474*
breakwaters, *402,* **492,** *498*
breccia
    due to meteor, *605–607, 606*
    volcanic, *129, 130*
British Columbia, heat wave in, *514*
brittle deformation, **62,** *64*
bromine, *616*

Bubonic Plague, 10
building codes
    for earthquake mitigation, 25, 113, 118–119
    for hurricane damage mitigation, 417
building damage from earthquakes
    due to ground shaking, 97, 97
    in Haiti, 102, 102–103
    mitigation of, 25, 26, 114, 114–119, 116–119
building permits and water shortage, 263
built environment, 441
Bull Shoals Dam, 461
buoyancy, 303
buoyancy force, 445, 445, A3
Bureau of Land Management (BLM), 22
Burin Peninsula tsunami, 235, 236
burns and burning, 522
    controlled (intentional, prescribed), 529, 529, 537, 538, 538
bushfires, 503, 503
business-as-usual scenario, 566, 566
Byron (Lord), 160

C

calamities, 14
Calbuco, Chile, volcano, 155
calcite, **38**
    and sinkholes, 266
calcium carbonate in groundwater, 261
caldera, **139**, 140, 147, 147–149
California
    delta plain subsidence in, 275, 276
    drought in, 509, 509–511, 510
    wildfires in, 521, 522, 528, 528–532, 529, 531, 532
California Current, 468
calories, **A4**
Camp Fire, 530, 531
canals, diversion, 462
Canary Current, 468
canopy, 524, 525
canopy fires, 524, 525, 535
canopy fuel, 525
Canyon Diablo meteorites, 608
capacity of stream, **424**
cap-and-trade policies, **580**
Cape Hatteras, N.C., 485
Cape Hatteras Lighthouse, 491, 491
Cape Town, South Africa, water crisis in, 251, 251, 263
Cape Verde Islands, flank collapse on, 238, 238
carbon budget, 563–565, **563–565**
carbon capture and sequestration, **579–580**
carbon credits, 580
carbon cycle, **68, 563**
    and climate change, 563–565, 563–565
carbon dioxide (CO_2)
    in air, 33, 291, 292, 292
    evolution of life and, 551
    as greenhouse gas, 545
    and infrared radiation, 544
    ocean formation and, 551
    rising atmospheric concentrations of, 561–565, 561–565
    volcanic eruptions and, 551
carbon emissions, reduction of, 578–580, 579, 580
carbon footprint, **578–580**, 579, 580
carbon-neutral energy sources, **579**
carbon sinks, **564**, 564
carbon sources, **564**, 564, 565

Caribbean hurricanes, 403–405, 403–405
Caribbean Sea, 468
Carrington event, **595**, 595
Cascade Volcanic Arc, 162
Cascadia
    ghost forests of, 230, 230–231
    megathrust hazard of, 112, **113**, 113
Caspian Sea, 468
casualties, **7**
    by disaster type, 16, 16
    from most deadly natural disasters, 11
catastrophes, **14**
catchment, 429, 429
cave networks, **266–267**
cavern, **266**
cellulose, 522
Celsius scale, 295, 296, A4
cement, **39**, 39
Center for Near-Earth Object Studies, 617
centigrade scale, 295, 296
Central American hurricanes, 405, 408
Central Arizona Project Canal, 255, 256
Centralia, Penn., coal-bed fire in, 277, 277
central uplift, 605, 606, 606
centrifugal force in hurricane formation, 392, 393
Ceres, 597, 597
CFCs (chlorofluorocarbons), 591, 591
CH_4. See methane (CH_4)
chain link fencing, 212, 213
channel, 422, 422, 424
channel currents, 482, 483
Channeled Scablands, 440, 441
charge, A1
charge separation, **340**
Charleston, S.C., earthquake, 87, 95
Chelyabinsk event, 604, 604
chemical(s), organic, 38
chemical bond, **A1**
chemical energy, **A3**
chemical formula, A1
chemical reaction, **A3**
chemical weathering, 40, 41
Chennai, water crisis in, 264, 264
Chesapeake Bay crater, 610, 611
Chicago
    heat wave in, 514, 514
    rising water levels in, 487
Chi Chi earthquake, 96
Chicxulub crater, 240, 240, 611, **611**–612, 613–617, 616
Chile
    earthquake in, 93
    tsunami in, 228
China
    and COVID-19 pandemic, 10
    drought in, 505
    earthquakes in, 11
    flooding in, 5, 6, 11, 433, 444, 447–450, 451, 462, 463
    hazardous winter weather in, 372, 373, 373
chinook winds, 528
chloride (Cl^-) in groundwater, 257
chlorine, 616
chlorofluorocarbons (CFCs), 591, 591
chondrites, 599, 599
chondrules, 599, 599
Christchurch, New Zealand, earthquake, 98, 99
chronic health disaster, **11**, 11
cinder(s), **138**, 138

cinder cones, 139, **141**, 141, 151, 167, 171
cirrocumulus clouds, 304
cirrostratus clouds, 304
cirrus clouds, 304, 304
Clark, William, 423
clast(s), **40–41**, 41, 479
clastic rocks, **39**, 39
clay, **40**, 41
    quick (sensitive), 99, 201, 201, 206
cliff, 176, 177
    sea, 477, **478**
cliff retreat, **489**, 489, 498
climate(s)
    classification of, 545–546, 545–547
    controls on Earth's, 543–546, 544–547
    defined, **34, 295, 545**
    polar, **317**
    and slope strength, 176
    temperate, **317**
    variation with location of, 545–546, 545–547
    volcanoes and, 160, 160–161
climate change, 542–583
    adapting to, 580–581, 581
    biological indicators of, 558–559, 560
    causes of, 550–553, 552, 553, 561–565
    changes in arctic due to, 569–572, 572
    changes in biosphere due to, 572–573, 573, 574, 574
    due to changes in CO_2 and CH_4 concentrations, 561–565, 561–565
    changes in global weather patterns due to, 559–561, 560
    changes in hazardous weather due to, 577
    changes in ocean due to, 573–577, 575, 576
    classification of climates and, 545–546, 545–547
    controls on Earth's climate and, 543–545, 544
    coral bleaching due to, 575, 575
    defined, 546–**547**
    degradation of fisheries due to, 576, 576
    and disaster risk, 20
    effects of, 11
    evidence of recent, 554–560, 554–561
    feedbacks affecting, 553
    global climate models to predict, 562–563, 563, 566, 566–567, 567
    impacts and hazards of, 566–578
    increased in heat waves and droughts due to, 577–578, 578
    international accords on, 581
    interruption of thermohaline circulation due to, 577
    long-term, 550–552
    melting glacial and sea ice due to, 557–558, 557–560
    melting of methane hydrate due to, 576, 576–577
    mitigation of impact of, 578–581, 579–581
    ocean acidification due to, 575, 575
    over course of history, 548–550, 550, 551
    overview of, 432, 543
    and paleoclimate indicators, 547–548, 548, 549
    recent, **554**
    redistribution of water resources due to, 567–569, 569
    reducing emissions for, 578–580, 579, 580
    removing greenhouse gases from atmosphere for, 580
    sea-level change due to, 556, 556–557, 567, 567, 568, 568, 570–571

climate change (continued)
  short-term, 552, 552–553, 553
  temperature change due to, 554, 554–555, 555
  visualization of, 570–571
climate feedback, **553**
climate groups, 506, 506, 546
climate types, 505, **506**, 506
climate zone(s), 505, **506**, 506
climate zone migration, 572–573, 573
climatological average, 516
climatological disaster, 7, 9
climatological drought, 504, 505
Cl⁻ (chloride) in groundwater, 257
cloud(s), 32, 33, 34
  anvil, **334**, 334
  classification of, 304, 304
  cold, 305
  composition of, 292, 293
  convective, 145, 145
  defined, **302**
  and energy, 305, 305–306
  formation of, 302–304, 303
  funnel, 349
  shelf, **335**, 335
  view from space of, 302, 303
  wall, **338**, 339, 350
  warm, 304–305
cloud cover, 295, **299**
cloud-to-cloud lightning, **340**, 340
cloud-to-ground lightning, 340–344, **341**–343
CME (coronal mass ejection), **588**–589, 589, 594, 595
CO_2. See carbon dioxide (CO_2)
coal ash, 444, 444
coal-bed fires, 277, 277
coast(s), 466–501
  currents along, 467–469, 468
  defined, **467**
  organic, 477, 478–482, 481
  overview of, 467
  protection of, 491–496, 492–496
  rocky, 475, 477, 478, 480
  sedimentary (sandy), 475, 477, 477–478, 479
  structure of, 475, 475–477, 476
  tides along, 469, 469–471, 470
  waves along, 471–474, 471–475
coastal challenge(s), 487–497
  beach erosion as, 490, 490–491, 491
  nuisance tides as, 487, 487, 488, 488
  sea-cliff retreat as, 489, 489
coastal disaster, 7, 8, 9
coastal flooding, 498
coastal hazards, 482–486
  biological, 497, 497
  due to currents, 482, 483
  due to hurricanes, 467, 467
  in southern Portugal, 500
  due to tides, 483–484, 485
  visualization of, 498–499
  due to waves, 484–486, 484–486
coastal plains, 274, 475, 476
  subsidence in, 274–276, 276
coastal properties, protection of, 491–496, 492–496
coastal sand dunes, 475, 477, 478
coastal vulnerability index, 491
coastal wetlands, 478, 481
Coast Guard, 22

coesite, 607, 613
coherent material, 176
cohesion, **182**
cold, 360–361, 361–363
cold air damming, 370, 370
cold cloud(s), 305
cold-cloud precipitation, 305
cold front, **318**, 319
cold pool, 330, 336, 336
cold waves, **362**, 363
collapse sinkholes, **267**–268, 269
collision, **56**–59, 58
collisional mountain belt, 57, 58, 94
Colorado
  flash floods in, 438–439, 439, 445, 455, 455
  wildfires in, 521
Colorado River
  depletion of, 255, 255–256
  salinification of, 257
Columbia River flood basalts, 150
Columbia River Plateau, 150
column failure, 114, 116
coma, 597–598, 598
combustion
  flaming, 523
  smoldering, 523
  spontaneous, 524
combustion stage, 522–523, 523
comet(s), 596, **597**–598, 598, 614
compaction, 41, 274
competence of stream, **424**
complex craters, 605, **606**, 606
complex landslides, **197**
complex slope, 178
composite volcanoes, 141, 141–142, 142
compositional banding, 43, 43
compound, **A2**
compression, 63
concave up slope, 176, 178
condensation and energy, 305, 306
condensation nucleus, 302
conditionally unstable air, 333, 333
conduction, 523, 545, **A4**, A5
conduit of volcano, 139
cone of depression, 260, 260
cone of uncertainty of hurricane tracks, 413, 413
conservation of energy, law of, A2
conservation of water, 263–266, 265
constructive interference, **472**, 473
contact force, A2, A3
containment of wildfire, **533**
contaminant plume, 262, 263
contamination
  of groundwater, 11, 261–263, 263
  of soil, 280
continent(s), **35**, 35
  and climate change, 552
continental air masses, 318, 318
continental arc, **56**, 56, 132, 132
continental collision, 56, 57
continental collision zones, earthquakes in, 94, 94
continental crust, 44–45, 45, 52
continental divide, **429**, 430
continental drift, 47, 47–**48**, 48
continental margins
  active vs. passive, 49, **52**
  submarine slumps along, 235–236, 235–237, 237

continental mid-latitude cyclones and winter weather, 366, 366–368
continental rifts, igneous activity in, 130, 132, 132
continental shelves, 34, 35, **49**, 49, 51, 60, 467
continental slope, 51
continental transform faults, 90–91, 92, 94
controlled burns, 529, 529, 537, 538, 538
convection
  defined, 36, **36**–**37**
  in energy exchange between Earth and atmosphere, 545
  heat transfer via, **A4**–**A5**, A5
  in ignition of wildfire, **526**, 526, 527
  within mantle, 45
convective cells, **312**, **A5**
convective cloud, 145, 145
convective column, 145, 166
convective zone, 586, 586
convergence lifting, 330
  in hurricane formation, **389**–390, 390
convergent boundary(ies), 52, 53, **54**–**56**, 55, 56
  igneous activity at, 130, 131–132, 132
convergent-boundary seismicity, 92–93, 93
convergent wind, 526, 527
convex-concave slope, 178
convex up slope, 176, 178
cooling and surface air pressure, 309, 309–310
cooling centers, 516
coral bleaching, 481, 482, 499, **575**, 575
coral reefs, 475, 480, **481**–472
Coral Sea, 468
core, 37, 37, 43, 44, 45, **46**
core body temperature, 361
Coriolis effect, **308**, 308
Coriolis force, 307–**309**, 308
corona, 586
coronal mass ejection (CME), **588**–589, 589, 594, 595
coronavirus, 7, 10, 10–11
  preparedness planning during, 25–26
corrosion and collapse of condominium complex in Florida, 484
cosmic speed, **600**
costs
  of disaster, 26, 26–27
  of pandemics, 11
cover-subsidence sinkholes, 267
COVID-19 pandemic, 7, 10, 10–11
  preparedness planning during, 25–26
cracks
  radial, 186
  transverse, 186
crater(s)
  complex, 605, **606**, 606
  defined, **138**, **605**
  formation of, 605, 605–606, 606
  simple, 605, **605**–606, 606
  transient, 605, 606
  visible, 608, 609
Crater Lake, 146, 147, 147, 149
Craters of the Moon, 162
creep, **184**, 184, 185
creeping disaster, 9–11
crest
  of ocean wave, 471, 471
  of seismic wave, 80, 82
  of wave train, 220
critical zone, **251**

crop damage in hazardous winter weather, 375–378
crossbeams in earthquake engineering, 117
cross-braces in earthquake engineering, 117, *117*
crown, of trees, 524, *525*
crown fires, **524**, *525*, 535
crust, 37, *37*, 43, *44*, **44–45**, *45*
crust-mantle boundary, 44, *45*
cryosphere, 31, **34**, *34*
crystal, **38**, *38*
    platy, *38*
crystalline rocks, **39**, *39*
crystalline texture of igneous rock, **128–129**, *129*
crystallization, fractional, **126**
crystal structure, 38
culverts, 442
cumulative probability, 457
cumulonimbus clouds, 304, *304*
cumulus clouds, 304, *304*
cumulus congestus clouds, *304*
current(s), 467–469
    channel, 482, *483*
    and climate change, 552
    dangerous, 482, *483*
    deep-ocean, 467
    defined, **467**
    longshore, **482**, *483*, 499
    nearshore regional (tidal), 482, *483*
    rip, **482**, *483*
    structure-related, 482, *483*
    surface, **467–469**
    warm *vs.* cold, *468*
    and water circulation, 218
curtain of lava, 139, *140*
cut bank, 428, *428*
cyclone(s)
    changes in severity of, 560–561
    deaths and economic impact of, 16, *16–17*, *17*
    defined, **383**
    of Indian Ocean and Southern Hemisphere, 411, 411–412, *412*
    mid-latitude (extra-tropical). *See* mid-latitude cyclones
    naming of, 384
    tropical, 16, *17*, *17*, **383**, 451, *452*
Cyclone Amphan, 415–416
Cyclone Bhola, 411, *411*, 415
Cyclone Idai, 411–412, *412*
Cyclone Kenneth, 411–412, *412*
Cyclone Tracy, 411
cyclonic flow, 392, *392*

D

dam(s)
    defined, 439
    failures of constructed, 442–443
    flash floods due to breaching of natural, 439–441, *440*, *441*
    for flood control, 460, *461*
    negative consequences of, 460–461
danger, language of, 7–13
DART (Deep-Ocean Assessment and Reporting of Tsunamis) buoys, **244**, *245*
dart leaders, 341, *342*
Day Zero in Cape Town, South Africa, 251, *251*, 263
"dead pool," 256
dead zones, 497, *497*
death toll, *15*, 15–16

debris avalanche, 197
debris falls, 184, 194–196, *195*, *196*
debris fans, 198
debris flows, 187–192
    defined, 184, **187**
    examples of, *187*, *189*, 189–192, *206*
    speed of, 189
    submarine, 198, *198*
debris slides, 184, 187, 193–194, *208*
decompression melting, *124*, **125**
deep-focus earthquake, 78, *93*
Deep-Ocean Assessment and Reporting of Tsunamis (DART) buoys, **244**, *245*
deep-ocean currents, 467
deep-sea trenches, 35, **49**, *49*, *52*, 54, *55*
deep time, 66, *66*, *67*
deflection of near-Earth objects, 619–620, *621*
deforestation
    changing slope strength due to, 204, *205*
    defined, **21**, *21*
    soil degradation due to, 280, *280*
deformation, **62**, *63*, *64*, 73
degree, 295, *296*
de-icing, 374
delta(s), 274, **426**, **426–427**
    swampy, *475*
delta plains, 274–276, *275*, 426, **427**, 434
Denali fault system, 112, 118
dengue fever, 574
density, A2
deposition, **41**
    and energy, 305
    by streams, 424–427, *425*, *426*
derecho, **336**, *337*
desalination, 253
desert air and heat waves, 519, 520–521
desertification, 287
desiccation cracks, 285
developing stage of thunderstorm, *334*, 334–335
Diablo winds, 528, 529, 530
diagonal crossbeams in earthquake engineering, 117
Did You Feel It? map, 87, *87*
differentiation, 37
diffusion, solid-state, 38
dikes, 54, *54*, 128, *128*, 139
diorite, *129*
dip, *445*
disaster(s), natural. *See* natural disasters
disaster-risk assessment, 27
disaster-risk reduction, 25
discharge, 256, **430–431**, *431*
    and flood stage, *432*, 433
discharge area, 258
disease
    due to earthquakes, 103
    global poleward spread of, 574, *574*
displaced people, **14**, *15*
displacement, **64**, *65*, 78–79, *79*
disrupted bedding, 200, *201*
dissipating stage
    of hurricane, 383, 393
    of thunderstorm, *334*, 335
dissolution sinkholes, 267
dissolved load, 424, *425*
distant tsunamis, 220–221, 243
distributaries, 426, 427

divergence, **320–321**
    high-altitude, 392, *392*
    low-altitude, 393
divergent boundary(ies), **52–54**, *52–55*
    igneous activity at, *130*, 130–131, *131*
divergent-boundary seismicity, 90, *91*
diversion canals, 462
divide, drainage, **429**, *429*, 430
Dixie Fire Complex, 528
Dominica and Hurricane Maria, 404, *404*
Donner Pass, 372
Doppler radar, **300**
dormant volcano, **161**, *162*
downbursts, **344**, 344–345
downcutting, 422, *422*
downdrafts
    defined, 299, **328**
    in thunderstorm formation, 330, 335, *336*
    in tornado formation, *348*, 348–350, *349*
downgoing plate, 54, *55*
downgrading of hurricane, 393
downslope stress, changing, 201–203, *202*, *203*
downslope winds
    defined, **528**
    and droughts, 511
    and heat waves, 521
    and wildfires, 528, *529*, 530–531, *531*
downstream floods, 421, 432–434, 434–437, *436–437*
drainage basin, **429**, *429*
drainage divide, **429**, *429*, 430
drainage networks, **429**, *429*–430, *430*
drawback, **223**, 227, *227*
drawdown
    of groundwater, 260–261, *260–262*
    of tsunami, 223, 227, *227*
drip irrigation, 265
drip torches, **536**, *537*
driving in hazardous winter weather, 374
drizzle, 292
dropsondes, 413
drought(s), 504–514
    agricultural, 504, **505**
    and climate zones, 505, *506*, 506–512
    deaths and economic impact of, 16, *17*
    defined, **6**, *8*, 503, **505**
    future increases in, 577–578, *578*
    and heat waves, 521
    hydrological, 504, **505**
    in India, 264, *264*
    in Mediterranean climate zones, 509–511, *509–511*
    mega-, **505–506**
    meteorological (climatological), 504, **505**
    monitoring and predicting, 512–513, *513*, *514*
    in monsoonal climate zones, 506–508, *507–509*
    in other temperate climate zones, 511, *511*–512
    overview of, 504, 504–505
    positive feedbacks that amplify, 512, *512*
    socioeconomic, 504, **505**
    in Somalia, *15*
    types of, 505
    and water shortage, 256
drowned river valleys, 475, *476*, 476
dry adiabatic lapse rate, **332**
dry air, 292
dry avalanche, 196, *197*
dry-land farming, 282
dry lightning, 524

dry wash, **427**, *427*
dry well, 260, *260*
dunes, *475, 477, 478*
Dust Bowl, **282**–283, *283*, 504, *504*–505
dust storms, 282, **282**–283, *283*, 284
dust tail, 598, *598*
dwarf planets, 594, *596*
dynamic pressure, 157
    of floodwaters, 445, *445*
    of hurricane, 397, *397*
    of tsunami, **226**, *227*
    of wind, 299

E

early warning systems for earthquakes, 110, *110*
Earth (planet)
    energy exchange between atmosphere and, *544*, 544–545
    formation of, 37, *37*–38
    layers of, 43–46, *44–46*
    materials of, 37–43
    orbital shape of, 552, *552*
    protective shields of, 589–592, *590–593*
    tilt of axis of, 552, *552*
    topography and structures of, 61–65
    view from space of, 31, *31*, 32, *32*
earthflow. *See* mudflows
earthquake(s), 72–121
    adjectives for describing, 89, *89*
    building damage due to, 25, *26*, 97, *97*, 102, *102*–103, 114, *114*–119, *116–119*
    casualties due to, 11, 119, *119*
    causes of, 72–79
    clusters of, 107, *107*
    comparing size of, 13
    in continental collision zones, 94, *94*
    damage due to, 96–105, *96–106*
    deaths and economic impact of, *16*, 17
    defined, **3**, **73**
    disease due to, 103
    distribution of, 50, *50*
    early warning systems for, 110, *110*
    energy release by, 90, *90*
    epicenter of, **78**, *78*, 85, *85*–86
    faults and faulting and, 73–74, *74–76*
    fire due to, 73, 100, *101*
    focus or hypocenter of, **78**, *78*, 89, *89*, 93
    foreshocks, mainshocks, and aftershocks of, **78**, 79
    ground rupture due to, 96, *96*
    ground shaking due to, 73, 86, 96–98, *97*
    in Haiti, 11, 102, *102*–103
    hazard maps for, *107*, 110–111, *111*
    due to induced seismicity, 95, *95*
    intensity of, 86, **86**–87, *87*, 89, *89*–90
    intraplate, 94, **94**–95, *95*
    landslides due to, 98, *98*
    location of, *78*, 78, 90–9691–95
    magnitude of, **87**–90, *88*, 89
    measuring, 73, 83–86, *84*, 85
    megathrust, **93**, *93*, 112, 113, *113*, 228–232, *229–232*, 243
    micro, 89
    not at plate boundaries, 93–95, *94, 95*
    overview of, 72, *72*
    paleoseismicity studies of, 107–108, *108*
    at plate boundaries, 90–93, *91–93*
    predicting, 106–112, *107–111*
    prevention of damage and casualties due to, 112–119, *114, 116–117*
    primary, secondary, and tertiary disasters from, 12, *12*
    probability of, 106–107, *107*
    recurrence interval of, 106–107
    regions of seismic risk for, 111–112, 113, *113*
    resonance disasters due to, 100, *101*
    in rifts, 94
    satellite data on, 109–110, *110*
    sediment liquefaction due to, 98–100, *99, 100*
    seismic gaps and, 108, *109*
    seismic waves and, **73**, 76–78, *77*, 79–83, *80–83*
    size of, 86–90, *86–90*
    slip from, 73, 78–79, *79*
    slow, **108**
    tsunamis due to, 73, 100–103, *103*, 217, *217*, 218, *218*, 228–233, *228–233*
    visualization of, 104–105
earthquake early warning systems, 110, *110*
earthquake engineering, **115**–118, *116–119*
earthquake hazard maps, *107*, 110–111, *111*
Earthquake Lake, 182, *183*
earthquake potential map, 24
earthquake preparedness drills, 119
earthquake-resistant designs, **115**–118, *116–119*
earthquake zoning, **118**–119, *119*
Earthrise (photograph), 31, *31*
Earth System, 32–37
    atmosphere in, 31, **33**–34, *34*
    biosphere in, 31, **33**, *33*
    cryosphere in, 31, **34**, *34*
    defined, **7**, **31**
    energy in, 35–37, *36*
    geosphere in, 31, **34**–35, *35*
    hydrosphere in, 31, **34**
    magnetic fields in, **32**–33, *33*
    overview of, 31
    realms of, **31**, *32*
    solar wind in, 33
East African Rift, 59, *59*, 94, *94*, 132
East Australian Current, 468
East coast of United States, seismic risk in, 112
easterly wave in hurricane formation, **390**–391, *391*, 392
East Greenland Current, 468
East Pacific Rise, 49
ebb tide, 470–471
eccentricity and climate change, 552, *552*
economic impact of disasters, **14**–17, *15–17*
economic scarcity, 263
eddy(ies), *432*, **469**
Edenville Dam, 442
effusive eruptions, **142**, *142–143*
EF (Enhanced Fujita) scale, 13, **351**–354, *351–355*
ejecta, **605**, *605*
ejecta blanket, **606**, *606, 609*, 613
ejecta curtain, 606
elastic deformation, **76**–77, *77*
elastic-rebound theory, **77**, *77*
El Capitan, Yosemite National Park, Calif., 3, *4*
Eldfell volcano, 169, *169*
electrical grid, 585, *585*, 595, *595*
electrical insulator, 340
electromagnetic energy, **A3**
electromagnetic radiation, 543, *544*
electromagnetic spectrum, 543, *544*
electron(s), A1, *A1*
electronic equipment, damage to, 594
electrostatic discharge, 340, *340*
electrostatic potential, 340
element, **A1**
elevation and surface temperature, 545
Ellicott City, Md., flash flooding in, 434–435, *435*
Elm, Switzerland, rockfall in, 196, *196*, 203
El Niño, **314**, *315*, 315, 508
El Niño/Southern Oscillation (ENSO), **315**
El Reno, Okla., tornado, 347, *349*
Embarras River, annual peak discharge for, 456, *456*
embayments, **478**, *480*
embers, 526
emergency action plans, 354
emergency response, 354
emergency response team, 327
emissions, reduction of, 578–580, *579, 580*
energy
    alternative sources of, 579, *580*
    chemical, **A3**
    clouds and, 305, *305*–306
    in Earth System, 35–37, *36*
    electromagnetic, **A3**
    external, 32, **36**, *36*
    for hurricane, **388**, 388–389, *389*
    internal, 32, **36**, *36*
    kinetic, 295, 613, **A3**
    law of conservation of, A2
    nuclear, **A3**–A4, *A4*
    potential, **A3**
    thermal, 295, **A3**, A4
energy density, 579
energy exchange between Earth and atmosphere, *544*, 544–545
energy release by earthquakes, 90, *90*
engineered structures, 115
engineering, earthquake, **115**–118, *116–119*
Enhanced Fujita (EF) scale, 13, **351**–354, *351–355*
ensemble simulations of hurricane tracks, 413, *413*
ENSO (El Niño/Southern Oscillation), **315**
environment
    of air parcel, 331
    built, 441
environmental lapse rate, **302**, 331–332
eons, 66, *67*
ephemeral streams, **427**, *427*
epicenter, **78**, *78*, 85, *85*–86
epidemics, 7
epochs, 66, *67*
Equatorial Counter Current, 468
equilibrium level
    of water, 219, *219*, 220
    of waves, 471, *472*, 472
eras, 66, *67*
Eros, 597
erosion, 280–284
    causes of, 280–284, *281*, 286–287
    defined, 11, **35**, **176**, **280**
    in Dust Bowl, 282, *282*–283, *283*
    and mass wasting, **176**
    rates of, 280, *281*
    by streams, 424–427, *425, 426*
erosion-control mats, 495, *496*
eruptive jet, **145**, *145*, 148, *149*, 166
eruptive style, 124, **150**–154, *151–155*

estuaries, *475*, **476**
Europe
　flooding from mid-latitude cyclones in, 453–455, *454*
　heat waves in, 514, *515*
eutrophication, **257**, *257*
evacuation(s)
　defined, 14, *15*, 21
　for hurricanes, 414–416, *416*
　for tsunamis, 245, *247*
evacuees, 14, *15*, 21
evaporation and energy, *305*, 306
evaporite sinkhole, *270*, 270
Everglades National Park, 273, *274*
evolution and climate change, 551
excavation stage of crater formation, 605
excessive heat warning, 517
exercise and heat illnesses, 515
exfoliation joints, 181, *181*, *183*
exothermic reactions, 522
expansive soil, *284*, 284–**285**
explosive eruptions, 142, 143–147
　classification of, 145–147, *146*
　defined, **143**
　hazards due to, 158–160
　due to interaction with water, 143, *143*–145, *144*
　of Mt. St. Helens, *148*, 148–149, *149*
　due to trapped gas bubbles, 145, *145*, *146*
　tsunamis due to, 238–239, *239*, *240*, *243*
exposure
　defined, **17**
　and risk, 18–19
external energy, *32*, **36**, *36*
extinction(s)
　defined, 612
　mass, *612*, 612–617, *616*
　sixth, **613**
extinction-level events, *596*, *612*, 612–617, *616*
extinct volcanoes, 60, **161**, *162*
extraction wells, 263
extraterrestrial disaster, 7, *8*, *9*
extra-tropical cyclones, 317, 320–322, *321*–*323*
extrusive igneous rocks, **128**, *128*
extrusive rocks, 40, *40*
eye of hurricane, **385**–**386**, *385*–*387*, *394*–*395*, *395*
eye wall of hurricane, **385**–**386**, *385*–*387*
Eyjafjallajökull, 158–159, *159*

F

Fahrenheit scale, 295, *296*, A4
failure, strength and, **176**
failure surfaces, **182**–**183**, *183*, *184*, *207*
faint young Sun, 551–552
Falling Rock Zone signs, 196
famine
　in Africa, 507, *507*
　death toll from, 16
far-field tsunamis, **220**–**221**, *243*
fatalities, number of, *15*, 15–16
fault(s) and faulting, 73–74
　active vs. inactive, 64, **74**
　in astroblemes, *607*, 607
　classification of, *74*, 74
　consequences and appearance of, 74, *75*, *76*
　defined, **50**, **73**
　development of, 73–74
　in geosphere, 35

normal, **74**, *74*
reverse, **74**, *74*
strike-slip, **74**, *74*, *75*, *76*, *96*
structure of, *64*, *65*
thrust, **74**, *74*, *96*
fault creep, 108
fault plane, 73–74
fault scarp, **74**, *76*
favelas, 20, *20*
Federal Emergency Management Agency (FEMA), **21**, *23*
　and flooding, 457, *459*
　and hurricanes, 406, 414
feedbacks affecting climate change, 553
felsic melts, 125, *126*, *129*, 135
felt report, 87
fetch, **472**, *473*
FFD (forward-flank downdraft), 338–339
　in tornado formation, 348, *348*
field force, A2
fine particulate matter, 295
fire(s)
　backing, **528**, *528*
　coal-bed, *277*, 277
　crown (canopy), **524**, *525*, 535
　defined, **522**, *523*
　due to earthquake, 73, *100*, *101*
　flank, 528
　ground, **524**, *525*, *526*
　heading, *527*, **528**
　ladder, **524**, *525*
　peat, 524
　perimeter of, **527**, *528*
　production of, 522, *523*
　severity and intensity of, 524, *526*
　spot, **526**
　surface, **524**, *525*
　wild- (forest). *See* wildfire(s)
fireball, **601**, *602*, 615
firebrands, **526**, *526*, *527*, **538**, *538*
firebreaks, **533**, *534*, *536*, *537*
fire complex, **528**, *528*, *529*
fire retardants, **533**–**536**, *536*
firestorm, 12, *100*, *101*, **527**, *527*
fire tornado, 527
fire triangle, **523**, *523*
fire whirl, **527**, *527*
first-order stream, 429–430, *430*
fisheries, degradation of, *576*, 576
fission reactions, *A4*, **A4**
fissure(s), 96, **131**, *140*, *206*
fissure eruptions, **139**, *140*
fjords, 234
　glacial, *475*, *476*
flame(s), **522**
flame height, 524, *526*
flame length, 524, *526*
flaming combustion, 523
flank(s), of volcano, 139
flank collapse of volcanic island, *236*, 236–**238**, *238*
flank fire, 528
flank vents, 138, *141*, *152*
flash flood(s) and flash flooding, 434–441
　due to breaching of natural dams, 439–441, *440*, *441*
　damage due to, *438*, 438
　defined, **421**, *435*

in Ellicott City, Md., 434–435, *435*
　due to extreme rainfall, 438–439, *439*
　due to rapid snow or ice melt, 439
flash flood warning, 460
flash flood watch, 460
Flint, Mich., water crisis in, 257
flood(s) and flooding
　areal, *441*, 441
　building codes for, 25, *26*
　coastal, 498
　damage due to rising water in, 445–447, *446*
　danger due to moving water in, 445, *445*, *446*
　deaths and economic impact of, 16, *17*
　defined, **5**, *6*, **421**
　estimating probability of, *456*, 456–457, *458*, 458–459
　flash, **421**, **434**–**441**, *435*, *438*–*441*
　due to hurricanes, *402*, 402, *407*, 407, 442
　inland. *See* inland flood(s) and flooding
　insurance for, 27, *27*
　mitigation of, 460–463, *461*–*463*
　outburst, **440**, *440*
　payment for damage due to, 462
　recurrence interval of, *456*, 456, *458*, 458–459, *459*
　response, recovery, and restoration after, 23
　risk of, 455–460, *456*–*460*
　shore, 421
　slow-onset (downstream, regional), **421**, **432**–**434**, *434*–*437*, *436*–*437*
　sunny-day, 484, 487, *487*, 488, *488*
　due to thunderstorm training, 452–453, *453*, 455
　urban, **441**–**442**, *442*, 568, *568*
　vulnerability for, 462–463
flood basalts, *134*, **134**–**135**
flood control, 460–462, *461*–*463*
Flood Control Act (1928), 460
flood-control channels, 442
flood crest, **432**, *433*
flooded river valleys, *475*, *476*, 476
flood-frequency curve, **457**, *457*
flood-frequency graphs, *457*, 457
flood-hazard map, 457–460, *459*
floodplain(s), *423*, **424**, *424*
　in Bangladesh, 434
　defined, 274, **424**
　and meandering streams, 428
　subsidence in, 274–276, *275*
floodplain deposits, *426*, 428
flood potential map, 24
flood stage, **432**, *433*
flood tide, 470, 499
floodwalls, 443, *461*, 461–462, *463*
flood warnings, 460, *460*
flood watches, 460, *460*
floodwaters
　damage due to rising, 445–447, *446*
　danger due to moving, 445, *445*, *446*
floodway, **462**, *463*
Florence, Italy, flooding in, *454*
Florida, collapse of condominium complex in, 484
Florida Current, *468*
flux melting, **125**, *125*
focus of earthquake, **78**, *78*, 89, *89*, *93*
fog, **304**
Fogo, Cape Verde Islands, flank collapse on, *238*, 238
folds, **64**, *65*

foliation, **42–43**, *43*, **64**, *65*, 182
foliation planes, *183*
food insecurity, 507
fool's gold, 262
footwall, *74*, *74*
force
    buoyancy, *445*, *445*, **A3**
    defined, **A1**
    examples of, *A3*
    field (noncontact), **A2**
    gravity as, *A3*, **A3**
    magnetism as, *A3*, **A3**
    mechanical (contact), **A2**, *A3*
forecasts, 22–25
    defined, **22**
    long-term, 23
    short-term, 23
foreign-aid offices, 22
foreshocks, **78**
forest fires. *See* wildfire(s)
forest litter, *524*, *525*
Forest Service, 22
Fort Tejon earthquake, 92
forward flank, of supercell, 338
forward-flank downdraft (FFD), 338–339
    in tornado formation, *348*, *348*
forward motion of hurricane, 386
fossil(s) as paleoclimate indicators, *548*, *548*
fossil fuels, decrease in use of, 579, *579*, *580*
foundation in earthquake engineering, 115
fracking, 95
fractional crystallization, **126**
fracture zones, **49**, *49*, *56*, *57*
fracturing in volcano, *139*
fragmental texture of igneous rock, *129*, **129–130**
Frankenstein (Shelley), 160
Franklin, Benjamin, 160
freeze and freezing
    deaths and economic impact of, 16
    and energy, 305
freeze watch, 359
freezing rain, *365*, **365–366**, *366*
freshwater, 34
    contamination and salinification of, *257*, *257*
    defined, **251**
    depletion of, *255*, **255–256**, *256*
friction, 77
    and mass wasting, *181*, **182**
frictional force and wind, 307
front
    cold, **318**, *319*
    defined, **318**, *319*, 330
    gust, **330**, *331*
    lifting due to movement of, *330*, *330*
    occluded, **318**, *319*
    polar, 318–**319**, *319*, *320*, *362*
    stationary, **318**, *319*
    warm, **318**, *319*
frontal lifting, 318
frontal squall lines, **336**, *337*
frost advisory, 359
frostbite, **361**
Juan de Fuca Plate, *113*, *113*
fuel, **522**, *523*, *524*, *525*, **526–527**
Fujita, Ted, 351
Fujita scale, 351
fumaroles, **163**, *167*, *171*

funnel cloud, 349
fusion reactions, *A4*, **A4**

G

gabbro, *40*, *40*, *129*, *131*
Galveston, Texas, hurricane, 405, *467*, *467*
gamma rays, 544
Ganges River delta
    cyclone over, *411*, *411*
    groundwater contamination of, 262
gas(es), *A2*, **A2**
    greenhouse, **544–545**
    trace, 33, 292
gas bubbles, explosions due to trapped, *145*, *145*, *146*, *148–149*
gas extraction, subsidence due to, 272
gauging station, *431*, *431*
GCMs (global climate models), **562–563**, *563*
    to predict future, *566*, 566–567
Geneseo, N.Y., salt-mine collapse in, *278*, *278*
Genoa Lows, *454*, *455*
gentle slope, *176*, *177*
geological disaster, 7, 8, *9*
geologic structures, **62–64**, *63–65*
geologic time, 31, **66**, *67*
geologic time scale, **66**, *67*
geologists, **37**
Great Lisbon Earthquake, *73*, *73*
geomagnetic storms, *586*, 595
GeoNET, *165*
geophysical disaster, 7, 8, *9*
geosphere
    defined, 31, **34–35**, *35*
    geologic time and global change in, 66–69
    layers of, 43–46
    materials of, 37–43
    plate tectonics in. *See* plate tectonics
    topography and structures of, 61–65
geostrophic balance, **310–311**, *311*
geostrophic wind, **311**, *311*
geysers, 161
ghost forests of Cascadia, *230*, 230–231
glacial fjords, *475*, *476*
Glacial Lake Missoula, flash flooding from, *440*, 440–441
glacial torrents, *440*, **440**
glaciations, *550*, *551*
glacier(s), **34**, *34*
    melting of, *557*, 557–558, *558*, *560*
glass, *vs.* mineral, *38*, *38*
glassy texture of igneous rock, **128**, *129*, *129*
global change, 66–69, *68*, *69*
    defined, **67**
global climate models (GCMs), **562–563**, *563*
    to predict future, *566*, 566–567
global cooling, **547**
global pandemics
    casualties of, 11
    COVID-19, 7, *10*, 10–11
    defined, **7**
    and economic collapse, 10
    economic costs of, 11
    history of, 10
    vs. physical natural disasters, *10*, 10–11
global positioning system (GPS), **61**, *61*, 593
global warming, 20, **547**
    from asteroid impact, 616–617
    mitigation of, *580*–581, *581*

prediction of, *566*, *566*
    See also climate change
global water crisis. *See* water crisis
Good Friday earthquake, 93, 99–100, *100*, 192
    tsunami due to, *229*, *229*, 231
GPS (global positioning system), **61**, *61*, 593
GRACE-FO satellite, 557
GRACE satellite, 557
grade of slope, *176*, *177*
grains, **38**, *39*
Grand Banks earthquake, 235
granite, *39*, **40**, *40*, 129
The Grapes of Wrath (Steinbeck), 504
grassland destruction, **21**
graupel, *305*, *305*
Graveyard of the Atlantic, 485
gravitational field, *A3*
gravity, **36**, *A3*, **A3**
gravity survey, 268
gravity wave, *219*, 219–220
graywater reclamation, 265
"great climate migration era," 581
"the Great Dying," 161
Great Lakes
    and lake-effect snowstorms, 370–371, *371*
    and rising water levels, 487
Great Lisbon Earthquake, *73*, *73*
Great Mississippi Flood of 1927, *433*–434, *435*, 460
Great Missoula Floods, *440*, *440*–441
Great Plains Committee, 282–283
"Great Wall of Japan," *246*, *247*
Greece, heat officer in, 581
greenhouse effect, 302, **544–545**
greenhouse gas emissions
    reduction of, 578–580, *579*, *580*
    removal from atmosphere of, 580
greenhouse gases, **544–545**
    and climate change, 550–551
Greenland, ice sheet covering, *555*, 557–558, *558*
green roofs, *518*, *518*
groins, **491**, *492*, 499
Gros Ventre Slide, 201–202, *202*
ground blizzard, **364**
ground fires, **524**, *525*, *526*
ground fuel, *525*
ground-penetrating radar, *268*, *269*
ground rupture due to earthquakes, *96*, *96*
ground shaking due to earthquake, *73*, *86*, **96–98**, *97*
groundwater
    behavior of, *258*–260, *258*–*260*
    challenges with, 258–263
    contamination of, 11, *261*–263, *263*
    defined, **9**, 34, **253**, *258*, *258*
    formation of, *258*, *258*
    as non-renewable resource, 261
    and permeable *vs.* impermeable material, *258*, *258*
    and pores, **258**, *258*
    and recharge and discharge areas, 258
    remediation of, 263
    and volcanic eruption, 164
groundwater depletion, *260*–**261**, *260*–262
    subsidence due to, *271*–272, *272*, *273*
    timeframe of, 9–11
growth rings, **548**, *549*
Guatemala City, sinkhole in, 270
Guinea Current, *468*
Gulf Coast hurricanes, *405*–408, *407*

Gulf of Mexico, 468
Gulf Stream, 468, 468
Gun-Yu Great Flood, 421
gust(s), 335
gust front, **330**, 331
gyres, 468, **469**

H

habitable zone, 543
haboobs, 282, 282–283, 283, 284
Hadley, George, 312–313
Hadley cells, 312, 312–313
    defined, **313**
    and drought, 506, 508
    expansion of, 578
    in hurricane formation, 390, 390
hail, 345, **345**–346, 346
hail nets, 346, 346
hailstone, **305**, 305, 345, 345, 346
hailstorm, **345**–346, 346
hail streak, 345, 345
hail swath, 345, 345
Haiti
    earthquakes in, 11, 102, 102–103
    and Hurricane Matthew, 403–405
Haiyuan earthquake, 11
Hamilton, William, 170
hanging wall, 74, 74
hard barriers, **491**–492, 492, 493
hard protection, 491–492, 492, 493
hard-rock mines, 277
hard water, 261
harmful algal blooms, **497**, 497, 545, 545
Hawaii
    earthquake risk in, 112
    submarine slumps around, 236, 236–238
    volcanic eruption in, 4, 60, 60, 133, 133, 162
Hawaiian eruptions, 150, 151–153, 152–153
Hawai'i Volcanoes National Park, 152
Hawai'i Volcano Observatory, 165
hazard(s)
    anthropogenic, **7**
    biological, **7**
    classification by source of, 7–9
    defined, 18
    vs. disaster, 7
    natural, **7**
    physical, **7**
    and risk, 18–19
    socio-, **7**
hazard maps
    earthquake, 107, 110–111, 111
    flood, **457**–460, 459
hazard mitigation, **25**
hazardous weather, future changes in, 577
hazard potential map, 24, **24**–25
hazard preparedness, **25**–26, 26
haze, **292**, 293
heading fire, 527, 528
headlands, 476, **478**, 480
head scarp, 184, **186**, 187, 188, 207
headward erosion, 422, 422
headwaters, 423, 423
heat
    defined, **295**, **A4**
    and drought, 512, 512
    in ignition of wildfire, 522, 523, 523, 524, 525, 526

    latent, **306**, 388–389, 389
    measurement of, A4
    metabolic, 514
    vs. temperature, A4
    transfer of, A4–A5, A5
heat advisory, 517
heat capacity, 368, 523
    of oceans, 546
heat dome, 519
heat exhaustion, 515
heat flow
    defined, **49**
    variations in, 49–50, 50
    and volcanic eruption, 163
heat-health warning systems, 517
heat illnesses, 515–516
heat index, **516**, 517
heat index table, 516, 517
heating and surface air pressure, 309, 309–310
heat islands, urban, **517**–518, 518
heat officer, 581
heat sources in atmosphere, 301, 301–302
heat stress, 515, 516
heat stroke, 515, 515
heat transfer, **A4–A5**, A5
    latent, 545
heat-transfer melting, **125**, 125
heat wave(s), 514–521
    danger of, 514, 514–516, 515
    defined, **6**, 503, **514**, 516–517, 517
    future increases in, 577–578, 578
    temperate-climate, 519, 519–521
    tropical-climate, 520, 521
    and urban heat islands, 517–518, 518
    and water shortage, 256
    weather patterns associated with, 519–521, 519–521
hectopascals (hPa), 296, 296
Heezen, Bruce, 48, 50
Heimaey volcano, 169, 169
Herculaneum, 171
Hess, Harry, 50, 51
heterosphere, 301, 302
high-altitude divergence in hurricane formation, 392, 392
high-emissions scenario, 566, 566
high latitudes, atmosphere of, 317–320, 318–320
high pressure at Earth's surface, 309, 309–310
high-pressure systems, **318**
high tide, 469, 469–470, 470
Hilina fault system, 236, 236
Hilo, Hawai'i, tsunami in, 217, 217, 228
Himalayas, 58, 62, 94
hockey-stick diagram, **554**, 554
H_2O (hydrogen dioxide) in air, 292, 292, 293
homosphere, 301, 302
Honduras and Hurricane Mitch, 451
Hong Kong, urban flooding in, 442
hook echo, 339, 339, 355, 355
Hoover Dam, 255
horizontal-motion seismograph, 83, 84
horizontal wind shear, 350, 351
hothouse intervals, 548, 550
hotshots, 536
hot spot(s), 52, 60, **60–61**, 162
    igneous activity at, 130, 133, 133–134
hot-spot track, 60, **61**, 133, 133

hot-spot volcanoes, 130, 133, 133–134
    and plate tectonics, 52, 60, 60–61
hot springs, **161–163**, 171
hot zones, **577**
hPa (hectopascals), 296, 296
Hudson Bay, 468
human influence on disaster risk, 20–21, 21
humanitarian crisis, 15
human toll of disasters, 14–17, 15–17
Humboldt Current, 468
humidity, relative, 295, 297–**298**, 298, 516, 517
hummocky surface, **186**, 188, 188, 285
hurricane(s), 382–419
    Atlantic and Gulf Coast, 405–408, 407
    Caribbean and Central American, 403–405, 403–405, 408
    categories of, 384, 384
    changes in severity of, 560–561
    comparing size of, 13, 13
    and cyclones of Indian Ocean and Southern Hemisphere, 411, 411–412, 412
    deaths and economic impact of, 16, 16, 17
    defined, 5, 5, **383**
    effect on coastline of, 291
    energy source of, 388, 388–389, 389
    evacuations for, 414–416, 416
    fatalities due to, 405, 406
    financial cost of, 405
    flooding due to, 402, 402, 407, 407, 442
    formation and evolution of, 389–395, 390–396
    Galveston, Texas, 405, 467, 467
    inland flooding due to, 451, 452
    intensity of, 383, 384, 384, 395
    life cycle of, 391–393, 391–393
    locations of, 388, 388
    major, 412
    mitigating damage from, 417, 417
    naming of, 384
    overview of, 383, 383
    Pacific, 408, 409, 410
    power outages due to, 397, 404, 406
    predicting tracks of, 413, 413–414, 414
    preparing for arrival of, 414–415, 415
    rain, inland flooding, and landslides from, 402, 402
    response, recovery, and restoration after, 23
    rotation of, 387
    societal disruptions due to, 412
    storm surge from, 400, 400–401
    structure of, 384–387, 385–387
    tracks and dimensions of, 395–397, 396
    visualization of, 398–399
    wave damage from, 401, 401–402
    wind damage from, 397, 397–400, 400
Hurricane Agnes, 402
Hurricane Andrew, 396
Hurricane Dorian, 386–387
Hurricane Florence, 387, 401, 413, 414
Hurricane Gaston, 393
Hurricane Harvey, 21, 383, 383, 451, 452
Hurricane Ida, 397, 402, 407, 442, 568
Hurricane Ike, 402, 402
Hurricane Irma, 383, 383, 385, 403, 405
Hurricane Isaias, 387, 414
Hurricane Katrina, 406–407, 407
    cost of, 405
    evacuation due to, 416

Hurricane Katrina (*continued*)
   size of, 13, *13*, *396*
   storm surge of, 401
   wind speeds of, *396*, *400*
Hurricane Maria, 383, *383*, 402, 404, *404*
Hurricane Matthew, 403–405, *490*
Hurricane Mitch, 405, 408, 451
Hurricane Patricia, 397
Hurricane Sandy
   beach restoration after, 494, *494*–495, *495*
   storm surge of, 401
   urban flooding due to, 568, *568*
   wave damage due to, *486*
Hurricane Sandy Relief Bill, 494–495
hurricane season, **388**
hurricane track, 395–397, *396*
   predicting, *413*, 413–414, *414*
hurricane warning, 406, **413**, *414*
hurricane watch, **413**
hydrofracturing, 95
hydrogen dioxide (H_2O) in air, 292, *292*, 293
hydrograph, **431**, *431*
hydrological disaster, 7, *9*
hydrological drought, 504, **505**
hydrologic cycle, **68**, **252**–253, *252*–254
hydrologists, 421
hydrosphere, 31, **34**
hydrothermal vents, *32*, 131, *131*
hygrometers, 300
hypocenter of earthquake, **78**, *78*
hypothermia, **361**
hypothesis, 50

I

ice, 34, *34*
   air bubbles in, 548
   sea, 558, *559*, *560*
ice age, **550**, *550*, 551
ice crystals, 293, 305, *305*
icehouse intervals, 548, 550
ice jams, 375, *375*, **441**, *441*
Iceland
   flash flooding in, 440, *440*
   hot-spot volcanoes in, 133, *134*
   lava flow from effusive eruption in, *142*
   smoke and ash from volcanic eruption in, 156–157, *157*, 160
ice melt, flash floods due to rapid, 439
ice nuclei, 303, 305, *305*
ice pellets, 366
ice sheet(s), 49
   melting of, 555, 557–558, *558*
ice storm(s), 291, 364–366, *365*, *366*
   in China, 373, *373*
   Great Ice Storm of 1998, 366, *367*
ice storm warning, 359
igneous activity, 130, 130–135
   in continental rifts, *130*, 132, *132*
   at convergent boundaries, *130*, 131–132, *132*
   defined, **124**
   at divergent boundaries, *130*, 130–131, *131*
   at hot spots, *130*, 133, *133*–134
   in large igneous provinces, *134*, 134–135
igneous provinces, large, *134*, **134–135**, 161, 551
igneous rocks, **39–40**, *40*
   extrusive, **128**, *128*
   formation of, 127–129, *127*–130

intrusive, **128**, *128*
   texture of, 128–130, *129*
   variety of, 128, *128*–130, *129*
ignimbrite, **145**, *170*
ignition of wildfire, 522–524, *523*, *524*
ignition point, 525, *528*, 534
impact breccia, 605–607, *606*
impact ejecta, **605**, *605*
impact melt, 605, *605*, *606*
impact velocity, **601**
impermeable material, 258, *258*
   and urban flooding, 441
impulsive techniques for deflection of near-Earth objects, 620
inactive faults, 64, 74
inches of mercury, 296
India
   famine in, 16
   smog in, 294
   water crisis in, 264, *264*
Indian Ocean, 468
   cyclones of, *411*, 411–412, *412*
   earthquake and tsunami in, 11, 25, 231, 231–232, *232*, 241, 244
Indian Ocean Dipole, *316*, 508
Indian Ocean Gyre, *468*
Indonesia earthquake, 99
induced seismicity, **95**, *95*
inert gas, 292
inertia, **83**, **308**
   in generation of tsunami, 219, *219*
influenza pandemic of 1918, 10
infrared radiation, 544, *544*
infrasound emissions and volcanic eruption, **165**
infrastructure, 14
   earthquake resistance of, 118
injection wells, 95, *95*, 263
inland flood(s) and flooding, 420–465
   areal, 441, *441*
   damage due to rising water in, 445–447, *446*
   danger due to moving water in, 445, *445*, *446*
   defined, 421
   estimating probability of, 456, *456*–457, *458*, 458–459
   due to failures of constructed dams and levees, 442–444, *443*, *444*
   flash, 421, **434–441**, *435*, *438*–*441*
   flood stages and crests in, 432, *433*
   from hurricane, 402, *402*
   due to mid-latitude cyclones, 451–455, *453*, *454*
   mitigation of, 460–463, *461*–*463*
   due to monsoons, 447, *447*–450, *449*, *450*
   due to mountain thunderstorms, 455, *455*
   overview of, 421, *421*
   payment for damage due to, 462
   recurrence interval of, 456, *456*, *458*, 458–459, *459*
   risk of, 455–460, *456*–*460*
   in rivers and streams, 432–441
   slow-onset (downstream, regional), 421, **432**–434, *434*–437, *436*–*437*
   due to tropical cyclones, 451, *452*
   urban, 441–442, *442*
   visualization of, *448*–*449*
   vulnerability for, 462–463
inner core
   of Earth, 44, *45*, 46
   of hurricane, 386, *387*

inner electron shell, A1
inner planets, 37
InSAR (Interferometic Synthetic Aperture Radar), 109–110, *110*, 164, 210
insolation, **545**, *546*
instrumental measurements, 547
insulation, 361
insurance, **26–27**, *27*
intact rock, 180
intensity
   of earthquakes, 86, **86–87**, *87*, 89, 89–90
   of hurricanes, 383, 384, *384*, 395
intensity maps, 87, *87*
intentional wildfires, 529, *529*, 537, 538, *538*
Interferometic Synthetic Aperture Radar (InSAR), 109–110, *110*, 164, 210
interglacials, 550, *551*
   and atmospheric CO_2 concentrations, 562, *562*
Intergovernmental Panel on Climate Change (IPCC), **554**
intermediate-focus earthquake, 78, 93
intermediate melts, 125, *126*, 135
intermountain seismic belt, 112
internal energy, 32, **36**, *36*
internally displaced people, 14, 15
international accords on climate change, 581
international aid agencies, 22
International Red Cross and Red Crescent, 22
intertidal zone, 470, **471**
intertropical convergence zone (ITCZ), 313, *313*–314
   defined, **313**
   and drought, 506–508, *509*
   in hurricane formation, 390, *390*
   and inland flooding, 447, *447*
   rainfall in, 546, *547*
intracontinental regions, seismic risk in, 112
intraplate earthquakes, 94, **94–95**, *95*
intrusive igneous rocks, **128**, *128*
intrusive rocks, 39–40, *40*
inundation depth, 223, *223*
inundation distance, **223**, *223*
inundation limit, 223, *223*, 227
inversion, 519, **520**
ion, **A1**
ionosphere, 301, *302*, **590**, 593
ion tail, 598, *598*
IPCC (Intergovernmental Panel on Climate Change), **554**
Iran
   earthquake in, 115
   water crisis in, 253
iridium, 607, 610–611, *611*, 613
iron, in Earth layers, 37
iron alloy, 46
iron meteorites, 599, *599*
iron oxides, 40
   in groundwater, 261
island(s), 35, *35*, 49
   barrier, **478**, *479*
island arc, **56**, *56*, 132, *132*
isobar(s), **306–307**, *307*
isobaric surface, 309, *309*
isotherm, 365
isotope(s), 548, *549*, **A1**
isotopic dating, **66**
isotopic studies to identify carbon sources, 565
Istanbul gap, 108, *109*

ITCZ. *See* intertropical convergence zone (ITCZ)
Itokawa asteroid, 621

J

Japan, earthquake and tsunamis in, 3, 3
Japan Current, 468
jet stream, **306**
  and drought, 510, 510, 511, 512
  and mid-latitude cyclones, 320–321, 321
  polar-front, 319–320, 320
jetty, **491**, 492
Johnston, David, 148–149
Johnstown, Penn., flood, 421, 421
joints, 64, 65, 180, **180**–182, 181
jökulhaup, 440, 440
Joplin, Missouri, tornado, 354, 354
Joso, Japan, flash flooding in, 438

K

Kaali Crater, 608, 609
Ka'oki fault system, 236
karst terrain, **266–267**, 267, 268
Keeling, Charles, 561
Keeling curve, 561, **561**–562
Kelvin scale, 295–296, A4
Kentland impact structure, 607, 610
Kilauea, 150, 152, 152–153, 162
  explosivity index for, 146
  InSAR image of, 164, 164
  mitigating risk of eruption of, 168, 168
  predicting eruption of, 165
  pyroclastic debris from, 140, 144
kinetic energy, 295, 613, **A3**
kinetic impactor, 620
king tides, **484**
K-Pg boundary, 613
K-Pg mass extinction, 613
Krakatau (Krakatoa), 144, 144–145
  hazards due to explosion of, 158–160
  pyroclastic debris from, 146
  tsunami due to eruption of, 238, 239
  volcanic island from, 149
Kuiper Belt, 596, 598
Kuroshio Current, 468
Kyoto Protocol, 581

L

Labrador Current, 468
La Conchita (Calif.) mudflows, 191, 191–192, 201
ladder fires, **524**, 525
ladder fuel, 525
lagoon, 477, **478**, 498
lag time, **432**, 433
lahars, 139, 166
  defined, **157**, **192**
  examples of, 184, 192, 192, 207
  flash floods due to, 439
  hazard potential map for, 24
  hazards due to, 157–158, 158
  mitigating risk from, 168
  on Mt. St. Helens, 149
lake(s)
  oxbow, 428, 428
  rising water levels in, 487
lake-effect snowstorms, **370–371**, 371
Lake Mead, 255, 256
Lake Nyos, volcanic gases in, 158, 159
Lake Okeechobee, 273, 274
Lake Pontchartrain and Hurricane Katrina, 406, 407
land, 35
landforms, **62**
landscapes for hurricane damage mitigation, 417
landslide(s), 175
  complex, **197**
  deaths and economic impact of, 17
  defined, **3**, **184**
  due to earthquakes, 98, 98
  example of, 175, 175
  flash floods due to, 440
  from hurricane, 402, 402
  in shantytowns, 20, 20
  subaerial, 234, 234–235, 243
  submarine, 198, 198–199, 199, 243
  tsunamis generated by, 234–238, 234–238, 243
landslide lake, 207
landslide potential maps, 25, **208**, 209
land subsidence, 271–278
  defined, **61**, **176**, 251, **271**
  in delta plains, floodplains, and coastal plains, 274–276, 275, 276
  due to groundwater depletion, 271–272, 272, 273
  due to mine collapse, 277, 277–278, 278
  due to oil and gas extraction, 272
  due to pore collapse, 271–272, 272, 273
  tectonic, 271
  visualization of, 286–287
  water loss and, 287
  in wetlands, 273–274, 274
La Niña, 314, **315**, 315, 508
lapilli, **138**, 138, 145, 145
large igneous provinces (LIPs), 134, **134–135**, 161, 551, 613
large-scale disaster, 14
Larsen B Ice Shelf, 558, 559
latent heat, **306**
  and hurricanes, 388–389, 389
latent heat transfer, 545
lateral blast, 148, 149, 166
lateral spreading, **192–193**, 193
latitude and surface temperature, 545, 545
lava
  in building of volcano, 135, 135–136, 136
  composition of, 125, 125–126, 126
  curtain of, 139, 140
  defined, **3**, **4**, **40**, 40, **124**
  hazards due to, 155, 155–156
  movement of, 164
  solidification of, 127, 127
lava dome, **135**, 135
  resurgent, 149
lava flows
  defined, **128**, 128, 135
  diversion or stoppage of, 168–169, 169
  hazard potential map for, 24
  rocks formed from, 40, 40
  in volcanic edifice, 139
lava fountains, 135, **136**, 136, 152, 167
lava lake, **135**–136, 136, 152
lava tubes, **135**, 152, 169
law of conservation of energy, A2
Law of the River, 256
laze, **158**, 159, 167
leaching, 261–262
  zone of, 279, 279
leishmaniasis, 574
Leonids meteor storm, 601
levees, 275
  alternatives to, 462
  building of, 461, 461–462
  failures of artificial, 443, 443–444, 444
  and Hurricane Katrina, 406, 407
  natural, **426**, 426, 428
  reinforcement of, 461, 462
Lewis, Meriwether, 423
lidar, 268, 269
lifting, 303, 303
  convergence, 330, **389–390**, 390
  orographic, **330**
lifting mechanisms
  in hurricane formation, 389–391, 390, 391
  and thunderstorms, 329, **330–331**, 330–332
lightning, 340–343
  cloud-to-cloud, **340**, 340
  cloud-to-ground, **340–344**, **341**–343
  defined, 5, 5, 340
  dry, 524
  examples of, 328
  formation of, 340, 340
  hazards of, 342–343, 343
  ignition of wildfire by, 342, 524, 524, 534
lightning bolt, 340
lightning rods, 343, 344
lightning stroke, **340**
limestone, 41, 42
LIPs (large igneous provinces), 134, **134–135**, 161, 551, 613
liquefaction hazard, 98–100, 99, 119, 119
liquefaction potential maps, 208
liquefaction sinkhole, 270, 270
liquid, A2, **A2**
liquidus, 127
lithification, **41**
lithosphere, 46, **46**
lithosphere plates, **51**, 53
lithospheric mantle, 45
  in plate tectonics, 51, 51, 52
Little Ice Age, 554, 554–555
Lituya Bay tsunami, 234, 234–235
local base level of stream, 423, 424
local disaster, 13–14
local magnitude (M_L), 88
local relief, 62, 62
local tsunamis, 220, 243
locusts, 8
Lō'ihi, 151, 162
Loma Prieta earthquake, 80, 89, 92, 118
Loma Prieta gap, 108, 109
London smog event, 294
lone meteor, 615
lone meteoroids, 601
longitudinal profile of stream, **423**, 423
long-period comets, 598
longshore bar, 477, **478**
longshore currents, **482**, 483, 499
longshore drift, **478**, 479, 492
long-term forecasting, 23
long-term predictions
  of earthquakes, 106–107, 107
  of volcanic eruptions, 163
Long Valley caldera, 150
longwall-advance mining, 277, 277

Loop Current, 406, 407
Los Angeles Aqueduct, 256
Louisiana
    delta plain and floodplain subsidence in, 274, 275
    Hurricane Ida and, 397, 407
    Hurricane Katrina and. *See* Hurricane Katrina
Lovell, Jim, 31
Love waves (L-waves), 80, *80*, *81*
    ground shaking due to, 96, *97*
    seismogram of, *85*
low-altitude divergence in hurricane formation, 393
low-emissions scenario, 566, *566*
lower mantle, 44, *45*
Lower Puna eruption, 153
low-latitude aurorae, 593, 595, *595*, 614
low-permeability material, 258, *258*
low pressure at Earth's surface, 309, 309–310
low-pressure center, **321**
low-pressure systems, **318**
low tide, 469, 469–470, *470*
lung cancer, 11
Lutetia (asteroid), 597
L-waves. *See* Love waves (L-waves)
Lyme disease, 574

M

Madagascar, cyclones in, 411
Madison Canyon Slide, 182–183, *183*
mafic melts
    appearance of, *137*
    defined, 125, *126*, 129
    flow of, *135*
    hazards due to, 155, *155*–156
magma
    composition of, 125, 125–126, *126*
    defined, **39**, **124**
    rise of, 127, *127*
    solidification of, 127, *127*
    types of, 125, *126*
magma chamber, 54, *54*, **127**, *127*, *128*
magnetic field(s), **32**–**33**, *33*, **A3**
    of Sun, 587, 587–588, *588*
magnetic polarity, 588
magnetic poles, 33, *33*
magnetism, A3, **A3**
magnetometers, 595, *595*
magnetosphere, **590**–592, *592*, *593*
magnetotail, 590
magnitude of earthquakes, **87**–90, *88*, *89*
magnitude scales, 87–89, *88*, *89*
mainshock, **78**
major air masses, 317–318, *318*
major disaster, 14
major disaster declaration, **22**
malaria, 574, *574*
mangrove swamp, 475, 478, 480
Manicougan Crater, 608, *610*
mantle, 37, *37*, 43, 44, **45**, *45*
mantle plume, 60, **60**–61, 130
map(s)
    bathymetric, 48–49, *49*, 198, *199*
    Did You Feel It?, 87, *87*
    earthquake hazard, 107, 110–111, *111*
    earthquake potential, 24
    of Earth's seismicity, *91*
    flood-hazard, 457–460, *459*

flood potential, 24
landslide potential, 25, **208**, *209*
liquefaction potential, 208
surface-wind streamline, 390, *390*
tornado potential, 25
tsunami inundation, 245, *246*
volcanic hazard assessment, **168**, *168*, *170*
wildfire hazard, 538–539, *539*
wind streaming, 387
maritime air masses, 318, *318*
maritime disasters, 7, *9*, 401, 484, **485**
marsh(es), salt, 478, *481*, 499
Marshall Islands, nuisance tides in, 488, *488*
masonry buildings, failure of, 114, *115*, 116
mass extinctions, 612, **612**–617, *616*
mass movement. *See* mass wasting
mass wasting, 174–215
    avalanches as, 196–197, *197*
    due to changing downslope stress or resistance stress, 201–203, *202*, *203*
    due to changing slope angle, 203, *204*
    due to changing slope strength, 204, *205*
    complex landslides as, 197
    creep as, 184, *184*, *185*
    debris flows as, *184*, 187, 187–192, *189*
    defined, **175**, *176*
    detecting start of, 210, *211*
    erosion and, 176
    factors that trigger, 200–205, *200*–*205*
    failure surfaces and, 182–183, *183*
    identifying regions at risk for, 205–208, *208*–*210*
    lahars as, *184*, 192, *192*
    lateral spreading as, 192–193, *193*
    mudflows as, *184*, 187, 187–192, *190*, *191*
    overview of, 175, *175*
    prevention and mitigation of, 211–212, *212*, *213*
    protection against, 205–212, *208*–*213*
    relief and, 176, *177*
    resistance to, 180–182, *180*–*182*
    rock and debris slides and falls as, *184*, 193–196, *194*–*196*
    safety factor and, 178–180, *179*
    due to sediment liquefaction, 200–201, *200*–*202*
    setting the stage for, 175–178, *176*–*178*
    due to shocks and vibrations, 200, *200*
    slopes and, 176, *177*
    slope stability and, 178–183, *179*–*183*
    slope steepness and shape and, 176, *177*, *178*
    slumps as, *184*, 186–187, *186*–*188*, 188
    solifluction as, *184*, 184–186, *186*
    subaerial, 198
    submarine, *198*, 198–199, *199*
    subsidence and, 176
    types of, *184*, 184–197
    uplift and, 175–176
    visualization of, 206–207
matter
    defined, **A1**
    nature of, *A1*, A1–A2
    states of, *A2*, **A2**
maturity stage of thunderstorm, *334*, 335
Mauna Loa, 162
    atmospheric CO_2 concentrations on, 561
maximum catch potential, 576, *576*
mb (millibars), 296, *296*
M_B (body-wave magnitude) scale, 88

meander, **427**, *428*
meandering stream, **427**–**428**, *428*
meander neck, 428, *428*
mean sea level, 400, *470*
mechanical force, **A2**, *A3*
mechanical seismograph, 83, *84*
median valley, 48
Medieval Warm Period, 554, *554*
Mediterranean climate, 505, 506, *506*, 509
    drought in, 509–511, *509*–*511*
Mediterranean Sea, 468
medium-emissions scenario, 566, *566*
megadisaster, **14**
megadroughts, **505**–**506**
megafires, **528**, *528*, 529
megathrust earthquakes, **93**, *93*
    of Cascadia, 112, *113*, 113
    tsunamis due to, 228–232, *229*–*232*, 243
megatsunami, **235**
Mei Yu front, 450, *451*
melt(s) and melting
    due to addition of volatiles, **125**, *125*
    decompression, 124, **125**
    due to decrease in pressure, 124
    and energy, 305
    felsic, 125, *126*, 129, *135*
    flux, **125**, *125*
    in igneous rock formation, 40
    intermediate, 125, *126*, *135*
    mafic, 125, *126*, 129, *135*, *137*
    partial, 125–126
    solidification of, 127, *127*
    source rock of, 125
    ultramafic, 125, *126*
melting rock, 123, 124–126, *124*–*126*
    solidification of, 127, *127*
meltwater, 557–558, *558*
Mendocino Fire Complex, 528
Mendoza, Argentina, hailstorms of, 346, *346*
Mercalli, Guiseppe, 86
mesocyclones, 338, *348*, 349
mesopause, *301*, 302, 593
mesosphere, *301*, **302**, 590, 593
metabolic heat, 514
metamorphic foliation, **42**–**43**, *43*
metamorphic rocks, **41**–**43**, *43*
metamorphism, 41–43, 607, *607*
meteor(s), 600–604, **600**–605, 615
Meteor Crater, 608, *609*, 619
meteorite(s), 599, **599**–600, *600*
meteorite craters
    complex, 605, **606**, *606*
    defined, **605**
    formation of, 605, 605–606, *606*
    simple, 605, **605**–606, *606*
    transient, 605, *606*
    visible, 608, *609*
meteorite falls, 599, *600*
meteorite finds, 599–600, *600*
meteorite impacts, 6, *8*
    effects of, 11
    tsunamis due to, 239–240, *240*
meteoroids, 596, **598**–605, *600*, *601*, 615
meteorological disaster, 7, *8*, *9*
meteorological drought, 504, **505**
meteorologists, **295**

meteor shower, **601**, *601*, 615
meteor storm, 601
methane (CH$_4$)
    in air, 33
    as greenhouse gas, 545
    and infrared radiation, 544
    rising atmospheric concentrations of, 562, *562*, *563–565*
methane hydrate, melting of, 576, *576–577*
Mexico City earthquake, 100
mica, 43, 182
microbursts, *344*, 344–345
microearthquakes, 89
microfossils, 548, *548*
microplates, 51
microtektites, 607, 610–611, *611*, 616
microwaves, 544
Mid-Atlantic Ridge, 48, *49*, 50
Midland, Mich., flooding of, 442–443
mid-latitude(s), **317**
    atmosphere of, 317–320, *318–320*
mid-latitude cyclones, 320–322, *321–323*
    continental, 366, *366–368*
    defined, 317, **321**
    inland flooding due to, 451–455
    oceanic, 368–370, **368–370**
mid-ocean ridges, 34–35, **49**, *49*, 51, 52
Milanković, Milutin, 552
Milankovitch cycles, **552**, *552*
millibars (mb), 296, *296*
mine(s) and mining
    hard-rock *vs.* soft-rock, 277
    longwall-advance, 277, *277*
    room-and-pillar, 277
    salt, 277–278, *278*
mine collapse, subsidence due to, 277, *277–278*, *278*
mineral(s), **38**, *38*
Mineral, Virg., earthquake, 95
minor disaster, 13–14
Mississippi, Hurricane Ida and, 397
Mississippi delta, 274, *275*, 426
Mississippi River
    discharge and watershed of, 431, *433*
    flooding of, 433–434, *435–437*, 436–437, 460
Mitchell, John, 73
mitigation
    of climate change impact, 578–581, *579–581*
    hazard, **25**
    for hurricanes, 417
    of impact risk from near-Earth objects, 617–621, *617–621*
    of mass wasting, 211–212, *212*, *213*
    of tsunamis, 244–246, *245–247*
    of wildfires, 537–539, *537–539*
M$_L$ (local magnitude), 88
moderate slope, 176, *177*
modification stage of crater formation, 605–606
Modified Mercalli Intensity (MMI) scale, 13, **86**, *86*
Moho, **44**, *45*
moist adiabatic lapse rate, **332–333**
moisture of air, *297*, 297–298, *298*, 313
    and thunderstorms, 329–330
molecule, **A1–A2**
molten rock, solidification of, 127, *127*
moment magnitude scale, 13, **89**
monitoring of droughts, 512–513, *513*, *514*

monsoon(s), *316*, 316–317, *317*
    defined, **316**, **447**
    flooding due to, 447, *447–450*, *449*, *450*
    in United States and Mexico, 450, *451*
monsoonal climate, 506, *506*
    drought in, 506–508, *507–509*
Monsoon Drift, 468
Montecito, Calif., mudflows in, 204, *205*
Monte Nuovo, 171
Montreal Protocol on Substances That Deplete the Ozone Layer, 591, *591*
Moon, craters on, 612, *612*
moon(s), 594, *596*
Moore, Okla., tornado, 5
moraine, 476
Morganza Spillway, 461
Morse, Bradford, 507
motion dampers, 117, *117*
mountain(s), thunderstorms in, 330–331, *332*
mountain belts, **35**, *35*, 52
mountain building, **62**, *62*
mountainous areas, flash floods in, 438–439, *439*
mountain range, and cold air damming, 370, *370*
mountain snowpacks, shrinking of, **569**, *569*
mountain snowstorms, 371–**372**, *372*
mountain thunderstorms, flooding due to, 455, *455*
mountain uplift, 32
mouth of stream, 423, *423*
Mozambique, cyclones in, 411–412, *412*
M$_S$ (surface-wave magnitude), 88
M$_S$ (surface-wave magnitude) scale, 88
Mt. Etna, 168–169, *169*
Mt. Fuji, Japan, *139*, 141
Mt. Kilimanjaro, 142
Mt. Mazama, 147
Mt. Nyiragongo, 155, *156*
Mt. Pelée, 153–154, *154*, 157
Mt. Pinatubo eruption, 142
    ash from, 156
    climate consequences of, 160, *160*
    explosivity index of, 147
    lahars following, 192, *192*
    predictions of, 163
    pyroclastic debris from, 146
Mt. Ranier, 168
Mt. Redoubt, 146, 158
Mt. Ruapehu, 158
Mt. Shasta, 142
Mt. St. Helens eruption, 3, *4*, *148*, 148–149, *149*
    explosivity index of, 146
    lahars from, 149, 158, 192, *192*
    pyroclastic debris from, 146
Mt. Vesuvius eruption, 123, *123*
    evacuation prior to, 168, *170*, 170–171
    explosivity index of, 146
    as Plinian eruption, 154
    pyroclastic debris from, 146, 157
    size of, 142
MT Phoenix, 486
mudcracks, 504, *505*
mudflows, 187–192
    defined, *184*, **187**
    examples of, *187*, *190*, 190, *191*, 191–192, 206
    hazard potential map for, 24
    speed of, 189
mudpots, 163

mudslides. *See* mudflows
Muir Glacier, 557
multicell storms, 335–336, *335–337*
multiple-vortex tornadoes, 349, *350*
MV Prestige, 484

N

N$_2$ (nitrogen) in air, 33, 292, *292*
Naples, Italy, 168, *170*, 170–171
National Aeronautics and Space Agency (NASA)
    on impact risk of near-Earth objects, 617, 619–620
    landslide potential predictions by, 208, *209*
National Flood Insurance Program, 27, 417, 462
National Guard, 22, *23*, 168, 414
National Hurricane Center (NHC), 406, 413
National Interagency Fire Center (NIFC), 537
National Oceanic and Atmospheric Administration (NOAA), 22
National Tsunami Warning Center, 244
National Weather Service (NWS), **21**
    watches and warnings from, 23, *24*, 327, 355, 359, *359*
National Wildfire Coordinating Group (NWCG), 537
natural dams, flash floods due to breaching of, 439–441, *440*, *441*
natural disasters, 2–29
    biological, 7, 8, *9*
    characterizing exposure and vulnerability to, 17–18
    chronic health, **11**, *11*
    classification by source of, 7–9
    climatological, 7, *9*
    coastal, 7, 8, *9*
    comparing frequency of, 14
    comparing sizes of, 13–14
    comparing time frames of, 9–11
    costs of, 26–27
    dealing with, 21–27
    defined, 7
    extraterrestrial, 7, 8, *9*
    geophysical or geological, 7, 8, *9*
    hazard *vs.*, 7
    human toll and economic impact of, 14–17
    hydrological, 7, *9*
    introduction to, 3–7
    and language of danger, 7–13
    local, minor, or small-scale, 13–14
    major or large-scale, 14
    maritime, 7, *9*
    meteorological, 7, 8, *9*
    most deadly, 10–11
    physical, 7
    predictions, forecasts, and warnings of, 22–25
    preparedness, mitigation, and planning for, 25–26
    primary, secondary, and tertiary, **12**, *12*
    rapid-onset, **9**, *9*
    responding to, 21–22
    risk of, 18–21, *19–21*
    slow-onset, **9**, *9*
    socio-, 7, 16
    stages of, 21–22, *22*, *23*
    stealth (creeping), **9–11**
natural hazard, 7
natural hazardous events, 7
natural levees, **426**, *426*, 428

neap tides, 470, *470*
near-Earth objects (NEOs), 614, *617*, **617**–618, *618*
near-field tsunamis, **220**, 243
nearshore regional currents, 482, *483*
nebula, 37, *37*
negative feedback and climate change, 553
negative polarity cloud-to-ground stroke, 342, *342*
NEOs (near-Earth objects), 614, *617*, **617**–618, *618*
Nepal earthquake, *94*
neutral, 575
neutrons, A1
Nevada wildfires, 528
Nevado del Ruiz volcano, 158, 192
New England, Hurricane Ida and, 402
Newfoundland tsunami, 235, *236*
New Jersey, Hurricane Ida and, 402
New Madrid earthquake, *94*, 95, *95*, 106
New Madrid zone, 112
New Orleans
   and Hurricane Ida, 407
   and Hurricane Katrina. *See* Hurricane Katrina
Newton, Isaac, A2
New York City
   flooding in, 568, *568*
   Hurricane Ida and, 402, *568*
New Zealand, 165–168
NGOs (Non-Governmental Organizations), 414
NHC (National Hurricane Center), 406, 413
Nicaragua and Hurricane Mitch, 405, 408, 451
NIFC (National Interagency Fire Center), 537
Niigata, Japan, earthquake, 98, *99*
Nisqually, Wash., earthquake, *89*
nitrogen (N_2) in air, 33, 291, *292*
nitrous oxide (N_2O) as greenhouse gas, 545
NOAA (National Oceanic and Atmospheric Administration), 22
noncontact force, A2
non-engineered buildings, 115
Non-Governmental Organizations (NGOs), 414
non-renewable resource, 261
nonstructural components in earthquake engineering, 115
non-supercell tornadoes, 350, *351*
nor'easters, 368–370, *368–370*
normal fault, **74**, *74*
normal shoreline, 223, *223*
normal stress, **179**, *179*
North American Plate, 113, *113*
North Anatolian fault, 90–91, 108, *109*
North Atlantic Current, *468*
North Atlantic Gyre, *468*
North Atlantic Ocean, *468*
North Equatorial Current, *468*
North geographic pole, *33*
North Island, New Zealand, 165–168
North magnetic pole, 33, *33*
North Pacific Gyre, *468*
North Pacific Ocean, *468*
Northridge earthquake, 79, *89*, 89, 98, *98*
North Sea, *468*
Norwegian Sea, *468*
no-till farming, 284
nuclear bond, **A1**
nuclear energy, **A3**–**A4**, *A4*
nuclear fusion, **586**, *586*
nuclear reaction, **A4**, *A4*
nuclear winter, 616

nucleus
   of atom, A1, *A1*
   of comet, 598, *598*
nuées ardentes, 154
nuisance tides, **484**, 487, *487*, 488, *488*
numerical age, **66**
nutrients in topsoil, 279
NWCG (National Wildfire Coordinating Group), 537
NWS (National Weather Service), **21**
   watches and warnings from, 23, *24*, 327, 355, 359, *359*

O

O_2 (oxygen)
   in air, 33, 291, *292*, 292
   in ignition of wildfire, 523, *523*
O_3. *See* ozone (O_3)
Oakland Firestorm, 530
obsidian, 129, *129*
occluded front, **318**, *319*
ocean(s)
   acidification of, **575**, *575*
   and climate change, 551, 573–577, *575*, *576*
   defined, **34**
   future changes in, 573–577, *575*, *576*
   heat capacity of, 546
   names of, 467, *468*
   warming of, 543, *543*
ocean basin, 34, *35*
oceanic crust, 44–45, *45*
oceanic cyclones and winter weather, 368–370, **368–370**
oceanic lithosphere, 52–54, *54*, *55*
oceanographers, 467
ocean waves, **471**, *471*
offshore sandbars, 223, *475*, *477*, **478**, *492*, *498*
Ogallala aquifer, 261, *262*
oil extraction, subsidence due to, 272
100-year flood, 456
Oort Cloud, *596*, 598
open pit mine, 203
orbital shape and climate change, 552, *552*
ordinary thunderstorms, **333**–335, *334*
ordinary wells, 260, *260*
ore deposits, 608
Oregon
   heat wave in, 514
   wildfires in, 521, *522*
organic chemicals, 38
organic coasts, 477, **478**–482, *481*
orographic lifting, **330**
orphan tsunami, 231
Oso (Wash.) landslide, 208, *210*
outburst flood, **440**, *440*
outcrops, **39**, *39*
outer core, 44, *45*, 46
outer electron shell, A1
outer planets, 37
outer rainbands of hurricane, 386, *387*
outer zone of hurricane, 386
outgassing, 597, *598*
overgrazing, 280, *280*
overhang, 176, *177*
overharvesting, 280, *280*
overland flow, 422, *422*
overriding plate, 54, *55*

overtopping, **443**, *443*
Owens, Mae, 266, *267*
Owens Lake, 256
oxbow lake, 428, *428*
oxidation, 274
oxides in magma, 125, *125*
oxygen (O_2)
   in air, 33, 291, *292*, 292
   in ignition of wildfire, 523, *523*
oxygen-isotope ratios, 548, *549*
ozone (O_3)
   in air, 33, 292, 294
   as greenhouse gas, 545
   as shield against UV radiation, 590, *590*, 591, *591*
ozone (O_3) hole, 590, **591**, *591*
ozone (O_3) layer, 161, 590, *590*, 591, *591*

P

Pacific hurricanes and typhoons, 408, *409*, 410
Pacific Ocean, *468*
Pacific Plate, 54, *55*
Pacific Plate–North American Plate boundary, 92, *93*
Pacific Tsunami Warning Center, 244
pāhoehoe, **135**, *136*
Pakistan, flooding in, 447, 450, *450*
Paleocene-Eocene climatic optimum, 550
paleoclimate(s), **547**
paleoclimate indicators, 547–**548**, *548*, *549*
paleoclimate proxies, 547–548, *548*, *549*
paleomagnetism, 50–51
paleoseismicity, **107**–108, *108*
Palmer Drought Severity Index (PDSI), **512**, *513*, 514
Palu, Indonesia
   lateral spreading in, 201, *202*
   tsunami in, 224
pancaking, 114, *116*
pandemics
   casualties of, 11
   COVID-19, 7, 10, 10–11
   defined, **7**
   economic costs of, 10, 11
   history of, 10
   vs. physical natural disasters, 11
Pangaea, *47*, **47**–48, *48*
Panic of 1857, 485
Pan-STARRS telescope, *617*
Papua New Guinea tsunami, 237, *237*
Paricutín volcano, 136, *137*
Paris Agreement, 581
partial melting, 125–126
passages, 266, *267*
passive margins, *49*, **52**
passive-margin slumps, 198
PDSI (Palmer Drought Severity Index), **512**, *513*, 514
peak discharge, 432, *433*
peak ground acceleration (PGA), **97**, 110–111, *111*
peat, 273–274
peat fire, 524
pebble beach, 477
Peléan eruptions, 153–154, *154*
peridotite, 45, *45*, 130–131
period(s)
   geologic, 66, *67*
   of seismic waves, 80, *82*
   of wave train, 220
permafrost, **184**–186, *186*
   thawing of, **569**–572, *572*

permanent streams, **427**, *427*
permeability, **258**
permeable material, 258, *258*
Permian mass extinction, 613
Perseids meteor shower, 601, *601*
Persian Gulf, *468*
personal safety in hazardous winter weather, 364, 374
Peru Current, *468*
Peshtigo Fire, 532
pests, global poleward spread of, 574, *574*
PGA (peak ground acceleration), **97**, 110–111, *111*
pH, 575
phase change, 305, **305**–306
Philippines
    nuisance tides in, 488, *488*
    tropical cyclones and, 408, *409*
Philippine Sea, *468*
Phlegraean Fields, *146*, 170, *171*
photochemical smog, 294
phreatic eruptions, *143*, **143**–144
phreatomagmatic eruptions, **143**, *143*
physical activity and heat illnesses, 515
physical disaster, 7
    vs. global pandemic, 11
physical scarcity, 263
physical weathering, 40, *41*
phytoplankton, 497, *497*
pillows, 131, *131*, 136, *137*
Pingualuit crater, 608, *609*
pipelines, earthquake resistance of, 118
planar slope, 176, *178*
plane(s) of weakness, 176, 182–183, *183*, 188
planet(s), 594
    defined, **37**, *37*, 596
    dwarf, 594, *596*
    formation of, *37*, 37–38
    inner and outer, 37
Planetary Defense Coordination Office, 617
planetesimals, 37, *37*
plant hardiness zones, 559, *560*
plasma, **586**, *587*, **A2**
plastic deformation, **62**, *64*
plastic flow, **45**, *45*
plate(s), 31, **51**, *53*
plate boundary(ies)
    birth and death of, 56–60, *58*, *59*
    convergent, *52*, *53*, **54–56**, *55*, *56*
    defined, **51**, *53*
    divergent, **52–54**, *52*–*55*
    earthquakes at, 90–93, *91*–*93*
    earthquakes not at, 93–95, *94*, *95*
    and hot spots, 60, **60–61**
    nature of, 52–61
    and plate motion, 61, *61*
    transform, *52*, *53*, **56**, *57*
plate-boundary volcanoes, 60
plate interior, **51**
plate motion, 61, *61*
plate tectonics
    and continental drift, *47*, 47–**48**, *48*
    discovery of, 46–52
    and distribution of earthquakes, 50, *50*
    proposal of, 50–51, *51*
    and seafloor bathymetry, 48–49, *49*
    theory of, 31, **51–52**, *52*, *53*
    and variations in heat flow and seafloor sediment thickness, 49–50, *50*

platforms, 191
Pleistocene Ice Age, 550, *550*, *551*
Plinian eruptions, 154, *155*
plutons, 128, *128*, 131–132
point bars, 428, *428*
polar air masses, 318, *318*
polar amplification, **569**–572, *572*
polar bears, 573, *573*
polar climate, **317**, *505*, *506*
polar front, 318–**319**, *319*, *320*, 362
polar-front jet stream, **319**–320, *320*
polar vortex, **362**, *363*
pollen, 293
pollutants, primary *vs.* secondary, 294, *294*
pollution
    air, 294, **294**–295
    water, 257, *257*
Pompeii, 123, 152, 154, 157, 170, *170*
ponds, settling, 444, *444*
population explosion, 19, **19–20**, *20*
population growth and disaster risk, 19, 19–20, *20*
pore(s), 182
    and groundwater, **258**, *258*
pore collapse, subsidence due to, **271**–272, *272*, *273*
porosity, **258**
Portugal, coastal hazards in, 500
Portuguese Bend slump, 188, *188*
positive feedback
    and climate change, 553
    in growth of hurricane, **389**
positive polarity cloud-to-ground stroke, 342, *342*
positive streamer, 341, *341*
potential energy, **A3**
pounds per square inch, 296, *296*
power grids, disruption of, 593–594
power outages
    due to hazardous winter weather, 367, 375, 378–379
    due to hurricanes, 397, 404, 406
power surges, 595, *595*
precession and climate change, 552, *552*
precipitation, 304–306
    changes in global patterns of, 559–561, *560*, 569, *569*
    cold-cloud, 305
    defined, 5, 32, 33, 291, 295, **299**
    formation of, 304–305, *305*
    in mineral formation, 38
    and surface temperature, 546, *547*
    total, 299–300
    warm-cloud, 304–305, *305*
precipitation rate, 299
precursors, **24**
prediction(s), 22–25
    defined, **22**
    of droughts, 512–513, *513*, *514*
    of earthquakes, 106–112, 107–111
    of tsunamis, 244–246, *245*–*247*
    of volcanic eruptions, *163*, 163–164, *164*
preheating stage, 522, *523*, 526
preparation stage, **22**, *22*
preparedness, hazard, **25**
preparedness planning, 25–26, *26*
prescribed burns, 529, *529*, 537, **538**, *538*
pressure, **62**, *63*
    air, 33
    dynamic. *See* dynamic pressure
    melting due to decrease in, 124

pressure gradient and wind, **306**–307, *307*
pressure-gradient force and wind, 306–307, *307*
pressure ridge, *76*
prevention
    of earthquake damage and casualties, 112–119, *114*, *116*–*117*
    of mass wasting, 211–212, *212*, *213*
primary disaster, **12**, *12*
primary pollutants, 294, *294*
primary waves (P-waves), **80**, *80*, *81*
    ground shaking due to, 96, *97*
    seismogram of, 84, *85*, 85
principal rainband of hurricane, 386, *387*
probabilistic earthquake hazard map, 110–111, *111*
probabilistic seismic-risk assessments, 106
producing well, 260, *260*
property buyout programs, 462
property damage, 16–17
    in hazardous winter weather, 375, *375*
protection of coastal properties, 491–496, *492*–*496*
protective landscapes for hurricane damage mitigation, 417
protolith, 41, *43*
protons, A1
protoplanets, 37, *37*
pseudoscience, 106
pseudotachylyte, 606, *606*, **607**, *607*
Puerto Rico and Hurricane Maria, 383, *383*, 402, 404, *404*
Pulido, Dionisio, 136
pumice, **129**, *129*
Puʻu ʻŌʻō, 152, *152*
P-waves. *See* primary waves (P-waves)
pyrite, 262
pyroclastic debris, 136–138, *137*–*139*
    defined, **128**, *128*
    igneous rock formation from, 40
    in volcanic explosivity index, 147, *147*
pyroclastic flow, 139, **145**, *145*, 166
    hazards due to, 157, *157*
pyroclastic shield, 142, *142*
pyrolysis stage, 522, *523*, 526

Q

quake lakes, 98
quarrying, rockfalls due to, 196, *196*
quartz, **38**
    shocked, 607, 613
Quebec, electrical grid of, 585, *585*, 595, *595*
quick clay, 99, 201, *201*, 206
Qushan, China, 200

R

radar, weather, 300, *300*
radar echo, 300, *300*
radar reflectivity, 300
    of hurricane, 386
radial cracks, 186
radiation, 36
    electromagnetic, 543, *544*
    heat transfer via, **A4**, A5
    in ignition of wildfire, 526, *526*, *527*
    solar, 543–544
radiation hardened equipment, 594
radiative zone, 586, *586*
radioactive atoms, 36
radiometric dating, 66

radio transmissions, disruption of, 593
radiowaves, 544
rain, **304–305**, 305
    acid, 294
    freezing, 365, **365–366**, 366
    from hurricane, 402, 402
rainbands
    outer, 386, 387
    principal, 386, 387
    spiral, 385, **385–386**, 386
raindrops, 292, 292, **304**, 305
rainfall
    flash floods due to extreme, 438–439, 439
    total, 299–300
rainforest destruction, 565, 565
raised rim of crater, 606, 606
ram compression, **600**
rapid-onset disaster, **9**, 9
rawinsondes, 300
Rayleigh waves (R-waves), 80, 80, 81
    ground shaking due to, 96, 97
    seismogram of, 85
reactive barriers, 263
realms, 7, **31**, 32, 32–35
rear flank, of supercell, 338
rear-flank downdraft (rFD), 349
    in tornado formation, 348, 348
rebar, 114
rebound, 76–77, 77
recent climate change, **554**
recharge, 261
recharge area, 258, 261
reclamation of water, 263–266, 265
recovery stage, **22**, 22, 23
recurrence interval (RI)
    defined, **14**, 14
    of earthquakes, 106–107, 111
    of flooding, **456**, 456, 458, 458–459, 459
    of volcanic eruptions, 163
redbeds, 42
Red Sea, 468
red tides, **497**, 497, 499
reef, **478**
    coral, 471, 475, **481–482**
reef bleaching, 481, 482, 499
reflection
    of seismic waves, **83**, 83
    in tsunami, 221
refraction, 43
    of ocean waves, **473–475**, 474
    of seismic waves, **83**, 83
    in tsunami, 221
refugees, 14
regional floods, 421, 432–434, 434–437, 436–437
regolith, **175**, 176
reinsurance, **27**
relative ages, 66
relative humidity (RH), 295, 297–**298**, 298
    and heat illnesses, 516
    and heat wave, 516, 517
relief, **35**, **176**, 177
    lifting due to high, 330
rescue, 21
reservoirs, 251
    dams and, 461, 461
    in hydrologic cycle, 252–253, 253, 254
residence time, 553

resistance, to mass wasting, 180–182, 180–182
resistance stress, 179, 179
    changing, 201–203, 202, 203
resistivity survey, 268
resilience, **22**
resonance, **100**
resonance disasters, **100**, 101
resonant period, 100, 101
response stage, **22**, 22, 23
restoration stage, **22**, 22, 23
restoring force, 219
resurgent lava done, 149
retaining wall, 212, 213
retreats, 550, 551
return stroke, 341, 341–342
revegetation, 211, 212
reverse capacity, 379
reverse fault, **74**, 74
rFD (rear-flank downdraft), 349
    in tornado formation, 348, 348
RH. *See* relative humidity (RH)
rhyolite, 129
RI. *See* recurrence interval (RI)
Richter, Charles, 88
Richter magnitude, 88, 88
Richter scale, **88**, 88
ridge, **320**, 320
    and drought, **510**, 510
    and heat waves, 519, 519
ridge(s), transverse, 186
Ridgecrest, Calif., earthquake, 87
ridge-push force, 61
rift(s) and rifting, 52, 59, **59–60**
    earthquakes in, 94
rift basins, 59, 59
rills, 281, 281
rime, 305, 305
Ring of Fire, **131**
Rio de Janeiro, mudflows in, 190, 190
Rio Grande Rift, 94, 162
rip currents, **482**, 483
ripples, 472
riprap, 211, 212, 491
rip tide, 498
risk
    acceptable, **19**
    defining, **18–19**
    effect of climate change on, 20
    effect of population growth on, 19, 19–20, 20
    factors influencing, 19–21
    human influence on, 20–21, 21
risk mitigation for volcanic eruptions, 168, 168–169, 169
river(s)
    atmospheric, **453**, 454
    defined, **422**
    flash floods of, 434–441, 435, 438–441
    flood stages and crests in, 432
    natural flooding in, 432–441
    slow-onset floods of, 432–434, 434–437, 436–437
river flood warning, 460
river valleys, drowned (flooded), 475, 476, 476
rock(s)
    clastic, **39**, 39
    crystalline, **39**, 39
    defined, **38–39**, 39
    extrusive, 40, 40

    igneous, **39–40**, 40, 127–129, 127–130
    intact, 180
    intrusive, 39–40, 40
    melting (molten), 123, 124–126, 124–126
    metamorphic, **41–43**, 43
    sedimentary, **40–41**, 41, 42
    texture of, **39**
    weathering of, **40**, 41
rock bolts, 212, 213
rock cycle, **68**, 68
rockfalls, 3, 4, 184, **194–196**, 195, 196, 206
rockslides, 184, **193–194**, 194, 195, 206
rocky coasts, 475, 477, **478**, 480
Rocky Mountain Arsenal, 95
rogue waves, 485, **485–486**, 486, 499
roofs, green, 518, 518
room-and-pillar mining, 277
rotation, of thunderstorm, 338
rotational slump, 186, 187, 207
rotation axis, 33
runoff, **422**
run-up elevation, **223**, 223
Russia
    famine in, 16
    heat wave in, 514, 515
R-waves. *See* Rayleigh waves (R-waves)

S

safety factor for slope, 178–**180**, 179
Saffir-Simpson scale, 13, **384**, 384
sag pond, 74, 76
Saint-Pierre, Martinique, 153–154, 154, 157
salinification, **257**
salinity, 257
salt domes, 278
salt flat, 256
salt marshes, 478, 481, 499
salt mine, collapse of, 277–278, 278
saltwater, **252**
saltwater intrusion, 261
San Andreas fault
    and Loma Prieta gap, 108, 109
    seismicity due to, 90–92, 92–94
    seismic risk due to, 112
    as strike-slip fault, 75, 76
    and transform boundaries, 56, 57
San Antonio, Texas, sinkhole in, 270
sandbars, 475, 477, **478**, 498
    due to jetties, 492
    and tsunamis, 223
sand beach, 477
sand dunes, coastal, 475, 477, 478
sand fencing, 495, 496
sandflies, 574
sand liquefaction, mass wasting due to, 200–201, 200–202
sand spit, **478**, 479, 499
sandstone, 39, 41, 42
sand volcanoes, 98, 99, 108, 200, 444
sandy shores, 475, 477, 477–478, 479
San Fernando, Calif., earthquake, 118–119
Sanford Dam, 442
San Francisco earthquake (1906), 76, 79, 91–92, 92
San Francisco Peaks, 162
San Joaquin delta, subsidence in, 275, 276
Santa Ana winds, 528, 529, 530, 531, 535
Santorini caldera, 147, 147–149

Santorini eruption, 239, 240
saprolite, 190, 190
satellite(s), electrical discharges on, 593
satellite data on earthquakes, 109–110, 110
saturated air, 332
saturated zone, 258–259, 259
saturation, 182
saturation vapor pressure (SVP), 297, **297**–298
scale, 257
scarp(s)
    head, 184, **186**, 187, 188, 207
    transverse, 186
schist, 43, 43
scientific revolution, 48
scoria, 129, 129
sea(s), names of, 467, 468
sea arch, 478, 480, 498
sea cliff, 477, **478**
sea-cliff retreat, **489**, 489, 498
seafloor, **34–35**, 35
seafloor bathymetry, **34**, 48–49, 49
seafloor sediment thickness, variations in, 49–50, 50
seafloor spreading, **50–51**, 51, 52
sea ice, 558, 559, 560
sea level, **34**
    mean, 400, 470
sea-level cycle, **68**, 69
sea-level rise
    and nuisance tides, 487, 487
    predicting future, 567, 567
    recent, 556, 556–557
    and urban flooding, 568, 568
    visualizing consequence of, 570–571
seamounts, **49**, 49, 51
Sea of Japan, 468
Sea of Okhotsk, 468
sea salt, 293
sea spray, 389
sea stacks, 30–31, **478**, 480, 498
sea-surface temperature and drought, 508
Seattle, Wash.
    heat wave in, 516
    oceanic cyclones in, 369, 369
    sinkhole in, 270
seawalls, 246, 247, **492**, 492, 581
seawater, 467
secondary disaster, **12**, 12
secondary pollutants, 294
secondary waves (S-waves), **80**, 80, 81
    ground shaking due to, 96, 97
    seismogram of, 84, 85, 85
second-order stream, 430, 430
sediment, **41**
    deposition of, 424–427, 425, 426
sedimentary basins, 271
sedimentary bedding, 41, 42
sedimentary coasts, 477, 477–478, 479
sedimentary rocks, **40–41**, 41, 42
sediment budget, 476, 477
sediment liquefaction
    due to earthquakes, **98**–100, 99, 119, 119
    lateral spreading due to, 193, 193, 201, 202
    mass wasting due to, **200**–201, 200–202
    sinkholes due to, 270, 270
sediment load, 424, 425
seismic belts, 24–25, **50**, 50, **89**, 90
seismic gaps, **108**, 109

seismic hazard maps, 107, **110**–111, 111
seismicity
    within continents, 93–95, 95
    convergent-boundary, 92–93, 93
    defined, **73**
    divergent-boundary, 90, 91
    induced, **95**, 95
    map of Earth's, 91
    transform-boundary, 90–92, 92, 93
    and volcanic eruption, 163, 163
seismic moment, 88–89
seismic ray, 80, 82
seismic-reflection profiling, 268
seismic retrofitting, **118**, 119
seismic risk
    probabilistic assessments of, 106
    US regions of, 111–112
seismic waves
    amplitude of, **80**, 82, 88
    attenuation and amplification of, 80–82, 82, 96
    crest of, 80, 82
    defined, **43**, **73**
    generation of, 76–78, 77
    nature of, 79–83
    period of, 80, 82
    propagation of, 80, 81, 82
    reflection and refraction of, **83**, 83
    trough of, 80, 82
    types of, 79–80, 80, 81
    velocity of, 80
    wave frequency of, 80
    wave front of, 80, 82
    wave height of, 80, 82
    wavelength of, 80, 82
    wave train of, 80, 82
seismogram, 84, **85**
seismograph, **83**, 84
seismologists, 73
seismology, **73**
sensitive clay, 99, 201, 201, 206
sequestration, **564**
settling ponds, 444, 444
severe storm, deaths and economic impact of, 16, 16, 17
severe thunderstorm, 327, **328**
severe thunderstorm warning, 327
ShakeAlert system, 110, 110
ShakeMap, 87, 87
shale, 41, 42, 43, 43
shallow-focus earthquake, 78, 93
shantytowns in Brazil, 20, 20
shatter cones, 607, 607
shear, 63
shear stress, **179**, 179, 181
shear walls in earthquake engineering, 115, 117–118, 118
shed over road, 212, 213
sheetwash, 281, 322
shelf cloud, 335, 335
Shelley, Mary, 160
shelter in place for hurricane, 415
shield volcanoes, **139**, 140, 142
shivering, 361, 515
shoaling, 222
shock(s), mass wasting due to, 200, 200
shocked quartz, 607, 613
shock metamorphism, 607, 607

shock waves, **601**, 606, 615
Shoemaker-Levy 9, 603, 603, 617
shooting stars, 600–604, 600–605
shore(s)
    contrasting types of, 475–477, 475–482
    defined, **467**
    floods and flooding of, 421
    organic, 477, 478–482, 481
    rocky, 475, 477, 478, 480
    sedimentary (sandy), 475, 477, 477–478, 479
    waves approaching, 473–475, 474
    See also coast(s)
shoreline, **470–471**
    normal, 223, 223
short-period comets, 598
short-term forecasting, 23
short-term predictions
    of earthquakes, 108–109
    of volcanic eruptions, 163, 163–164, 164
shotcrete, 212, 213
shrinkage, 274
Siberia, wildfires in, 521
Siberian air mass, 372, 373, 373
Siberian Traps, 161
Sichuan earthquake, 98
significant wave height, **485**, 485
silica (SiO_2) in igneous rock, 40
silicosis, 11, 156, 282
sills, 128, 128
    basalt, 161
simple craters, 605, **605**–606, 606
single-cell thunderstorms, 333–335, 334
sink(s), in carbon budget, **564**, 564
sinkhole(s), 266–271
    collapse, **267**–268, 269
    consequences of, 268
    cover-subsidence, 267
    dealing with risk of, 258–271, 269
    defined, 251, **266**, 266
    dissolution, 267
    evaporite, 270, 270
    filling of, 271
    formation of, 267–268, 269, 270, 270
    and karst terrain, **266–267**, 267, 268
    liquefaction, 99, 99, 270, 270
    water loss and, 287
SiO_2 (silica) in igneous rock, 40
sixth extinction, **613**
slab avalanche, 196, 197
slab-pull force, 61
slash-and-burn agriculture, 280
slate, 43
slaty cleavage, 43
sleet, 365, 366
slides
    land-. *See* landslide(s)
    mud-. *See* mudflows
    submarine, 198, 198–199, 199
slip
    and cause of earthquake, 73–74
    and distribution of earthquakes, 50
    extent of, 78–79, 79
    and transform boundaries, 56
slope(s)
    complex (convex-concave), 178
    concave up, 176, 178
    convex up, 176, 178

slope(s) (continued)
  defined, **176**, 177
  drainage of, 211, 213
  failure of, **180**
  formation of, 176
  gentle, 176, 177
  grade of, 176, 177
  moderate, 176, 177
  planar, 176, 178
  safety factor and, 178–180, 179
  shape of, 176, 178
  stability of, 178–183, 179–183
  stable, **179**, 179
  stair-step profile, 176, 178
  steep, 176, 177
  steepness of, 176, 177, 178
  submarine, 177
  talus, 207
  unstable, 175, 179, **179–180**
  vertical, 176, 178
slope angle, changing, 203, 204
slope erosion due to wildfires, 532–533, 533
slope-parallel joints, 181
slope strength, changing, 204, 205
slow earthquakes, **108**
slow-onset disaster, **9**, 9
slow-onset floods, 421, **432–434**, 434–437, 436–437
slow-push techniques for deflection of near-Earth objects, 620
slump(s), 184, 186–187, 186–188, 188, 206
  defined, **186**
  detecting start of, 210, 211
  identifying areas of, 205, 208
  nonrotational, 206
  passive-margin, 198
  rotational, 186, 187, 207
  submarine, 198, 198, 206, 235–236, 235–237, 237
  translational, 186, 187, 188
slump block, 184, 186, 187
small-scale disaster, 13–14
smog, 294
smoke, **522**
smoke inhalation, 532, 533
smoke jumpers, 536
Smokey Bear, 537, 537, 538
smoldering combustion, 523
Snake River Plain hot-spot track, 162
snow
  dry, 364
  light, moderate, and heavy, 363, 364
snow avalanche, **196**, 197, 372, 372
snowball Earth, 548
snowdrifts, 364
snowfall, 363, 364
  intensity of, 363
  total, 300
snowflake, **305**, 305
snow line, 569, 569
snow melt, flash floods due to rapid, 439
snowpack(s), **363**, 372
  shrinking of, **569**, 569
snowstorm(s), 358–381
  in 2021, 378, 378–379
  agricultural damage due to, 375–378
  aircraft hazards due to, 364, 374, 374–375
  in Asia, 372, 373, 373
  blizzards as, 363–364, 364

causes of, 366–374
cold air damming and, 370, 370
continental mid-latitude cyclones and, 366, 366–368
dealing with, 374–379
hazards of, 363, 364
lake-effect, **370–371**, 371
mountain, **371–372**, 372
nor'easters and oceanic cyclones and, 368–370, 368–370
personal safety hazards due to, 364, 374
preparing for, 379
property damage due to, 375, 375
visualization of, 376–377
SO₂ (sulfur dioxide) in air, 292
social distancing, 25–26
societal disruptions due to hurricanes, 412
socioeconomic drought, 504, **505**
socio-natural disasters, 7, 16
soft protection, 492–496, 493–496
soft-rock mines, 277
soil(s)
  composition of, 278
  conservation of, 284
  and construction of impervious surfaces, 284
  contamination of, **280**
  defined, **40**, 42, 252, **278**
  degradation of, **279**–284, 280, 287
  depletion of, 279–**280**
  destruction of, 252
  diversity in, 279
  erosion of. See erosion
  expansive, 284, **284–285**
  formation of, 278, 278–279, 279
  hazards related to, 278–285
  layers of, 278, 278–279, 279
  modified texture of, 280, 282
  sub-, 278, 279, 279
  top-, 278, 279, 279
soil horizons, 279, 279
soil moisture, **253**, 258
solar atmosphere, 586, 587
solar core, 586, 586
solar cycle, **588**, 588
solar energy, 579, 580
solar flare, **588**, 589, 593, 615
solar prominence, **588**, 589
solar radiation, 543–544, 544
  changes in intensity of, 551–552
  and heat illnesses, 515–516
  and heat waves, 519–520
solar storms, **6**, 585, 585, **588–589**, 589
solar wind, 33, **586**, 587, 592, 592
solid, A2, **A2**
solidification, 38, 40
  of melts, 127, 127
solid-state diffusion, 38
solidus, 127
solifluction, 184, 184–**186**, 186, 206
Somalia, drought in, 15, 505
soot, 293, **522**
Soufrière Hills volcano (Montserrat), 156
source(s) in carbon budget, **564**, 564, 565
source rock of melt, 125
South Africa, rogue waves south of, 485
South Atlantic Gyre, 468
South Atlantic Ocean, 468

South China Sea, 468
South Equatorial Current, 468
Southern Hemisphere cyclones, 411–412, 412
Southern Ocean, 468
Southern Oscillation, 314, **315**, 315
South geographic pole, 33
South magnetic pole, 33, 33
South Pacific Gyre, 468
South Pacific Ocean, 468
Spaceguard Survey Report, 617
space junk, 593, 614
space object(s), 594–621, 596
  asteroids as, 585, 585, 596, 597, 597
  comets as, 596, 597–598, 598
  dwarf planets as, 594, 596
  extinction-level impacts of, 612, 612–617, 616
  impact effect on crust of, 605–612, 605–612
  Kuiper belt as, 596
  meteoroids and meteorites as, 596, 598–600, 599
  meteors, fireballs, and bolides as, 600–604, 601–604
  mitigating impact risk of, 617–621, 617–621
  moons as, 594, 596
  Oort cloud as, 596
  planets as, 594, 596
  visualizing impact of, 614–615
space weather
  consequences of, 593–594
  defined, **585–586**
  and earth's protective shields, 589–592, 590–593
  intense past, 595, 595
  mitigating consequences of, 594, 594, 595, 595
  overview of, 585, 585–586
  and solar storms, 588–589, 589
  sun's energy source and structure and, 586, 586, 587
  sun's magnetic field and sunspots and, 587, 587–588, 588
space weathering, 618
Space Weather Prediction Center, 594, 594
spalling, 114
sparks, 526, 534
SPC (Storm Prediction Center), 327, 355
SPEI (standardized precipitation-evapotranspiration index), 511
speleothems, 267
spillway, 182, 442, 461, 461
spiral rainbands, 385, **385–386**, 386
Spirit Lake, 148, 149
sponge zones, 462
spontaneous combustion, 524
spot fires, **526**
spring(s), **259**, 259
spring tides, 469–470, 470
S–P time, 85, 85, 88
sputtering, 600, 601
squall lines, 335
  frontal, **336**, 337
squall-line thunderstorms, **335–336**, 335–337
SS Central America, 484, 484–485
St. Elmo's fire, 156, 167
stable air, 331
stable slope, **179**, 179
stage of stream, **431**, 431
stair-step profile slope, 176, 178
stalactites, 267, 267
stalagmites, 267

standardized precipitation-evapotranspiration index (SPEI), 511
stars, shooting, 600–604, 600–605
state of emergency declaration, **22**
states of matter, A2, **A2**
stationary front, **318**, 319
steady-state condition, 68
stealth disasters, **9–11**, 252, 278
steam, volcanic explosions involving, 143, 143
steep slope, 176, 177
Steinbeck, John, 504
steppe(s), 312
stepped leader, 341, 341
stick-slip behavior, **77**
stick-slip model, 107, 107
stilts, beach houses on, 496, 496
sting jets, 369–370, 370
stishovite, 607, 613, 616
stony-iron meteorites, 599, 599
stony meteorites, 599, 599
Storegga slide, 198
    tsunami due to, 235, 235
storm(s)
    changes in severity of, 560–561
    deaths and economic impact of, 16, 16, 17
    defined, **5**, **33**, 291
    dust, 282, **282**–**283**, 283, 284
    fire-, 12, 100, 101, **527**, 527
    geomagnetic, 586, 595
    hail-, **345–346**, 346
    ice, 291, 366, 367, 373, 373, 394–396
    increase in intensity and duration of, 577
    meteor, 601
    multicell, 335–336, 335–337
    snow-. *See* snowstorm(s)
    solar, **6**, 585, **585**, **588–589**, 589
    thunder. *See* thunderstorm(s)
    tropical, **384**
storm drains, 265
Storm Prediction Center (SPC), 327, 355
storm sewers, 442
storm surge, 13, **400**, 498
    from hurricane, 400, 400–401
    from Hurricane Katrina, 406, 407
storm-surge barrier, **492**, 493
storm-surge warnings, 413–414
straight-line winds, **335**, 336
stratocumulus clouds, 304
stratopause, 301, 302, 593
stratosphere, 301, **302**, 590, 593
stratospheric haze, 145
stratovolcanoes, 141, **141–142**, 142, 148
stratus clouds, 304, 304
stream(s), 422–432
    braided, **428–429**, 429
    capacity of, **424**
    competence of, **424**
    defined, **422**
    describing, 422–424, 423, 424
    discharge and stage of, 430–431, 431, 432
    and drainage networks and watersheds, 429, 429–430, 430
    ephemeral, **427**, 427
    erosion, transport, and deposition by, 424–427, 425, 426
    flash floods of, 434–441, 438–441
    flood stages and crests in, 432, 433
    gradient of, 423, 423
    longitudinal profile of, 423, 423
    meandering, **427–428**, 428
    natural flooding in, 432–441
    permanent, **427**, 427
    runoff and formation of, 422, 422
    slow-onset floods of, 432–434, 434, 435, 436–437
    trunk, 423, **429**, 429
    types of, 427–429, 427–429
    velocity of, 431, 432
streambed, 422, 422, 431, 432
stream flow, 422, 422
stream order, **429–430**, 430
strength of material, **176**
stress, 62
    and atmospheric pressure, 296
    downslope, 201–203, 202, 203
    and faulting, 73
    normal, **179**, 179
    resistance, 179, 179, 201–203, 202, 203
    shear, **179**, 179, 181
    and slope stability, **179**, 179
strike line, 74
strike-slip fault, **74**, 74, 75, 76, 96
Strombolian eruptions, 151–153, 153
structural components in earthquake engineering, 115
structural failure of levees, 443, **444**
structure-related currents, 482, 483
subaerial landslides, tsunamis due to, 234, 234–235, 243
subaerial mass wasting, 198
subaerial volcanoes, 60, **131**
subatomic particles, A1
subduction, **54**
subduction zones, 51, 52, **54**
sublimation and energy, 305
submarine debris flows, 198, 198
submarine mass wasting, 198, **198–199**, 199
submarine slides, 198, 198–199, 199
    tsunamis due to, 243
submarine slope, 177
submarine slumps, 198, 198, 206
    tsunamis due to, 235–236, 235–237, 237
submarine volcanoes, 60
subsidence. *See* land subsidence
subsoil, 278, 279, 279
substrate, 86
subtropics, **311**
    atmosphere of, 311–317
suction vortices, 349, 350
Sudan, drought in, 505
Sudbury Basin, 607, 608, 610
sulfate aerosols, 616
sulfur dioxide (SO_2) in air, 292
sulfurous gases from volcanic eruptions, 158, 159
Sumatra earthquake, 89, 93, 98, 100
    tsunami due to, 218, 218, 220, 231, 231–232, 232
summit eruption, 138, 140
summit vent, 138
Sun
    energy source and structure of, 586, 586, 587
    faint young, 551–552
    magnetic field and sunspots of, 587, 587–588, 588
Sunda Trench, 231
sunny-day floods, 484, 487, 487, 488, 488
Sunset Crater, 142
sunspots, 587, **588**, 588
supercell(s). *See* supercell thunderstorms
supercell thunderstorms, 336–339, 338, 339
    defined, **337**
supercell tornadoes, 348–349, 348–350
supercontinent cycle, **68**, 69
supercooled droplets, 305, 305, 365
Superfund sites, 263
supersaturation, 302–304
super typhoon(s), **408**
Super Typhoon Haiyan, 408, 409
supervolcanoes, 11, 142, **149–150**, 150
support columns
    in earthquake engineering, 117, 117
    failure of, 114, 116
surface air pressure, effect of heating and cooling on, 309, 309–310
surface currents, **467–469**
surface fires, **524**, 525
surface freshwater
    contamination and salinification of, 257, 257
    depletion of, 255, 255–256, 256
surface temperatures, average monthly, 545, 545
surface tension, 181, 472
surface water, 34, **253**
    challenges with, 255–257, 255–257
surface wave(s), **80**, 84, 85
surface-wave magnitude (M_S), 88
surface-wave magnitude (M_S) scale, 88
surface-wind streamline maps, 390, 390
surf zone, 473, 474, 477, 485, 486
Surtsey volcano, 150–151, 151
Surtseyan eruptions, 150–151, 151
suspended load, 424, 425
sustained winds, 383, 384, 384, 397, 397–400
suture, 58, 59
SVP (saturation vapor pressure), 297, **297–298**
swamps, 478
    mangrove, 475, 478, 481
swampy delta, 475
swash, 473, 474, 479
S-waves. *See* secondary waves (S-waves)
sweating, 515
swells, **473**, 473
system, **A2**

T

tailings, 444, 444
Taiwan, typhoons in, 408, 410
talus piles, **182**, 182
talus slope, 207
Tambora, Indonesia volcano, 146, 147, 160
Tangshan earthquake, 11, 115
target of meteorite, 605
Tasman Sea, 468
TDS (tornado debris signature), 355, 355
tectonic activity, **35**
tectonic subsidence, 271
teletsunamis, 220–221, 243
temperate climate, **317**, 505, 506
    drought in, 511, 511–512
temperate-climate heat waves, 519, 519–521
temperature
    adapting to increase in global, 580–581, 581
    air, **295–296**, 296
    and altitude, 301, 301–302
    apparent, **516**

temperature (continued)
　average monthly, 545, 545
　body, 514–515
　core body, 361
　defined, **A4**
　within Earth, 45, 45
　latitude and, 545, 545
　predicting changes in, 566, 566
　recent changes in, 554, 554–555, 555
temperature anomalies, **548**, 550, 555, 555
temperature extremes, deaths and economic impact of, 17
temperature regulation, 515
temperature scales, 295–296, 296
10-year flood, 456, 456
Tennessee, thunderstorm training in, 452–453
tension, 63
tephra, **136**, 139
　in volcanic explosivity index, 147, 147
terraces, 191, 191
　uplifted, 475, 476
tertiary disaster, **12**, 12
Texas
　and Hurricane Harvey, 451, 452
　winter-weather disaster in, 378, 378–379
texture of rock, **39**
Tharp, Marie, 48–49, 49, 50
theory, 50
Thera eruption, 147, 239, 240
thermal energy, 295, **A3**, A4
thermal expansion, 557
thermohaline circulation
　and climate change, 553, **553**
　interruption of, 577
thermometers, **295**–296, 296, 300
thermosphere, 301, **302**, 593
　heating of, 593
third-order stream, 430, 430
30-year flood, 456
Thistle, Utah, debris flow, 189, 189
Thomas Fire, 530, 531
thrust fault, 74, 74, 93, 96
thunder, 343–**344**
thunderstorm(s), 328–348
　atmospheric instability and, 329, 331–333, 332, 333
　conditions that produce, 328–333
　defined, **5**, 313, 323, **328**, 328
　developing stage of, 334, 334–335
　dissipating stage of, 334, 335
　downbursts and microbursts in, 344, 344–345
　global distribution of, 328–329, 329
　hail in, 345, 345–346, 346
　hazards of, 340–348
　life cycle of, 334, 334–335
　lifting mechanisms and, 329, **330**–331, 330–332
　lightning in, 328, **340**–343, 340–344
　maturity stage of, 334, 335
　moist air and, 329–330
　mountain, 330–331, 332, 455, 455
　ordinary (single-cell, air-mass), **333**–335, 334
　rotation of, 338
　severe, 327, **328**
　size of, 328
　squall-line (multicell), **335**–336, 335–337
　supercell, 336–339, 338, 339
　thunder in, 343–**344**
　timing of, 328
　training of, 452–453, 453, 455
　types of, 333–339
　updraft and downdrafts in, 328
thunderstorm days, 329, 329
tidal bore, **484**, 484
tidal bulges, 469, 469, 470
tidal currents, 482, 483
tidal flats, **478**, 479, 484, 499
tidal gauges, 223, 225
tidal range, **470**, 470, 483–484
"tidal waves," 218–219
tide(s), 469, 469–471
　causes of, 469, 469–470
　dangerous, 483–484, 484
　defined, **469**
　flood vs. ebb, 470–471, 499
　king, **484**
　low vs. high, 469, 469–470, 470
　nuisance, **484**, 487, 487, 488, 488
　red, **497**, 497, 499
　rip, 498
　spring vs. neap, 469–470, 470
　and tsunamis, 218
tide-generating force, 469, 469
tiltmeters, 148, 164
tilt of Earth's axis and climate change, 552, 552
tinder, 525, 526
Toba volcano, Sumatra, 146, 150
toe of slump, 184, 186, 187, 207
Tōhoku, Japan earthquake, 3
　fire due to, 100, 101
　magnitude of, 89, 93
　slip during, 79, 79
Tōhoku, Japan tsunami, 3, 3, 103, 103, 232–233, 233
　relief and recovery from, 241
　satellite imagery of, 221
　and tsunami warnings, 244–245, 246, 247
Tokyo earthquake, 100
topography, **35**, 49
　development of, 61–62, 62
toppling, 194, 195
topsoil, 278, 279, 279
Torino scale, 619, 621
tornado(es), 347–355
　comparing size of, 13
　damage from, 291, 327, 327
　defined, **5**, 5, **347**, 347
　destruction by, 355, 355
　detection of, 355, 355
　fire, 527
　formation of, 347–350, 348–351
　forward speed of, 347
　due to hurricane, 386
　intensity of, 347, 350–355, 351
　life cycle of, 352–353
　location of, 347, 347
　multiple-vortex, 349, 350
　non-supercell, 350, 351
　safety from, 355
　size of, 347
　supercell, 348–349, 348–350
　threat to urban centers by, 354, 354
Tornado Alley, 347, 347
tornado debris signature (TDS), 355, 355
tornado drills, 355
tornado outbreak, **347**
tornado potential map, 25
tornado shelters, 25, 25, 327
tornado track, **347**, 348
tornado warning, 24, 327, 355
tornado warning sirens, 25, 25
tornado watch, 327, 355
total people affected, **14**, 15
total precipitation, 299–300
total water withdrawal, 263
trace gases, 33, 292
trade winds, **312**, 312
　and drought, 506–508
　and heat waves, 520, 520
training of thunderstorms, 452–453, 453, 455
Trans Alaska Pipeline, 118
transform, **56**, 57
transform boundary, 52, 53, **56**, 57
transform-boundary seismicity, 90–92, 92, 93
transform fault, **56**, 57
transient crater, 605, 606
transition zone, 44, 45, 55
translational slump, 186, 187, 188
transport by streams, 424–427, 425, 426
transverse cracks, 186
Transverse Ranges, 92, 93
transverse ridges, 186
transverse scarps, 186
travel time graph, 85, 85–86
tree rings, 548, 549
trenches, 35, **49**, 49, 52, 54, 55
tributary, 423, **424**, 429
triple junction, 52
Tri-State Tornado of 1925, 347
tropical air masses, 318, 318
　and heat waves, 520, 520
tropical climate, 505, 506
tropical-climate heat waves, 520, 521
tropical cyclones, 16, **383**
　deaths and economic impact of, 16, 16, 17
　inland flooding due to, 451, 452
tropical depression, **384**
tropical diseases, 574, 574
tropical disturbance, **384**
tropical storm, **384**
Tropical Storm Lee, 402, 402
tropical storm warning, **413**, 414
tropical storm watch, **413**
tropics, **311**
　atmosphere of, 311–317
tropopause, 301, **302**, 593
troposphere, 301, **302**, 590, 593
trough(s)
　and drought, 512
　and heat waves, 519, 519
　in jet stream, **320**, 320
　of ocean wave, 471, 471
　of seismic wave, 80, 82
　of wave train, 220
trunk stream, 423, **429**, 429
Tsar Bomba explosion, 602
tsunami(s), 216–249
　from asteroid impact, 613
　coming ashore by, 223, 223–225
　in deep ocean, 222, 222
　defined, **3**, 3, 217, **218–219**
　destruction by, 226–228, 227
　drawback of, **223**, 227, 227

due to earthquakes, 73, 100–103, *103*, 217, *217*, 218, *218*, 228–233, *228–233*, 243
 far-field (distant, tele-), **220–221**, *243*
 generation of, *219*, 219–220, *220*
 landslide-generated, 234–238, *234–238*, 243
 mega-, **235**
 due to meteorite impacts, 239–240, *240*
 near-field (local), **220**, 243
 orphan, 231
 overview of, *217*, 217–218, *218*
 predication and mitigation of, 244–246, *245–247*
 recurrence interval of, 218
 relief and recovery from, 241, *244*
 secondary disasters related to, 240–241, *241*
 in shallow water, 222, *222*
 shoaling of, 222, *222*
 signs of impending, 245
 variation in height of, 220–221, *221*
 visualization of, 242–243
 due to volcanic eruptions, 238–239, *239*, *240*, 243
 vs. wind-driven wave, 223–226, *225*, *226*
tsunami beaming, 221, *221*, 231
tsunami elevation, **223**, *223*
tsunami inundation maps, 245, *246*
tsunami stones, 244–245, *246*
tsunami warning, **244**
tsunami watch, **244**
tsunami wave train, 220, **220–221**, *221*
tuff, *129*, **130**, 147
 air-fall or ash-fall, 145
 ash-flow, 145
 welded, 145
Tunguska event, **604**, *604*
turbidity currents, 198, *198*, 206
turbulence, 431, *432*
Turnagain Heights disaster, 99–100, *100*, 192
Twain, Mark, 460
Twelve Apostles (sea stacks), *30–31*
typhoon(s)
 deaths and economic impact of, 16, *16*, *17*
 defined, 13, **383**
 formation of, 408, *409*
 impact of, 408, *409*, *410*
 naming of, 384
 super-, **408**
Typhoon Morakot, 408, *410*
Typhoon Tip, 13, *13*, *396*

U

ultimate base level of stream, *423*, 424
ultramafic melts, 125, *126*
ultraviolet A (UVA), *590*, 590
ultraviolet B (UVB), *590*, 590
ultraviolet C (UVC), *590*, 590
ultraviolet (UV) light, 544
 from asteroid impact, 616
ultraviolet (UV) radiation, atmosphere as shield against, *590*, 590, *591*, 591
unconsolidated material, 176, *180*, 182
undercutting, **202**–203, *203*, 208
 prevention of, 211, *212*
undergrowth, 524, *525*
undermining, *443*, 443–444
underseeping, *443*, **443–444**
undertow, **486**, *486*
uniformitarianism, **66**, *66*
United Nations, 22
United Nations Disaster Relief Organization (UNDRO), 115
unsaturated air, 332
unsaturated zone, 258–259, *259*
unstable air, 331
 conditionally, 333, *333*
 in hurricane formation, 389
unstable slope, 175, *179*, **180**
updraft
 defined, 299, **328**
 in thunderstorm, 328, 333, 335, 338
 in tornado formation, 348, *348*, *349*, 350
uplift, **61**, **175–176**
 and climate change, 551
uplifted terraces, *475*, 476
upper mantle, *44*, 45
urban flooding, **441–442**, *442*, 568, *568*
urban heat islands, **517**–518, *518*
urbanization, 21, *21*
urban planning, 25
US Army Corps of Engineers, 22, 406, 434, 460–461
US Drought Monitor, 512
US Geological Survey (USGS), 87, *87*, 148, 163, 165, 165–168, *168*, 491
US National Integrated Drought Information System, 512
Utah, wildfires in, 528
UV. *See* ultraviolet (UV)

V

Vaiont Dam disaster, 193, *194*
Valdivia megathrust earthquake, 228
Valles Caldera, *162*
VAN (Volcano Activity Notice), 165
Van Allen radiation belts, *592*, 592
Van de Graaff generator, *343*
vapor plume, **603**, 613, 616
vapor pressure, *297*, **297–298**
Vargas (Venezuela) tragedy, 189
vector, **179**
vegetation cover, changing slope strength due to removal of, 204, *205*
VEI (volcanic explosivity index), **145–147**, *146*, 168
velocity
 of object, **308–309**
 of wave train, 220
Venice, coastal plain subsidence in, 276, *276*
vent(s), 123, *141*
 flank, 138, *141*, *152*
 hydrothermal, 32, 131, *131*
 summit, 138
vertical-motion seismograph, 83, *84*
vertical slope, 176, *178*
vertical structure in atmosphere, 301, *301*–302
vertical wind shear, 337
vesicles, 129
vibrations, mass wasting due to, 200, *200*
viscosity of melt, **126**
visibility, 295, **299**
visible light, 544
vog, **158**, *159*, 167
volatiles
 in magma, 126, *126*
 melting due to addition of, **125**, *125*
volcanic agglomerate, 130
volcanic arc, *51*, **55**, *56*, 130, **132**, *132*
volcanic bombs, **138**, *138*, *139*
volcanic breccia, 129, 130
volcanic edifices, **138–142**, *139–142*
volcanic eruptions, 122–173
 ash cloud from, 35, *35*
 and building of volcano, 135–142, *135–142*
 and climate, 160, *160*–161
 and climate change, 551
 deaths and economic impact of, *17*
 defined, **3**, **4**, **8**, **123**
 effusive, *142*, **142**–143
 explosive, 142, 143–146, **143–147**
 fissure, **139**, *140*
 flash floods due to, 439
 Hawaiian, 150, *151*–153, *152–153*
 hazards of, 24, 155–160, *155–161*, 168, 168–169, *169*
 igneous activity and, **124**, 130–134, *130–135*
 and igneous rock formation, 127–129, *127–130*
 large caldera, 147, *147–149*
 lava and, **124**, *125*, 125–126, *126*
 magma and, **124**, 125–126, *125–127*, *127*
 melting rock and, 123, 124–126, *124–126*
 mitigating risk from, 168, *168*–169, *169*
 overview of, 123, *123*–124
 Peléan, 153–154, *154*
 phreatic, *143*, **143–144**
 phreatomagmatic, **143**, *143*
 Plinian, 154, *155*
 prediction of, 163, *163*–164, *164*
 protection from, 161–169, *162–165*, *168*, *169*
 and society, 148, *148*–149, *149*, 170, *170*–171, *171*
 Strombolian, 151–153, *153*
 style of, 124, **150**–154, *151–155*
 summit, 138, *140*
 super-, 11, *142*, 149–150, *150*
 Surtseyan, 150–151, *151*
 tsunamis due to, 238–239, *239*, *240*, 243
 visualization of, 166–167
 Vulcanian, 153, *153*
 warnings of, 165, *165*–168
volcanic explosivity index (VEI), **145–147**, *146*, 168
volcanic gases
 hazards due to, 158, *159*
 increases in, 164
volcanic hazard assessment map, **168**, *168*, *170*
volcanic island, flank collapse of, *236*, 236–**238**, *238*
volcanic island arc, **56**, *56*, 132, *132*
volcanic unrest, 163, **163–164**
volcano(es)
 active, *161*, *162*
 building of, 135–142, *135–142*
 bulging of, 136, 148, *148*, *149*, 164
 and climate, 160, *160*–161
 computer models of, 164
 defined, *3*, 35, *35*
 dormant, **161**, *162*
 extinct, 60, **161**, *162*
 hot-spot, *52*, 60, **60–61**, *130*, 133, *133–134*
 location of, 130, *130*
 plate-boundary, 60
 pyroclastic shield, 142, *142*
 sand, 98, *99*, 108, 200, 444
 and seafloor spreading, *51*
 shield, **139**, *140*
 strato- (composite), *141*, **141–142**, *142*, 148
 subaerial, 60, **131**
 submarine, 60
 summit of, 138, *139*

volcano(es) (continued)
super-, 11, 142, 149–150, 150
throat of, 138
Volcano Activity Notice (VAN), 165
volcano observatories, **165,** 165
Volcano Observatory Notifications for Aviation (VONA), 165
vortex(ices), 362
polar, **362,** 363
suction, 349, 350
of tornado, 347, 348–349, 349
Vredefort Dome, 608, 610
V-shaped valleys, 423, 424
Vulcanian eruptions, 153, 153
vulnerability, **17–18,** 18
and risk, 18–19

W

Wadati-Benioff zone, 55, **93,** 93
wadi, 427, 427
Walker circulation, 314, **314**–316
wall cloud, **338,** 339, 350
wall rock, assimilation of, 126
warm cloud, 304
warm-cloud precipitation, 304–305, 305
warm front, **318,** 319
warning, **23,** 24
Washington, heat wave in, 514, 516
watch, **23**
water
in air, 292, 292, 293
conservation and reclamation of, 263–266, 265
Earth's supply of, 252–253, 252–254
explosions due to interaction with, 143, 143–145, 144
forms of, 292, 293
fresh-, 34, **251,** 255–257, **255**–257
ground-. See groundwater
hard, 261
as non-renewable resource, 261
society's use of, 253, 254
surface, 34
water content, changing slope strength due to, 204
water crisis, 251–266
in Cape Town, South Africa, 251, 251, 263
defined, **253,** 254
in Flint, Mich., 257
groundwater challenges in, 258–263, 258–263
in India, 264, 264
overview of, 251, 251–252, 252
responding to, 263–266, 264, 265
surface-water challenges in, 255–257, 255–257
visualization of, 286–287
in western US, 256, 263
water demand, 263, 265
water droplets, 292, 292, 293
supercooled, 305, 305, 365
water equivalent of snow, 300
water erosion, soil degradation due to, 280, 281, 281
water footprint, 263
water pipes, frozen, 375
water pollution, **257,** 257
water resources, redistribution of, 567–569, 569
water security, **263**–266, 265
watershed, **255,** 255, **429,** 429, 431
water shortage, 256, 263
waterspouts, 347

water stockpiling, 263
water stress, **263**–266, 265, 507
water subsidies, 265
water table, 200, 258–**259,** 259
lowering of, 211, 212, 260, 260–261, 261
water theft, 263
water treatment plant, 265
water vapor, 33, **292,** 293
as greenhouse gas, 545
water withdrawal, total, 263
wave(s)
approaching shore, 473–475, 474
body, **80**
along coast, 471–474, 471–475
cold, **362,** 363
constructive interference between, **472,** 473
easterly, **390**–391, 391, 392
formation of, 471, 471–472
gravity, 219, 219–220
hazards of, 484–486, 484–486
heat. See heat wave(s)
high storm, 8
motion of, 472, 472–473, 473
ocean, **471,** 471
radio-, 544
rogue, **485,** 485–486, 486, 499
seismic. See seismic waves
shock, **601,** 606, 615
"tidal," 218–219
wind-driven, 218, 223–226, 225, 226, 471–475
wave amplitude
of ocean waves, 401, 471
of wave train, 220
wave base, 226, **472,** 472, 473
wave-cut benches, **476,** 476, 480, 489, 489, 498
wave-cut notch, 480, **489,** 489
wave damage from hurricane, 401, 401–402
wave drift, 472
wave erosion, **477**
wave frequency of seismic waves, 80
wave front of seismic wave, 80, 82
wave height
of ocean wave, 471, 471, 473
of seismic waves, 80, 82
significant, **485,** 485
of wave train, 220
wavelength
of ocean wave, 471, 471–473
of seismic waves, 80, 82
of wave train, 220
wave period
of ocean waves, 471–472
of seismic waves, 80, 82
of wave train, 220
wave refraction, **473–475,** 474
wave speed, 471, 472
wave train, **80,** 82, 471, **471**–472
tsunami, 220, **220**–221, 221
wave velocity of wave train, 220
weather, 33, 292, **295,** 545
future changes in hazardous, 577
weathering, **40,** 41
changing slope strength due to, 204
weather patterns, changes in global, 559–561, 560
weather radar, 300, 300
weather satellites, 300, 300–301, 301
weather stations, **300**

Wegener, Alfred, 47, 47–48, 48, 50
Weibull equation, 458
weight, A3
welded tuff, 145
wells, **260,** 260
artesian, 260
dry, 260, 260
extraction, 263
injection, 95, 95, 263
ordinary, 260, 260
producing, 260, 260
western United States
water crisis in, 256, 263
wildfires in, 521
West Greenland Current, 468
West Wind Drift, 468
wet avalanche, 196, 197
wetlands, subsidence in, 273–274, 274
Whakaari, New Zealand, volcano, 143, 165–168
whirlpool, 432
whitecaps, 472
White Island, New Zealand, 143, 165–168
whiteout conditions, 291, 363
wildfire(s), 521–539
from asteroid impact, 616, 616
climate controls on, 529, 529–532, 532
cost of, 532
crown (canopy), **524,** 525, 535
defined, **6,** 6, 9, **503,** 521
distribution of, 529–532, 532
downslope winds and, 528, 529, 530–531, 531
fighting, 533–537, 536, 537
ground, **524,** 525, 526
ignition of, 522–524, 523, 524
impacts of, 16, 17, 532–533, 533
intentional (controlled), 529, 529, 537, 538, 538
ladder, **524,** 525
due to lightning, 342, 524, 524, 534
mitigation of, 537–539, 537–539
naming of, 527
overview of, 503, 503, 521–522, 522
parts of, 527–528, 528
peat, 524
secondary disasters due to, 532–533, 533
severity and intensity of, 524, 526
spot, **526**
spread of, 524–528, 526–528
surface, **524,** 525
triggers for, 524, 524
types of, 524, 525, 526
visualization of, 534–535
vulnerability to, 539, 539
in western United States, 521
wildfire complex, **528,** 528, 529
wildfire hazard maps, 538–539, 539
wildfire suppression, **537,** 537
Wilson, J. Tuzo, 56
wind(s), 306–311
background, 526
convergent, 526, 527
defined, 32, 33, 291, **299,** 306
downslope, 511, 521, **528,** 529, 530–531, 531
and effect of heating and cooling on surface air pressure, 309, 309–310
and forces acting on air, 306–309, 307, 308
geostrophic, **311,** 311
in jet streams, 306

solar, **33**, **586**, 587, 592, *592*
straight line, **335**, *336*
sustained, 383, *384*, **384**, 397, *397–400*
trade, **312**, *312*
wind barbs, 299, *299*
wind chill advisory, 359
wind chill factor, **361**, *362*
wind chill temperature, **361**, *362*
wind chill warning, 359
wind damage from hurricane, *397*, 397–400, *400*
wind direction, 295, **298–299**, *299*
control of, *310*, 310–311, *311*
wind-driven waves, 218, 471–475
approaching shore, 473–475, *474*
defined, **472**
formation of, *471*, 471–472
motion of, *472*, 472–473, *473*
vs. tsunamis, 223–226, *225*, *226*
wind energy, 579, *580*
wind erosion, soil degradation due to, 280, 281–284, *282*, *283*
wind gusts, 335
wind shear
horizontal, 350, *351*
in hurricane, 394
vertical, 337
wind speed, 295, **298–299**, *299*
and heat illnesses, 516
and heat wave, 516, *517*
of hurricane, 383, *384*, **384**, 397, *397–400*, *400*
wind streaming map, 387
wind streamline maps, 390, *390*

wind strength and drought, 506–508
wind turbines, *580*
windward side, 371
winter, nuclear, 616
Winter Park sinkhole, 266, *266*
winter storm warning, 359, *359*
winter storm watch, 359, *359*
winter weather, 358–381
in 2021, *378*, 378–379
agricultural damage due to, 375–378
in Asia, 372, *373*, 373
causes of, 366–374
cold air damming and, *370*, 370
cold as, 360–363, *361*–*363*
continental mid-latitude cyclones and, *366*, 366–368
dealing with, 374–379
hazardous, 359
hazards to aircraft due to, *374*, 374–375
hazards to personal safety due to, 374
ice storms as, 364–366, *365*–*367*, 367
lake-effect, 370–371, *371*
mountain, 371–372, *372*
nor'easters and oceanic cyclones and, 368–370, *368*–*370*
overview of, 359–360
preparing for, 379
property damage due to, *375*, 375
snowstorm as, *360*, 363–364, *364*
terminology for, 359
visualization of, 376–377

winter weather advisory, 359, *359*
Wizard Island, *147*, 149
work, **A2**
World Health Organization and COVID-19 pandemic, 10
World Heritage Sites, 503
Wuhan, China, and COVID-19 pandemic, 10

X
Xenia, Ohio, tornado, 349
x-rays, *544*

Y
Yangtze River, China, flooding, 5, 6, 11
yellow fever, 574
Yellowstone National Park
caldera in, 150
hot spots in, 60, *133*, 133–134, *162*
pyroclastic debris in, *146*
pyroclastic shield in, *142*
wildfires in, *537*, 538
Yousafzai, Malala, 450
Yungay, Peru landslide, *175*, 175

Z
Zika virus, 574
zone of accumulation, 279, *279*
zone of leaching, 279, *279*
zoning, 25
earthquake, **118**–119, *119*
for hurricane damage mitigation, 417